Photo-initiated Quantum Molecular Dynamics

University of Nottingham, UK
15–17 April 2013

FARADAY DISCUSSIONS

Volume 163, 2013

RSC Publishing

The Faraday Division of the Royal Society of Chemistry, previously the Faraday Society, founded in 1903 to promote the study of sciences lying between Chemistry, Physics and Biology.

EDITORIAL STAFF

Editor
Philip Earis

Deputy editor
Heather Montgomery

Development editor
Rowan Frame

Senior publishing editor
Susan Weatherby

Publishing editors
Tanya Smekal, Jessica Cocker

Publishing assistants
Emily Barbour, Sian Gordon, Emily Porter

Publisher
Niamh O'Connor

Faraday Discussions (Print ISSN 1359-6640, Electronic ISSN 1364-5498) is published 8 times a year by the Royal Society of Chemistry, Thomas Graham House, Science Park, Milton Road, Cambridge, UK CB4 0WF. Volume 163 ISBN-13: 978-1-84973-690-9

2013 annual subscription price: print+electronic £765, US $1428; electronic only £727, US $1356. Customers in Canada will be subject to a surcharge to cover GST. Customers in the EU subscribing to the electronic version only will be charged VAT. All orders, with cheques made payable to the Royal Society of Chemistry, should be sent to RSC Order Department, Royal Society of Chemistry, Thomas Graham House, Science Park, Milton Road, Cambridge, CB4 0WF, UK. Tel +44 (0) 1223 432398; E-mail orders@rsc.org

If you take an institutional subscription to any RSC journal you are entitled to free, site-wide web access to that journal. You can arrange access *via* Internet Protocol (IP) address at www.rsc.org/ip. Customers should make payments by cheque in sterling payable on a UK clearing bank or in US dollars payable on a US clearing bank.

PRINTED IN THE UK

MIX
Paper from responsible sources
FSC www.fsc.org FSC® C013604

Faraday Discussions documents a long-established series of *Faraday Discussion* meetings which provide a unique international forum for the exchange of views and newly acquired results in developing areas of physical chemistry, biophysical chemistry and chemical physics.

Photo-initiated Quantum Molecular Dynamics

Faraday Discussions
www.rsc.org/faraday_d

A General Discussion on Photo-initiated Quantum Molecular Dynamics was held in Nottingham, UK on 15th, 16th and 17th April 2013.

CONTENTS

ISSN 1359-6640; ISBN 978-1-84973-690-9

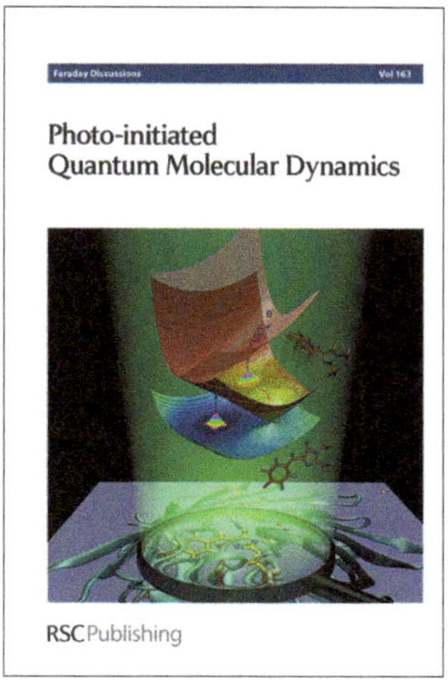

Cover
See Bochenkova and Andersen, *Faraday Discuss.*, 2013, **163**, 297–319.

When taken out of the Green Fluorescent Protein, its anionic light-absorbing molecular unit shows a remarkable fundamental interplay between electronic and nuclear excited-state decay channels in the ultrafast photo-initiated dynamics occurring on a pile of potential energy surfaces.

Image reproduced by permission of Anastasia Bochenkova from *Faraday Discuss.*, 2013, **163**, 297.

INTRODUCTORY LECTURE

PAPERS AND DISCUSSIONS

Co-sponsor

IOP | Institute of Physics
Biological Physics Group

CONCLUDING REMARKS

ADDITIONAL INFORMATION

Faraday Discussions

RSC Publishing

PAPER

The three pillars of photo-initiated quantum molecular dynamics

Albert Stolow*

Received 3rd July 2013, Accepted 4th July 2013
DOI: 10.1039/c3fd90021e

Studies of the interaction of light with matter, based on spectroscopy and diffraction, yielded a microscopic view of nature. The detailed structure of molecules and solids emerged from this effort and led to many of the advances of the 20th century, with the 'structure-function' paradigm being perhaps paramount. This static view of Nature is a necessary but insufficient advance and, in the 21st century, we will need to develop microscopic yet dynamical views of Nature. We consider here issues in Photo-initiated Quantum Molecular Dynamics raised by Faraday Discussions 163, in the context of three main categories or 'pillars' of light-matter interaction: energy/time, phase/coherence, intensity.

1 Introduction

The subject of Faraday Discussion 163 is 'Photo-initiated Quantum Molecular Dynamics'. The topics under discussion range very broadly, from high resolution spectroscopy,[1,2] to ultrafast dynamics[3–9] and coherence[10,11] quantum control[12–15] and strong field attosecond physics.[16–18] Yet underlying this seemingly disparate collection of topics is the unifying theme of light-matter interactions. Why is this an important subject? One hundred years ago, it was studies of the interaction of light with matter which led scientists to a microscopic view of nature. It was spectroscopy and diffraction, combined with quantum mechanics, which led us to understand the detailed structure of matter—the shapes of molecules, biomolecules and solids. It is hard to overstate how significant a step this was. One might fairly call it the 'great leap forward' of the 20th century. From the point of view of molecular sciences, this led to the central paradigm of 'structure-function relationships', the basis of much of our understanding of chemistry, biochemistry and material science. The canonical example is the structure of double helix DNA, the shape of which clearly determines its function as the template for the storage and transfer of genetic information. As powerful and successful as 'structure-function relations' may be, they remain a necessary but insufficient condition for our understanding of the world around and within us.

National Research Council, 100 Sussex Drive, Ottawa, ON K1A 0R6, Canada. E-mail: albert.stolow@nrc.ca; Fax: +01-613-991-3437; Tel: +01-613-993-7388

Simply put, this is because Nature is not static. There are dynamical, often fast competing processes which determine the outcomes of many natural phenomena and these cannot be discerned from a static, structural perspective alone.

It is perhaps useful to consider an example. A compelling case is the process of Vision which involves the light-sensitive *trans*-membrane protein rhodopsin, comprised of an opsin protein 'shell' surrounding a photoactive chromophore. The chromophore is the molecular sub-unit retinal which contains a polyene chain that acts, in essence, as the 'light collector'. The first step in Vision, following absorption of a visible photon, is the *cis-trans* isomerisation of *11-cis* retinal about the C11–12 double bond and occurs on the 10^{-13} s time scale.[19] Given that the overall visual response is much slower (10^{-2} s), one could reasonably ask whether this ultrafast time scale is a curiosity or, in fact, is essential for the process of Vision. I argue that it is essential and that if this first step was much slower, Vision would be impossible. The rapid *cis-trans* isomerisation stresses and 'deforms' the *trans*-membrane opsin protein, permitting the now 're-shaped' cytoplasmic side of the protein to catalyze many hundreds of transducin proteins to an activated state.[20] This leads to a closing of cationic *trans*-membrane transport channels, hyperpolarizing the cell membrane and affecting a nearby synaptic terminal, thereby converting visual information to an electrical signal. However, the retinal chromophore is surrounded both by protein and by water—a highly dissipative environment. Dissipative processes are typically fast (ps time scale) and eventually lead to the conversion of photon energy to a simple, local temperature increase. In such a case, the energy associated with photon absorption would not lead to 're-shaping' of the protein: there would be no 'signal transduction'. Given that these fast dissipative mechanisms do exist, Nature was left with only one choice: the signal transduction pathway must be much faster than dissipation in order to compete. The first steps in Vision must be ultrafast, otherwise it would be impossible. We can see that a purely static 'structure-function' perspective would not lead to any understanding of the ultrafast functioning of the rhodopsin protein. In this example, and in many others, dynamics is central to the function. I suggest that

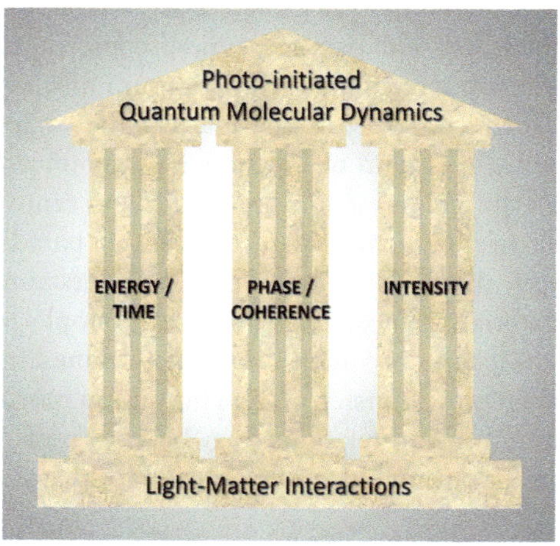

Fig. 1 The three pillars of photo-initiated quantum molecular dynamics. The light-matter interactions which underlie much of this subject can be broadly categorized into these three themes.

in the 21st century we will need to develop 'dynamics–function' relationships to augment and amplify our 'structure–function' view of Nature.

In Faraday 163, we are concerned with light-matter interactions. In the following, I will make use of the narrative device illustrated in Fig. 1. I will try to broadly (but perhaps somewhat incorrectly) classify light-matter interactions into three categories or 'pillars': (1) energy/time; (2) phase/coherence; (3) intensity. Modern optics allows for the creation of coherent pulses of light which are only a few cycles in duration. This allows for probing of the fastest events in chemistry. The other side of this coin is the implicit control over optical phase that comes with short pulses. The transfer of this phase control to molecular systems underlies all control scenarios. Finally, modern optics allows for the creation of very precise but high intensity laser pulses which allow the application of strong electric fields, opening up new regimes of light-matter interaction. Hopefully, the device suggested by Fig. 1 will allow us to consider these broad themes which underlie the subject of Faraday Discussions 163.

2 Energy/time

The first pillar of 'photo-initiated quantum molecular dynamics' is that of energy and/or time. Our studies of molecular dynamics benefit enormously from understanding light-matter interactions. Spectroscopy is the study of such interactions from a frequency domain point of view. This subject is so well established and so broadly studied that it needs no further discussion here. Conveniently, modern optical techniques allow for the creation of optical pulses which are on the time scales of dynamical processes within molecules. This has led to a time domain approach to spectroscopy and dynamics. In the following, I will try to connect time and frequency domain forms of spectroscopy, showing their correspondence. Pump–probe spectroscopy involves by definition the use of at least two photons. As such, it is a double resonance technique and, therefore, can be compared and contrasted with frequency domain double resonance spectroscopies.

Time domain techniques make use of spectrally broad, coherent laser pulses, as illustrated in Fig. 2. A laser pulse is described by the set of frequencies which comprise it. For example, two colours ω_j and ω_k within the pulse may have amplitudes A_j and A_k and a fixed phase difference $\phi_k - \phi_j$. A spectrometer measures only the intensities $|A_j|^2$ and $|A_k|^2$, therefore revealing nothing of these

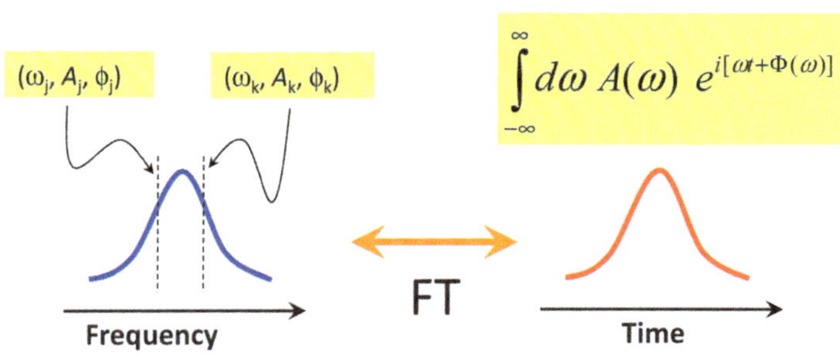

Fig. 2 Ultrashort laser pulses are understood in terms of the Fourier components which comprise them. The amplitudes $A(\omega)$ and phases $\Phi(\omega)$ determine the temporal shape of the pulse. Active control over these functions is now nearly routine.

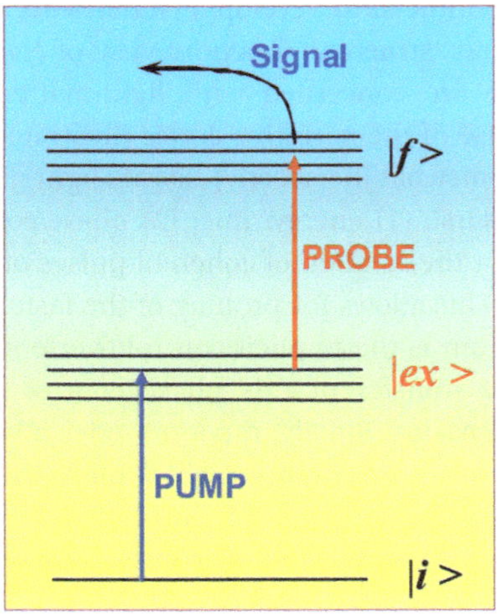

Fig. 3 Time-resolved experiments involve preparing a non-stationary state (a wavepacket) with an ultrashort pump pulse. Its field-free dynamics is probed by a time-delayed probe pulse which projects it onto a suitable final state. The signal is proportional to the population in the final state at the end of the pump–probe sequence. The choice of final state is critically important.

relative phases. The Fourier synthesis of all frequencies ω having amplitudes $A(\omega)$ and phases $\Phi(\omega)$ determines the temporal pulse shape. For example, when $\Phi(\omega)$ is a constant, the pulse assumes its shortest duration, the so-called transform-limit pulse. The transfer of the coherent properties of light to material systems, *via* dipole interactions, is what underlies all pump–probe and quantum control experiments.

From a time domain perspective, illustrated in Fig. 3, a short pump pulse prepares a non-stationary excited state |*ex*>, generally called a wavepacket, which evolves as a function of time. The wavepacket dynamics are projected onto a final state |*f*> as a function of time *via* a suitable probe pulse. The measured signal is proportional to the population in the final state at the end of the two-pulse sequence. The utility of time domain approaches to spectroscopy and dynamics is that the superposition of exact molecular eigenstates—which in general are extremely complicated—constructs, for a short period of time, a zeroth order state

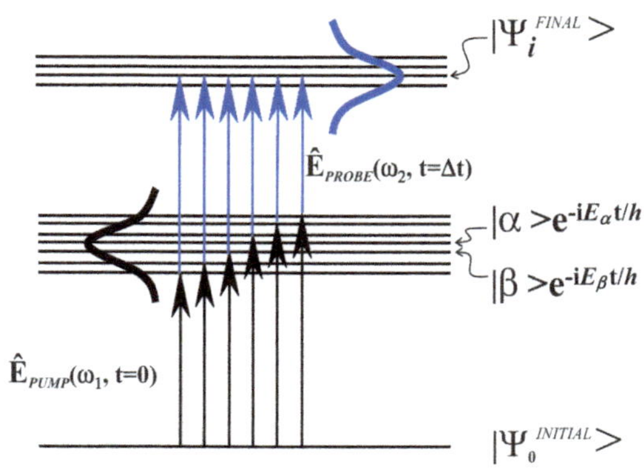

Fig. 4 A frequency domain picture of the preparation and detection of wavepacket dynamics.

which is a solution to a simpler Hamiltonian. One example would be the coherent superposition of exact non-Born-Oppeneheimer molecular eigenstates which, for a short period of time, can behave like a Born-Oppenehimer (BO) state. The latter will, of course, evolve due to the non-adiabatic coupling terms which were neglected in the BO approximation. This leads to the coupled evolution of electronic and vibrational character within the excited molecule, generally termed non-adiabatic processes. It is the time evolution of these zeroth order states which yields information about the coupling terms of interest. Importantly, this perspective applies even at 'chemical' levels of excitation, when bonds are being broken/formed and true continuum channels are open, and no high resolution spectrum exists.

In Fig. 4, we present a frequency domain perspective on such wavepacket measurements. The wavepacket is prepared by an ultrashort pump pulse (black) at time $t = 0$. The E_n are the excited state eigenenergies. (N.B. This discussion of wavepacket dynamics can be readily extended to include continuum states). The wavepacket $\chi(\Delta t)$, given by eqn (1), subsequently evolves in a field-free manner, governed by relative energy phase factors in the superposition.

$$|\chi(\Delta t)\rangle = \sum_n \tilde{a}_n |\Psi_n > \mathrm{e}^{-iE_n\Delta t/\hbar} \qquad \text{where} \qquad |\Psi_n\rangle \in |\alpha\rangle, |\beta\rangle.... \tag{1}$$

The a_n complex coefficients determine both the amplitudes and initial phases of the exact (non-BO) molecular eigenstates $|\Psi_n>$ which were prepared by the pump laser $\mathbf{E}_{\mathrm{PUMP}}(\omega_1, t = 0)$. The probe laser $\mathbf{E}_{\mathrm{PROBE}}(\omega_2, t = \Delta t)$, shown in blue, projects the wavepacket onto a specific final state $|\Psi_i>$ at time delay Δt. The time dependence of the differential signal, $S_i(\Delta t)$, the projection onto a *single* final state $|\Psi_i>$, is:

$$S_i(\Delta t) = \left|\left\langle \Psi_i \right| \vec{\mu}\,(\vec{r}) \cdot \vec{E}_{\mathrm{PROBE}} \left| \chi(\Delta t) \right\rangle\right|^2 = \left| \sum_n \tilde{b}_n \, \mathrm{e}^{-iE_n\Delta t/\hbar} \right|^2 \tag{2}$$

$$\textit{where } \tilde{b}_n = \tilde{a}_n \langle \Psi_i | \vec{\mu}(\vec{r}) \cdot \vec{E}_{\mathrm{PROBE}}(\omega_2, \Delta t) | \Psi_n \rangle$$

$$S_i(\Delta t) = \sum_n \sum_{m \leq n} \left|\tilde{b}_n\right| \left|\tilde{b}_m\right| \cos\{(E_n - E_m)\,\Delta t\,/\,\hbar + \Phi_{nm}\}$$

The complex coefficients b_n contain both the a_n from eqn (1) and the probe transition dipole moments to the final state $|\Psi_i>$. The differential signal $S_i(\Delta t)$ is the coherent sum over all two-photon transition amplitudes consistent with the pump and probe laser bandwidths and contains interferences between all degenerate two-photon transitions. Therefore, the signal as a function of Δt contains modulations at frequencies $(E_n - E_m)$, the set of excited state level spacings. The wavepacket signal is, in essence, a multi-level quantum beat. The Fourier power spectrum of this signal yields frequencies, related to the set of level spacings in the problem, and their amplitudes (*i.e.* modulation depths), related to the respective overlaps of the wavepacket states with the specific, chosen final state $|\Psi_i\rangle$. Importantly, different final states $|\Psi_i\rangle$ will generally have differing overlaps with the wavepacket, leading to differing amplitudes in the power spectrum. Choosing an appropriate final state is therefore critical. Finally, in order to avoid losing information due to averaging over final states, differential detection techniques such as dispersed fluorescence or photoelectron spectroscopy may be important.

Excited state dynamics in polyatomic systems generally do not obey the BO approximation. This is because excited state potential energy surfaces of polyatomics intersect, leading to ultrafast radiationless processes such as internal conversion. These electronic degeneracies are termed conical intersections, 3N-8 dimensional hypersurfaces of intersection between potential energy surfaces.[21] The remaining two dimensions locally lift the electronic degeneracy in linear fashion and therefore have the topography of a double cone, hence the name. Conical intersections play the same role in excited state dynamics as do transition states in ground state dynamics. As such, we might be able to gain some insights into vibrational dynamics at conical intersections by making analogies to vibrational motions through transition states. J. C. Polanyi developed simple rules to help guide our thinking about vibrational motions at transition states for A + BC collisions.[22] In particular, the location of the barrier (*i.e.* 'early', 'middle' or 'late') with respect to the col (saddle point) of the potential significantly governs which types of vibrational motions lead to chemical reaction. The effects of the particle masses are also important, determining the skew angle of the potential in a mass-weighted coordinate system. Will there be analogous 'Polanyi rules' for dynamics at conical intersections? The higher dimensionality of the problem renders this non-obvious. An experimentalist's initial approach to this problem might be *via* phenomenology. Much was learned about 'structure-function' relationships through the use of systematic chemical substitution. Can systematic variation of molecular properties reveal dynamical information about vibrational motions at conical intersections?

One of the most innocuous of chemical substitutions is the replacement of a hydrogen atom by a methyl group. By systematically moving a methyl

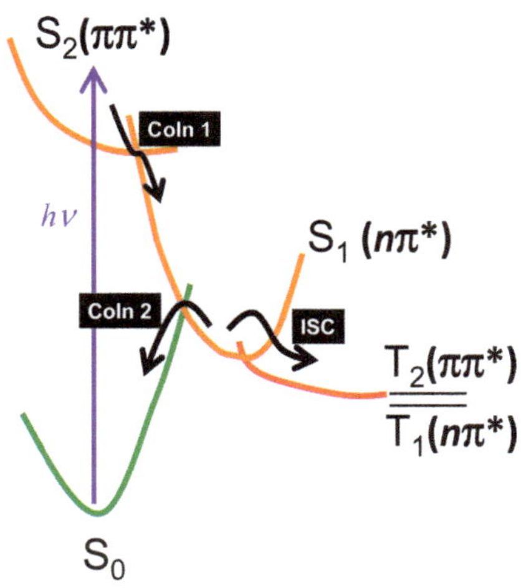

Fig. 5 Excited state photophysics of the unsaturated aldehyde acrolien (propenal). Ultrafast non-adiabatic dynamics involve conical intersections between S_2 and S_1 (CoIn 1) and between S_1 and S_0 (CoIn 2).

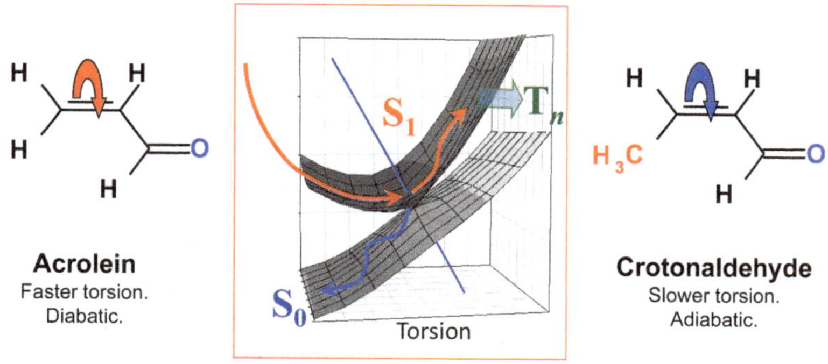

Fig. 6 The competition between intersystem crossing (ISC) and internal conversion to the S_0 ground state depends on vibrational dynamics at the S_1–S_0 conical intersection. The speed of twisting (torsion) of the terminal methylene group is related to diabatic *versus* adiabatic passage through the conical intersection.

substituent to different locations on the molecular frame, one can alter the relative inertia of certain vibrations with respect to others, while keeping the overall density of states roughly constant. We have been trying to use this approach to discern vibrational dynamics at conical intersections in unsaturated hydrocarbons.[23–26] An example is the Norrish Type 1 photoreaction of the unsaturated aldehydes and ketones, the simplest of which is acrolein (propenal). The excited state photo-initiated dynamics of acrolein is depicted in Fig. 5. The UV (200nm) absorbing state is the $S_2(\pi\pi^*)$ state, analogous that in ethylene. The 'bright' S_2 state has a conical intersection with the lower lying but 'dark' $S_1(n\pi^*)$ state. The excited state lifetime of the S_2 state is extremely short and, as in ethylene, is determined by passage through the S_2–S_1 conical intersection *via* vibrational motions involving large amplitude torsion of the terminal methylene group. Motion on the steep $S_1(n\pi^*)$ potential leads to a second conical intersection, in this case between the S_1 state and S_0 ground state. Diabatic *versus* adiabatic passage though this second conical intersection governs the fate of the excited molecule. Diabatic passage preserves the $S_1(n\pi^*)$ electronic character, leading to triplet formation and subsequent α-cleavage photochemistry (HCO and C_2H_3 radical products), typical of aldehydes and ketones. By contrast, adiabatic passage leads to formation of the hot ground state molecule and subsequent singlet state photochemistry and molecular products (*e.g.* CO, C_2H_4, H_2). We have found that the location of a single methyl substituent greatly affects the excited state lifetime and singlet-triple branching.[23] Locating the methyl group on the central carbon atom (*i.e.* methacrolein) or carbonyl group carbon atom (*i.e.* methyl vinyl ketone) has only a small effect. Surprisingly, locating the methyl group on the ethylenic end (*i.e.* crotonaldehyde) simultaneously reduces the excited state lifetime and, amazingly, completely turns off the triplet photochemistry channel. These results cannot be rationalized in terms of slight energetic shifts due to methylation, changes in conical intersection topography or density of states arguments. An explanation for this remarkable result is depicted in Fig. 6. The conical intersection between S_1 and S_0 is achieved *via* large amplitude torsional motions of the terminal methylene group, originating at S_2–S_1 conical intersection. When this terminal methylene group contains only hydrogen atoms (acrolein), the 'torsional velocity' is high. This rapid motion leads to a more diabatic passage through the S_1–S_0 conical intersection. Remaining on the $S_1(n\pi^*)$ potential, spin-orbit coupling leads to triplet state formation and subsequent α-cleavage photochemistry (typical of aldehydes and ketones). In contrast, replacing an ethylenic hydrogen atom with a methyl group adds significant inertia (mass 15 *vs.* mass 1) to this torsional motion. This slows down the 'torsional velocity' and leads to a more adiabatic passage through the S_1–S_0 conical intersection. This simultaneously reduces the excited state lifetime and turns off the triplet photochemistry channel. These types of studies give some first glimpses of the important role of vibrational motions at conical intersections. From a control point of view, it is remarkable that the location of a methyl group has such a dramatic effect on the excited state dynamics and photochemical branching ratios, a consequence not predictable from a 'structure–function' perspective alone.

As reflected by the topics covered in Faraday 163, an excited state photodynamical process of current interest is photodamage to DNA.[3,7,27] The photostability (*i.e.* protection against damage) of biomolecules such as DNA is generally determined by a competition between ultrafast excited state electronic relaxation processes. In DNA, UV excitation can lead to singlet excited state photochemistry and potential mutagenic damage. Although intracellular DNA repair mechanisms exist, these are slow and energetically inefficient. Therefore, molecular features that led to any intrinsic photostability will have had an evolutionary advantage. What determines this intrinsic photostability? Generally, excited states of molecules are much more reactive than are ground states. Therefore, electronically excited states are 'dangerous' for biomolecules because their rapid photochemistry can alter their structure and ability to function. Therefore, ultrafast excited state mechanisms which can 'quench' or 'relax' the electronically excited state, thereby competing effectively with excited state chemistry, are important photostabilization mechanisms. Generally, such 'electronic relaxation' processes are

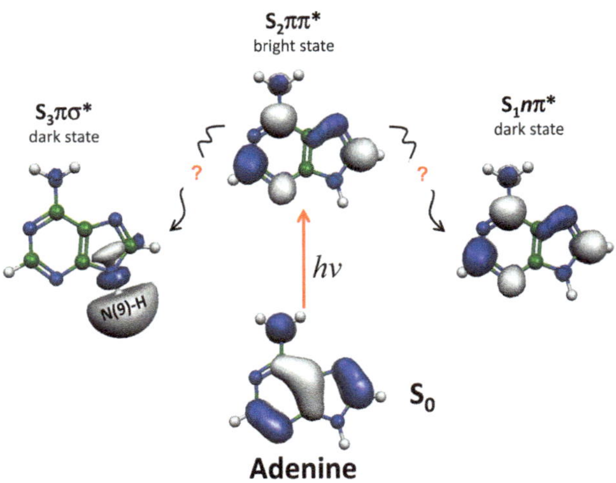

Fig. 7 The UV $\pi\pi^*$ state of the DNA nucleobase adenine is protected against photodamage by ultrafast internal conversion processes involving several excited states. Internal conversion to the ground state converts 'dangerous' electronic excitations to 'less dangerous' vibrational excitations which may then be rapidly cooled in aqueous environments.

achieved *via* ultrafast internal conversion to the ground state, typically involving conical intersections. This process converts 'dangerous' electronic energy into 'less dangerous' vibrational energy, since the latter will be cooled efficiently in aqueous solution. The UV chromophores in DNA are the aromatic nucleobases themselves and it is their photophysics which determines the outcome of UV absorption.

One nucleobase which has attracted considerable experimental and theoretical attention is adenine. Some of the relevant excited states of isolated adenine are shown in Fig. 7. The UV absorbing state is the $S_2(\pi\pi^*)$ state. The S_2 state has conical intersections with two lower lying excited states, the $S_1(n\pi^*)$ state and the $S_3(\pi\sigma^*)$ state, leading to a very rapid quenching of the S_2 state electronic excitation. Although details of the vibrational motions which govern branching between these competing pathways have yet to be finalized, it seems clear that very rapid non-adiabatic processes can lead to stabilization against excited state photochemistry. Once again, these important dynamical effects are not predictable from a 'structure–function' perspective.

In general, scattering processes are vectorial in nature. Whether they involve photon or particle interactions, scattering or absorption depends on how the molecule is oriented with respect to the direction of approach. For example, transition dipole moments generally differ along different axes (*i.e.* $\mu_x \neq \mu_y \neq \mu_z$), as does chemical reactivity (*i.e.* steric effects). Therefore, the most detailed views of molecular dynamics will obtain from the molecular frame (MF) of reference, avoiding the loss of information due to lab frame (LF) orientational averaging. For time-resolved dynamics in isolated gas phase molecules, there are two general approaches to such MF studies: (i) determine the MF orientation *post facto* on an event-by-event basis using coincidence techniques;[28] (ii) pre-align the MF to the LF using laser-induced alignment techniques.[29] An example of the first which is based on photoionization detection is the Coincidence Imaging Spectroscopy (CIS) technique, illustrated in Fig. 8. This approach is based on dissociation events which 'label' the MF direction in space. For example, a given

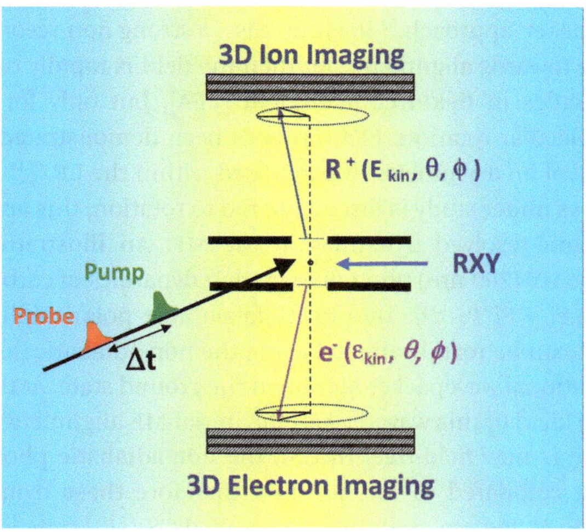

Fig. 8 Coincidence imaging spectroscopy (CIS) measures kinematically complete 3D recoil momentum vectors of both ions and electrons, in coincidence and as a function of time.

photo-initiated reaction R–XY + $h\nu \rightarrow$ R + XY is vectorial in nature, the photofragment angular distribution being referred to the MF. Any LF orientational averaging will blur this angular information, since the parent molecules were randomly oriented before excitation. However, in the so-called axial recoil limit where the emission direction of the R photofragment is along the LF direction of the original R-XY bond, measurement of the 3D fragment recoil vector yields information on the orientation of the MF within the LF. Using time-resolved photoelectron spectroscopy (TRPES), the time-evolution of vibrational and electronic dynamics may be observed.[30] In particular, measuring the time evolution of the 3D photoelectron angular distribution (PAD) in coincidence with the 3D photofragment recoil distribution allows the PAD to be transformed from the LF to the MF on an event-by-event basis. The time-resolved MF-PAD yields detailed information on the rearrangement of electronic structure (bonding) within the molecule during a photo-initiated dynamical process. As one example, the CIS approach to photodissociation dynamics studies was applied to the non-adiabatic photodissociation of the nitric oxide dimer:[28] $(NO)_2 + h\nu \rightarrow NO(A, 3s) + NO(X)$.

The second MF approach to time-resolved studies of dynamics in isolated gas phase molecules is based on laser pre-alignment of the MF within the LF. Importantly, strong (but non-ionizing) non-resonant laser fields can induce degrees of alignment ($\cos^n$) much greater than the $\cos^2$ alignment typically obtainable *via* single photon absorption. A polarizable molecule in an electric field will have a dipole moment induced by the field. A dipole in a field will feel a torque which induces the dipole to align with the field. This is the origin of the non-resonant laser-induced alignment of molecules. This effect can be used to fix the MF alignment within the LF, thereby minimizing the loss of information due to orientational averaging. While the strong laser field remains on, the induced dipole will feel a pendular potential, maintaining the axis alignment of the MF along the laser polarization direction. However, for the purposes of studying native dynamics in molecules, the presence of a strong laser field will strongly perturb such dynamics. Ideally, the molecule should be aligned but field-free during the dynamical evolution. The field-free alignment of molecules can be achieved *via* two means: the non-adiabatic or 'kick' approach,[31] or the adiabatic 'switched wavepacket' approach.[32] In either case, a strong non-resonant laser field induces a torque towards alignment, but then the field is rapidly turned off. This leaves the molecules in field-free alignment (FFA), but only for a time short compared to molecular rotation. FFA has even been demonstrated in 3D, where all three MF axes of an asymmetric top are fixed within the LF.[33,34] As long as the dynamical process under study is fast compared to rotation, this approach can be used to study time-resolved dynamics in the MF. An illustrative example is the non-adiabatic UV (200 nm) photodissociation dynamics of carbon disulfide:[29] $CS_2 + h\nu \rightarrow CS(X) + S(^3P, {}^1D)$. Due to its favourable polarizability anisotropy, ground state CS_2 can be readily aligned using the non-adiabatic (kick) approach. This induces rotational wavepacket motion in the ground state. At the revival time (~75 ps after the kick) of this wavepacket, the initial MF alignment is reproduced but the molecule is now field-free. In CS_2, the non-adiabatic photodissociation dynamics is fast compared to rotation and therefore these dynamics may be studied in the MF. At 200 nm excitation, the excited $CS_2(^1B_2)$ state is quasi-bound, undergoing compound stretch-bend vibrational motions perpendicular to the dissociation coordinate, leading to vibrational structure in the UV absorption

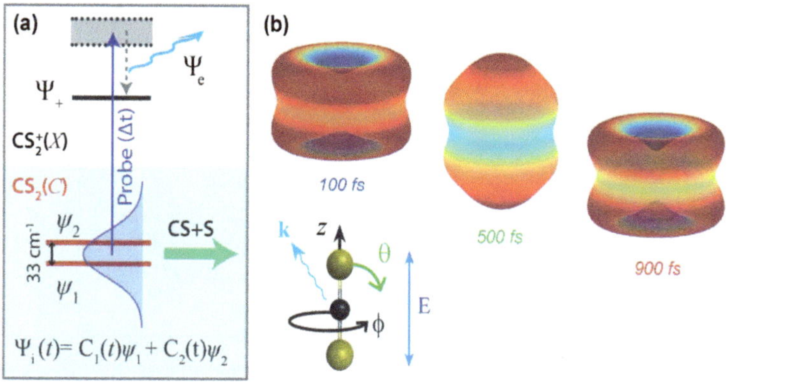

Fig. 9 Probing excited state wavepacket dynamics in the CS_2 (^1B_u) state from the molecule's point of view. A strong, polarized laser field 'kicks' the molecular frame (MF) into alignment. Time-resolved MF photoelectron angular distributions (MFPADs), right, recorded at the time of alignment, are able to directly probe the purely electronic dynamics of the valence electrons.

spectrum. As illustrated in Fig. 9(a), an ultrashort pulse can be used to prepare a coherent superposition of these compound stretch-bend predissocation resonances (here spaced by 33 cm^{-1}). The superposition of compound stretch-bend resonances leads to a quasi-bound vibrational motions which oscillate from stretch to bend during dissociation. Using photoionization as a probe, the MFPAD may be measured as a function of time, as shown in Fig. 9(b). In the $CS_2(^1B_2)$ state, bending mixes Π_g electronic character into the initial Σ_u electronic wavefunction, due to Renner–Teller interactions in the next-higher-lying state. This means that vibrational and electronic dynamics are coupled during dissociation. This leads to the observed changes in the MFPAD seen at 500 fs, compared to 100 and 900 fs. Importantly, the MFPAD is able to project out the purely valence electronic dynamics[35] even though these are strongly coupled to the nuclear motions.

The most familiar forms of light-matter interactions used in the study of photo-initiated quantum molecular dynamics are those of spectroscopy and photodissociation dynamics. These interactions are typically perturbative in terms of the strength of light-matter coupling. This represents the majority of the cases presented here in Faraday 163. In the next sections, we will also consider what happens as laser fields become stronger and the conventional perturbative picture of spectroscopy fails.

3 Phase/coherence

The second pillar of 'photo-initiated quantum molecular dynamics' is that of phase and/or coherence. As indicated by Fig. 2, laser pulses by definition have a certain degree of phase coherence between the frequencies which make up the pulse. The ability to control these phases and to transfer them to molecular systems *via* dipole interactions is the basis of the field of quantum control. Since electric forces underpin the dynamics of all chemical processes, the use of externally applied electric fields is a natural approach. The catalysis of reactions by metals is an example of externally applied electric fields (due to the surface) being used to suppress the activation barrier along the reaction co-ordinate.

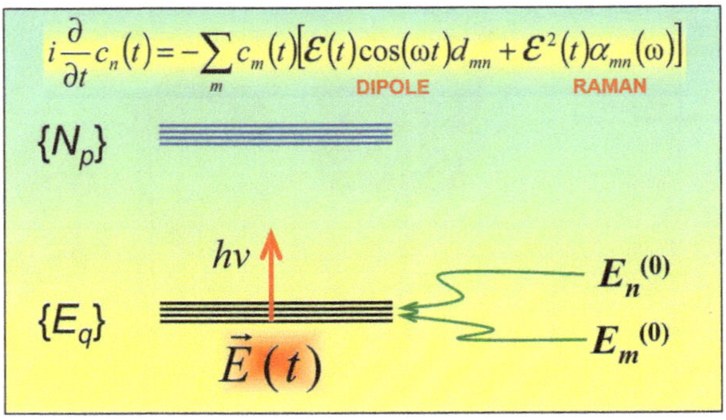

Fig. 10 Quantum dynamics of the non-resonant dynamic Stark effect (NRDSE) in a strong laser field. The time evolution of the superposition in the field is driven by two source terms, one dipole-type and one Raman-type. For details, see the text.

Amplified lasers allow for the application of very precise electric fields, even at strengths which exceed the fields that bind matter. In the perturbative limit, quantum control schemes are based on the interference between degenerate pathways. For examples, interference between coherent 1- and 3-photon transitions, controlling total cross sections, or between coherent 1- and 2-photon transitions, controlling differential cross sections (*i.e.* angular distributions), have been implemented. The interference between coherent 2-photon transitions is even more common and, as can be seen from Fig. 4, is the basis of ultrafast pump–probe experiments. Commonly, control over the amplitudes $A(\omega)$ and phases $\phi(\omega)$ of a laser field, indicated within Fig. 2, is achieved *via* Fourier domain spatial light modulators (*i.e.* pulse shapers). These can be coupled to computer-controlled feedback mechanisms which vary $A(\omega)$ and $\phi(\omega)$ so as to optimize a specific outcome (*e.g.* product branching ratio). As long as perturbation theory applies, this type of quantum control can be understood in terms of interfering transitions between field-free states of the molecule. This picture will hold until perturbation theory fails—*i.e.* when the laser electric field significantly perturbs the eigenstates and the optical transitions can no longer be understood in terms of field-free transitions. One of the first non-perturbative interactions to emerge is the dynamic Stark effect,[36] discussed in more detail below.

In Fig. 10, we consider a molecular Hamiltonian which has the typical band structure of well-separated electronic states, with each electronic state having the smaller level spacing structures of vibrational states. The set of ground electronic states are labelled $\{E_q\}$ whereas the excited states are labelled $\{N_p\}$. Here we are interested in controlling dynamics within the set of electronic ground eigenstates $E_n^{(0)}$ *via* the application of a strong but non-resonant electric field, indicated by the red arrow. By non-resonant, we mean that the laser frequency is large compared to the small (vibrational) level spacings but small compared to the large (electronic) level spacings, as shown in the figure. We now wish to consider the quantum dynamics of these levels in a strong laser field.

In the optical domain ($\lambda \sim$ 100–1000 nm) where the wavelength of light is much larger than the molecule (~1 nm) and, therefore, its spatial variation across the molecule can be neglected, the fundamental interaction between light and matter is *via* dipole coupling:

$$V(t) = -\vec{\mu}\cdot\vec{E}(t) \tag{3}$$

where

$$E(t) = 1/2\ \varepsilon\,(t)\,\mathrm{e}^{i\omega t} + c.c.$$

Here $\varepsilon(t)$ is the laser pulse envelope (*e.g.* a Gaussian) and ω is the frequency of light. In order to develop the quantum dynamics of the ground state in a strong laser field, we expand the states in the field free basis but with time dependent coefficients:

$$|\Psi(t)\rangle = \sum_n c_n(t)\mathrm{e}^{-i\omega_n t}|\,n\rangle \tag{4}$$

The problem is now to determine the time varying coefficients $c_n(t)$. To do this, we plug the ansatz of eqn (4) into the Schrödinger equation to get the coupled equations of motion for the state coefficients:

$$ih\frac{\partial c_n}{\partial t} = \sum_m \mathrm{e}^{-i\omega_{mn}t}V_{mn}(t)c_m(t) \tag{5}$$

This is the formal solution to the problem and involves a sum over all states: all dipole-coupled states m, whether near or far away, affect the time evolution of the state n. Anything else is an approximation. For the case of strong but non-ionizing non-resonant laser fields of interest here, we can treat the near and far away states differently. We will keep the full quantum dynamics of the near $\{E_q\}$ states but approximate the effect of the far away $\{N_p\}$ states on the nearby states by a derived effective interaction, namely the polarizability. In this case, the equations of motion of the nearby state coefficients take the form:

$$i\frac{\partial}{\partial t}c_n(t) = -\sum_m c_m(t)\left[\varepsilon(t)\cos(\omega t)d_{mn} + \varepsilon^2(t)\,\alpha_{mn}(\omega)\right] \tag{6}$$

It can be seen that there are two source terms driving the state coefficients $c_n(t)$. These represent the two limiting cases for the non-resonant dynamic Stark effect (NRDSE). If the neighbouring states $E_m^{(0)}$ and $E_n^{(0)}$ have dipole allowed transitions between them ($d_{mn} \neq 0$), then the dominant term in eqn (6) is the first term, the dipole coupling. In this case, the superposition in the field will oscillates at the frequency of light due to the cos(ωt) factor. In essence, the state $E_m^{(0)}$ makes a transition to neighbouring state $E_n^{(0)}$ with an off-resonant photon, the large detuning being made up for by the field strength. It is the Rabi cycling of these types of transitions which builds the superposition in the field. In contrast, if the neighbouring states $E_m^{(0)}$ and $E_n^{(0)}$ do not have dipole allowed transitions between them ($d_{mn} = 0$), then the dominant term becomes the second one, the polarizability coupling. In this case, the state $E_m^{(0)}$ tries to makes a dipole transition to the far away $\{N_p\}$ states but cannot do so since they are too far away. Instead, a stimulated Raman transition back down to the $E_n^{(0)}$ state occurs and it is the Rabi cycling of this process

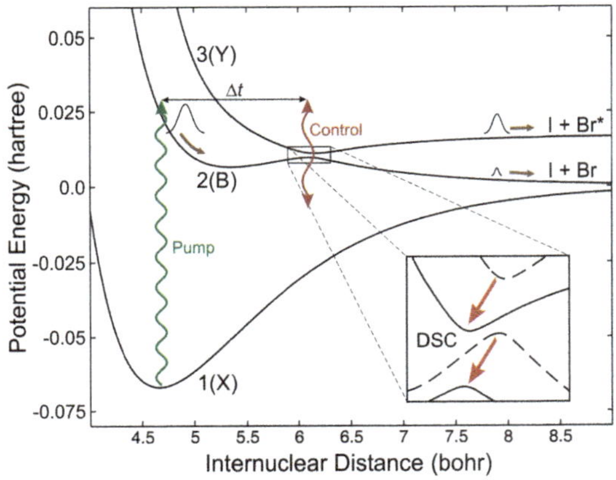

Fig. 11 Dynamic Stark Control (DSC) of the IBr → I + Br/Br* non-adiabatic photodissociation. A strong non-resonant IR field modifies the curve crossing at the time of wavepacket passage, thereby controlling the branching ratio.

which builds the superposition in the field. Since this 'up and down process' represents a difference frequency within the laser pulse, the carrier frequency ω of the light falls out and only the laser pulse envelope $\varepsilon(t)$ remains important. Thus, the superposition in the field follows only the slower pulse envelope and does not oscillate at the frequency of light itself. This polarizability-dominated NRDSE regime is what underlies the laser-induced alignment of molecules (in which case the states $E_m^{(0)}$ are rotational states). The polarizability NRDSE can also be used to control photo-initiated quantum dynamics (*e.g.* photodissociation) in molecular excited states,[13] and discussed in more detail below.

In Fig. 11, we consider the ultrafast non-adiabatic photodissociation of the diatomic molecule iodine bromide:[37] IBr + $h\nu$ → I($^2P_{3/2}$) + Br($^2P_{3/2}$)/Br*($^2P_{1/2}$). The optically bright, bound B-state correlates to I + Br* excited state products. It is crossed by the dark, dissociative Y-state which correlates with I + Br ground state products. The coupling between the B and Y states is in the intermediate strength coupling regime and, hence, neither the diabatic nor adiabatic potential curves apply. This is seen in the field-free Br/Br* branching ratio, which is around 50% at visible wavelengths. Since the B and Y states correlate to products of differing spin multiplicity, the dipole coupling between the B and Y states is small. Therefore, in a strong laser field, the dominant NRDSE is of the polarizability type. Imagine that we excite the B-state with a short visible pump pulse (green), creating a wave-packet. If left to evolve field-free, the expected Br/Br* branching ratio would obtain. However, if we applied a strong, non-resonant field (red) at time delay Δt, as the wavepacket crossed the region of non-adiabatic coupling, we might be able to modify the hopping probability. What type of control should we expect? We can understand the basic effect by considering the Stark effect on the diabatic states. Since the B-state is bound, it has continuous electron density between the atoms. Therefore, we might expect it to have a significant polarizability. In contrast, the dissociative Y-state has a node in the electronic wavefunction between the two atoms and, therefore, it should have a lower polarizability. Therefore, in a strong non-resonant field, the B and Y states should have different Stark shifts, leading to a field-induced change in the crossing point (inset) between the two states. This

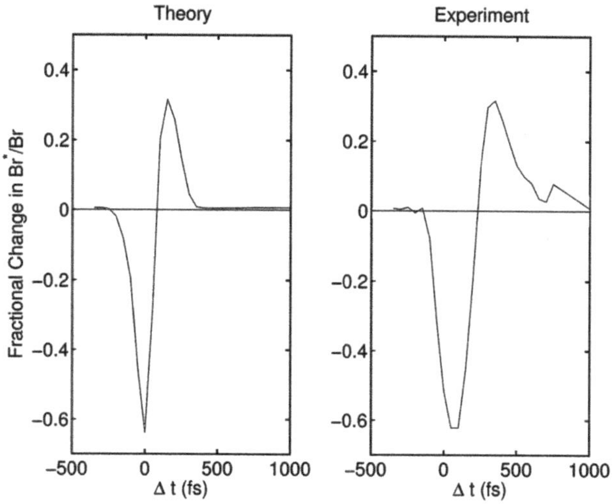

Fig. 12 Using the dynamic Stark control scheme of Fig. 11, the Br/Br* branching ratio may be controlled by a strong non-resonant laser field. Importantly, there is no net absorption of photons from the strong control field.

would lead to a change in the surface hopping probability and, therefore, Dynamic Stark Control (DSC) over the Br/Br* branching ratio. This scenario was experimentally demonstrated[38] and, as shown in Fig. 12, large fractional changes in the Br/Br* branching ratio were obtained, depending on the timing between the creation of the wavepacket and the application of the NRDSE control pulse. It is important to realize that, in NRDSE control, there is no net consumption of photons. This means that the dynamics remain on the coupled potential energy surfaces to be controlled, rather than making electronic transitions to other states. In this sense, the NRDSE control pulse acts like a catalyst rather than as a reagent. It can also be seen that the field-free Br/Br* branching ratio obtains when the NRDSE pulse is applied too early or too late. This demonstrates that the NRDSE reversibly follows the control pulse envelope, leaving the molecule in its field-free configuration after the Stark interaction is over.

As researcher's control over photo-initiated quantum molecular dynamics continues to improve, we may hope to see the range of applications of the NRDSE expand. However, as non-resonant laser fields become only somewhat stronger, ionization becomes the dominant response. This is the subject of the following section.

4 Intensity

The third pillar of 'photo-initiated quantum molecular dynamics' is that of intensity. At laser intensities above 10^{13} W cm^{-2}, laser electric fields applied to matter start to become comparable to the fields which bind electrons to matter itself. In this regime, a new physics emerges, that of strong field ionization (SFI), tunnel ionization and strongly driven electrons.

A simple quasi-static model[39] applies to the SFI of rare gas atoms whereby, for sufficiently long wavelengths, tunnelling of the single most weakly bound electron (of binding energy I_p) occurs through an essentially static barrier formed by the superposition of the core potential Coulomb electric field with the laser electric field. This model ignores any electronic excitations inside the Coulomb potential, as well as any multi-electron effects. The subsequent evolution of the SFI liberated electron in the strong field leads to a range of important new phenomena. Most significant is that the strongly driven electron can recollide with its parent ion core, emitting a burst of XUV radiation. This process is known as high harmonic generation (HHG). Controlling the recollision electron leads to the formation of attosecond pulses[40] and the birth of a new field of research: attosecond science.

Implicit in the quasi-static model[39] is an approximation that the bound electrons can adiabatically follow the 1–2 fs time scale oscillations of the laser field. This is analogous the BO approximation wherein the bound valence electrons adiabatically follow the time dependent electric field due to the vibrational motions of the charged nuclei. In the rare gas atoms, the electronic level spacings are very large (~10 eV) and, therefore, the time scale of electronic motion (the inverse of the level spacing) is very fast compared to optical periods. The adiabatic approximation holds very well for rare gas atoms. A second important approximation in the quasi-static model of SFI is that only a single active electron (SAE) responds to the strong field. This may seem odd for the rare gas atoms such as Ne, Ar, Kr *etc.* where there are six equivalent outer shell electrons. However, in the rare gas atoms, the first excited state is typically about 2/3 of the way to the ionization potential. This means that there exist no doubly excited states below the first I_p. Therefore, only single excitations contribute effectively to the strong field polarizability: the rare gas atoms behave in SFI as if they have only a SAE. Finally, we must remember that the inert gases have a pathological electronic structure. Will these approximations hold for polyatomic molecules which generally have extended lengths and doubly excited states existing below the first I_p?

Laser driven electrons in an extended molecular potential suggests a rough analogy with the splashing of water in a periodically tipping bathtub of length L. An oscillating electric field causes the potential to periodically tip left and right at the laser frequency ω_L, forcing the electron density to oscillate within the well.

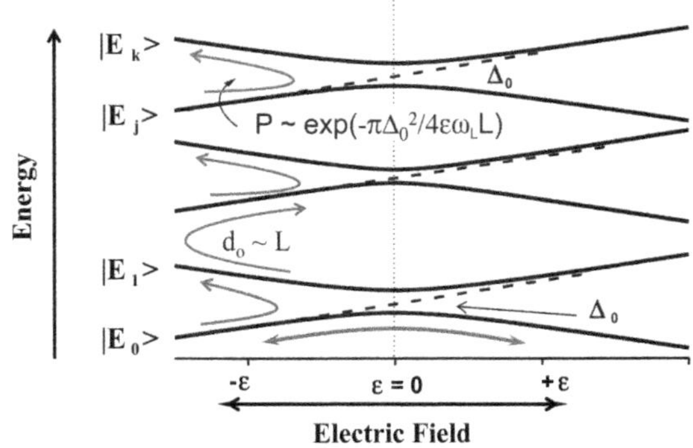

Fig. 13 In a strong ionizing laser field ε, the driven electrons within the molecular potential may adiabatically follow (solid black lines) the applied laser field. However, depending on level spacings, dipole coupling strengths and laser frequency, the system may also make non-adiabatic transitions (dashed black lines) at the zero crossings of the laser field. In such a case, the tunnelling picture of strong field ionization will not apply.

In the low frequency limit (adiabatic dynamics), the 'bathtub' tips very slowly and the 'surface of the water' remains flat at all times: this is the quasi-static picture. As the frequency of the tipping motion increases, the water no longer remains flat—splashing (non-adiabatic dynamics) occurs. As the amplitude increases, water is splashed out of the bathtub. This is non-adiabatic multielectron ionization dynamics, which we have termed NME.[41,42]

A simple model can be used to estimate when NME becomes important. Referring to Fig. 13, we consider a Hückel-type molecular potential of characteristic length L and field-free electronic level spacings Δ_0 subjected to a strong oscillating electric field $E = \varepsilon \sin\omega_L t$. We allow for a typical dipole coupling between states to have value μ. For laser frequencies $\omega_L \ll \Delta_0$, the level shifts adiabatically follow the field (solid black lines) and populations remain in their Stark shifted states at all times. As the laser frequency increases, however, the probability of Landau–Zener (LZ) type non-adiabatic transitions (dashed black lines) significantly increases. These transitions will occur when the laser field is changing most rapidly (near the zero crossings $\varepsilon = 0$), rather than when the field is strongest (near the maxima $\pm\varepsilon$). The strongest transition dipoles will have the greatest influence. In molecules, these would likely be the charge transfer transitions, where the dipole $\mu \equiv d_0 \approx L$. The probability P_{LZ} of non-resonant LZ transition during one-half of a laser cycle will be:

$$P_{LZ} \approx e^{-\pi\Delta_0^2/4\omega_L\varepsilon\mu} \tag{7}$$

When $\omega_L \varepsilon L \sim {\Delta_0}^2$, there will be strong non-resonant absorption by all delocalized electrons. This non-adiabatic electronic 'ladder climbing' ionization

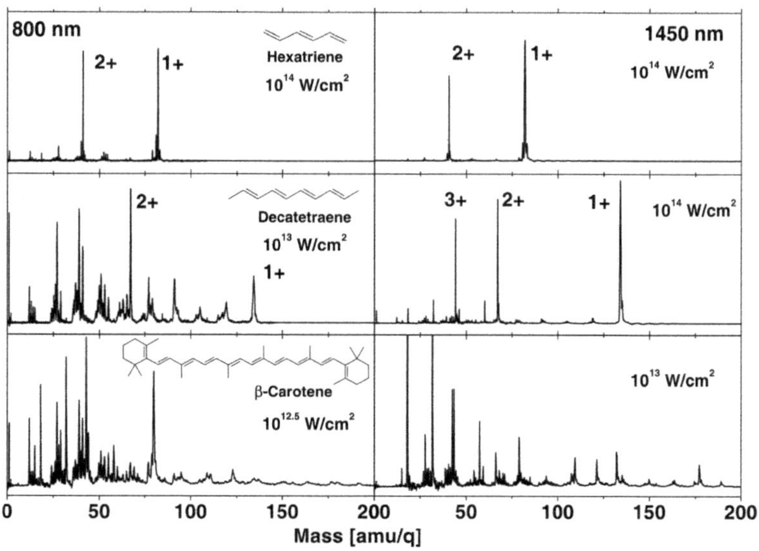

Fig. 14 Failure of the strong field adiabatic approximation. SFI mass spectra of linear, conjugated hydrocarbons of increasing length, hexatriene, decatetraene and β-carotene, at wavelengths of 800 and 1450 nm. Hexatriene shows characteristics of quasi-static tunnel ionization. The longer decatetraene has a fragmentation pattern which varies with wavelength, exhibiting a transition to complete fragmentation by 800 nm. This is a failure of the adiabatic approximation. The much longer β-carotene shows extensive fragmentation at all wavelengths studied.

mechanism (NME) represents a complete failure of the quasi-static picture. As a specific example, we consider a linear polyene-type molecule with $L = 13.5$ Å, $\Delta_0 = 4$ eV, and $\lambda = 700$ nm. In this case, $\omega_L \varepsilon L \sim {\Delta_0}^2$ is achieved at an intensity $I \sim 5 \times 10^{12}$ W cm^{-2}, a situation which occurs in many strong field experiments on molecules.

In Fig. 14 we show the SFI mass spectra of linear, fully conjugated *all trans* hydrocarbons of increasing length, hexatriene, decatetraene and β-carotene, using intense 40 fs pulses at the wavelengths of 800 and 1450 nm.[41] (We remind that the quasi-static model is frequency independent). Hexatriene (upper) shows the formation of several charge states of the parent molecule, with little fragmentation (no excited ionic states) and little variation with wavelength (frequency independence). This is characteristic of quasi-static tunnel ionization. The longer decatetraene (middle) shows a similar result at 1450 nm, forming singly, doubly and triply charged parent ions. However, its fragmentation pattern varies greatly with wavelength, exhibiting a transition to explosive fragmentation by 800 nm. This wavelength dependence indicates a complete failure of the quasi-static picture and is due to the emergence of NME. The longer β-carotene (bottom) shows extensive fragmentation at all wavelengths. These results suggest that, for polyatomic systems, the adiabatic approximation upon which the quasi-static model rests can fail dramatically, depending on both molecular and laser properties.

Even in the longer wavelength adiabatic limit, there can still be multielectron effects in SFI. The SAE picture of SFI assumes that only a single electron is driven by the strong laser field and that all other bound electrons remain unaffected by this field. This would be strictly true only if the ion core had zero polarizability. Were ion core polarizable, the remaining bound electrons would displace in the field, affecting the Coulomb potential barrier for the active electron. One can get a sense for this issue by considering a classical, perfectly conducting shell in a static external electric field. The electrons on the surface of the shell will arrange in such a way so as to counteract (screen) the applied field. Due to the exponential dependence of tunnelling rates on the properties of the barrier, dynamic multielectron screening will have significant effects on ionization rates.

In order to characterize dynamic screening effects in SFI, we studied the ionization rates of transition metal atoms[43] and clusters[44] in strong infrared (λ =

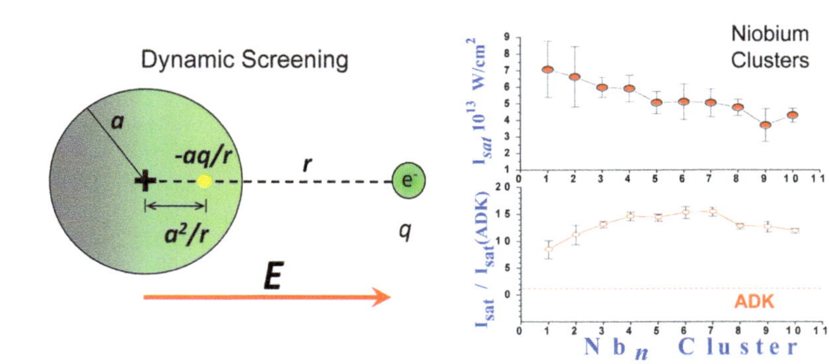

Fig. 15 Failure of the single active electron (SAE) approximation. The strong field ionization response of transition metals clusters, here of niobium, reveals a dramatic suppression of ionization as compared to the expectations of SAE theories such as ADK. This is due to the dynamic screening of the applied field by the polarisable electrons of the ionic core.

1.5 μm) laser fields. In Fig. 15, we show results for the case of Niobium clusters. The saturation intensity, I_{sat}, a measure of the SFI response, is plotted (upper panel) as a function of the Nb_n cluster size n. In order to compare with the expectations of a non-polarizable core model, we used the well established Ammosov–Delone–Krainov (ADK) theory[45] which considers only the I_p of the species in question and neglects any electronic dynamics inside the potential. ADK theory accurately describes the SFI rates of the rare gas atoms. The ratio of the experimental I_{sat} to the I_{sat}(ADK) is plotted in the lower panel. If the Niobium clusters experienced SFI as do rare gas atoms, this ratio would have a value equal to one. Instead, it can be seen that this ratio can be an order of magnitude larger. This represents a several order of magnitude suppression of SFI rates for the Niobium clusters as compared to an artificial 'rare gas atom' of the same I_p. The origin of this suppression is the dynamic polarization of the core, screening the applied laser field felt by the 'active' electron. A simple model (left) which treats the cluster as a metal sphere by including the polarizability, the image charge and laser field interactions, is largely able to capture the essential physics.[44] Although metal clusters are expected to be more polarizable than many molecules, these results show that even in the quasi-static limit, there are important multielectron effects in the SFI of polyatomic molecules.

The failures of the adiabatic and SAE approximations have important consequences for the study of polyatomic molecular dynamics using strong laser fields. In SFI, the driven multielectron response occurs on the sub-cycle (attosecond) time scale. There is considerable interest in probing coherent electron dynamics in atomic and molecular systems, using both strong and weak laser fields. This is the domain of the emerging field of attosecond science.[46]

Attosecond science is based on the strong field process of high harmonic generation (HHG). Since its origins[40] in 1994, it was understood that attosecond time scale measurements could probe the dynamics of electrons in a manner analogous to the femtosecond probing of vibrational and vibronic dynamics. Isolated attosecond pulses as short as 80 as have been created. In terms of time-resolved measurements, both strong and weak field approaches have been proposed, although to date all attosecond measurements have involved the use of at least one strong laser field. As discussed above, the physics of multi-electron polyatomic molecules in strong laser fields is complex. As such, SFI-based processes such as HHG may not be the most direct probes of native electronic dynamics in molecules. However, by analogy with femtosecond wavepacket experiments, the generation of isolated attosecond pulses will allow for attosecond pump–probe studies of electronic wavepacket dynamics. For reference, a Gaussian 100 as laser pulse has a bandwidth of ~18 eV, spanning many electronic states, and a central frequency in the XUV range. What are the consequences of this? Well, 'wavepackets are wavepackets' and much of the intuition that we have developed in understanding wavepacket dynamics using narrower bandwidth femtosecond pulses should be translatable, adjusting level spacing and time scales, to attosecond wavepacket dynamics. This idea is discussed a little further, below.

On sub-femtosecond time scales, the atoms are essentially frozen and the dynamics is purely electronic. (N.B. The fastest molecular vibration, that of H_2, has a frequency of 4155 cm^{-1}, corresponding to a period of ~8 fs). What kind of electronic wavepackets will very broad attosecond pulses prepare? For simplicity, we will restrict ourselves to bound states, the extension to electronic continua being straightforward. With a frozen core electronic Hamiltonian, the exact field-free eigenstates are stationary, many-body solutions to the Schrödinger equation—*i.e.* the exact n-body electronic wavefunction for all electrons. These wavefunctions implicitly contain all static and dynamic electron correlations. (Artificial single particle functions such as molecular orbitals are certainly not solutions to this Schrödinger equation.) We may not be able to calculate exact electronic wavefunctions, but the molecule knows them. One may be tempted to think of an electronic state as composed of electrons whizzing about the molecular core at near-relativistic velocities. However, this is not what the Schrödinger equation tells us. The eigenstates are stationary solutions: the n-body electronic density is time-independent. Field-free dynamics (time dependence) therefore obtains only *via* the coherent superposition of eigenstates—in other words, formation of a wavepacket.

Using the broad bandwidths of attosecond pulses, we could create a coherent superposition of exact n-body electronic wavefunctions. This superposition may, for a short time, appear like the solution to a simpler Hamiltonian wherein certain couplings (*e.g.* electron correlation) are neglected. For example, a specific coherent superposition of exact n-body wavefunctions could behave, for a short period of time, like a molecular orbital—a single particle excitation. Electron correlation would, of course, eventually convert this single particle excitation into many-body excitations. Analogous to the fs wavepacket dynamics case discussed above, the time-resolved study of this process would yield information about the electronic couplings which evolve charge distributions on the molecular frame.

Using the principle that "the same equations have the same solutions", I speculate that we can map the problem of attosecond electronic wavepackets in molecules onto a well known femtosecond time scale problem related to intramolecular vibrational energy redistribution (IVR). The problem I have in mind is that of local modes, an example of which is the C–H stretch in benzene, C_6H_6. Remarkably, the C–H overtone absorption spectrum of benzene looks like that of a diatomic Morse oscillator. It seems as if an H atom were vibrating against a structureless particle of mass C_6H_5. However, the linewidths of the overtones increases with C–H vibrational quantum number. This is because the 'C_6H_5 particle' is not really structureless and, as the energy in the C–H stretch increases, it couples ever more strongly into the internal modes of the benzene ring. This flow of vibrational energy from a C–H local mode to the set of ring modes is called IVR.

We can construct a wavepacket picture of these local mode dynamics, referring to the tier model diagram in Fig. 16. The exact molecular vibrational eigenstates (which are generally not the harmonic normal modes) in principle exist, even if we cannot calculate these very well. These necessarily involve the coupled motions of all C and H atoms within the molecule. As such, they are exact n-body stationary solutions to the vibrational Schrödinger equation. These exact states are shown in the middle panel. In frequency domain language, the oscillator strength for these vibrational transitions comes from the C–H stretch local oscillator mode, shown left. The vibrational spectrum, shown right, would have this oscillator strength distributed over all coupled modes. A spectrally broad, coherent laser pulse could prepare a superposition of exact n-body states such that, for a short time, they behave like a localized zeroth order state – in this case, the 'one-particle' C–H stretch. This phased superposition would initially look like a single particle excitation - an H atom vibrating against a structureless particle of mass C_6H_5.

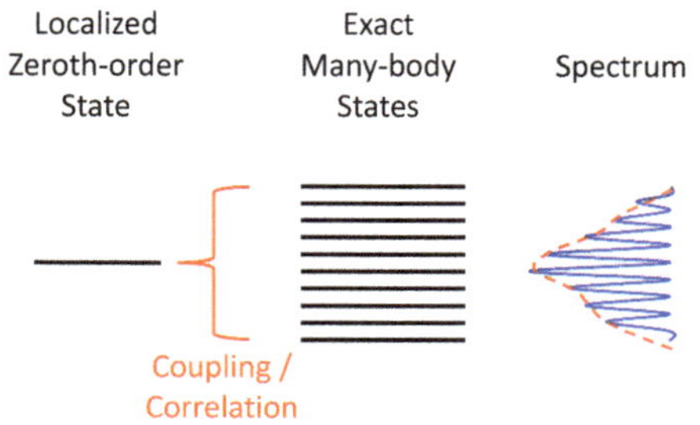

Fig. 16 A localized zeroth order state such as a local mode or molecular orbital can be created by the coherent superposition of exact many-body states. Energy or charge redistribution is understood in terms of the dephasing of this superposition. For details, see the text.

However, due to their differing energy phase factors, this superposition will evolve and the interference which initially cancelled out the ring modes will be spoiled. This dephasing of the vibrational wavepacket will lead to energy flow into the ring (a many-body excitation).

This same picture can be applied to attosecond electronic dynamics in molecules. If we make the mapping "local mode" ↔ "molecular orbital" and "exact vibrational wavefunction" ↔ "exact n-body electronic wavefunction", then a very similar discussion of wavepacket dynamics ensues, provided one translates the time scales and level spacings involved. Referring again to Fig. 16, the middle panel now corresponds to the exact n-body electronic wavefunctions. The localized zeroth order state is now the single particle molecular orbital. (The coupling would now be electron correlation rather than anharmonic vibrational couplings). Once again, a spectrally broad coherent (attosecond in this case) laser pulse could prepare a coherent superposition of exact n-body electronic states which, for a short time, could look like a single particle excitation—a molecular orbital. Again, the phased superposition will evolve and the localized electronic motion will become a more generalized many-body excitation. Energy flows from the single excited electron to all other coupled electrons in the molecule. This attosecond charge redistribution would represent the real-time observation of electron correlation in molecules.

Finally, I would like to address the issue of "attosecond chemistry". A coherent superposition of many bound electronic states of a molecule, such as illustrated in Fig. 16, would indeed exhibit attosecond time scale electronic motions and one might discern electron correlation in molecules. Chemistry, however, requires that the atoms move: each electronic state in the superposition represents a different potential energy surface for these motions and likely correlates to different product channels. Conventionally, we would call these different chemical reactions (*e.g.* ground state *vs.* excited state reactions). What would be seen in an attosecond wavepacket experiment once the atoms start to move? In general, molecular potential energy surfaces are not parallel to each other and their electronic structure will vary with nuclear coordinates in differing ways. Due to this variation of both electronic level spacings and structure with coordinates, the electronic coherences prepared on attosecond time scales will appear to irreversibly dephase once the atoms start to move.

A given chemical reaction occurs on a given potential energy surface and, by definition, involves the making and breaking of chemical bonds. This requires

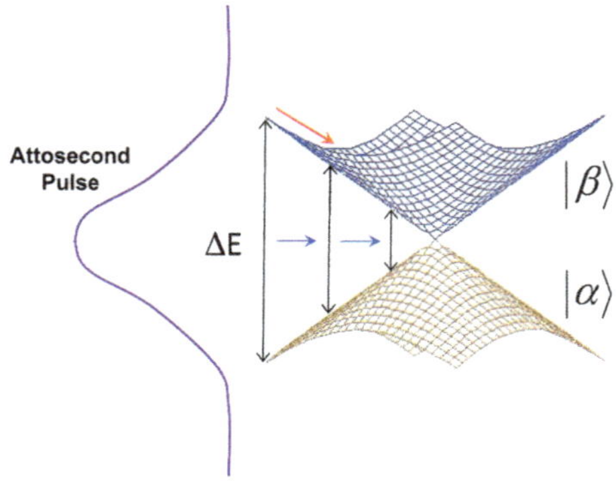

Fig. 17 Time scales of electronic dynamics during chemical reaction. One of the fastest processes in chemistry is passage through a conical intersection. An attosecond pulse creates a coherent superposition of two electronic states |α> and |β>. This oscillating electronic coherence will slow down as the nuclei move towards the conical intersection, where they finally achieve the time scales of nuclear motions.

the concerted motions of both atoms and electrons. This concerted motion may be either adiabatic or non-adiabatic, depending on whether or not the electrons can 'follow' the nuclei. In the adiabatic case, the time scales of electron motion during a chemical reaction are clearly the same as the time scale of atomic motions (femtosecond). The non-adiabatic case, passage through a conical intersection for example, is more interesting because it is one of the fastest processes in chemistry and requires a bit more analysis. For this purpose, we refer to Fig. 17. We consider two electronic states, |α> and |β>, of a polyatomic molecule which exhibit a conical intersection at some geometry. We imagine that, at the Franck–Condon geometry of the ground state (not shown), the states |α> and |β> are well separated with spacing ΔE, as shown on the left side of the figure. An attosecond pulse would prepare a coherent superposition of these two states. At short times, the atoms are stationary and a purely electronic two-level quantum beat is created wherein the charge distribution on the molecular frame oscillates between the sum and difference of the two electronic wavefunctions, at a time scale given by the inverse level spacing ΔE. Until the atoms move, nothing else can happen. Let us now let the atoms move, in this case towards the conical intersection, and ask what happens to the time scale of electron dynamics. As the two surfaces approach, the electronic energy level spacing diminishes, meaning that the electron dynamics slows down. Close to the conical intersection, the electronic level spacings become comparable to the vibrational ones and, therefore, the time scale of electron motion has slowed down to the time scale of vibrational motions. In fact, this is the reason why the BO separation fails. Therefore, at conical intersections, the dynamics of electron motions becomes comparable to the time scale of vibrational motions. The above simple arguments seem to suggest, perhaps more generally, that the time scale of electron dynamics during a chemical reaction is femtoseconds, the time scale of the motions of the atoms.

5 Conclusion

The interaction of light with matter underlies many important areas of research, ranging from fundamental physics and chemistry to medicine and engineering.

As new optical tools continue to become available, our ability to apply them will depend on our understanding of the nature and range of light-matter interactions. Using the narrative device of the 'three pillars' shown in Fig. 1, we have attempted to classify light-matter interactions as they apply to photo-initiated quantum molecular dynamics, the subject of Faraday Discussion 163.

New light sources will lead to new regimes of resolution in both the time and frequency domains. In particular, non-linear optics will become an increasingly useful tool due to the commercial availability of appropriate lasers. The consequences of this are that understanding and controlling optical phase, important only in non-linear optics, will open broad new avenues for studying molecular dynamics, life sciences and engineering. The ability to apply precise yet strong laser fields offers novel routes to controlling or modifying molecular systems. The three pillars of photo-initiated quantum molecular dynamics will continue to offer new opportunities for science and technology for the foreseeable future.

References

1 K. Prozument, R. G. Shaver, M. Ciuba, J. S. Muenter, G. B. Park, J. F. Stanton, H. Guo, B. M. Wong, D. S. Perry and R. W. Field, *Faraday Discuss.*, 2013, **163**, DOI: 10.1039/C3FD20160K.
2 E. M. M. Tan, S. Amirjalayer, B. H. Bakker and W. J. Buma, *Faraday Discuss.*, 2013, **163**, DOI: 10.1039/C2FD20139A.
3 S. B. King, M. A. Yandell and D. M. Neumark, *Faraday Discuss.*, 2013, **163**, DOI: 10.1039/C3FD20158A.
4 G. M. Roberts, C. A. Williams, H. Yu, A. S. Chatterley, J. D. Young, S. Ullrich and V. G. Stavros, *Faraday Discuss.*, 2013, **163**, DOI: 10.1039/C2FD20140B.
5 E. Riedle, M. Bradler, M. Wenninger, C. F. Sailer and I. Pugliesi, *Faraday Discuss.*, 2013, **163**, DOI: 10.1039/C3FD00010A.
6 B. C. Arruda, B. Smith, K. G. Spears and R. J. Sension, *Faraday Discuss.*, 2013, **163**, DOI: 10.1039/C3FD20149J.
7 M. C. Stuhldreier and F. Temps, *Faraday Discuss.*, 2013, **163**, DOI: 10.1039/C3FD00003F.
8 S. Meech, K. Addison, I. Heisler, J. Convard, T. Dixon and P. Page, *Faraday Discuss.*, 2013, **163**, DOI: 10.1039/C3FD00019B.
9 A. V. Bochenkova and L. H. Andersen, *Faraday Discuss.*, 2013, **163**, DOI: 10.1039/C3FD20150C.
10 R. Binder, J. Wahl, S. Römer and I. Burghardt, *Faraday Discuss.*, 2013, **163**, DOI: 10.1039/C3FD20148A.
11 F. Fassioli, D. G. Oblinsky and G. D. Scholes, *Faraday Discuss.*, 2013, **163**, DOI: 10.1039/C3FD00009E.
12 D. Sharma, A. Huijser, J. Savolainen, G. Steen and J. Herek, *Faraday Discuss.*, 2013, **163**, DOI: 10.1039/C3FD20156B.
13 M. E. Corrales, G. Balerdi, V. Loriot, R. de Nalda and L. Bañares, *Faraday Discuss.*, 2013, **163**, DOI: 10.1039/C2FD20143G.
14 L. A. Pachón, L. Yu and P. Brumer, *Faraday Discuss.*, 2013, **163**, DOI: 10.1039/C3FD20144A.
15 S. Ruiz-Barragan and L. Blancafort, *Faraday Discuss.*, 2013, **163**, DOI: 10.1039/C3FD20155D.
16 R. Pazourek, S. Nagele and J. Burgdörfer, *Faraday Discuss.*, 2013, **163**, DOI: 10.1039/C3FD00004D.
17 N. Schirmel, N. Reusch, P. Horsch and K.-M. Weitzel, *Faraday Discuss.*, 2013, **163**, DOI: 10.1039/C3FD20152J.
18 P. H. Bucksbaum and V. Petrovic, *Faraday Discuss.*, 2013, **163**, DOI: 10.1039/C3FD20154F.
19 R. W. Schoenlein, L. A. Peteanu, R. A. Mathies and C. V. Shank, *Science*, 1991, **254**, 412.
20 D. Baylor, *Proc. Natl. Acad. Sci. U. S. A.*, 1996, **93**, 560.
21 *Conical Intersections: Theory, Computation and Experiment.*, ed. W. Domcke, D. Yarkony and H. Köppel, World Scientific, Singapore, 2011.
22 J. C. Polanyi, *Acc. Chem. Res.*, 1972, **5**, 161.
23 A. M. D. Lee, J. D. Coe, S. Ullrich, M.-L. Ho, S.-J. Lee, B.-M. Cheng, M. Z. Zgierski, I.-C. Chen, T. J. Martinez and A. Stolow, *J. Phys. Chem. A*, 2007, **111**, 11948.

24 O. Schalk, A. Boguslavskiy and A. Stolow, *J.Phys. Chem. A*, 2010, **114**, 4058.
25 O. Schalk, A. E. Boguslavskiy, A. Stolow and M. S. Schuurman, *J. Am. Chem. Soc.*, 2011, **133**, 16451.
26 G. Wu, A. E. Boguslavskiy, O. Schalk, M. S. Schuurman and A. Stolow, *J. Chem. Phys.*, 2011, **135**, 164309.
27 D. Picconi, F. J. Avila Ferrer, R. Improta, A. Lami and F. Santoro, *Faraday Discuss.*, 2013, **163**, DOI: 10.1039/C3FD20147C.
28 O. Geßner, A. M. D. Lee, J. P. Shaffer, H. Reisler, S. V. Levchenko, A. I. Krylov, J. G. Underwood, H. Shi, A. L. L. East, D. M. Wardlaw, Et-H. Chrysostom, C. C. Hayden and A. Stolow, *Science*, 2006, **311**, 219.
29 C. Z. Bisgaard, O. J. Clarkin, G. Wu, A. M. D. Lee, O. Geßner, C. C. Hayden and A. Stolow, *Science*, 2009, **323**, 1464.
30 A. Stolow and J. G. Underwood, *Adv. Chem. Phys.*, 2008, **139**, 497.
31 F. Rosca-Pruna and M. J. J. Vrakking, *Phys. Rev. Lett.*, 2001, **87**, 153902.
32 J. G. Underwood, M. Spanner, M. Yu. Ivanov, J. Mottershead, B. J. Sussman and A. Stolow, *Phys. Rev. Lett.*, 2003, **90**, 223001.
33 J. G. Underwood, B. J. Sussman and A. Stolow, *Phys. Rev. Lett.*, 2005, **94**, 143002.
34 K. F. Lee, D. M. Villeneuve, P. B. Corkum, A. Stolow and J. G. Underwood, *Phys. Rev. Lett.*, 2006, **97**, 173001.
35 P. Hockett, C. Z. Bisgaard, O. J. Clarkin and A. Stolow, *Nat. Phys.*, 2011, **7**, 612.
36 B. J. Sussman, J. G. Underwood, R. Lausten, M. Yu. Ivanov and A. Stolow, *Phys. Rev. A: At., Mol., Opt. Phys.*, 2006, **73**, 053403.
37 M. J. J. Vrakking, D. M. Villeneuve and A. Stolow, *J. Chem. Phys.*, 1996, **105**, 5647.
38 B. J. Sussman, D. Townsend, M. Yu. Ivanov and A. Stolow, *Science*, 2006, **314**, 278.
39 P. B. Corkum, *Phys. Rev. Lett.*, 1993, **71**, 1994.
40 P. B. Corkum, M. Yu. Ivanov and N. H. Burnett, *Opt. Lett.*, 1994, **19**, 1870.
41 M. Lezius, V. Blanchet, D. M. Rayner, D. M. Villeneuve, A. Stolow and M. Yu. Ivanov, *Phys. Rev. Lett.*, 2001, **86**, 51.
42 M. Lezius, V. Blanchet, M. Yu. Ivanov and A. Stolow, *J. Chem. Phys.*, 2002, **117**, 1575.
43 M. Smits, C. A. de Lange, A. Stolow and D. M. Rayner, *Phys. Rev. Lett.*, 2004, **93**, 213003.
44 M. Smits, C. A. de Lange, A. Stolow and D. M. Rayner, *Phys. Rev. Lett.*, 2004, **93**, 203402.
45 M. V. Ammosov, N. B. Delone and V. P. Krainov, *Sov. Phys. JETP*, 1986, **64**, 1191.
46 P. B. Corkum and F. Krausz, *Nat. Phys.*, 2007, **3**, 381.

Faraday Discussions

RSC Publishing

PAPER

A new approach toward transition state spectroscopy

Kirill Prozument,[a] Rachel Glyn Shaver,[a] Monika A. Ciuba,[b] John S. Muenter,[c] G. Barratt Park,[a] John F. Stanton,[d] Hua Guo,[e] Bryan M. Wong,[f] David S. Perry[g] and Robert W. Field*[a]

Received 29th December 2012, Accepted 1st February 2013
DOI: 10.1039/c3fd20160k

Chirped-Pulse millimetre-Wave (CPmmW) rotational spectroscopy provides a new class of information about photolysis transition state(s). Measured intensities in rotational spectra determine species-isomer-vibrational populations, provided that the rotational populations can be thermalized. The formation and detection of S_0 vinylidene is discussed in the limits of low and high initial rotational excitation. CPmmW spectra of 193 nm photolysis of vinyl cyanide (acrylonitrile) contain $J = 0$–1 transitions in more than 20 vibrational levels of HCN and HNC, but no transitions in vinylidene or highly excited local-bender vibrational levels of acetylene. Reasons for the non-observation of the vinylidene co-product of HCN are discussed.

A Introduction

Chirped-Pulse millimetre-Wave (CPmmW) spectroscopy[1–3] is capable of determining the relative species-conformer-vibrational level populations of all polar products of a photolysis reaction. These experimentally determined populations encode the structures of the transition states that are most important at each photolysis wavelength or combination of wavelengths.[4] The isomer-conformer-vibrational level population information obtained from a CPmmW spectrum is more complete than what is obtainable by mass spectrometry[5] and free of the need for the transition strength and quantum yield determinations that are

[a]*MIT, Department of Chemistry, Cambridge, MA 02139. E-mail: kuyanov@mit.edu; rshaver@mit.edu; barratt@mit.edu; rwfield@mit.edu*
[b]*Arizona State University, Department of Chemistry and Biochemistry, Tempe, AZ 85287-5601. E-mail: monika.ciuba@asu.edu*
[c]*University of Rochester, Department of Chemistry, Rochester, NY 14627. E-mail: muenter@chem.rochester.edu*
[d]*University of Texas at Austin, Department of Chemistry and Biochemistry, Austin, TX 78712. E-mail: jfstanton@mail.utexas.edu*
[e]*University of New Mexico, Department of Chemistry, Albuquerque, NM 87131. E-mail: hguo@unm.edu*
[f]*Sandia National Laboratories, Materials Chemistry Department, Livermore, CA 94550. E-mail: usagi@alum.mit.edu*
[g]*The University of Akron, Department of Chemistry, Akron, OH 44325-3601. E-mail: dperry@uakron.edu*

required for most laser-based population measurements. However, difficulties exist in the use of populations determined by CPmmW spectroscopy to characterize transition states.

This paper is a case study of the application of CPmmW spectroscopy to the 193 nm photolysis of vinyl cyanide (acrylonitrile): $CH_2{=}CHCN$.[5] The most important results are the observation of many vibrationally excited levels of HCN[6,7] and HNC,[7,8] the comparison of the observed isomer and vibrational population distribution when $CH_2{=}CHCN$ is replaced by $CH_2{=}CDCN$, the dependence of the population distribution on the amount of post-photolysis rotational cooling, and the failure to observe either S_0 vinylidene or extreme local-bender vibrationally excited S_0 acetylene. Both vinylidene and local-bender acetylene have large electric dipole moments along the inertial a-axis, which are expected to give rise to strong $J = 0\text{–}1$, $K_a = 0\text{–}0$ transitions in the 70–90 GHz region of our CPmmW spectrometer.

B Chirped pulse spectroscopy

Chirped Pulse Fourier Transform Microwave Spectroscopy is a revolutionary technique developed in the research group of Brooks Pate at the University of Virginia.[2,3] In its initial implementation, the frequency of a microwave pulse is chirped linearly in time over several GHz. This microwave pulse is broadcast into a gas phase molecular sample, polarizing all two-level systems with frequencies within the spectral interval of the chirped pulse. These polarizations relax by emitting Free Induction Decay (FID) radiation, which is collected, down-converted by mixing with a local oscillator (heterodyne detection), and recorded as a time-domain signal on a fast oscilloscope. All components of the experiment are synchronized to a 10 MHz master clock, thus the FIDs from all chirps are perfectly phase coherent and may be averaged in oscilloscope memory. The detected signal scales as[9]

$$P_{1,2} \propto \frac{\mu^2 \varepsilon_0 \Delta N_{1,2}}{\alpha^{1/2}} A^{1/2} L \tag{1}$$

where μ is the electric dipole moment, ε_0 is the peak microwave electric field, $\Delta N_{1,2}$ is the population density difference between levels 1 and 2, α is the chirp rate, A is the cross-sectional area of the microwave beam, and L is the length of the irradiated region. A chirp over 10 GHz can be completed in a time between 10 ns and 10 μs. The advantage of a slow chirp (large $\alpha^{-1/2}$) is limited by the necessity that the CP terminate before the FID has decayed significantly. The FID is collected over a time interval from ~1 μs to 20 μs, where the decay rate of the FID is determined by inelastic collisions (T_1, typically 10 MHz/Torr) or Doppler dephasing ($T_2 \propto \Delta E_{1,2}{}^{-1} T^{-1/2}$). Usually, CP spectra are recorded at sufficiently low pressure that T_1 is slightly longer than T_2.

When the time-domain signal is transformed into the frequency-domain, linewidths in the 100 to 500 kHz range are routinely obtained. A single chirp can contain $\sim 10^5$ resolution elements.[1–3] Since all two-level systems within the chirp frequency region are sampled in each chirp, relative transition intensities may be measured with accuracies approaching the 1% level, even for transition frequencies separated by 10 GHz.

The microwave electric field, ε_0, scales as $I^{1/2}$, where I is the intensity in units of power/area. If the diameter of the microwave beam is increased by a factor of 10, ε_0 decreases by a factor of 10 but $\Delta N_{1,2} A^{1/2}$ increases by an exactly cancelling factor of 10. Therefore there is no penalty in choosing the cross-section of the microwave beam to be as large as that of the molecular sample. However, optimum CP signals are obtained when the active volume is as long as possible. In the $\Delta E_{1,2} < k_B T$ limit, the Boltzmann contribution to the CP signal scales as $\Delta E_{1,2}/T^2$. For pure rotational CP spectroscopy of photolysis products, rotational thermalization to $T \approx 5$K is very important.

In NMR there is a simple relationship between the integrated intensity of a single resolved transition and the number of nuclei in the chemical environment that corresponds to the selected chemical shift.[10] In microwave spectroscopy there is a similarly simple relationship between the transition intensity and the population difference between the upper and lower levels connected by a selected transition frequency.[11] One needs to know the rotational temperature, the permanent electric dipole moment, and the transition width. These three quantities are routinely measured by microwave spectroscopy. However, two fundamental problems complicate the use of CP spectroscopy to measure populations of very small molecules, especially when the goal is to determine the nascent, non-thermalized, species-conformer-vibrational level populations of reaction products. (i) The rotational constants of small (near prolate top, *a*-dipole) molecules are typically so large that *only* the lowest $J \leftrightarrow J + 1$ transition lies within the operation range of the spectrometer. Thus, unless one assumes that rotation of the target molecule is thermalized at a temperature determined by exploiting the measured rotational population distribution of a heavier "thermometer molecule" or from the translational temperature determined from a Doppler linewidth, one cannot derive vibrational populations from a single measured rotational population difference. (ii) Since rotation–thermalizing transitions occur typically at a rate 10^3 times that of vibration–thermalizing transitions,[12] it is plausible that in a supersonic jet expansion, rotational populations can be thermalized, yet vibrational populations will be near-nascent. However, when an enormous amount of reaction exoergicity is deposited in rotation, it is dangerous to make any kind of assumption about rotational thermalization.

C S_0 vinylidene: $C{=}CH_2$

On the acetylene electronic ground state potential energy surface (S_0), linear acetylene is separated by an ~2 eV barrier from the C_{2v} symmetry vinylidene isomer.[13,14] Although vinylidene is a widely invoked reaction intermediate, vinylidene has never been chemically isolated. There have been definitive spectroscopic observations of S_0 vinylidene,[15–17] but these observations have often been misinterpreted. From the vinylidene side, the barrier to isomerization to acetylene is very low, $\lesssim 0.09$ eV. However, the acetylene$\leftrightarrow$vinylidene saddle point lies well below the lowest dissociation limit at 5.6 eV,[18] thus all of the information about isomerization rates and mechanisms is encoded in energy-resolved eigenstates,[16] in particular in the distribution of acetylene and vinylidene basis state characters in each eigenstate. It is *not* a textbook style bound$\leftrightarrow$free barrier-tunnelling problem.

As shown in Fig. 1 and 2, the minimum energy isomerization path from linear acetylene to vinylidene is almost exclusively a large amplitude acetylene local-bend (*i.e.* one CCH bend, not one of the two bending normal modes, ν_4-trans bend or ν_5-cis bend).[19,20] This suggests that, in the absence of strong rotation-induced "Coriolis" or "rotational ℓ-resonance" mixing[21–24] [two different forms of rotationally

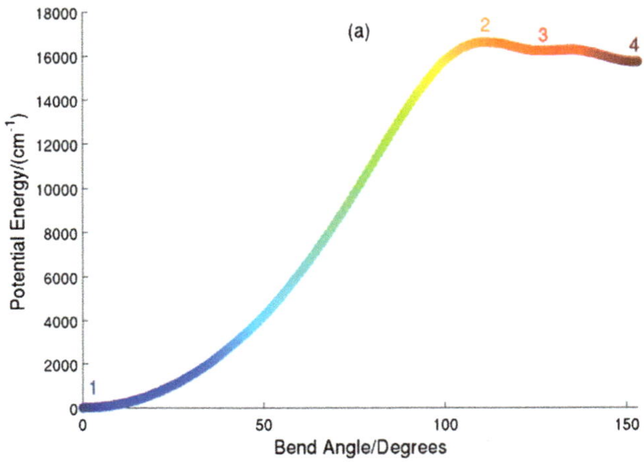

Fig. 1 The Minimum Energy Isomerization Path between acetylene and vinylidene. The one-dimensional path is colour-coded to correspond to the dependence of the molecular geometry, shown in Fig. 2, on the progress along the isomerization path, which is predominantly a local CCH bend. The numbers 1–4 correspond to the locations of the two H atoms along the isomerization path, the structure of which is also labelled on Fig. 2.[19]

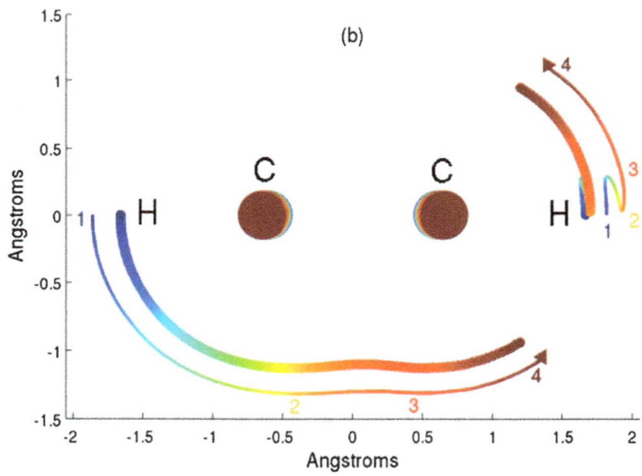

Fig. 2 Geometric structure of S_0 HCCH as a function of the (colour-coded) progress along the isomerization path shown in Fig. 1.[19] The agreement between the local-bender ω_{eff} frequency *vs.* N_{bend} obtained from *ab initio* calculations and the $\mathbf{H}^{eff}$ model is displayed in Fig. 3.

mediated interactions, referred to hereafter as Coriolis] of highly excited acetylene basis states, one expects that vibrationally unexcited vinylidene will interact exclusively with the very sparse manifold of highly excited local-bender levels of acetylene. Local-bender levels correspond to roughly 0.5% of the total vibrational density of states at the vibrational excitation energy of the barrier. Thus nonrotating C_2H_2 will encode isomerization as few-level vinylidene↔local-bender acetylene quantum beats, but highly rotating C_2H_2 might present to each vinylidene basis state the full (symmetry-sorted) density of acetylene eigenstates (about one vibrational level per 3 cm^{-1}), each with a small fractional local-bender character. Thus,

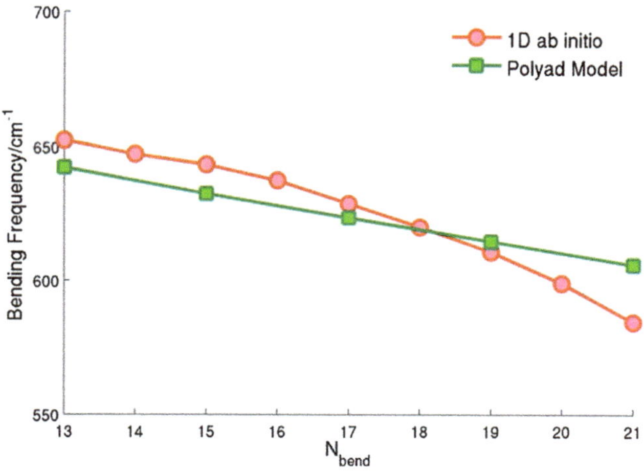

Fig. 3 Comparison of the frequency of the *ab initio* one-dimensional local-bending vibration with the frequency obtained from the low-J $\mathbf{H}^{eff}$ fit model.[21] The agreement is satisfactory in the region between the emergence of the local-bend at $N_{bend} \approx 14$ and at the top of the isomerization barrier near $N_{bend} \approx 24$.[19,20]

profoundly different limiting behaviours might be expected, dependent on whether the energized species is formed rotationally hot or rotationally cold.

There are two classes of experiments that seem to have sampled these two limits. The rotationally cold limit is realized by **photodetachment** of the extra electron from the vinylidene negative ion.[15–17] $C{=}CH_2^-$ has a geometric and vibrational structure almost identical to that of S_0 vinylidene. Ejection of a light electron cannot be accompanied by significant rotational excitation of the heavy molecule. In contrast, formation of vinylidene by **photolysis** of vinyl cyanide,[5] vinyl bromide, vinyl chloride,[25] or propene creates a highly rotationally excited C_2H_2 fragment. Even if, at high-J, the lowest energy vibrational levels of vinylidene are not affected by Coriolis mixing, the highly excited vibrational levels of acetylene formed by photolysis of vinyl-X are profoundly Coriolis-mixed. The acetylene local-bender basis state character, which serves to vinylidene as the unique doorway into acetylene, is admixed at high-J into the entirety of the dense manifold of vibrationally excited acetylene levels. This rotation-induced mixing (exclusively on the acetylene side of the barrier) serves as the on-switch for the more familiar mechanism of isomerization into a dense and ergodic manifold of vibrational states.

The spectrum and dynamics of S_0 vinylidene became a subject of great excitement and controversy when the Lineberger research group observed and assigned the lowest few vibrational levels of S_0 vinylidene in the photoelectron spectrum (PES) of the $H_2C{=}C^-$ negative ion (Fig. 4).[15,16] A fixed frequency UV line of an Argon ion laser was used to detach the extra electron from $H_2C{=}C^-$ and the difference between the energy of the UV photon and the kinetic energy of the detached electron provided information about the energies and widths of the S_0 vinylidene vibrational "resonances." The resolution of the PES was insufficient to resolve the J, K_a rotational structure of these vibrational resonances in this near-prolate asymmetric top. In the original paper, *most* of each observed linewidth was explicitly and correctly interpreted as inhomogeneous broadening due to unresolved rotational structure.[16] Despite this carefully crafted explanation, many papers from other

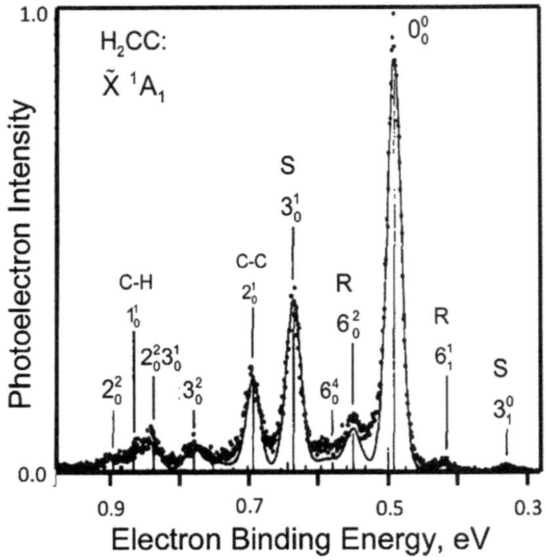

Fig. 4 The photoelectron spectrum of the vinylidene negative ion. The widths of each vibrational band feature are only slightly larger than the computed unresolved rotational structure at T_{rot} equal to T_{trans} of the helium carrier gas, upper right panel of Fig. 5 of Ervin *et al.*[16] Reprinted with permission from K. M. Ervin, J. Ho, and W. C. Lineberger, *J. Chem. Phys.*, 1989, **91**, 5974. Copyright 1989, American Institute of Physics.

research groups were quick to interpret *all* of the resonance width to homogeneous broadening due to tunnelling through a low barrier into either a continuum or vibrational quasi-continuum of S_0 acetylene. Lifetimes of S_0 vinylidene vibrational levels ranging from a few picoseconds to as short as ~100 femtoseconds have been cited as having been "experimentally measured" in the PES spectrum.[16] There is a third possible explanation, strongly supported by Coulomb explosion experiments,[17] that the excess width is due to few-level mixing between one zero-order vinylidene vibrational basis state and one or two near-degenerate acetylene vibrational basis states that have some extreme local-bender character.

Like the Lineberger group negative ion PES spectra,[15,16] the Coulomb explosion experiments[17] start with the $H_2C{=}C^-$ negative ion. This negative ion is accelerated to very high kinetic energy and impacted onto a thin foil. This impact results in the removal of all electrons. After impact (the Coulomb explosion), the positions and arrival times of the resultant two H^+ and two C^{6+} cations (*all four ions* from *each* parent molecule) are recorded, event by event, on a multi-particle detector. From the arrival times and positions, the structure of the parent molecule is back-calculated at the instant of collision with the foil. The extraordinary thing about the Coulomb explosion experiments is that, even when the extra electron on the $H_2C{=}C^-$ negative ion is detached 3.5 μs before the parent molecule collides with the foil, 50% of the reconstructed images are of the vinylidene structure.[17] One explanation that accounts for the extra width in the negative ion PES and the long-time persistence of the vinylidene structure is a few-level quantum beating system where the long-lived eigenstates are mixtures of one vinylidene and one or two acetylene basis states that have admixed local-bender character. Interestingly, the *initial* decay of vinylidene character in this quantum-beating system could be on the subpicosecond time-scale, but the dynamics must be periodic rather than irreversible.

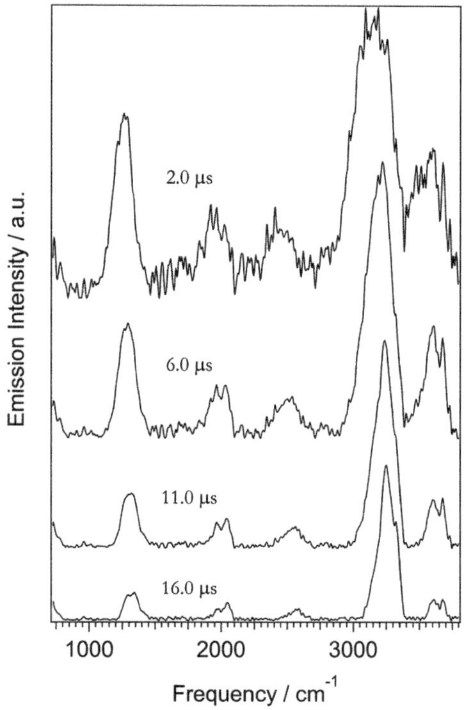

Fig. 5 The time-resolved Fourier transform infrared fluorescence spectrum (TR-FTIR) resulting from the 193 nm photolysis of vinyl cyanide. The strong early-time feature near 1350 cm^{-1} is assigned here as superimposed acetylene $\Delta v = -2$ local-bender (parallel type) transitions and in the original publication as $\nu_4 + \nu_5$ (perpendicular type) combination bands. This is Fig. 1 from Wilhelm *et al.*[26] Reprinted with permission from M. J. Wilhelm, M. Nikow, L. Letendre, and H.-L. Dai, *J. Chem. Phys.*, 2009, **130**, 044307. Copyright 2009, American Institute of Physics.

Detachment of an electron from the $H_2C{=}C^-$ negative ion results in much less vibrational and rotational excitation than ultraviolet photolysis of vinyl–*X* (*X* = Cl, Br, CN, CH_3). For photolysis, this is partly a kinematic effect and partly a Franck–Condon mediated energetic effect. There is an enormous difference in nuclear geometry between the S_0 zero-point vibrational level of the parent molecule and the geometry of the conical intersection that feeds the photofragmentation transition states. This ensures, due to Franck–Condon restrictions, that UV photolysis occurs at energies far above the thermodynamic threshold. The photofragments are formed significantly rotationally and vibrationally excited. This is bad news for high resolution spectroscopic measures of energy partitioning, because population is distributed over many rotation-vibration levels, individual rovibronically resolved transitions will be weak, and the spectrum will be complex and congested. Time-resolved FTIR spectroscopy is well suited to the characterization of photolysis product distributions.[26] The relatively low resolution of TR-FTIR makes it very difficult to determine rotational population distributions. However, very strong $\Delta v = -1$ vibrational propensity rules for infrared active normal modes often usefully simplify the low-resolution vibrational spectrum. For example, in S_0 acetylene, only ν_3 (antisymmetric CH stretch, parallel type) and ν_5 (*cis*-bend, perpendicular type) are infrared active and have recognizably different rotational band contours.

The Hai-Lung Dai group has applied Time-Resolved-FTIR spectroscopy to the 193 nm photolysis of vinyl–*X* molecules.[26] They observe spectrally distinct vibrational features that enable measurement of the HCN : HNC product ratio. Surprisingly, one of the strongest vibrational features in their early-time spectrum cannot be assigned as one of the known IR active fundamentals of acetylene, vinylidene, HCN, or HNC (Fig. 5). They assign it as the acetylene $\nu_4 + \nu_5$ combination band (perpendicular type) and attribute its presence in the spectrum as due to collision-induced decay of vinylidene into high vibrational levels of acetylene.[5,26,27] Combination bands are usually weak relative to IR active fundamentals, even when the combination band originates from a high-(ν_4,ν_5) level. These features are more likely assignable as $\Delta\nu = -2$, $\Delta\ell = 0$ local-bender overtone bands (parallel type), which arise directly from the decay of vinylidene along the isomerization path[28] into highly excited acetylene local-bender levels. The choice between these competing assignments might eventually be settled by observation of the rotational band contour, which is very different for parallel (no Q branch) *vs.* perpendicular (strong Q branch) type bands.

The low-resolution of the TR-FTIR spectra will preserve the intensity associated with emission from highly excited local-bender levels, even in the presence of strong Coriolis mixing of the local-bender character into a quasi-continuum of highly excited vibrational levels. The mixing does not cause the local-bender character to vanish. It dilutes it into many vibrational levels. The local-bender character will light up low-resolution spectra, but it will be diluted into unobservability in an eigenstate-resolved, high-resolution spectrum.

D Photolysis of vinyl cyanide. Mass spectrometry and photofragment translational spectroscopy.

When vinyl cyanide is photolyzed by 193 nm radiation, the following photofragmentation products are energetically accessible:[5]

$$H_2C{=}C{-}C{\equiv}N\ (m/e\ 52) + H\ (m/e\ 1)$$

$$HC{=}CHC{\equiv}N\ (m/e\ 52) + H\ (m/e\ 1)$$

$$H_2C{=}CH\ (m/e\ 27) + C{\equiv}N\ (m/e\ 26)$$

$$HC{=}C{-}C{\equiv}N\ (m/e\ 51) + H_2\ (m/e\ 2)$$

$$^1(C{=}CHC{\equiv}N)\ (m/e\ 51) + H_2\ (m/e\ 2)$$

$$^3(C{=}CHC{\equiv}N)\ (m/e\ 51) + H_2\ (m/e\ 2)$$

$$HC{\equiv}CH\ (m/e\ 26) + HCN \text{ or } HNC\ (m/e\ 27)$$

$$^1(H_2C{=}C)\ (m/e\ 26) + HCN \text{ or } HNC\ (m/e\ 27)$$

$$^3(H_2C{=}C)\ (m/e\ 26) + HCN \text{ or } HNC\ (m/e\ 27)$$

Provided that vacuum ultraviolet radiation at sufficiently short wavelength is available, each of these photofragments may be ionized, mass selected, and detected. The time-of-flight distribution from photodissociation to ionization provides the translational energy distribution, $P(E_{Trans})$, for each mass-selected photofragment. The $P(E_{Trans})$ contains information about the nascent vibrational energy distribution (unaffected by post-photolysis collisions) within each photofragment. Mass spectrometry alone is not sufficient to distinguish two different chemical species (*e.g.* $H_2C{=}CH$ *vs.* HCN or $HC{\equiv}CH$ *vs.* $C{\equiv}N$) [except at extremely high mass resolution, which is starting to be available at national user facilities] or different isomers (*e.g.* HCN *vs.* HNC or $HC{\equiv}CH$ *vs.* $H_2C{=}C$) with the same *m*/*e*. However, different chemical species can usually be distinguished *via* their distinct ionization thresholds.[5] Whether isomers can be distinguished depends on whether the unstable isomer decays on the time-scale of the photofragmentation half-collision. HNC is sufficiently stable that its contribution to the *m*/*e* 27 $P(E_{Trans})$ can be distinguished from that of HCN. However, the isomerization rate of vinylidene has been a subject of controversy and also the motivating factor behind the long-term effort in the Field group to spectroscopically characterize vinylidene.

There is an enormous difference in the amount of rotational excitation of vinylidene formed by the photodetachment of an electron from the vinylidene negative ion *vs.* vinylidene formed by the photolysis of vinyl cyanide. Rotation matters! As demonstrated by the $\mathbf{H}^{\text{eff}}$ of Perry *et al.*,[24] the exceptional stability of low-*J* acetylene extreme local-bender states[21–23] is destroyed at high *J* (see Fig. 9 and 10 in Section E). At high-*J*, the local-benders undergo rapid Intramolecular Vibrational Redistribution (IVR) into the dense and ergodic manifold of highly excited vibrational levels. The minimum energy isomerization path between acetylene and vinylidene lies along the local-bend from the acetylene side[19] and the in-plane wag from the vinylidene side[16] of the isomerization barrier. Nonrotating vinylidene interacts exclusively with isoenergetic local-bender levels of acetylene. The density of acetylene local-bender vibrational states is very low, typical of a *diatomic* molecule. The Coulomb explosion experiments (that also form low-*J* vinylidene by the photodetachment of an electron from the vinylidene negative ion[17]) show that the acetylene–vinylidene interaction is that of few-level quantum beats. At low-*J*, there is no irreversible isomerization from vinylidene into a dense manifold of highly excited vibrational levels of acetylene.

The negative ion PES experiments of the Lineberger group[16] reveal vinylidene vibrational resonances for which most of the width is due to rotational inhomogeneous broadening (Fig. 4). The extra width could well be due to unresolved splittings between pairs of locally-perturbing levels rather than a Fermi Golden Rule broadening due to mixing into a vibrational quasi-continuum. The fast decay of vinylidene character could well be the initial (reversible) decay of a two-level quantum beating system. Yet Ervin *et al.*[16] write "The width of the singlet transition yields a lower limit to the vinylidene lifetime $\tau > 0.027$ ps. From an analysis of line shapes including instrumental and rotational broadening, we estimate the lifetime of the singlet to be $\tau \sim 0.04$–0.2 ps." This corresponds to FWHM of 250–50 cm^{-1}. [Earlier in their paper, Ervin *et al.*[16] presciently state "Since the acetylene–vinylidene system is bound, in a much higher resolution experiment there would be no 'lifetime broadening' as such. Rather, the zero-order vibrational levels we identify with vinylidene would be split into many eigenlevels of the complete system."] This lifetime estimate is echoed by Blank *et al.*,[5] "Based on negative ion photodetachment spectral linewidths, Lineberger and co-workers estimated a lifetime of 40–200 fs for the isomerization of vinylidene to acetylene."

Everyone is correct. At low-J, as in photodetachment experiments, vinylidene lives longer than 3.5 μs. Its lifetime is probably on the same time-scale as a typical acetylene local-bender $\nu_{bend} = 24 \rightarrow 22$ vibrational radiative decay, ~1 ms. At high-J, as in photolysis experiments, the vinylidene lifetime is acetylene IVR limited. 40–200 fs is probably an excellent estimate.

Zou *et al.*[13] report full 6-dimensional calculations of the $J = 0$ vibrational levels in the energy region from ~500 cm^{-1} below the vinylidene zero-point level to ~1800 cm^{-1} above the acetylene–vinylidene saddle point. They find *isolated* vibrational eigenstates that have predominant vinylidene character (shown in their Fig. 2, 3 and 6) at an ~1 per 3 cm^{-1} symmetry-sorted (A_1' in the Complete Nuclear Permutation Inversion Group, G_8) vibrational density of states. The total vibrational density of states is four times larger than 1/3 cm^{-1} because only four of the eight CNPI symmetry species have a nonzero nuclear statistical weight. Since the tunneling interactions between acetylene and vinylidene vibrational basis states are diagonal within each CNPI symmetry species, 1/3 cm^{-1} is the density of states that is relevant to a Fermi Golden Rule estimate of the interaction matrix elements between acetylene and vinylidene basis states.

The FWHM associated with a state that has 0.2 ps lifetime is 50 cm^{-1}. There are only ~50/3 A_1' vibrational eigenstates within this FWHM. Although each of these ~17 eigenstates is long-lived (longer than 3.5 μs), together they encode the decay of a single vinylidene basis state formed by the photodetachment event at $t = 0$. Using the Fermi Golden Rule in reverse to estimate the average tunneling matrix element between one vinylidene basis state and each of the nearby acetylene basis states,

$$w = (2\pi/\hbar)\langle\phi_{\text{vinylidene}}|H|\phi_{\text{acetylene}}\rangle^2\rho, \quad (2)$$

where $w = 1/0.2$ ps is the tunneling rate, H is the interaction matrix element, and $\rho = 1/3$ cm^{-1} is the vibrational density of states, then the average matrix element, in cm^{-1} units, is 2.5 cm^{-1}. With such a small average tunneling matrix element, significant vinylidene~acetylene mixing (approaching 50 : 50) only occurs between nearest neighbour eigenstates. This is quite consistent with the Coulomb explosion result,[17] in which images are observed consisting typically of 50% acetylene and 50% vinylidene structures.

E Spectroscopy of acetylene

The vibration-rotation energy levels of the acetylene S_0 electronic state have been exhaustively examined in high-resolution (~0.001 cm^{-1}) infrared and Raman,[29] moderate resolution (~0.03 cm^{-1}) $S_1 - S_0$ Stimulated Emission Pumping (SEP),[30,31] and low-resolution (~5 cm^{-1}) $S_1 - S_0$ Dispersed Fluorescence (DF) spectra.[30–33]

As for any linear four-atom molecule, there are seven vibrational degrees of freedom, three nondegenerate stretching modes (ν_1 symmetric CH stretch, ν_2 CC stretch, ν_3 antisymmetric CH stretch) and two doubly-degenerate bends (ν_4 *trans*-bend and ν_5 *cis*-bend). All of the normal mode vibrational frequencies are in approximate integer multiple ratios $[(\nu_1 : \nu_2 : \nu_3 : \nu_4 : \nu_5) = (5 : 3 : 5 : 1 : 1)]$.[34,35] Whenever vibrational frequencies are in near integer multiple ratios, second-order perturbation theory breaks down and quasi-degenerate perturbation theory requires diagonalization of an approximately block diagonal effective Hamiltonian matrix, $\mathbf{H}^{\text{eff}}$.[34] Each block of this $\mathbf{H}^{\text{eff}}$ is called a **polyad**. The vast majority of vibrational levels are members of a polyad, and all of the normal mode vibrational quantum numbers $(v_1,v_2,v_3,v_4,\ell_4,v_5,\ell_5)$ are destroyed by anharmonic interactions between quasi-degenerate basis states. The result of these inter-mode anharmonic interactions in S_0 acetylene is that the only surviving approximately good vibrational quantum numbers are the polyad quantum numbers: $N_{\text{resonance}} = 5v_1 + 3v_2 + 5v_3 + v_4 + v_5$, $N_{\text{stretch}} = v_1 + v_2 + v_3$, and $\ell_{\text{total}} = \ell_4 + \ell_5$.[34]

The total vibrational Hamiltonian is approximately block-diagonal and each block is a polyad that is denoted by its three polyad quantum numbers $[N_{\text{res}},N_{\text{s}},\ell_{\text{tot}}]$. The intrapolyad matrix elements follow standard harmonic oscillator matrix element selection and scaling rules.[34] For example, the intrapolyad anharmonic matrix element $k_{1133}\mathbf{Q}_1{}^2\mathbf{Q}_3{}^2$ has intra-polyad selection rules $\Delta v_1 = -\Delta v_3 = \pm 2$ and 0 and the values of such matrix elements follow the scaling rule

$$\begin{aligned} \left\langle v_1 v_3 \middle| k_{1133}\mathbf{Q}_1^2\mathbf{Q}_3^2 \middle| v_1 v_3 \right\rangle &= \frac{1}{2}K_{1133} v_1 v_3 \\ \left\langle v_1 v_3 \middle| k_{1133}\mathbf{Q}_1^2\mathbf{Q}_3^2 \middle| v_1+2, v_3-2 \right\rangle &= \frac{1}{4}K_{1133}\left[(v_1+2)(v_1+1)(v_3)(v_1-1)\right]^{1/2} \end{aligned} \tag{3}$$

where k_{1133} is the force constant for quartic interaction between modes 1 and 3 and K_{1133} is the v_1,v_3-independent scaling parameter. The $k_{1133}\mathbf{Q}_1{}^2\mathbf{Q}_3{}^2$ term also gives rise to inter-polyad interactions, for example

$$\left\langle v_1 v_3 \middle| k_{1133}\mathbf{Q}_1^2\mathbf{Q}_3^2 \middle| v_1+2, v_3+2 \right\rangle = \frac{1}{4}K_{1133}\left[(v_1+2)(v_1+1)(v_3+2)(v_3+1)\right]^{1/2} \tag{4}$$

but the energy denominator, $2\omega_1 + 2\omega_2$, permits treatment of such interpolyad effects by second-order perturbation theory. This makes it possible to replace the exact Hamiltonian by a lower-dimension block-diagonal $\mathbf{H}^{\text{eff}}$, where the membership and matrix elements in each polyad block are explicitly related to those in all other blocks.

In the special case that $N_{\text{s}} = 0$, then $N_{\text{res}} = N_{\text{bend}} = v_4 + v_5$. The $[N_{\text{bend}} = 10, \ell_{\text{tot}} = 0]$ polyad, for example factors into four blocks, [10,0,*g*,e] (12 members), [10,0,*g*,f] (6 members), [10,0,*u*,e] (9 members), and [10,0,*u*, f] (9 members).

The $\mathbf{H}^{\text{eff}}$ is a powerful and convenient fit-model, because it requires far fewer adjustable parameters than a complete, full-dimensional Hamiltonian, $\mathbf{H}$. The $\mathbf{H}^{\text{eff}}$ provides accurate predictions of the energies of unobserved vibrational levels, the eigenvector for each eigenstate expressed as a linear combination of normal mode product basis states that are members of the $[N_{\text{res}},N_{\text{s}},\ell_{\text{tot}}]$ polyad,

$$\left|[N_{\text{res}}, N_{\text{s}}, l_{\text{tot}}], \text{energy rank } i\right\rangle = \sum_{\substack{\text{basis state} \\ j}} c_j^i \left|[N_{\text{res}}, N_{\text{s}}, l_{\text{tot}}], j\right\rangle \tag{5}$$

predictions of transition propensity rules and relative intensities, for example when the *k*-th basis state is "bright" and all of the other basis states in the polyad are "dark",

$$I_{[N_{\text{res}},N_{\text{s}},N_{\text{tot}}],i} = \left(c_k^i\right)^2 \tag{6}$$

and predictions of inter-isomer interaction matrix elements, for example when only the *k*-th polyad basis state (*e.g.* an extreme local-bender) has significant overlap with the other-isomer basis state,

$$\langle [N_{\text{res}}, N_s, N_{\text{tot}}], i | \mathbf{H} | \text{other isomer} \rangle \propto c_k^i. \tag{7}$$

In addition to the complete Hamiltonian, $\mathbf{H}$, there are two classes of $\mathbf{H}^{\text{eff}}$ models for the vibration-rotation levels of acetylene: (i) the low-J $\mathbf{H}^{\text{eff}}$ that is determined primarily by fits to low-resolution SEP and Dispersed Fluorescence (DF) data mostly obtained from polyads with significant excitation in the bending modes,[21–23] and (ii) the all-J $\mathbf{H}^{\text{eff}}$ that is determined by global fits to high resolution spectra, accurate to measurement uncertainty, of all observed rotation-vibration energy levels.[24]

The complete $\mathbf{H}$, derived from a high-level *ab initio* potential energy surface, can be an excellent approximation to "the truth," but it is not organized in a way that directly identifies the almost-good quantum numbers and all of the relevant matrix element propensity and scaling rules. The low-J $\mathbf{H}^{\text{eff}}$ is parametrically parsimonious, compact, and it explicitly reveals the scalable and mechanistically explicit forms of all spectroscopic and dynamical parameters.[21] However, since the low-J $\mathbf{H}^{\text{eff}}$ is determined from low-J data, it does not include J-dependent interaction mechanisms (in particular the rotational ℓ-resonance term, the matrix elements for which scale as $J(J+1)$) and cannot be expected to scale reliably to high-J levels. The all-J $\mathbf{H}^{\text{eff}}$ is expressed in terms of an enormous number of fit parameters and requires numerical experimentation to reveal the goodness of approximately conserved observables, in particular how well quantities that are well conserved at low-J continue to be conserved at high-J.[24]

A staggering amount of information is contained in the set of eigenvectors of the complete or an effective Hamiltonian. Two ways of looking for the emergence or disappearance of a class of localization in state space are the "participation number"[36] and a form of correlation diagram known as a "spaghetti diagram".[21] The participation number is a measure of how many basis states have a significant presence in a specified eigenstate. For the i-th eigenstate the participation number, P_i, is

$$P_i = \left[\sum_j \left(c_j^i \right)^4 \right]^{-1}, \tag{8}$$

where the mixing coefficients, c_j^i, are defined in Eq. (5). A P_i value near 1 means that the i-th eigenstate is localized in a single basis state, whereas a P_i value near 10 means that 10 basis states contribute significantly to the i-th eigenstate. [For example, if there are 10 basis states that contribute equally to the i-th eigenstate, then each mixing coefficient is $10^{-1/2}$ and $P_i = 10$.] A complementary quantity, the basis state "dilution factor," D_j, may be defined similarly,

$$D_j = \sum_i \left(c_j^i \right)^4. \tag{9}$$

When the j-th basis state is diluted effectively into 10 eigenstates, $D_j = 1/10$. A fundamental objection is that the P_i and D_j, as defined in Eqs. (8) and (9), depend on the choice of basis set. Starting from the complete normal mode basis set, it is possible to define a variety of physically relevant basis sets by applying a unitary transformation that diagonalizes $\mathbf{H}^{(0)}$ plus one or more of the anharmonic resonance terms in $\mathbf{H}^{(1)}$.[24] One choice is to transform from a normal mode to a local-mode Hamiltonian.[21] Each choice of basis set is capable of conveying different classes of insight into the nature of the dynamics in various regions of state space.[24,37]

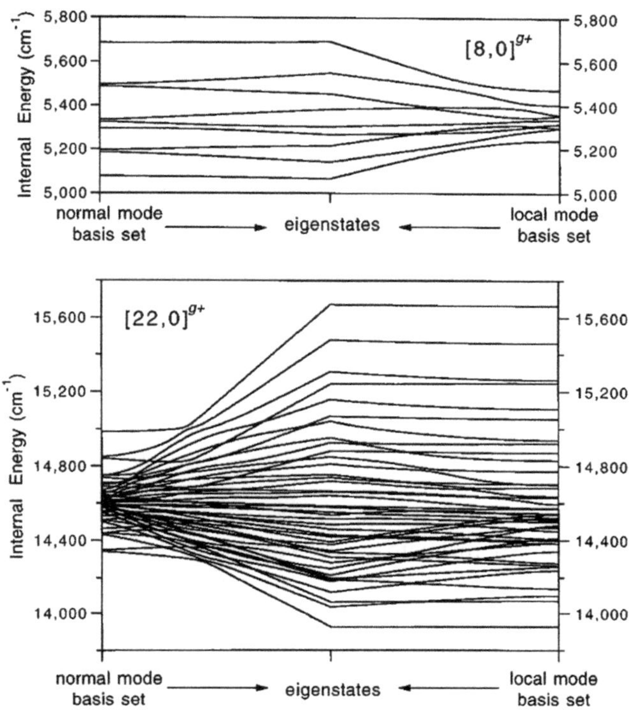

Fig. 6 Correlation diagrams for the normal and local mode basis sets for the [$N_{\text{bend}} = 8, \ell_{\text{total}} = 0$, *gerade*, e-symmetry] (top) and [22,0,*g*,e] (bottom) polyads. The lines in the diagram represent the state energies within the polyad in different limits. In the middle of the diagram are the eigenenergies, and at the far left and right are the zero-order energies of the normal mode and local mode basis sets, respectively. At positions intermediate between the eigenstate and basis state extremes, the energies are calculated by diagonalizing the effective Hamiltonian with the off-diagonal elements multiplied by a scaling factor between 0 (the unperturbed basis set limit) and 1 (the eigenstate limit). An eigenstate is likely to be assignable in terms of quantum numbers associated with a given basis set if the line which passes from the eigenstate to a zero-order basis state does so with minimal deviations, or can be followed through avoided crossings. It is clear from this diagram that many more eigenstates in the [22,0,*g*,e] polyad are assignable in the local mode basis set than in the traditional normal mode basis set, but that the normal mode basis set provides a better zero-order description of the eigenstates in the [8,0,*g*,e] polyad. This is Fig. 8 from Jacobson *et al.*[22] Reprinted with permission from M. P. Jacobson, R. J. Silbey, and R. W. Field, *J. Chem. Phys.*, 1999, **110**, 845. Copyright 1999, American Institute of Physics.

Figure 6 is a correlation (spaghetti) diagram that displays the goodness of the basis state quantum numbers in different energy regions of a polyad and also for high- *vs.* low-energy polyads.[21] The polyads shown are "pure- bending" polyads. They are pure-bending because, when $N_{\text{str}} = 0$, then $N_{\text{res}} = N_{\text{bend}} = \nu_4 + \nu_5$.

At the extreme left and right respectively are the normal mode and local mode basis state energies. In the middle are the eigen-energies. The connecting lines are the energies obtained with *all of the off-diagonal matrix elements* of $\mathbf{H}^{\text{eff}}$ multiplied by a scaling factor that is varied from zero at the extremes to 1 at the centre. Connection between the extreme and the centre without avoided crossings is a sign that the basis state quantum numbers remain good and that the P_i and D_j values remain near 1.

There are two spaghetti diagrams in Fig. 6, one for $N_{\text{bend}} = 8$ and one for $N_{\text{bend}} = 22$ (very close to the top of the acetylene–vinylidene barrier, which is located just above $N_{bend} = 24$). At low N_{bend}, the normal mode basis set is best, especially at the top and bottom of the polyad. At high N_{bend}, the local mode basis set is a good basis, except for the tangle of levels near the middle of the polyad. The top (counter-rotators) and bottom (local-benders) eigenstates are nearly pure basis states,[38] with $P_i \approx 1$ and $D_j \approx 1$. A dramatic bifurcation occurs near $N_{bend} = 14$. Pure-bending eigenstates emerge and become increasingly pure as N_{bend} increases. These extreme local-benders have *gerade*, *ungerade* inversion symmetry, as is required for any molecule, such as HCCH, which has a centre of symmetry. For example, at $N_{bend} = 22$, there is a g,u pair of eigenstates

$$|N_{bend} = 22,g> = 2^{-1/2}[|v_{left} = 22, v_{right} = 0> + |v_{left} = 0, v_{right} = 22>]$$

$$|N_{bend} = 22,u> = 2^{-1/2}[|v_{left} = 22, v_{right} = 0> - |v_{left} = 0, v_{right} = 22>]. \quad (10)$$

The emergence of pure, extreme local-bender eigenstates is a clearly demonstrated property of an approximate, reduced-dimension $\mathbf{H}^{eff}$. However, Hua Guo's high-level *ab initio* calculations of low-J $g \leftrightarrow u$ vibrational transition strengths for the complete $\mathbf{H}$, illustrated by Fig. 7 and 8, show that the emergence of local-benders is not an artefact that arises from the use of an approximate $\mathbf{H}^{eff}$.[39]

This emergence of local-benders is part of a global reorganization of the structure and dynamics of S_0 acetylene. At low energy, vibrational transition probabilities are small and apparently indiscriminately distributed over many g,u pairs of rotation–vibration energy levels. At higher energy, all of the transition strength seems to coalesce into one strongly allowed $\Delta N_{bend} = 0$, $g \leftrightarrow u$, $\Delta J = \pm 1$ pair of vibration–rotation transitions.[39]

Owing to the near perfect degeneracy between the same-J, same-N_{bend} pair of g,u local-bender vibrational levels,[21–23] this g–u allowed vibration-rotation

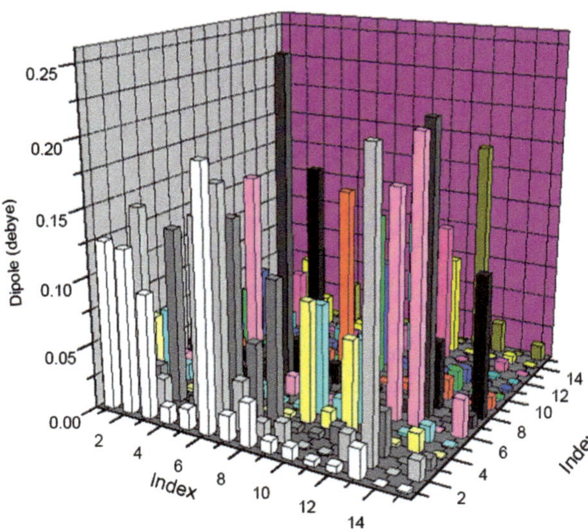

Fig. 7 Acetylene S_0 g–u $\ell = 0$–0 computed transition moments between low energy vibrational levels in the normal mode limit.[39] Notice that there are many weakly allowed transitions and most transition moments are smaller than 0.1 Debye.

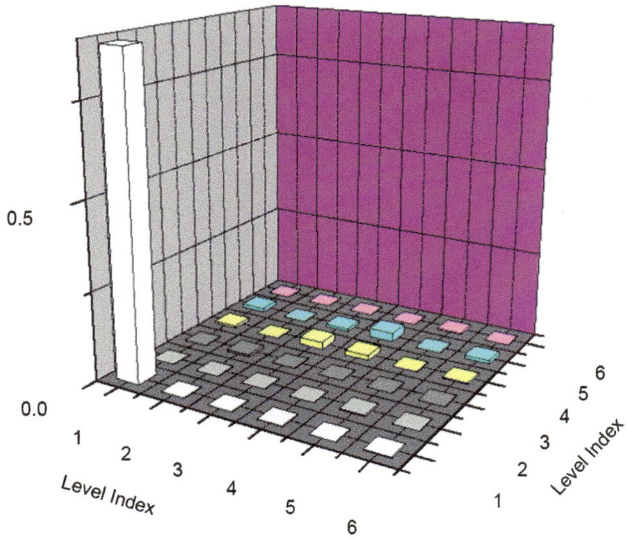

Fig. 8 Computed vibrational transition moments between vibrational levels that are well into the local mode limit.[39] Notice that there is only one transition, [$N_{bend} = 20, \ell_{total} = 0, g$] to [$N_{bend} = 20, \ell_{total} = 0, u$], that has appreciable transition strength, ~1 Debye, and all of the other transition moments are ≪0.1 Debye. Contrary to normal expectations of greater complexity at high vibrational energy, a dramatic simplification of the transition structure emerges.

transition occurs at exactly the frequency of what would be a *g*–*g* forbidden pure-rotation transition in a local-bender state. Pure rotation transitions in S_0 acetylene are forbidden, but these local-bender vibration–rotation transitions are strongly allowed, with electric dipole transition moments in excess of 1 Debye.

The Field group designed its CPmmW spectrometer to observe *a*-dipole $J = 0$–1 transitions in *both* S_0 local-bender acetylene and S_0 vinylidene, *both* expected to occur in the 70–85 GHz frequency region. We expected that the 193 nm photolysis of vinyl cyanide would produce vinylidene, local-bender acetylene, HNC, and HCN, all of which have $J = 0$–1 transitions within the tuning range of the spectrometer. We also expected that, even if vinylidene were produced at very high *J*, rotational cooling in the supersonic jet expansion would cool the rotations to $T_{rot} \sim 5$K. We were confident that the extreme purity of the local-bender levels of acetylene would ensure that redistribution of the local-bender character into the dense manifold of non-local-bender states would not occur. This confidence was based on the $\mathbf{H}^{eff}$ determined from fits to low-*J* levels.[21–23]

The high-*J* $\mathbf{H}^{eff}$ fitted by Michel Herman's research group and subjected to dynamical analysis by David Perry and Michel Herman[24] reveals a dramatic destruction of the stability of extreme local-bender states at very high *J*. For example, the participation numbers for the extreme local- bender level in the $N_{bend} = 16$ polyad are respectively 1.0034, 7.23, and 43.7 for $J = 2$, 30, and 100. This turning on of Intramolecular Vibrational Redistribution at high-*J* is illustrated more globally in Fig. 9 and 10.

The emergence of transition mode vibrational states, as exemplified by the local-bender states at high-ν_{bend} of acetylene, is a general phenomenon. Whenever a molecular potential energy surface supports two isomeric forms separated by a barrier (saddle point), the vibrational eigenstates must rearrange in a way

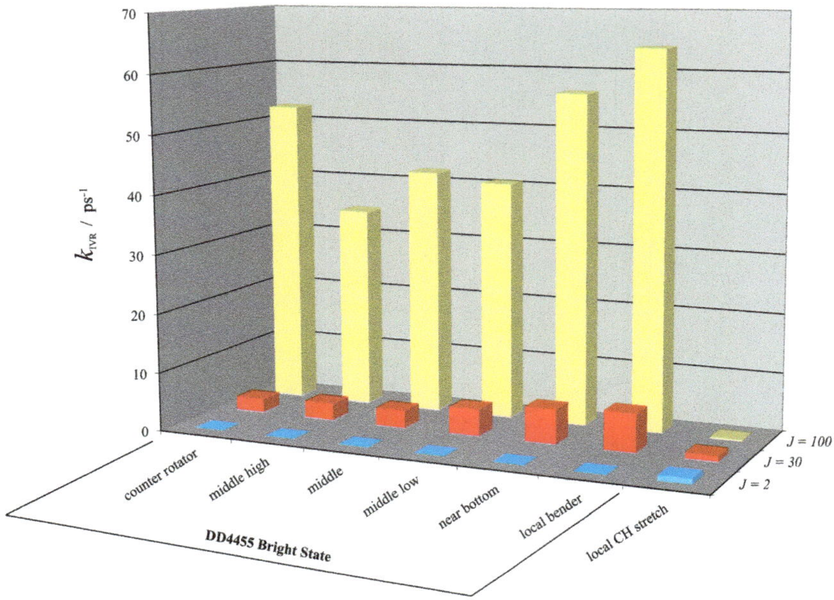

Fig. 9 Initial rates of IVR in acetylene for bending bright states ($\ell_{total} = 0$, $N_{stretch} = 0$) for various polyad members, expressed in the DD4455 basis set, and for the local CH stretch bright state. Data is for $J = 2$, 30, and 100 in the [$N_{bend} = 16, g, e$] polyad. In the DD4455 basis, all of the vibrational interaction matrix elements at $J = 0$ have been prediagonalized; therefore, the plotted rates reflect only the rotationally mediated interactions, including the rotational *l*-type resonance and Coriolis interactions. This figure is Figure 10 from Perry *et al.*[24] The rates are derived from a spectroscopic Hamiltonian fitted to 19 582 vibration–rotation transitions between vibrational states with energies up to 13 000 cm^{-1}. Reprinted with permission from David Perry.

that encodes the isomerization barrier. Some of the vibrational wavefunctions become localized along the minimum energy isomerization path. The potential along the isomerization path flattens out as the barrier is approached, causing the energy spacings in this series of transition mode states (*i.e.* their effective vibrational frequency) to decrease. This means that, within a polyad that spans the isomerization coordinate (*e.g.* the *trans*-bend plus *cis*-bend normal modes or the local-bender modes of S_0 acetylene), an isomerization mode must emerge and its energy will fall far below those of the other members of the polyad. Figure 6 illustrates this reorganization of the (nonrotating) vibrational states in acetylene. It is interesting to note that the mechanism of isomerization is implicit in the normal mode frequencies (ω_i) and anharmonicities (x_{ii}, x_{ij}, and x_{jj}) that are determined by fits exclusively to the low-energy vibrational levels. For example in acetylene, the zero-order energy of the all-*cis* level in the N_{bend} polyad, $E_{cis}(N_{bend}) = \omega_5 N_{bend} + x_{55} N_{bend}^2$, becomes degenerate with that of the all-*trans* level, $E_{trans}(N_{bend}) = \omega_4 N_{bend} + x_{44} N_{bend}^2$, at $N_{bend} = (729.1-608.7)/(2.389 + 3.483) = 20.5$. This causes the polyad to turn itself inside-out, which results in non-healing qualitative changes in the eigenvector characters, especially for the highest (counter-rotator) and lowest (local-bender) energy eigenstates. The actual appearance of new classes of eigenstates occurs at a lower value of N_{bend} than this naïve prediction.

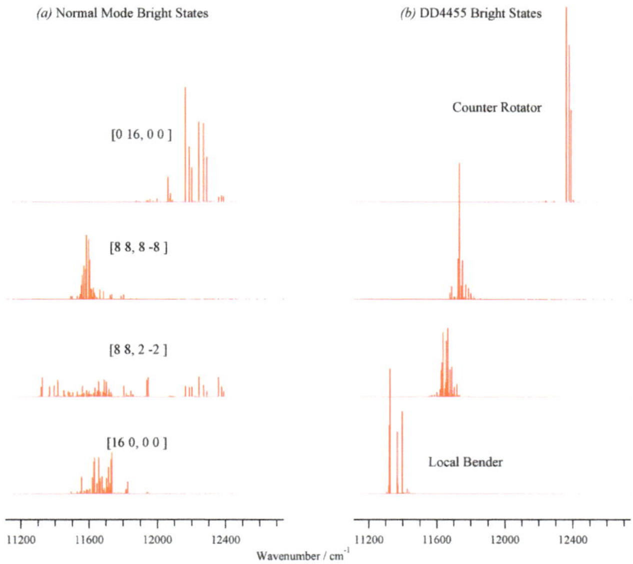

Fig. 10 Simulated spectra of different bright states for $J = 30$ in the [16,*g*,*e*] polyad of acetylene. (a) Normal mode bright states. (b) Bright states in the DD4455 basis set. The indicated wavenumbers include the vibrational energies plus about 1120 cm^{-1} of rotational energy. The fractionation in (a) results from the combination of anharmonic vibrational interactions plus the rotationally mediated interactions; whereas in (b) the fractionation arises entirely from the rotationally mediated interactions. This figure is Figure 8 from Perry *et al.*[24] Reprinted with permission from David Perry.

F CPmmW spectroscopy of the 193 nm photolysis of vinyl cyanide

C_2H_2 (S_0 acetylene and vinylidene), S_0 HCN, and S_0 HNC are dominant products of the 193 nm photolysis of vinyl cyanide. Pure rotation $J = 0$–1 *a*-dipole spectra of local-bender vibrationally excited acetylene, vinylidene, HCN, and HNC are each in principle detectable in the Field group's CPmmW spectrometer. The $J = 0$–1 transitions in many vibrational levels of each of these species are expected to be well resolved and, from the intensities of these transitions, we expected that it would be possible to determine a set of relative species-vibrational populations. However, in order to extract species-vibrational populations from our pure rotation CPmmW spectra, it is necessary to cool the nascent rotational distribution to a $T_{rot} \approx 5$ K *thermal* distribution in a supersonic expansion without significantly affecting the nascent species-vibrational population distribution. Our expectation, based on the low-J $\mathbf{H}^{eff}$ [21] and the *ab initio* results that reveal the dramatic emergence of strong local-bender vibrational transitions,[39] was that the three-centre reaction that forms HCN + vinylidene co-products populates low-lying vibrational levels of S_0 vinylidene. We expected that these vinylidene resonances would in-turn funnel population into nearby extreme local-bender vibrational levels of S_0 acetylene.

1 Experimental details

Supersonic jet cooling of the rotational degrees of freedom of photolysis products is essential. Since vibrational relaxation occurs on a much slower time-scale compared to that of rotational relaxation, it should be possible to control the amount of collisional cooling of the photolysis products by precisely locating the photolysis laser beam within the collisional region of the adiabatic expansion. In an axial jet expansion from a 1 mm diameter nozzle, however, the number density decreases as the square of the distance from the nozzle, and the entire collisional region, in which rotational cooling occurs, is contained within a distance of only a few nozzle diameters from the nozzle orifice. In previous Field group experiments it was difficult to control the amount of rotational cooling by adjusting the location of the photolysis beam within the expansion region.

A slit jet,[40–42] designed following drawings supplied by Terry Miller's group,[42] was used in the experiments described here. Two general valve solenoid drivers operate the poppet with an attached rubber cord, which seals the 5 cm long, 0.2 mm wide slit from inside of the slit jet body. Adiabatic expansion of the molecular beam produced by a slit jet is mostly one-dimensional with number density in the expanding gas decreasing linearly with the distance from the slit. The collisional region in our experiment extends 2–3 centimetres from the slit, is 3–4 cm wide, and can accommodate the unfocused photolysis laser beam with a cross section of 1 × 2.5 cm. Another advantage of the slit jet relative to the axial jet is a reduction of Doppler dephasing of the FID collected in a direction parallel to the slit. We observe a 150 kHz Doppler linewidth (in the 63–100 GHz region), which is a 2–3 fold improvement over what we could achieve with the axial jet. Clustering of the molecules expanded from a slit jet is more problematic than in an axial jet, thus greater dilution in carrier gas was required.

Vinyl cyanide (VCN) at ≥ 99% purity and its isotopologues at ≥ 98% purity were purchased from Sigma Aldrich and mixed, without further purification, with argon gas in a sample cylinder at 0.25% VCN in Ar. We found no difference in cyanovinyl radical signal strength when nitrogen was used instead of Ar as a carrier gas, but we obtained an order of magnitude weaker signal with helium, which we attribute to less efficient rotational cooling. The mixture is expanded from the slit jet, at a stagnation pressure of 150 mbar, into a vacuum chamber with approximately 60 L volume, which was pumped by a 2400 L/s turbo-molecular pump (Osaka TG2400MBWC), situated opposite to the slit jet. The turbo pump is backed by a 45 cfm two-stage rotary vane pump (Alcatel 2063). The pressure in the chamber, measured by a 0.05 Torr scale capacitance manometer (MKS 627DU5TDD1B), was 5×10^{-4} mbar when the pulsed slit jet operates at a 20 Hz repetition rate. The gas pulse duration was estimated to be 700 μs when the on-time of the multi-channel IOTA ONE driver was set to 350 μs. A 193 nm ArF excimer laser (Lambda Physik COMPex 102) with 25–50 mJ/pulse energy photolyzes the VCN molecules 1.5 cm downstream from the slit. The mm-waves were loosely focused to an ~7 cm diameter beam by two 50 cm focal length Teflon lenses located inside the vacuum chamber with their optical axis 7.5 cm downstream from the slit. Two 2.5 cm thick Teflon windows, located adjacent to the lenses transmit the chirped mm-wave pulse into and the FID out of the vacuum system.

2 Spectroscopic results

The 70–102 GHz operating region of our current CPmmW spectrometer is sufficient to capture *only* the $J = 0$–1 transitions of *all two-heavy-atom* polar molecular

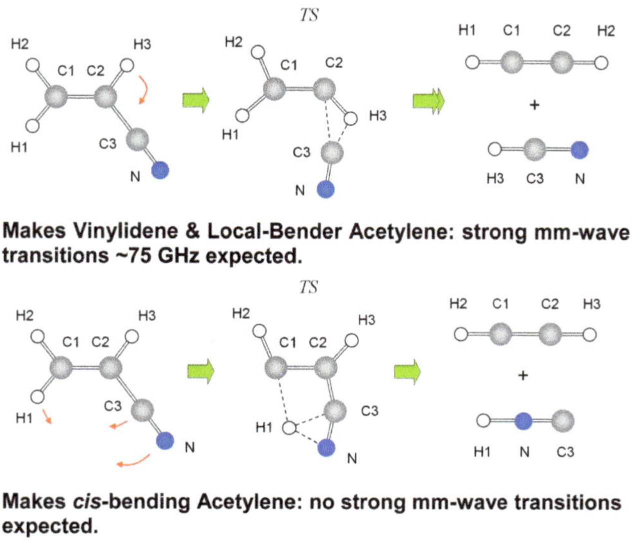

Fig. 11 A simple model for the relationships between the C_2H_2 and (HCN, HNC) photolysis co-products.[27]

products (except CN) of 193 nm photolysis of VCN. The $J = 1$–2 transitions of all two-heavy-atom molecules occur at too high frequency for our spectrometer, thus there cannot presently be any direct measurement of T_{rot}, or even a determination of whether the rotational populations are thermalized. However, there are several different polar molecules (parent VCN, photofragments, and known impurities in our VCN sample) that have more than two heavy atoms. For these molecules, several consecutive J–$J+1$ transitions are observed. The rotational populations for these larger molecules appear to be thermalized and can be used to estimate T_{rot} for the two-heavy-atom molecules. However, if a two-heavy-atom photofragment is formed at extremely high rotational excitation, there is no assurance that the intensity of a $J = 0$–1 transition bears any relationship to a nascent vibrational population. In fact, rotational cooling might be too slow to transfer population into $J = 0$ and 1 at a rate that is sufficient to yield a detectable $J = 0$–1 transition at our operating conditions.

CPmmW spectra of HCN and HNC are shown in Fig. 12. The $J = 0$–1 transitions in HCN and HNC, respectively, are labelled with downward-pointing blue and red arrows. The HCN *vs.* HNC line assignments are largely based on the presence of much larger eqQ quadrupolar hyperfine splittings in HCN than HNC. The magnitude of the eqQ constant is strongly dependent on the amount of excitation in the bending mode (ν_2) and weakly dependent on excitation in the CN stretch mode (ν_3).[7] Vibrational assignments are based on the combination of the rotational constant, B, and eqQ. Globally, excitation of the bending mode increases B, while excitation of the stretches decrease B. Some vibrational assignments are shown on Fig. 14 and 15, but with few exceptions, vibrational assignments are unambiguous. Two strong transitions near 88.48 GHz are assigned as belonging to the (0,2,1) and (1,2,0) vibrational levels.[43] The intensities of the $J = 0$–1 rotational transitions in these vibrational levels correspond

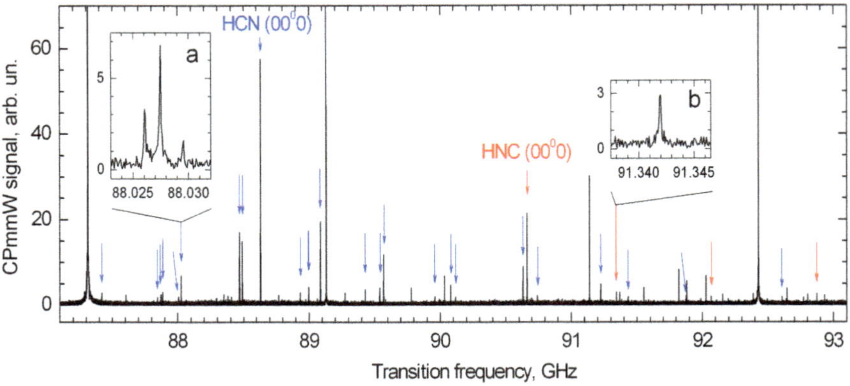

Fig. 12 The CPmmW spectrum of products of 193 nm photolysis of VCN. The $J = 0$–1 transitions in more than 25 HCN, HNC vibrational levels are identified. The insets show the characteristic eqQ hyperfine structure of HCN *vs.* HNC.

to a much higher vibrational temperature than most of the bend overtone $(0,\nu_2,0)$ levels sampled in our CPmmW spectra.

Vinyl cyanide photolysis: 3 *vs.* 4 centre transition state. There is vastly more information about the HCN and HNC products of 193 nm photolysis of VCN than was available from the mass spectrometric plus photofragment translational spectroscopy results.[6] Figure 11 illustrates the simple model that had been used to organize the previous state-of-the-art experimental results.[27] The primary co-products of the three-centre transition state (1,1 elimination) are expected to be S_0 vinylidene, local-bend excited S_0 acetylene, and bend-excited HCN, while those of the four-centre transition state (1,2 elimination) are expected to be HNC and *cis*-bend excited acetylene. All three products of the three-centre mechanism are expected to be detectable in our CPmmW spectrometer. In contrast, only the HNC product of the four-centre mechanism is

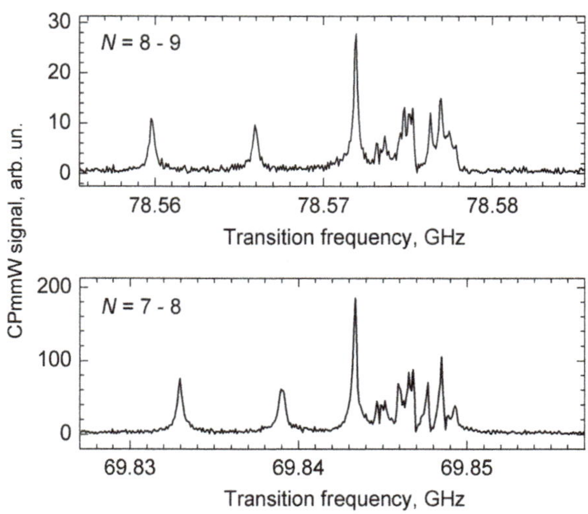

Fig. 13 $N = 8$–9 rotational transitions (top) and $N = 7$–8 transitions (bottom) in the cyanovinyl radical.[46,47]

expected to be detectable. The *cis*-bend excited acetylene is expected to be anharmonically admixed into a dense manifold of essentially ergodic vibrational levels of acetylene, which will have undetectably weak $J = 0$–1 transitions. This weakness is due to both small electric dipole transition moments and the distribution of population over too many vibrational levels.[30]

Encouraged by the very strong HCN spectra, we searched very long and hard for the $J = 0$–1 transitions in vinylidene and local-bender acetylene. At the bottom of Figure 13 is a spectrum that we initially believed was that of vinylidene, perhaps split into many components *via* interactions of an energetically remote acetylene local-bender doorway state with near degenerate ergodic acetylene vibrational levels.

How does one prove the molecular identity of a group of previously unobserved transitions in a mm-wave spectrum? High-level quantum chemical calculations of $(B + C)/2 = \bar{B}$ rotational constants for vinylidene and local-bender acetylene[14] lead to predictions of $J = 0$–1 transitions near 75 and 67 GHz, respectively. The observed cluster of lines near 70 GHz is unacceptably discrepant from the predictions. Use of $H_2{}^{13}C_2H^{13}CN$ ^{13}C isotopically substituted VCN results in a 2.55 GHz red-shift of the 69.84 GHz cluster that is compatible with vinylidene. The Zeeman effect for S_0 vinylidene should be very small, whereas that for a radical should be ~1 MHz/Gauss. Application of a few Gauss inhomogeneous magnetic field causes a few of the lines to vanish into baseline, but several of the lines appear to be resistant to shifting and broadening. This mixed verdict led us to the final test: ^{15}N substitution of the VCN. There is no N in vinylidene, yet the entire cluster of lines shifts 2.02 GHz to the red, implying that the molecule contains one N atom. Utter failure! The final nail in the coffin is the rational multiple test. For *a*-dipole transitions in a near prolate top, $J - (J + 1)$ transitions occur at $(2\bar{B})(J + 1)$. This means that if one transition is $J - (J + 1)$ the $(J + 1) - (J + 2)$ transition will occur at a frequency larger by the factor $(J + 2)/(J + 1)$. Figure 13 shows the near-perfect replication of the cluster of transitions at frequencies consistent with $N = 8$–9 (top) and $N = 7$–8 (bottom).[46,47]

Figure 14 shows the effects of a change in the amount of rotational cooling on the relative intensities of HCN and HNC $J = 0$–1 transitions. When the 193 nm laser beam passes through the VCN beam at 1.5 cm downstream from the slit jet orifice, significant rotational cooling of the photofragments occurs. However, when the laser beam is shifted to 3 cm downstream, the expansion is nearly complete and very little rotational cooling occurs. It appears that reduced rotational cooling causes the $J = 0$–1 transitions in the HCN stretch-bend combination vibrational levels and in all of the HNC vibrational levels to disappear. This suggests that the HCN stretch-bend combination levels and all of the HNC levels are formed rotationally much hotter than the HCN bending levels. Perhaps the degree of rotational excitation is a qualitative marker for fragmentation *via* distinct transition states? Degree of rotational excitation is an important class of dynamical variable that is uniquely accessible to CPmmW spectroscopy. However, we have not had an opportunity to explore systematically variations in the extent of rotational cooling or to obtain spectra with S : N improved sufficiently to reveal the transitions that are unobservably weak in the presently obtained spectra. Systematic study of this rotational signature requires, at minimum, the ability to observe the relative intensities of $J = 0$–1 *and* $J = 1$–2 transitions.

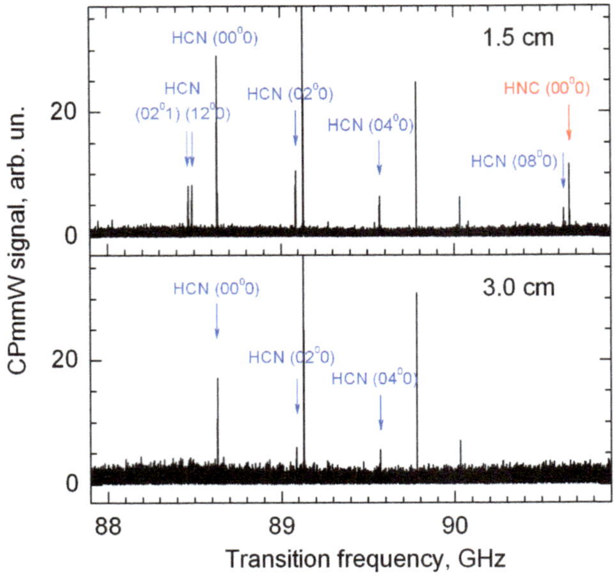

Fig. 14 The effect of rotational cooling on the CPmmW spectrum of 193 nm photolysis products. When photolysis occurs 1.5 cm from the slit jet (top panel) significant rotational cooling occurs in the supersonic jet expansion. When photolysis occurs 3 cm downstream of the slit jet (bottom panel), rotational cooling is negligible. Notice the disappearance of several transitions in the 3 cm downstream spectrum owing to the failure of the population to be relaxed all the way down to $J = 0$ and 1.

Figure 15 shows the effect of partial deuteration of the VCN. If HCN is primarily formed *via* the 3-centre transition state (1,1 elimination) then H$\rightarrow$D replacement on C-1 should significantly reduce the intensity of all HCN $J = 0$–1

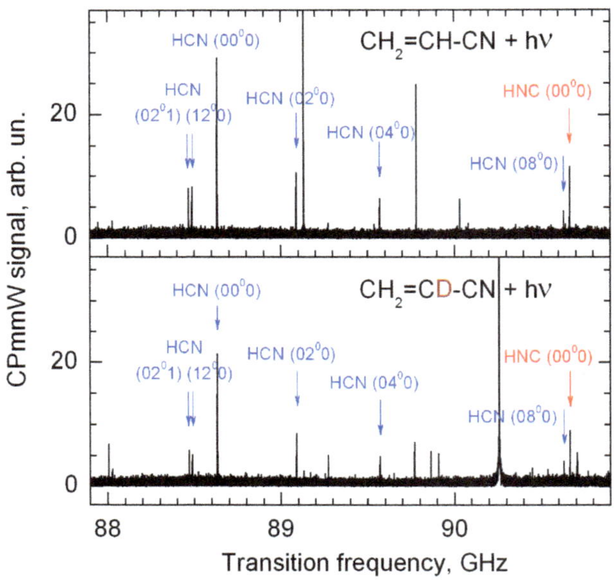

Fig. 15 The effect of deuteration of the VCN parent molecule. Deuteration at the C-1 position should shut down the three-centre transition state for production of HCN. The small change in transition intensities upon deuteration implies that the simple three-centre picture is not valid.

transitions, because DCN rather than HCN will be the primary photolysis product. DCN transitions (not shown) are detected, but at very different mm-wave frequencies. Figure 15 shows that photolysis of the D-substituted VCN parent molecule yields a reduction of only ~30% in the HCN transition intensities. It also appears that the HNC intensity (believed to be primarily a product of the four-centre transition state) is also reduced by ~30%. These are preliminary results, but they suggest that the three *vs.* four-centre picture is over-simplified.[27,44] Perhaps there is H-scrambling at the transition state? Perhaps there is CN roaming? The important point is that pure rotational spectroscopy introduces important new diagnostics for dynamics.

G Future needs

The essential weakness of pure rotational spectroscopy as a diagnostic for chemical dynamics is that, in order to provide a measure of nascent vibrational populations, it is necessary that the rotation be thermalized. Although it seems reasonable to use T_{rot}, measured for parent or fragment "thermometer molecules" that contain more than two heavy atoms, as a measure of T_{rot} for a two-heavy-atom photofragment, it is never possible to be sure that a molecule formed rotationally hot is rotationally cooled in the supersonic expansion sufficiently for the $J = 0$–1 transition to be detectable in the CPmmW spectrum. At minimum, it is necessary to observe at least two rotational transitions in every populated vibrational level. Then the T_{rot} inferred from the measured population ratio can be compared to the T_{rot} of the thermometer molecule. If the two values of T_{rot} are approximately equal, one can probably assume that rotational thermalization has occurred.

We failed to see the pure rotation spectrum of vinylidene. There are two likely reasons for this failure. One is that Franck–Condon restrictions invariably result in excitation of the parent molecule to energies well above the energy of the lowest photofragmentation transition state. Vibrationally mediated photodissociation[4] offers possibilities for excitation nearer threshold or for starting the initial wavepacket close to a selected conical intersection on the electronically excited potential energy surface. However, in the case of vinylidene, extreme rotational excitation is the enemy. We believe that, at low-J, vinylidene sees only a sparse manifold of extreme local-bender states of acetylene. At high-J, owing to strong Coriolis interactions between each extreme local-bender state and the dense manifold of "ergodic" acetylene states, the interaction between vinylidene and nearby acetylene states is promiscuous. It is Fermi Golden Rule statistical rather than selective. The relatively slow variation of vinylidene vibrational resonance widths in the PES[16] suggests that the interaction between vinylidene and vibrationally highly excited acetylene is not dependent on accidental near degeneracy of each vinylidene basis state with a local- bender basis state. The interaction is mediated by energetically remote local-bender doorway states.[30] Proof of this hypothesis will require an experimental scheme for the photolysis of a vinyl-X molecule that forms vinylidene at minimal initial rotational excitation. Infrared Multi-Photon Dissociation (IRMPD) using a CO_2 laser has been suggested.[45] This scheme could be optimized by vibrationally pre-heating the vinyl-X in a pyrolysis nozzle or by selection of an X as a leaving group that is an exceptionally good absorber of CO_2 laser photons.

H Power of CPmmW in photolysis that produces triatomic co-products

The CPmmW pure rotation spectra of HCN/HNC are extremely rich in information about the photolysis transition state(s). The vibrational and isomer specificity is unrivalled by other classes of chemical dynamics diagnostics. It is likely that all polar triatomic photofragments will carry previously unimaginable information about photolysis transition states, especially information about photolysis co-product(s).

Conclusions

CPmmW spectroscopy captures a new class of chemical dynamics information about the 193 nm photolysis of vinyl cyanide. It is clear that the photolysis mechanisms are more complicated than the simple 3-centre *vs.* 4-centre model depicted in Fig. 11.[5,27,44] Relative populations in many HCN, HNC vibrational levels are observed, but S_0 vinylidene has escaped detection. Rotation plays a critical role in the "stability" and high-resolution observability of vinylidene. When vinylidene is formed in low-J rotational levels, as in photodetachment, it is as stable as a highly excited vibrational level of a diatomic molecule. However, when vinylidene is formed rotationally highly excited, as in photolysis of V–X (X = Cl, Br, CN), it mixes into the acetylene vibrational quasi-continuum. Additional schemes for enhancing the capabilities of CPmmW spectroscopy and for observing the pure rotation spectrum of S_0 vinylidene are discussed.

Acknowledgements

We are grateful to Dr. Gabriel Just and to Prof. T. A. Miller at The Ohio State University for sharing their expertise on the slit jet design, and to Dr. Michel Costes at the University of Bordeaux for providing us with the multi-channel IOTA ONE driver. RWF thanks the Department of Energy (grant #DEFG0287ER13671) for primary support of this work (equipment and personnel support for KP, RGS, GBP, BMW), the Petroleum Research Fund (grant #50650-ND6) for the support of KP, the Air Force Office of Scientific Research (contract #FA9550-09-1-0330) for the support of MC, and the National Science Foundation (grant #1126380) for equipment and for partial support of GBP. We thank Joshua Baraban for many experimental and theoretical contributions.

Notes and references

1 G. Barratt Park, Adam H. Steeves, Kirill Kuyanov-Prozument, Justin L. Neill and Robert W. Field, *J. Chem. Phys.*, 2011, **135**, 024202.
2 G. G. Brown, B. C. Dian, K. O. Douglass, S. M. Geyer, S. T. Shipman and Brooks H. Pate, *Rev. Sci. Instrum.*, 2008, **79**, 053103.
3 B. C. Dian, G. G. Brown, K. O. Douglass and Brooks H. Pate, *Science*, 2008, **320**, 924.
4 F. Fleming Crim, *Annu. Rev. Phys. Chem.*, 1993, **44**, 397.
5 David A. Blank, Arthur G. Suits, Yuan T. Lee, Simon W. North and Gregory E. Hall, *J. Chem. Phys.*, 1998, **108**, 5784.
6 Frank C. De Lucia and Paul A. Helminger, *J. Chem. Phys.*, 1977, **67**, 4262.
7 Hans A. Bechtel, Adam H. Steeves, Bryan M. Wong and Robert W. Field, *Angew. Chem., Int. Ed.*, 2008, **47**, 2969.

8 R. J. Saykally, P. G. Szanto, T. G. Anderson and R. C. Woods, *International Symposium on Molecular Spectroscopy*, 1976, Talk TS-1, The Ohio State University; G. L. Blackman, R. D. Brown, P. D. Godfrey and H. I. Gunn, *Nature*, 1976, **261**, 395.
9 J. C. McGurk, T. G. Schmalz and W. H. Flygare, *J. Chem. Phys.*, 1974, **60**, 4181.
10 C. P. Slichter, "*Principles of Magnetic Resonance*", 1990, Springer.
11 C. H. Townes and A. L. Schawlow, "*Microwave Spectroscopy*", 1955, McGraw Hill.
12 E. Weitz and G. W. Flynn, *Annu. Rev. Phys. Chem.*, 1974, **25**, 275.
13 S. Zhou, J. M. Bowman and A. Brown, *J. Chem. Phys.*, 2003, **118**, 10012.
14 Hyunwoo Lee, Joshua H. Baraban, Robert W. Field, and John F. Stanton, *J. Chem. Phys.*, Submitted.
15 S. M. Burnett, A. E. Stevens, C. S. Feigerle and W. C. Lineberger, *Chem. Phys. Lett.*, 1983, **100**, 124.
16 Kent M. Ervin, J. Ho and W. Carl Lineberger, *J. Chem. Phys.*, 1989, **91**, 5974.
17 J. Levin, H. Feldman, A. Baer, D. Ben-Hamu, O. Heber, D. Zajfman and Z. Vager, *Phys. Rev. Lett.*, 1998, **81**, 3347.
18 David H. Mordaunt and Michael N. R. Ashfold, *J. Chem. Phys.*, 1994, **101**, 2630.
19 Bryan W. Wong, Adam H. Steeves and Robert W. Field, *J. Phys. Chem. B*, 2006, **110**, 18912.
20 Bryan M. Wong, Ryan L. Thom and Robert W. Field, *J. Phys. Chem. A*, 2006, **110**, 7406.
21 Mathew P. Jacobson and Robert W. Field, *J. Phys. Chem.*, 2000, **104**, 3086.
22 Matthew P. Jacobson, Robert J. Silbey and Robert W. Field, *J. Chem. Phys.*, 1999, **110**, 845.
23 Matthew P. Jacobson, Jonathan P. O'Brien, Robert J. Silbey and Robert W. Field, *J. Chem. Phys.*, 1998, **109**, 121.
24 David S. Perry, J. Martens, B. Amyay and Michel Herman, *Mol. Phys.*, 2012, **110**, 2687.
25 Y. Huang, Y. Yang, G. He, S. Hashimoto and Robert J. Gordon, *J. Chem. Phys.*, 1995, **103**, 5476.
26 Michael J. Wilhelm, Matthew Nikow, Laura Letendre and Hai-Lung Dai, *J. Chem. Phys.*, 2009, **130**, 044307.
27 A. Derecskei-Kovacs and S. W. North, *J. Chem. Phys.*, 1999, **110**, 2862.
28 T. Carrington, L. M. Hubbard, H. F. Schaefer III and W. H. Miller, *J. Chem. Phys.*, 1989, **80**, 4347.
29 Michel Herman, in "*Handbook of High Resolution Spectroscopies*", Martin Quack and Frederic Merkt (editors), John Wiley and Sons, 2010.
30 K. L. Bittinger and Robert W. Field, *J. Chem. Phys.*, 2010, **132**, 134302.
31 Adam H. Steeves, Ph. D. Thesis, 2009, Department of Chemistry, M. I. T.
32 David M. Jonas, Ph.D. Thesis, 1992, Department of Chemistry, M. I. T.
33 M. P. Jacobson, S. L. Coy and R. W. Field, *J. Chem. Phys.*, 1997, **107**, 8349–8357.
34 Robert W. Field, Joshua H. Baraban, Samuel H. Lipoff, and Annelise R. Beck, in "*Handbook of High-Resolution Spectroscopies*", Martin Quack and Frederic Merkt (editors), John Wiley and Sons, 2010.
35 M. P. Jacobson, C. Jung, H. S. Taylor and R. W. Field, *J. Chem. Phys.*, 1999, **111**, 600.
36 A. E. Douglas, *J. Chem. Phys.*, 1966, **45**, 1007; G. M. Stewart and J. D. McDonald, *J. Chem. Phys.*, 1983, **78**, 3907; G. A. Bethardy, X. Wang and D. S. Perry, *Can. J. Chem.*, 1994, **72**, 652.
37 Matthew P. Jacobson and Robert W. Field, *Chem. Phys. Lett.*, 2000, **320**, 553.
38 M. E. Kellman, *Annu. Rev. Phys. Chem.*, 1995, **46**, 395, http://darkwing.uoregon.edu/~meklab/migration2/index.html.
39 Hua Guo, unpublished private communication.
40 C. M. Lovejoy and D. J. Nesbitt, *Rev. Sci. Instrum.*, 1987, **58**, 807.
41 K. Liu, R. S. Fellers, M. R. Viant, R. P. McLaughlin, M. G. Brown and R. J. Saykally, *Rev. Sci. Instrum.*, 1996, **67**, 410.
42 S. H. Wu, P. Dupre and T. A. Miller, *Phys. Chem. Chem. Phys.*, 2006, **8**, 1682.
43 Z. Zelinger, T. Amano, V. Ahrens, S. Brunken, F. Lewen, H. S. P. Muller and G. Winnewisser, *J. Mol. Spectrosc.*, 2003, **220**, 223.
44 Zahra Homayoon, Saulo A. Vazquez, Roberto Rodrigues-Fernandez and Emilio Martinez-Nunez, *J. Phys. Chem. A*, 2011, **115**, 979.
45 P. A. Schulz, Aa. S. Sudbo, D. J. Krajnovich, H. S. Kwok, Y. R. Shen and Y. T. Lee, *Annu. Rev. Phys. Chem.*, 1979, **30**, 379.
46 K. Seiki, Y. Sumiyoshi, Y. Oshima, Y. Endo and J. Tang, Detection of the H_2CCCN radical by FTMW spectroscopy and observation of the tunneling inversion motion, *Bunshi Kozo Sogo Toronkai Koen Yoshishu*, 2000, **2000**, 11 (in Japanese).
47 L. Letendre and H. L. Dai, *J. Phys. Chem. A*, 2002, **106**, 12035.

Faraday Discussions RSC Publishing

PAPER

Time-resolved photoelectron imaging of the iodide–thymine and iodide–uracil binary cluster systems

Sarah B. King,[a] Margaret A. Yandell[a] and Daniel M. Neumark*[ab]

Received 22nd December 2012, Accepted 15th February 2013
DOI: 10.1039/c3fd20158a

The energetics and dynamics of thymine and uracil transient negative ions were examined using femtosecond time-resolved photoelectron imaging. The vertical detachment energies (VDEs) of these systems were found to be 4.05 eV and 4.11 eV for iodide–thymine ($I^-\cdot T$) and iodide–uracil ($I^-\cdot U$) clusters, respectively. An ultraviolet pump pulse was used to promote intracluster charge transfer from iodide to the nucleobase. Subsequent electron detachment using an infrared probe pulse monitored the dynamics of the resulting transient negative ion. Photoelectron spectra reveal two primary features: a near-zero electron kinetic energy signal attributed to autodetachment and a transient feature representing photodetachment from the excited anion state. The transient state exhibits biexponential decay in both thymine and uracil complexes with short and long decay time constants ranging from 150–600 fs and 1–50 ps, respectively, depending on the excitation energy. However, both time constants are systematically shorter for $I^-\cdot T$. Vibrational autodetachment and iodine loss are identified as the primary decay mechanisms of the transient negative ions of thymine and uracil.

1 Introduction

Ionizing radiation is a major cause of damage to DNA.[1,2] Though experiments and theory estimate the lowest ionization potentials for DNA components to be between 8–11 eV,[3–7] electrons with less than 3 eV of kinetic energy have been demonstrated to induce both single and double strand breaks in DNA.[8] The mechanism of strand cleavage by ionizing radiation below the relevant ionization potentials has been the subject of much recent study. Theoretical work by Simons[9] and Schaefer[10] posits that these strand breaks result from electron attachment to the empty π^* orbital of the nucleobase followed by energy transfer and cleavage of the C–O σ bonds in the sugar phosphate backbone of the DNA strand. Cleavage of the N–C bond linking the sugar to the nucleobase at the N1 position (Fig. 1) and N–H cleavage at the N3 position may also occur following electron transfer from the π^* orbital of the

[a]*Department of Chemistry, University of California, Berkeley, CA 94720, USA. E-mail: dneumark@berkeley.edu*
[b]*Chemical Sciences Division, Lawrence Berkeley National Laboratory, Berkeley, CA 94702, USA*

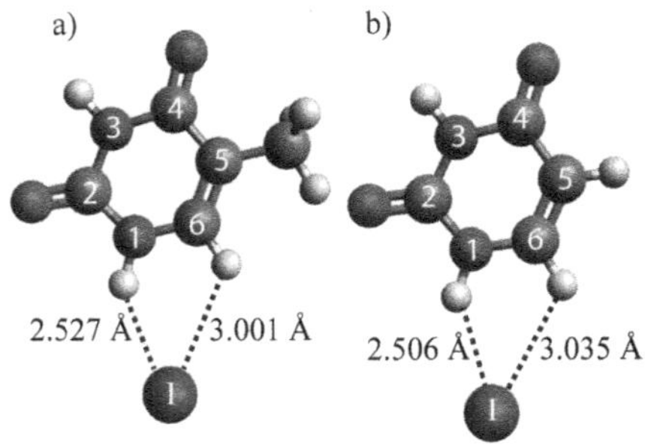

Fig. 1 Structures of a) iodide–thymine and b) iodide–uracil clusters with calculated bond distances indicated.

nucleobase. Understanding the mechanisms driving relaxation and fragmentation of nucleobases subsequent to charge transfer will provide insight into how low energy electrons cause DNA damage. In this article, we explore time-resolved radiation chemistry by performing one-photon photoelectron (PE) spectroscopy and two-color time-resolved PE spectroscopy of iodide–thymine ($I^- \cdot T$) and iodide-uracil ($I^- \cdot U$) complexes, expanding on our previous report of $I^- \cdot U$.[11]

Extensive experimental and theoretical work has been carried out on the nucleobases of DNA and RNA to gain insight into the mechanisms of DNA damage. Dissociative electron attachment (DEA) experiments examining dissociation pathways of isolated nucleobases as a function of incident electron collision energy[12–16] find that the dominant channel at low collision energy is production of the deprotonated DNA base $(B-H)^-$ *via* hydrogen loss from the N1 position. Sharp resonances in the DEA spectra at and just below 1 eV have been attributed to vibrational Feshbach resonances resulting from the coupling of dipole-bound states of the isolated nucleobase to the σ^* orbital of the N1–H bond, leading to fragmentation and hydrogen loss.[17,18]

PE spectroscopy and Rydberg electron transfer (RET) experiments demonstrate that the anions of the isolated nucleobases are preferentially formed in dipole-bound states rather than conventional valence-bound states.[19–22] Dipole-bound anions readily form for molecules with dipole moments greater than 2.5 Debye,[23] and isolated nucleobases have dipole moments on the order of 4 D.[24,25] Bowen and co-workers[20] recorded PE spectra characteristic of dipole-bound states for thymine and uracil anions, yielding adiabatic electron affinities (AEAs) for uracil and thymine of 93 $\pm$ 7 meV and 69 $\pm$ 7 meV, respectively. These spectra are dominated by a single narrow peak representing the transition between the anion and neutral ground vibrational states due to very similar anion and neutral geometries. There is, however, experimental and theoretical evidence of low-lying valence states of nucleobase anions. These species are expected to have relatively broad PE spectra and higher vertical detachment energies (VDEs, defined as the difference in energy between the anion and neutral at the geometry of the anion) than dipole-bound states.[26] RET experiments on nucleobases by Schermann and co-workers[21,22] suggested that valence-bound states of uracil could be formed by electron attachment to $U(Ar)_n$ clusters, followed by Ar evaporation.

Experiments by Bowen[19] on uracil-rare gas binary cluster anions found that the uracil anion remains a dipole-bound state upon complexation with Ar or Kr. However, the PE spectrum of the uracil-Xe anion complex showed evidence of both a dipole-bound and valence-bound uracil anion, with the latter having a VDE of ~0.6 eV. Only the valence-bound state was seen in the binary complex with water. Schiedt *et al.*[27] reported PE spectra of anionic DNA bases solvated by one or more water molecules. By extrapolating to zero water molecules, the electron affinities of uracil and thymine to form valence-bound states were estimated to be 150 ± 120 meV and 120 ± 120 meV respectively, within the range of the electron affinity to form dipole-bound states. Calculations report VDEs of the valence-bound anions of 0.5 eV for thymine[28] and 0.6 eV for uracil[29,30] and suggest that a stable valence-bound anion should exist in a puckered ring geometry.[26]

Here we apply conventional PE spectroscopy and femtosecond time-resolved PE spectroscopy[31,32] to iodide–thymine ($I^-\cdot T$) and iodide–uracil ($I^-\cdot U$) anion complexes in order to study the energetics and dynamics of nucleobase transient negative ions. Calculated structures, see section 3.3, for both complexes are shown in Fig. 1. Using an ultraviolet pump pulse, the excess electron photodetached from the iodide can interact with the nucleobase, forming a transient negative ion $I\cdots B^{*-}$ in analogy to electron scattering studies. In the time-resolved experiments, a probe pulse detaches the electron from the transient negative ion of thymine or uracil, yielding the decay times of these species and providing insight into their dynamics. Through these experiments we hope to understand the energetics and dynamics of the thymine and uracil transient negative ions.

2 Experimental details

The experimental apparatus has been described in detail elsewhere.[33,34] Clusters of $I^-\cdot T$ and $I^-\cdot U$ are produced by passing argon gas at 50 psig over a reservoir containing methyl iodide. This gas mixture is expanded through an Even-Lavie pulsed solenoid valve operating at 500 Hz with a cartridge containing the sample heated to 205 °C. The pulsed beam passes through a ring filament ionizer creating iodide–nucleobase anionic clusters. The anionic clusters are accelerated perpendicularly and separated in time using a Wiley–McLaren time-of-flight (TOF) mass spectrometer.[35] The mass spectrum is calibrated using the characteristic peak progression of $I^-(Ar)_n$ clusters. The cluster size of interest is mass-selected and crossed with one or two laser beams perpendicular to the TOF axis. The resulting photoelectrons are accelerated collinearly by a set of velocity map imaging plates[36] onto a position sensitive detector consisting of two chevron stacked microchannel plates, a phosphor screen, and a charge-coupled device camera.

The femtosecond laser pulses used in our experiment are generated from a Ti:Sapphire oscillator and multipass amplifier (KM Labs Griffin Oscillator and Dragon Amplifier). The laser operates at a 1 kHz repetition rate and produces 1.8 mJ per pulse centered at 790 nm. An optical chopper reduces the repetition rate of the laser to 500 Hz to match the repetition rate of the pulsed solenoid valve. In order to generate the UV pump wavelengths used in the experiment, a portion of the 790 nm pulse is either frequency-tripled to generate 265 nm or sent into an optical parametric amplifier (Light Conversion TOPAS-C). The output of the TOPAS can be frequency-doubled to produce nearly continuous wavelengths from 235 nm to 330 nm (5.27–3.75 eV), with energies of approximately 10 μJ/pulse. Part of the remaining fundamental at 790 nm (1.57 eV) serves as the probe pulse. Measurements of the cross correlation between the pump and the probe directly before the laser pulses enter the experimental apparatus yield a cross correlation of approximately 150 fs.

The photoelectron images obtained on our detector are reconstructed following four-way symmetrization using the basis-set expansion method (BASEX).[37] Both electron kinetic energy (eKE) distributions as well as photoelectron angular distributions (PADs), which will not be considered in this paper, can be obtained from the reconstructed images. The electron kinetic energy distributions were calibrated using the well-known spectrum of iodide.

3 Results

3.1 One-photon photoelectron spectra

Fig. 2a and 2b show one-photon photoelectron spectra for $I^-{\cdot}T$ and $I^-{\cdot}U$ measured at excitation energies ranging from 4.10–5.30 eV. For ease of comparison at different photon energies, the spectra are plotted as a function of electron binding energy, *eBE*, where $eBE = h\nu - eKE$. Both the iodide–thymine and iodide–uracil spectra show electron signal due to direct detachment from the anion to the neutral iodine–nucleobase complex (feature A) as well as photoelectrons with nearly zero kinetic energy (feature C), found where $eBE \approx h\nu$. At 5.30 eV, an additional signal (feature B) is seen. This feature represents detachment to the $^2P_{1/2}$ state of iodine complexed to the nucleobase. Vertical detachment energies of both anions can be derived by fitting feature A with a Gaussian function to find the peak center, yielding values of 4.05 ± 0.05 eV and 4.11 ± 0.05 eV for $I^-{\cdot}T$ and $I^-{\cdot}U$, respectively.

For both anion complexes, feature C is observed as a distinct feature from 5.30 eV to 4.20 eV, a range of 0.90 eV. Near-zero kinetic energy photoelectron signal is observed for excitation energies as low as 3.76 eV for both complexes, but below 4.20 eV only a single peak or rising edge is seen that cannot definitively be assigned as feature A or feature C. Table 1 gives the ratio of intensities of features C and A for both species at excitation energies where both features are seen and clearly defined. This ratio is highest at 4.77 eV for both complexes, although $I^-{\cdot}T$ complexes have consistently higher C/A ratios than $I^-{\cdot}U$ complexes at all photon energies.

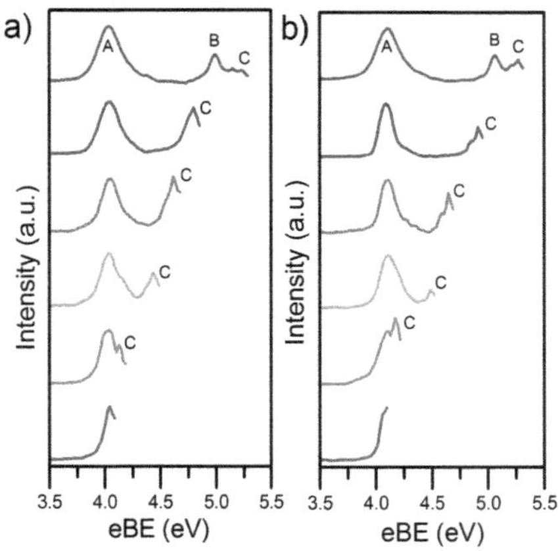

Fig. 2 One-photon photoelectron spectra at different photon energies of iodide–thymine a), and iodide–uracil b), where A indicates photoelectrons from direct detachment to the $^2P_{3/2}$ neutral state and, and B to the $^2P_{1/2}$ neutral state, and C zero kinetic energy electrons. From top to bottom photon energies are as follows in a) 5.30 eV, 4.87 eV, 4.70 eV, 4.51 eV, 4.20 eV, and 4.10 eV and in b) 5.30 eV, 4.92 eV, 4.68 eV, 4.51 eV, 4.20 eV, and 4.10 eV.

Table 1 Ratios of feature C to feature A

Pump energy (eV)	Thymine	Uracil
4.92	—	0.36
4.87	0.56	—
4.77	0.62	0.42
4.70	0.57	—
4.68	—	0.36
4.51	0.33	0.13

3.2 Two-photon time-resolved photoelectron spectra of $I^-\cdot T$ and $I^-\cdot U$

Fig. 3 shows a representative time-resolved photoelectron spectrum of $I^-\cdot T$ excited at 4.79 eV and probed at 1.57 eV, plotting photoelectron signal *vs.* eKE at a series of pump-probe delays. Three features are apparent in the spectra: the direct detachment feature A, near-zero kinetic energy electrons (feature C), and a broad, low intensity signal (feature D) with eKE ranging between 1.0–1.6 eV. Features A and C appear at all pump-probe delays, while the transient feature D appears at zero pump-probe delay, t_0, and persists for tens of picoseconds. The rise and decline of the transient feature is mirrored by depletion and subsequent recovery in the intensity of feature C.

Fig. 4 displays the time-dependent integrated intensities of features C and D for $I^-\cdot T$ measured with an excitation energy of 4.69 eV at short and long time delays. The open circles and squares represent measured intensities for features C and D, respectively, while the curves are obtained from the fitting procedure described in the next section. At all pump energies, feature D rises abruptly at t_0, then decays monotonically over a time scale of 100–300 fs. A more detailed analysis in the next section shows that feature D decays biexponentially at all excitation energies studied. Features C and D show complementary dynamics: feature C has non-zero intensity at negative time delays, is abruptly depleted at t_0, then recovers on time scales similar to the decay of D. Both signals reach their asymptotic values by 5 ps.

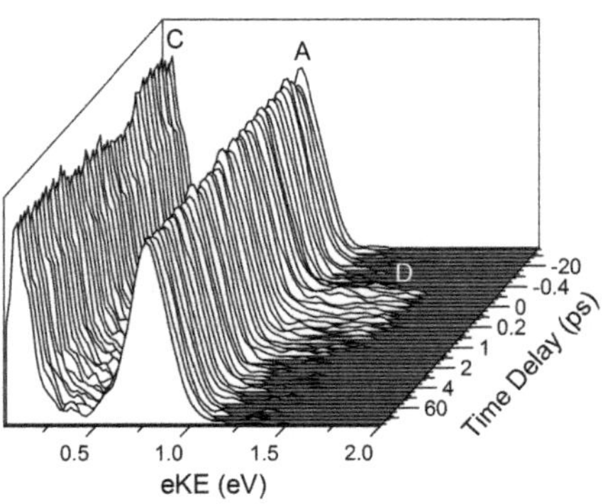

Fig. 3 Time-resolved photoelectron spectrum of the iodide–thymine cluster excited at 4.79 eV and probed at 1.57 eV. Features A, C, and D represent direct detachment of the anion to form the corresponding neutral species, vibrational autodetachment, and photodetachment of the transient negative ion, respectively.

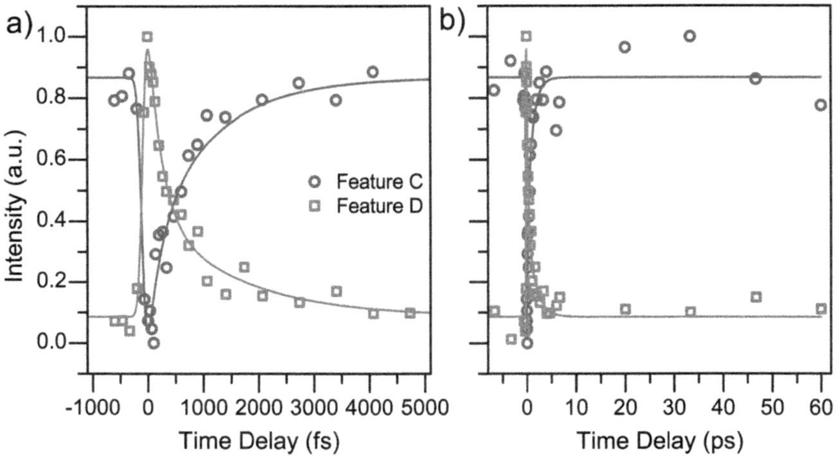

Fig. 4 Normalized integrated intensities for features C and D in $I^-\cdot T$ clusters excited at 4.69 eV and probed at 1.57 eV, where a) shows short time dynamics and b) shows long time dynamics.

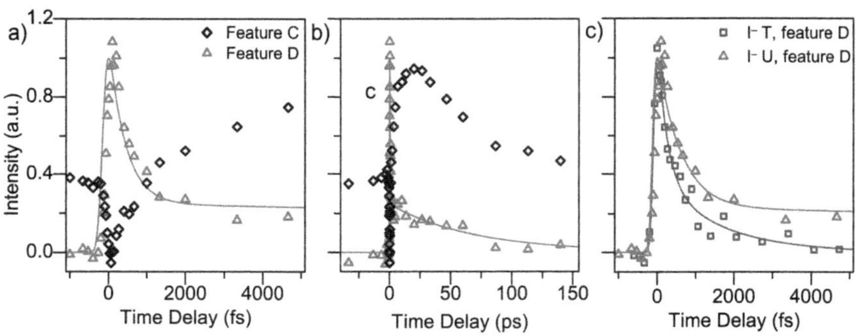

Fig. 5 Normalized integrated intensities for $I^-\cdot U$ clusters excited at 4.69 eV and probed at 1.57 eV where a) shows the short time dynamics of features C and D, b) the long time dynamics and c) a comparison between the short time dynamics of feature D in $I^-\cdot U$ with that of $I^-\cdot T$ excited at 4.69 eV.

Analogous plots for iodide–uracil clusters based on our earlier work[11] are presented in Fig. 5a and 5b. As in $I^-\cdot T$, the intensity of feature D sharply increases at t_0 and then undergoes biexponential decay. Fig. 5c compares the short-time dynamics of feature D for $I^-\cdot T$ and $I^-\cdot U$ at the same pump energy, showing slower decay at short times and more persistent signal at longer times for $I^-\cdot U$. Feature C is also quite different in $I^-\cdot U$. While this feature shows similar depletion and recovery for both complexes at early times, it subsequently recovers past its initial intensity in $I^-\cdot U$. Maximum intensity is achieved after about 25 ps, followed by decay to its initial value over tens of picoseconds.

3.3 Electronic structure calculations

The structures of the iodide–nucleobase complexes were optimized using second order perturbation theory in the Gaussian 09 software package[38] with an augmented double zeta Dunning-type basis and an aug-cc-pVDZ-pp pseudopotential[39] for iodide. Calculated structures of the $I^-\cdot T$ and $I^-\cdot U$ complexes are shown in Fig. 1, with calculated bond distances indicated between iodide and the

N1 and C6 protons. The excess charge is localized on the iodine atom. The iodide resides between the hydrogen bound to N1 and the neighboring vinyl hydrogen in both clusters and appears to interact somewhat more with the hydrogen bound to C6 in the thymine complex than in the uracil complex. The presence of iodide leads to slight elongation of the N1–H and C5–C6 bonds relative to un-complexed species,[26,40] as well as bowing of both the N1 and C6 hydrogen atoms toward the halide. Single point energy calculations carried out using CCSD with the same basis sets estimate VDEs as 4.09 and 4.13 eV for thymine- and uracil-containing complexes, respectively. The calculated geometry and VDE for $I^-\cdot U$ agree well with a recent calculation by Ortiz[41] using density functional theory.

4 Analysis

The integrated intensity of feature D and, for $I^-\cdot T$, feature C, can be fit to eqn (1), representing a convolution of a Gaussian instrumental response function, with a FWHM of the measured cross correlation of 150 fs, and a step function at $t = t_0$ that then evolves bi-exponentially:

$$I(t) = e^{-t^2/\sigma^2} * \begin{cases} I_0 & t < t_0, \\ I_0 + A_0\delta(t - t_0) + A_1 e^{-(t-t_0)/\tau_1} + A_2 e^{-(t-t_0)/\tau_2} & t \geq t_0 \end{cases} \quad (1)$$

Table 2 Time constants of feature D in iodide–thymine and iodide–uracil clusters

Nucleobase	Pump energy (eV)	A_1/A_2	τ_1 (fs)	τ_2 (ps)
Thymine	4.60	3.8	300 ± 50	1.9 ± 1.0
	4.69	2.4	200 ± 40	1.5 ± 0.4
	4.79	1.5	160 ± 60	0.9 ± 0.2
Uracil[11]	4.69	7.0	620 ± 50	52 ± 20
	4.79	5.6	390 ± 80	37 ± 20
	4.90	5.6	300 ± 50	12 ± 6

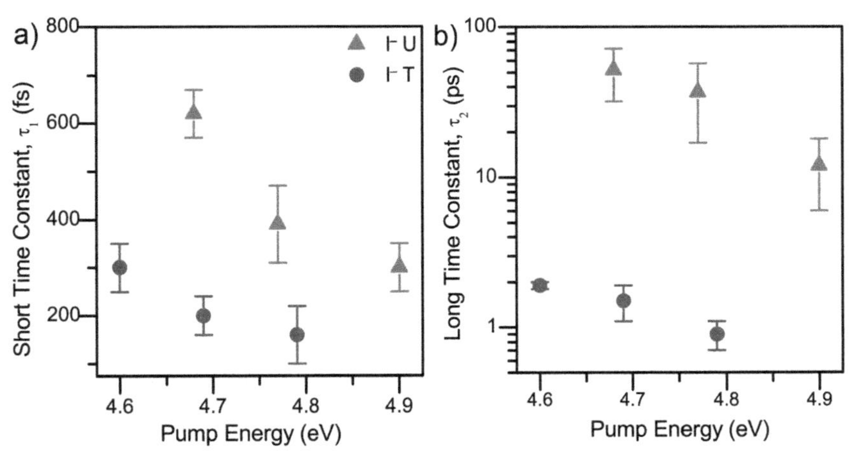

Fig. 6 a) Short and b) long time constants b) in Table 2 *versus* excitation energy for $I^-\cdot T$ and $I^-\cdot U$ clusters.

Table 2 lists the time constants τ_1 and τ_2 and the amplitude ratio A_1/A_2 for $I^-\cdot T$ and $I^-\cdot U$ at the various excitation energies used in our experiment, while Fig. 6a and 6b plot the two time constants against excitation energy. For both systems, τ_1 and τ_2 decrease with increasing excitation energy. The time constants τ_1 and τ_2 are systematically shorter for $I^-\cdot T$. This trend is particularly striking for τ_2 values, which are over an order of magnitude shorter for $I^-\cdot T$ than for $I^-\cdot U$. Moreover, while the time-dependence of feature C for $I^-\cdot T$ can be completely described by eqn (1) with the same constants that fit feature D (except that A_1 and A_2 are negative), this functional form cannot capture the long-time dynamics of feature C in $I^-\cdot U$.

5 Discussion

5.1 Nature of iodide–nucleobase clusters

In one-photon PE spectra of both $I^-\cdot T$ and $I^-\cdot U$ at 5.30 eV, the two direct detachment features A and B are split by the spin orbit splitting of iodine, indicating the anion has the electronic character of iodide, and the excess electron in the ground state of these complexes is localized on the iodine atom. This interpretation is fully supported by the electronic structure calculations reported here and elsewhere.[41]

The one-photon photoelectron spectra in Fig. 2 show clear evidence of very slow electrons, feature C, over a wide range of excitation energies. This feature is assigned to vibrational autodetachment from the transient negative ion (TNI), $I\cdots B^{*-}$, formed after charge transfer from iodide to the nucleobase, as follows:

$$I^-B \xrightarrow{h\nu_{\text{pump}}} I\cdots B^{*-} \xrightarrow{k_1} I\cdots B + e^- \tag{2}$$

The resulting photoelectrons have nearly zero kinetic energy. A similar slow electron signal, seen in our previous work on iodide–solvent clusters, was also ascribed to autodetachment.[42–45] The TNI is also the source of pump–probe signal observed as feature D in the time-resolved experiments according to:

$$I^-B \xrightarrow{h\nu_{\text{pump}}} I\cdots B^{*-} \xrightarrow{h\nu_{\text{probe}}} I\cdots B + e^- \tag{3}$$

The resulting photoelectrons are ejected with a kinetic energy given by $eKE = h\nu_{\text{probe}} - VDE(I\cdots B^{*-})$ where the last term is the VDE of the TNI in eqn (3). The eKE of feature D ranges from 1.0–1.6 eV. Owing to the diffuse nature of this transient signal, we cannot definitively assign the VDE of the TNI but can place it in the range of 0–0.6 eV, which includes the calculated VDEs of 0.5 and 0.6 eV for the valence bound anions of thymine and uracil.[28,26] Notably, we can see no spectroscopic evidence of the sharp peak characteristic of dipole-bound nucleobases at any pump-probe delay. We therefore conclude that the TNI formed in our experiment upon UV initiated charge transfer is the valence-bound state of the nucleobase. Anion photoelectron spectroscopy[19] and Rydberg electron transfer[22] experiments show that complexation to a polarizable species stabilizes valence-bound anions relative to dipole-bound states, and we may be seeing a manifestation of that effect here owing to the nearby iodine atom.

Our experiment can be considered in terms of intracluster electron scattering, where the electron produced by photodetachment from the iodide moiety is either ejected into free space, resulting in the direct detachment features A and B, or is indirectly scattered from the nucleobase forming the TNI leading to the time-dependent features C and D. A similar mechanism was also proposed by Continetti and co-workers[46] in their dissociative photodetachment experiments on the iodide–aniline complex.

Since thymine and uracil absorb at the excitation energies used in our experiment,[47,48] we must also consider the possibility that UV absorption by the nucleobase could excite the nucleobase to a $\pi \rightarrow \pi^*$ state, followed by transfer of an electron from iodide into the π hole of the nucleobase and forming the same $I\cdots B^{*-}$ state as in in eqn (2) and (3). There are several arguments against this mechanism, however. First, Fig. 4 and 5 show formation of the TNI immediately at t_0, with an appearance rate limited by the Gaussian shape of our laser pulses, whereas electron transfer from iodide to an electronically excited nucleobase would likely result in delayed formation of the TNI. Secondly, the gas phase absorption cross section for uracil increases by more than a factor of ten from 4.2 to 5.0 eV,[48] and the gas phase spectrum of thymine also rises steeply above its onset around 4.4 eV.[49] The autodetachment signal in Fig. 2 does not show a corresponding increase with energy, however, indicating that it is not correlated with UV absorption by the nucleobase.

5.2 Time-resolved dynamics

The transient feature D in the time-resolved photoelectron spectra decays bi-exponentially for both $I^-\cdot T$ and $I^-\cdot U$ clusters. Key trends are shown in Table 2 and Fig. 6a and 6b. For both complexes, the two time constants τ_1 and τ_2 decrease with increasing excitation energy. This result is consistent with a statistical decay process such as vibrational autodetachment in eqn (2), as shown in our recent work on other iodide–solvent clusters.[50] The fast time constant τ_1 ranges from 150–600 fs. Fig. 6a shows that at comparable excitation energies, τ_1 is more than a factor of two smaller for thymine than uracil complexes. For example, at 4.69 eV, $\tau_1 = 200$ fs for $I^-\cdot T$ but is 620 fs for $I^-\cdot U$. The slow time constant τ_2 varies from 1–2 ps for $I^-\cdot T$ and is considerably longer (12–52 ps) for $I^-\cdot U$.

The bi-exponential decay of feature D indicates that the $I\cdots B^{*-}$ state created by the pump pulse does not decay solely by autodetachment. Instead it must also decay to a longer-lived anion state that can be detached with a 1.57 eV probe photon. In $I^-\cdot T$, recovery of the autodetachment signal C is bi-exponential and is fit well with the same time constants as those used to fit feature D. This correspondence indicates that in $I^-\cdot T$, both the initial $I\cdots B^{*-}$ state and its anionic decay product undergo autodetachment. The situation for autodetachment from the uracil complex is more complicated and is considered in more detail below. An overall mechanism consistent with our results is as follows:

$$I^-B \xrightarrow{h\nu_{\text{pump}}} I\cdots B^{*-} \xrightarrow{h\nu_{\text{probe}}} I\cdots B + e^- \quad (4a)$$
$$I\cdots B^{*-} \xrightarrow{k_1} I\cdots B + e^- \quad (4b)$$
$$I\cdots B^{*-} \xrightarrow{k_2} I + B^{*-} \quad (4c)$$
$$B^{*-} \xrightarrow{h\nu_{\text{probe}}} B + e^- \quad (4d)$$
$$B^{*-} \xrightarrow{k_3} B + e^- \quad (4e)$$

Here, the initially formed $I\cdots B^{*-}$ species can be detached by a probe photon (4a), autodetach (4b), or lose an iodine atom (4c). The B^{*-} fragment from (4c) can also be photodetached by a probe photon (4d) or autodetach (4e). The iodine atom should be bound to the nucleobase by ~50 meV,[42] so the B^{*-}fragment should have less vibrational energy than the initially formed $I\cdots B^{*-}$ resulting in an autodetachment rate k_3 slower than k_1. According to this scheme, the two time constants τ_1 and τ_2 are related to the three rate constants in (4) by:

$$\tau_1 = 1/(k_1 + k_2),\ \tau_2 = 1/k_3 \tag{5}$$

The overall mechanism in eqn (4) explains the decay of feature D in both uracil and thymine and the dynamics of feature C in thymine. However, it does not explain the overshoot in feature C observed in uracil. As has been argued previously,[11,51] this overshoot indicates enhancement of the autodetachment signal by the probe pulse. Fig. 5b shows that the time interval during which feature C exceeds its initial level roughly coincides with the long-time decay of feature D. Hence, the overshoot may occur through probe absorption by the U^{*-} fragment, produced in step (4c), to an autodetaching state. In other words, the probe pulse may not only induce direct detachment (4d) but may also enhance autodetachment. Once the U^{*-} fragment has decayed by (4e) the probe pulse can no longer be absorbed and the overshoot disappears. Note that because τ_2 is considerably shorter in $I^-\cdot T$, the lifetime of the fragment that absorbs the probe pulse may be too short to see significant probe-enhanced autodetachment, which might explain the absence of this effect in $I^-\cdot T$ complexes.

We should also consider other possible contributions to the time-resolved dynamics seen in our experiments, summarized in Fig. 7 along with previously discussed iodine loss. DEA experiments show that low-energy electron collisions with uracil[12,52] and thymine[15] lead to formation of the deprotonated $(U\text{-}H)^-$ and $(T\text{-}H)^-$ anions *via* hydrogen atom loss from the N1 position. In the low energy regime, the yield for this channel shows an onset of 0.6 eV, a small peak at 0.7 eV

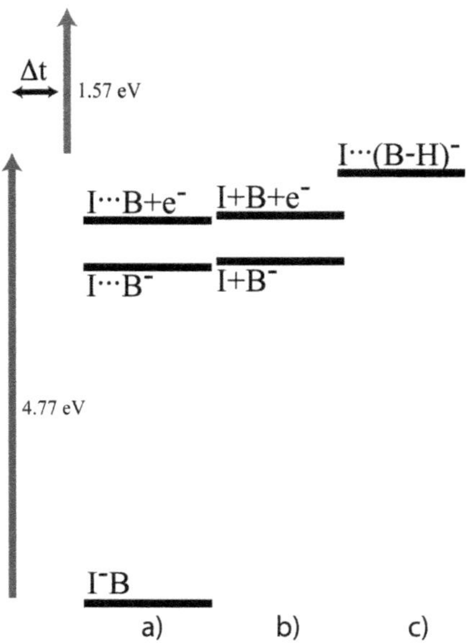

Fig. 7 Energy diagram showing approximate relative energies of possible fragmentation pathways of thymine and uracil subsequent to UV initiated charge transfer from iodide where a) shows UV initiated charge transfer, b) iodine loss, and c) hydrogen loss.

and a much larger resonance at 1 eV.[17] These results raise the question of whether the analogous channel is energetically accessible in our experiments, whereby electron transfer to the nucleobase from the iodide leads to H-atom loss. Calculations by Martínez *et al.*[41] find that removal of a hydrogen atom from the N1 position in $I^- \cdot U$ requires 4.6 eV, which falls in the range of excitation energies used in our experiments. Alternatively, from the intracluster electron scattering perspective, the electron produced by photodetachment of the iodide moiety collides with the nucleobase at an effective collision energy $E_c = h\nu - VDE(I^- \cdot B)$. The VDEs, as discussed above, are 4.05 eV for $I^- \cdot T$ and 4.11 eV for $I^- \cdot U$. Based on the pump energies in Table 2, the range of E_c is 0.55–0.74 eV and 0.57–0.79 eV for $I^- \cdot T$ and $I^- \cdot U$, respectively, overlapping the threshold for H atom loss in both cases.

It thus appears that H-atom loss is a viable decay channel that can occur in parallel with autodetachment, affecting the decay constants τ_1 and τ_2. Unfortunately, we are unable to observe this channel directly in our current experimental configuration. Photoelectron spectroscopy experiments on deprotonated thymine find that the electron affinity of the T-H radical is 3.25 eV,[53] and one can reasonably assume a similar value for U-H. Therefore the probe energy in our experiment, 1.57 eV, is not sufficient to detach these species and the excess electron is too tightly bound to autodetach. Experiments at higher probe energies will need to be performed in order to assess the importance of this channel. A more indirect approach would be to perform the same experiments reported here on deuterated thymine and uracil. DEA experiments on deuterated thymine show a 40-fold drop in deuterium loss at low energies, compared to hydrogen loss in native thymine.[16] A similar effect is predicted in calculations on uracil.[18] This dramatic isotope effect is attributed to a lower tunneling probability for D-atom loss, whereas autodetachment is unaffected. Hence, if H-atom loss is significant in our experiments, we should see considerably slower decay of the transient feature D upon deuteration as the hydrogen loss channel would be largely turned off.

The possible role of anion tautomerization should also be considered. PES experiments by Bowen and co-workers[29] using a laser ablation ionization source found a uracil anion with a VDE of 2.5 eV. Comparison to electronic structure calculations suggests this species is a tautomer of the uracil anion in which a hydrogen atom has migrated from N3 to C5. Several other tautomers have been identified in calculations on uracil and thymine anions.[54,55] All hydrogen migration tautomers have higher VDEs than the canonical tautomer, and one (the N1-C5 tautomer) is predicted to be more stable than the canonical tautomer by 1.4 kcal mol^{-1} for thymine and 2.6 kcal mol^{-1} for uracil. One can envision a scenario in our experiment in which electron transfer from the iodide to the nucleobase is followed by tautomerization to a species with a higher VDE than the initially formed TNI. Such a process would contribute to the decay of the transient feature D. However, the barriers to tautomerization are calculated to be 40 kcal mol^{-1},[54,55] and are not accessible in our experiment. Therefore, absent a lower energy path, it is unlikely that tautomerization can occur under the conditions of our experiment. As with H-atom loss, further experiments with higher photon energies should be able to verify the absence of tautomerization.

Given the similarities between uracil and thymine, the observation of considerably faster decay dynamics in $I^- \cdot T$ complexes compared to $I^- \cdot U$ is striking. The electronic structure of the two nucleobases is similar, as are the DEA cross sections for H atom loss. The additional methyl group in thymine might be expected to slow, rather than accelerate, the autodetachment rate owing to a larger density of reactant states in the TNI. However, our previous work on iodide–solvent clusters showed that relatively small increases in the energy available for autodetachment can significantly raise the autodetachment rate.[50] Such an effect may be operative here as well. The VDEs of the valence-bound states of thymine and uracil are calculated to be 0.5 and 0.6 eV, respectively,[26,28,30] while the VDE of $I^-\cdot T$ is 0.06 eV lower than that of $I^-\cdot U$. Therefore at the same excitation energy, there is 0.16 eV more energy available for autodetachment in $I^-\cdot T$ than in $I^-\cdot U$, possibly leading to more rapid autodetachment and faster decay in $I^-\cdot T$.

6 Conclusion

We have investigated the energetics and dynamics of the $I^-\cdot T$ and $I^-\cdot U$ binary clusters using one-photon and two-photon time-resolved photoelectron spectroscopy. The one-photon experiments show a stabilization of iodide by the nucleobase, increasing the VDE from 3.05 eV for bare iodide to 4.11 eV for $I^-\cdot U$ clusters and 4.05 eV for $I^-\cdot T$ clusters. In addition to direct detachment from the anionic cluster to the neutral manifold, we observe a wavelength-independent zero electron kinetic energy feature that we attribute to charge transfer from iodide to the nucleobase, forming a transient negative ion that decays by vibrational autodetachment.

Two-photon time resolved experiments on the $I^-\cdot U$ and $I^-\cdot T$ clusters show that the TNI formed by charge transfer has a VDE between 0–0.6 eV. In both thymine and uracil clusters, this transient feature decays bi-exponentially at all pump energies studied, 4.60–4.90 eV. The short time constants range between 160–300 fs in thymine and 300–600 fs in uracil, while the long time constants range between 1–2 ps in thymine and 12–52 ps in uracil. The bi-exponential decay is attributed to changing rates of autodetachment upon iodine loss from the TNI. The shorter time constants in $I^-\cdot T$ clusters may reflect small differences in the iodide–nucleobase binding energy and the VDE of the nucleobase valence anion. The role of hydrogen atom loss and tautomerization merit further exploration. In addition, more detailed theoretical work into the nature of the iodide–nucleobase interaction both in the ground and charge transfer state would be of considerable interest both for understanding the dynamics observed in this experiment as well as how these experiments can be used to provide insight into the interactions of excess electrons with DNA.

Acknowledgements

This research is supported by the National Science Foundation (NSF) under grant CHE-1011819. S.B.K. and M.A.Y. are grateful for funding through NSF Graduate Research Fellowships.

References

1 C. von Sonntag, *Chemical Basis of Radiation Biology*, Taylor and Francis, London, 1987.
2 H. Schussler, S. Navaratnam and L. Distel, *Radiat. Phys. Chem.*, 2005, **73**, 163–168.
3 N. S. Hush and A. S. Cheung, *Chem. Phys. Lett.*, 1975, **34**, 11–13.

4 A. O. Colson, B. Besler and M. D. Sevilla, *J. Phys. Chem.*, 1993, **97**, 8092–8097.
5 K.-W. Choi, J.-H. Lee and S. K. Kim, *J. Am. Chem. Soc.*, 2005, **127**, 15674–15675.
6 K. B. Bravaya, O. Kostko, S. Dolgikh, A. Landau, M. Ahmed and A. I. Krylov, *J. Phys. Chem. A*, 2010, **114**, 12305–12317.
7 D. Ghosh, A. Golan, L. K. Takahashi, A. I. Krylov and M. Ahmed, *J. Phys. Chem. Lett.*, 2011, **3**, 97–101.
8 B. Boudaiffa, P. Cloutier, D. Hunting, M. A. Huels and L. Sanche, *Science*, 2000, **287**, 1658–1660.
9 J. Simons, *Acc. Chem. Res.*, 2006, **39**, 772–779.
10 J. Gu, Y. Xie and H. F. Schaefer, *J. Am. Chem. Soc.*, 2006, **128**, 1250–1252.
11 M. A. Yandell, S. B. King and D. M. Neumark, *J. Am. Chem. Soc.*, 2013, **135**, 2128–2131.
12 G. Hanel, B. Gstir, S. Denifl, P. Scheier, M. Probst, B. Farizon, M. Farizon, E. Illenberger and T. D. Märk, *Phys. Rev. Lett.*, 2003, **90**, 188104.
13 A. M. Scheer, C. Silvernail, J. A. Belot, K. Aflatooni, G. A. Gallup and P. D. Burrow, *Chem. Phys. Lett.*, 2005, **411**, 46–50.
14 S. Ptasińska, S. Denifl, V. Grill, T. D. Märk, E. Illenberger and P. Scheier, *Phys. Rev. Lett.*, 2005, **95**, 093201.
15 S. Ptasińska, S. Denifl, B. Mróz, M. Probst, V. Grill, E. Illenberger, P. Scheier and T. D. Märk, *J. Chem. Phys.*, 2005, **123**, 124302.
16 S. Denifl, P. Sulzer, F. Zappa, S. Moser, B. Kräutler, O. Echt, D. K. Bohme, T. D. Märk and P. Scheier, *Int. J. Mass Spectrom.*, 2008, **277**, 296–299.
17 P. D. Burrow, G. A. Gallup, A. M. Scheer, S. Denifl, S. Ptasińska, T. Märk and P. Scheier, *J. Chem. Phys.*, 2006, **124**, 124310.
18 G. A. Gallup and I. I. Fabrikant, *Phys. Rev. A: At., Mol., Opt. Phys.*, 2011, **83**, 012706.
19 J. H. Hendricks, S. A. Lyapustina, H. L. de Clercq and K. H. Bowen, *J. Chem. Phys.*, 1998, **108**, 8–11.
20 J. H. Hendricks, S. A. Lyapustina, H. L. de Clercq, J. T. Snodgrass and K. H. Bowen, *J. Chem. Phys.*, 1996, **104**, 7788–7791.
21 C. Desfrançois, H. Abdoul-Carime and J. P. Schermann, *J. Chem. Phys.*, 1996, **104**, 7792–7794.
22 C. Desfrançois, V. Periquet, Y. Bouteiller and J. Schermann, *J. Phys. Chem. A*, 1998, **102**, 1274–1278.
23 O. H. Crawford, *Mol. Phys.*, 1971, **20**, 585–591.
24 R. D. Brown, P. D. Godfrey, D. McNaughton and A. P. Pierlot, *J. Am. Chem. Soc.*, 1988, **110**, 2329–2330.
25 N. A. Oyler and L. Adamowicz, *J. Phys. Chem.*, 1993, **97**, 11122–11123.
26 R. A. Bachorz, W. Klopper and M. Gutowski, *J. Chem. Phys.*, 2007, **126**, 085101.
27 J. Schiedt, R. Weinkauf, D. M. Neumark and E. Schlag, *Chem. Phys.*, 1998, **239**, 511–524.
28 D. Svozil, T. Frigato, Z. k. Havlas and P. Jungwirth, *Phys. Chem. Chem. Phys.*, 2005, **7**, 840–845.
29 R. A. Bachorz, W. Klopper, M. Gutowski, X. Li and K. H. Bowen, *J. Chem. Phys.*, 2008, **129**, 054309.
30 P. Dedíková, L. Demovič, M. Pitoňák, P. Neogrády and M. Urban, *Chem. Phys. Lett.*, 2009, **481**, 107–111.
31 T. Suzuki and B. J. Whitaker, *Int. Rev. Phys. Chem.*, 2001, **20**, 313.
32 A. Stolow, A. E. Bragg and D. M. Neumark, *Chem. Rev.*, 2004, **104**, 1719–1757.
33 A. V. Davis, R. Wester, A. E. Bragg and D. M. Neumark, *J. Chem. Phys.*, 2003, **118**, 999–1002.
34 A. E. Bragg, J. R. R. Verlet, A. Kammrath, O. Cheshnovsky and D. M. Neumark, *J. Am. Chem. Soc.*, 2005, **127**, 15283–15295.
35 W. C. Wiley and I. H. McLaren, *Rev. Sci. Instrum.*, 1955, **26**, 1150–1157.
36 A. T. J. B. Eppink and D. H. Parker, *Rev. Sci. Instrum.*, 1997, **68**, 3477–3484.
37 V. Dribinski, A. Ossadtchi, V. Mandelshtam and H. Reisler, *Rev. Sci. Instrum.*, 2002, **73**, 2634–2642.
38 M. J. Frisch, G. W. Trucks, H. B. Schlegel, G. E. Scuseria, M. A. Robb, J. R. Cheeseman, G. Scalmani, V. Barone, B. Mennucci, G. A. Petersson, H. Nakatsuji, M. Caricato, X. Li, H. P. Hratchian, A. F. Izmaylov, J. Bloino, G. Zheng, J. L. Sonnenberg, M. Hada, M. Ehara, K. Toyota, R. Fukuda, J. Hasegawa, M. Ishida, T. Nakajima, Y. Honda, O. Kitao, H. Nakai, T. Vreven, J. A. Montgomery, Jr., J. E. Peralta, F. Ogliaro, M. Bearpark, J. J. Heyd, E. Brothers, K. N. Kudin, V. N. Staroverov, R. Kobayashi, J. Normand, K. Raghavachari, A. Rendell, J. C. Burant, S. S. Iyengar, J. Tomasi, M. Cossi, N. Rega, J. M. Millam, M. Klene, J. E. Knox, J. B. Cross, V. Bakken, C. Adamo, J. Jaramillo, R. Gomperts, R. E. Stratmann, O. Yazyev, A. J. Austin, R. Cammi, C. Pomelli, J. W. Ochterski, R. L. Martin, K. Morokuma, V. G. Zakrzewski, G. A. Voth, P. Salvador, J. J. Dannenberg, S. Dapprich, A. D. Daniels, Farkas,

J. B. Foresman, J. V. Ortiz, J. Cioslowski and D. J. Fox, *Gaussian 09*, Gaussian Inc.Wallingford CT2009.
39 K. A. Peterson, B. C. Shepler, D. Figgen and H. Stoll, *J. Phys. Chem. A*, 2006, **110**, 13877–13883.
40 L. T. M. Profeta, J. D. Larkin and H. F. Schaefer, *Mol. Phys.*, 2003, **101**, 3277–3284.
41 A. Martínez, O. Dolgounitcheva, V. G. Zakrzewski and J. V. Ortiz, *J. Phys. Chem. A*, 2008, **112**, 10399–10404.
42 A. Kammrath, J. R. R. Verlet, A. E. Bragg, G. B. Griffin and D. M. Neumark, *J. Phys. Chem. A*, 2005, **109**, 11475–11483.
43 O. T. Ehrler, G. B. Griffin, R. M. Young and D. M. Neumark, *J. Phys. Chem. B*, 2009, **113**, 4031–4037.
44 R. M. Young, M. A. Yandell and D. M. Neumark, *J. Chem. Phys.*, 2011, **134**, 124311.
45 R. M. Young, R. J. Azar, M. A. Yandell, S. B. King, M. Head-Gordon and D. M. Neumark, *Mol. Phys.*, 2012, **110**, 1787–1799.
46 M. S. Bowen, M. Becucci and R. E. Continetti, *J. Chem. Phys.*, 2006, **125**, 133309.
47 L. B. Clark, G. G. Peschel and I. Tinoco, *J. Phys. Chem.*, 1965, **69**, 3615–3618.
48 M. Barbatti, A. J. A. Aquino and H. Lischka, *Phys. Chem. Chem. Phys.*, 2010, **12**, 4959–4967.
49 B. B. Brady, L. A. Peteanu and D. H. Levy, *Chem. Phys. Lett.*, 1988, **147**, 538–543.
50 M. A. Yandell, R. M. Young, S. B. King and D. M. Neumark, *J. Phys. Chem. A*, 2012, **116**, 2750–2757.
51 O. Kornilov, O. Bunermann, D. J. Haxton, S. R. Leone, D. M. Neumark and O. Gessner, *J. Phys. Chem. A*, 2011, **115**, 7891–7900.
52 S. Denifl, S. Ptasińska, G. Hanel, B. Gstir, M. Probst, P. Scheier and T. D. Märk, *J. Chem. Phys.*, 2004, **120**, 6557–6565.
53 B. F. Parsons, S. M. Sheehan, T. A. Yen, D. M. Neumark, N. Wehres and R. Weinkauf, *Phys. Chem. Chem. Phys.*, 2007, **9**, 3291–3297.
54 R. A. Bachorz, J. Rak and M. Gutowski, *Phys. Chem. Chem. Phys.*, 2005, **7**, 2116–2125.
55 K. Mazurkiewicz, R. A. Bachorz, M. Gutowski and J. Rak, *J. Phys. Chem. B*, 2006, **110**, 24696–24707.

Faraday Discussions RSCPublishing

PAPER

A multi-sheeted three-dimensional potential-energy surface for the H-atom photodissociation of phenol

Sai G. Ramesh[a] and Wolfgang Domcke[b]

Received 18th January 2013, Accepted 11th February 2013
DOI: 10.1039/c3fd00006k

Extending the previous work of Lan *et al.* [*J. Chem. Phys.*, **122**, 224315 (2005)], a multi-state potential model for the H atom photodissociation is presented. All three "disappearing coordinates" of the departing H atom have been considered. *Ab initio* CASSCF computations have been carried out for the linear COH geometry of C_{2v} symmetry, and for several COH angles with the OH group in the ring plane and also perpendicular to the ring plane. By keeping the C_6H_5O fragment frozen in a C_{2v}-constrained geometry throughout, we have been able to apply symmetry-based simplifications in the constructions of a diabatic model. This model is able to capture the overall trends of twelve adiabats at both torsional limits for a wide range of COH bend angles.

I Introduction

The photophysics and photochemistry of simple aromatic or heteroaromatic molecules with acidic groups, such as pyrrole, indole and phenol, have attracted considerable interest in recent years. Around the turn of the century, *ab initio* electronic-structure calculations for indole, pyrrole and phenol predicted the existence of optically dark and photochemically reactive singlet excited states, the so-called $^1\pi\sigma^*$ states of azoles and aromatic enols.[1-3] The calculations of scans of the electronic potential-energy (PE) surfaces along NH or OH stretching coordinates revealed the generically dissociative character of these states. Since the $^1\pi\sigma^*$ PE surfaces form conical intersections with the electronic ground state, they can give rise to either ultrafast internal conversion to the electronic ground state or to the direct photodetachment of fast hydrogen atoms.[1-3] The combination of advanced spectroscopic methodologies, which provide detailed energy-resolved or time-resolved information on the reaction dynamics, with *ab initio* computational methods, which provide convincing interpretations of the experimental data, has promoted the H-atom detachment photochemistry in aromatic

[a]*Department of Inorganic and Physical Chemistry, Indian Institute of Science, Bangalore 560 012, India. E-mail: sai.ramesh@ipc.iisc.ernet.in*
[b]*Department of Chemistry, Technische Universität München, D-85748, Garching, Germany. E-mail: wolfgang.domcke@ch.tum.de*

molecules to a very active field of research.[4–31] The experiments have provided strong evidence that the mechanisms of sub-picosecond H-atom photodetachment processes in these systems are in sharp contrast to the textbook mechanism of unimolecular photodissociation of organic molecules, in which excited electronic states undergo rapid internal conversion to the electronic ground state, followed by slow statistical fragmentation.[13,22,28] Moreover, the photophysics of these simple organic chromophores is of general interest, since they can serve as prototypes of important biological molecules, such as the DNA/RNA bases or the aromatic amino acids tryptophan and tyrosine.[32]

The $^1\pi\sigma^*$ excited states also play an essential role in electron/proton-transfer processes in clusters of pyrrole, indole, phenol and related chromophores with amphoteric solvent molecules, such as water or ammonia.[33–36] For example, it has been found that the photoinduced dynamics of the S_1 ($^1\pi\pi^*$) state of phenol–water or phenol–ammonia complexes is not, as previously thought, governed by the coupling of the $^1\pi\pi^*$ state with ion-pair states leading to proton transfer, but rather by the coupling with $^1\pi\sigma^*$ states, which leads to hydrogen transfer from phenol to the solvent molecules *via* a concerted electron/proton-transfer process.[3,32]

The photoinduced H-atom detachment process in phenol was first observed by Tseng *et al.* with multimass ion imaging with excitation at 248 and 193 nm.[7] Ashfold and coworkers applied high-resolution photofragment translational spectroscopy (H-atom Rydberg tagging) and obtained vibrationally resolved kinetic-energy spectra for a multitude of excitation wavelengths for phenol, substituted phenols and related molecules.[8–13,17–19,22,24,26,28] Stavros and collaborators used time-resolved velocity map ion imaging to clock the H-atom appearance in indole, phenol and deuterated phenol, revealing that the fast as well as the slow H-atoms appear on an ultrafast time scale (<200 fs).[21,25,27] Experiments by Crim and coworkers revealed a significant increase of the fraction of slow H-atoms upon excitation of the OH-stretching mode of phenol, which reflects the preferred formation of excited-state phenoxyl fragments.[20] All these observations were qualitatively rationalized with reference to *ab initio* calculated one-dimensional PE profiles for H-atom dissociation in the electronic ground state and the lowest $^1\pi\pi^*$ and $^1\pi\sigma^*$ states.[1–3,16,24,26,28,32]

The electronic-structure calculations for phenol reveal that the H-atom photodissociation process essentially involves the three electronic states S_0, $^1\pi\pi^*$ and $^1\pi\sigma^*$.[3,24,28,32] The second $^1\pi\pi^*$ state becomes relevant for excitation wavelengths shorter than about 200 nm.[24] Owing to its dissociative character along the OH stretching coordinate, the $^1\pi\sigma^*$ PE function intersects the bound PE function of the allowed $^1\pi\pi^*$ state as well as that of the electronic ground state. When the appropriate out-of-plane vibrational modes are taken into account, the two curve crossings become conical intersections.[3,32] Lan *et al.* identified the OH torsion coordinate as the dominant coupling mode at the $^1\pi\pi^*$ – $^1\pi\sigma^*$ and $^1\pi\sigma^*$ – S_0 conical intersections of phenol, based on complete-active-space self-consistent-field (CASSCF) calculations.[37] This finding implies that both coupling and tuning modes[38] at the two conical intersections are "disappearing modes" in the photodissociation reaction, that is, they are lacking in the phenoxyl fragment. Ashfold and collaborators, on the other hand, concluded from the analysis of internal energy spectra of phenoxyl that two out-of-plane deformation modes of the phenyl ring (normal modes ν_{16a} and ν_{16b}) are the primary coupling modes at the conical intersections.[12,13,17,19,22,24,28] A CASSCF computational survey of all symmetry-allowed coupling modes confirmed, however, that the OH torsion is by far the strongest coupling mode at the $^1\pi\pi^*$ – $^1\pi\pi^*$ and $^1\pi\sigma^*$ – S_0 conical intersections.[39]

While numerous *ab initio* calculations of PE functions for H-atom photodissociation have been reported for phenol, substituted phenols, thiophenol and related molecules, only a few investigations have addressed the dynamics of the photodissociation reaction. Lan *et al.*, Abe *et al.*, An and Baeck, as well as Venkatesan *et al.* performed time-dependent quantum wave-packet calculations for two-dimensional (OH stretching and OH torsion) models of the three lowest PE surfaces of phenol[37,40,41] and thiophenol.[42] Nix *et al.* and Dixon *et al.* explored geometric-phase effects and H-atom tunneling effects with two-dimensional wave-packet simulations, including the OH stretching mode and a conserved out-of-plane mode of phenol.[43,44] The nonadiabatic photoinduced dynamics of pyrrole, for which the dissociative $^1\pi\sigma^*$ state is the lowest excited state, has been explored with quantum wave-packet calculations as well as with classical trajectory surface-hopping calculations.[45–49]

The aim of the present work is the extension of the two-dimensional (2D) PE surface of Lan *et al.*[37] to include all three disappearing modes of phenol (OH stretching, OH torsion, OH bending), thus providing a full-dimensional description of the dissociating H-atom. The inclusion of the C_{2v}-symmetric linear COH configuration provides important insight into the crossing patterns of the diabatic electronic states involved in the photodissociation of phenol. It turns out that additional electronic configurations and additional conical intersections enter the picture in the three-dimensional description of the H-atom dissociation. Multi-state adiabatic PE profiles have been computed as functions of the OH stretching coordinate for a wide range of the COH bending angle and for two values of the OH torsional angle (in-plane as well as perpendicular to the phenyl ring). A preliminary diabatic PE model has been constructed *via* a least-squares fitting of the adiabatic *ab initio* energy data.

II *Ab initio* methodology

A Reference structure and use of symmetry

The ground-state equilibrium geometry of phenol was optimized with the second-order Møller–Plesset (MP2) method using the aug-cc-pVDZ basis set. Phenol is a planar molecule with C_s symmetry. The CO bond is slightly in-plane tilted with respect to the six-membered ring. Instead of using the ground-state equilibrium geometry as the reference geometry, it is preferable to use a reference structure with a C_{2v} symmetry constraint on the C_6H_5O subsystem. This reference structure has been obtained by an MP2 geometry optimization. The choice of the axis system and the numbering of atoms is shown in Fig. 1. The position of the OH hydrogen atom is described by the polar coordinates (r, θ, ϕ). The geometry parameters of the equilibrium structure of phenol obtained with the C_6H_5O-constrained and unconstrained optimizations, respectively, are given in Table 1. With the exception of the OCC bond angle, the differences between the two geometries are minor.

Throughout the PE surface calculations, the C_6H_5O fragment is kept rigid, the geometry parameters being given by the left column of Table 1. The ground-state and excited-state energies are determined as functions of the coordinates (r, θ, ϕ) of the H-atom of the OH group. It follows from these definitions that the PEs as

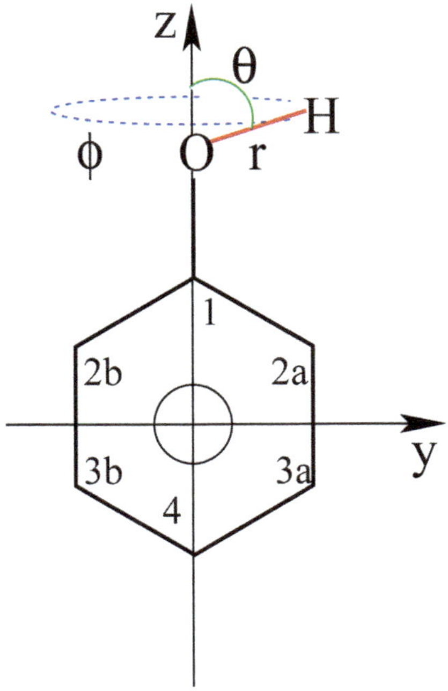

Fig. 1 Axis system and numbering of atoms for phenol.

Table 1 Structural parameters for phenol, obtained at the MP2/aug-cc-pVDZ level, with and without symmetry restrictions. The fully optimized structure exhibits C_s symmetry. It is compared with the structure obtained with C_{2v} constraint on the phenoxy fragment. The atom numbering is defined in Fig. 1

	C_{2v} (phenoxy)	C_s	Diff.
r	0.9674 Å	0.9680 Å	0.0006 Å
θ	71.91°	71.41°	0.50°
ϕ	0.00°	0.00°	0.00°
r_{OC}	1.3820 Å	1.3817 Å	0.0003 Å
$r_{C_1C_{2a}}$	1.4072 Å	1.4072 Å	<0.0001 Å
$r_{C_{2a}C_{3a}}$	1.4064 Å	1.4076 Å	0.0012 Å
$r_{C_{2a}H_{2a}}$	1.0946 Å	1.0958 Å	0.0012 Å
$r_{C_{3a}H_{3a}}$	1.0940 Å	1.0940 Å	<0.0001 Å
$\alpha_{OC_1C_{2a}}$	119.87°	122.71°	2.84°
$\alpha_{C_1C_{2a}C_{3a}}$	119.65°	119.63°	0.02°
$\alpha_{C_{3a}C_{2a}H_{2a}}$	120.73°	120.25°	0.48°
$\alpha_{C_{2a}C_{3a}H_{3a}}$	119.33°	119.36°	0.03°

functions of the azimuthal (torsional) angle ϕ exhibit two-fold periodicity. More generally, these definitions allow us to invoke C_{2v} symmetry for the symmetry analysis of the interactions among electronic states.

A key slice of the PE surface is that for $\theta = 0°$, *i.e.*, for linear COH geometry. For ease of identification, we label the C_{2v} planes of the C_6H_5O fragment as $\sigma_{\parallel}$ (molecular plane) and $\sigma_{\perp}$. The wavefunctions of A_2 states are antisymmetric with respect to both of these planes. B_1 states have characters +1 and −1 for $\sigma_{\perp}$ and $\sigma_{\parallel}$, respectively. These characters are the opposite for B_2 states. Away from the linearity of the COH group, the point group of phenol reduces to $C_s^{\parallel} = \{\hat{E}, \hat{\sigma}_{\parallel}\}$ for $\phi = 0°$. For $\phi = 90°$, the point group becomes $C_s^{\perp} = \{\hat{E}, \hat{\sigma}_{\perp}\}$. At these two values of the torsional angle, the following symmetry correlations apply:

$$\begin{aligned} C_{2v} \rightarrow C_s^{\parallel} &: \begin{cases} A_1, & B_2 \rightarrow A'_{\parallel} \\ A_2, & B_1 \rightarrow A''_{\parallel} \end{cases} \\ C_{2v} \rightarrow C_s^{\perp} &: \begin{cases} A_1, & B_1 \rightarrow A'_{\perp} \\ A_2, & B_2 \rightarrow A''_{\perp} \end{cases} \end{aligned} \tag{1}$$

Eqn (1) tells us which states are permitted to interact (mix) when the OH bond moves away from the C_2 axis of the C_6H_5O fragment.

States that correlate with A_1 and B_2 states at linearity ($\theta = 0°$) can interact when the H atom lies in the ring plane. These couplings vanish when the OH bond is perpendicular to the ring. Instead, new couplings emerge between states that correlate with A_1 and B_1 states at linearity. These (A_1, B_1) sets of states form conical intersection (CI) seams for $\phi = 0°$. Likewise, for $\phi = 90°$, the states of A_1 parentage form CIs with those of B_2 parentage. Note that couplings of B_1 and B_2 states vanish for planar as well as perpendicular OH bonds. This is true for A_1–A_2 state pairs as well.

The above symmetry arguments can only be employed if the C_{2v} restriction on the phenoxyl part of phenol is a good approximation. The geometries optimized with and without C_{2v} constraint are given in Table 1. All bond lengths are essentially the same. The bond angle differences are also small: The largest change is about 3° for the OC_1C_{2a} angle. The energy difference between the two structures is merely 134.3 cm^{-1}. It can thus be concluded that the model with a C_{2v}-constrained rigid phenoxy group is appropriate for the description of the PE surfaces of phenol.

The discussion above bears some semblance to a recent work of Dixon *et al.*[44] that analyses the role of the OH torsion at the two main CIs of phenol. To this end, they used the non-rigid G_4 group to describe the torsional tunneling, and made symmetry arguments about the vibrational modes that act as coupling modes at the CI. While the present work focuses on the H atom motions alone, it incorporates the two-fold symmetry associated with the torsional motion.

We reiterate that all PE surface computations discussed in what follows were performed with a fixed C_{2v}-symmetric geometry of the C_6H_5O fragment, whose bond lengths and angles are given in Table 1. The only geometrical variables that are allowed to vary are the coordinates (r, θ, ϕ) of the H-atom of the OH-group. These are the three "disappearing coordinates" in the H-atom detachment from phenol.

B CASSCF calculations and choice of the active space

The choice of the active space is a crucial and nontrivial issue if one wants to calculate excited-state PE surfaces for a wide range of the (r, θ, ϕ) coordinates of the H-atom. In the previous work of Lan *et al.*[37] and Vieuxmaire *et al.*,[39] an active space of 10 electrons in 9 orbitals, CAS (10, 9), was employed. It comprises the seven π-MOs of the phenoxy fragment and two orbitals that represent the σ_{OH} and σ^*_{OH} orbitals at large OH bond lengths. This active space, as well as several others built upon it, turn out to be inadequate when the COH angle is varied and higher excited states are considered. An active space comprising 12 electrons in 13 orbitals, CAS (12, 13), was found to be appropriate for an extended range of θ and

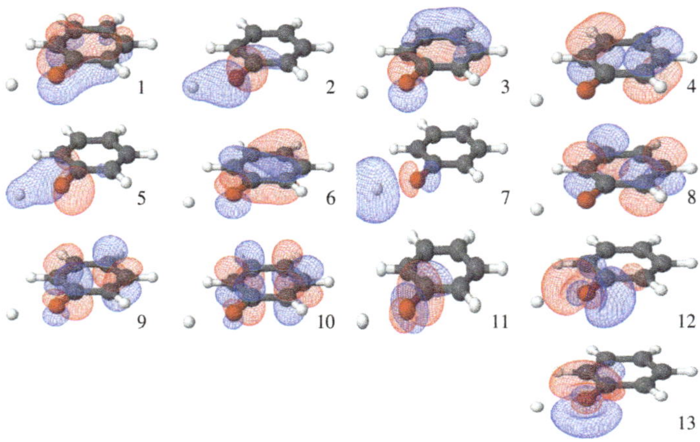

Fig. 2 The 13 orbitals of the active space for $\phi = 0°$ (planar molecule).

planar as well as perpendicular orientation of the OH group. The 13 orbitals of this active space are shown in Fig. 2 for $\phi = 0°$ (OH in ring plane) and in Fig. 3 for $\phi = 90°$ (OH perpendicular to ring plane), with $r = 1.80$ Å and $\theta = 40°$.

The additional orbitals in CAS (12,13) are the MOs 2, 11, 12, and 13 in both figures. The 13 MOs can be regarded as linear combinations of the following fragment orbitals: the 6 π-MOs of the benzene ring, the three p orbitals of the oxygen atom, the σ^* orbital of the OH bond, the σ^* orbital of the CO bond, and two $3p$ Rydberg orbitals on the oxygen atom. With respect to the various OH bond geometries, the first 10 orbitals are sufficient. However, the in-plane p_z orbital of the oxygen is tied up in σ_{CO} (MO 1 in the figures). In the absence of the corresponding σ^*_{CO}, the energy curves of the CAS orbitals and the resulting state energies exhibit discontinuities. The inclusion of the CO antibonding orbital and the energetically vicinal $3p_x$ and $3p_y$ Rydberg orbitals stabilizes the CASSCF calculations.

The orbitals of CAS (12,13) are found to be appropriate for the OH torsional angles $\phi = 0°$ and 90°, and are expected to be adequate for ϕ values in between as well. For $\phi = 0°$, the active space comprises 5 $a'_{||}$ and 8 $a''_{||}$ orbitals. For $\phi = 90°$,

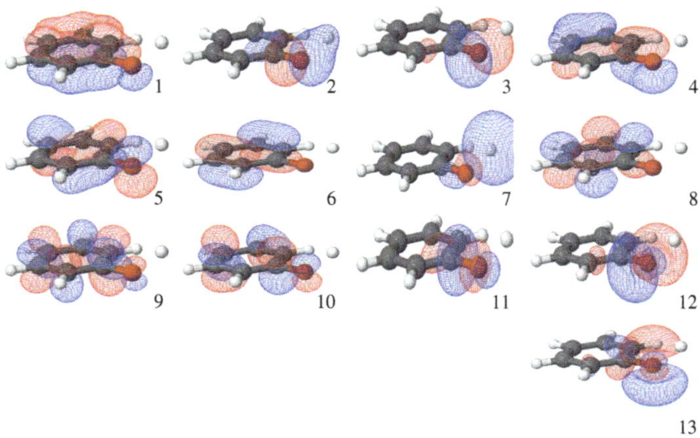

Fig. 3 The 13 orbitals of the active space for $\phi = 90°$ (perpendicular OH bond).

there are 9 $a'_{\perp}$ and 4 $a''_{\perp}$ orbitals. At $\theta = 0°$ (C_{2v} symmetry), the active orbitals are classified as 3 a_1, 6 b_1, 2 b_2, and 2 a_2.

In view of the large number of nuclear geometries to be sampled in the CASSCF calculations, the size of the basis set was reduced by the removal of the diffuse functions of the aug-cc-pVDZ basis set on the five H-atoms of the phenyl ring. The MOLPRO program package[50] was used for all *ab initio* electronic-structure calculations of this work.

C Computation of the *ab initio* PE surfaces

Ab initio CASSCF energy data was generated as one-dimensional cuts along the radial coordinate r for select values of the polar angle θ and the azimuthal angle ϕ. The range of r extends from 0.7 Å to 4.0 Å. The grid spacing has been adjusted as required for the characterization of avoided crossings and other sharp variations of the PEs. The polar angle θ was varied from 0° to 100° in steps of 10°. For these values of θ, cuts along the radial coordinate were computed for $\phi = 0°$ and $\phi = 90°$.

At each (r, θ, ϕ) point, we have carried out state-averaged CASSCF calculations including various states of a given symmetry species in the energy functional. A two-stage procedure was adopted. In the first stage, both the orbitals and the configuration-interaction coefficients were optimized. For given values of θ and ϕ, the number of states of each symmetry species and their individual weights were varied with the OH bond length (through standard options in the CASSCF routines of MOLPRO). This is necessary to ensure that the wavefunctions and orbitals change smoothly across symmetry-allowed curve crossings or avoided crossings. Thereby, energy discontinuities are eliminated.

Once satisfactory energy data had been obtained for the whole relevant range of r, the optimized orbitals were used in the stage-2 calculations, in which only the configuration-interaction coefficients were optimized (also available in MOLPRO). In this stage, a fixed number of states of a given symmetry species were included in the state averaging. The CASSCF energies computed in the first stage are recovered and the energies of additional (higher lying) electronic states are obtained. We defer the specification of the number of included states to later sections. In the case of $\theta = 0°$, stage-2 calculations were not carried out, since there was sufficient information available from the stage-1 calculations.

III *Ab initio* results and discussion

A Reference PE curves at $\theta = 0°$

The PEs as a function of r at the linear COH geometry serve as a reference for the construction of a three-dimensional PE surface of phenol. Fig. 4 shows the result of (stage-1) state-averaged CASSCF calculations of the four symmetry species : A_1 (red), B_1 (blue), B_2 (green), and A_2 (salmon). The number of states included in the state-averaging was varied as needed for each symmetry species. At least four states of each symmetry were computed at each r value. For certain regions, the number of included states is as high as 8 for B_2 and 7 for A_1 symmetry.

At large r, the PE curves dissociate towards the $C_6H_5O^\cdot$ + $H^\cdot$ radical pair. The lowest dissociation limit defines the zero of the energy scale. The lowest two energy levels of the phenoxyl radical are of B_1 and B_2 symmetry, in this order. The B_1 state has the unpaired electron in the oxygen p_x orbital, that is, in a π-orbital. The B_2 state

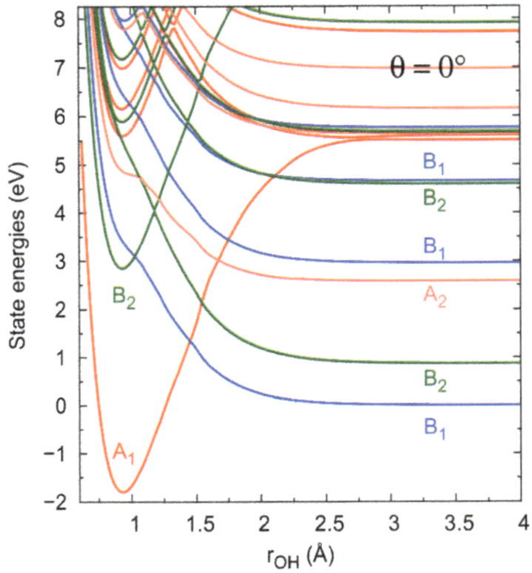

Fig. 4 CASSCF PE functions for phenol at linear COH geometry. The lowest dissociation limit defines the zero of the energy scale. The C_{2v} symmetry labels are specified for the states with lower energies. A_1 energies are shown in red, B_1 energies in blue, B_2 energies in green, and A_2 energies in salmon. While the set of states is complete at short r up to 8 eV, not all states with dissociation limits above 6 eV were computed.

has the electron in the oxygen p_y orbital, that is, in a σ-orbital that lies in the ring plane. As r decreases, the energies of these two states rise monotonically. Since the H-atom approaches the doubly-occupied oxygen p_z (that is, σ_{CO}) orbital, no bonding can result. Both these curves are particularly important for the photochemistry of phenol, since they are the final states of the two main photodissociation channels.

The next two dissociative PE functions are also potentially accessible dissociation channels in the case of sub-200 nm excitation. In the A_2 state of the phenoxyl radical, the unpaired electron occupies a π orbital of a_2 symmetry. The structure of the slightly higher lying second B_1 state is similar to that of the lowest B_1 state. Several other dissociative states of various symmetries are found at higher energies – a cluster of them is visible between 5.5 and 6 eV.

Notably, there exists a stable OH bond in linear phenol: Fig. 4 shows that the PE function of A_1 symmetry exhibits a deep minimum at $r \approx 0.93$ Å. In this state, a pair of electrons occupies the lowest a_1 orbital, which is a CO-bonding as well as an OH-bonding σ orbital. At large r, this state leads to a homolytic cleavage of the OH bond, the unpaired electron of the phenoxyl radical being in the σ_{CO} orbital. This dissociation limit lies about 5.5 eV above the $^2B_1$ ground state of the radical. Evidently, the A_1 state is the lowest bound electronic state of phenol with a linear COH group. The second bound state is of B_2 symmetry. It is about 5 eV higher in energy, and is composed of π excitations with respect to the A_1 state. This B_2 state represents the bright $^1\pi\pi^*$ state of phenol. The next group of bound states is too high in energy to be experimentally relevant, but these states are needed for the model we construct. Note that the lowest bound B_2 state does not exhibit an avoided crossing with the lowest dissociative B_2 state. The analysis of the wavefunctions reveals that the leading determinants of the two states differ by more than two spin orbitals and hence they are non-interacting in first order. Interestingly, this behaviour persists for bent COH geometries.

The PE functions of Fig. 4 are extremely useful for the construction of a multi-dimensional multi-sheeted PE surface for the H-atom photodissociation of phenol. Due to the high symmetry (C_{2v}), most of the curve crossings seen are symmetry allowed crossings, which become avoided crossings when the COH polar angle θ deviates from zero. The curves of Fig. 4 thus serve as diabats for the construction of the three-dimensional adiabatic PE surface. For energies up to as high as 6–7 eV above the lowest dissociation limit, one finds a few bound curves with several dissociative curves cutting through them. When the symmetry is lowered to C_s, a subset of these states may interact. The dissociative curves are of all symmetries, while the low-lying bound curves are of A_1 or B_2 symmetry (there are two high-lying bound curves of B_1 and A_2 symmetry). In $C_s^{||}$ symmetry, this means that we need only consider couplings within the states of $A'_{||}$ symmetry. In $C_s^{\perp}$ symmetry, the coupling structure is expected to be somewhat more complex, since the A_1 and B_2 states become $A'_{\perp}$ and $A''_{\perp}$ states, respectively. These observations are useful for the understanding of the topography of the energy surfaces away from linearity.

B PE curves for $\theta > 0°$ and $\phi = 0°$ as well as 90°

The symmetry species of $C_s^{||}$ and $C_s^{\perp}$ are the appropriate state labels for $\phi = 0°$ and 90°, respectively, at bent COH geometries ($\theta > 0°$). However, we will frequently refer to them with their parent C_{2v} symmetry labels, since this helps to rationalize the curve-crossing patterns. We also seek to limit the total number of states included in stage-2 of our computations. Two criteria help in the selection. First, we observe that the inclusion of the electronic states associated with the A_1 PE function of Fig. 4 is essential for obtaining smooth PE curves up to sufficiently high energies for a wide θ range. This criterion implies the inclusion of the lowest 12 states at the dissociation limit in Fig. 4, all of which have energies below 6 eV at large r. This set comprises 3 A_1, 4 B_1, 3 B_2, and 2 A_2 states. These states become 6 $A'_{||}$ states and 6 $A''_{||}$ states for $\theta > 0°$ and $\phi = 0°$, and 7 $A'_{\perp}$ states and 5 $A''_{\perp}$ states for $\theta > 0°$ and $\phi = 90°$. All of these states were included for all grid points in the stage-2 calculations.† The second criterion is that sufficiently many states have to be included in order that the lowest two dissociative diabats (of B_1 and B_2 symmetry in C_{2v}) can be associated, to the extent possible, with the adiabatic PEs at short r. For the most part, the inclusion of the 12 states given above is sufficient for this purpose as well.

1 Analysis of the PE functions for $\theta = 20°$. When the COH angle is 20° from linearity, various inter-state couplings become symmetry allowed, but are still small in magnitude. This provides a clear picture of state mixing with respect to the scenario at linearity. The state energies at $\theta = 20°$ are shown in Fig. 5 for $\phi =$ 0° (left) and 90° (right). The dots are the results of the stage-1 computations, while the solid lines are from the stage-2 calculations.

For $\phi = 0°$, the energies of the bound $A'_{||}$ (A_1) state and the lowest dissociative $A'_{||}$ (B_2) state evidently form a strongly avoided crossing. Note that the $A'_{||}$ adiabats are pushed apart by ≈2 eV for a deviation from linearity of merely 20°. The lower adiabat is what has been computed in several earlier works on phenol, as it leads to the B_2 ($^2\sigma$) dissociation channel. The upper $A'_{||}$ (A_1) adiabat continues smoothly to

† In the stage-1 calculations, the number of computed states was mostly smaller. However, in some regions of the PE slices, as many as 9–10 states of A' symmetry were needed in the stage-1 calculations in order that important lower-energy curves were reasonably smooth.

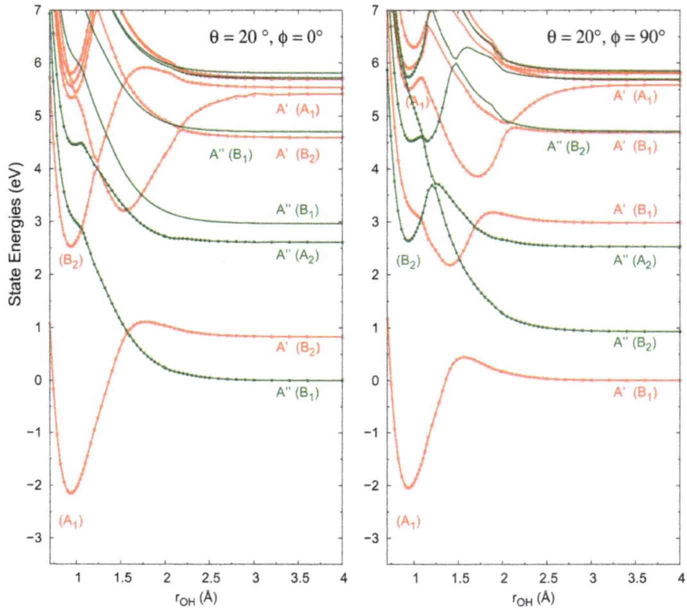

Fig. 5 CASSCF PE functions for $\theta = 20°$ and $\phi = 0°$(left) and $\phi = 90°$ (right). The points were obtained with stage-1 computations, while the lines are from stage-2 calculations; see text. The energies of A' states are shown in red, energies of A'' states in green.

the dissociation limit without significant interaction with another $A'_{||}$ (B_2) state that crosses its path. We can rationalize the latter observation through the configuration-interaction coefficients at linearity: the leading determinants of the parent states differ by several spin orbitals, implying a negligible coupling matrix element.

A second avoided crossing among the $A'_{||}$ states for $\phi = 0°$ exists between the lowest bound $A'_{||}$ (B_2) state and the lowest dissociative $A'_{||}$ (A_1) state. (The next $A'_{||}$ (A_1) dissociative state lies a fraction of an eV higher, but appears non-interacting.) Note that there are no visible avoided crossings among the $A''_{||}$ states for $\phi = 0°$. As discussed in Section III A, there are no low-lying bound $A''_{||}$ curves in the considered energy range.

We now turn our attention to the $\phi = 90°$ panel in Fig. 5. As anticipated in Section III A, there exist more avoided crossings for $\phi = 90°$ than for $\phi = 0°$. The bound $A'_{\perp}$ (A_1) curve forms avoided crossings with two dissociative $A'_{\perp}$ (B_1) curves. Note the remarkable strength of these interactions: For a mere 20° deviation from linearity, the lowest two $A'_{\perp}$ states are pushed apart by about 2 eV, similar to the avoided crossing of the lowest $A'_{||}$ states in the left panel of Fig. 5.

As the bound $A'_{\perp}$ (A_1) state continues to dissociation, the curve shows a weak interaction with the third repulsive $A'_{\perp}$ (B_1) curve. Among the $A''_{\perp}$ set of states, the bound $A''_{\perp}$ (B_2) curve exhibits avoided crossings with two $A''_{\perp}$ (A_2) dissociative curves. The lowest dissociative $A''_{\perp}$ (B_2) curve crosses the former $A''_{\perp}$ curves without significant interaction.

Although the COH angle of 20° is far from that of the ground-state minimum geometry of phenol, we can make some useful observations in connection with the photoinduced dynamics in phenol. It is evident from Fig. 4 and 5 that the OH torsion, in conjunction with the COH bending, significantly alters the crossing and interaction pattern of the radial PE curves. In particular, the number of accessible conical interactions changes with the torsional angle. The lowest $A'_{||}$–$A''_{||}$ conical intersection of the planar molecule vanishes, by symmetry, for a torsional angle of 90°. Two new conical intersections appear instead at somewhat higher energies, leading to the second ($A''_{\perp}$ (B_2)) and third ($A''_{\perp}$ (A_2)) dissociation limits. These observations suggest that the couplings between the diabats may have simple sinusoidal forms as functions of ϕ.

2 **Analysis of the PE functions for $\theta = 40°$, 70°, and 90°.** As the COH bending angle increases, the states that interact remain the same, but the strength of the coupling between them grows. This is evidenced in the PE surfaces in Fig. 6, 7, and 8 for the angles $\theta = 40°$, 70°, and 90°, respectively. The PE curves for $\theta = 70°$ plots should be similar to those for the equilibrium bond angle $\theta_{min} \approx 72°$.

An evident aspect of the series of PE functions for $\phi = 0°$ is that the repulsion of the bound $A'_{||}$ (A_1) and lowest repulsive $A'_{||}$ (B_2) curves increases markedly with decreasing COH bond angle. As a result, the barrier of the lowest $A'_{||}$ PE curve at $r \approx$ 1.7 Å disappears for $\theta > 40°$ (see left panels of Fig. 6–8). The same effect is observed for the lowest avoided crossing among the $A'_{\perp}$ states at $\phi = 90°$: The barrier of the lowest $A'_{\perp}$ PE function also disappears for $\theta > 40°$ (see right panels of Fig. 6–8).

One state, however, remains essentially unaffected by the bending angle, namely the lowest $A''_{\perp}$ (B_2) state for $\phi = 90°$. The explanation is that the unpaired electron of the phenoxy radical resides in the oxygen $2p_y$ orbital. The H-atom thus approaches the radical along the nodal plane of the singly occupied orbital.

We now consider the evolution of the CIs with the bending angle. For $\phi = 0°$, the first excited $A'_{||}$ state is intersected by the two lowest $A''_{||}$ states. The positions of these intersections move slightly outward as θ increases. The position of the intersection of

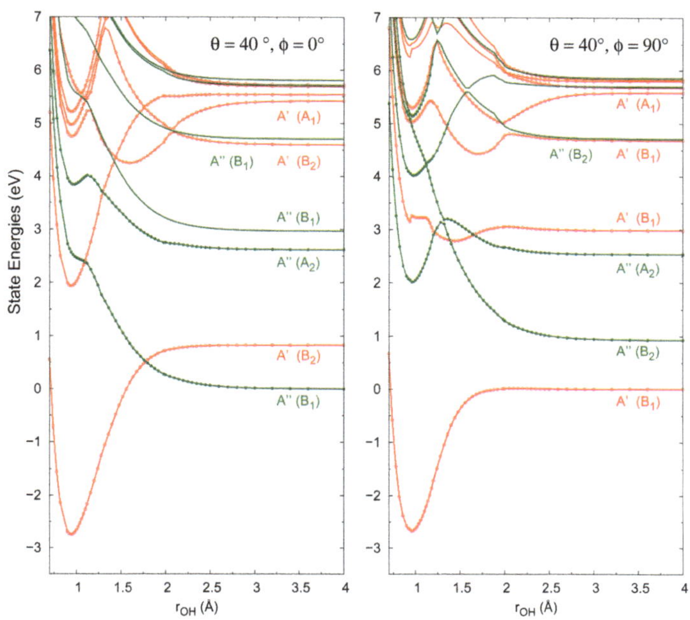

Fig. 6 CASSCF PE functions for $\theta = 40°$ and $\phi = 0°$ (left) and $\phi = 90°$ (right). The points were obtained with stage-1 computations, while the lines are from stage-2 calculations. The energies of A' states are shown in red, and energies of A'' states are in green.

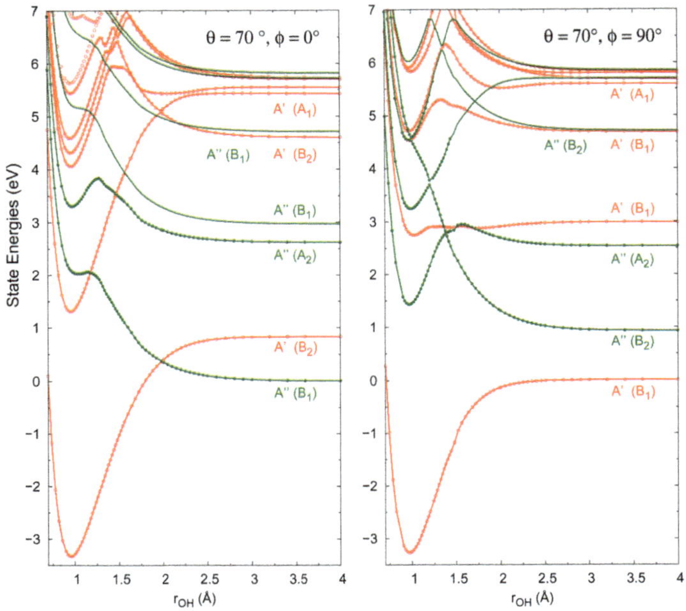

Fig. 7 CASSCF PE functions for $\theta = 70°$ and $\phi = 0°$ (left) and $\phi = 90°$ (right). The points were obtained with stage-1 computations, while the lines are from stage-2 calculations. The energies of A' states are shown in red, and energies of A'' states are in green.

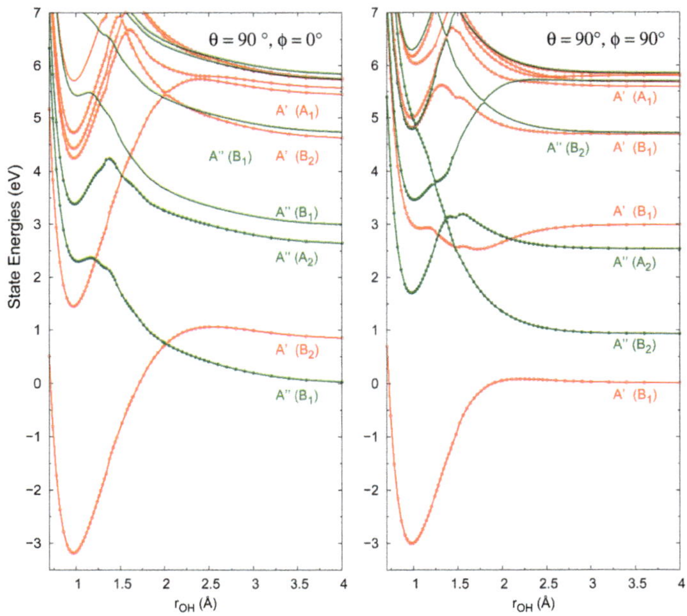

Fig. 8 CASSCF PE functions for $\theta = 90°$ and $\phi = 0°$ (left) and $\phi = 90°$ (right). The points were obtained with stage-1 computations, while the lines are from stage-2 calculations. The energies of A' states are shown in red, and energies of A'' states are in green.

the lowest $A''_{||}$ state with the $A'_{||}$ ground state also moves outward with decreasing COH angle.

For $\phi = 90°$, the well of the second $A'_{\perp}$ state (of B_1 parentage) is very shallow close to θ_{min}, but becomes deeper in either direction. It gives rise to three intersections in close proximity. These intersections are symmetry-allowed at $\phi = 90°$, but become avoided crossings for intermediate values of ϕ. This will result in very complex PE surfaces in this energy range and thus complex wavepacket-dynamics evolving through this region.

We note that the flat regions of the PE curves become repulsive for $\theta \gtrsim 90°$. This effect is particularly visible for $\phi = 0°$, see left panel of Fig. 8, and is due to the repulsion of the mobile H-atom by the *ortho* H-atom of the benzene ring.

Finally, we compare the vertical excitation energies in this work with those in prior studies. At the lowest energy point in our PE slices [r = 0.96 Å, θ = 70° ($\approx \theta_{min}$), ϕ = 0°], the S_0 – $\pi\pi^*$ and S_0 – $\pi\sigma^*$ vertical energies are 4.64 eV and 5.43 eV. These compare well with the data compiled in Table 1 of Dixon *et al.*,[44] which reports the experimental (theoretical) energies are 4.51 eV (4.82–4.86 eV) and 5.12 eV (5.37–5.7 eV) for the two transitions, respectively.

3 **Some remarks on the results.** We make note of an important aspect of the stage-1 computations. The state counts for each symmetry species are different for the C_{2v}, $C_s^{||}$, and $C_s^{\perp}$ geometries during this stage. One direct consequence is that the $r \to \infty$ limits are somewhat different for the three sets of geometries. For instance, the B_2 channel has a final energy of about 0.82 eV for the $C_s^{||}$ geometries while it is obtained as about 0.93 eV for the $C_s^{\perp}$ geometries. However, for all θ values within the $C_s^{||}$ or $C_s^{\perp}$ set of geometries, the dissociation limits are the same.

The energy difference between the $\phi = 0°$ and $\phi = 90°$ limits, though modest on the energy scale of the plots discussed in the last few sections, affects the comparability of the PE surfaces of the two torsional limits. Energy gradients and barriers in the ϕ coordinate are important parameters for the photodynamics. The torsional S_0 barrier height in the present work is about 510 cm^{-1}, which is significantly lower than the value of 1200 cm^{-1} obtained in earlier works. Both values are of the same order of magnitude as the aforementioned difference in the $\phi = 0°$–$\phi = 90°$ energy limits. The removal of these quantitative deficiencies will require CASSCF potential scans for $0° \leq \phi \leq 90°$ with consistent state counts throughout.

IV Construction of a multi-sheeted diabatic PE surface

We now attempt to extract an analytic diabatic PE model from the CASSCF *ab initio* data that can be used for a quantum wave-packet description of the photodissociation of phenol. The diagonal terms of the potential model, the diabats, will be based upon the results obtained at the linear COH geometry. As we saw in the previous sections, only a few specific diabats are coupled at the planar and perpendicular geometries of the OH bond, which greatly simplifies the construction of the three-dimensional PE surface. The fits of model functions to the *ab initio* data were obtained with the Levenberg–Marquardt nonlinear fitting procedure.[51] The present attempt is similar in spirit to the recent works of Gatti and coworkers, who have developed a 17-state vibronic model for ethylene.[52,53]

A Identification of the diabats

We return to Fig. 4, which shows the radial PE functions for linear COH geometry and their C_{2v} symmetry labels. Apart from the symmetry labels, we classify the PE functions in Fig. 4 as well-type potentials (W) and repulsive potentials (R).

At the dissociation limit, there are 12 PEs below 6 eV, comprising 3 A_1, 4 B_1, 3 B_2, and 2 A_2 states. One of them is the lowest bound A_1 curve at its dissociation limit, which we shall henceforth refer to as $W_1(A_1)$. The remaining 11 states are all dissociative states. The lowest four will be denoted as $R_1(B_1)$, $R_1(B_2)$, $R_1(A_2)$, and $R_2(B_1)$. In addition, there are two dissociative A_1 states, which we label as $R_1(A_1)$ and $R_2(A_1)$, among the cluster of states with 5.5–6.0 eV dissociation energies. With this labelling scheme, the diabats at the dissociation limit for all geometries are the following:

$$\begin{array}{cccc} W_1(A_1) & R_1(B_1) & R_1(B_2) & R_1(A_2) \\ R_1(A_1) & R_2(B_1) & R_2(B_2) & R_2(A_2) \\ R_2(A_1) & R_3(B_1) & R_3(B_2) & \\ & R_4(B_1) & & \end{array}$$

It remains to identify the diabats for these 12 PEs at short r. For $\phi = 0°$, we have a total of 6 $A'_{||}$ (A_1, B_2) states and 6 $A''_{||}$ (B_1, A_2) states. Likewise, for $\phi = 90°$, there are 7 $A'_{\perp}$ (A_1, B_1) states and 5 $A''_{\perp}$ (B_1, A_2) states. With these symmetry correlations in mind, the following diabats are suitable at short r:

$\phi = 0°$				$\phi = 90°$			
$A'_{\|\|}$	$A''_{\|\|}$	$A'_{\|\|}$	$A''_{\|\|}$	$A'_{\perp}$	$A'_{\perp}$	$A''_{\perp}$	$A''_{\perp}$
$W_1(A_1)$	$R_1(B_1)$	$W_1(B_2)$	$R_1(A_2)$	$W_1(A_1)$	$R_1(B_1)$	$W_1(B_2)$	$R_1(A_2)$
$W_2(A_1)$	$R_2(B_1)$	$R_1(B_2)$	$W_1(A_2)$	$W_2(A_1)$	$R_2(B_1)$	$R_1(B_2)$	
$W_3(A_1)$	$R_3(B_1)$	$W_2(B_2)$		$W_3(A_1)$	$R_3(B_1)$	$W_2(B_2)$	
	$W_1(B_1)$			$W_4(A_1)$	$W_1(B_1)$	$W_3(B_2)$	

All bound states up to 8 eV at $r = r_{min}$ are included in these lists. Note that the lists are slightly different for planar and perpendicular OH bond geometries.

Putting the diabats at long and short r together, the potential model has 20 diabats. In terms of symmetry, they are classified as $W_{1\text{-}4}(A_1)$, $R_{1\text{-}2}(A_1)$, $W_1(B_1)$, $R_{1\text{-}4}(B_1)$, $W_{1\text{-}3}(B_2)$, $R_{1\text{-}3}(B_2)$, $W_1(A_2)$, and $R_{1\text{-}2}(A_2)$. For $\theta = 0°$, the energies of the diabats can be fitted to the *ab initio* data in a controlled manner.

B Fit to curves at $\theta = 0°$

At linear COH geometry, there are no significant couplings between any of the states in Fig. 4. The bound curves are approximated by the following Morse or Varshni[54] forms:

$$\begin{aligned} f^W_{morse}(r) &= D_0 + D_2\left[\left(1 - e^{-a(r-r_e)}\right)^2 - 1\right], \\ f^W_{varshni}(r) &= D_0 + D_2\left[\left(1 - \frac{r_e}{r}e^{-a\left(r^2-r_e^2\right)}\right)^2 - 1\right]. \end{aligned} \tag{2a}$$

For $W_{1\text{-}3}(A_1)$ and $W_{1\text{-}2}(B_2)$, the Varshni form is found to be more suitable. The standard Morse shape is adopted for $W_1(A_2)$ and $W_1(B_1)$. The dissociative potential functions are modelled using one of the following expressions

$$f^R_{antimorse}(r) = D_0 + D_2\left[\left(1+e^{-a(r-r_0)}\right)^2 - 1\right],$$

$$f^R_{comb}(r) = \frac{1}{2}\left[f^W_{morse}(r) + f^R_{antimorse}(r)\right] - \frac{1}{2}\sqrt{\left[f^W_{morse}(r) - f^R_{antimorse}(r)\right]^2 + 4\left[f^G(r)\right]^2}. \tag{2b}$$

In the last equation, $f^G(r) = A^G e^{-b^G(r-r_0^G)^2}$ is a Gaussian function. For the low-lying dissociative states like $R_{1\text{-}2}(B_1)$, $R_1(B_2)$, and $R_1(A_2)$, we use the 'comb' function that exhibits an anti-Morse behaviour at long r but has a region of flatness close to r_{min}.[37] For some of the higher dissociative states, such as $R_1(A_1)$, sufficiently detailed information is not available. Hence, we model them as anti-Morse functions.

Although the total set of diabats of the model is large, implying a rather large set of adjustable parameters, a few observations help control the fitting procedure. First of all, the D_0 values of all the dissociative curves as well as $W_1(A_1)$ are held fixed at their values at $r = 10.0$ Å throughout. Next, whenever possible, potential curves of similar shape are considered as clones of each other. For example, among the W diabats, only the parameters of $W_1(A_1)$ can unambiguously be determined. For the rest of the W diabats, the dissociation limits are not known from our computations. Therefore, they are treated as clones of $W_1(A_1)$ with adjustable vertical energy offsets. ($W_1(A_2)$ and $W_1(B_1)$ are treated separately, since their minima are slightly different from the minimum of $W_1(A_1)$.) A few of the repulsive states, such as $R_1(B_1)$ and $R_2(B_2)$, can be described by a common parameter set, except for a vertical shift in energy.

The result of the fitting of the PE functions at $\theta = 0°$ is shown in Fig. 9. The colour coding of the *ab initio* the data is the same as in Fig. 4. The fitted functions are shown as thin black lines. The 20-function model traces the computed potentials very satisfactorily for the whole energy range of interest. The fitted parameters, which form the basis for the subsequent fits, are listed in Table 2.

C Preliminary fits at $\theta > 0°$

We now introduce the off-diagonal terms of the diabatic potential matrix. The functional form we choose for both $\phi = 0°$ and $\phi = 90°$ is[55,56]

$$f^C(r) = V^C r^n e^{-b^C r}, n = 0 \text{ or } 2. \tag{3}$$

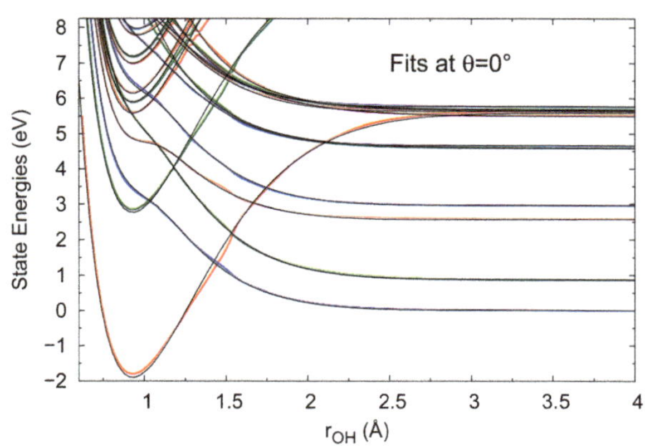

Fig. 9 Fits to the state-averaged CASSCF energies at linear COH geometry. The colour coding for data is the same as that of Fig. 4. The solid black lines are the fitted functions.

Table 2 Parameters for the fitted functions at the linear ($\theta = 0°$) COH geometry

Diabat	Shape	Parameters
$W_1(A_1)$	Varshni	$D_0 = 5.5957, D_2 = 7.4984, a = 0.5692, r_0 = 0.93$
$W_2(A_1)$	Varshni	Clone of $W_1(A_1)$; $\delta D_0 = 13.0765$
$W_3(A_1)$	Varshni	Clone of $W_1(A_1)$; $\delta D_0 = 13.6389$
$W_4(A_1)$	Varshni	Clone of $W_1(A_1)$; $\delta D_0 = 14.4957$
$R_1(A_1)$	AntiMorse	$D_0 = 5.4896, D_2 = 13.2830, a = 2.4813, r_0 = 0.25$
$R_2(A_1)$	AntiMorse	$D_0 = 5.6742, D_2 = 30.5098, a = 3.0904, r_0 = 0.25$
$R_1(B_1)$	Comb	$D_0^R = 0.0, D_2^R = 2.2041, a^R = 2.0822, r_0^R = 1.00$ $D_0^W = 8.4146, D_2^W = 3.8897, a^W = 2.7873, r_0^W = 0.93$ $A^G = 2.2120, b^G = 0.3059, r_0^G = 1.05$
$R_2(B_1)$	Comb	Clone of B_1R_1; $\delta D_0^R=2.9529, \delta D_0^W = 2.9529$
$R_3(B_1)$	Comb	Clone of A_2R_2; $\delta D_0^R = -0.9742, \delta D_0^W=-0.9742$
$R_4(B_1)$	Comb	Clone of A_2R_2; $\delta D_0^R = 0.1292, \delta D_0^W = 0.1292$
$W_1(B_1)$	Morse	Clone of A_2W_1; $\delta D_0 = 0.1820$
$W_1(B_2)$	Varshni	$D_0 = 10.2672, D_2 = 7.4984, a = 0.5692, r_0 = 0.93$
$W_2(B_2)$	Varshni	Clone of $W_1(A_1)$; $\delta D_0 = 13.3963$
$W_3(B_2)$	Varshni	Clone of $W_1(A_1)$; $\delta D_0 = 14.6957$
$R_1(B_2)$	Comb	$D_0^R=0.8608, D_2^R=2.8824, a^R = 1.9453, r_0^R=1.00$ $D_0^W=12.1889, D_2^W=4.9076, a^W = 2.5976, r_0^W=0.93$ $A^G = 3.1859, b^G = 0.3036, r_0^G=1.05$
$R_2(B_2)$	AntiMorse	$D_0 = 4.5898, D_2 = 15.9226, a = 2.7456, r_0 = 0.25$
$R_3(B_2)$	AntiMorse	$D_0 = 5.6746, D_2 = 12.4271, a = 2.6170, r_0 = 0.25$
$R_1(A_2)$	Comb	$D_0^R=2.5740, D_2^R=1.6245, {}_a{}^R = 2.2934, r_0^R=1.00$ $D_0^W=8.9778, D_2^W=3.5983, {}_a{}^W = 2.8502, r_0^W=0.93$ $A^G = 1.4049, b^G = 0.2756, r_0^G=1.05$
$R_2(A_2)$	Comb	$D_0^R=5.6260, D_2^R=1.9713, a^R = 1.8709, r_0^R=1.00$ $D_0^W=13.0000, D_2^W=2.5645, a^W = 3.2429, r_0^W=0.93$ $A^G = 2.4553, b^G = 0.3115, r_0^G=1.05$
$W_1(A_2)$	Morse	$D_0 = 11.2420, D_2 = 3.4526, a = 2.6978, r_0 = 0.96$

In the sense that the couplings between the states are expected to vanish as the H atom moves away from C_6H_5O, such a form is suitable. We carry out the fits only for $\theta \leq 80°$, since the repulsive effect of the *ortho*-H atom becomes significant thereafter, the description of which would need additional distance-dependent terms in the model.

1 Parametrization of the $\phi = 0°$ PE curves. The starting parameters of the $\phi = 0°$ diabats are taken from those at $\theta = 0°$. Of course, the $r \rightarrow \infty$ limits (the D_0 values) for $\theta > 0°$ are appropriately set. (Recall that these are a little different from the limits at $\theta = 0°$; see Section II B 3.) As θ increases, two avoided crossings within the $A'_{||}$ manifold become prominent, namely those for the $W_1(A_1)$ – $R_1(B_2)$ and $W_1(B_2) - R_1(A_1)$ pairs; compare Fig. 4 and 5. For the former pair, the coupled states repel each other strongly with increasing θ. Save for a vertical energy gap, the lower adiabats of both pairs (the first two $A'_{||}$ wells) are found to be very nearly of the same shape, suggesting that the coupling is strong for the second pair as well. Therefore, we choose two $f^C(r)$ functions with $n = 2$ for these avoided crossings.‡

‡ As for the linear COH geometry, the $W_1(B_2)$ diabat is modelled as an energy-upshifted version of $W_1(A_1)$. However, the $R_1(B_2)$ and $R_1(A_1)$ were modelled as a 'comb' and an Antimorse function, respectively, at the linear geometry. On this account, the couplings within the diabat pairs are taken to be different.

The other $A'_{||}$ wells of A_1 or B_2 origin are of nearly the same shape as the lowest $A'_{||}$ adiabat (within the r-range that they appear in the calculations). We model them as the lower energy curves resulting from upshifted clones of $W_1(A_1)$ being coupled to upshifted clones of $R_1(B_2)$. The coupling between these clones is taken to be identical to the $W_1(A_1)$ - $R_1(B_2)$ coupling. By construction, this procedure does not add new parameters to the model.

The left panels of Fig. 5–8 show that there are no other interactions between diabats for all $\phi = 0°$ configurations. Those we have just discussed reside entirely within the $A'_{||}$ manifold of states. There are no avoided crossings between the $A''_{||}$ states in the energy range presently explored.

Apart from the parameters of the coupling functions just described, we also optimize a select subset of parameters of the diabats themselves at each θ value. The most important choice we make is that for $W_1(A_1)$, which controls the shape of all the wells in the PE surfaces only the a (exponential) parameter of this diabat (see eqn (2a)) is allowed to change.

The parameter optimizations were carried out sequentially with increasing θ. The results for $\theta = 20°$ and $70°$ are shown in the left panels of Fig. 10 and 11, the latter being nearly at θ_{min}. The fitted results are the solid black traces, while the dashed curves are the two key diabats, $W_1(A_1)$ and $R_1(B_2)$. The model performs surprisingly well, in view of the energy and coordinate range it spans. However, it is unable to reproduce all details with high accuracy, particularly the depths of the potential wells.

2 Parametrization of the $\phi = 90°$ PE curves. For the $C_s^{\perp}$ configurations, there are more avoided crossings to be accounted for, which occur in both the $A'_{\perp}$ and the $A''_{\perp}$ states. In the $A'_{\perp}$ manifold, $W_1(A_1)$ crosses the $R_{1\text{-}3}(B_1)$ states at various energies. Its interactions with $R_1(B_1)$ and $R_2(B_1)$ lead to two important avoided

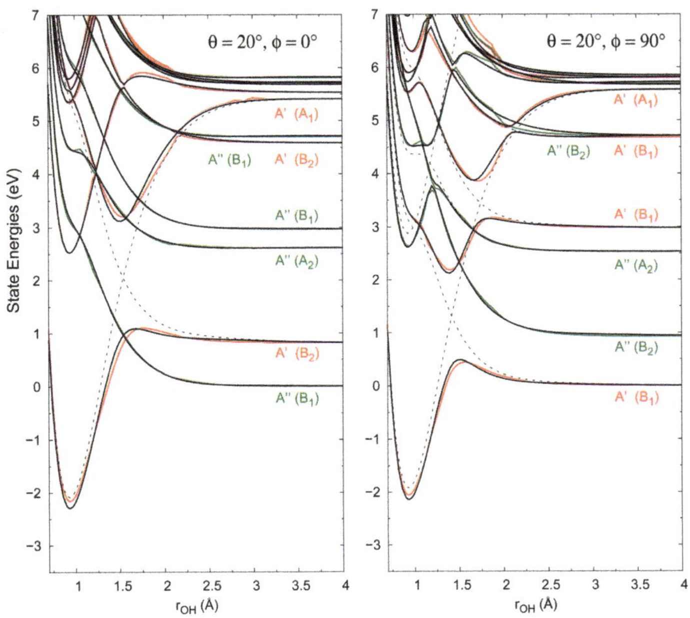

Fig. 10 Fits to the state-averaged CASSCF energies at $\theta = 20°$ and $\phi = 0°$ (left panel) and $\phi = 90°$ (right panel). The line colours are the same as in Fig. 5. The solid black lines are the result of the fits, while the dashed lines show the most important of the diabats with optimized parameters.

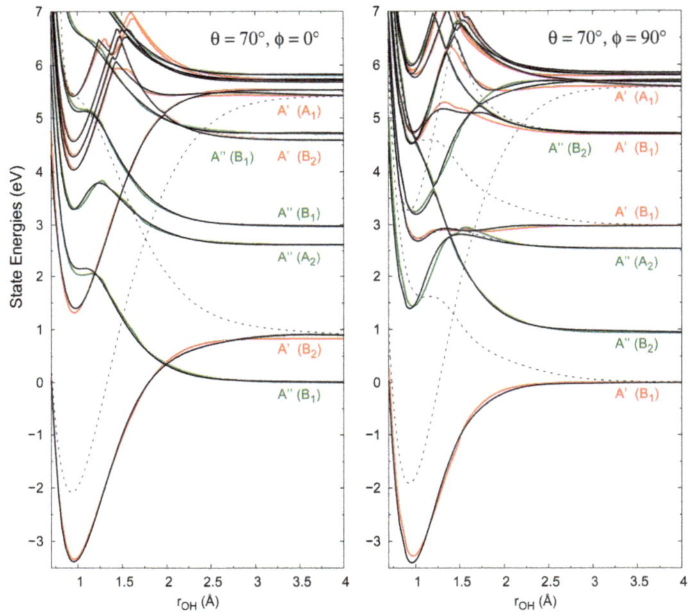

Fig. 11 Fits to the state-averaged CASSCF energies at $\theta = 70°$ and $\phi = 0°$ (left panel) and $\phi = 90°$ (right panel). The line colours are the same as Fig. 5. The solid black lines are the results of the fits, while the dashed lines show the most important of the diabats with optimized parameters.

crossings, which we model with two $f^C(r)$, $n = 2$ functions. The avoided crossing with $R_3(B_1)$ is more visible closer to $\theta_{\rm min}$, and we model it with a simple exponential function. $W_2(A_1)$ and $W_3(A_1)$ show avoided crossings with B_1R_2 at distances close to $r_{\rm min}$. It was found sufficient to model them with exponential functions.

Turning to the $A''_\perp$ set of states, the key player is $W_1(B_2)$, which shows avoided crossings with $R_1(A_2)$ at short r as well as with $R_2(A_2)$ at large r. The former is modelled with a simple exponential, while the latter is modelled with yet another $f^C(r)$, $n = 2$ function.

As for the $\phi = 0°$ case, the other diabats of well-type, *viz.* $W_{2\text{-}4}(A_1)$ and $W_{2,3}(B_2)$, may be given high energy clones of $R_1(B_1)$ to interact with. The coupling within each such clone-clone pair is an exact copy of that for the $W_1(A_1) - R_1(B_1)$ pair. As before, this procedure introduces no new parameters, but increases the size of the diabatic potential matrix.

For the reparametrization of the diabats as functions of θ, we found the prescription applied to the $\phi = 0°$ fits to be suitable here as well. That is, only a select set of parameters were adjusted in the fitting. We also held the parameters of $W_1(A_1)$ frozen at their $\theta = 0°$ values throughout, except for the exponential parameter.

The results of the fits are shown in the right hand panels of Fig. 10 and 11. The black curves show the lowest 12 adiabats resulting from the model, while the dashed lines represent some key diabats. The accuracy of the fits is reasonable, although slightly poorer than those for $\phi = 0°$.

D Some remarks on the fits

It is difficult to make an *a priori* guess of the functional dependence of the parameters of the $f^C(r)$ functions on θ. Analysing the diabats, only a few of them

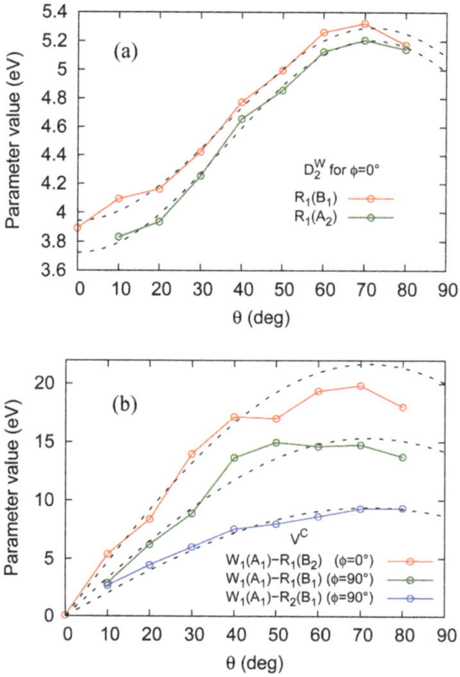

Fig. 12 (a) Fits of the D_2^W parameter of the $R_1(B_1)$ (red) and $R_1(A_2)$ (green) diabats as a function of θ at $\phi = 0°$. The dashed lines represent the fits to sinusoidal forms given in the text. (b) Fits as a function of θ of the V^C parameter of the $W_1(A_1) - R_1(B_2)$ coupling at $\phi = 0°$ (red) and $W_1(A_1) - R_1(B_1)$ (green) and $W_1(A_1) - R_2(B_1)$ (blue) couplings at $\phi = 90°$. The dashed lines are fits to sinusoidal forms given in the text.

show an identifiable trend with θ that can be modelled with simple functional forms. For example, the well-depths (D_2^W) of the Morse portions of $R_1(B_1)$ and $R_1(A_2)$ at $\phi = 0°$ appear to be adequately described by the function $\alpha + \beta\sin^2\left(\frac{\theta}{\theta_o}\frac{\pi}{2}\right)$, where $\theta_o = 72° \approx \theta_{min}$. This is depicted in Fig. 12a. For $\phi = 90°$, the same parameters for $R_1(A_2)$ shows a similar trend (albeit a little weaker), but $R_1(B_1)$ does not. Certain others, such as some exponential parameters, appear to change roughly linearly with θ.

We now analyse the θ dependence of a few coupling elements, namely those for the $W_1(A_1) - R_1(B_2)$ pair at $\phi = 0°$ and the $W_1(A_1) - R_1(B_1)$ and $W_1(A_1) - R_2(B_1)$ pairs at $\phi = 90°$. The pre-exponential (V^C) parameters of all three functions appear to be reasonably approximated by the form $\beta\sin\left(\frac{\theta}{\theta_o}\frac{\pi}{2}\right)$, as shown in Fig. 12b. The exponential parameters (b^C) do not show a very clear trend (save for that of the third pair), and it might be sufficient to treat all of them as linear functions of θ.

A persistent difficulty is the accurate description of the well-depths of the adiabats. In particular, the torsional barrier of the S_0 PE surface is poorly recovered. For the $\theta = 70° \approx \theta_{min}$ PE slice, the model's ground-state minimum energy for $\phi = 90°$ is slightly lower in energy than that for $\phi = 0°$. This underlines the need to improve the fidelity of the model for the description of the various barriers and gradients. Some of this improvement may result from refinements of the *ab initio* data. Additional improvements may be achieved by more sophisticated and more flexible model functions.

The dependence of the fitting parameters on the torsional angle ϕ may be modelled with $\cos(n\phi)$ or $\sin(n\phi)$ functions, where $n = 1$ or 2. The couplings between states of B_1 and B_2 parentage vanish at both torsional limits. Therefore, a $\sin(2\phi)$ dependence appears suitable. (We note that this is in agreement with the analysis of Dixon *et al.*:[44] $\sin(2\phi)$ is an a_2-symmetric function.) States of A_1 and B_1 parentage are coupled at $\phi = 90°$, but belong to different symmetry species at $\phi = 0°$. This suggests a $\sin(\phi)$ torsional dependence. This is, in fact, the expected situation for the S_0 - $^1\pi\sigma^*$ conical intersection that occurs at 1.98 Å, $\theta = 70°$ and $\phi = 0°$. We may presently estimate the size of the coupling as simply the size of the $W_1(A_1) - R_1(B_1)$ coupling term at the same r and θ but at $\phi = 90°$. The value obtained with the present model is 1.685 eV, which is comparable to the parameter value of 1.476 eV determined in the work of Lan *et al.*[37] ($\lambda_{12,max}$ in Table IV). A refinement of this result, as well as the verification of the sinusiodal torsional modulation of the coupling functions, requires calculations of additional PE slices for $0° < \phi < 90°$.

V Summary and concluding remarks

In this work, we have constructed a diabatic multi-state potential model for phenol OH photodissociation that includes all three disappearing modes, namely the OH stretch, CCOH torsion, and the COH bend. We have computed an extensive set of energies at the CASSCF level of theory with a carefully tailored active space. A key element is the application of C_{2v} constraint to the C_6H_5O fragment, which allows us to exploit C_{2v} symmetry selection rules in the construction of the model. Three geometric limits are considered in detail, namely linear COH geometry, bent COH group with the H atom in the ring plane, and bent COH group with the H atom in the bisector plane vertical to the ring.

The linear COH geometry, where C_{2v} symmetry applies, serves as the starting point of the analysis. Most of the potential curves turn out to be uncoupled at this geometry by symmetry or by the structure of their wave functions. As the COH angle bends away from linearity, several inter-state couplings are activated, but in a mutually exclusive manner at planar and perpendicular torsional angles. Thanks to these simplifications, a diabatic potential model can be constructed which is able to describe the twelve adiabats computed at both torsional limits for an extended range of COH bending angles.

With the parameters determined by a nonlinear least-squares fit of the large set of *ab initio* data, the model is able to capture the overall trends of the surfaces with remarkable accuracy. At the present stage, the model has some shortcomings which need to be removed before the surface is used for quantum dynamics simulations following the photoexcitation of phenol. Once refined, the model will allow us to explore the ultrafast H-atom dissociation dynamics of phenol on multiple coupled electronic states.

Acknowledgements

SGR thanks the Deutsche Forschungsgemeinschaft (DFG) for support under the SFB 749 programme. The authors would like to thank Sergy Grebenshchikov and Raffaele Borrelli for helpful discussions.

References

1 A. L. Sobolewski and W. Domcke, *Chem. Phys. Lett.*, 1999, **315**, 293.
2 A. L. Sobolewski and W. Domcke, *Chem. Phys.*, 2000, **259**, 181.
3 A. L. Sobolewski and W. Domcke, *J. Phys. Chem. A*, 2001, **105**, 9275.
4 D. A. Blank, S. W. North and Y. T. Lee, *Chem. Phys.*, 1994, **187**, 35.
5 J. Wei, A. Kuczman, J. Riedel, F. Renth and F. Temps, *Phys. Chem. Chem. Phys.*, 2003, **5**, 315.
6 H. Lippert, H.-H. Ritze, I. V. Hertel and W. Radloff, *ChemPhysChem*, 2004, **5**, 1423.
7 C.-M. Tseng, Y. T. Lee and C.-K. Ni, *J. Chem. Phys.*, 2004, **121**, 2459.
8 B. Cronin, M. G. D. Nix, R. H. Qadiri and M. N. R. Ashfold, *Phys. Chem. Chem. Phys.*, 2004, **6**, 5031.
9 M. G. D. Nix, A. L. Devine, B. Cronin and M. N. R. Ashfold, *Phys. Chem. Chem. Phys.*, 2006, **8**, 2610.
10 A. L. Devine, B. Cronin, M. G. D. Nix and M. N. R. Ashfold, *J. Chem. Phys.*, 2006, **125**, 184302.
11 B. Cronin, A. L. Devine, M. G. D. Nix and M. N. R. Ashfold, *Phys. Chem. Chem. Phys.*, 2006, **8**, 3440.
12 M. G. D. Nix, A. L. Devine, B. Cronin, R. N. Dixon and M. N. R. Ashfold, *J. Chem. Phys.*, 2006, **125**, 133318.
13 M. N. R. Ashfold, B. Cronin, A. L. Devine, R. N. Dixon and M. G. D. Nix, *Science*, 2006, **312**, 1637.
14 C.-M. Tseng, Y. T. Lee, C.-K. Ni and J.-L. Chang, *J. Phys. Chem. A*, 2007, **111**, 6674.
15 C.-M. Tseng, Y. T. Lee, M.-F. Lin, C.-K. Ni, S.-Y. Liu, Y.-P. Lee, Z. F. Xu and M. C. Lin, *J. Phys. Chem. A*, 2007, **111**, 9463.
16 I. S. Lim, J. S. Lim, Y. S. Lee and S. K. Kim, *J. Chem. Phys.*, 2007, **126**, 034306.
17 A. L. Devine, M. G. D. Nix, B. Cronin and M. N. R. Ashfold, *Phys. Chem. Chem. Phys.*, 2007, **9**, 3749.
18 A. L. Devine, M. G. D. Nix, R. N. Dixon and M. N. R. Ashfold, *J. Phys. Chem. A*, 2008, **112**, 9563.
19 G. A. King, A. L. Devine, M. G. D. Nix, D. E. Kelly and M. N. R. Ashfold, *Phys. Chem. Chem. Phys.*, 2008, **10**, 6417.
20 M. L. Hause, Y. H. Yoon, A. S. Case and F. F. Crim, *J. Chem. Phys.*, 2008, **128**, 104307.
21 A. Iqbal, L.-P. Pegg and V. G. Stavros, *J. Phys. Chem. A*, 2008, **112**, 9531.
22 M. N. R. Ashfold, A. L. Devine, R. N. Dixon, G. A. King, M. G. D. Nix and T. A. A. Oliver, *Proc. Natl. Acad. Sci. U. S. A.*, 2008, **105**, 12701.
23 J. S. Lim, H. Choi, I. S. Lim, S. B. Park, Y. S. Lee and S. K. Kim, *J. Phys. Chem. A*, 2009, **113**, 10410.
24 G. A. King, T. A. A. Oliver, M. G. D. Nix and M. N. R. Ashfold, *J. Phys. Chem. A*, 2009, **113**, 7984.
25 A. Iqbal, M. S. Y. Cheung, M. G. D. Nix and V. G. Stavros, *J. Phys. Chem. A*, 2009, **113**, 8157.
26 G. A. King, T. A. A. Oliver, M. G. D. Nix and M. N. R. Ashfold, *J. Chem. Phys.*, 2010, **132**, 064305.
27 A. Iqbal and V. G. Stavros, *J. Phys. Chem. A*, 2010, **114**, 68.
28 M. N. R. Ashfold, G. A. King, D. Murdock, M. G. D. Nix, T. A. A. Oliver and A. G. Sage, *Phys. Chem. Chem. Phys.*, 2010, **12**, 1218.
29 D. J. Hadden, K. L. Wells, G. M. Roberts, L. T. Bergendahl, M. J. Paterson and V. G. Stavros, *Phys. Chem. Chem. Phys.*, 2011, **13**, 10342.
30 R. Crespo-Otero, M. Barbatti, H. Yu, N. L. Evans and S. Ullrich, *ChemPhysChem*, 2011, **17**, 3365.
31 G. A. Williams, G. M. Roberts, H. Yu, N. L. Evans, S. Ullrich and V. G. Stavros, *J. Phys. Chem. A*, 2012, **116**, 2600.
32 A. L. Sobolewski, W. Domcke, C. Dedonder-Lardeux and C. Jouvet, *Phys. Chem. Chem. Phys.*, 2002, **4**, 1093.
33 G. A. Pino, G. Gregoire, C. Dedonder-Lardeux, C. Jouvet, S. Martrenchard and G. Solgadi, *Phys. Chem. Chem. Phys.*, 2000, **2**, 893.
34 S.-I. Ishiuchi, K. Daigoku, M. Saeki, M. Sakai, K. Hashimoto and M. Fujii, *J. Chem. Phys.*, 2002, **117**, 7077.
35 C. Dedonder-Lardeux, D. Grosswasser, C. Jouvet and S. Martrenchard, *PhysChemComm*, 2001, **4**, 1.
36 H. Lippert, V. Stert, L. Hesse, C. P. Schultz, I. Hertel and W. Radloff, *J. Phys. Chem. A*, 2003, **107**, 8239.
37 Z. Lan, W. Domcke, V. Vallet, A. L. Sobolewski and S. Mahapatra, *J. Chem. Phys.*, 2005, **122**, 224315.

38 H. Köppel, W. Domcke and L. S. Cederbaum, *Adv. Chem. Phys.*, 1984, **57**, 59.
39 O. P. J. Vieuxmaire, Z. Lan, A. L. Sobolewski and W. Domcke, *J. Chem. Phys.*, 2008, **129**, 224307.
40 M. Abe, Y. Ohtsuki, Y. Fujimura, Z. Lan and W. Domcke, *J. Chem. Phys.*, 2006, **124**, 224316.
41 H. An and K. K. Baeck, *J. Phys. Chem. A*, 2011, **115**, 13309.
42 T. S. Venkatesan, S. G. Ramesh, Z. Lan and W. Domcke, *J. Chem. Phys.*, 2012, **136**, 174312.
43 M. G. D. Nix, A. L. Devine, R. N. Dixon and M. N. R. Ashfold, *Chem. Phys. Lett.*, 2008, **463**, 305.
44 R. N. Dixon, T. A. A. Oliver and M. N. R. Ashfold, *J. Chem. Phys.*, 2011, **134**, 194303.
45 V. Vallet, Z. Lan, S. Mahapatra, A. L. Sobolewski and W. Domcke, *J. Chem. Phys.*, 2005, **123**, 144307.
46 I. Frank and K. Damianos, *J. Chem. Phys.*, 2007, **126**, 125105.
47 M. Vazdar, M. Eckert-Maksic, M. Barbatti and H. Lischka, *Mol. Phys.*, 2009, **107**, 845.
48 M. Barbatti, J. Pittner, M. Pederzoli, U. Werner, R. Mitric, V. Bonacic-Koutecky and H. Lischka, *Chem. Phys.*, 2010, **375**, 26.
49 S. Faraji, M. Vazdar, V. S. Reddy, M. Eckert-Maksic, H. Lischka and H. Köppel, *J. Chem. Phys.*, 2011, **135**, 154310.
50 H.-J. Werner, P. J. Knowles, R. Lindh, F. R. Manby, M. Schütz, P. Celani, T. Korona, G. Rauhut, R. D. Amos, A. Bernhardsson, *et al.*, *MOLPRO, version 2006.1, a package of ab initio programs*. See http://www.molpro.net.
51 K. Madsen, H. B. Nielsen, and O. Tingleff, *"Methods for non-linear least squares problems"* (2004), Electronically available at http://www2.imm.dtu.dk/pubdb/.
52 B. Lasorne, M. A. Robb, H.-D. Meyer and F. Gatti, *Chem. Phys.*, 2010, **377**, 30.
53 J. Jornet-Somoza, B. Lasorne, M. A. Robb, H.-D. Meyer, D. Lauvergnat and F. Gatti, *J. Chem. Phys.*, 2012, **137**, 084304.
54 Y. P. Varshni and R. C. Shukla, *J. Chem. Phys.*, 1964, **40**, 250.
55 R. S. Mulliken, C. A. Rieke, D. Orloff and H. Orloff, *J. Chem. Phys.*, 1949, **17**, 1248.
56 R. H. McKenzie, *Chem. Phys. Lett.*, 2012, **535**, 196.

Faraday Discussions RSC Publishing

PAPER

Probing ultrafast dynamics in photoexcited pyrrole: timescales for $^1\pi\sigma^*$ mediated H-atom elimination†

Gareth M. Roberts,[a] Craig A. Williams,[a] Hui Yu,[b] Adam S. Chatterley,[ac] Jamie D. Young,[a] Susanne Ullrich[b] and Vasilios G. Stavros*[a]

Received 5th December 2012, Accepted 7th January 2013
DOI: 10.1039/c2fd20140b

The heteroaromatic ultraviolet chromophore pyrrole is found as a subunit in a number of important biomolecules: it is present in heme, the non-protein component of hemoglobin, and in the amino acid tryptophan. To date there have been several experimental studies, in both the time- and frequency-domains, which have interrogated the excited state dynamics of pyrrole. In this work, we specifically aim to unravel any differences in the H-atom elimination dynamics from pyrrole across an excitation wavelength range of 250–200 nm, which encompasses: (*i*) direct excitation to the (formally electric dipole forbidden) $1^1\pi\sigma^*$ ($^1A_2$) state; and (*ii*) initial photoexcitation to the higher energy $^1\pi\pi^*$ ($^1B_2$) state. This is achieved by using a combination of ultrafast time-resolved ion yield and time-resolved velocity map ion imaging techniques in the gas phase. Following direct excitation to $1^1\pi\sigma^*$ ($^1A_2$) at 250 nm, we observe a single time-constant of 126 ± 28 fs for N–H bond fission. We assign this to tunnelling out of the quasi-bound 3s Rydberg component of the $1^1\pi\sigma^*$ ($^1A_2$) surface in the vertical Franck–Condon region, followed by non-adiabatic coupling through a $1^1\pi\sigma^*/S_0$ conical intersection to yield pyrrolyl radicals in their electronic ground state ($C_4H_4N(\tilde{X})$) together with H-atoms. At 238 nm, direct excitation to, and N–H dissociation along, the $1^1\pi\sigma^*$ ($^1A_2$) surface is observed to occur with a time-constant of 46 ± 22 fs. Upon initial population of the $^1\pi\pi^*$ ($^1B_2$) state at 200 nm, a rapid $^1\pi\pi^*$ ($^1B_2$) → $1^1\pi\sigma^*$ ($^1A_2$) → N–H fission process takes place within 52 ± 12 fs. In addition to ultrafast N–H bond cleavage at 200 nm, we also observe the onset of statistical unimolecular H-atom elimination from vibrationally hot S_0 ground state species, formed after the relaxation of excited electronic states, with a time-constant of 1.0 ± 0.4 ns. Analogous measurements on pyrrole-d_1 reveal that these statistical H-atoms are released only through C–H bond cleavage.

[a]*Department of Chemistry, University of Warwick, Library Road, Coventry, CV4 7AL, UK. E-mail: v.stavros@warwick.ac.uk*
[b]*Department of Physics and Astronomy, University of Georgia, Athens, GA 30602, USA*
[c]*Department of Chemistry, Durham University, South Road, Durham, DH1 3LE, UK*

† Electronic supplementary information (ESI) available. See DOI: 10.1039/c2fd20140b

1 Introduction

The gas phase spectroscopy and dynamics of heteroaromatic subunits of biomolecules have received considerable interest in the recent literature.[1–3] Notable examples include studies on DNA nucleobases[4–9] and aromatic amino acid chromophores[10–17] following the absorption of ultraviolet (UV) radiation. Whilst these studies lack the realistic environment to which these molecules ultimately function in—for example, surrounding solvent molecules or the hydrophobic cavity of an extended protein—such a 'bottom-up' approach has nonetheless enabled considerable insight into many of the key aspects of their photochemistry and photophysics. Recent work has indeed begun to demonstrate that these gas phase studies can be used as a benchmark for understanding how this behaviour may evolve in the condensed phase.[18–21]

In this discussion paper, we focus on the excited state photochemistry of pyrrole (molecular structure shown in inset in Fig. 1(a)). A five-membered aromatic heterocycle, pyrrole appears in a number of naturally occurring biomolecules—most notably it is present in heme, the non-protein component of hemoglobin, and as a subunit within the amino acid tryptophan. The prevalence

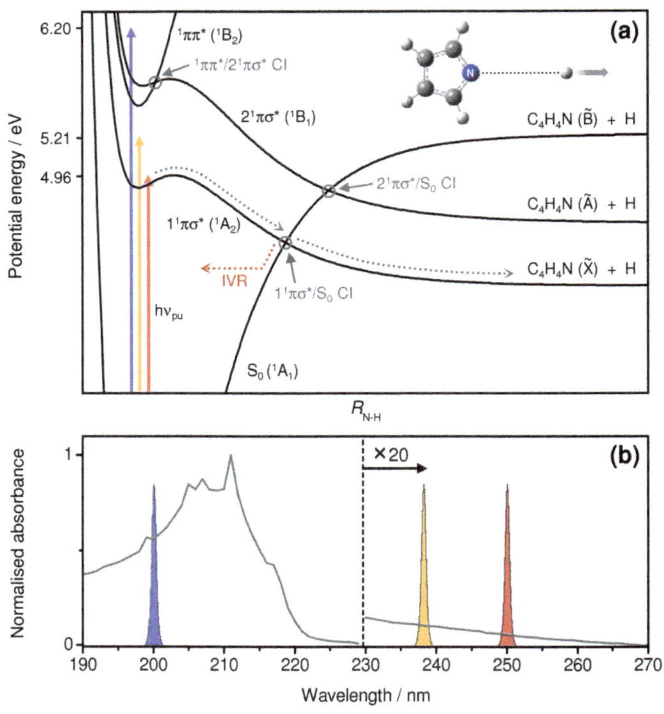

Fig. 1 (a) Schematic 1-D potential energy cuts of the S_0 ($^1A_1$), $1^1\pi\sigma^*$ ($^1A_2$), $^1\pi\pi^*$ ($^1B_2$) and $2^1\pi\sigma^*$ ($^1B_1$) electronic states of pyrrole (molecular structure inset), as a function of N–H bond length (R_{N-H}). Excitations ($h\nu_{pu}$) at pump wavelengths of 250 (red), 238 (yellow) and 200 nm (blue) are represented by vertical arrows. Conical intersections (CIs) along the R_{N-H} coordinate are labelled in grey, while intramolecular vibrational energy redistribution (IVR) after coupling back on to S_0 is represented by the dotted red line. (b) Vapour phase UV absorption spectrum of pyrrole between 270–190 nm (grey). Approximate spectral profiles of the fs pump pulses at 250 (red), 238 (yellow) and 200 nm (blue) are also shown.

of pyrrole in nature has triggered numerous spectroscopic studies, in both the time-[22,23] and frequency-domains,[24–28] as well as a number of theoretical investigations,[1,29–33] all aimed at unravelling the excited state dynamics which ensue after photoexcitation with UV radiation.

The UV absorption spectrum of pyrrole is shown in Fig. 1(b). Over past decades, there have been a number of theoretical studies conducted in an attempt to characterise the intense peak centred at ~210 nm.[34–39] Although the precise ordering of the electronic excited states in pyrrole still remains controversial, it is generally accepted that the majority of absorbance in this band arises from a $^1\pi\pi^*$ ($^1B_2$) ← S_0 ($^1A_1$) transition[26,29,39] (~15% has also been attributed to a $^1\pi\pi^*$ ($^1A_1$) ← S_0 ($^1A_1$) transition,[26,39] not shown in Fig. 1(a)). The pronounced resonances superimposed on top of this broad feature have been attributed to excitation to Rydberg states, with the strongest resonance at ~211 nm assigned by Roos *et al.*[39] to the origin band of the $^1\pi 3p_y$ ($^1B_1$) ← S_0 ($^1A_1$) transition. This broad absorption feature decreases significantly in intensity by ~220 nm, although a very weak tail of absorbance persists down to ~270 nm. This weak tail in the absorption spectrum has been assigned to at least three "optically forbidden" excited electronic states,[27] the most significant of which is the $1^1\pi\sigma^*$ ($^1A_2$) ← S_0 ($^1A_1$) transition. In the vertical Franck–Condon region this $1^1\pi\sigma^*$ ($^1A_2$) state possesses notable 3s Rydberg character associated with the N atom.[1,39] Although formally electric dipole forbidden in C_{2v} symmetry, excitation to $1^1\pi\sigma^*$ ($^1A_2$) can be vibronically induced *via* intensity borrowing from higher lying electronic states, *e.g.* $^1\pi\pi^*$ ($^1B_2$).[23,25–28]

In recent years there has been considerable interest in H-atom elimination dynamics from the N–H bond in pyrrole, mediated by the $1^1\pi\sigma^*$ ($^1A_2$) state.[22–28] This has largely been driven by the seminal theoretical work by Domcke, Sobolewski and co-workers.[1] These authors proposed that the highly efficient ultrafast relaxation dynamics of heteroaromatic molecules can be due, in part, to dissociative $^1\pi\sigma^*$ states localised along X–H (where X = N or O) bond coordinates. With specific reference to pyrrole and the corresponding schematic 1-D potential energy cuts (PECs) shown along its N–H bond coordinate (R_{N-H}) in Fig. 1(a), population directly imparted to the dissociative $1^1\pi\sigma^*$ ($^1A_2$) surface may evolve towards a $1^1\pi\sigma^*/S_0$ conical intersection (CI) and undergo non-adiabatic coupling. This can lead to either repopulation of the electronic ground state, S_0, or direct N–H bond cleavage to yield H-atoms with large amounts of kinetic energy (KE) and pyrrolyl radical co-fragments in their electronic ground state, $C_4H_4N(\tilde{X})$. Alternatively, when excitation occurs at higher photon energies, internal conversion (IC) from the strongly absorbing $^1\pi\pi^*$ ($^1B_2$) state can lead to subsequent population of the $1^1\pi\sigma^*$ ($^1A_2$) surface, also giving rise to rapid N–H bond scission or repopulation of S_0.[24,26–28]

Experimental determination of the exact branching ratio of population bifurcating into S_0 and radical photoproducts at the $1^1\pi\sigma^*/S_0$ CI is non-trivial.[40] However, one is able to measure the KE of the H-atoms eliminated following photoexcitation and then correlate features in the KE spectrum to H-atoms eliminated along the $1^1\pi\sigma^*$ ($^1A_2$) state, providing a signature for the participation of this state in the excited state dynamics;[24–28] ref. 3 provides a more general discussion on this approach to studying $^1\pi\sigma^*$ mediated dynamics in heteroaromatic molecules. For pyrrole, this was first demonstrated using photofragment translational spectroscopy (PTS) techniques by Blank *et al.*[24] upon excitation at 193 and 248 nm, and later with velocity map ion imaging (VMI) techniques by Wei *et al.*,[25,26] following excitation at 217 and 243.1 nm. In the latter studies, H-atoms with high KE ($\sim$7500 cm^{-1}) were attributed to direct dissociation mediated by the $1^1\pi\sigma^*$ ($^1A_2$) state, whilst H-atoms with low KE were attributed to unimolecular dissociation from the S_0 ground state. Based on the measured KE of the H-atoms eliminated, the persistence of high KE D-atoms observed following deuteration of the N–H coordinate (pyrrole-d_1), and the near-limiting recoil anisotropy parameter ($\beta_2 = -0.95$), Wei *et al.* concluded that at 217 nm a $^1\pi\pi^*$ ($^1B_2$) → $1^1\pi\sigma^*$ ($^1A_2$) → N–H fission pathway occurred in <300 fs.[26] The wavelength range for studying this behaviour was extended in H(Rydberg) atom PTS (HRA-PTS) measurements by Cronin *et al.*[27] from $193.3 < \lambda < 254.0$ nm, enabling these authors to identify the onset for the $^1\pi\pi^*$ ($^1B_2$) state absorption channel ($\sim$218 nm) and also determine the N–H bond strength to be D_0(N–H) = 32850 ± 40 cm^{-1}. At wavelengths greater than 254 nm, no significant H-atom signal was observed by these authors.[27]

The first ultrafast time-resolved measurements of H-atom elimination from pyrrole (and indeed, any aromatic heterocycle) were reported in the seminal paper by Lippert *et al.*[22] In this work, the $1^1\pi\sigma^*$ ($^1A_2$) state was directly photoexcited at 250 nm and the H-atoms were probed as a function of pump–probe time delay through 2+1 resonance enhanced multiphoton ionisation (REMPI). The measured H^+ signal, as a function of increasing pump–probe delay, exhibited a bi-exponential rise profile, to which kinetic fits returned two time-constants, τ_1 and τ_2, with values of 110 ± 80 fs and 1.1 ± 0.5 ps, respectively.[22] Once again, with reference to the PECs along R_{N-H} in Fig. 1(a), τ_1 (110 fs) was attributed to the timescale for direct $1^1\pi\sigma^*$ ($^1A_2$) mediated N–H bond fission, proceeding *via* non-adiabatic passage through the $1^1\pi\sigma^*/S_0$ CI, yielding H-atoms with large amounts of KE in coincidence with $C_4H_4N(\tilde{X})$ radical co-fragments. The second, slower time constant, τ_2 (1.1 ps), was attributed to statistical unimolecular H-atom elimination from the hot S_0 ground state of pyrrole, following IC back onto S_0 at the $1^1\pi\sigma^*/S_0$ CI.[22] Most recently, Montero *et al.*[23] have performed ultrafast time-resolved ion yield (TR-IY) measurements on the pyrrole cation signal (pyrrole$^+$) to study the lifetimes of the initial photoexcited states of pyrrole at a number of wavelengths. From these studies, they infer a timescale of 15 ± 3 fs for the initially prepared wavepacket on the $1^1\pi\sigma^*$ ($^1A_2$) surface to evolve out of the vertical ionisation window, while at higher photon energies they report a timescale of 19 ± 3 fs for the $^1\pi\pi^*$ ($^1B_2$) → $1^1\pi\sigma^*$ ($^1A_2$) IC process to occur.[23] The dynamical picture arising from these combined experimental studies, in both the time- and frequency-domains, is also qualitatively supported by a number of *ab initio* studies, which include mixed quantum-classical trajectory[30–33] and fully quantum mechanical wavepacket calculations.[29]

To establish the timescales for H-atom elimination dynamics in pyrrole, we probe dissociated H-atoms using a combination of ultrafast TR-IY and time-resolved velocity map ion imaging (TR-VMI) following excitation at a number of wavelengths (250, 238 and 200 nm). The ability to obtain both energy *and* time information using TR-VMI enables us to determine the timescales for direct dissociation following excitation to the $1^1\pi\sigma^*$ ($^1A_2$) state and infer a timeframe for $^1\pi\pi^*$ ($^1B_2$) → $1^1\pi\sigma^*$ ($^1A_2$) coupling. Complementary TR-IY and TR-VMI measurements of D-atom elimination from pyrrole-d_1 (which is selectively deuterated at the N–H site), also enable us to: (*i*) investigate the role of both statistical unimolecular N–H *and* C–H bond fission from pyrrole molecules in a vibrationally hot S_0 ground state, depending on the initially prepared excited state ($1^1\pi\sigma^*$ ($^1A_2$) or $^1\pi\pi^*$ ($^1B_2$)); and (*ii*) reveal the presence of tunnelling through a small exit barrier in the $1^1\pi\sigma^*$ ($^1A_2$) surface, prior to rapid N–H bond fission.

2 Experimental methods

The reader is directed to previous work for further experimental details of the TR-VMI apparatus[41,42] and therefore only a brief overview is provided here. TR-VMI experiments utilise femtosecond (fs) laser pulses derived from a commercial Ti:Sapphire oscillator and regenerative amplifier system (Spectra-Physics, Tsunami and Spitfire XP, respectively). The regenerative amplifier is set at 125 Hz for these experiments, delivering 3 mJ/pulse centred at 800 nm with a temporal full width at half maximum (FWHM) of ~40 fs. The fundamental output is subsequently split into three beams of equal energy. Of these three beams, one is up-converted into a 200 nm pump beam, which is achieved *via* frequency doubling the fundamental in a type-I β-barium borate (BBO) crystal to generate 400 nm, followed by successive frequency mixing (type-II and type-I BBO crystals, respectively) with the residual 800 nm pulses to generate third harmonic (267 nm) and finally the desired 200 nm fourth harmonic (~1 μJ/pulse). The remaining two beams are used to pump two optical parametric amplifiers (Light Conversion, TOPAS-C). One of these produces a tunable pump in the range 238–250 nm (3–4 μJ/pulse) whilst the second is used to generate the probe, set at 243.1 nm (~7 μJ/pulse) to selectively ionise any H-atoms *via* a 2+1 REMPI process, resonant with the two-photon allowed 2s ← 1s transition. The time-delay between pump and probe pulses (Δt) is controlled by passing the 800 nm seed beam for pump generation through a protected-gold hollow retroreflector mounted on a computer controlled delay stage (Physik-Instrumente, M-521.DD). The pump and probe pulses are focused through $f = 50$ cm MgF_2 and CaF_2 lenses, respectively, and recombined prior to entering the interaction region of a VMI spectrometer[43] where they perpendicularly intersect a molecular beam pulse seeded with pyrrole. The power densities of the pump and probe pulses are $\sim10^{11}$–10^{12} and $\sim10^{13}$ W cm^{-2}, respectively. Non-resonant multiphoton ionisation of methanol is employed as a method for determining both the delay position corresponding to the temporal overlap of the pump and probe ($\Delta t = 0$), to within an accuracy of ±15 fs, and the associated cross-correlation/instrument response function (IRF). For these experiments a Gaussian IRF is measured with a FWHM of ~120 fs for pump wavelengths in the range 250–238 nm and ~160 fs for the 200 nm pump. The polarisation of both beams is set parallel to the plane of the imaging detector.

Molecular beam pulses are generated by seeding pyrrole (Sigma-Aldrich, >98%) molecules in helium (~2 bar) which are transferred into vacuum using an Even-Lavie pulsed solenoid valve[44] operating at 125 Hz, which is synchronized to the laser system. The valve temperature is maintained at ~50 °C, while the helium carrier gas pressure and opening time of the valve are optimized to reduce the degree of clustering of the parent ion relative to the parent ion monomer. A typical opening time for the valve is ~13 μs. The molecular beam machine consists of differentially pumped source and interaction chambers, partitioned by a skimmer with a 2 mm aperture. The pulsed valve is housed within the source chamber, whilst the interaction chamber contains VMI optics, replicating the arrangement described by Eppink and Parker.[43] Generated cations are then accelerated towards a position sensitive detector (under VMI focusing conditions) which consists of a pair of 40 mm diameter chevron microchannel plates (MCPs) coupled to a P-43 phosphor screen (Photek, VID-240). A timed voltage pulse is applied to the rear MCP to gate the detector for measuring only the H^+ or D^+ signal. A 2-D projection of the 3-D ion distribution is subsequently obtained by capturing the light emitted from the phosphor screen using a charge-coupled device (Basler, A-312f). 1-D radial/velocity spectra of the REMPI probed H or D atoms (H^+ or D^+) are obtained through deconvolution of the recorded H^+ or D^+ images by incorporating a polar onion-peeling method[45] into our data acquisition program coded in LabVIEW. Finally, total kinetic energy release (TKER) spectra are obtained from the derived 1-D radial (r) spectra by using an appropriate Jacobian ($r^2 \propto$ KE), assuming the mass of the radical co-fragment is C_4H_4N (66 amu), and a KE calibration factor (obtained by photolysis of HBr at 200 nm[46]).

Alternatively, TR-IY measurements of the total H^+ or D^+ signal, as a function of Δt, can be recorded by monitoring the current output directly from the phosphor screen using an oscilloscope,[47] which is interfaced to our data acquisition program with a general purpose interface bus. This enables a time-of-flight mass spectrum to be recorded at each Δt. Individual mass spectra at each Δt are acquired over ~1250 laser shots, and the measured total H^+ or D^+ signal is then integrated. Resultant TR-IY transients are the average of 20 individual time-resolved scans.

Pyrrole-d_1 was synthesised using a method similar to that described in ref. 26, by repeated stirring of undeuterated pyrrole in excess D_2O in a light sealed vessel for ~48 h. The pyrrole-d_1 product was then separated from the D_2O, and dried over Na_2CO_3. ^{1}H-NMR spectroscopy was used to check for deuteration (>95%).

In addition to the TR-IY and TR-VMI experiments, a vapour phase UV absorption spectrum was recorded with a commercial UV-Visible absorption spectrometer (Perkin Elmer, Lambda 25, 1 nm resolution) by bubbling helium through a sample of pyrrole and collecting the gas in a fused silica sample cell.

3 Results and discussion

3.1 Excited state dynamics following direct excitation to $1^1\pi\sigma^*$ ($^1A_2$)

a H^+ velocity map image, total kinetic energy release spectrum and recoil anisotropy parameters at 250 nm. The inset in Fig. 2(a) depicts a velocity map image of the H-atoms generated following direct excitation to the $1^1\pi\sigma^*$ ($^1A_2$) state of pyrrole at 250 nm, which are subsequently ionised through 2 + 1 REMPI at 243.1 nm with a time-delayed probe pulse set at $\Delta t = 1.2$ ns. The left half displays the recorded image while the right half corresponds to a reconstructed slice through the centre of the original 3-D velocity distribution, extracted using the polar onion-peeling method.[45] A one-colour (243.1 nm probe only) background image has been subtracted from the presented image, such that the signal corresponds exclusively to H-atoms formed and probed after photolysis at 250 nm. The image contains information about both speed and angular recoil distributions of the photodissociated H-atoms. The radial distance of signal from the centre of the image is proportional to the velocity of the H-atoms, with a clear anisotropic ring appearing at large radius, associated with H-atoms carrying large amounts of KE following photodissociation. The enhanced signal intensity

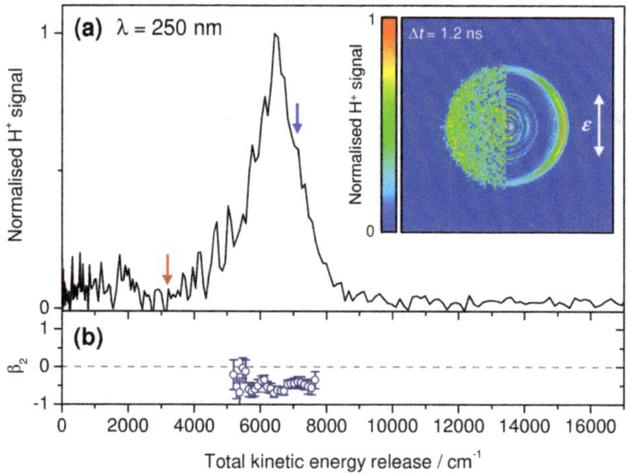

Fig. 2 (a) TKER spectrum and corresponding H$^+$ velocity map image (inset), obtained following photoexcitation of pyrrole at 250 nm and at a pump–probe delay of $\Delta t = 1.2$ ns. The left half presents the recorded image, while the right half presents a reconstructed slice through the centre of the original 3-D ion distribution (the vertical white arrow indicates the electric field polarisation, ε, of the pump pulse, $h\nu_{pu}$). Vertical blue and red arrows indicate the predicted TKER$_{max}$ values for photodissociation into the $C_4H_4N(\tilde{X})$ + H and $C_4H_4N(\tilde{A})$ + H photoproducts channels, respectively (see Fig. 1(a) and eqn (1)). (b) β_2 recoil anisotropy parameters as a function of TKER between ~5000–8000 cm^{-1} at $\Delta t = 1.2$ ns.

perpendicular to the electric field vector of $h\nu_{pu}$, ε, exhibited by this feature is worthy of note, and indicates that these H-atoms are eliminated on a timescale faster than the rotational timescale of the parent pyrrole molecules; this recoil anisotropy will be discussed further in the ensuing paragraphs.

The TKER spectrum derived from the image is also shown in Fig. 2(a) by the solid black line. This spectrum is dominated by a broad peak centred at ~6500 cm^{-1}, the location of which is in reasonable accord with previous HRA-PTS data by Cronin *et al.*,[27] as well as slice-imaging measurements by Rubio-Lago *et al.*[28] at the same wavelength. However, we concede that the inherently poorer resolution of our ultrafast TR-VMI setup does not allow us to resolve any vibrational mode structure in the TKER distribution. Despite this, we can determine the maximum TKER for dissociation into $C_4H_4N(\tilde{X})$ + H photofragments, TKER$_{max}$, according to the general relationship:

$$\mathrm{TKER_{max}} = h\nu_{pu} - D_0(\mathrm{N{-}H}) - E_{int} \quad (1)$$

In eqn (1) $h\nu_{pu}$ corresponds to the pump photon energy (40000 cm^{-1}), D_0(N–H) is the adiabatic N–H dissociation energy (32850 cm^{-1}),[27] and E_{int} is the total internal (electronic and vibrational) energy of the radical co-fragments. In the case of $C_4H_4N(\tilde{X})$ + H photoproducts with $E_{int} = 0$ cm^{-1}, this returns a value for TKER$_{max}$ of ~7150 cm^{-1}, correlated to a scenario where the $C_4H_4N(\tilde{X})$ co-fragments are formed in their respective zero point vibrational energy level ($v = 0$). This TKER$_{max}$ value is represented by the vertical blue arrow in Fig. 2(a). Within the resolution of our VMI spectrometer,[48] this is in reasonable agreement with the high energy 'tail' of the observed peak centred at ~6500 cm^{-1}, and as such we assign this feature to the formation of $C_4H_4N(\tilde{X})$ + H *via* direct dissociation along $1^1\pi\sigma^*$ ($^1A_2$). In principle, dissociation into C_4H_4N radicals in their first electronic excited state, $C_4H_4N(\tilde{A})$, together with H, is an energetically open channel at 250 nm. The predicted $TKER_{max}$ for such a channel, using eqn (1) and a calculated $\tilde{A} \leftarrow \tilde{X}$ energy separation of $\sim$3950 cm^{-1},[49] is represented by the vertical red arrow in Fig. 2(a). Based on the very low signal levels observed around 3200 cm^{-1}, we conclude it is unlikely that the $C_4H_4N(\tilde{A})$ + H dissociation channel is accessed at 250 nm, in line with previous observations at the same excitation wavelength.[27,28]

To obtain insight into which modes are promoting the optically forbidden $1^1\pi\sigma^*$ ($^1A_2$) $\leftarrow$ S_0 ($^1A_1$) transition, one can turn to the measured anisotropy parameters, β_2, shown in Fig. 2(b) between $\sim$5000–8000 cm^{-1}. An electronic transition will have greatest intensity when its transition dipole moment ($\boldsymbol{\mu}$) is aligned parallel to ε (in accordance with a $\cos^2\theta$ distribution, where θ is the angle between $\boldsymbol{\mu}$ and ε[50]). Within the limit of axial recoil and assuming dissociation occurs on a timeframe much faster than molecular rotation (as one anticipates here, given that dissociation is fast following excitation to a dissociative state such as $1^1\pi\sigma^*$ ($^1A_2$)), the distribution in the H-atom photoproducts allows us to ascertain the alignment of the dissociating N–H bond relative to $\boldsymbol{\mu}$. Formally, β_2 can take values between $-1 \leq \beta_2 \leq 2$, where these limiting values of -1 and 2 correspond to scenarios in which the dissociating bond is aligned perpendicular or parallel to $\boldsymbol{\mu}$, respectively. The measured non-limiting (average) value of $\beta_2 \approx -0.5$ in Fig. 2(b) is in good agreement with previous observations,[27,28] which largely report β_2 values of < 0. This suggests that the dominant modes responsible for promoting the transition possess either b_2 or b_1 symmetry, Γ_ν, inducing vibronic transition moments of either b_1 or b_2 symmetry, Γ_μ, (transforming as x and y, respectively, in C_{2v} symmetry) both of which lie perpendicular to the dissociating N–H bond. Such a conclusion is formed on the basis that $A_1 \otimes \Gamma_\nu \otimes \Gamma_\mu \otimes A_2 \supseteq A_1$. However, we note that modes of a_2 symmetry may also promote the $1^1\pi\sigma^*$ ($^1A_2$) $\leftarrow$ S_0 ($^1A_1$) transition, inducing a vibronic transition moment of a_1 symmetry (parallel to the N–H bond), leading to $\beta_2 > 0$ for peaks associated with a_2 vibrations, as observed in the high-resolution TKER spectra reported by Cronin *et al.*[27] The reduced resolution in our broadband ($\sim$500 cm^{-1}) measurements prohibits a similar observation here, although contributions from a_2 vibrations may account for the non-limiting β_2 value of -0.5.

b H-atom elimination dynamics at 250 nm. A normalised TR-IY transient of the *total* H^+ signal as a function of pump–probe delay, between $\Delta t = -1$ ps and 2.5 ps, following photoexcitation at 250 nm and probing at 243.1 nm is shown in Fig. 3(a) (open circles). A 243.1 nm probe only background signal has been subtracted from the displayed transient, noting that negligible 250 nm pump-only signal was observed, in line with 250 nm being non-resonant with the 2s $\leftarrow$ 1s transition in H. At negative pump–probe delays ($\Delta t < 0$), an appreciable 'probe-pump' H^+ signal is observed. This concurs with observations in earlier TR-IY measurements conducted at 250 nm by Lippert *et al.*,[22] and we discuss the origins of this signal in greater detail below. Around $\Delta t = 0$, a sharp rise in signal is observed, which then decays within our IRF (<120 fs). Upon increasing the time delay to positive pump–probe delays ($\Delta t > 0$), the signal then steadily rises before plateauing around $\sim$500 fs.

Unlike previous studies by Lippert *et al.*[22] however, we have additionally performed TR-VMI measurements to generate a series of complementary time-resolved TKER spectra at a series of Δt between -1 ps and 1.2 ns, presented in Fig. 3(b). At $\Delta t = -1$ ps a small feature centred at $\sim$7000 cm^{-1} is observed, which

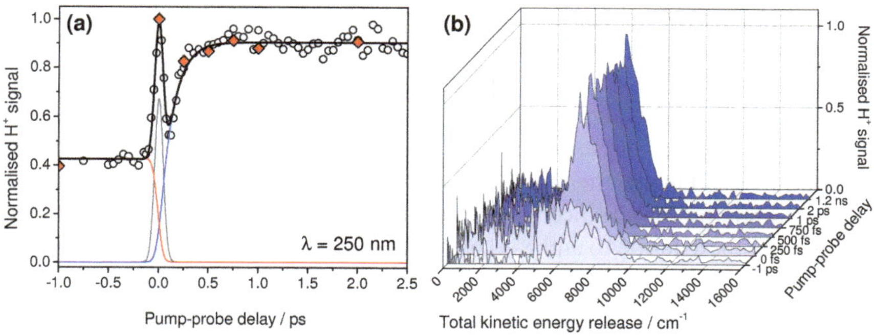

Fig. 3 (a) Normalised H^+ signal transient obtained using TR-IY (open circles), with a 250 nm pump and 243.1 nm probe combination, together with the resultant kinetic fit (black line). A one-colour 243.1 nm background signal has been subtracted from the displayed transient. Individual fit components associated with $\tau_f = 126$ fs (blue), $\tau_r < 30$ fs (red) and $\tau_d < 30$ fs (grey) are also shown (see main text for details). (b) Background (243.1 nm probe only) subtracted time-resolved TKER spectra, at a series Δt, following excitation at 250 nm. The total integrated H^+ signal in each time-resolved TKER spectrum is shown superimposed on (a) as red diamonds.

then moves towards lower TKER (~6500 cm^{-1}) and significantly broadens in width at $\Delta t = 0$. In addition, there is also a notable increase in signal at low TKER (<4000 cm^{-1}) at this pump–probe delay, which exhibits a broad profile. By $\Delta t =$ 500 fs, a new narrower feature centred at ~6500 cm^{-1} has appeared, correlated to probed H-atoms formed following N–H dissociation along the $1^1\pi\sigma^*$ ($^1A_2$) surface at 250 nm, as discussed above (*cf.* Fig. 2(a)). This feature exhibits a constant intensity out to $\Delta t = 1.2$ ns, with no significant signal at low TKER (<4000 cm^{-1}) in any spectra recorded at $\Delta t \geq 250$ fs. This latter observation instils confidence that the rapid (sub-500 fs) rise in signal observed at positive pump–probe delays in Fig. 3(a), correlates (almost) exclusively to direct $1^1\pi\sigma^*$ ($^1A_2$) mediated N–H bond scission, with negligible contributions from slower statistical H-atom elimination from hot S_0 species (by 1.2 ns), following IC back to S_0 at the $1^1\pi\sigma^*/S_0$ CI (*vide infra*). Agreement between the TR-VMI and TR-IY measurements is confirmed by overlaying the normalised total integrated H^+ signal in each of the time-resolved TKER spectra on top of the normalised TR-IY trace (red diamonds, Fig. 3(a)), showing excellent agreement with each other.

In order to extract a time-constant for the appearance of H-atoms eliminated along the $1^1\pi\sigma^*$ ($^1A_2$) at 250 nm, the H^+ signal transient in Fig. 3(a) is fitted to the sum of an exponential rise (τ_f) and decay (τ_d) function at positive Δt, and a single exponential rise function (τ_r) at negative Δt. All three functions are convolved with our Gaussian IRF of 120 fs. The reader is referred to the electronic supplementary information (ESI†) online for further details of the kinetic fitting procedure. The result of this fit is shown as the solid black line through the experimental data points in Fig. 3(a). The time-constants returned from the kinetic fit have values of $\tau_f = 126 \pm 28$ fs, $\tau_d < 30$ fs and $\tau_r < 30$ fs; these individual fit components are indicated by the blue, grey and red lines in Fig. 3(a), respectively. The minimum temporal resolution of our measurements is ~30 fs (~25% of our 120 fs Gaussian IRF)[7] and as a result we quote values for both τ_d and τ_r as <30 fs. τ_r is attributed to 'reverse dynamics', discussed further below. τ_d corresponds to the decay of the peak present in the H^+ transient around $\Delta t = 0$, and arises due to multiphoton processes mediated by the increased photon flux when the pump and probe pulses are temporally overlapped, which may yield H^+ directly (*e.g.* dissociative ionisation), as seen in our previous studies on other aromatic heterocycles.[6,10,51,52] The rapid decay of this signal after $\Delta t = 0$ is merely a reflection of the extremely short lived nature of the resonant intermediate state that promotes this multi-photon process. Finally, the remaining component of the fit, $\tau_f = 126 \pm 28$ fs, is attributed to the appearance timescale of (REMPI probed) H-atoms formed through photoexcitation to and N–H dissociation along the $1^1\pi\sigma^*$ ($^1A_2$) surface at 250 nm.

It is worth noting the disparity in the time-constants for the forward ($\tau_f = 126 \pm 28$ fs) and reverse ($\tau_r < 30$ fs) dynamics. The forward dynamics result from single photon excitation with 250 nm to the $1^1\pi\sigma^*$ ($^1A_2$) state which subsequently dissociates to yield H-atoms probed through 2+1 REMPI with 243.1 nm. In contrast, the reverse dynamics correspond to single photon excitation with 243.1 nm to the $1^1\pi\sigma^*$ ($^1A_2$) state, forming H-atoms which are then two-photon excited with 243.1 nm to the 2s level. These excited H(2s)-atoms are then ionised with a single photon of 250 nm to yield H^+. Such an interpretation of these reverse dynamics has also previously been put forward by Lippert *et al.*[22] The disparity between the values of τ_f and τ_r therefore most likely reflects the different mechanisms *via* which H^+ is generated at $\Delta t < 0$ and $\Delta t > 0$: for the reverse dynamics, photodissociation of the N–H bond *and* two-photon excitation of the 2s ← 1s transition all needs to occur within the temporal width of the 243.1 nm pulse (~80 fs), hence giving rise to the very fast time-constant of $\tau_r < 30$ fs – this is not the case for the forward dynamics. In further support of the 'probe-pump' mechanism proposed for forming H^+ at $\Delta t < 0$, we recall the fact that the feature observed in the TKER spectrum at $\Delta t = -1$ ps is centred at ~7000 cm^{-1}, whereas the pronounced peak in the spectra recorded at $\Delta t \geq 250$ fs is centred at ~6500 cm^{-1}. The blue-shift in central TKER of the feature observed at $\Delta t = -1$ ps, relative to the spectra at $\Delta t \geq 250$ fs in Fig. 3(b), is a direct reflection of the increased photon energy imparted to the $1^1\pi\sigma^*$ ($^1A_2$) state at 243.1 nm (compared to 250 nm), enabling more of the available energy to be partitioned into TKER during the N–H fission process.

As this stage, we highlight one key difference between the H^+ transient presented here and those reported earlier by Lippert *et al.*[22] Unlike the results presented in Fig. 3(a), these authors measured a H^+ signal transient which exhibited a bi-exponential rise profile with two time-constants, τ_1 and τ_2, with values of 110 ± 80 fs and 1.1 ± 0.5 ps, respectively.[22] Although τ_1 compares very favourably with our measured value of $\tau_f = 126 \pm 28$ fs, we do not observe any evidence of a second slower rise in our H^+ signal correlating to τ_2 (~1.1 ps). Lippert *et al.* speculated that the latter time-constant of 1.1 ps, arose through some fraction of the dissociating wavepacket non-adiabatically coupling back into a vibrationally hot S_0 state, which may then undergo N–H dissociation to yield a 'statistical' distribution of slower H-atoms. We observe negligible H-atoms with a statistical distribution in our time-resolved TKER spectra out to $\Delta t = 1.2$ ns (Fig. 3(b)). This also accords with the high-resolution TKER spectrum measured by Cronin *et al.* at 250 nm ($\Delta t \sim 5$ ns),[27] although we acknowledge that Rubio-Lago *et al.* do observe a reasonable fraction of H-atoms with a statistical profile at the same wavelength.[28] Such a distribution can, however, arise from undesired multiphoton dissociative ionisation events, which may appear on an ultrafast (<200 fs) timeframe.[14,51,53,54]

Additionally, in previous TR-VMI studies on related aromatic heterocycles (imidazole and pyrazole) at 200 nm,[11] as well as in the present work on pyrrole (see section 3.2), we have indeed observed evidence for the onset of statistical H-atom elimination from hot S_0 species, but taking place on a much longer timeframe of hundreds of picoseconds to nanoseconds. Ultimately, the preceding arguments and our present findings open a debate as to what role, if any, does a statistical H-atom elimination pathway play in the photophysics of pyrrole at 250 nm within ~1 ns? – at present this remains somewhat unclear based on the conflicting experimental data.

c The role of tunnelling on the $1^1\pi\sigma^*$ ($^1A_2$) surface at 250 nm. The value of $\tau_f = 126 \pm 28$ fs, while still ultrafast, is perhaps longer than one might anticipate for the appearance of H-atoms, following direct excitation to a state such as $1^1\pi\sigma^*$ ($^1A_2$) in pyrrole, given its repulsive profile with respect to R_{N-H}. Lippert *et al.*[22] also briefly acknowledged this fact at 250 nm, and suggested that this may be due to the dissociating wavepacket being impeded by a small exit barrier to N–H fission, which lies close to the vertical Franck–Condon region in $1^1\pi\sigma^*$ ($^1A_2$) state. With reference to Fig. 4(a), this exit barrier arises from the fact that the effective 'adiabatic' $1^1\pi\sigma^*$ ($^1A_2$) surface, shown by the solid black line, corresponds to the lower adiabat formed through an avoided-crossing between the diabatic $^1\pi$3s Rydberg and $^1\pi\sigma^*$ valence states (dotted grey lines).[55,56] Previous theoretical calculations predict the energy (height) of the exit barrier to be between ~240–3200 cm^{-1},[29,33] although to date, no direct experimental evidence for this exit barrier has been reported. Nonetheless, by analogy with σ* mediated N–H bond fission dynamics observed in related systems, such as ammonia[57] and aniline (aminobenzene),[56] following excitation into the quasi-bound 3s Rydberg well of the $1^1\pi\sigma^*$ ($^1A_2$) state, it could be possible for population to escape *via* tunnelling through this exit barrier and evolve towards N–H dissociation. Fig. 4(a) shows a reduced 1-D picture for this tunnelling mechanism at 250 nm along R_{N-H}, originating from the zero-point of the N–H stretch ($\nu_{NH} = 0$) in a (vibrationally excited)

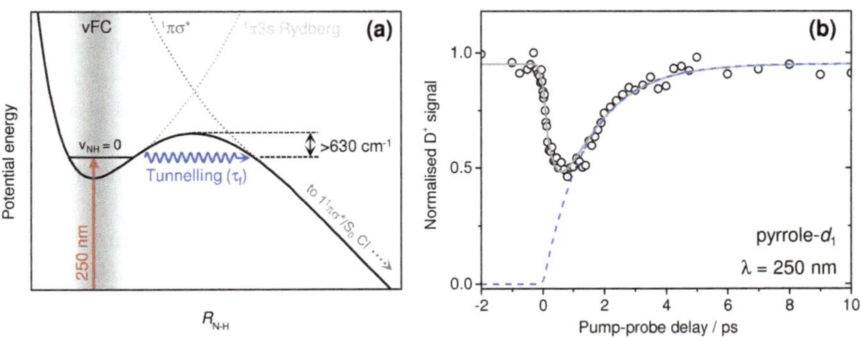

Fig. 4 (a) Enlarged schematic of the 'adiabatic' $1^1\pi\sigma^*$ ($^1A_2$) surface (solid black line) in pyrrole as a function of R_{N-H}. The diabatic $^1\pi$3s Rydberg and $^1\pi\sigma^*$ valence surfaces are also shown by the dotted grey lines. Excitation into the quasi-bound 3s Rydberg well (in the vertical Franck–Condon (vFC) region; grey shading) of a (vibrationally excited) adiabatic $1^1\pi\sigma^*$ ($^1A_2$) surface at 250 nm is shown by the red arrow. Tunnelling through the exit barrier onto the dissociative region of this potential is shown by the blue arrow. (b) Normalised D$^+$ signal transient obtained using TR-IY (open circles), following photolysis of pyrrole-d_1 with a 250 nm pump and 243.1 nm probe combination. The $\tau_f = 1.4$ ps fit (dashed blue line) is also shown (see ESI† for details). The grey line is provided as a visual guide.

adiabatic $1^1\pi\sigma^*$ ($^1A_2$) potential. Such a tunnelling mechanism would most likely impede N–H fission dynamics, and account for the slightly longer value of $\tau_f = 126 \pm 28$ fs we measure here at 250 nm.

To investigate whether such a tunnelling mechanism is active following excitation to the $1^1\pi\sigma^*$ ($^1A_2$) surface at 250 nm, we have also performed TR-IY measurements on the *total* D^+ signal as a function of Δt, after photolysis of pyrrole-d_1 at the same wavelength. The results of these measurements are presented in Fig. 4(b) (open circles). The D^+ signal transient bears some similarities to the H^+ transient in Fig. 3(a), although there is no obvious sharp 'spike' in D^+ signal around $\Delta t = 0$, suggesting that the spike observed at $\Delta t = 0$ in Fig. 3(a) has significant contributions from multiphoton-driven C–H fission. However, perhaps the most striking difference is the significantly slower rise in the D^+ signal at $\Delta t > 0$, which does not plateau until ~6 ps. A kinetic fit, obtained by fitting to the *rising* D^+ signal at $\Delta t \geq 1$ ps only (see ESI† for details), is shown by the dashed blue line in Fig. 4(b). This fit returns a value of $\tau_f = 1.4 \pm 0.3$ ps, and in turn yields a kinetic isotope effect (k_H/k_D) of ~11. Such a large kinetic isotope effect strongly supports the idea that following excitation at 250 nm, N–H bond fission proceeds *via* tunnelling through a small exit barrier in the $1^1\pi\sigma^*$ ($^1A_2$) surface. This is, to the best of our knowledge, the first experimental evidence that the $1^1\pi\sigma^*$ ($^1A_2$) surface in pyrrole does indeed possess a quasi-bound well in the vertical Franck–Condon region, and that population can evolve out of it *via* tunnelling. Based on previous HRA-PTS measurements by Cronin *et al.*,[27] who report that the origin band of the $1^1\pi\sigma^*$ ($^1A_2$) state must lie at >254 nm, and our measurements here at 250 nm, we conclude that the exit barrier to N–H dissociation in this state must be >630 cm^{-1}.

d H-atom elimination dynamics at 238 nm: effects of excess internal energy. We now consider the effect of increasing excitation energy on the observed dynamics, after direct excitation to $1^1\pi\sigma^*$ ($^1A_2$). The inset in Fig. 5(a) contains a velocity map image of H-atoms generated following excitation at 238 nm to the $1^1\pi\sigma^*$ ($^1A_2$) state, which is probed with a time-delayed 243.1 nm probe ($\Delta t = 1.2$ ns). The left half of the image once again displays the collected image (after one-colour 243.1 nm background subtraction) while the right half presents a reconstructed slice through the centre of the original 3-D H^+ distribution. As with photoexcitation at 250 nm, there is a clear anisotropic feature ($\beta_2 \approx -0.4$, see Fig. S1 of the ESI†) at large radius, corresponding to photodissociated H-atoms born with large amounts of KE. The similarities between both images in Fig. 3(a) and 5(a) indicate that these H-atoms are indeed formed *via* similar mechanisms – *i.e.* $1^1\pi\sigma^*$ ($^1A_2$) mediated N–H bond fission.

Further similarities/differences between excitation at 250 and 238 nm can be gleaned through the corresponding time-resolved TKER spectra and H^+ signal transient, which are shown in Fig. 5(a) and 5(b), respectively. As with the measurements at 250 nm, the TKER spectrum at $\Delta t = -5$ ps displays a small amount of signal peaked at ~7000 cm^{-1}, and can analogously be assigned to N–H bond cleavage along $1^1\pi\sigma^*$ ($^1A_2$) at 243.1 nm, where the time-delayed 238 nm photons are involved in the H-atom ionisation step. Cursory inspection of the TKER spectrum in Fig. 2(a) and those at $\Delta t = 1$ ps and 1.2 ns in Fig. 5(a), clearly reveals a difference in the peak position of the feature at high TKER when excited at 250 and 238 nm, respectively. For excitation at 250 nm the feature appears centred at ~6500 cm^{-1}, whereas at 238 nm it appears at ~7200 cm^{-1} due to the

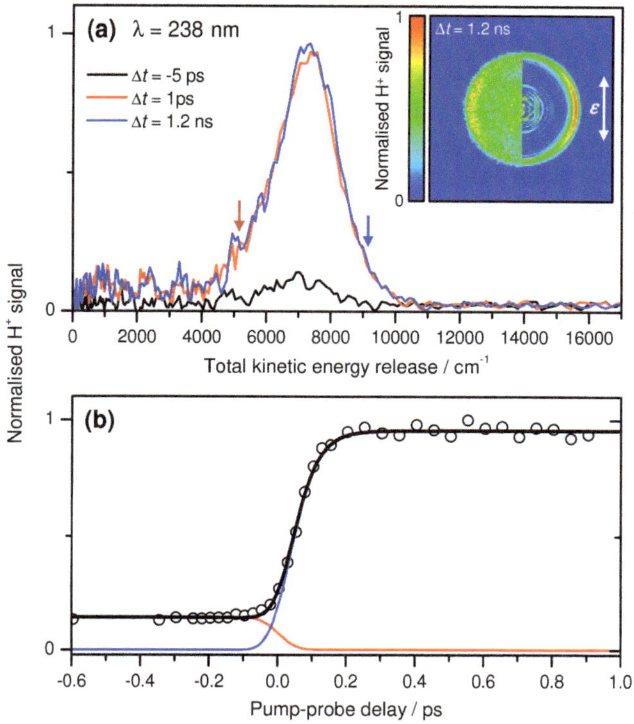

Fig. 5 (a) Time-resolved TKER spectra recorded following photoexcitation of pyrrole at 238 nm, and probing H-atoms with a time-delayed 243.1 nm pulse. Vertical arrows indicate the predicted $TKER_{max}$ values for N–H dissociation into the $C_4H_4N(\tilde{X})$ + H (blue) and $C_4H_4N(\tilde{A})$ + H (red) photoproduct channels (see Fig. 1(a) and eqn (1)). Inset: H^+ velocity map image (left) and reconstructed slice through the centre of the original 3-D ion distribution (right), recorded at $\Delta t = 1.2$ ns (white arrow indicates the polarisation of $h\nu_{pu}$). (b) Integrated H^+signal transient (6000–8500 cm^{-1}), following excitation at 238 nm (open circles), together with an associated kinetic fit (black line). Individual fit components associated with τ_f = 46 fs (blue) and $\tau_r < 30$ fs (red) are also shown (see main text for details).

increased photon energy. Similarly, the feature at ~7200 cm^{-1} is assigned to dissociation into $C_4H_4N(\tilde{X})$ + H photoproducts *via* the $1^1\pi\sigma^*$ ($^1A_2$) surface at 238 nm, confirmed through agreement between the predicted $TKER_{max}$ for this channel (~9170 cm^{-1} at 238 nm) and the location of the high energy tail of this feature (vertical blue arrow, Fig. 5(a)). No significant signal associated with the higher energy $C_4H_4N(\tilde{A})$ + H photoproduct channel is observed in Fig. 5(a). We also note that N–H dissociation dynamics along the $1^1\pi\sigma^*$ ($^1A_2$) surface appear complete by $\Delta t = 1$ ps, as there is no further evolution of the amplitude of the high TKER feature between 1 ps and 1.2 ns, in line with direct excitation to a dissociative state.

The H^+ signal transient in Fig. 5(b) (open circles) is obtained by integrating the high TKER feature (6000–8500 cm^{-1}) in TKER spectra recorded at many Δt values between −0.6 to 1 ps, and enables a more quantitative analysis of the timescale for H-atom elimination along $1^1\pi\sigma^*$ ($^1A_2$) at 238 nm. Using a similar approach to that described in section 3.1b, time-constants of $\tau_f = 46 \pm 22$ fs and $\tau_r < 30$ fs, are extracted *via* a kinetic fit (black line); the τ_f and τ_r components of the fit are shown by the blue and red lines in Fig. 5(b), respectively. Unlike the data at 250 nm

(Fig. 3(a)), there is no noticeable 'spike' in the measured H^+ signal around $\Delta t = 0$. This may suggest that an appropriate intermediate state, which can undergo dissociative ionisation to yield H^+, is not accessible *via* multiphoton excitation at this combination of pump and probe wavelengths, although unlike TR-IY measurements at 250 nm (Fig. 3(a)), the transient in Fig. 5(b) is obtained by integrating *only* signal associated with $1^1\pi\sigma^*$ ($^1A_2$) driven N–H fission. Reassuringly, τ_r once again yields a time-constant that is faster than our IRF (<30 fs), and can be reconciled with the mechanism outlined above (see section 3.1b).

We draw this section to a close by pointing the reader's attention to the notable difference in the τ_f values extracted following excitation at 250 and 238 nm; $\tau_f = 126 \pm 28$ fs and 46 ± 22 fs, respectively. At 250 nm we concluded that N–H bond fission proceeds by population tunnelling out of the small quasi-bound 3s Rydberg well in the vertical Franck–Condon region of the $1^1\pi\sigma^*$ ($^1A_2$) surface, giving rise to a H-atom appearance time-constant which is slower than one might traditionally anticipate if N–H fission occurred in a purely 'ballistic' fashion along a highly repulsive surface. At 238 nm, the time-constant of $\tau_f = 46 \pm 22$ fs is ~3 times faster than at 250 nm, suggesting that direct excitation to $1^1\pi\sigma^*$ ($^1A_2$) at this wavelength is above the exit barrier and N–H fission occurs directly, rather than *via* tunnelling. Indeed, TR-IY measurements on pyrrole-d_1, which monitor the *total* D^+ signal as a function of Δt at 238 nm, return a time-constant of $\tau_f = 136 \pm 38$ fs for the appearance of D-atoms formed *via* N–D fission along $1^1\pi\sigma^*$ ($^1A_2$) (see Fig. S2 of the ESI†). This returns a much smaller kinetic isotope effect of ~2.9, which indicates that tunnelling plays a negligible role in the H-atom elimination dynamics at 238 nm in pyrrole, resulting in very rapid N–H bond fission within 46 ± 22 fs.

3.2 Excited state dynamics following excitation to $^1\pi\pi^*$ ($^1B_2$) at 200 nm

a Time-resolved total kinetic energy release spectra. We now turn our attention to dynamics occurring following photoexcitation of the $^1\pi\pi^*$ ($^1B_2$) state at 200 nm. Time-resolved TKER spectra recorded following H-atom elimination from pyrrole at 200 nm are shown in Fig. 6(a) at $\Delta t = -5$ ps (black), 1 ps (blue) and 1.2 ns (red). Reconstructed slices through the centre of the original 3-D H^+ distributions at $\Delta t = 1$ ps (left) and 1.2 ns (right) are also shown in inset. At 200 nm, no 'probe-pump' signal is observed at $\Delta t = -5$ ps, indicating an absence of reverse dynamics. As with measurements at 250 and 238 nm, both TKER spectra at $\Delta t > 0$ once again clearly display an intense feature at high TKER, peaked around ~7200 cm^{-1}, correlated with the anisotropic signal ($\beta_2 \approx -0.5$, see Fig. S3 of the ESI†) at large radii in the inset velocity map images. Unlike measurements at longer wavelengths however (section 3.1), this feature at 200 nm exhibits a pronounced tail which decays in intensity towards higher TKER, returning to the baseline signal level between ~16000–17000 cm^{-1}. The predicted $TKER_{max}$ values for dissociation into the three possible photoproduct channels (see Fig. 1(a)) after N–H bond scission at 200 nm are shown by the blue, red and light brown arrows, respectively, in Fig. 6(a); $TKER_{max}$ for the higher energy $C_4H_4N(\tilde{B})$ + H photoproduct channel (light brown arrow) is determined using eqn (1) and a calculated $\tilde{B} \leftarrow \tilde{X}$ energy separation of ~10400 cm^{-1}.[58] Unfortunately, given the spectral bandwidth of the fs pump pulse (~500 cm^{-1}) and the restricted resolution of our VMI spectrometer,[48] it is challenging to identify which electronic

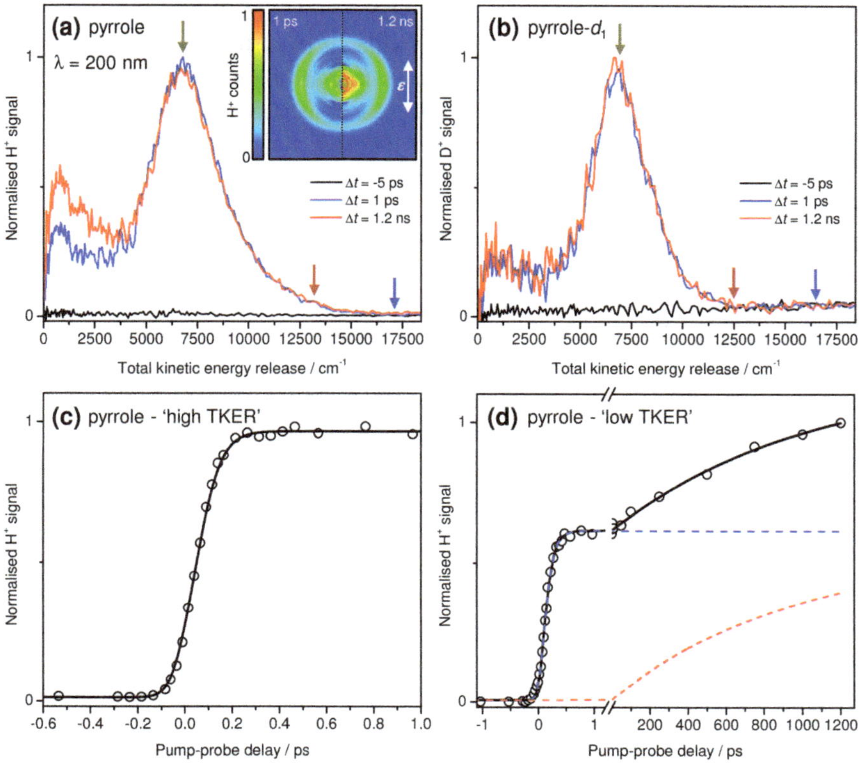

Fig. 6 (a), (b) Time-resolved TKER spectra recorded following photoexcitation of pyrrole and pyrrole-d_1 at 200 nm, and probing H/D-atoms with a time-delayed 243.1 nm pulse, respectively. Vertical arrows indicate the predicted $TKER_{max}$ values for N–H dissociation into the $C_4H_4N(\tilde{X})$ + H (blue), $C_4H_4N(\tilde{A})$ + H (red) and $C_4H_4N(\tilde{B})$ + H (light brown) photoproduct channels (see Fig. 1(a) and eqn (1)). Inset in (a): Reconstructed slices through the centre of the original 3-D H^+ distribution at Δt = 1 ps (left) and 1.2 ns (right). The white arrow indicates the polarisation of $h\nu_{pu}$. (c) Integrated 'high TKER' H^+ signal transient (5000–10000 cm^{-1}), following excitation of pyrrole at 200 nm (open circles), together with an associated kinetic fit with a single component of τ_f = 52 fs (black line). (d) Integrated 'low TKER' H^+ signal transient (200–3000 cm^{-1}), following excitation of pyrrole at 200 nm (open circles), together with an associated kinetic fit (black line). Individual fit components associated with τ_L = 108 fs (dashed blue line) and τ_{stat} = 1.0 ns (dashed red line) are also shown (see main text for details).

state the pyrrolyl co-fragments are formed in. Previous high-resolution HRA-PTS measurements at 193.3 nm by Cronin *et al.*[27] do, however, allow an excellent comparative insight into the branching between the three possible photoproduct channels at 200 nm, and strongly suggests that only the $C_4H_4N(\tilde{X})$ + H channel is populated. With reference to Fig. 1(a), this implies that flux does not couple directly from the $^1\pi\pi^*$ ($^1B_2$) state onto the higher lying $2^1\pi\sigma^*$ ($^1B_1$) state (*via* a $^1\pi\pi^*/2^1\pi\sigma^*$ CI), the diabatic asymptote of which is correlated with the $C_4H_4N(\tilde{A})$ + H channel, even though such a pathway is proposed to be energetically open by 200 nm.[58] We therefore propose that the high TKER feature *predominantly* arises through flux coupling onto the $1^1\pi\sigma^*$ ($^1A_2$) state from the initially populated $^1\pi\pi^*$ ($^1B_2$) state, *via* an appropriate $^1\pi\pi^*/1^1\pi\sigma^*$ CI (located along a 'ring-distortion' coordinate orthogonal to R_{N-H}),[30–33] followed by subsequent N–H bond cleavage to yield $C_4H_4N(\tilde{X})$ + H photoproducts. We acknowledge however, that it is not possible to definitively rule out initial population of other electronic excited states at 200 nm, particularly the $^1\pi\pi^*$ ($^1A_1$) state, which is thought to contribute ~15% of the total observed absorbance in pyrrole's first strong absorption band (see Fig. 1(b)).[26,39] The non-limiting β_2 value of −0.5 associated with the high TKER feature in Fig. 6(a), provides some evidence that this may indeed be the case: for the stronger $^1\pi\pi^*$ ($^1B_2$) ← S_0 ($^1A_1$) transition, μ_y (b_2 symmetry) is aligned perpendicular to the N–H bond, while for $^1\pi\pi^*$ ($^1A_1$) ← S_0 ($^1A_1$) excitation μ_z (a_1 symmetry) is aligned parallel to the N–H bond, a superposition of which may yield a non-limiting negative β_2 value of −0.5.[59]

In contrast to the TKER spectra recorded after direct excitation to $1^1\pi\sigma^*$ ($^1A_2$), the spectra at $\Delta t > 0$ in Fig. 6(a) contain a significant fraction of signal at <4000 cm^{-1}, with a 'Boltzmann-like' profile peaked at ~1000 cm^{-1}. At 250 and 238 nm such signal contributions are minimal (Fig. 2(a) and 5(a)). Furthermore, what is perhaps most striking is the clear time dependence of the intensity of this 'low TKER' feature, as evidenced by the change in signal level below ~4000 cm^{-1} between $\Delta t = 1$ ps (blue) and 1.2 ns (red), which is clearly absent in the high TKER feature. This is also clearly seen in the associated time-resolved velocity map images (Fig. 6(a), inset) as an enhanced number of H^+ counts near to the centre of the image. In order to ascertain the origins to this slower rise in signal at low TKER, we have performed comparative studies on pyrrole-d_1. These measurements enable a clear determination of whether these H-atoms at low TKER are born through either N–H or C–H bond fission, or both. Time-resolved TKER spectra recorded at $\Delta t = -5$ ps (black), 1 ps (blue) and 1.2 ns (red) after 200 nm excitation of pyrrole-d_1 are shown in Fig. 6(b). These spectra were obtained by imaging D^+, rather than H^+, following 2+1 REMPI of any eliminated D-atoms at 243.1 nm. The feature at high TKER is peaked at slightly lower energy (~7000 cm^{-1}), in line with the stronger N–D bond strength (D_0(N–D) ~33600 cm^{-1}).[60,61] Additionally, the high energy tail of this feature only extends to ~12500 cm^{-1} some ~3500–4500 cm^{-1} less than that observed in undeuterated pyrrole (Fig. 6(a)). This indicates that, on average, the $C_4H_4N(\tilde{X})$ co-fragments retain more of the available excess energy as internal (vibrational) energy following N–D bond fission, relative to N–H bond fission in undeuterated pyrrole. However, what is most revealing is the apparent invariance in the amplitude of the low TKER component as a function of Δt, compared to measurements on the undeuterated counterpart. Unlike the spectra presented in Fig. 6(a), which may, in principle, possess signal components from both N–H *and* C–H bond dissociation, the signal in Fig. 6(b) can only arise from N–D bond scission. Therefore, given that the low TKER component in Fig. 6(b) does not increase between $\Delta t = 1$ ps and 1.2 ns we can, with reasonable confidence, exclusively attribute the rise in signal at low TKER when $\Delta t > 1$ ps in Fig. 6(a) to a mechanism involving C–H bond cleavage. We discuss the precise nature of this C–H fission mechanism in greater detail below (section 3.2c).

b H-atom elimination dynamics *via* the $1^1\pi\sigma^*$ ($^1A_2$) surface. Timescales for H-atom elimination dynamics at 200 nm are extracted from normalised integrated H^+ signal transients for both the high TKER (5000–10000 cm^{-1}) and low TKER (200–3000 cm^{-1}) features, presented in Fig. 6(c) and 6(d), respectively (open circles). We initially consider the high TKER H^+ signal transient in Fig. 6(c), presented between $\Delta t = -0.6$ to 1 ps. Beyond $\Delta t = 1$ ps the signal associated with the high TKER component plateaus, as evidenced by the time-resolved TKER spectra in Fig. 6(a). Unlike the H^+ signal transients obtained at 238 and 250 nm excitation, no reverse dynamics are observed at 200 nm and only a single time-constant of $\tau_f = 52 \pm 12$ fs is required to fit the data (black line), which is assigned as the overall timescale for a $^1\pi\pi^*$ ($^1B_2$) → $1^1\pi\sigma^*$ ($^1A_2$) → N–H fission mechanism to take place. Such a mechanism is based on/supported by conclusions drawn in previous experimental work.[24,26,27] This value for τ_f is similar to that obtained following 238 nm excitation (46 ± 22 fs), and the implications of this are discussed below.

From a theoretical perspective, mixed quantum-classical dynamics simulations by Barbatti and co-workers[32,33] predict that ~80% of the initial $^1\pi\pi^*$ ($^1B_2$) state population is deactivated *via* $^1\pi\pi^*$ ($^1B_2$) → $1^1\pi\sigma^*$ ($^1A_2$) coupling, and return an *average* timescale of 54 fs for the onset of N–H bond dissociation along $1^1\pi\sigma^*$ ($^1A_2$),[32] in excellent accord with our reported appearance timescale of 52 ± 12 fs for high TKER H-atoms at 200 nm. Given the measured value for τ_f at 200 nm, we can conclude that population transfer in the initial $^1\pi\pi^*$ ($^1B_2$) → $1^1\pi\sigma^*$ ($^1A_2$) IC step must take place within a sub-50 fs timeframe. Once again, theory[32] predicts that coupling through appropriate CIs, located along ring deformation coordinates,[30–33] occurs within ~20 fs. Furthermore, we also draw on recent ultrafast TR-IY studies on the pyrrole cation (pyrrole$^+$) signal by Montero *et al.*,[23] who report a timescale of 19 ± 3 fs for the initial $^1\pi\pi^*$ ($^1B_2$) → $1^1\pi\sigma^*$ ($^1A_2$) IC process, in good agreement with both our current observations and previous dynamics simulations. However, after direct excitation of $1^1\pi\sigma^*$ ($^1A_2$) at 238 nm we measure a value for τ_f of 46 ± 22 fs, which is very comparable to the value of 52 ± 12 fs extracted at 200 nm. In principle, one may anticipate that H-atoms born through a $^1\pi\pi^*$ ($^1B_2$) → $1^1\pi\sigma^*$ ($^1A_2$) → N–H fission pathway at 200 nm will possess a longer appearance timescale than those formed after direct excitation to $1^1\pi\sigma^*$ ($^1A_2$), given the additional time (~20 fs) required for the initial $^1\pi\pi^*$ ($^1B_2$) → $1^1\pi\sigma^*$ ($^1A_2$) step. Such a timescale is within the statistical uncertainty (±22 fs) of the extracted τ_f value at 238 nm, and we postulate that this, together with the increased kinetic energy of the dissociating wavepacket, is the most likely origin of such a discrepancy.

c Statistical H-atom elimination dynamics. We now return to discuss the low TKER H^+ signal transient depicted in Fig. 6(d). In stark contrast to the high TKER H^+ signal transient, the temporal profile of the H^+ signal at low TKER clearly exhibits a bi-exponential rise, occurring within two very different time-constants. Moreover, the slower rise in signal at low TKER has not plateaued by the maximum temporal delay achievable in our current experiments (Δt = 1.2 ns), suggesting that these dynamics are incomplete by 1.2 ns. This H^+ signal transient can be fitted using two exponential rise functions (black line), which returns time-constants of $\tau_L = 108 \pm 20$ fs (dashed blue line) and $\tau_{stat} = 1.0 \pm 0.4$ ns (dashed red line). Analogous to conclusions drawn in previous work on related aromatic heterocycles,[10,51,53,54] we attribute the τ_L time-constant to the appearance timescale for H-atoms formed *via* undesired multiphoton induced pathways; such a result is not a central discussion point in this paper and the reader is referred to the aforementioned references for further details. We now move on to discuss the origins of the $\tau_{stat} = 1.0 \pm 0.4$ ns time-constant in the ensuing paragraphs.

Theoretical calculations can provide an excellent starting point for discussing the $\tau_{stat} = 1.0 \pm 0.4$ ns time-constant extracted from the low TKER H^+ signal transient in Fig. 6(d). As mentioned above in section 3.2b, ~80% of the initial population imparted to ${}^1\pi\pi^*$ (${}^1B_2$) is predicted to decay *via* coupling onto the lower lying $1^1\pi\sigma^*$ (${}^1A_2$) surface.[32,33] However, the remaining population can directly couple back onto S_0 at appropriate ${}^1\pi\pi^*/S_0$ CIs,[30–33] or evolve towards alternative photofragmentation pathways (*e.g.* HCN + C_3H_4), the latter of which has been experimentally observed by Blank *et al.* at 193 nm;[24] we focus on the former as a pathway for forming H-atoms at low TKER. It was postulated in these theoretical studies,[32,33] that low TKER H-atoms could be generated through population of vibrationally hot S_0 pyrrole molecules, which subsequently undergo statistical unimolecular dissociation on a timeframe significantly longer than ultrafast N–H bond fission along the $1^1\pi\sigma^*$ (${}^1A_2$) surface. IC from ${}^1\pi\pi^*$ (${}^1B_2$) → S_0, mediated at an appropriate CI along a ring-distortion coordinate, offers one viable pathway for forming hot S_0 species, which can then eject statistical H-atoms with low TKER. Alternatively, after IC from ${}^1\pi\pi^*$ (${}^1B_2$) → $1^1\pi\sigma^*$ (${}^1A_2$), population can access the $1^1\pi\sigma^*/S_0$ CI at extended R_{N-H} distances, also providing a route to forming hot S_0 molecules, ultimately yielding low TKER H-atoms (see PECs in Fig. 1(a), dashed red arrow). Moreover, earlier experiments (at 217 nm) suggest that a large fraction of these H-atoms with low TKER occur as a result of C–H bond dissociation in hot S_0 molecules, rather than N–H bond cleavage.[26] These earlier observations directly concur with our findings presented in Fig. 6(a) and 6(b): upon selectively deuterating at the N–H site we almost exclusively observe low TKER H-atoms eliminated through C–H bond dissociation. Guided by all of these previous works, we therefore propose that $\tau_{stat} = 1.0 \pm 0.4$ ns correlates to the timescale for statistical unimolecular C–H bond dissociation from vibrationally hot S_0 molecules.

As alluded to above, the statistical C–H fission dynamics from hot S_0 species have not terminated by the temporal limit of our experiments ($\Delta t = 1.2$ ns). We thus suggest that the value for τ_{stat} extracted from the kinetic fit to the low TKER H^+ signal transient in Fig. 6(d) should be regarded as a conservative 'lower-limit' for the timescale of such a process (*i.e.* $\tau_{stat} \geq 1$ ns). Such a consideration is also guided by previous TKER spectra reported in previous frequency-resolved experiments,[27,28] recorded at much longer time-delays of $\Delta t \sim 5$ to 10 ns. In particular, the TKER spectra recorded at 210.7 and 193 nm by Cronin *et al.*[27] ($\Delta t \sim 5$ ns) clearly possess a signal component at low TKER which has a significantly greater amplitude than the high TKER signal. In line with our above comments, this further suggests that by 1.2 ns, statistical H-atom elimination dynamics from hot S_0 are not complete in pyrrole. Based on our current measurements, it is unclear whether or not this additional signal rise after 5 ns comes about solely through C–H fission from S_0. It may be possible that analogous statistical N–H bond fission from hot S_0 species will take longer than C–H fission, and begin to contribute to the large low TKER signal observed by Cronin *et al.*[27] and Rubio-Lago *et al.*[28] by $\Delta t = 5$–10 ns.

To close this section, we briefly consider the notable contribution of statistical H-atom elimination dynamics after populating ${}^1\pi\pi^*$ (${}^1B_2$) at 200 nm, relative to the apparently negligible participation of this behaviour (by 1.2 ns) following direct excitation to $1^1\pi\sigma^*$ (${}^1A_2$) at 250 and 238 nm. Whilst we cannot provide a definitive answer to this observation, we postulate that, unlike direct excitation to $1^1\pi\sigma^*$ (${}^1A_2$) where S_0 can only be populated *via* IC from $1^1\pi\sigma^*$ (${}^1A_2$) → S_0, excitation to ${}^1\pi\pi^*$ (${}^1B_2$) offers a number of possible pathways through which S_0 can be re-populated: for example, direct IC from ${}^1\pi\pi^*$ (${}^1B_2$) → S_0 mediated by appropriate CIs as well as sequential IC from $^1\pi\pi^*$ ($^1B_2$) → $1^1\pi\sigma^*$ ($^1A_2$) → S_0. Coupled to the increased internal energy following excitation to $^1\pi\pi^*$ ($^1B_2$), this may enable access to higher energy dissociation channels from hot S_0 species, such as C–H fission, potentially rationalising the notable difference in statistical H-atom elimination dynamics at the excitation wavelengths studied here.

4 Conclusions

Using a combination of ultrafast time-resolved ion-yield and time-resolved velocity map ion imaging techniques we have directly probed H-atom elimination dynamics in pyrrole following: (*i*) direct excitation to the $1^1\pi\sigma^*$ ($^1A_2$) state at both 250 and 238 nm; and (*ii*) initial photoexcitation to the higher energy $^1\pi\pi^*$ ($^1B_2$) state at 200 nm.

After direct excitation to the dissociative $1^1\pi\sigma^*$ ($^1A_2$) surface we almost exclusively observe H-atoms born with large amounts of kinetic energy release (6500–7200 cm^{-1}), which are attributed to rapid dissociation of the N–H bond and the formation of $C_4H_4N(\tilde{X})$ + H photoproducts. At 250 nm, we measure an ultrafast appearance time of 126 ± 28 fs for these high kinetic energy H-atoms. With the aid of complementary measurements of D-atom elimination from pyrrole-d_1 molecules, we conclude that excitation at 250 nm imparts population into the quasi-bound 3s Rydberg well of the $1^1\pi\sigma^*$ ($^1A_2$) surface in the vertical Franck–Condon region, which then rapidly tunnels through a small exit barrier (>630 cm^{-1}), subsequently leading to ultrafast N–H bond fission. At 238 nm, excitation to $1^1\pi\sigma^*$ ($^1A_2$) occurs above the exit barrier to dissociation, leading to extremely rapid N–H bond fission within 46 ± 22 fs. Measurements on pyrrole-d_1 indicate that tunnelling plays a negligible role in the N–H bond cleavage dynamics at 238 nm.

Excitation at a wavelength of 200 nm initially accesses the higher lying $^1\pi\pi^*$ ($^1B_2$) state, which yields a bi-modal distribution of H-atoms with high amounts of kinetic energy as well as low kinetic energies (with a 'statistical' profile), in contrast to the measurements at 250 and 238 nm. The H-atoms possessing large amounts of kinetic energy are once again attributed to the formation of $C_4H_4N(\tilde{X})$ + H photoproducts, except this time generated though a sequential $^1\pi\pi^*$ ($^1B_2$) → $1^1\pi\sigma^*$ ($^1A_2$) → N–H fission mechanism. We extract an appearance time-constant of 52 ± 12 fs for H-atoms formed *via* this process. Based on this timescale, this implies that the initial $^1\pi\pi^*$ ($^1B_2$) → $1^1\pi\sigma^*$ ($^1A_2$) internal conversion step (*via* a $^1\pi\pi^*/1^1\pi\sigma^*$ CI located along a ring-deformation coordinate[30–33]) takes place on a sub-50 fs timeframe, which is in good accord with findings from previous complementary experimental[23] and theoretical work.[32,33]

The appearance of low kinetic energy H-atoms at 200 nm is characterised by a bi-exponential rise, with associated time-constants of 108 ± 20 fs and 1.0 ± 0.4 ns. The former ultrafast time-constant of 108 ± 20 fs is attributed to H-atoms formed through undesired multiphoton processes. However, the latter slower rise in low kinetic energy H-atom signal is assigned to the onset of statistical unimolecular H-atom elimination from vibrationally hot S_0 ground state species, after initial re-coupling onto S_0 through appropriate CIs (*e.g.* $1^1\pi\sigma^*/S_0$ or $^1\pi\pi^*/S_0$ CIs). Specifically, probing D-atom elimination from pyrrole-d_1 at 200 nm reveals that, up to 1.2 ns after excitation, statistical H-atom elimination only results from C–H bond cleavage on S_0. Moreover, based on our measurements here and by comparison with earlier frequency-domain studies,[27,28] these statistical C–H fission dynamics appear to be incomplete by 1.2 ns, indicating that the extracted time-constant of ~1.0 ns should be considered as a 'lower-limit' for the timescale of such a process in pyrrole.

More generally, the ultrafast time-resolved studies presented here provide new and complementary insights into the dynamical processes which occur following photoexcitation of pyrrole between 250 to 200 nm. In particular, these studies for the first time highlight: (*i*) the role of tunnelling after direct excitation to the $1^1\pi\sigma^*$ ($^1A_2$) surface at 250 nm, providing strong evidence that this state does indeed possess a quasi-bound 3s Rydberg well in the vertical Franck–Condon region (similar to $^1n\sigma^*$ and $^1\pi\sigma^*$ surfaces in other related species, such as ammonia[57] and aniline,[56] respectively); and (*ii*) the onset of slower statistical C–H bond fission after (re-)coupling onto a vibrationally hot S_0 ground state following initial excitation to $^1\pi\pi^*$ ($^1B_2$), and the timescales on which this behaviour starts to occurs.

Acknowledgements

The authors gratefully thank Dr Dave Hadden and Dr Michael Staniforth for experimental assistance and helpful discussions and Professor Mike Ashfold for helpful discussions. G.M.R. and A.S.C. thank the Leverhulme Trust for a post-doctoral research fellowship and doctoral funding, respectively. C.A.W. thanks the EPSRC for a doctoral research fellowship. J.D.Y. thanks the University of Warwick for a doctoral training award. H.Y. thanks the NSF for a graduate research assistantship. S.U. and V.G.S. would like to thank the NSF and EPSRC for grants (NSF-CHE-0924456 and EP/H003401) which have been awarded through the International Collaboration in Chemistry Program. V.G.S. would also like to thank the Royal Society for a University Research Fellowship.

References

1 A. L. Sobolewski, W. Domcke, C. Dedonder-Lardeux and C. Jouvet, *Phys. Chem. Chem. Phys.*, 2002, **4**, 1093–1100.
2 M. N. R. Ashfold, B. Cronin, A. L. Devine, R. N. Dixon and M. G. D. Nix, *Science*, 2006, **312**, 1637–1640.
3 M. N. R. Ashfold, G. A. King, D. Murdock, M. G. D. Nix, T. A. A. Oliver and A. G. Sage, *Phys. Chem. Chem. Phys.*, 2010, **12**, 1218–1238.
4 N. L. Evans and S. Ullrich, *J. Phys. Chem. A*, 2010, **114**, 11225–11230.
5 M. G. D. Nix, A. L. Devine, B. Cronin and M. N. R. Ashfold, *J. Chem. Phys.*, 2007, **126**, 124312.
6 K. L. Wells, D. J. Hadden, M. G. D. Nix and V. G. Stavros, *J. Phys. Chem. Lett.*, 2010, **1**, 993–996.
7 S. Ullrich, T. Schultz, M. Z. Zgierski and A. Stolow, *Phys. Chem. Chem. Phys.*, 2004, **6**, 2796–2801.
8 C. T. Middleton, K. de La Harpe, C. Su, Y. K. Law, C. E. Crespo-Hernandez and B. Kohler, *Annu. Rev. Phys. Chem.*, 2009, **60**, 217–239.
9 H. Satzger, D. Townsend, M. Z. Zgierski, S. Patchkovskii, S. Ullrich and A. Stolow, *Proc. Natl. Acad. Sci. U. S. A.*, 2006, **103**, 10196–10201.
10 A. Iqbal and V. G. Stavros, *J. Phys. Chem. Lett.*, 2010, **1**, 2274–2278.
11 G. M. Roberts, C. A. Williams, M. J. Paterson, S. Ullrich and V. G. Stavros, *Chem. Sci.*, 2012, **3**, 1192–1199.
12 A. L. Devine, B. Cronin, M. G. D. Nix and M. N. R. Ashfold, *J. Chem. Phys.*, 2006, **125**, 184302.
13 R. Montero, A. P. Conde, V. Ovejas, F. Castano and A. Longarte, *J. Phys. Chem. A*, 2012, **116**, 2698–2703.

14 A. Iqbal and V. G. Stavros, *J. Phys. Chem. A*, 2010, **114**, 68–72.
15 G. M. Roberts, A. S. Chatterley, J. D. Young and V. G. Stavros, *J. Phys. Chem. Lett.*, 2012, **3**, 348–352.
16 M. N. R. Ashfold, A. L. Devine, R. N. Dixon, G. A. King, M. G. D. Nix and T. A. A. Oliver, *Proc. Natl. Acad. Sci. U. S. A.*, 2008, **105**, 12701–12706.
17 M. G. D. Nix, A. L. Devine, B. Cronin and M. N. R. Ashfold, *Phys. Chem. Chem. Phys.*, 2006, **8**, 2610–2618.
18 Y. Zhang, T. A. A. Oliver, M. N. R. Ashfold and S. E. Bradforth, *Faraday Discuss.*, 2012, **157**, 141–163.
19 T. A. A. Oliver, Y. Zhang, M. N. R. Ashfold and S. E. Bradforth, *Faraday Discuss.*, 2011, **150**, 439–458.
20 F. F. Crim, *Faraday Discuss.*, 2012, **157**, 9–26.
21 D. Murdock, S. J. Harris, T. N. V. Karsili, G. M. Greetham, I. P. Clark, M. Towrie, A. J. Orr-Ewing and M. N. R. Ashfold, *J. Phys. Chem. Lett.*, 2012, **3**, 3715–3720.
22 H. Lippert, H. H. Ritze, I. V. Hertel and W. Radloff, *ChemPhysChem*, 2004, **5**, 1423–1427.
23 R. Montero, A. P. Conde, V. Ovejas, M. Fernandez-Fernandez, F. Castano, J. R. V. de Aldana and A. Longarte, *J. Chem. Phys.*, 2012, **137**, 064317.
24 D. A. Blank, S. W. North and Y. T. Lee, *Chem. Phys.*, 1994, **187**, 35–47.
25 J. Wei, A. Kuczmann, J. Riedel, F. Renth and F. Temps, *Phys. Chem. Chem. Phys.*, 2003, **5**, 315–320.
26 J. Wei, J. Riedel, A. Kuczmann, F. Renth and F. Temps, *Faraday Discuss.*, 2004, **127**, 267–282.
27 B. Cronin, M. G. D. Nix, R. H. Qadiri and M. N. R. Ashfold, *Phys. Chem. Chem. Phys.*, 2004, **6**, 5031–5041.
28 L. Rubio-Lago, D. Zaouris, Y. Sakellariou, D. Sofikitis, T. N. Kitsopoulos, F. Wang, X. Yang, B. Cronin, A. L. Devine, G. A. King, M. G. D. Nix, M. N. R. Ashfold and S. S. Xantheas, *J. Chem. Phys.*, 2007, **127**, 064306.
29 V. Vallet, Z. G. Lan, S. Mahapatra, A. L. Sobolewski and W. Domcke, *J. Chem. Phys.*, 2005, **123**, 144307.
30 M. Barbatti, M. Vazdar, A. J. A. Aquino, M. Eckert-Maksic and H. Lischka, *J. Chem. Phys.*, 2006, **125**, 164323.
31 B. Sellner, M. Barbatti and H. Lischka, *J. Chem. Phys.*, 2009, **131**, 024312.
32 M. Vazdar, M. Eckert-Maksic, M. Barbatti and H. Lischka, *Mol. Phys.*, 2009, **107**, 845–854.
33 M. Barbatti, J. Pittner, M. Pederzoli, U. Werner, R. Mitric, V. Bonacic-Koutecky and H. Lischka, *Chem. Phys.*, 2010, **375**, 26–34.
34 H. Nakatsuji, O. Kitao and T. Yonezawa, *J. Chem. Phys.*, 1985, **83**, 723–734.
35 H. Nakano, T. Tsuneda, T. Hashimoto and K. Hirao, *J. Chem. Phys.*, 1996, **104**, 2312–2320.
36 M. H. Palmer, I. C. Walker and M. F. Guest, *Chem. Phys.*, 1998, **238**, 179–199.
37 O. Christiansen, J. Gauss, J. F. Stanton and P. Jorgensen, *J. Chem. Phys.*, 1999, **111**, 525–537.
38 J. Wan, J. Meller, M. Hada, M. Ehara and H. Nakatsuji, *J. Chem. Phys.*, 2000, **113**, 7853–7866.
39 B. O. Roos, P. A. Malmqvist, V. Molina, L. Serrano-Andres and M. Merchan, *J. Chem. Phys.*, 2002, **116**, 7526–7536.
40 G. M. Roberts and V. G. Stavros, in *Ultrafast Phenomena in Molecular Sciences*, ed. L. Banares, Springer, in press.
41 K. L. Wells, G. Perriam and V. G. Stavros, *J. Chem. Phys.*, 2009, **130**, 074304.
42 G. M. Roberts, C. A. Williams, J. D. Young, S. Ullrich, M. J. Paterson and V. G. Stavros, *J. Am. Chem. Soc.*, 2012, **134**, 12578–12589.
43 A. T. J. B. Eppink and D. H. Parker, *Rev. Sci. Instrum.*, 1997, **68**, 3477–3484.
44 U. Even, J. Jortner, D. Noy, N. Lavie and C. Cossart-Magos, *J. Chem. Phys.*, 2000, **112**, 8068–8071.
45 G. M. Roberts, J. L. Nixon, J. Lecointre, E. Wrede and J. R. R. Verlet, *Rev. Sci. Instrum.*, 2009, **80**, 053104.
46 P. M. Regan, S. R. Langford, A. J. Orr-Ewing and M. N. R. Ashfold, *J. Chem. Phys.*, 1999, **110**, 281–288.
47 A. Iqbal, L. J. Pegg and V. G. Stavros, *J. Phys. Chem. A*, 2008, **112**, 9531–9534.
48 Below $\sim$7000 cm^{-1} we note that the VMI spectrometer has a restricted resolution of $\Delta E/E$ $\sim$ 15–20%.
49 A. J. Gianola, T. Ichino, R. L. Hoenigman, S. Kato, V. M. Bierbaum and W. C. Lineberger, *J. Phys. Chem. A*, 2004, **108**, 10326–10335.
50 P. L. Houston, in *Imaging in Molecular Dynamics: Technology and Applications*, ed. B. J. Whitaker, Cambridge University Press, Cambridge, 2003, ch. 1, pp. 3–19.
51 C. A. Williams, G. M. Roberts, H. Yu, N. L. Evans, S. Ullrich and V. G. Stavros, *J. Phys. Chem. A*, 2012, **116**, 2600–2609.

52 D. J. Hadden, C. A. Williams, G. M. Roberts and V. G. Stavros, *Phys. Chem. Chem. Phys.*, 2011, **13**, 4494–4499.
53 D. J. Hadden, K. L. Wells, G. M. Roberts, L. T. Bergendahl, M. J. Paterson and V. G. Stavros, *Phys. Chem. Chem. Phys.*, 2011, **13**, 10342–10349.
54 A. Iqbal, M. S. Y. Cheung, M. G. D. Nix and V. G. Stavros, *J. Phys. Chem. A*, 2009, **113**, 8157–8163.
55 H. Reisler and A. I. Krylov, *Int. Rev. Phys. Chem.*, 2009, **28**, 267–308.
56 G. A. King, T. A. A. Oliver and M. N. R. Ashfold, *J. Chem. Phys.*, 2010, **132**, 214307.
57 M. N. R. Ashfold, R. N. Dixon, M. Kono, D. H. Mordaunt and C. L. Reed, *Philos. Trans. R. Soc. London, Ser. A*, 1997, **355**, 1659–1674.
58 G. A. King, T. A. A. Oliver, M. G. D. Nix and M. N. R. Ashfold, *J. Chem. Phys.*, 2010, **132**, 064305.
59 At 200 nm we acknowledge that the ‘low TKER’ and ‘high TKER’ signal components in the TKER spectra in Fig. 6(a) are partially convoluted. This may, in principle, also provide an alternative explanation for the ‘effective’ non-limiting β_2 value extracted for the ‘high TKER’ feature between 6000–10000 cm^{-1}, given the near-isotropic distribution of the underlying ‘low TKER’ signal component ($\beta_2 \approx 0$); this explanation was previously adopted by Wei *et al.* at 217 nm (see ref. 26). A fit to the profile of the TKER spectrum at $\Delta t = 1.2$ ns (Fig. S3) indicates that the low TKER feature only contributes ~7% of the total signal between 6000–10000 cm^{-1}, which would in turn result in a ‘corrected’ β_2 value of −0.54 (see ref. 26). As such, we favour the arguments put forward in the main text (section 3.2a) to explain the non-limiting value of $\beta_2 \approx -0.5$ for the high TKER feature at 200 nm.
60 B. Cronin, A. L. Devine, M. G. D. Nix and M. N. R. Ashfold, *Phys. Chem. Chem. Phys.*, 2006, **8**, 3440–3445.
61 This is a conservative estimate based on the value given in ref. 60, which corresponds to the fully deuterated pyrrole, *i.e.*, pyrrole-d_5.

DISCUSSIONS

General discussion

DOI: 10.1039/c3fd90017g

Dr Worth opened the discussion of the paper by Professor Albert Stolow: You classify the control of molecules by a laser field as 'coherent control'. Do you think that all control experiments use coherence as a mechanism, or are other control mechanisms possible?

Professor Stolow responded: By 'coherent control' one usually means control over the relative optical phases of an incident electric field. These phases are transferred to a material system *via* dipole coupling, thereby exerting control over the material response. Alternatively, varying the frequency of the incident electric field (*i.e.* spectroscopy) is a very well established method for controlling the outcome of molecular dynamics. There are, of course, other types of control such as those based on non-perturbative laser intensities which modify the material Hamiltonian as a function of time. An example is the use of the Dynamic Stark Effect. We note, however, that all electromagnetic fields have certain coherence properties which can be fully quantified, even if thought to be 'random'.

Professor Duxbury opened the discussion of the paper by Professor Robert W. Field: In this paper you describe the development of a new chirped pulse millimetre wave spectrometer to observe a-dipole, $J = 1$ transitions in both S_0 low bender acetylene and in vinylidene. You also note the problems associated with the accessibility of $J = 1$–2 transitions and higher.

My question concerns the potential use of another type of spectrometer which is complementary to the one which you have developed. Chirped pulse quantum cascade (CPQC) laser spectrometers have some similar characteristics to those employing chirped pulse Fourier transform microwave spectroscopy in the millimetre wave region. They have been designed based upon the concepts of McGurk, Schmaltz and Flygare, *J. Chem. Phys.*, **60**, 4181 (1974), your reference (9). This describes the use a numerical solution of the coupled Maxwell Bloch equations, including both homogeneous and heterogeneous broadening. The resolution of CPQC instruments is not as high as those of the Pate type instruments, CPmmW, (your ref. 2 and 3) operating in the mm region, as they operate between ca 11 μm to 4 μm. However, they can cover a wider range of quantum numbers, and also allow the direct measurement of some of the vibrational and collisional properties to be carried out, as well as the determination of the vibrational transition dipole moment. Examples are given in the following papers.

1. G. Duxbury, N. Langford, K. Hay and N. Tasinato, "Quantum cascade laser spectroscopy: diagnostics to non-linear optics", *J. Mod. Optics*, **56**, 2034–2048 (2009).

2. N. Tasinato, K. G. Hay, N. Langford. G. Duxbury and D. Wilson, "Time dependent measurements of nitrous oxide and carbon dioxide collisional relaxation processes by frequency down-chirped quantum cascade laser: Rapid passage signals and the time dependence of collisional processes", *J. Chem. Phys.*, **132**, 164301 (2010).
3. G. Duxbury, J. F. Kelly, T. A. Blake, and N. Langford, "Sub-Doppler spectra of infrared hyperfine transitions of nitric oxide using a pulse modulated quantum cascade laser: Rapid passage, free induction decay, and the ac Stark effect", *J. Chem. Phys.*, **136**, 174319 (2012).

Keywords: chirped frequency quantum cascade laser, Maxwell-Bloch calculations, molecular alignment in the infrared and microwave regions.

Professor Field replied: CPQC laser spectrometers are a valuable tool for spectroscopy and dynamics. However, pure rotation spectroscopy using a Chirped Pulse millimeter Wave spectrometer offers some important advantages. The spectra are simpler, consisting of only one or two rotational transitions per vibrational level. Fig. 12 of my paper shows $J = 0$–1 transitions in more than 30 assigned vibrational levels of HCN/HNC. The relationship between an observed transition intensity and the population difference between upper and lower levels depends only on the permanent electric dipole moment in each sampled vibrational level, rather than the level-specific dipole derivative. It is in principle possible to observe more than one rotational transition per vibrational level, but populations in extremely high-J rotational levels are presently unobservable by CPmmW spectroscopy. High-J level populations are routinely observable by CPQC spectroscopy.

Professor Ashfold said: The various transition states for HCN(HNC) elimination from vinyl cyanide correspond to local maxima on the ground state potential energy surface (PES).[1] The difference in energy between the transition state and the asymptotic products of interest will be released dynamically (*i.e.* will be determined by the forces arising from the topology of the PES in this region). Is there not a danger that these forces will lead to rotationally excited vinylidene products, the lifetimes of which are too short to benefit from the collisional cooling that would be required for their observation in the present experiment, or even in a future incarnation using infrared multiple photon dissociation to drive the elimination?

1. Z. Homayoon, S.A. Vazquez, R-Rodrigues-Fernandez and E- Martinez-Nunez, *J. Phys. Chem. A*, 2011, **115**, 979.

Professor Field answered: The partitioning of energy between product vibration and rotation should depend on the above-threshold energy at the transition state. 193 nm photolysis of vinyl cyanide is 22 000 cm^{-1} above threshold. A near-threshold scheme for photolysis of vinyl cyanide, such as CO_2 laser infrared multiphoton dissociation, might produce rotationally cooler vinylidene.

Professor Neumark asked: As you are aware, one of the cleanest ways to generate vinylidene is by photodetachment of the vinylidene anion. Is it possible to use your method to probe vinylidene produced by photodetachment? What number densities are needed for the CPmmW experiment to work?

Professor Field responded: Thank you for this exciting suggestion. It appears that the sensitivity of our CPmmW spectrometer may be sufficient to record mm–wave spectra of positive and negative ions and of neutral molecules obtained by photodetachment of negative ions. Two crucial factors weigh heavily in favor of extending CPmmW spectroscopy to ions:

- the $\gtrsim 100\ \text{cm}^3$ volume of the active region;
- the ~ 0.5 MHz resolution of the spectrometer, which makes pre-resonance mass and species selection unnecessary.

Our CPmmW spectrometer operates in the 70–100 GHz region. Consider a linear ion with a B-value of $0.75\ \text{cm}^{-1}$. The $J = N = 1$ - 2 transition occurs at $3\ \text{cm}^{-1}$. The rotational partition function at $T_{\text{rot}} = 20$ K is $q_{\text{rot}} = 18.5$ and the $N =$ 1 - N = 2 population dierence is

$$\frac{N_{\text{tot}}(3)(0.336)}{18.5} = 0.054\, N_{\text{tot}}$$

This population difference for a transition near $3\ \text{cm}^{-1}$ is approximately $\propto B/T^2$.

We have observed Rydberg–Rydberg $\Delta n = \pm 1$ transitions for Ca atoms with principal quantum numbers in the $40 \lesssim n \lesssim 50$ range. Laser excitation is used to populate a *single* n, l Rydberg state, in contrast to the Boltzmann rotational *distribution* expected for molecular-ion sources. The active region (irradiated by the CPmmW pulse) of our Rydberg spectrometer is $5 \times 5 \times 10\ \text{cm}^3$ and contains $\sim 10^6$ atoms in the single selected $n, l, m_l = 0$ quantum state. The transition moment for a $\Delta n = \pm 1$ Rydberg–Rydberg transition is $\sim$ 1000D. The polarization achieved is typically 1000 D/molecule. We typically obtain $S : N = 10$ in a single chip.

How many ions in the $\sim 250\ \text{cm}^3$ active volume of our spectrometer would be required to achieve S/N = 10 after 1 min of signal averaging at a chirped pulse repetition rate of 10 Hz (600 chirps)? If the permanent electric dipole moment of the ion is 1 Debye and, owing to the $\sim$ 30mW of mm-wave power available to us, then we can achieve only 1% polarization or $\sim$ 0.01 D/molecule. This is inferior by a factor of 10^5 to the polarization that can be achieved for Rydberg–Rydberg transitions. The N_{tot} required would be

$$N_{\text{tot}} = \left(\frac{10^6}{0.054}\right)(10^5)\frac{1}{(600)^{1/2}} = 8 \times 10^{10}\ \text{ions}$$

or a $250\ \text{cm}^3$ volume filled with 20 K ions at $3 \times 10^8\ \text{ions/cm}^3$. This does not seem to be an impossibly large ion density.

The sensitivity of the CPmmW spectrometer is degraded by dephasing of the FID caused by line broadening due to collisions, effects of inhomogenous electric fields, and Doppler broadening. The most serious limiting factor is probably the Doppler dephasing of an unskimmed supersonic jet.

Professor Orr-Ewing asked: The challenge you face appears to be to find a method to generate rotationally cold vinylidene in sufficient quantity that you can observe the CPmmW spectrum. In experiments using a flowing afterglow ion source and various further stages to produce a mass-selected ion beam, Lineberger and coworkers showed that electron detachment from the vinylidene negative ion H_2CC^- produces the desired rotationally cold neutral vinylidene.[1] The densities in such an ion beam after mass and velocity selection are unlikely to be sufficient for your purposes, but might you be able to combine CPmmW spectroscopy with the flowing afterglow source and an electron detachment laser to observe H_2CC? Your spectroscopic resolution should be sufficient to observe the spectral lines of the vinylidene against a background mixture of other gases, and the density should be considerably higher in the source than downstream in the ion beam in a Lineberger type experiment.

1. K. M. Ervin, J. Ho and W. C. Lineberger, *J. Chem. Phys.*, 1989, **91**, 5974.

Professor Field answered: Thank you for the suggestion. Photodetachment of a rotationally cold negative ion beam would produce rotationally cold vinylidene that could only decay into a sparse manifold of extreme local-bender states of acetylene. Rotation-dependent mixing on the acetylene side of the isomerization barrier would be effectively turned off. This would facilitate CPmmW detection of both the nominal vinylidene and nominal local-bender acetylene states. The sub-MHz resolution of CPmmW spectroscopy would make mass-selection unnecessary. The crucial issue is whether the number of rotationally cold ions produced within the 100 cm^3 active volume will be sufficient for CPmmW spectroscopy: $\sim 10^{12}$.

Professor Orr-Ewing enquired: Towards the end of your paper, you speculate about the possibility of a roaming pathway producing HCN (or HNC) following ultraviolet photoexcitation of vinyl cyanide. Is there any experimental evidence of this type of roaming pathway competing with the 3- and 4-centre transition states postulated to eliminate HCN and HNC?

Professor Field answered: We have no evidence of roaming. We do have decisive proof that the 3-center (1,1 elimination) mechanism, which had previously been thought to be the dominant mechanism, accounts for a negligible fraction of the HCN products. Deuteration of the 1-carbon results in only a 20–30% reduction in the CPmmW-detected HCN products. This might be evidence of a statistical H,D scrambling process, or for CN roaming, or for the blindness of the CPmmW experiment to HCN formed at *J*-values too high to be cooled in the supersonic expansion.

Professor Stolow asked: When the mean time between collisions is shorter than the emission time (inverse of the A-factor), then rotations can be easily relaxed, as you have stated. This problem has roots in the use of infrared chemiluminescence by J. C. Polanyi and co-workers. I recall that they expended considerable efforts on understanding the nature of the rotational relaxation process and developed techniques such as 'arrested relaxation' which were meant to assist in these studies. I wonder if these existing methods could be applied to chirped pulse microwave probing of chemical reaction dynamics?

Professor Field responded: The problem faced with the use of an ArF excimer laser to produce the photofragments is that some method must be found to remove an enormous amount of rotational energy from the C_2H_2 photofragment without significantly altering the nascent vibrational excitation. I believe that the only photolytic path to weakly rotating vinylidene must involve photolysis or thermolysis nearer threshold and finding a parent molecule that spits out nonrotating vinylidene by a concerted rather than sequential breaking of the two bonds attaching the vinylidene moiety to the cleverly designed leaving group.

Professor Reid remarked: Is it possible to extend the microwave experiment to enable the study of transitions other than $J = 0$ to 1? Would this help in the study of the vinylidene isomerization?

Professor Field responded: It would be straightforward and desirable to extend the capability of our CPmmW spectrometer to capture at least one higher-J transition. To do so would require two frequency doublers, one for the excitation and one for the detection functions of the spectrometer. The ability to measure the absolute intensities of $J = 0$–1 and one other higher-J transition would yield a rotational temperature, assuming that rotation is thermalized in the supersonic expansion. Further evidence in support of rotational thermalization would be a finding that the rotational temperature inferred from two transitions in a relatively light target molecule is equal to the rotational temperature of a heavy "thermometer molecule." However, it is currently not possible to observe very high-J transitions. Lacking such measurements, it would be difficult to rule out a mechanism that exclusively produces molecules at such high-J that cooling in a supersonic expansion is ineffective.

Professor Dr Weitzel enquired: You discuss the intrinsic difficulty of detecting vinylidene radicals as a transient species in the 193 nm photodissociation of vinylcyanide. In recent years several groups including my own have demonstrated that femtosecond laser ionization is a viable tool for distinguishing structural isomers, see *e.g.* G. Urbasch, H. G. Breunig and K.-M. Weitzel, "Distinction of *ortho*- and *para*-Xylene by Femtosecond-Laser Mass Spectrometry", *ChemPhysChem*, **8**, 2185–2188, (2007). Can you comment on the prospect of applying fs-laser ionization to detect vinylidene radicals and distinguish it from acetylene?

Professor Field answered: The vinylidene species in its S_0 electronic state has been detected and distinguished from acetylene in the Coulomb Explosion experiments discussed briefly in my paper. The use of femtosecond photolytic techniques to prepare vinylidene for detection by CPmmW spectroscopy is impractical owing to the smaller number of photons available from such a pulse (assuming one vinylidene molecule per photolysis photon) relative to those from the 50 mJ ArF excimer laser pulses that were available for our experiments. There have been femtosecond experiments on the C_2H_2 di-cation, which distinguish vinylidene from acetylene structures, but the di-cation is not the same species as the neutral molecule. I cannot comment on the feasibility using fs-laser ionization for CPmmW-free detection of vinylidene.

Mr Chatterley opened the discussion of the paper by Professor Daniel M. Neumark: We have recently acquired time resolved photoelectron spectra of the gas phase nucleotide anions (bases with the sugar/phosphate backbone attached), using our electrospray photoelectron imaging apparatus. The demonstration that gas phase dynamics of nucleotides can be measured is especially relevant here, as in biological systems the bonds which break upon electron attachment are located on the backbone.[1] Our work is not on electron attachment, but on the related topic of UV induced photodamage, which was also brought up in Prof. Stolow's introductory talk,[2] as well as Prof. Temps' paper on solution phase dynamics. The key point for the current discussions is that our measurements provide some confidence that the dynamics observed in these larger gas-phase systems are directly comparable to those observed in the condensed phase and that the prospect for building up in complexity and biological relevance are very good indeed.

1. B. Boudaiffa, P. Cloutier, D. Hunting, M. A. Huels and L. Sanche, *Science*, 2000, **287**, 1658–1660.
2. S. Ullrich, T. Schultz, M. Z. Zgierski and A. Stolow, *Phys. Chem. Chem. Phys.*, 2004, **6**, 2796–2801.

Professor Neumark responded: This is a great result. We hope to extend our time-resolved electron attachment experiments to nucleosides and nucleotides in the near future.

Professor Dr Bucksbaum asked: The advantage of your experiment appears to be that you can study the effect of selective population of the π^* orbital without removing an electron from any bonding orbitals. But photoprotection from uv light involves an excitation of π–π^*, that is, a state where there is a vacancy in a bonding π orbital in addition to the presence of an antibonding π^* orbital. Recent work on the mechanisms for photoprotection by McFarland *et al.* (arXiv:1301.3104 [physics.chem-ph]) points to a rapid recovery of the π–π bond directly from relaxation of π–π^*. Can your results help of to separate the effects of π^* and π orbitals in photoprotection?

Professor Neumark answered: As you point out, the electron scattering experiment described in our paper fills an empty π^* without creating a π hole, so it does not bear directly on the dynamics subsequent to $\pi\pi^*$ photoexcitation. However, as we discuss, the wavelengths at which we photodetach the iodide overlap the $\pi\pi^*$ transition in the nucleobase. We would expect the resulting excited state to relax eventually and produce a vibrationally excited nucleobase (NB) in its ground electronic state complexed to the iodide anion. Such a complex would dissociate to I^- + NB by vibrational predissociation. We could measure the overall time constant for this process with time-resolved photoelectron spectroscopy at a probe wavelength energetic enough to photodetach the atomic iodide product (EA(I) = 3.06 eV, so iodide cannot be detached at the 800 nm probe wavelength used in our paper). Such an experiment would provide, at the very least, an upper bound for the relaxation time of the $\pi\pi^*$ state to the ground state. Preliminary results in our laboratory suggest that iodide production is indeed occuring, and we hope to have more quantitative data on the dynamics of this process very soon.

Professor Domcke commented: This is a comment on Phil Bucksbaum's question to Albert Stolow concerning the role of π and π^* orbitals for the out-of-plane deformation of aromatic rings upon excitation to $\pi\pi^*$ states. According to *ab initio* calculations, the out-of-plane deformation of $\pi\pi^*$ states arises primarily from the π^* orbital rather than the π orbital. This is in agreement with the experimental observation that cations (with a hole in the π orbital) are less likely to be out-of-plane deformed than anions (with an additional electron in the π^* orbital).

Dr Verlet enquired: Electron transfer from the iodide to thymine leads to the observation of low energy electrons in the photoelectron spectra which is typical of a statistical electron emission process as described in the paper. However, the time-resolved spectra suggest that autodetachment proceeds on a few 100 fs timescale. Does this relatively fast time-scale not contradict the low energy electron observed as there is insufficient time for the energy to redistribute leading to this statistical electron distribution?

Professor Neumark replied: This is a very interesting question. Recall that the fast time constant τ_1 is given by $1/(k_1 + k_2)$. So if $k_2 \gg k_1$, *i.e.* iodine loss is considerably faster than autodetachment, then the fast decay at early times would reflect the rate of iodine loss, not the autodetachment rate. If this is indeed the case, then our results do not imply that autodetachment occurs on a 100 fs timescale.

Professor Orr-Ewing asked: There is a striking difference between your measured values of τ_2 for the thymine and uracil systems. Your kinetic model associates τ_2 with the lifetime for autodetachment from the anion of the internally excited nucleobase following loss of an iodine atom from the binary cluster. Why might these lifetimes be so different for thymine and uracil anions?

Professor Neumark answered: We consider the difference in the dynamics between the two systems to be a striking and unexpected result. As mentioned in the paper, we estimate that there can be as much as 0.16 eV more available energy in IT than in IU at the same excitation energy, which could lead to faster dynamics overall in IT. An interesting possibility is that a parallel channel such as H atom loss occurs in IT but not IU because of these energetic considerations. We do not yet have direct evidence of such a channel but experiments at higher probe photon energies are planned in the near future; such experiments would detect the deprotonated nucleobase anion produced by H atom loss.

Professor Field said: The processes, autodetachment of an electron from a negative ion and autoionization of neutral Rydberg states, are superficially similar in that an electron is spontaneously ejected from a parent species. Why do these two processes require qualitatively distinct models? For Rydberg states there are powerful quantum number scaling rules that describe the rates of large families of related processes and identify the fundamental cause for each family of processes. The simple answer, that the processes are fundamentally distinct because the autodetaching negative ion state is not a member of an infinite series of related states, seems over-simple. One crucial difference between autoionization and autodetachment is that the Rydberg state lives on a potential surface that differs by an amount that scales as $1/n^{*3}$ (n^* is the effective principal quantum number) from that of the cation. This sort of small distortion might be appropriate for a dipole bound anion, but not for an anion where the extra electron is in a valence orbital. Another reason is that the electronic matrix elements that govern Rydberg processes accumulate in the spatial region near the nuclei, where a stationary phase relationship exists between the initial and final electronic wavefunctions. This stationary phase localization results in another $1/n^{*3}$ scaling of the interaction matrix element. None of these simplifying factors appears to be relevant to autodetachment. So my question is asked and answered.

Professor Reid enquired: The I^- anion is used as the source of electrons in your experiments. Does the presence of the iodine atom influence the observations? Would it be possible to use other anions in order to investigate this?

Professor Neumark answered: We have proposed that loss of the iodine atom is responsible for the observed bi-exponential decay, so in that sense its presence influences our observations. While other anionic sources of electrons could be used, iodide is probably the simplest and least perturbative. Of all the halide anions, we would expect iodide to bind the most weakly to the nucleobase, for example. Molecular anionic partners add to the complexity of the experiment since photodetachment will most likely produce a range of electron kinetic energies owing to Franck–Condon considerations. Atomic anions with lower electron binding energies such as sodide (Na^-) might be of interest, as these would enable us to access a wider range of electron collision energies, but the neutral sodium atom formed by photodetachment could easily affect subsequent observations.

Professor Fielding opened the discussion of the paper by Professor Wolfgang Domcke: Professor Domcke has presented a detailed theoretical description of a large number of electronic potential energy surfaces relevant to H atom loss in phenol following the absorption of UV radiation. May I check which excited electronic states have been included? In our earlier work investigating the dynamics of aniline following the absorption of a UV photon,[1,2] we find that the 3s component of the $\pi 3s/\pi\sigma^*$ state, and possibly also the 3p Rydberg states, transfer population back down to the electronic ground state. I would be interested to hear Professor Domcke's views on the possible role of p Rydberg states on the excited state dynamics in phenol.

1. R. Spesyvtsev, O. M. Kirkby and H. H. Fielding, *Faraday Discuss.*, 2012, **157**, 165–179.
2. R. Spesyvtsev, O. M. Kirkby, M. Vacher and H. H. Fielding, *Phys. Chem. Chem. Phys.*, 2012, **14**, 9942–9947.

Professor Domcke replied: The dissociative electronic excited states included, at least the low lying ones, are primarily of $\pi 3s$ character. The second and higher dissociative states have additional π excitations. An inspection of the relevant σ^* orbital as a function of geometry reveals that the 3p Rydberg orbitals don't play a role for linear C–O–H geometry, but become increasingly mixed with the 3s Rydberg orbital for increasing bending angle, both with in-plane bending and out-of-plane bending of the OH group. This observation seems to agree with your findings for aniline. Note that the above statements do not pertain to or indicate any direct and significant contributions of determinants containing the 3p Rydbergs. In fact, such an involvement is not seen for any of the 12 states we include in our work at near-minimum geometries. However, as noted in section II.B of the paper, the presence of the 3p Rydberg orbitals stabilizes the active space in our CASSCF calculations.

Miss Perveaux asked: I would have expected the ring to have some impact on the dynamics of the H dissociation, especially at the conical intersection where it may act as a dissipative bath coupled to the system undergoing internal conversion.[1,2] However, in this paper, the ring is kept entirely frozen. Do you know if the role of the ring in the kinetics of the reaction can be neglected, and if not, do you know how you could treat this?

This would be a perfect case of application for using the recent development of polyspherical coordinates in quantum dynamics.[3–10]

1. Axel Kühl and Wolfgang Domcke, *J. Chem. Phys.*, 2002, **116**, 263.
2. Axel Kühl and Wolfgang Domcke, Chemical Physics 259 (2000) 227-236.
3. E. V. Gromov, A. B. Trofimov, F. Gatti, and H. Köppel, *J. Chem. Phys.*, 2010, **133**, 164309.
4. Mamadou Ndong, Loic Joubert-Doriol, Hans-Dieter Meyer, André Nauts, Fabien Gatti *et al.*, *J. Chem. Phys.*, 2012, **136**, 034107.
5. Falk Richter, Fabien Gatti, Céline Léonard, Frédéric Le Quéré, and Hans-Dieter Meyer, *J. Chem. Phys.*, 2007, **127**, 164315.
6. loic Joubert-Doriol, Benjamin Lasorne, Fabien Gatti, Markus Schröder, Oriol Vendrel and Hans-Dieter Meyer, *Computat. Theor. Chem.*, 2012, **990**, 75–89.
7. Fabien Gatti and Christophe Iung, *Phys. Rep.*, 2009, **484**, 1–69.
8. Gauthier Pasin, Fabien Gatti, Christophe Iung, and Hans-Dieter Meyer, *J. Chem. Phys.*, 2006, **124**, 194304.
9. B. Lasorne, Michael A. Robb, Hans-Dieter Meyer and F. Gatti, *Chem. Phys.*, 2010, **377**, 30–45.
10. Joaquim Jornet-Somoza, Benjamin Lasorne, Michael A. Robb, Hans-Dieter Meyer, David Lauvergnat, and Fabien Gatti, *J. Chem. Phys.*, 2012, **137**, 084304.

Professor Domcke replied: I agree that the consideration of more than three internal degrees of freedom in the theoretical description of the photochemistry of phenol is highly desirable. While the quantum dynamics of nuclear motion could presumably be treated with the powerful methods mentioned in this comment, the bottleneck is the construction of an appropriate multi-sheeted multi-dimensional potential-energy surface of phenol (ultimately 33 vibrational degrees of freedom). A possible strategy could be the combination of the three-dimensional large-amplitude surface for the hydrogen atom with a small-amplitude surface for the vibrations of the phenoxyl fragment, for example, a fourth-order expansion in the normal modes of phenoxyl. A preliminary attempt in such a direction has recently been made for pyrrole.[1] The experimental data for phenol indicate that the coupling of the dissociating H-atom with the phenoxyl modes is, with the exception of the CO in-plane bending mode, relatively weak.[2–5] The likely reason is the very fast timescale of the H-atom dissociation.

1. Z. Lan and W. Domcke, *Chem. Phys.*, 2008, **350**, 125.
2. C.-M.Tseng, Y. T. Lee and C.-K. Ni, *J. Chem. Phys.*, 2004, **121**, 2459.
3. G. A. King, T. A. A. Oliver, M. G. D. Nix and M. N. R. Ashfold, *J. Phys. Chem. A*, 2009, **113**, 7984.
4. M. G. D. Nix, A. L. Devine, B. Cronin, R. N. Dixon and M. N. R. Ashfold, *J. Chem. Phys.*, 2006, **125**, 133318.
5. M. N. R. Ashfold, G. A. King, D. Murdock, M. G. D. Nix, T. A. A. Oliver and A. G. Sage, *Phys. Chem. Chem. Phys.*, 2010, **12**, 1218.

Dr Santoro said: Beyond being essential for quantum-dynamical (QD) simulations, I think that diabatic representations are very important to understand the chemistry behind complex ultrafast nonadiabatic phenomena. In this sense, I am fascinated by the big challenge of trying to work out a model comprising 12 different diabatic surfaces for the dissociative process you deal with, and I would like to ask some additional details. I see that you exploit symmetry arguments. Would you describe your diabatization strategy as "energy-based", "property-based" or are other definitions more appropriate? In principle one would like to have diabatic states with very small kinetic nonadiabatic couplings (since they are then neglected in QD calculations). Did you check whether this is the case in the representation you built? Do you think it would be helpful and feasible to check the coordinate dependence of matrix elements of some properly-chosen operator, *e.g.* transition dipoles, to ensure that the diabatic states have the desired behaviour?

Professor Domcke responded: In this particular application, we primarily relied on the high symmetry (C_{2v}) of an idealized phenol molecule with a linear C–O–H group. The diabatic surfaces in the three-dimensional space of the H-atom coordinates were obtained *via* "diabatization by ansatz", that is, a physically motivated diabatic potential model was constructed and the parameters of the model were determined by a least-squares fit of the eigenvalues of the 12 × 12 model Hamiltonian to the adiabatic *ab initio* data. The diabatization strategy may thus be classified as "energy-based". The *ab initio* electronic wave functions were not considered at all in this diabatization procedure. Having the 12 × 12 adiabatic-to-diabatic transformation matrix, we could transform *ab initio* property surfaces, such as transition dipole moment surfaces, from the adiabatic to the diabatic representation. Their smoothness near conical intersections would verify *a posteriori* the success of the diabatization procedure. It may be worthwhile to perform this test in the future for the most relevant conical intersections in phenol.

Dr Roberts asked: In the introduction to Prof. Domcke's discussion paper, it is rightly noted that a detailed *ab initio* survey of all symmetry allowed coupling modes (within C_s symmetry) at the $^1\pi\pi^*/^1\pi\sigma^*$ and $^1\pi\sigma^*/S_0$ conical intersections (CIs) in phenol indicate that the O–H torsion motion (τ_{OH}) has by far the strongest coupling strength at both CIs.[1] However, it is also acknowledged that analysis of high resolution H(Rydberg) atom photofragment translational spectroscopy experiments led Ashfold and co-workers to suggest that out-of-plane ring deformation modes, ν_{16a} and ν_{16b}, act as the dominant vibronic coupling motions at the CIs during O–H bond fission[2,3]—no experimental evidence for τ_{OH} acting as the coupling mode was found, in apparent contradiction to *ab initio* studies. Guided by earlier spectroscopic studies on phenol,[4] the most recent interpretation of these experimental data by Dixon *et al.*,[3] invokes non-rigid G_4 molecular symmetry (isomorphous with the C_{2v} point group) to explain the dominance of ν_{16a} for promoting vibronic coupling between the $^1\pi\pi^*$ and $^1\pi\sigma^*$ states, *via* tunnelling under the inner $^1\pi\pi^*/^1\pi\sigma^*$ CI. In G_4 symmetry ν_{16a} is the lowest frequency mode possessing the correct a_2 symmetry to promote $^1\pi\pi^*$ ($^1B_2$) → $^1\pi\sigma^*$ ($^1B_1$) population transfer; τ_{OH} however, becomes a b_1 mode and no longer possesses the correct symmetry to promote coupling.[3]

To qualitatively investigate the effects of higher-order C_{2v} (isomorphous with G_4) symmetry on the coupling between the $^1\pi\pi^*$ and $^1\pi\sigma^*$ states, in collaboration with Prof. Martin Paterson, we have performed preliminary CASSCF calculations of the branching space[5] for the $^1\pi\pi^*/^1\pi\sigma^*$ CI at a linear C–O–H geometry (labelled $\theta = 0°$ in Prof. Domcke's paper), which conforms to C_{2v} symmetry. The calculated branching space for the $^1\pi\pi^*/^1\pi\sigma^*$ CI at $\theta = 0°$ is shown in Fig. 1. While the gradient difference (**g**) logically remains as the O–H stretch, we find that the derivative coupling motion (**h**) does not display any out-of-plane O–H deformation (which would be akin to τ_{OH} at the equilibrium geometry), but only exhibits out-of-plane motions on the phenyl ring, which bear semblance to normal mode ν_{16a} (shown inset); for reference the calculated branching space at the $^1\pi\pi^*/^1\pi\sigma^*$ minimum energy crossing point (MECP) with C_s symmetry is also shown in Fig. 1 and compares very favourably with findings from previous calculations,[1] where $\mathbf{h} = \tau_{OH}$. Although we acknowledge $\theta = 0°$ is a maximum on the $^1\pi\pi^*/^1\pi\sigma^*$ CI seam along θ (and hence unlikely to play a major role in the actual O–H fission dynamics), our findings nonetheless appear to qualitatively align with the idea that upon applying non-rigid G_4 molecular symmetry to the entire $^1\pi\pi^*/^1\pi\sigma^*$ CI seam, mode ν_{16a} would become the dominant symmetry allowed coupling motion.[3] In light of this, I wonder if there is any prospect of rigorously treating the O–H fission dynamics in phenol within G_4 molecular symmetry, rather than C_s?—perhaps in a similar manner to a recent treatment of excited state dynamics in ethylene, which invokes G_{16} molecular symmetry.[6,7]

1 O. P. J. Vieuxmaire, Z. Lan, A. L. Sobowleski and W. Domcke, *J. Chem. Phys.*, 2008, **129**, 224307.
2 M. G. D. Nix, A. L. Devine, B. Cronin, R. N. Dixon and M. N. R. Ashfold, *J. Chem. Phys.*, 2006, **125**, 133318.
3 R. N. Dixon, T. A. A. Oliver and M. N. R. Ashfold, *J. Chem. Phys.*, 2011, **134**, 194303.
4 G. Berden, W. L. Meerts, M. Schmitt and K. Kleinermanns, *J. Chem. Phys.*, 1996, **104**, 972–982.
5 I. N. Ragazos, M. A. Robb, F. Bernardi and M. Olivucci, *Chem. Phys. Lett.*, 1992, **197**, 217–223.
6 B. Lasorne, M. A. Robb, H. Meyer and F. Gatti, *Chem. Phys.*, 2010, **377**, 30–45.
7 J. Jornet-Somoza, B. Lasorne, M. A. Robb, H. Meyer, D. Lauvergnat and F. Gatti, *J. Chem. Phys.*, 2012, **137**, 084304.

Professor Domcke replied: This comment and the reported preliminary results address an important open question concerning the $^1\pi\pi^*/^1\pi\sigma^*$ conical intersection in phenol. When a reference geometry of C_{2v} symmetry (isomorphic with G_4 symmetry) is chosen as in the presented paper, pure torsion indeed does not have the right symmetry to serve as a coupling mode at the $^1\pi\pi^*/^1\pi\sigma^*$ conical intersection. The ring deformation mode ν_{16a} does have the right symmetry, but the combination of in-plane CO bending (toward the S_0 equilibrium geometry)

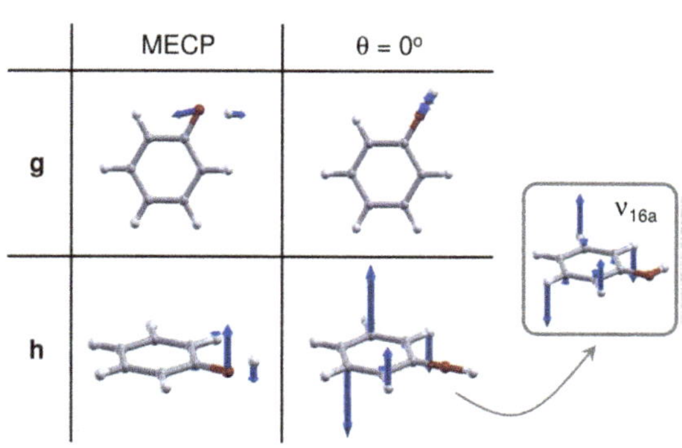

Fig. 1 Branching space motions of the $^1\pi\pi^*/^1\pi\sigma^*$ CI in phenol, calculated at the CASSCF(10,9)/6-311G* level of theory for the minimum energy crossing point (C_s symmetry) and $\theta = 0°$ (C_{2v} symmetry) geometries. **g** = gradient difference, **h** = derivative coupling. For comparison, normal mode ν_{16a} is also given.

and OH torsion (bilinear coupling in C_{2v} symmetry) also has the appropriate a_2 symmetry. The question of which internal coordinates primarily contribute to the $^1\pi\pi^*/^1\pi\sigma^*$ coupling in phenol indeed is not settled at the present stage.

Dr Oliver remarked: In your introduction, you referred to phenol as "spectroscopic water" because it is an excellent source of photoinduced solvated electrons and hydronium ions (*via* proton coupled electron transfer). Recent analysis[1] and time resolved measurements[2,3] have shown that excitation to the S_1 ($1^1\pi\pi^*$) state of phenol in the gas phase and cyclohexane solution, results in H atom tunnelling under the (S_1/S_2 ($1^1\pi\pi^*/1^1\pi\sigma^*$)) conical intersection on a nanosecond timescale resulting in O–H bond fission. Subsequent measurements of phenol in bulk water show a nanosecond rise time of charged products in contrast to homolytic bond fission.[4] Are these latest observations, especially the appearance timescale of the products, consistent with your prior theoretical predictions of PCET?[5]

1. R. N. Dixon, T. A. A. Oliver and M. N. R. Ashfold, *J. Chem. Phys.*, 2011, **134**, 194303.
2. Y. Zhang, T. A. A. Oliver, M. N. R. Ashfold and S. E. B. Bradforth, *Faraday Discuss.*, 2012, **157**, 141.
3. G. M. Roberts, A. S. Chatterely, J. D. Young and V. G. Stavros, *J. Phys. Chem. Lett.*, 2012, **3**, 348.
4. Y. Zhang, T. A. A. Oliver, M. N. R. Ashfold and S. E. Bradforth, unpublished results.
5. A. L. Sobolewski and W. Domcke, *J. Phys. Chem. A*, 2001, **105**, 9275.

Professor Domcke responded: I believe that our earlier calculations can be in agreement with the recent experimental observations referred to in the question, albeit we did not explicitly calculate appearance timescales of radicals or ions. Our calculations for phenol in the gas phase and for small phenol–water clusters clearly indicate that homolytic bond fission occurs after excitation of the $^1\pi\pi^*$ state of phenol.[1–3] The charge separation in aqueous solution, leading to solvated ions, most likely is a secondary process which is not taken into account in our calculations. For illustration, I refer to a recent paper by Riedle, Mayr and coworkers in which the homolytic generation of benzhydryl radicals and the subsequent charge separation in solution have been fully resolved in time.[4] Measurements at this technical level would be highly welcome for the investigation of the photolysis of phenol in liquid water.

1. A. L. Sobolewki and W. Domcke, *J. Phys. Chem. A*, 2001, **105**, 9275.
2. A. L. Sobolewski, W. Domcke, C. Dedonder-Lardeux and C. Jouvet, *Phys. Chem. Chem. Phys.*, 2002, **4**, 1093.
3. W. Domcke and A. L. Sobolewski, *Science*, 2003, **302**, 1693.
4. C. F. Sailer, B. P. Fingerhut, S. Thallmair, C. Nolte, J. Ammer, H. Mayr, I. Pugliesi, R. de Vivie-Riedle and E. Riedle, *ChemPhysChem*, 2013, **14**, 1423.

Dr Trevitt asked: Is there a "prefulvene" type deactivation channel for electronically excited phenol in this UV energy range—similar to the "channel 3" deactivation pathway of benzene? And, if so, could this lead to hot ground-state phenol that could then subsequently fragment to produce H atoms (or other products)?

Professor Domcke answered: Yes, one expects the existence of an internal-conversion pathway analogous to the "channel three" in the S_1 state of benzene also in substituted benzenes like phenol. For phenol, the existence of a low-barrier reaction path to an S_1–S_0 conical intersection of "prefulvenic" character has been verified by CASSCF/MRCI calculations.[1] The channel-three mechanism should thus be present in phenol with a threshold that is similar to the 3000 cm^{-1} channel-three threshold in benzene. The available experiments indicate, however, that the hydrogen atoms detached from phenol appear within a few hundred femtoseconds at all excitation energies.[2–5] This finding indicates that the direct photodetachment of the H-atom *via* the $\pi\sigma^*$ state outperforms the channel-three decay in phenol at all excitation energies.

1. O. P. J. Vieuxmaire, Z. Lan, A. L. Sobolewski and W. Domcke, *J. Chem. Phys.*, 2008, **129**, 224307.
2. C.-M. Tseng, Y. T. Lee and C.-K. Ni, *J. Chem. Phys.*, 2004, **121**, 2459.
3. A. Iqbal, L.-J. Pegg and V. G. Stavros, *J. Phys. Chem. A*, 2008, **112**, 9531.
4. A. Iqbal, S. Y. Cheung, M. G. D. Nix and V. G. Stavros, *J. Phys. Chem. A*, 2009, **113**, 8157.
5. M. N. R. Ashfold, G. A. King, D. Murdock, M. G. D. Nix, T. A. A. Oliver and A. G. Sage, *Phys. Chem. Chem. Phys.*, 2010, **12**, 1218.

Professor Neumark opened the discussion of the paper by Dr Vasilios G. Stavros: In your paper, you claim that excitation of pyrrole at 200 nm leads to H atom production *via* statistical decay, and propose that "N–H bond fission from hot S_0 species will take longer than C-H fission". This seems unlikely, since the N–H bond is weaker than the C–H bond. Can you explain your results in light of this consideration?

Dr Stavros replied: Professor Neumark is absolutely correct in that the N–H bond is weaker than the C–H bond and in light of this comment and ensuing discussions at the meeting, we acknowledge that the rise in the low KE feature of our TKER spectra in pyrrole would be very difficult to reconcile based solely on statistical unimolecular C–H bond elimination. Additional processes that may contribute to the rise in the low KE feature shown in Fig. 6(a) of our discussion paper could involve alternative fragmentation pathways such as HCN + C_3H_4 following 200 nm excitation, fragments which have been detected by Blank *et al.*[1] following excitation at 193 nm. These fragments could then potentially undergo multiphoton dissociative ionization with our 243 nm probe to generate H-atoms, which are subsequently ionized through REMPI to generate H^+. One way to potentially establish the extent to which these additional fragmentation pathways are contributing to the rise in the low KE feature is to carry out time-resolved ion-yield measurements. These are currently underway in one of our laboratories (Ullrich).

1. D. A. Blank, S. W. North and Y. T. Lee, *Chem. Phys.*, 1994, **187**, 35–47.

Professor Dr Temps said: Pyrroles may also undergo intramolecular rearrangement and/or ring opening reactions after photoexcitation. This may lead to scrambling of the H and D atoms. Have the authors observed any experimental evidence for such processes?

Dr Stavros replied: This is a very interesting question and one that we have also considered after reading your excellent paper from an earlier Faraday Discussion.[1] Unfortunately our present studies cannot provide us with information regarding intramolecular H- and D-atom scrambling. However, one way we could establish whether such scrambling is indeed operative is to carry out time-resolved ion-yield measurements on the deuterated isotopomer pyrrole-d1. If we detect fragments containing D such as C_3H_3D (Blank *et al.*[2] observed C_3H_4) this could be the first step towards establishing whether intramolecular D/H exchange in the excited state is participating in the excited state dynamics of pyrrole.

1. D. A. Blank, S. W. North and Y. T. Lee, *Chem. Phys.*, 1994, **187**, 35–47.
2. J. Wei, J. Riedel, A. Kuczmann, F. Renth and F. Temps, *Faraday Discuss.*, 2004, **127**, 267–282.

Professor Dr Temps said: In ultrafast dissociation reactions like that of pyrrole, products may not be observable immediately but only after some delay, depending on the time a fragment needs to traverse inner regions of the potential energy surface *en route* to the separated fragments. This may have to be accounted for somehow in the data analysis, as fitting exponential rise times without a possible initial delay time may not be appropriate. Also, very precise determinations of time-zero are necessary in these cases. Have any delay times been observed?

Dr Stavros responded: In our original data analysis, we explored fitting the experimental data by sampling a larger parameter space in the fitting algorithm. As such, we did attempt to fit our transients by allowing the time-zero delay in our fit functions to vary. However, our fitting algorithm would always converge to a scenario where there was essentially no delay added to our time-zero (within our experimental accuracy). Based on this, the agreement between our extracted time-constants with those of Montero *et al.*[1] and the theoretical calculations by Vazdar *et al.*,[2] we are confident that our analysis is as good as our instrument response function allows it to be. Time-resolved photoelectron spectroscopy measurements would certainly provide further insight into the dynamics along the inner regions of the potential energy surface *en route* to dissociation and are currently underway in Professor Fielding's laboratory (see comment by Mr Oliver Kirby).

1. R. Montero, A. P. Conde, V. Ovejas, M. Fernandez-Fernandez, F. Castano, J. R. V. de Aldana and A. Longarte, *J. Chem. Phys.*, 2012, **137**, 064317.
2. M. Vazdar, M. Eckert-Maksic, M. Barbatti and H. Lischka, *Mol. Phys.*, 2009, **107**, 845–854.

Mr Kirkby commented: Dr Stavros has presented a really interesting and thorough study of how relaxation dynamics of pyrrole can be probed and unravelled by H-atom elimination methods, building upon recent work on similar hetero-aromatic systems.[1–5] We have also recently studied the ultrafast non-radiative decay dynamics in pyrrole,[6] using time-resolved photoelectron imaging. Our work in this area has stimulated two questions:

First, we notice that your results are in remarkable agreement with Lippert *et al.*,[7] who also employ femtosecond H-atom detection. In contrast, we measure decay times of < 50 fs for all excitation wavelengths from 250–200 nm, in agreement with measurements of Castano *et al.*[7]

Perhaps Dr Stavros could comment on whether or not he feels it is possible that dynamics are occurring on the excited potential energy surface between the Franck–Condon region (where our measurements are made) and H-atom elimination (where his measurements are made). Second, our experiments have shown several pressure dependent processes, with their own distinct lifetimes (see Fig. 2, the rising peak at 2.4 eV). We observe these effects even when the monomer to dimer ratio is small (5% at 0.9 bar, 10% at 1.8 bar). Do you think it is possible that you have dimers in your experiment?

1. G. M. Roberts, C .A. Williams, J. D. Young, S. Ullrich, M. J. Paterson and V. G. Stavros, *J. Am. Chem. Soc.*, 2012, **134**, 12578–12589.
2. R. Spesyvtsev, O. M. Kirkby, M. Vacher and H. H. Fielding, *Phys. Chem. Chem. Phys.*, 2012, **14**, 9942–9947.
3. R. Spesyvtsev, O. M. Kirkby and H. H. Fielding, *Faraday Discuss.*, 2012, **157**, 165–179.
4. T. N. V. Karsili, A. M. Wenge, S. J. Harris, D. Murdock, J. N. Harvey, R. N. Dixon and M. N. R. Ashfold, *Chem. Sci.*, 2013, **4**, 2434–2446.
5. M. N. R. Ashfold, G. A. King, D. Murdock, M. G. D. Nix, T. A. A. Oliver and A. G. Sage, *Phys. Chem. Chem. Phys.*, 2010, **12**, 1218–1238.
6. O. M. Kirkby and H. H. Fielding, publication forthcoming 2013.
7. H. Lippert, H.-H. Ritze, I.V. Hertel and W. Radloff, *ChemPhysChem*, 2004, **5**, 1423.
8. R. Montero, A. P. Conde, V. Ovejas, M. Fernández-Fernández, F. Castaño, J. R. Vázquez de Aldana and A. Longarte, *J. Chem. Phys.*, 2012, **137**, 064317.

Dr Stavros answered: The kinetic isotope effect we measure (k_H/k_D) of ~11 strongly supports the idea that N–H bond fission proceeds *via* tunnelling through a small exit barrier along the $1^1\pi\sigma^*$ ($^1A_2$) state. As such, it is very difficult for us to reconcile any alternative dynamics, such as excited state flux that is trapped on the upper cone of the $1^1\pi\sigma^*/S_0$ CI that may re-sample this CI and proceed to N–H dissociation.[1] Our rationale for this is simply that we would not anticipate this pathway to show such a notable kinetic isotope effect as is observed here. We can seek further confirmation for such tunnelling mechanism by carrying out measurements in 2,4-dimethylpyrrole. As the comment by Professor Mike Ashfold shows, electronic structure calculations on 2,4-dimethylpyrrole clearly show an enhanced barrier to tunnelling which should manifest itself in slower H-atom dissociation timescales in TR-VMI of H^+.

It is important to note however that when we photoexcite above this small exit barrier, within the uncertainty of our measurements, there seems to be excellent agreement between our measured time-constants (46 ± 22 fs at 238 nm and 52 ± 12 fs at 200 nm) and the time-constants that you measure (<50 fs for all excitation wavelengths).

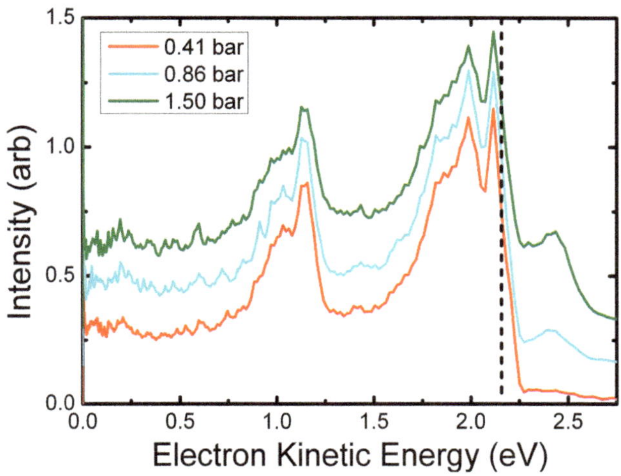

Fig. 2 Photoelectron spectrum of pyrrole, by (1 + 1) ionisation with 240 nm radiation, at various He backing pressures.

Regarding whether we have dimers or not in our molecular beam, we have recorded the TOF-MS at these excitation wavelengths and see no measurable signal at the appropriate *m*/*z* to indicate that we have dimers in our molecular beam. Having said that though, even if dimers were present in our molecular beam with a monomer/dimer ratio similar to that measured in your PES, we very much doubt that their contribution to the total transient H^+ signal would be significant so as to influence the extracted time-constants from our transients at 250, 238 and 200 nm.

1. H. Lippert, H. H. Ritze, I. V. Hertel and W. Radloff, *ChemPhysChem*, 2004, **5**, 1423–1427.

Professor Stolow commented: In pump–probe photoionization spectroscopies, there can be two pathways to the same ionic fragment. One pathway, the one typically under study, is a neutral fragmentation process followed by ionization of the free neutral fragment. The time evolution of this signal represents the time evolution of the chemical dynamic process under investigation. There exists, however, another pathway which equally yields a time-dependent ionic fragment. This pathway involves ionization with the pump laser and subsequent fragmentation of that cation with the probe laser. This too can be time dependent, but in this case probes dynamics in the cation, not the neutral. One way of dealing with this difficult situation is to use coincidence spectroscopies where one correlates the outgoing photoelectron with the fragment ion. If one doesn't make use of coincidences, how can we be sure that your measured time constants are not a blend of the two dynamic processes described above?

Dr Stavros answered: Professor Stolow raises a very important question regarding pathways for H-atom production. Indeed, following multi-photon excitation, the cation can potentially undergo dissociative ionisation to yield H-atoms, which can be probed with 243 nm. The measured transients are then a reflection of the dynamics occurring in the cation rather than the neutral. However, our confidence in our assignment that these H-atoms are borne from a neutral fragmentation process and not an ion fragmentation process largely comes from the general agreement between our measured TKER spectra and those obtained by Professor Ashfold and co-workers using HRA-PTS.[1] In addition, the transients we measure for both pyrrole and pyrrole-d_1 at 250 nm and 238 nm, in conjunction with the HRA-PTS data, suggest that a tunnelling mechanism is operative with a small energy barrier (> 630 cm^{-1}). We believe it would be very unlikely that at the 3-photon limit (H atoms have been observed at appearance energies of 12.85 eV [2]), we were photoexciting pyrrole$^+$ to such a localized energy region, that N–H dissociation mediated by H-atom tunnelling was occurring. Admittedly, we cannot definitively rule this out and as Professor Stolow quite correctly points out, coincidence measurements would be ideal to pinpoint the exact origin of these H-atoms.

1. B. Cronin, M. G. D. Nix, R. H. Qadiri and M. N. R. Ashfold, *Phys. Chem. Chem. Phys.*, 2004, **6**, 5031.
2. G. D. Willet and T. Baer, *J. Am. Chem. Soc.*, 1980, **102**, 6774.

Professor Ashfold commented: The finding of a longer time constant for forming H atom photoproducts when exciting pyrrole at ~250 nm (*i.e.* at the long wavelength edge of its UV absorption) contradicts conclusions reached in the experimental study of Montero *et al.*[2] but is plausibly consistent with electronic structure calculations which reveal a small energy barrier in the N–H stretch coordinate (R_{N-H}) of the $S_1(^1\pi\sigma^*)$ potential.[3] This is clearly evident in Fig. 3(a), which shows *ab initio* potential energy cuts (PECs) along R_{N-H} for the ground (S_0), S_1 and S_2 (both $^1\pi\sigma^*$), and S_3 and S_4 (both $^1\pi\pi^*$) excited states of pyrrole calculated using: (i) complete active space with second order perturbation theory (CASPT2) based on a fully state averaged complete active space self-consistent field (SA-CASSCF) reference wavefunction, with the remainder of the nuclear framework held fixed at the ground state equilibrium (*i.e.* planar) geometry, (ii) Dunning's augmented correlation consistent basis set of triple ζ quality: aug-cc-pVTZ with extra even tempered sets of *s* and *p* diffuse functions on the N atom to describe the Rydberg-valence coupling more effectively and (iii) an active space comprising 10 electrons in the following 10 orbitals: the 3 ring-centred π bonding orbitals and the two π* antibonding orbitals, the N centred 3s Rydberg orbital and two occupied σ and two virtual σ* orbitals, one of each of which is localised around the N–H bond.[4] An expanded view of barrier in the $S_1(\pi\sigma^*)$ PEC is shown in Fig. 2(b), along with a second (red) curve showing the effective potential (*i.e.* a vibrationally adiabatic potential) for molecules in the zero-point level of the S_1 state. The latter was obtained by estimating the zero-point energy associated with all normal modes of the S_1 molecule at a range of R_{N-H} values and of the ground state pyrrolyl radical (*i.e.* ZPE($R_{N-H} = \infty$)), and then raising the *ab initio* PEC by the appropriate difference [ZPE(R_{N-H}) − ZPE($R_{N-H} = \infty$)] at each R_{N-H}. The effective

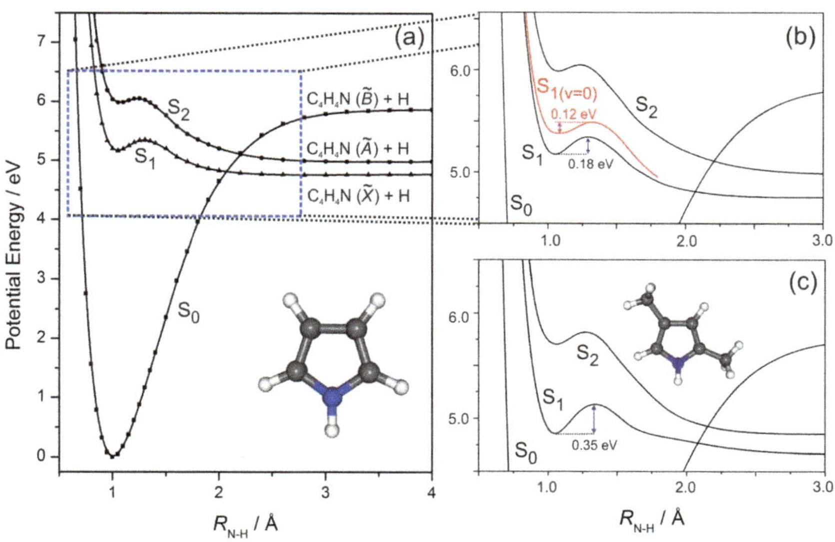

Fig. 3 (a) PECs along R_{N-H} for the S_0 and first two excited singlet states of pyrrole calculated at the CASPT2(10/10)/aug-cc-pVTZ level, with extra even tempered sets of *s* and *p* diffuse functions on the N atom. (b) Expanded view of the 4.5–6.5 eV / 0.5–3.0 Å region, showing the *ab initio* S_1 and vibrationally adiabatic $S_1(v = 0)$ PECs (black and red curves, respectively). (c) Similarly expanded view of the corresponding *ab initio* PECs for 2,4-DMP, that serves to highlight the greater barrier in the S_1 PEC (*cf.* pyrrole)

barrier to N–H bond fission by tunnelling so derived (~0.12 eV) is in good accord with that estimated by Roberts *et al.*[1]

Our previous H Rydberg atom photofragment translational spectroscopy (PTS) studies of jet-cooled pyrrole molecules at many wavelengths in the range $254 \geq \lambda \geq 230$ nm revealed structure attributable to population of specific vibrational (v) levels of the pyrrolyl partner fragment.[5] Different H + pyrrolyl(v) product channels show different recoil anisotropies—a finding that can be explained by recognising that the electric dipole forbidden S_1–S_0 ($^1A_2$–X^1A_1) transition gains oscillator strength by vibronic coupling with higher lying excited states and that the (ring-centred) parent vibrations that facilitate such coupling are orthogonal to the N–H dissociation coordinate and map through, essentially unchanged, into the radical product. The 'mean' recoil anisotropy obtained by integrating over all H + pyrrolyl products formed in the photolysis of pyrrole at $\lambda \sim 250$ nm ($\beta \sim -0.5$) is in good accord with that reported by Roberts *et al.*[1]

Tunnelling effects should be easier to investigate in an asymmetrically substituted system like 2-ethylpyrrole or 2,4-dimethylpyrrole (2,4-DMP). Such alkyl substitution will have two consequences. First, they break the (C_{2v}) symmetry of pyrrole, increasing the S_1–S_0 absorption cross-section. Second, though alkyl substituents are generally viewed as relatively mild perturbers of the σ-framework (+*I* inductive effect), alkyl substitution in the 2-position nonetheless stabilises the ground state radical (*cf.* pyrrolyl) and thereby reduces the N–H bond strength (and the S_1–S_0 energy separation).[4] As Fig. 3(c) shows, the calculated energy barrier to N–H bond fission in the *ab initio* S_1 PEC for 2,4-DMP is roughly twice that in pyrrole, which should reveal itself *via* a reduced tunnelling rate for 2,4-DMP(S_1, v = 0) molecules.

1. G. M. Roberts, *et al.*, *Phys. Chem. Chem. Phys.*, 2013, **163**, DOI: 10.1039/c2fd20140b.
2. R. Montero *et al.*, *J. Chem. Phys.*, 2012, **137**, 35 (ref. 23 in the paper by Roberts *et al.*).
3. A. L. Sobolewski, W. Domcke, C. Dedonder-Lardeux and C. Jouvet, *Phys. Chem. Chem. Phys.*, 2002, **4**, 1093.
4. T. N. V. Karsili, B. Marchetti, R. Moca and M. N. R. Ashfold, *J. Phys. Chem. A* (submitted).
5. B. Cronin, M. G. D. Nix, R. H. Qadiri and M. N. R. Ashfold, *Phys. Chem. Chem. Phys.*, 2004, **4**, 5031.

Dr Stavros replied: We appreciate the invaluable contribution by Professor Mike Ashfold and agree that TR-VMI measurements on 2,4-dimethylpyrrole would make for a highly interesting comparative study into the tunnelling dynamics between pyrrole and 2,4-dimethylpyrrole, given the enhanced S_1–S_0 absorption cross-section and increase in the barrier height towards N–H dissociation.

Professor Dr Buma commented: I actually have two questions. The first question concerns your conclusion that after excitation of pyrrole at 250 nm tunneling occurs to the πσ* state. You come to this conclusion based on experiments on the deuterated compound. Wouldn't it be much more direct to do experiments at slightly lower wavelengths so that you would be closer to the barrier?

The second question concerns the results you show in Fig. 3. You have indicated why you think that the TKER is larger for for negative pump–probe delays but this explanation does not explain why the width of the peak is significantly larger. Might this indicate that another state becomes involved when excitation occurs at 243.1 nm instead of 250 nm?

Dr Stavros responded: These are both very interesting questions. Regarding the first question, given the spectral bandwidth of our femtosecond laser pulses (approximately 500 cm^{-1}), the amount of control one would have in tuning the wavelength is limited, as the barrier height we predict is comparable to our spectral bandwidth. Coupled to this, N–H dissociation proceeds *via* the zero-point in the N–H stretch, as we have observed for O–H dissociation in the related heteroaromatic phenol.[1] Therefore, any excess energy excites orthogonal modes to the N–H stretch, which are largely carried through adiabatically into to the pyrrolyl co-fragment (although there are examples of vibrational non-adiabaticity, as noted by Professor Ashfold and co-workers following photoexcitation at 254 nm in pyrrole[2]). Regarding the apparent increase in width of the peak at ~7000 cm^{-1} in the TKER at negative ($\Delta t = -1$ ps) time delays, this is a noteworthy observation, and one which we admittedly overlooked. This is merely a reflection of the amount of internal energy that is carried off by the partner fragment (pyrrolyl) as we increase the pump photon energy. This is more clearly seen in the TKER spectra recorded using the HRA-TOF method by Professor Ashfold and co-workers (see Fig. 2 and 3 in ref. 2).

1. G. M. Roberts, A. S. Chatterley, J. D. Young and V. G. Stavros, *J. Phys. Chem. Lett.*, 2012, **3**, 348–352.
2. B. Cronin, M. G. D. Nix, R. H. Qadiri and M. N. R. Ashfold, *Phys. Chem. Chem. Phys.*, 2004, **6**, 5031–5041.

Professor Bañares asked: Related with the paper by Stavros and co-workers on the ultrafast photodynamics of pyrrole, we have a comment and a question. The comment is that we have carried out pump–probe experiments using nanosecond laser pulses and velocity map imaging to study the photodynamics of pyrrole-ammonia clusters, $PyH\cdot(NH_3)_n$, with $n = 2$–6, in the wavelength range from 236 nm (threshold for cluster absorption) down to 199 nm.[1,2] In those experiments, we excite the clusters by the pump laser and after a delay we probe the $NH_4(NH_3)_m$, $m = n - 1$, products by multiphoton ionization at 333.5 nm. Special care was taken to avoid contributions from cluster evaporation in all cases by carefully choosing the pump and probe laser intensities.[2] All our results are compatible with an excited state hydrogen transfer (ESHT) process between the chromophore and the solvent cage, where the eliminated H atom from the pyrrole acts as a projectile towards the closest ammonia molecule of the solvent cage in the cluster. Thus, the ESHT process can be understood as a bimolecular process between the H-atom from the pyrrole chromophore and the solvent cage. The effect of microsolvation on the pyrrole potential energy surfaces became apparent in four experimental facts when compared to the bare molecule case: First, the spectroscopic range where the hydrogen atom transfer was observed is blue shifted by ~15 nm; second, the coupling between the S_2 and S_1 surfaces involves considerably higher vibrational modes; third, no evidence of statistical deactivation was found, even at the shortest excitation wavelengths, in contrast with bare pyrrole; and fourth, no evidence of tunnelling effect at the larger excitation wavelengths as that reported by Stavros and co-workers for bare pyrrole was observed in the clusters.

The question is for Prof. Domcke. How do you think all these experimental facts related with the effect of microsolvation on the chromophore potential energy surfaces can be rationalized from a theoretical point of view? In your experience from *ab initio* calculations on similar systems, do you expect to find such effects under clusterization of the chromophore?

1. L. Rubio-Lago, G. A. Amaral, A. N. Oldani, J. D. Rodriguez, M. G. González, G. A. Pino and L. Bañares, *Phys. Chem. Chem. Phys.*, 2011, **13**, 1082.
2. J. D. Rodríguez, M. G. González, L. Rubio-Lago, L. Bañares, *J. Chem. Phys.*, 2012 **137**, 094305.

Professor Domcke answered: While there is general agreement that an excited-state hydrogen-transfer process takes place after photoexcitation in pyrrole–ammonia clusters, the actual mechanism is a matter of discussion. While David *et al.* proposed a relatively slow (adiabatic) H-atom transfer mechanism on the basis of their experimental findings and preliminary potential-energy surface calculations,[1] Rubio-Lago *et al.* suggested an impulsive mechanism, in which a fast photodissociated H-atom from pyrrole "collides" with the solvent cluster.[2] Recently, Zilberg and Haas showed by CASSCF calculations that two hydrogen-bonded isomers coexist on the $S_1(\pi\sigma^*)$ potential-energy surface of the pyrrole-ammonia cluster, separated by a low barrier.[3] This picture seems to favor the adiabatic process. The final answer can only come from femtosecond time-resolved experiments or first-principles nonadiabatic dynamics calculations. First attempts towards *ab initio* dynamics calculations have been reported for pyrrole–water clusters and indicate an electron/proton transfer timescale of about 100 fs which would be longer than the timescale expected for impulsive H-atom transfer.[4]

1. O. David, C. Dedonder-Lardeux, C. Jouvet, H. Kang, S. Martrenchard, T. Ebata and A. L. Sobolewski, *J. Chem. Phys.*, 2004, **120**, 10101.
2. L. Rubio-Lago, G. A. Amaral, A. N. Oldani, J. Rodriguez, M. G. Gonzalez, G. A. Pino and L. Bañares, *Phys. Chem. Chem. Phys.*, 2011, **13**, 1082.
3. S. Zilberg and Y. Haas, *Phys. Chem. Chem. Phys.*, 2012, **14**, 8836.
4. A. Kumar, M. Kolaski and K. S. Kim, *J. Chem. Phys.*, 2008, **128**, 034304.

Mr Picconi enquired: The question refers to the kinetic energy distributions Dr Stavros observes at the different excitation wavelength.

When you decrease the excitation wavelength from 250 to 238 nm you observe a shift of the peak of kinetic energy toward higher values (from 6500 to 7200 cm^{-1}). When you excite with 200 nm light you do not see any further shift, altough the system has gained 1 eV of excess energy more.

Thus the excess energy flows almost completely into the internal degrees of freedom of the pyrrolyl fragment.

In your experience, is this result surprising or expected? Is it somewhat related to the presence of the conical intersection?

Dr Stavros answered: This is an excellent question and an observation that is entirely expected, as shown in a number of studies by Professor Mike Ashfold and co-workers on related heteroaromatic hydrides.[1] In the specific case of pyrrole, this mean TKER corresponds to the energy difference between the "plateau" region of the $1^1\pi\sigma^*$ state and the diabatic asymptote of the $1^1\pi\sigma^*$ state (see Fig. 1 in ref. 2).

1. M. N. R. Ashfold, G. A. King, D. Murdock, M. G. D. Nix, T. A. A. Oliver and A. G. Sage, *Phys. Chem. Chem. Phys.*, 2010, **12**, 1218–1238.
2. M. N. R. Ashfold, B. Cronin, A. L. Devine, R. N. Dixon and M. G. D. Nix, *Science*, 2006, **312**, 1637–1640.

Professor Dr Weitzel commented: You have addressed the aspect that for excitation of pyrrole at 200 nm you observe the onset of "statistical H loss". Is this claim solely based on the observation of H-atoms with low kinetic energy together with a slow time constant? Are there any competing reaction channels accessible at this energy? Is the claim of statistical decay supported by statistical rate theory?

Dr Stavros responded: Indeed our initial interpretation of statistical H-atom loss was inspired by the observation of H-atoms with low KE and the much longer time-constant we determined for these low KE H-atoms. However, as noted earlier by Professor Neumark and as indicated by our response to his question, there are potentially other fragmentation channels that are likely in competition with statistical H-atom loss. As such, we are now of the opinion that the low KE signal is very likely multicomponent in nature and further measurements are currently underway to probe these additional channels. However, to address your final question regarding whether statistical decay is supported by statistical rate theory, the time-constant that we measure is significantly shorter than what one would predict using RRKM theory ($\sim$30 μs, obtained using a similar approach to that described in ref. 1 given below). In previous work on related heteroaromatics imidazole and pyrazole, we have rationalized this as follows:[1] RRKM theory assumes that the internal (vibrational) energy is fully equilibrated in the system. However, following repopulation of S_0 through vibrational couplings at appropriate CIs, this leads to vibrational flux, which is localized in specific nuclear degrees of freedom, which over time will undergo IVR. This opposes the main assumption of RRKM theory, which assumes that energy is fully equilibrated and as such may explain, in part, why statistical H-atom loss in these experiments occurs on a much shorter timescale that what one would predict using statistical rate theory.

1. G. M. Roberts, C. A. Williams, M. J. Paterson, S. Ullrich and V. G. Stavros, *Chem. Sci.*, 2012, 3, 1192–1199.

Professor Dr Riedle asked: In our analysis of the sub-100 fs excited state proton transfer in 2-(2'-hydroxyphenyl)benzothiazole we have found that the appearance of the keto signal cannot properly be modeled by an exponential function. This becomes particularly clear when the time-zero between pump and probe pulse is determined to better than 10 fs. Instead a model utilizing a delayed step-like appearance convoluted with the instrument response function of 35 fs models the signal very well (see Fig. 8 of ref. 1). This can be readily understood as an evolution of the optically prepared wavepacket towards the crossing point between the enol and keto potential energy surface. Similarly the bond cleavage we report in our contribution at this meeting displays a clear two-step signal increase (see Fig. 7 and 8). I am wondering whether the very fast times seen in your analysis also need such a wavepacket model and how significant the reported rate constants are.

1. S. Lochbrunner, A. J. Wurzer and E. Riedle, *J. Phys. Chem. A*, 2003, **107**, 10580.

Dr Stavros responded: We have used various approaches to model the measured transients presented in our discussion paper. Akin to what you describe above, one of our approaches was to sample a larger parameter space, such as freely varying our time-zero between our pump and probe pulse during the iterative procedure. Even with this approach, the fitting algorithm would converge at 'zero' for our time-zero. Further credence to our fitting procedure here is given from our more recent work modelling the photodissociation of ammonia following excitation to the first electronically excited state, which has explicitly required a step-wise model to describe the production of H-atoms (+ NH_2 (A-state) co-fragments), where dissociation flux is temporarily trapped around the upper cone of the CI between the ground and first excited electronic states prior to dissociation.[1]

1. A. S. Chatterley, G. M. Roberts and V. G. Stavros, in press.

Professor Dr Temps commented: The gas phase dinucleotide photoelectron spectra presented during the discussion by Mr Chatterley are highly intriguing. It is very surprising in view of the available solution phase data on the excited-state dynamics of oligonucleotides, which clearly show the existence of very long-lived excited states, to see no long-lived components in the data. Are there any plausible explanations?

Mr Chatterley answered: There are a number of plausible explanations for the lack of observation of long-lived excited states in dAA^- and $dAAA^-$.

Firstly, it could be that the bases are not stacked in the gas-phase and therefore, no excimer states can form. Ion mobility studies by Gidden and Bowers[1] showed that the dimer (dAA^-) favours a stacked configuration at lower temperatures (80 K), however opens up as the temperature is increased to room temperature. Our experiments were performed at around 300 K,[2] for which the unfolding of the oligonucleotides is entropically favoured.

Secondly, the probe energy of our experiments (3.1 eV) may be insufficient to detach from the excimer state. In a neutral hydrated stacked dimer of adenine, Smith *et al.* measured a vertical ionization energy of 9.16 eV, with a very small tail to an onset at 8.01 eV.[3] We do not know how the presence of the charge affects these particular states, and it may push them outside our ionization window.

Thirdly, in aqueous solution, the charge on the phosphate group is completely screened,[4] but not so in the gas-phase where it has a strong effect on the electronic structure of the system.[5] The excimer state may be destabilized by the charge such that it cannot be formed following excitation to the bright states at 266 nm.

Of the above possibilities, the first is perhaps the most likely and experiments are underway to reduce the internal temperature of the ions.

1. J. Gidden and M. T. Bowers, *Eur. Phys. J. D*, 2002, **20**, 409–419.
2. D. A. Horke, A. S. Chatterley and J. R. R. Verlet, *Phys. Chem. Lett.*, 2012, **3**, 834–838.
3. V. R. Smith, E. Samoylova, H.-H. Ritze, W. Radloff and T. Schultz, *Phys. Chem. Chem. Phys.*, 2010, **12**, 9632–9636.
4. P. Slavicek, B. Winter, M. Faubel, S. E. Bradforth and P. J. Jungwirth, *J. Am. Chem. Soc.*, 2009, **131**, 6460–6467.
5. A. S. Chatterley, A. S. Johns, V. G. Stavros, and J. R. R. Verlet, *J. Phys. Chem. A*, 2013, **117**, 5299–5305.

Faraday Discussions RSC Publishing

PAPER

Electronic transient spectroscopy from the deep UV to the NIR: unambiguous disentanglement of complex processes

Eberhard Riedle, Maximilian Bradler, Matthias Wenninger, Christian F. Sailer and Igor Pugliesi

Received 5th February 2013, Accepted 11th February 2013
DOI: 10.1039/c3fd00010a

Complex multi-stage relaxation and reaction pathways after the optical excitation of molecules makes the disentanglement of the underlying mechanisms challenging. We present four examples that a new transient spectrometer with excitation fully tunable from the deep UV to the IR and 225 to 1700 nm probing allows for an analysis with greatly reduced ambiguity. The temporal resolution of about 50 fs allows us to resolve all relevant processes. For each example there is a new twist in the sequence of relaxation steps that had previously been overlooked. In malachite green it appears that the importance of the phenyl twisting has been overemphasized and rather a charge transfer state should be considered. In TINUVIN-P the predicted twisting as the driving motion for the ultrafast IC is confirmed and leads to a resolution of the earlier puzzle that the sub-5 ps regime shows kinetics deviating from a pure cooling process despite the sub-ps proton transfer cycle. For the bond cleavage of Ph_2CH–Cl and Ph_2CH–Br the degree of electron transfer within the radical pair can now be determined quantitatively and leads to a profound understanding of the long term cation yield. For the first time coherent wavepacket motion in the photoproducts is reported. Last but not least the measurement of the GSB recovery in the deep UV allows for the surprising result, that even after S_2 excitation of cyclopentenones the triplet states are reached with near unity probability within a few picoseconds.

1 Introduction

Optical spectroscopy in a wide range of varieties is a prime source of microscopic information on molecular systems and processes. Simple continuous wave ultraviolet-visible (UV/vis) absorption allows the identification of the energies of excited electronic states and infrared (IR) measurements render typical vibrational modes that are used to identify the associated motives in the molecule. The

Lehrstuhl für BioMolekulare Optik, Ludwig-Maximilians-Universität München, Oettingenstr. 67, 80538 München, Germany. E-mail: Riedle@physik.uni-muenchen.de; Fax: (+49)089-2180-9202; Tel: (+49)089-2180-9210

two regimes are complementary in many ways and the skilful choice or combination is often needed.

Modern time resolved versions of UV/vis and IR spectroscopy permit the observation of the temporal evolution of the system with well below 100 fs precision. Particularly in the UV/vis practically all relevant processes can now be resolved. The interpretation of the raw measurements does, however, require the identification of unambiguous signatures of the reaction intermediates and the eventual products. This is by far not trivial, as only the spectrum of the reactants is typically well known and possibly that of the product, if it can be isolated in sufficient quantity and purity. Any short lived intermediates can only be characterized by the transient measurement itself and so there is often a "chicken and egg" dilemma that a model of the reaction is needed to assign the observed spectral signatures and these would be of great help to develop the model.

As the structural motives that determine the vibrational frequencies are quite localized, *e.g.*, N–H or C–C stretches or specific bending motions in the fingerprint region, at least first guesses for the assignment are readily made. However, this also means that only moderate changes result in many regions of the IR spectrum due to the chemical transformation. Identifying the significant ones and isolating them from solvent signals can then become quite challenging. The electronic bands do change quite largely on the contrary, both in position and in strength. The problem there is the large homogenous and inhomogeneous width of each electronic transition. Sometimes even the decision of what is a separate electronic band or just a vibronic structure can become difficult.

Many of the seminal experiments in femtochemistry were done with fixed frequency pulses[1] and this - in retrospect - required a fair amount of chemical intuition for the analysis. Broadband transient measurements have, however, been developed continuously and the change from liquid based continuum generation to crystal based continua has allowed an ever widening application. Due to the intrinsically broad spectrum of a femtosecond IR pulse - relative to the typical 15 cm^{-1} width of vibrational transitions - somewhat of an overview was originally easier to gain in this spectral range.[2] The coverage of a wide spectral range requires tuning the probe pulse repeatedly and performing the time scan at each setting. This is where UV/vis transient spectroscopy has an intrinsic advantage, as extremely broad continua are now available[3,4] and can be combined with detectors of a thousand or more pixels.

In this contribution we want to demonstrate the potential of a cutting edge broadband UV/vis/NIR transient spectrometer that will be briefly described in Sec. 2. We then present and discuss measurements on four sample systems ranging from ultrafast nonradiative decay, excited state proton transfer, and bond cleavage to ultrafast intersystem crossing. For all systems we find surprising new details that had been overlooked by the earlier investigations with limited spectral coverage. We thus can demonstrate that transient electronic spectroscopy is now becoming extremely powerful and the redundancy of information obtained in the measurements gives an unambiguous and clear picture of the complex dynamics. Combined with high level computations the energy flow in molecules after absorption of a photon and leading eventually to photochemistry can be traced most reliably.

2 Ultra-broadband transient spectroscopy with 50 fs resolution

The transient spectrometer used for our measurements relies on a fully tunable pump source and a broadband continuum for wavelength selective probing. The general setup has been described in detail[4] and is based on a kHz Ti:sapphire amplifier (CPA2001; Clark-MXR, Inc.). The pump pulses are derived from a blue pumped noncollinear optical parametric amplifier (NOPA) frequency doubled in a BBO crystal with typically 100 μm thickness. The visible NOPA pulses are over-compressed in a fused silica prism compressor to precompensate for all dispersing elements after the doubling stage. With various mixing strategies and a possible intermediate continuum stage the range from 189 nm to 5 μm can be covered without any gaps.[5,6] The pump pulse lengths can be adjusted to 30 fs or below from the deep UV to 1200 nm and are only slightly longer at longer wavelengths.

In the original setup we used a 775 nm pumped single filament super-continuum that spans from 290 to 720 nm as the probe. Such a wide continuum is generated in a selected 5 mm CaF_2 plate that is translated in a circular fashion to avoid optically induced damage. The plate is carefully oriented to avoid depolarization of the continuum.[7,8] The broadband probe light is dispersed in a fused silica prism and detected with a multichannel detector. The continuum chirp is corrected in the data postprocessing. The pump–probe delay up to 2 ns is adjusted with the help of a computer controlled delay line accurate to better than 10 fs. The overall temporal resolution is as good as 50 fs, mainly limited by residual group velocity mismatch between pump and probe in the 120 μm solution sample. Transient spectrometers with parameters similar or close to ours are in use in a number of laboratories around the world.[9–12]

To access an even wider probe spectral range we alternatively pump the CaF_2 based continuum generation by the frequency doubled or tripled Ti:sapphire

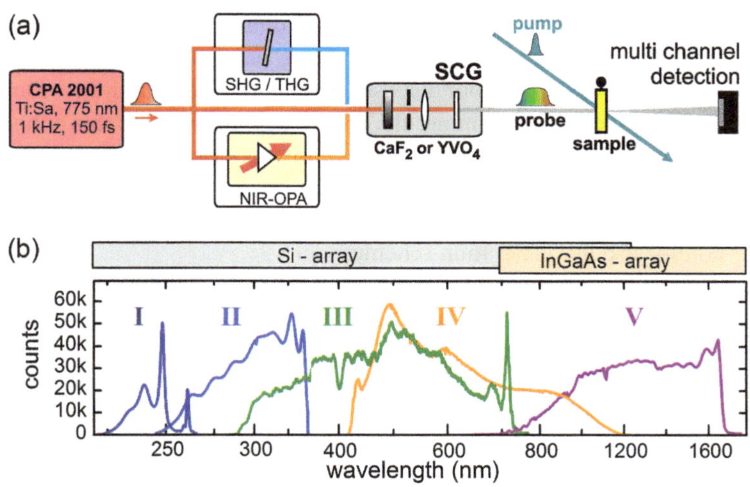

Fig. 1 (a) Scheme of the supercontinuum generation with various harmonics of the Ti:sapphire amplifier or the output of a NIR OPA. (b) Spectra of the continua as seen on the spectroscopic multi-channel detector.

output (see Fig. 1(a)). We find that significantly less pump energy is needed at these shorter wavelengths (387 and 258 nm) and therefore no damage is found even in prolonged operation. The resulting spectra are shown in Fig. 1(b). They span from close to the pump wavelengths down to 245 and 225 nm. The long wavelength cut-off is given by the dielectric filter used to suppress the high spectral power near the pump wavelength. Optimized filters with a smooth roll off could add some more spectral range in this vicinity to the probe pulse.

To access the near infrared (NIR) probe range we use a single-stage collinear optical parametric amplifier (OPA) to generate a red shifted continuum.[3] With the OPA signal at 1200 nm and the CaF_2 plate the useful continuum spans from 420 to 1100 nm. In this range the silicon based CCD or diode array can still be used. For the 750 to 1700 nm range we use the idler at 2100 nm to generate the continuum in a 4 mm YVO_4 crystal. As detector an InGaAs based 256 pixel detector (Series 2000CV2 camera with uncooled Hamamatsu G9203-256D detector; Entwicklungsbüro Stresing) was used.

All continua are single filament and can therefore be well collimated and focused into the sample cell. Their spectra are summarized in Fig. 1(b). They overlap sufficiently and in total span from 225 nm to 1.7 μm. Typically two or three of the ranges are needed to cover the region of interest for a particular molecule. They are measured sequentially and then stitched together in the postprocessing to render the comprehensive manifold of transient spectra for all desired delay times.

3 Selection of sample molecules and processes

The development of the transient spectrometer described above has been motivated by our various investigations of ultrafast molecular and chemical dynamics. These range from photophysical process, *via* excited state electron and proton transfer to complex chemical reaction sequences. For this discussion we have chosen four examples that serve to demonstrate the newly available capabilities from the application point of view.

The first decision that has to be made in an ultrafast molecular experiment is the choice of the pump wavelength, λ_{pump}. As we have ultrashort pulses at all possibly needed wavelengths to our disposal, we choose λ_{pump} purely according to the energy level structure of the molecule. We strongly believe that this enhances the value of the results as we do not have to worry about some initial electronic relaxation that is inevitably connected to some vibrational excitation. As shortly discussed above, such a choice of pump wavelength is always possible with the existing nonlinear pulse generation schemes.

The choice of λ_{pump} can be made quite readily by inspecting the linear absorption spectrum of the sample. The spectra of the four substances are shown in Fig. 2. For 2,3,4,5-tetramethyl-2-cyclopentenone (TMCPO) in cyclohexane there is an extremely weak $S_1 \leftarrow S_0$ transition around 320 nm and a quite strong $S_2 \leftarrow S_0$ transition below 250 nm. We excite TMCPO at 228 nm to specifically investigate the relaxation from the S_2 state. Diphenylmethylbromide (Ph_2CH–Br) can be excited through the weak benzoic absorption at 270 nm. In 2-(2′-hydroxy-5′-methylphenyl)-benzotriazole (commercial name TINUVIN-P) ultrafast excited state intramolecular proton transfer (ESIPT) can be triggered by π–π^* absorption at 350 nm. The cationic dye malachite green (4-[(4-dimethylaminophenyl)phenyl-methyl]-

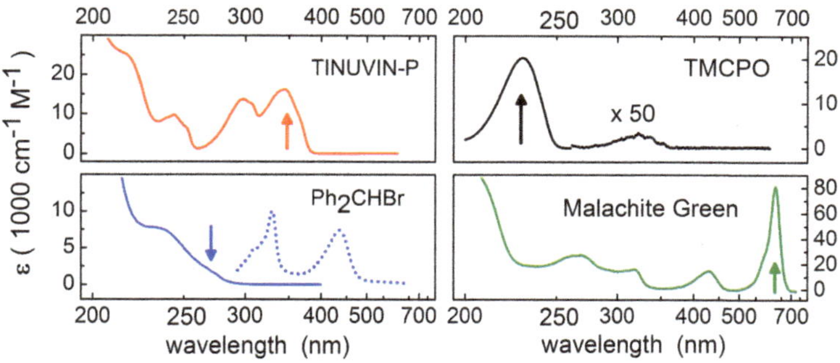

Fig. 2 Absorption spectra of the molecules studied in this work. TMCPO: 2,3,4,5-tetramethyl-2-cyclopentenone. In the panel for Ph_2CHBr also the spectra of the benzhydryl radical and cation are shown (dotted line). The arrows indicate the pump wavelength.

N,*N*-dimethylaniline; MG) is known to laser spectroscopists as it made the first real femtosecond laser system, the colliding pulse mode locked laser (CPM) possible due to its saturability at the 620 nm emission wavelength of Rhodamine 6G. In the present report we reinvestigate the ultrafast dynamics after 620 nm excitation.

4 Putting the analysis of the ultrafast dynamics in malachite green on a sound basis

Malachite green (MG, see Fig. 3(a)) is a cationic triarylmethane dye that absorbs strongly in the visible region. Since this absorption overlaps perfectly with the output of the CPM laser - it is actually used as a saturable absorber in this device - it was one of the first molecules investigated for its ultrafast dynamics. The early studies concentrated on the ground state bleach (GSB) recovery. Rather complex kinetics with a main time constant of 2.1 ps in methanol and a strong solvent dependence were found.[13,14] The solvent dependence is attributed to the viscosity of the solvent and it is postulated that an intermediate state S_x is reached after

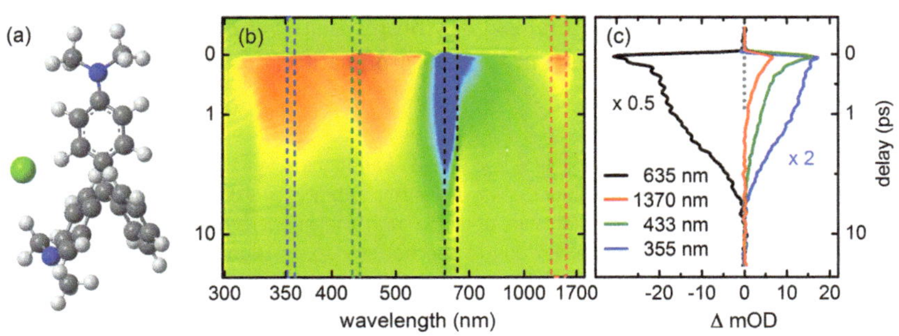

Fig. 3 Transient spectroscopy of malachite green dissolved in ethanol. (a) Structure of the dye. (b) Manifold of transient spectra. Red colours indicate an increase in absorption, blue colours a decrease. The wavelength scale is roughly reciprocal to pronounce different energetic parts equally. The time scale is linear in the range of −1 to +1 ps and logarithmic beyond. (c) Selected kinetic traces at the maximum of the NIR absorption of the LE state (1420 nm), the broad excited state absorption in the UV (350 nm) and the GSB (620 nm).

excitation through torsion of the phenyl rings.[14–19] After relaxation from the S_x state to the ground state the geometrical relaxation and cooling of the hot S_0 is thought to complete the cycle.

We investigated the dynamics of MG in ethanol after 620 nm pumping and utilizing the probe ranges III and V to cover the range from 300 to 1700 nm. The resulting manifold of the transient spectra is shown in Fig. 3(b). A pump–probe cross correlation of 35 fs in the NIR and slightly longer in the UV is reached and matches the resolution of previous one-[14,20] or two-colour pump–probe[15] and fluorescence up-conversion[18] measurements.

The strongest feature in the transient spectrum is the region of negative optical density around 635 nm (blue colours; Fig. 3(b))), where GSB and stimulated emission (SE) overlap. This is also the spectral region where most of the investigations were performed. In addition, the whole range from 300 nm to the GSB region and the NIR displays strong excited state absorption (ESA). A preliminary global fit analysis shows a 100 fs component with an increase of signal in the SE region and a decrease in the GSB region. This can be interpreted as the inertial motion of the optically prepared wavepacket away from the Franck–Condon point towards the equilibrium geometry of the excited electronic state S_1 that leads to a time dependent red shift of the SE. Such an extremely fast dynamics was already reported previously.[14,20]

A second decay component of about 440 fs has major peaks in the decay associated spectrum (DADS) at 433 nm (with 33 nm width) and at 1400 nm (430 nm width). These peaks can be readily assigned to the ESA following S_1 excitation and leading to the S_2 and S_4 excited electronic states. In the absorption spectrum, transitions to these two states are found at the matching positions of 428 and 255 nm (compared in Table 1). A third spectral feature with a positive amplitude in the DADS is found to be nearly coinciding with the $S_1 \leftarrow S_0$ absorption band, but is significantly different in shape. If this feature was due to a change in the GSB, it would indicate an increase in GSB with time. There is no argument known to us that would predict such a change in the hundreds of fs range. Therefore we have to assign this feature to the $S_3 \leftarrow S_1$ transient absorption (ESA). Finally a decrease in SE with a 440 fs time constant is found that reaches from 635 nm to 1100 nm. This matches the spontaneous emission spectrum within the limitation that certainly the stimulated emission in the sub-ps range does not originate from a thermalized S_1 vibronic distribution. In summary, this means that the strong transient signal around the usual 620 nm pump

Table 1 Absorption energies for the $S_n \leftarrow S_0$ and $S_n \leftarrow S_1$ transitions obtained from the cw spectrum of MG together with the $S_n \leftarrow S_1$ transitions energies (widths in brackets) extracted from the transient absorption (TA) of MG. The time constants are obtained from single wavelength fits. "d" signals where the time constant cannot be determined due to spectral overlap, "a" signals no amplitude for the associated time constant

	λ (nm) $S_n \leftarrow S_0$	λ (nm) $S_n \leftarrow S_1$	λ (nm) TA exp	τ_1 (ps)	τ_2 (ps)	τ_3 (ps)
S_1	622			0.10	0.44	2.2
S_2	428	1372	1400 (430)	a	0.44	a
S_3	315	638	~635	d	d	d
S_4	255	432	433 (33)	a	0.33	1.7
ESA			355	a	a	2.0

wavelength of most investigations is the sum of GSB, SE and ESA. It follows that the undiscussed averaging over the spectral width of the pump/probe pulse will not lead to any unambiguous interpretation of the dynamics of MG.

To display the transient absorption at selected times more clearly, we extracted the spectrum of absorption change (black lines in Fig. 4) from the complete manifold shown in Fig. 3(b). We then added a constant fraction of the ground state absorption spectrum (grey line in all panels of Fig. 4) to obtain the momentary spectrum (blue lines) of the sub-ensemble of molecules that was initially excited. The rapid decay of the 1400 and 433 nm peaks is clearly seen, as are the changes in the 635 nm region. At 1 ps, there is a broad absorption beyond the ground state absorption remaining that decays with about 2 ps. At 2 ps the 622 and 428 nm bands are the only remaining signals. The 622 nm $S_1 \leftarrow S_0$ absorption band has, however, not yet fully recovered. It is interesting to notice that the shape of the 622 nm band does not significantly change during the relaxation processes. This is surprising in view of the generally accepted interpretation that large amplitude torsional motions of the phenyl rings dominate the dynamics. We cannot rationalize how such motions do not lead to changes in the 622 nm band of more than the 3% upper limit we can deduce as maximum spectral change compared to the full GSB signal.

Kinetic traces at selected wavelengths are shown in Fig. 3(c). The fitted exponential decay constants are shown in Table 1. At 1400 nm we explicitly probe the S_1 population through the $S_2 \leftarrow S_1$ ESA. We find a relaxation time of 440 fs in good agreement with fluorescence measurements.[21] It should be noted that the largely increased probe range of our experiment allows for this unambiguous determination of the S_1 lifetime by transient spectroscopy alone. At 433 nm there is an overlap between the specific $S_4 \leftarrow S_1$ ESA and the very broad ESA that lasts into the ps range. The decay curve can be fitted with a first 330 fs exponential contribution that again signals the S_1 decay and a further 1.7 ps components. At 355 nm there is no ground state absorption and we do not have to expect any GSB. At this wavelength we can therefore exclusively probe the intermediate state S_x that the S_1 state relaxes to with the 440 fs time constant. The 2.0 ps decay is within experimental accuracy identical to the 1.7 ps component already found at 432 nm.

For the GSB recovery we need a 100 fs, a 440 fs and a 2.2 ps time constant. The first two most likely just result from the overlapping SE and ESA components. The

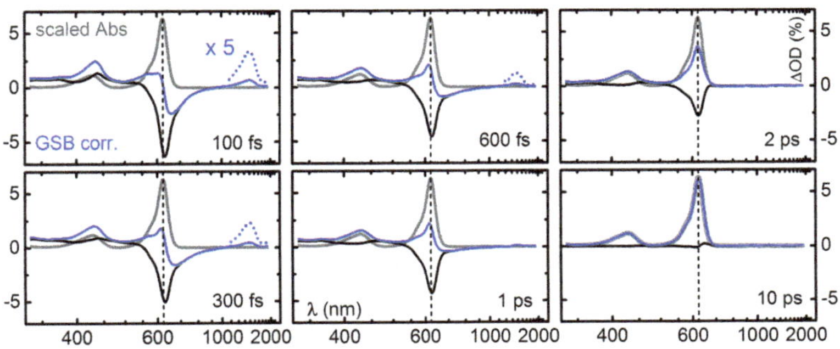

Fig. 4 Selected transient spectra at the indicated pump–probe delays (black lines). For visualization of the underlying changes in malachite green dissolved in ethanol, an equal amount of absorption spectrum is added to each transient spectrum (blue lines). In this way the recovery to the absorption spectrum (grey line) of the sub-ensemble excited by the pump pulse can be recognized.

slightly longer 2.2 ps recovery time of the GSB as compared to the S_x decay is quite reasonable in view of cooling processes. Temporal signal oscillations of about 226 cm^{-1} that can be attributed to a symmetric carbon–phenyl stretch vibration[14,22,23] are only found in the region from 605 to 635 nm. This must mean that a ground state wavepacket prepared by resonant stimulated Raman scattering is observed.

The extremely wide probe range of our measurements and the detailed analysis leads to the following conclusions. After S_1 excitation MG relaxes to a state S_x that neither fluoresces below 900 nm nor carries any oscillator strength to the known higher electronic states. Instead a nearly 2 eV broad ESA band decaying with a time constant very close to the GSB recovery is found. This all points to a charge transfer (CT) character of the S_x state. Indeed, recent high level calculations predict such a character already for the S_1 state[24] and for the related molecule crystal violet a relaxation mechanism involving charge redistribution has been proposed.[25,26] We suggest that the initially excited (local excited) state has an electronic character more similar to the ground state. This property would explain the very strong $S_1 \leftarrow S_0$ absorption. The often discussed torsion of the phenyl rings starts after the initial 100 fs geometric relaxation and leads to a simultaneous change in the electronic character towards a CT state on the half-ps time scale. Charge recombination then takes about 2 ps to repopulate the S_0 state with little optical signature from the "reverse" torsion of the phenyl rings.

From an extensive search of the literature it does not become clear, why repeatedly claims were made that the polarity of the solvent does not influence the dynamics of MG. It was always the viscosity that has been considered. If one, however, considers a series of alcohols with increasing size, not only the viscosity increases but also the polarity decreases. Future work should investigate this point by the reanalysis of existing data and solvent dependent transient spectroscopy. The solvation times of alcohols that are much longer than the ps times in the decay of excited MG are no intrinsic hindrance, as already a partial solvation can lead to efficient CT.[27] Given the capabilities demonstrated in this section, a final decision on the MG relaxation mechanisms should readily be possible.

5 An overlooked twisting isomerization in the proton transfer cycle of TINUVIN-P

TINUVIN-P is one of the best studied ESIPT systems.[28–31] After a 60–80 fs proton transfer the keto-type molecule relaxes to the electronic ground state with a 120–150 fs time constant. In the excited keto state a pronounced wavepacket motion is observed.[31] As the internal conversion (IC) was assumed to happen in the closed keto form originating from the optically excited enol form, the subsequent 600 fs decay found at probe wavelengths in the visible region was assigned to the proton back-transfer.[29,30] In the enol form a 180° twist around the central bond between the phenyl and the benzotriazole ring regenerates an identical conformation and a possible "isomerization" cannot be detected spectroscopically. In the keto-type form the same twist transfers TINUVIN-P from the closed form with the intramolecular hydrogen bond to the open form.[32]

The ultrafast proton transfer cycle leaves the TINUVIN-P in a highly vibrationally excited state. This is seen as a positive transient absorption below the $S_1 \leftarrow S_0$ absorption edge at 367 nm and a decrease in absorption at shorter

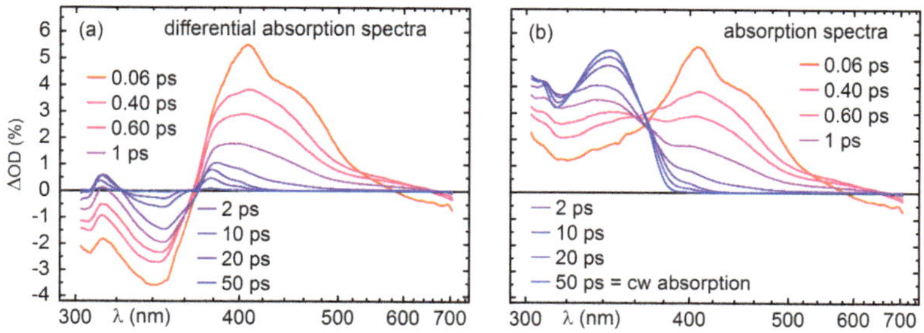

Fig. 5 (a) Transient absorption changes at selected delay times for TINUVIN-P in cyclohexane excited at 355 nm. (b) Absorption spectrum of the sub-ensemble excited by the pump pulse.

wavelengths.[33,34] For times from 5 ps onward the relaxation of this transmission change can be well modelled with a cooling process. However, it was already realized in the earlier investigations that the cooling picture was not sufficient to explain the spectral evolution in the first few picoseconds.

With our broadband spectrometer we investigated the dynamics of TINUVIN-P in cyclohexane. The differential absorption spectra shown in Fig. 5(a) display a rich variety of kinetics in the first 50 ps. To allow for an easier discussion of the underlying dynamics, we added a suitable fixed fraction of the ground state spectrum to all traces. The resulting time dependent absorption spectrum of the sub-ensemble excited by the pump pulse (see Fig. 5(b)) is used to discuss the observations.

At 60 fs after the excitation depletion of the ground state absorption is clearly seen below 367 nm. In the blue spectral region a strong transient absorption is found that initially displays some vibronic structure. This initial absorption must be due to ESA from the enol- or keto excited state. The absorption extends into the UV and overlaps with the ground state absorption. A weaker structure between 500 and 600 nm that develops with some delay is assigned to the closed keto ground state in accordance with ref. 30. Beyond 600 nm stimulated emission is found. The SE decays very rapidly and therefore can be assigned to the excited keto-type state.

Kinetic traces at selected wavelengths are shown in Fig. 6. For clarity the traces are shown on an expanded scale in part (a,c) and over the full relevant range of 30 ps in part (b,d). At 684 nm we find a delay/increase of the SE with about 80 fs and a decay with 154 fs. Within the experimental precision this ESIPT and IC behaviour matches the previous reports.[29,30] At 551 nm we find a 144 fs increase of the signal and a 588 fs decay. Therefore we can confirm that in the 500 to 600 nm region the suggested proton back-transfer is observed and takes 600 fs.

At exactly the vibrationless 0–0 transition of the optically pumped $S_1 \leftarrow S_0$ band (367 nm) we find an isosbestic point in the sub-ensemble spectra (Fig. 5(b)) at times beyond some ps. This is in close agreement with the earlier reports[33,34] and we can safely assign the decrease of absorption around 400 nm and the increase in the range around 350 nm to vibrational cooling. The spectral evolution of a cooling process does not render simple exponential decays with a wavelength independent decay time. Still, the value of 13 ps found at 344 nm near the band maximum gives a characteristic time constant of the cooling process. A very detailed resonance Raman study finds mode specific behaviour with time constants in the same range.[35,36] The traces at 362 and 375 nm, energetically

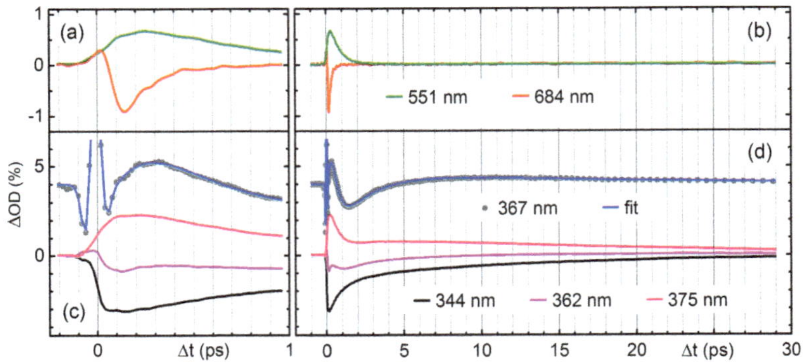

Fig. 6 Selected kinetic traces at the spectral positions, where the various conformers and relaxation processes of TINUVIN-P can be best differentiated. (a) and (c) show the traces within the first picosecond, while (b) and (d) show the range up to 30 ps. For clarity the coherent artefact due to the solvent cyclohexane has been subtracted from the traces except at 367 nm. The trace for 367 nm is multiplied by a factor of 4 and shifted upward.

slightly above and below the 0–0 transition consequently display quite complicated kinetics as many opposing effects contribute to the optical signal.

At exactly 367 nm the cooling does not influence the optical signal and the kinetics at this wavelength can therefore report the pure population evolution of the enol electronic ground state. We still have to assume that none of the other states involved in the full ESIPT and relaxation cycle contributes to the signal. Since the ESIPT itself proceeds in below 100 fs, the IC in 150 fs and the back-transfer from the closed keto-type form in 600 fs, this is a fairly safe assumption after about 1 ps. Inspection of the 367 nm trace (Fig. 6(c,d); grey filled circles for the experimental values and blue line for a multi-exponential fit) shows indeed an 80 fs decrease, a 150 fs increase and a 410 fs decrease. The vibrational cooling is seen as a very weak 15 ps decrease of the signal back to zero. Clearly there is another 2.2 ps contribution that was not seen in earlier work. In a global fit this time constant can be found as a prominent decay within most of the GSB range and the range around 450 nm. However, single wavelength fits show a systematic variation of the time constant due to the cooling and the overlap with the ESA.

To assign this 2.2 ps relaxation constant we return to the red region of the spectrum. The width of the spontaneous emission spectrum of the closed keto-type form of TINUVIN-P is known[31] and it is also established that the ground state closed keto-type form shows a mirror image to the fluorescence spectrum with a 0–0 transition at 625 nm.[30] Assuming a similar width and strength for emission and the absorption bands, only the region down to about 500 nm in the transient absorption spectrum can be assigned to the closed keto-type form. What is left to explain is the strong absorption in the range 380 to 500 nm and the 2.2 ps time constant.

It has been calculated, that the IC of the excited closed keto-type form proceeds *via* a conical intersection that involves a twisting around the central bond and a pyramidization.[32] A similar prediction backed up by experimental evidence was also made for the related 2-(2′-hydroxyphenyl)benzothiazole (HBT).[37] As a result of the twisting motion necessary for the extremely fast IC, parts of the molecules do not relax to the closed keto-type form but to the open keto-type form. This is directly corroborated by transient IR spectroscopy[38] and in line with early indications.[29,39] In view of all the evidence we assign the 2.2 ps time constant in TINUVIN-P to the back-isomerization of the open to the closed keto-type form that is then followed by an even faster back-transfer of the proton to lead to vibrationally excited enol molecules. These are subsequently cooled by the solvent in 10 to 20 ps. The majority of the strong transient absorption in the blue is consequently assigned to the open form.

These additional species populated with the 150 fs IC can readily explain the difficulties discussed previously to model the cooling kinetics at times before 5 ps.[34] It is not yet clear what this means for the interpretation of the resonance Raman studies.[35,36] In summary, the relaxation cycle in UV-excited TINUVIN-P, that takes as little as some ten picoseconds for completion, involves forward ESIPT ($\tau = 80$ fs), a branching in the IC to the closed and open keto-type form ($\tau = 150$ fs), a back-twisting ($\tau = 2.2$ ps), back-proton transfer ($\tau = 600$ fs) and finally cooling ($\tau = 15$ ps). One might ask oneself whether such a complex sequence is a rare exception or a more general occurrence. The ring opening of 1,3-cyclohexadiene comes to mind with the subsequent multi-step relaxation to the lowest energy tZt-conformer. In this system also highly complicated kinetics were reported.[40]

6 The generation and interconversion of benzhydryl radicals and cations by photolysis of diphenylmethylbromide

After UV excitation diarylmethylhalides undergo homolysis to the pair of a benzhydryl radical and halogen radical or heterolysis to the benzhydryl cation and the halogen anion. The overwhelming majority of investigations on this process and the subsequent interconversion between radical and ion pairs were performed without transient spectroscopy which renders spectra of the reactive intermediates, but rather by *a priori* chosen pump and probe pulse wavelengths. In addition, very intricate analysis with multiple exponential rates and a highly complex interpretation was deduced from these fixed colour recordings.[41] Last but not least, only ps resolution was employed, even though it had been shown that the initial processes proceed on the sub-ps time scale.[42]

We utilized 30 fs pulses at the first maximum of the absorption spectrum (270 nm; see Fig. 1) to excite either diphenylmethylchloride (Ph_2CH–Cl) or diphenylmethylbromide (Ph_2CH–Br). As the strong absorption features of the photolytic products Ph_2CH^+ and $Ph_2CH^\bullet$ are well separated from the absorption of the precursors, we can use the range III of our continuum generation for a full process analysis. The spectrum of $Ph_2CH^\bullet$ at 330 nm and the one of Ph_2CH^+ at 430 nm are shown as dotted lines in Fig. 1.

The manifold of transient spectra for Ph_2CH–Br in acetonitrile at various pump probe delays is shown in Fig. 7(a). It can be readily seen that the radical signal appears very early after the optical excitation and decreases in the ps range. The cation signal increases more slowly and then also decreases later. The kinetic traces shown in Fig. 7(b) give a more detailed view. They show the integrated signal over the full bands. The optical signal can be considered to be proportional to the species population within good approximation. The step in the rise of $Ph_2CH^\bullet$ (blue line) at about 50% amplitude happens at 100 fs. In agreement with

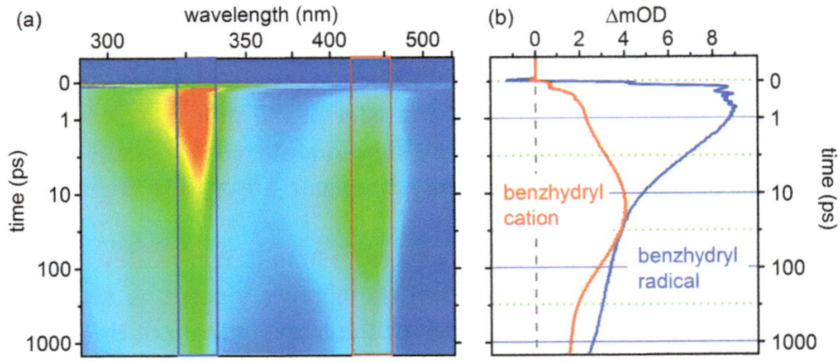

Fig. 7 (a) Transient spectra after 270 nm excitation of diphenylmethylbromide in acetonitrile. Red colour indicates a large transient absorption, green an intermediate one and blue colours little. (b) Signal for the resulting benzhydryl radical (blue line) and benzhydryl cation (red line) integrated over the respective band.

the observation for Ph_2CH–Cl, where we observe such a step at 80 fs,[43] we assign this to the delayed generation of the radical. The excited state wavepacket propagates from the Franck–Condon region to a first conical intersection that leads part of the wavepacket to the radical product. The remainder of the wavepacket stays on the excited potential hypersurface and reaches a second conical intersection leading to the cation product. This takes a total time of 125 fs in Ph_2CH–Cl and 200 fs in Ph_2CH–Br (see step in the red line). This means that the cation does not appear until this later time. No more radical and cation population is generated at later times. The slight increase in the optical signals is rather due to an increase in transition strength due to solvation and planarization of the products.[44]

A close inspection of the product signal curves and the full set of transient spectra show significant modulations up to about 2 ps. The frequency is about 300 cm^{-1} for the radical and 350 cm^{-1} for the cation. To analyze these modulations, we compare the signal at the blue side of the radical absorption to that at the red side (see Fig. 8(a)). Clearly a π phase shift is seen that proves that the modulation is due to a vibronic wavepacket. In addition, we determined the position of the maximum of the radical and cation absorption for each delay.[45,46] The resulting temporal dependence of the peak position shows the already mentioned ps relaxation due to solvation and planarization and also the oscillations. This confirms the interpretation as a vibronic wavepacket. An analysis of the normal modes of the radical and cation suggests that these wavepackets correspond to a symmetric stretch motion of the phenyl rings relative to the central carbon. This is very similar to the situation in MG (see above). However, in the photolysis of Ph_2CH–Br the wavepacket motion is in the photochemical products. The bond cleavage proceeds so fast that an impulsive excitation of the vibrational motion occurs. The stabilization of the radical and the cation leads to an increased electron density in the vicinity of the central carbon. Such an increase in electron density leads in turn to an impulsive shortening of the bonds and the suggested vibrational motion.

In the time up to about 10 ps the radical population decreases while the population of the cation increases. This is due to electron transfer between closely spaced benzhydryl and bromine radicals leading to the ion pair. We find that a

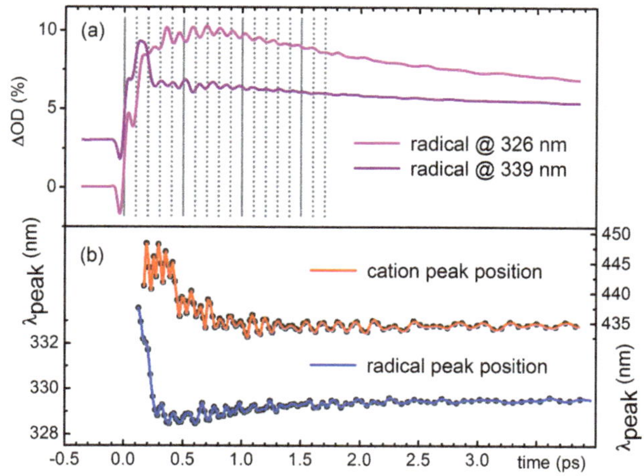

Fig. 8 (a) Transient absorption changes after 270 nm excitation of Ph_2CH–Br on the low wavelength (326 nm) and high wavelength (339 nm) edge of the radical product band. (b) Temporal dependence of the peak position of the benzhydryl radical (blue curve) and cation (red curve) band.

stretched exponential describes this process best, just as in our recent analysis of Ph_2CH–Cl.[46] The only difference is that for the chloride system an 11 ps time constant was found for this process compared to the shorter time constant of 2.1 ps in the bromide system. The decay of the radical population and the increase of the cation population with the identical temporal behaviour is direct proof of the interconversion by electron transfer. This finding is in contrast to the conclusions drawn from ps experiments on Ph_2CH–Br.[47,48] Clearly the very fast termination of the electron transfer in the bromide system did not allow the correct interpretation in the earlier experiments with 20 ps resolution. The electron transfer is stopped by the depletion of closely spaced radical pairs.[46]

After the end of the electron transfer the remaining radical pairs separate by diffusion to give a stable population of free radicals. The ion pairs stick together for a longer time due to the Coulomb attraction. Therefore they can effectively undergo recombination and the reduced population of free ions eventually evolves (94 ps time for Ph_2CH–Br and 138 ps for Ph_2CH–Cl).

The complete set of transient data cannot be modelled anymore with the usual kinetic scheme of well defined species like contact pairs, solvent separated pairs and free species. Instead we developed a combined model of Marcus-type electron transfer and Smoluchowski-diffusion.[46] We have to assume an initial radical and cation pair distance distribution that extends far beyond the contact distance considered explicitly in earlier work.[41,42,47,48] The various processes are treated as distance dependent. The time dependent populations can then be deduced by integration over the full radial distribution for each delay time. The time dependent rates result from the convolution of the distributions with the distance dependent rates.[46]

The observation of the full range of optical signals connected with the photolysis of diarylmethylhalides with 50 fs temporal resolution renders many hitherto covered details on the dynamics. Hypotheses that have been carried on for decades can now finally be scrutinized. The investigation of complex and multistep chemical process is now possible in a most direct fashion. Together with high level theoretical considerations the influence of conical intersections[49] and the microscopic arrangement can be investigated quantitatively. Since the breadth of optical signals renders a high degree of redundancy and cross checks, the ambiguity of many theories will be put to a serious test.

7 Direct determination of the triplet yield after S_2 excitation

One of the little mentioned understandings in ultrafast molecular dynamics seems to be that the rates connecting the originally excited state to the series of product states or chemical species are evaluated, presented and discussed, while the efficiencies and possible branching ratios are much less often put into the focus of the investigation. This stems from the fact that the yields of transient species are much less direct to determine compared to final product yields. For a hypothetical photochemical process leading from a precursor A to a stable product B the yield can readily be determined, as well established chemical means can be used to determine the product quantum yield. With modern illumination sources like high power LEDs this is extremely simple.[50] As a by-product one finds out which fraction returns to the reactant and effectively just undergoes a photophysical nonradiative relaxation.

For transient species or states the situation is much more difficult. Even if a time resolved measurement is available and able to uniquely identify the transient species - as exemplified in the sections above - one can still not rely on the Beer–Lambert law as the molar absorption coefficient ε of the species is generally not known. So only the product of ε and concentration can be determined. For the benzhydrylbromide photolysis discussed in Sec. 6 one can chemically stabilize the cation and actually measure the value of ε. If in addition the fluence of the pump pulses at the sample position and all other relevant geometric parameters are determined, quantitative transient spectroscopy results that actually determines the time dependent yield of all species absolutely.[46]

More frequently other methods have been employed to determine the yield of intermediate species, particularly triplet states. Triplets are of special interest to chemistry as they are long-lived compared to singlet states and therefore can allow for diffusion controlled photo-initiated bimolecular processes like the ones found in photocatalysis[51] and photo-initiated organic synthesis. One method uses triplet–triplet energy transfer to a system with a known triplet absorption coefficient and the measurement of the resulting quencher triplet by spectral photometry can be used. However this procedure requires very careful calibration and a certain degree of assumptions. The oxygen production by triplet-sensitized photooxidation of 1,3-diphenylisobenzofuran is an example of actually producing a physically measurable specific product that should be proportional in amount to the amount of triplet molecules produced over a selected period.[52] Another spectroscopic method that requires many assumptions and checks is the measurements of the ESR spectrum that specifically can report the triplet concentration.[53] To decouple the triplet yield and the triplet absorption cross section, Van Stryland and coworkers have introduced a very elaborate double-pump method that relies on the partial saturation of transitions. This approach still has limitations given mainly by the molecular energy level structure.[54] The most direct method is the determination of the ground state bleach recovery. This method relies on the ability to separate the time scales of the decay of the initially excited singlet and the resulting triplet in the GSB signal. This has been demonstrated by various single colour probe transient absorption measurements in the visible region.[55–57]

A more general application of the GSB recovery has so far been hampered by two issues. First, many systems of interest have their first strong absorption far in the UV. Second, a measurement at just one GSB wavelength still leaves the uncertainty that a superimposed ESA falsifies the interpretation. Both issues can now be solved with our extended probing range reaching as low as 225 nm. This is certainly sufficient to reach the strong absorption bands of almost any substance of chemical interest. By recording a full transient spectrum, the shape of the known ground state absorption spectrum can be used to unambiguously disentangle GSB and ESA or product contributions.

For an experimental demonstration we chose 2,3,4,5-tetramethyl-2-cyclopentenone (TMCPO) in cyclohexane. As seen from the absorption spectrum displayed in Fig. 2, there is a very weak $S_1 \leftarrow S_0$ absorption around 320 nm and a strong $S_2 \leftarrow S_0$ absorption centred at 229 nm. The assignment of the spectrum and the qualitative picture of the dynamics are made in analogy to the closely related acrolein.[58] We pump the S_2 state with 228 nm pulses of sub-60 fs duration. The transient spectra are measured with the three continuum ranges I–III (Fig. 1(b)). They display very broad structures (see Fig. 9(a)) and it becomes clear that such a wide probe range is needed for a clear interpretation. The spectrum at 0.15 ps pump–probe delay is dominated by the S_2 state through its ESA and possibly direct stimulated emission. The TMCPO relaxes to the S_1 state with a time constant of 0.25 ps, a quite reasonable time for an IC mediated by a conical intersection. Subsequently the molecule relaxes to the triplet manifold within 4 ps. This is quite fast and the underlying mechanisms are presently under study (compare the scheme in Fig. 9(b)).

To determine the S_1 and triplet yield, we evaluate kinetic traces at probe wavelengths that allow for a straightforward interpretation. A full analysis shows that a kinetic scheme purely based on population transfer cannot describe the observed dynamics. A more complex model incorporating solvation and vibrational relaxation has to be considered. At 400 nm we can see predominantly the S_2

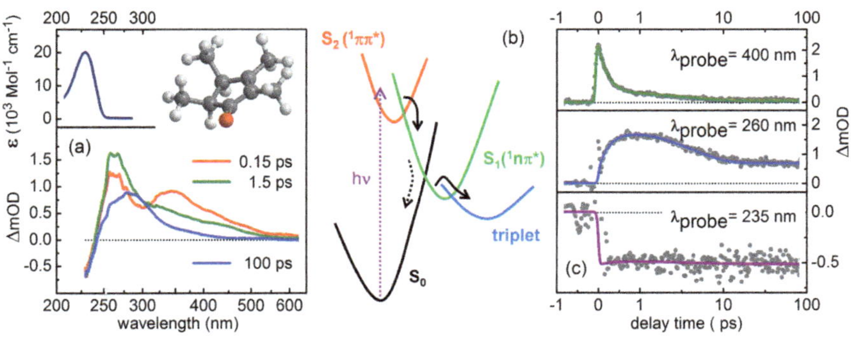

Fig. 9 (a) Selected transient spectra after 227 nm excitation of TMCPO in cyclohexane. The spectra were recorded in three parts and stitched afterwards. (b) Reaction scheme after S_2 excitation. (c) Selected kinetic traces monitoring mainly the S_2 population (400 nm), the S_1 population (260 nm) and the GSB (235 nm).

state relaxation with a 0.25 ps exponential decay. At 260 nm the signal reports the S_1 build-up and decay with 0.25 and 4 ps. Finally, at 235 nm the lack of any GSB recovery can be directly seen. Within the present experimental uncertainty given by signal fluctuations and the coherent contributions from the solvent and cell windows, we see no recovery of the GSB. This means that the whole S_2 population initially generated by optical excitation transfers to the triplet. We can compare this to the previously reported triplet yield of the parent molecule cyclopentenone of also near unity.[59] While an ultrafast intersystem crossing (ISC) proceeding with such a high yield is not unreasonable when the vibrationless S_1 state is excited, the finding of a unity yield after S_2 excitation is surprising. The vibrational excess energy released in the $S_2 \rightarrow S_1$ relaxation would easily suffice for IC to the electronic ground state as reported for acrolein.[58] High level quantum chemical calculations do indeed show a low lying conical intersection between the S_1 and the S_0 state. The fact that this process is highly inefficient while the ISC is highly efficient indicates that kinetic factors determine the dynamics and the branching ratio. That the triplet is accessed with near unity yield can be deduced from the transient spectra without any necessary assumptions as the GSB acts as an easily interpretable internal standard. The measurement is also corroborated by further recordings of TMCPO and 3-methyl-2-cyclopentenone in acetonitrile that also show near unity triplet quantum yield after S_2 excitation.

8 Summary and conclusions

The four molecules investigated in this work all display complex multi-stage relaxation after optical excitation. This has in the past lead to investigations with partially contradicting interpretations and conclusions. The extremely broadband probe investigations now possible with about 50 fs temporal resolution render a detailed picture that allows an interpretation with unprecedented unambiguity. For each example a new twist in the story of the relaxation results that had been overlooked. In malachite green it appears that the importance of the phenyl twisting has been overemphasized and rather a charge transfer state should be considered. In TINUVIN-P the predicted twisting as the driving motion for the ultrafast IC is confirmed and leads to a resolution of the earlier puzzle that the sub-5 ps regime shows kinetics deviating from a pure cooling process. For the bond cleavage of Ph_2CH–Cl and Ph_2CH–Br the degree of electron transfer within the radical pair can now be determined quantitatively. The earlier single-wavelengths measurements with ps resolution lead to strongly contradicting conclusions. For the first time coherent wavepacket motion in the photoproducts is reported. Last but not least the measurement of the GSB recovery in the deep UV allows for the surprising result, that even after S_2 excitation of cyclopentenones the triplet states are reached with near unity probability within a few picoseconds.

All these investigations were purely based on UV/vis/NIR transient spectroscopy. As similar problems have been tackled in the past with transient IR spectroscopy, we would like to ask ourselves how the two methods with state-of-the-art setups[4,10–12,60–62] compare. For the comparison a figure of merit is helpful. A sensible figure of merit seems to be the ratio of the broadest spectral probe range available to the smallest width found in the spectrum. In the IR vibrational transitions in solution are about 15 cm^{-1} wide and the whole spectrum covers about 3000 cm^{-1}. This gives a factor of 200. The spectral width in the electronic spectrum is at least as wide as the homogeneous broadening determined by the pure electronic dephasing on the time scale of tens of fs. This converts to a linewidth of 200 cm^{-1}. Our experimental setup can cover close to 40 000 cm^{-1} probe range and again a factor of 200 results. This means that transient electronic spectroscopy has finally caught up with transient IR spectroscopy with respect to detection width.

In both kinds of spectroscopies one has to identify the relevant spectral regions where the intermediates can be most unambiguously identified and tracked. The continua used in our setup (compare Fig. 1(b)) cover huge ranges and together with the highly resolving UV/vis/NIR detector arrays a first overview can be gained from a single time scan measurement requiring one or two hours. The temporal resolution is typically 50 to 100 fs, somewhat superior to the IR. Most importantly we find that the extension to the NIR allows for a qualitative new feature. We can always probe to the red of the GSB and SE region where only ESA or product bands can be located. Identifying the "last transition" in the NIR then allows to obtain the kinetics of the single species. This is nicely seen in the MG transient peak at 1400 nm assigned to the $S_2 \leftarrow S_1$ transition. Once the kinetics of this species are deduced from the measurement, the full associated spectral components can be extracted and used to decompose the set of transient spectra. In a second step the next to last transition is tackled and so on leading to a complete picture of photoinduced dynamics, however complex it might be.

The quantitative determination of transition strengths is readily possible by comparison with the intrinsic standard of the GSB of known strength. This is then often helpful to determine the electronic character of the intermediate species by comparison to quantum chemical calculations. What our setup also provides is sufficient spectral reproducibility and resolution to allow the observation of shifts as small as 0.1 nm or 10 cm^{-1}. These can be identified with solvation, ion or radical pairing,[46] and cooling. The temporal evolution of the shifts gives extra information on the dynamics.

It is also interesting to compare the ultra-broadband TA spectroscopy with fs-fluorescence spectroscopy (fs-FS) as it can be measured with up-conversion[63,64] or Kerr gating.[65] The reason that much effort is put into developing techniques of fs-FS is that the fluorescence is a very clean signature of the electronically excited molecule. There is no superposition in the signal by ESA or GSB as typically seen in TA spectroscopy with a limited detection range. As demonstrated in the examples of this work, ultra-broadband TA spectroscopy can fulfil the same condition by choosing the right spectral range, typically to the red of the GSB and stimulated emission. The well separated ESA transitions seen in this low energy part of the spectrum are also clean reporters of excited state dynamics. The great advantage is that the transient ESA signal can be measured with sub-50 fs temporal resolution, only reached by extremely well optimized fs-FS experiments.[64] Contrary to fs-FS, where an improvement in temporal resolution leads to a largely decreased signal strength and long averaging times, the TA spectroscopy can be performed quite quickly. We therefore expect that particularly transient NIR detection will be able to rival fs-Fs without the need for a dedicated setup.

Last but not least we want to address the detection and identification of vibronic coherences. In many one-colour pump–probe measurements much debate is needed to find out whether the observed modulations in the TA signal are due to vibrational wavepackets in the electronic ground state or to vibronic wavepackets in an excited state. The first are a nice confirmation of our understanding of ultrafast experiments and the underlying molecular level structure, but they do not reveal much information beyond that already known from Raman spectroscopy and resonance Raman spectroscopy. Only the latter render real new information on the structure of the excited molecule and the course of reactions. We were able to report earlier that in ESIPT those modes contribute to the wavepacket signature that are intimately connected with the proton transfer by shortening of the donor–acceptor distance.[66] Now we even find that in the bond cleavage discussed in Sec. 6 the photoproducts display long lasting vibrational coherences. This can only be understood if the whole cleavage process proceeds coherently. It gives great hope that coherent control, *e.g.*, by double pulses or shaped pulses, can be used to control the branching ratio between homolysis and heterolysis and the yield of individual products. Also incoherent control by time delayed pulses of largely differing wavelengths might be feasible as the electron transfer, the recombination and the diffusional separation are understood in detail.

The newly available experimental capabilities should be able to revitalize the use of electronic spectroscopy for the elucidation and quantitative analysis of complex molecular and chemical processes. Unlike the typical use of UV/vis spectroscopy for stable species or just processes in the ms range, our strategy comes with 50 fs temporal resolution. As a consequence, even the shortest lived intermediates can be looked for. Unlike the earlier fs experiments the ultrabroadband detection allows the full range of spectroscopic knowledge to be utilized. We therefore think that the best is still to come for electronic spectroscopy and we look forward to a bright, visible future.

Acknowledgements

We thank P. Lang and Oliver Schalk for experimental help, and R. de Vivie-Riedle, S. Thallmair and T. Tahara for stimulating discussions. Financial support of this work by the Deutsche Forschungsgemeinschaft through the SFB 749, the GRK 1626, and the excellence cluster 'Munich Center for Advanced Photonics' (MAP) is gratefully acknowledged.

References

1 A. H. Zewail, *Angew. Chem., Int. Ed.*, 2000, **39**, 2586.
2 P. Hamm, S. Wiemann, M. Zurek and W. Zinth, *Opt. Lett.*, 1994, **19**, 1642.
3 M. Bradler, P. Baum and E. Riedle, *Appl. Phys. B: Lasers Opt.*, 2009, **97**, 561.
4 U. Megerle, I. Pugliesi, C. Schriever, C. F. Sailer and E. Riedle, *Appl. Phys. B: Lasers Opt.*, 2009, **96**, 215.
5 M. Bradler, C. Homann and E. Riedle, *Opt. Lett.*, 2011, **36**, 4212.
6 C. Homann, P. Lang and E. Riedle, *J. Opt. Soc. Am. B*, 2012, **29**, 2765.
7 P. J. M. Johnson, V. I. Prokhorenko and R. J. D. Miller, *Opt. Express*, 2009, **17**, 21488.
8 I. Buchvarov, A. Trifonov and T. Fiebig, *Opt. Lett.*, 2007, **32**, 1539.
9 R. Huber, T. Köhler, M. O. Lenz, E. Bamberg, R. Kalmbach, M. Engelhard and J. Wachtveitl, *Biochemistry*, 2005, **44**, 1800.
10 A. N. Tarnovsky, W. Gawelda, M. Johnson, C. Bressler and M. Chergui, *J. Phys. Chem. B*, 2006, **110**, 26497.
11 T. A. Zeidan, Q. Wang, T. Fiebig and F. D. Lewis, *J. Am. Chem. Soc.*, 2007, **129**, 9848.
12 A. L. Dobryakov, S. A. Kovalenko, A. Weigel, J. L. Pérez-Lustres, J. Lange, A. Müller and N. P. Ernsting, *Rev. Sci. Instrum.*, 2010, **81**, 113106.
13 E. P. Ippen, C. V. Shank and A. Bergman, *Chem. Phys. Lett.*, 1976, **38**, 611.
14 Y. Nagasawa, Y. Ando and T. Okada, *Chem. Phys. Lett.*, 1999, **312**, 161.

15 Y. Nagasawa, Y. Ando, D. Kataoka, H. Matsuda, H. Miyasaka and T. Okada, *J. Phys. Chem. A*, 2002, **106**, 2024.
16 D. Ben-Amotz and C. B. Harris, *J. Chem. Phys.*, 1987, **86**, 4856.
17 M. D. Hirsch and H. Mahr, *Chem. Phys. Lett.*, 1979, **60**, 299.
18 A. Mokhtari, L. Fini and J. Chesnoy, *J. Chem. Phys.*, 1987, **87**, 3429; A. Mokhtari, A. Chebira and J. Chesnoy, *J. Opt. Soc. Am. B*, 1990, **8**, 1551.
19 F. Pellegrino, A. Dagen and R. R. Alfano, *Chem. Phys.*, 1982, **67**, 1982.
20 P. Sen, S. Yamaguchi and T. Tahara, *Faraday Discuss.*, 2010, **145**, 411.
21 N. H. Gottfried, B. Roither and P. O. J. Scherer, *Opt. Commun.*, 1997, **143**, 261.
22 M. J. Rosker, F. W. Wise and C. L. Tang, *Phys. Rev. Lett.*, 1986, **57**, 321.
23 K. A. Nelson and L. R. Williams, *Phys. Rev. Lett.*, 1987, **58**, 745.
24 A. Nakayama and T. Taketsugu, *J. Phys. Chem. A*, 2011, **115**, 8808.
25 M. M. Martin, E. Breheret, F. Nesa and Y. H. Meyer, *Chem. Phys.*, 1989, **130**, 279.
26 Z. R. Grabowski, K. Rotkiewicz and W. Rettig, *Chem. Rev.*, 2003, **103**, 3899.
27 U. Schmidhammer, U. Megerle, S. Lochbrunner, E. Riedle and J. Karpiuk, *J. Phys. Chem. A*, 2008, **112**, 8487.
28 S. R. Flom and P. F. Barbara, *Chem. Phys. Lett.*, 1983, **94**, 488.
29 M. Wiechmann, H. Port, W. Frey, F. Lärmer and T. Elsässer, *J. Phys. Chem.*, 1991, **95**, 1918.
30 C. Chudoba, S. Lutgen, T. Jentzsch, E. Riedle, M. Woerner and T. Elsaesser, *Chem. Phys. Lett.*, 1995, **240**, 35.
31 C. Chudoba, E. Riedle, M. Pfeiffer and T. Elsaesser, *Chem. Phys. Lett.*, 1996, **263**, 622.
32 A. L. Sobolewski, W. Domcke and C. Hättig, *J. Phys. Chem. A*, 2006, **110**, 6301.
33 W. Frey and T. Elsaesser, *Chem. Phys. Lett.*, 1992, **189**, 565.
34 K. Lenz, M. Pfeiffer, A. Lau and T. Elsaesser, *Chem. Phys. Lett.*, 1994, **229**, 340.
35 V. Kozich and W. Werncke, *J. Mol. Struct.*, 2005, **735**, 145.
36 V. Kozich, J. Dreyer and W. Werncke, *Chem. Phys. Lett.*, 2004, **399**, 484.
37 M. Barbatti, A. J. A. Aquino, H. Lischka, C. Schriever, S. Lochbrunner and E. Riedle, *Phys. Chem. Chem. Phys.*, 2009, **11**, 664.
38 O. F. Mohammed, S. Luber, V. S. Batista and E. T. J. Nibbering, *J. Phys. Chem. A*, 2011, **115**, 7550.
39 P. F. Barbara, L. E. Brus and P. M. Rentzepis, *J. Am. Chem. Soc.*, 1980, **102**, 5631.
40 S. Lochbrunner, W. Fuss, W. E. Schmid and K.-L. Kompa, *J. Phys. Chem. A*, 1998, **102**, 9334.
41 K. S. Peters, *Chem. Rev.*, 2007, **107**, 859.
42 M. Lipson, A. A. Deniz and K. S. Peters, *Chem. Phys. Lett.*, 1998, **288**, 781.
43 C. F. Sailer, N. Krebs, B. P. Fingerhut, R. de Vivie-Riedle, E. Riedle, *Phys. Rev. Lett.*, submitted for publication.
44 B. P. Fingerhut, C. F. Sailer, J. Ammer, E. Riedle and R. de Vivie-Riedle, *J. Phys. Chem. A*, 2012, **116**, 11064.
45 C. F. Sailer, R. B. Singh, J. Ammer, E. Riedle and I. Pugliesi, *Chem. Phys. Lett.*, 2011, **512**, 60.
46 C. F. Sailer, B. P. Fingerhut, S. Thallmair, C. Nolte, J. Ammer, H. Mayr, I. Pugliesi, R. de Vivie-Riedle and E. Riedle, *ChemPhysChem*, 2013 (in print).
47 J. Dreyer and K. S. Peters, *J. Phys. Chem.*, 1996, **100**, 15156.
48 J. Dreyer, M. Lipson and K. S. Peters, *J. Phys. Chem.*, 1996, **100**, 15162.
49 B. P. Fingerhut, D. Geppert and R. de Vivie-Riedle, *Chem. Phys.*, 2008, **343**, 329.
50 U. Megerle, R. Lechner, B. König and E. Riedle, *Photochem. Photobiol. Sci.*, 2010, **9**, 1400.
51 M. Wenninger and E. Riedle, in *Chemical Photocatalysis*, ed. B. König, DeGryuter, Berlin, 2013.
52 M. R. Detty and P. B. Merkel, *J. Am. Chem. Soc.*, 1990, **112**, 3845.
53 M. Guéron, J. Eisinger and R. G. Shulman, *Mol. Phys.*, 2006, **14**, 116.
54 D. Peceli, S. Webster, D. A. Fishman, C. M. Cirloganu, H. Hu, O. V. Przhonska, V. V. Kurdyukov, Y. L. Slominsky, A. I. Tolmachev, A. D. Kachkovski, R. R. Dasari, S. Barlow, S. R. Marder, D. J. Hagan and E. W. Van Stryland, *J. Phys. Chem. A*, 2012, **116**, 4833.
55 H. E. Lessing, A. von Jena and M. Reichert, *Chem. Phys. Lett.*, 1976, **42**, 218.
56 S. M. King, C. Rothe, D. Dai and A. P. Monkman, *J. Chem. Phys.*, 2006, **124**, 234903.
57 B. Lament, J. Karpiuk and J. Waluk, *Photochem. Photobiol. Sci.*, 2003, **2**, 267.
58 A. M. D. Lee, J. D. Coe, S. Ullrich, M.-L. Ho, S.-J. Lee, B.-M. Cheng, M. Z. Zgierski, I.-C. Chen, T. J. Martinez and A. Stolow, *J. Phys. Chem. A*, 2007, **111**, 11948.
59 R. Bonneau, *J. Am. Chem. Soc.*, 1980, **102**, 3816.
60 E. T. J. Nibbering and T. Elsaesser, *Chem. Rev.*, 2004, **104**, 1887.
61 E. T. J. Nibbering, H. Fidder and E. Pines, *Annu. Rev. Phys. Chem.*, 2005, **56**, 337.

62 M. L. Groot, L. J. G. W. van Wilderen and M. Di Donato, *Photochem. Photobiol. Sci.*, 2007, **6**, 501.
63 X.-X. Zhang, C. Würth, L. Zhao, U. Resch-Genger, N. P. Ernsting and M. Sajadi, *Rev. Sci. Instrum.*, 2011, **82**, 063108.
64 C. H. Kim and T. Joo, *Opt. Express*, 2008, **16**, 20742.
65 B. Schmidt, S. Laimgruber, W. Zinth and P. Gilch, *Appl. Phys. B: Lasers Opt.*, 2003, **76**, 809.
66 S. Lochbrunner, A. J. Wurzer and E. Riedle, *J. Phys. Chem. A*, 2003, **107**, 10580.

Faraday Discussions

RSC Publishing

PAPER

Ultrafast ring-opening reactions: a comparison of α-terpinene, α-phellandrene, and 7-dehydrocholesterol with 1,3-cyclohexadiene

Brenden C. Arruda, Broc Smith, Kenneth G. Spears and Roseanne J. Sension*

Received 13th December 2012, Accepted 9th January 2013
DOI: 10.1039/c3fd20149j

The excited and ground state dynamics of a series of 1,3-cyclohexadiene derivatives and their hexatriene photoproducts are studied using ultrafast broadband UV-visible transient absorption spectroscopy. The substitution pattern around the cyclohexadiene backbone alters the excited state potential energy surface in the Franck–Condon region as evidenced by changes in the excited state absorption and fluorescence properties of the systems. Unsubstituted 1,3-cyclohexadiene and α-phellandrene exhibit no excited state absorption while a strong excited state absorption in the visible spectral region is observed for both α-terpinene and Provitamin D_3. Steric factors introduced by the ring substitutions determine the dominant rotational isomer at equilibrium for the hexatriene photoproducts. Coupling to the solvent during the conformational relaxation from the initial helical all-*cis* hexatriene is unique to each photoproduct, but the relaxation process occurs on a 6–10 ps timescale regardless of the size or substitution pattern on the triene.

1 Introduction

Photochemical transformations in the condensed phase are at the forefront of physical chemistry and find application in molecular electronics, nano-scale optical switches,[1,2] molecular motors,[3] catalysis,[4] and photobiology.[5–7] The 1,3-cyclohexadiene (CHD) chromophore is one of the most well characterized photochemically active molecules, with an enormous literature on its gas and solution phase dynamics in the electronically excited state and the ground state (see for instance ref. 8 and the references therein). Substitution around the CHD backbone can alter the ground and excited state potential energy surfaces, but the photochemical transformation from diene to triene is pervasive (Fig. 1A). The electrocyclic ring opening as well as the reverse ring-closure reaction of CHD and similar dienes such as cyclobutadiene played a role in the development of the Woodward Hoffman rules for the conservation of orbital symmetry as well as the van der Lugt–Oosterhoff model which extended the theory to account for surface

Department of Chemistry, University of Michigan, Ann Arbor, Michigan 48109, USA

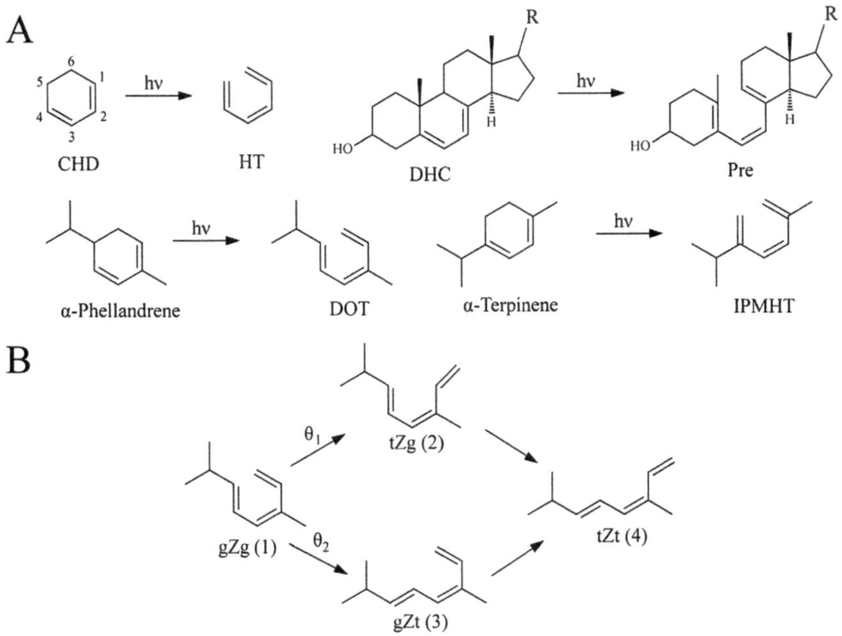

Fig. 1 A: Photochemical transformations of the cyclohexadienes under investigation here. The numbers indicate the carbon number on the CHD backbone and the other structures are drawn to be consistent with this numbering scheme. B: The energetically distinct isomerization pathways for DOT following rotation around the dihedrals θ_1 and θ_2.

crossings.[9–12] A number of theoretical studies have targeted the excited state potentials and motions that drive the molecule as it traverses the electronic states involved in the nonadiabatic ring opening process,[13–16] and the current model described below is accepted due to the large body of experimental work supporting these predictions.[17–27]

Excitation of CHD in the UV leads to a Franck–Condon wavepacket on the 1B electronic state which propagates towards a conical intersection with the spectroscopically "dark" 2A state on a *ca.* 50 fs timescale. Relaxation on this surface leads to a second conical intersection coupling the 2A state to the 1A ground state. Decay out of the excited state occurs on a *ca.* 150 fs timescale overall.[21,25] If the trajectory is reactive, this process results in the generation of the 1,3,5-hexatriene (HT) photoproduct. On return to the ground state, the molecule is left with a large excess of vibrational energy, which is redistributed among the vibrational degrees of freedom *via* solvent assisted intramolecular vibrational energy redistribution (IVR), and dissipated to the bath, cooling the internal temperature of the molecule from an estimated 2200 K to the ambient temperature of the surroundings.[19]

The general features of the excited state reaction pathway have been confirmed by a variety of experiments with ultrafast time resolution. Kosma *et al.* observed the ring opening process with 13 fs resolution in a transient ionization experiment, building off of a previous experiment with lower time resolution.[21,25] Observation of coherent oscillations in the transient ion signal confirmed the proposed symmetry-breaking in the radiationless 1B → 2A transition. In addition, Kim *et al.* demonstrated using three-pulse experiments on gas phase CHD that optical coupling between the ground and excited state is limited to the first ~50 fs of the reaction.[24]

The all-*cis* hexatriene photoproduct is generated initially in agreement with the principle of least motion, but is energetically less favourable than the "*trans*" rotamers. Approach to the equilibrium distribution for a HT population requires rotamer interconversion through the cZt intermediate to the tZt-HT conformation dominant at room temperature. The traditional naming scheme for the HT rotamers includes designations for the single and central double bonds, where the "c" and "t" index refer to the *cis* and *trans* orientations of the single bonds and "Z" or "E" refer to the orientation of the central double bond. Steric effects cause helicity in the *cis* forms, so for the substituted trienes under investigation here, *gauche* (g) is used instead of *cis* to describe the single bond conformation. This notation is also adopted for the all-"*cis*" or gZg-HT conformer. Previous studies in condensed phases have found the thermalization process for HT to occur on a picosecond timescale which depends only on solvent type (alcohols *vs.* alkanes) but not on specific solvents within a series.[19,22]

Interest in the photo-biological activity of CHD is driven by the inclusion of this chromophore in 7-dehydrocholesterol, or provitamin D_3 (Pro, DHC).[5–7] The ring-opening reaction in this system leads to formation of Previtamin D_3 (Pre), the thermal precursor to the essential human nutrient Vitamin D_3. Though aspects of the ring-opening of CHD can be used to describe the photochemistry of DHC, significant differences remain due to the vibrational degrees of freedom, steric effects of the steroid skeleton, and electronic contributions of the substituents to the potential energy surfaces of the chromophore. The influence of the substituents is observed experimentally as an increased excited state lifetime indicated by excited state absorption (ESA) as well as an increase in the fluorescence quantum yield by two orders of magnitude (2×10^{-6} for CHD and 2×10^{-4} for DHC).[7,18]

The decay of the ESA in DHC is well modelled by a fast exponential decay component of 0.4–0.6 ps and a slow component of 1–1.8 ps with relative amplitudes that depend on the solvent.[7] This has been explained by an asymmetric branching on the excited state surface to two distinct populations which decay independently. The 1–2 ps formation of Pre from DHC has been confirmed by Meyer-Ilse *et al.* using ultrafast circular dichroism spectroscopy.[28] The bulky steric groups on the hexatriene photoproduct also serve to shift the ground state equilibrium to the tZg rotamer at room temperature in the gas phase, although the ground state conformational relaxation is quite complex as shown by Tapavicza *et al.* using surface hopping molecular dynamics simulations.[16]

Table 1 Acronyms used in this manuscript and their corresponding definitions[a]

Acronym	Definition
CHD	1,3-cyclohexadiene
HT	1,3,5-hexatriene
IPMHT	(Z)-2-isopropyl-5-methyl-1,3,5-hexatriene
DOT	(3*Z*,5*E*)-3,7-dimethylocta-1,3,5-triene
DHC	7-dehydrocholesterol, Provitamin D_3
Pre	Previtamin D_3

[a] See Fig. 1 and ref. 31 for structures corresponding to the compounds discussed here.

The photochemistry and ground state conformational relaxation of α-terpinene[29] and α-phellandrene,[30] cyclic dienes with intermediate sizes between CHD and DHC, have recently been studied by Marzec *et al.* in matrix isolation experiments. These experiments revealed that the tZg rotamer of the α-terpinene photoproduct, (Z)-2-isopropyl-5-methyl-1,3,5-hexatriene (IPMHT), is the most populated at equilibrium. These studies, confirmed by our own calculations, find no stable minimum for the tZt conformer of IPMHT. The size of the isopropyl group precludes substantial occupation of the gZt rotamer. Transient absorption experiments in solution phase showed that approach to the equilibrated population was virtually solvent independent for a wide range of polarities, hydrogen bonding capacity, and bulk viscosity.[31]

For the α-phellandrene photoproduct, (3*Z*,5*E*)-3,7-dimethylocta-1,3,5-triene (DOT), Marzec *et al.* showed that in the matrix isolation environment, the tZg conformer once again dominates the equilibrium population.[30] However, for a non-static environment such as liquid solution, full exploration of the conformational space is expected, as the most energetically favorable minimum is found to be the tZt rotamer. Relaxation to this rotamer requires traversing one of two different reaction pathways which are energetically distinct (Fig. 1B). For the ease of the reader, a list of acronyms and their definitions are provided in Table 1.

In the work presented here, we have used broadband femtosecond transient absorption spectroscopy (TA spectroscopy) in the UV-Visible spectral region to compare the excited and ground state dynamics of CHD, α-terpinene, α-phellandrene, and DHC. Reid *et al.* performed picosecond time-resolved UV resonance Raman measurements on CHD and α-phellandrene in an effort to characterize the ring opening reaction.[17] The limitations on the time resolution in these experiments prohibited an investigation into the excited state ring opening reaction however, and the measurements likely probed only the conformational relaxation on the ground state. Previous work in our lab examined the similarities and differences between α-terpinene and DHC photochemistry.[31] This report presents transient absorption measurements of α-phellandrene and provides a global comparison of the ground and excited state behaviour of these four CHD containing compounds. Emphasis is placed on the effects of hydrocarbon substitution on the CHD backbone, which leads to a logical grouping of the behavior of the molecules based on the location of the steric groups, as previously postulated by Garavelli *et al.* for gas phase CHD, α-phellandrene, and α-terpinene.[21] This work connects the femtosecond timescale excited state behaviour observed by Garavelli *et al.*[21] to the picosecond timescale ground state relaxation observed by Reid *et al.*[17] This is the first examination of the ground state conformational relaxation of DOT in solution following photoexcitation of α-phellandrene with femtosecond time resolution and the first broadband spectral observation of the photoproduct conformational relaxation.

2 Experimental

Ultrafast transient absorption measurements were conducted under conditions described previously.[31] Briefly, the fundamental frequency of a Ti:sapphire femtosecond oscillator was amplified and frequency tripled in β-barium borate (BBO) nonlinear optical crystals to create a 266 pump beam. The 400 nm second harmonic was used to produce a broad band continuum in a translating CaF_2

window. The resulting continuum spanned the range from ~270 to 600 nm. The instrument response in the UV ($\lambda < 400$ nm) ranged from 150 to 300 fs. When looking for excited state absorption in the visible spectral region, care was taken to optimize the pulse duration to <100 fs. Multiphoton absorption and photolysis or photoionization in the solvent was minimized by attenuating the pulse energy to <250 nJ. The solvent signal was obtained with each sample measurement to ensure that the solvent signal remained small compared to the sample signal. All spectra presented here are solvent subtracted using the process described previously.[31]

Steady state photolysis experiments utilized a mercury arc lamp as the UV source. The spectra were collected on a Shimadzu UV-2401pc UV-VIS spectrometer with a wavelength resolution of 0.1 nm. When photolysis measurements were conducted in alcohols, the sample was deoxygenated to minimize secondary product formation. 1,3-Cyclohexadiene (97%), α-terpinene (95%), and 7-dehydrocholesterol (98%) were acquired from Sigma-Aldrich while α-phellandrene (>95%) was obtained from MP Bio. All chemicals were used without further purification. The transient absorption measurements were conducted at a concentration of ~1 mM.

Density functional theory (DFT) calculations utilized the Gaussian03 software package to conduct the vibrational frequency and electronic excited state

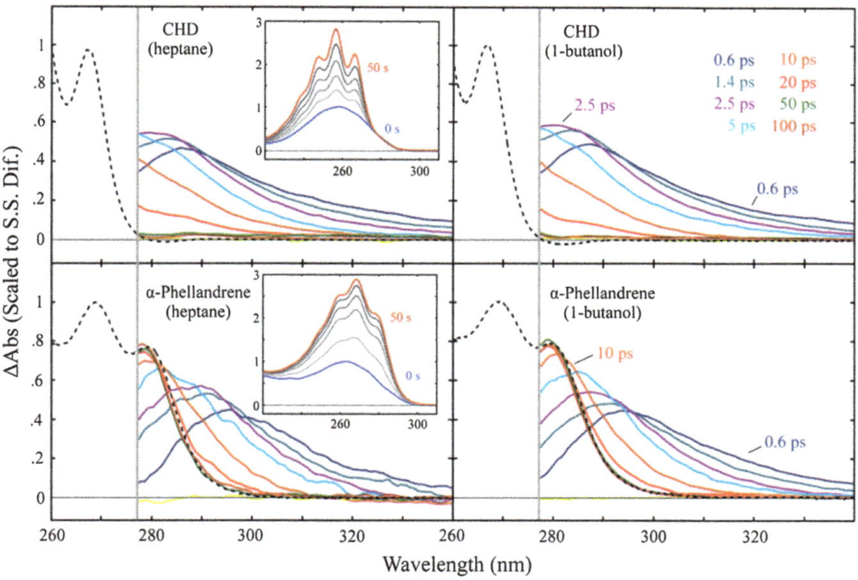

Fig. 2 Transient absorption spectra at various time delays of CHD and α-phellandrene. For ease of comparison the same time delays are presented in all four spectra. Note that the traces for CHD shift to the blue out of our spectral window, and appear to decay to 0. The black dashed lines are the steady state photolysis difference spectra (S.S. Dif.) and the trace around zero in α-phellandrene is for the time delay −1 ps. There is uncertainty in the intensity of the HT steady state spectrum relative to the transient absorption data due to the extent of the continuum in these measurements. The intensity shown is adapted from data reported earlier.[19,22] The insets in the left hand panels show the steady state photolysis of CHD and α-phellandrene in heptane from 0 to 50 s. UV irradiation was applied to the sample in 10 s increments, and the middle spectra emphasizing the growth of the product are the intermediate irradiation times between 0 and 50 s.

calculations.[32] Vertical electronic transitions were calculated using a time-dependent density functional theory (TD-DFT) approach. The 6-311++G(d,p) basis set and the B3LYP density functional were used in all calculations. The transition state structures were found using the Synchronous transit-guided quasi-Newton (STQN) method.[33] Frequency calculations were used to verify that one imaginary (negative) frequency was present at the transition state corresponding to the reactive coordinate.

3 Results

Steady state electronic spectroscopy of α-phellandrene

The solution-phase steady state UV-Vis absorption spectra of α-phellandrene and its photoproducts in heptane for various UV irradiation times are shown in Fig. 2 (inset). The spectrum evolves from a broad unstructured peak with a maximum at 263.2 nm into a more intense and vibronically structured spectrum characteristic of planar conjugated hexatrienes. The vibronic bands peak at 279.0, 267.6, and 260.2 nm. Further irradiation results in continued growth of the triene photoproduct, until at exceptionally long times (>3 min) the peak decreases in intensity as side reactions begin to dominate the photochemistry. Spectra of α-phellandrene and its photoproducts in all solvents used here were identical with only minor (~1–2 nm) shifts in the peak positions.

UV-Vis transient absorption spectroscopy of α-phellandrene

The ultrafast spectral dynamics of α-phellandrene were studied in four solvents of varying polarities and bulk viscosities: heptane, hexadecane, methanol and 1-butanol. All spectra show a coherent signal at $t = 0$ corresponding to two photon absorption of the sample. In contrast to both α-terpinene and DHC,[7,19,31] α-phellandrene shows no excited state absorption (ESA) in the visible region of the spectrum. As for CHD the only transient absorption signals observed are in the UV region of the spectrum. The early time spectra of α-phellandrene show a peak at $\lambda \sim 294$ nm. At longer time delays, the peak shifts to the blue resulting in a difference spectrum consistent with the formation of tZt-DOT in thermal

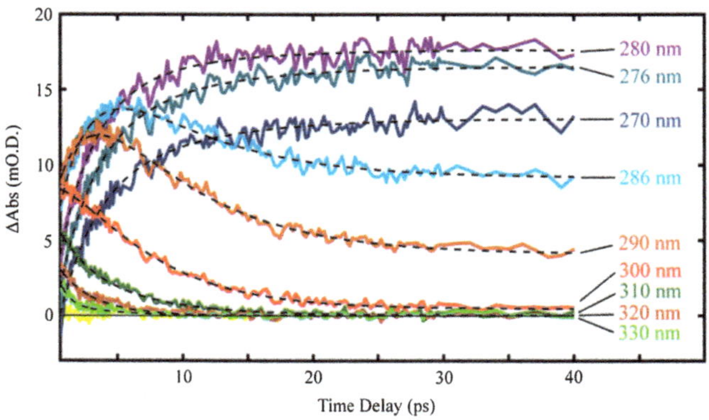

Fig. 3 Kinetic traces across the transient absorption spectrum of α-phellandrene. The yellow curve is the residual for the fits. There is uncertainty in the intensity of the 270 and 276 nm traces due to low white light intensity to the blue of 277 nm.

Table 2 Average fitting parameters for the ultrafast dynamics following excitation of CHD and α-phellandrene

Sample	Solvent	η (mPa s)[a]	τ_1[b]	τ_2
CHD	Heptane	0.387	1.4 ± 0.4	10.1 ± 0.4
	Hexadecane	3.03	1.4 ± 0.55	9.4 ± 0.55
	Methanol	0.544	1.0 ± 0.5	7.8 ± 0.5
	1-Butanol	2.54	1.4 ± 0.55	8.4 ± 0.55
α-Phellandrene	Heptane	0.387	2.5 ± 0.3	9.2 ± 0.3
	Hexadecane	3.03	2.9 ± 0.4	7.7 ± 0.4
	Methanol	0.544	2.8 ± 0.45	6.6 ± 0.45
	1-Butanol	2.54	2.6 ± 0.4	7.5 ± 0.4

[a] Viscosities are reported at 298 K.[34] [b] Average between 280 nm and 290 nm for CHD; between 270 nm and 300 nm for α-phellandrene.

equilibrium with the surrounding solvent within 50–100 ps. Fig. 2 shows the transient absorption spectra in n-heptane and 1-butanol.

Kinetic traces of α-phellandrene in 1-butanol at select wavelengths are shown in Fig. 3. The time domain behaviour was modelled using a biexponential decay with a third constant amplitude to account for the formation of a permanent photoproduct. The necessity for a model consisting of at least two exponential components is clear from the rise and decay seen in the wavelength region of 282–300 nm. For most solvents two exponential components were sufficient to fully characterize the behaviour within the signal to noise of the data.

The two components account for the decay of the initial peak and the rise of the peak absorbing to the blue. The time constant associated with the decay of the initial spectrum is fast relative to the rise component, with a decay constant of

Table 3 First singlet transition energies and the relative ground state free energies of the various conformations of HT and DOT. The ground state free energies for the HT and DOT conformers are given relative to the lowest energy tZt conformations respectively. Entries labelled 'a–b TS' are the transition states between conformer a and b

Conformer	Calculated oscillator strength	Singlet transition (nm)	ΔG (kJ mol^{-1})
CHD	0.1186	268.57	−35.9
gZg-HT	0.1455	292.75	38.9
gZg-cZt TS	—	—	44.3
cZt-HT	0.4186	274.71	16.8
cZt-tZt TS	—	—	27.4
tZt-HT	0.8855	267.41	0
α-Phellandrene	0.0862	272.28	−20.6
gZg-DOT	0.1717	290.13	35.5
gZg-tZg TS	—	—	41.3
gZg-gZt TS	—	—	44.5
tZg	0.4888	283.36	17.3
gZt	0.3906	284.23	12.4
tZg-tZt TS	—	—	28.1
gZt-tZt TS	—	—	21.7
tZt-DOT	0.9755	280.27	0

2.5–3 ps depending on the solvent, while the slower component ranges from 6.5–9 ps. The decay components are free fitting parameters at each wavelength, and the overall decay becomes somewhat faster at the reddest wavelengths. Between 300 and 270 nm the decay components are essentially wavelength independent. The faster decay observed on the red side of the spectrum reflects thermal cooling of the initial hot photoproduct that accompanies conformational relaxation. Table 2 shows the average time constant obtained from the blue portion of the spectrum where the fitting parameters were constant. The data in methanol had significantly better signal to noise, and required a third sub-picosecond time constant to fit the data well. This time constant ranged from 0.6 ± 0.2 ps at $\lambda < 280$ nm to 0.4 ± 0.2 ps at redder wavelengths.

Quantum chemical calculations on α-phellandrene and DOT

Optimized structures for α-phellandrene and the various conformers of its triene photoproduct were obtained at the B3LYP level of theory with the 6-311++G(d,p) basis set. These structures were used for ground and excited state calculations. TD-DFT calculations were used to predict the spectral pattern expected for the formation of an all-*cis* triene photoproduct (gZg-DOT) and the shifts expected from its subsequent relaxation to the steady state photoproduct (tZt-DOT). The purpose of this calculation was to aid in the qualitative assignment of the ultrafast and steady state UV-Vis absorption measurements. Quantitative predictions cannot be expected within the limitations of the TD-DFT method and the level of theory employed. Nevertheless, relatively good agreement is found in the predicted absorption pattern and the observed spectral evolution in the ultrafast and steady state time regimes.

The TD-DFT results predict that the initial gZg-DOT photoproduct will absorb at a longer wavelength than α-phellandrene. Isomerization through the spectrally overlapping gZt- and tZg-DOT conformers is predicted to result in a blue shift relative to gZg-DOT, and further isomerization to the steady state tZt-DOT rotamer should shift the spectrum bluer still. The tZt rotamer is predicted to absorb at the highest frequency of the DOT rotamers considered, but the photochemical conversion to DOT still results in a net red shift of the spectrum. These predictions are completely consistent with the observed spectral shifts in the ultrafast and steady state UV-Vis absorption experiments. The predicted singlet transitions are given in Table 3.

Vibrational frequency calculations were used to predict the barrier height for isomerization between the various conformers of DOT by optimizing to energy minima and transition state geometries. These calculations show that the gZg conformation has the highest energy, while interconversion between the conformers proceeds through two decay channels (see Fig. 1B), resulting in an overall decrease in free energy of $\sim$38 kJ mol^{-1} for tZt relative to gZg. The two pathways have similar, but not identical energy barriers. The results are compiled in Table 3.

UV-Vis transient absorption spectroscopy of CHD

Steady state photolysis and transient absorption spectra of CHD in heptane and 1-butanol are shown in Fig. 2 for comparison with the α-phellandrene measurements. The initial spectrum is a broad peak with a maximum at 286 nm in heptane and 287.5 nm in 1-butanol. At longer time delays, the peak shifts to the blue, and eventually out of the spectral window for our experiments. These results are consistent with those reported earlier, but the quality of the data is much improved.[19,26,27] The temporal behaviour across the spectrum required two exponentials, along with a constant amplitude for the permanent photoproduct, to properly fit the data at each wavelength. A strong wavelength dependence was found for all of the fitting parameters due to the competing processes of conformational relaxation and thermalization from the initial high internal molecular temperature. This wavelength dependent relaxation is consistent with earlier observations.[19,27] Although a complete description of the dynamics requires a more complex model, the spectral evolution can be characterized with a fast rise component and a slower decay component. The values shown in Table 2 are provided for comparison with α-phellandrene and represent the average single wavelength fitting parameters in the 280–290 nm range where a clear rise and decay is seen.

Quantum chemical calculations on CHD and HT

DFT calculations were conducted for CHD and for the gZg, cZt, and tZt rotamers of HT. TD-DFT calculations predict a spectral pattern for the first singlet transitions which are qualitatively consistent with the observed spectral shifts for CHD and the relaxation from gZg-HT to tZt-HT observed in the ultrafast experiments and with earlier calculations.[19,27,35,36] The cZc photoproduct is predicted to absorb to the red of CHD. Conformer interconversion leads to a blue shift with the tZt rotamer having a higher frequency transition than CHD. Isomerization on the ground state proceeds through the cZc → cZt barrier of 5.4 kJ mol^{-1} and the cZt → tZt barrier of 10.5 kJ mol^{-1}.

4 Discussion

Global comparison of excited state dynamics

The influence of steric substitution on the excited state dynamics of the CHD backbone was investigated by Garavelli *et al.* using *ab initio* calculations and transient ionization measurements.[21] It was found that the lifetime of the optically excited state was shortest for CHD and α-phellandrene and markedly longer for α-terpinene. These results led to the conclusion that the difference in the lifetime of the Franck–Condon wavepacket arose because of a mass effect on C1 and C4 (see numbering scheme in Fig. 1), which decreases the momentum in the reactive ethylenic-stretch coordinate for α-terpinene. Our measurements on α-terpinene and DHC are consistent with this mechanism. CHD leaves the Franck–Condon region rapidly, as is well established, resulting in a very low fluorescence quantum yield. Conversely, α-terpinene and DHC both have C1 and C4 substitution, and appear to traverse the Franck–Condon region on a longer time scale (~150 fs for α-terpinene and ~1 ps for DHC) as evidenced by fluorescence measurements and ESA in ultrafast TA experiments.[31]

The UV-visible transient absorption measurements of α-phellandrene provide no clear way to determine the excited state lifetime. While DHC and α-terpinene exhibit a strong excited state absorption in the visible, α-phellandrene exhibits no excited state absorption in the visible region of the spectrum and no clear signature of the excited state in the UV region. Measurements of the fluorescence quantum yield place an upper limit of *ca.* 1 ps on the lifetime of the initially excited state in heptane, but are at present insufficient to define the lifetime with greater precision. The appearance of the photoproduct absorption is consistent with ground state recovery on a time scale <200 fs, but is limited by the time resolution of the measurements. Femtosecond time-resolved fluorescence measurements may be required to determine the excited state lifetime with the necessary accuracy to fully test the model for excited state decay.

Although the excited state lifetime in α-phellandrene is uncertain, the absence of a detectable visible excited state absorption for α-phellandrene and CHD stands in contrast to the very strong visible excited state absorption of α-terpinene and DHC. This result points to a significant influence of substitution on the electronic structure of the cyclohexadiene chromophore. Accurate quantum chemical calculations should be able to elucidate the reasons for this significant difference.

Global comparison of ground state dynamics

The conformational relaxation in HT has been studied extensively as a function of solvent and initial conformation.[19,22,26] The data presented here are in general agreement with the trends observed previously. The initial peak at 286 nm represents the difference spectrum between gZg-HT and CHD. This assignment is supported by the quantum chemical calculations which place the absorption spectrum of the initial photoproduct to the red of the more stable conformations. It is also consistent with earlier measurements that allowed for the formulation of basis spectra corresponding to both gZg-HT and cZt-HT.[19,27,36] The relaxation from gZg-HT to cZt-HT and tZt-HT is faster in alcohols than in alkanes (Table 2 and Fig. 2) although the differences are smaller than deduced from measurements using single-wavelength probes.[19,22] The shoulder in the spectrum around 290 nm, most prominent in the 20 ps spectrum in Fig. 2, is evidence for sequential population of cZt-Ht and tZt-Ht as the hexatriene molecule thermalizes with the surrounding solvent. A small residual peak at 290 nm in the transient difference spectra in alkanes (heptane shown in Fig. 2) is evidence for trapping of a small population of the cZt-HT rotamer as discussed by Harris *et al.*[22]

The rotamer interconversion in DOT is qualitatively different than for HT, although it occurs on a similar time scale. The interconversion proceeds slightly faster in alcohols than in alkanes (Table 2). The differences are not large however. The initial peak in the TA spectra at 294 nm is assigned to the difference spectrum between gZg-DOT and α-phellandrene. The shift of the spectrum to shorter wavelengths is consistent with the spectral changes predicted by the TD-DFT calculations for isomerization to the tZt-DOT rotamer.

Despite the steric groups and comparable barriers for conformational isomerization, the overall relaxation of DOT is faster than the relaxation of HT. While one might expect that solvent interactions with the steric groups on DOT should lead to substantially more trapping of intermediate conformations, this is not observed. Notably, there is no spectral evidence for trapping of any residual gZt-DOT or tZg-DOT intermediate on short time scales or on longer time scales following excitation of α-phellandrene. Comparison of the 20 ps difference spectra in Fig. 2 (red lines) highlights these differences. The shoulder at 290 nm requires substantial persistent population of the cZt-HT conformer. In contrast the DOT spectra show no evidence for distinct population of the tZg or gZt conformations. The data for DOT shows a smooth transition from the initial photoproduct at 0.6 ps to the relaxed product at 50 ps. The calculations reported in Table 3 provide no clear explanation for the observed differences. The barriers for conformational isomerization are comparable for HT and DOT (~10 kJ mol^{-1}), and the oscillator strengths of the gZt/tZg-DOT and cZt-HT conformations are similar suggesting that gZt/tZg-DOT should be apparent in the spectrum if populated at a significant level.

An earlier study by Harris *et al.* investigated the cZt-HT → tZt-HT reaction as a function of solvent using excitation of tZt-HT to produce a small residual population of cZt-HT.[22] The decay of this trapped population was measured as a function of temperature and a barrier and Arrhenius prefactor for isomerization deduced. The barrier is higher than calculated and is dependent on the nature of the solvent, 17.4 kJ mol^{-1} in alcohols and 23.5 kJ mol^{-1} in alkanes. The data obtained for the isomerization of cZt-HT is not easily explained by simple models for condensed phase reaction dynamics.[22] The current study of DOT again points to an anomaly in the behaviour of cZt-HT. The smooth relaxation of DOT and the absence of trapped population in the gZt-DOT or tZg-DOT conformations are consistent with the calculated barriers and with simple theories for condensed phase reaction dynamics. More sophisticated molecular dynamics simulations are in progress to try to explain the influence of solvent on the isomerization dynamics of these simple polyenes in solution.

5 Conclusion

The aim of this study was to characterize the excited and ground state dynamics of α-phellandrene in solution and present this data in the broader context of the CHD chromophore and HT backbone dynamics. Quantum chemical calculations were used to predict the shifts in the UV-Vis spectra of the photoproducts to help in assignment of ultrafast TA and steady state photolysis measurements. The conformational relaxation of the DOT photoproduct of α-phellandrene was studied in multiple solvents with varying viscosity and hydrogen bonding capacity using UV transient absorption measurements.

Unlike the 1,4 substituted dienes, α-terpinene and DHC, no visible excited state absorption was detected in TA experiments conducted on α-phellandrene or CHD. Previous studies have attributed the longer lifetime of the Franck–Condon region of electronically excited α-terpinene to a mass effect due to the 1,4 substitution pattern.[21] Our observations are consistent with this mechanism. Rapid relaxation out of the Franck–Condon region and into a dark state in α-phellandrene could explain the lack of ESA in the ultrafast experiments compared with α-terpinene. It is also possible that no resonant state in the visible probe region exists or that a transition to the state is symmetry forbidden. Future work will seek to understand how the substitution on the HT backbone modifies the higher electronic states.

The ground state conformational relaxation of IPMHT and Pre formed following excitation of α-terpinene and DHC respectively is confined to the gZg → tZg. The gZg-IPMHT → tZg-IPMHT transition occurs on a time scale independent of bulk solvent properties like viscosity and polarity.[31] Similarly, the gZg → tZg isomerization of Pre shows a very weak dependence on solvent despite its size and the presence of a hydroxyl group.[31] For both of these compounds, the conformational relaxation occurs on a *ca.* 5–8 ps timescale. In contrast HT and DOT approach an equilibrium population dominated by the tZt conformer, with the relaxation gZg → tZg → tZt on a *ca.* 6–10 ps timescale. The overall relaxation is somewhat faster in alcohols compared with alkanes. The similarity in the relaxation timescale is surprising given the difference in size of the four trienes studied as well as the difference in the conformational minima available to the molecules. The contrast between the relaxation of cZt-HT and gZt/tZg-DOT again points to an anomaly in the behaviour of hexatriene.

Future studies will use molecular dynamics simulations accompanied by more extensive transient absorption measurements to explain the observed dynamics. Tunability of the pump wavelength and improvement in time resolution will allow for a more detailed examination of the rapid excited state decay of α-phellandrene and α-terpinene, while a deeper UV continuum probe will provide a more complete view of the photoproduct formation. Moving to the deep UV presents many experimental challenges, but will be necessary to enhance understanding of the conformational relaxation of the triene photoproducts studied here.

Acknowledgements

This work has been supported by the National Science Foundation through Grant No. CHE-0718219 and CHE-1150660.

References

1 D. Geppert, L. Seyfarth and R. de Vivie-Riedle, *Appl. Phys. B: Lasers Opt.*, 2004, **79**, 987–992.
2 N. L. Rangel, K. S. Williams and J. M. Seminario, *J. Phys. Chem. A*, 2009, **113**, 6740–6744.
3 J. Conyard, K. Addison, I. A. Heisler, A. Cnossen, W. R. Browne, B. L. Feringa and S. R. Meech, *Nat. Chem.*, 2012, **4**, 547–551.
4 J. B. Wang and B. L. Feringa, *Science*, 2011, **331**, 1429–1432.
5 N. A. Anderson, J. J. Shiang and R. J. Sension, *J. Phys. Chem. A*, 1999, **103**, 10730–10736.
6 W. Fuss, T. Hofer, P. Hering, K. L. Kompa, S. Lochbrunner, T. Schikarski and W. E. Schmid, *J. Phys. Chem.*, 1996, **100**, 921–927.
7 K.-C. Tang, A. Rury, M. B. Orozco, J. Egendorf, K. G. Spears and R. J. Sension, *J. Chem. Phys.*, 2011, **134**, 104503.
8 S. Deb and P. M. Weber, *Annu. Rev. Phys. Chem.*, 2011, **62**, 19–39.
9 R. B. Woodward and R. Hoffman, *The Conservation of Orbital Symmetry*, VCH, Weinheim, 1970.
10 R. B. Woodward and R. Hoffmann, *Angew. Chem., Int. Ed. Engl.*, 1969, **8**, 781–853.
11 W. T. A. M. van der Lugt and L. J. Oosterhoff, *Chem. Commun.*, 1968, 1235–1236.
12 W. T. A. M. Van der Lugt and L. J. Oosterhoff, *J. Am. Chem. Soc.*, 1969, **91**, 6042–6049.
13 A. Hofmann and R. de Vivie-Riedle, *J. Chem. Phys.*, 2000, **112**, 5054–5059.
14 A. Nenov, P. Kolle, M. A. Robb and R. de Vivie-Riedle, *J. Org. Chem.*, 2010, **75**, 123–129.
15 J. B. Schonborn, J. Sielk and B. Hartke, *J. Phys. Chem. A*, 2010, **114**, 4036–4044.
16 E. Tapavicza, A. M. Meyer and F. Furche, *Phys. Chem. Chem. Phys.*, 2011, **13**, 20986–20998.
17 P. J. Reid, S. J. Doig, S. D. Wickham and R. A. Mathies, *J. Am. Chem. Soc.*, 1993, **115**, 4754–4763.
18 M. O. Trulson, G. D. Dollinger and R. A. Mathies, *J. Chem. Phys.*, 1989, **90**, 4274–4281.
19 N. A. Anderson, S. H. Pullen, L. A. Walker, J. J. Shiang and R. J. Sension, *J. Phys. Chem. A*, 1998, **102**, 10588–10598.
20 E. C. Carroll, B. J. Pearson, A. C. Florean, P. H. Bucksbaum and R. J. Sension, *J. Chem. Phys.*, 2006, **124**, 114506.
21 M. Garavelli, C. S. Page, P. Celani, M. Olivucci, W. E. Schmid, S. A. Trushin and W. Fuss, *J. Phys. Chem. A*, 2001, **105**, 4458–4469.
22 D. A. Harris, M. B. Orozco and R. J. Sension, *J. Phys. Chem. A*, 2006, **110**, 9325–9333.

23 E. C. Carroll, J. L. White, A. C. Florean, P. H. Bucksbaum and R. J. Sension, *J. Phys. Chem. A*, 2008, **112**, 6811–6822.
24 J. Kim, H. L. Tao, J. L. White, V. S. Petrovic, T. J. Martinez and P. H. Bucksbaum, *J. Phys. Chem. A*, 2012, **116**, 2758–2763.
25 K. Kosma, S. A. Trushin, W. Fuss and W. E. Schmid, *Phys. Chem. Chem. Phys.*, 2009, **11**, 172–181.
26 S. Lochbrunner, W. Fuss, W. E. Schmid and K. L. Kompa, *J. Phys. Chem. A*, 1998, **102**, 9334–9344.
27 S. H. Pullen, N. A. Anderson, L. A. Walker and R. J. Sension, *J. Chem. Phys.*, 1998, **108**, 556–563.
28 J. Meyer-Ilse, D. Akimov and B. Dietzek, *J. Phys. Chem. Lett.*, 2012, **3**, 182–185.
29 K. M. Marzec, I. Reva, R. Fausto, K. Malek and L. M. Proniewicz, *J. Phys. Chem. A*, 2010, **114**, 5526–5536.
30 K. M. Marzec, I. Reva, R. Fausto and L. M. Proniewicz, *J. Phys. Chem. A*, 2011, **115**, 4342–4353.
31 B. Arruda, J. Peng, B. Smith, K. G. Spears and R. J. Sension, *J. Phys.Chem. B*, submitted.
32 M. J. Frisch, G. W. Trucks, H. B. Schlegel, G. E. Scuseria, M. A. Robb*et al.*, *Gaussian 03, Revision C.02.*
33 C. Y. Peng and H. B. Schlegel, *Isr. J. Chem.*, 1993, **33**, 449–454.
34 in *CRC Handbook of Chemistry and Physics*, ed. W. M. Haynes, CRC Press/Taylor and Francis, Boca Raton, FL.
35 S. Pullen, L. A. Walker, B. Donovan and R. J. Sension, *Chem. Phys. Lett.*, 1995, **242**, 415–420.
36 S. H. Pullen, N. A. Anderson, L. A. Walker and R. J. Sension, *J. Chem. Phys.*, 1997, **107**, 4985–4993.

Faraday Discussions RSC Publishing

PAPER

Ultrafast photo-initiated molecular quantum dynamics in the DNA dinucleotide d(ApG) revealed by broadband transient absorption spectroscopy

Mayra C. Stuhldreier and Friedrich Temps*

Received 8th January 2013, Accepted 16th January 2013
DOI: 10.1039/c3fd00003f

The ultrafast photo-initiated quantum dynamics of the adenine–guanine dinucleotide d(ApG) in aqueous solution (pH 7) has been studied by femtosecond time-resolved spectroscopy after excitation at $\lambda = 260$ nm. The results reveal a hierarchy of processes on time scales from $\tau < 100$ fs to $\tau > 100$ ps. Characteristic spectro-temporal signatures are observed indicating the transformation of the molecules in the electronic relaxation from the photo-excited state to a long-lived exciplex. In particular, broadband UV/VIS excited-state absorption (ESA) measurements detected a distinctive absorption by the excited dinucleotide around $\lambda = 335$ nm, $\approx$0.5 eV to the blue compared to the maximum of the broad and unstructured ESA spectrum after excitation of an equimolar mixture of the mononucleotides dAMP and dGMP. A similar feature has been identified as signature of the excimer in the dynamics of the adenine dinucleotide d(ApA). The lifetime of the d(ApG) exciplex was found to be $\tau = 124 \pm 4$ ps both from the ESA decay time and from the ground-state recovery time, far longer than the sub-picosecond lifetimes of excited dAMP or dGMP. Fluorescence–time profiles measured by the up-conversion technique indicate that the exciplex state is reached around $\approx$6 ps after excitation. Very weak residual fluorescence at longer times red-shifted to the emission from the photo-excited state shows that the exciplex is almost optically dark, but still has enough oscillator strength to give rise to the dual fluorescence of the dinucleotide in the static fluorescence spectrum.

1 Introduction

The understanding of photophysical and photochemical properties of the DNA and its building blocks is of fundamental scientific interest because of the sensitivity of the genetic information to UV light-induced chemical reactions which may result in mutagenic and potentially lethal lesions.[1-3] Stimulated by the huge advances in ultrafast time-resolved spectroscopy, the lifetimes and

Institute of Physical Chemistry, Christian-Albrechts-University Kiel, Olshausenstr. 40, D-24098 Kiel, Germany. E-mail: temps@phc.uni-kiel.de; Web: http://www.uni-kiel.de/phc/temps/; Fax: +49 880 7802; Tel: +49 880 7801

radiationless deactivation dynamics of the excited electronic states of the DNA bases and their nucleosides or nucleotides, isolated base pairs, and single- and double-stranded oligonucleotides have been measured in the last ten to fifteen years with the help of femtosecond pump–probe spectroscopy.[3,4] Following the measurement of highly efficient, ultrafast electronic deactivation processes in the free nucleobases on the sub-picosecond time scale,[3,5–7] experiments on DNA model strands revealed that the excited-state lifetimes in these large base assemblies can be some surprising three to four orders of magnitude longer than for the base monomers.[8] These striking observations created substantial controversy, whether the long-lived excited states of oligonucleotides originate from the formation of excimers or exciplexes[9] with (partial) charge transfer (CT) character between two neighbouring π-stacked bases of a strand[10] or whether they should be understood as (Frenkel) excitons delocalised over several bases.[10,11] Soon afterwards, the peculiar photo-initiated quantum molecular dynamics of the oligonucleotides turned out as the rule rather than exception.[4,12–18] Furthermore, first quantum chemical results became available for model systems consisting of two or more nucleobases with DNA-like orientation.[19–28] Both from the experimental and theoretical point of view, however, oligonucleotides with more than two bases are still too large for complete understanding of the excited-state structures and dynamics. Further, experiments on long oligonucleotides are hampered by structural inhomogeneities washing out more specific information that might be hidden in excited-state spectra.

In order to elucidate the photo-initiated quantum molecular dynamics in DNA oligonucleotides in more detail, we initiated a series of investigations by femtosecond time-resolved fluorescence and absorption spectroscopy of the ultrafast electronic relaxations of selected short di-, tri- and tetranucleotides.[29] These fragments are the smallest model systems showing the effect of π-stacking interactions between adjacent bases on the nature and dynamics of the excited electronic states. Although the molecules are dynamically disordered, circular dichroism (CD) and nuclear magnetic resonance (NMR) studies confirmed high fractions of stacked (B-DNA like) configurations.[30–39] Further, early static emission spectra suggested the formation of excimer/exciplex-like states upon UV excitation of stacked conformations for several dinucleotides.[1,40–46] Nevertheless, there have yet been few time-resolved experimental studies on dinucleotides.[47–50] Buchvarov *et al.* reported excited-state absorption (ESA) spectra for a series of $d(A)_n$ oligomers ($2 \leq n \leq 18$) and attributed the >100 ps lifetimes to delocalised, electronically relaxed excitons.[47] Takaya *et al.* determined electronic relaxation rates for several ribose dinucleotides after 267 nm excitation from measurements of the ground-state recovery (GSR) times at $\lambda = 252$ nm and identified the long-lived states as excimers/exciplexes.[48] More recently, Su *et al.* analysed the GSR time profiles of adenosine and 2′-deoxyadenosine oligonucleotides $(A)_n$ and $d(A)_n$ as function of length and correlated the quantum yields for excimer formation with the fractions of stacked bases in the strands.[49] In our laboratory, we investigated the fluorescence decay dynamics of the adenine dinucleotide.[50]

In this paper, we report on a comprehensive study of the ultrafast excited state dynamics of the adenine–guanine dinucleotide d(ApG) (see Fig. 1) in aqueous solution at pH 7 by femtosecond UV/VIS fluorescence up-conversion, broadband UV/VIS absorption and single-colour deep-UV absorption measurements.

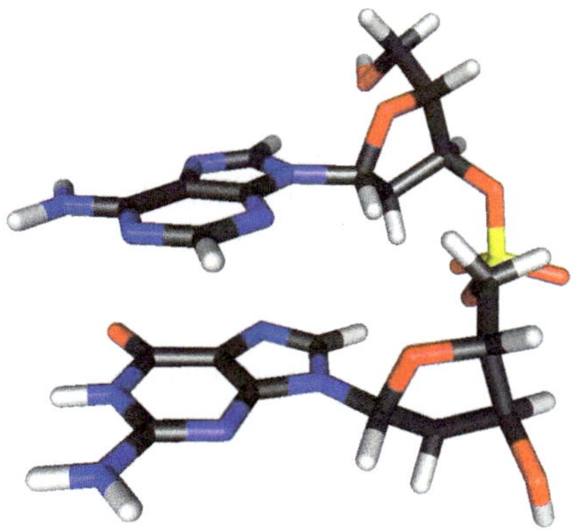

Fig. 1 Structure of d(ApG) in B-DNA conformation.

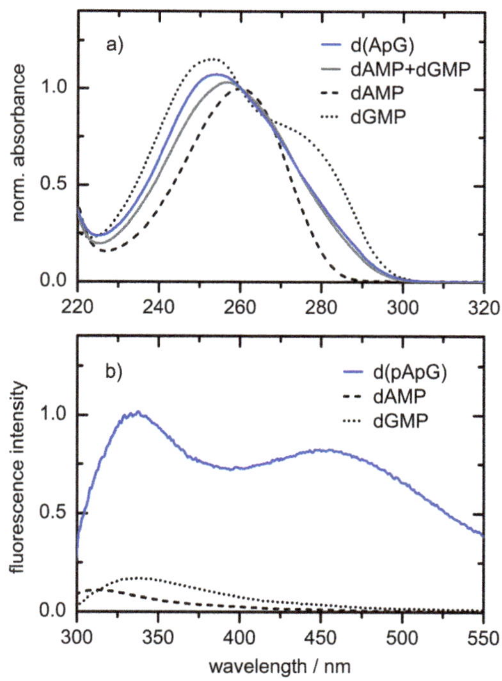

Fig. 2 a) UV absorption spectra of d(ApG), dAMP, dGMP and an equimolar mixture of dAMP and dGMP normalized at $\lambda = 260$ nm for better comparison of the band shapes. Relative to the mixture of dAMP and dGMP, the unnormalized spectrum of d(ApG) shows a normal hypochromic effect of $\approx 7\%$. b) Static fluorescence spectra of d(pApG), dAMP and dGMP after 260 nm excitation.[53] As noted in the text (*cf.* Experimental section), the extra phosphate of d(pApG) compared to d(ApG) does not alter the relevant dual fluorescence properties of the dinucleotide.[1,40–46]

In contrast to the extremely weak fluorescence in the UV region from the respective mononucleotides, d(ApG) exhibits a broad, red-shifted second emission maximum around $\lambda = 460$ nm typical for the alleged 'excimers' in addition to 'monomer-like' emission in the UV (*cf.* Fig. 2). Similar spectra have been observed for the adenine dinucleotide d(ApA), where the excimer state revealed itself by its exceptionally long (≈300–400 ps) fluorescence lifetime. The selection of the mixed dinucleotide d(ApG) for investigation was further motivated by the observed large effects on the fluorescence lifetimes upon substitution of an increasing number of adenines in a $d(A)_{20}$ strand by guanines.[12] The performed time-resolved measurements on d(ApG) are compared to data for the mononucleotides deoxyadenosine monophosphate (dAMP) and deoxyguanosine monophospate (dGMP) and to the recently obtained detailed picture for the excited-state dynamics of d(ApA).[29] The results reveal clear spectro-temporal signatures of the elusive excimers in the excited-state absorption spectra that allow us to assess their fast formation and subsequent slow relaxation.

2 Experimental

The femtosecond fluorescence and absorption experiments have been described in some detail elsewhere.[51,52] The required UV pump and probe pulses were derived from the output of a regeneratively amplified Ti:Sa laser system (Clark-MXR CPA2001, 775 nm, 150 fs, 1 mJ, 1 kHz) in home-built non-collinear optical parametric amplifiers with frequency doubling crystals. Broadband supercontinuum probe pulses (320–700 nm) were generated from a small fraction of the 775 nm beam in a CaF_2 plate. The investigated nucleotides were purchased from biomers.net (5′-d(ApG)-3′) and Sigma-Aldrich (dAMP, dGMP) and dissolved in phosphate buffered ultrapure water (16 mM NaH_2PO_4, 34 mM Na_2HPO_4, 117 mM NaCl).

All time-resolved experiments were carried out in flow cells with 0.2 mm fused silica windows. The excitation wavelength was $\lambda = 260$ nm with magic-angle polarisation in all cases. Fluorescence up-conversion measurements were performed on sample solutions of ≈0.2 mM concentration in 1 mm flow cells. The decay curves after photoexcitation were measured at three selected wavelengths ($\lambda_{FL} = 310$, 350 and 430 nm). The 775 nm gate pulses of the experiment limited the time resolution after deconvolution to $\Delta t \approx 150$ fs. Better time resolution ($\Delta t \approx 50$ fs) was reached by the transient absorption experiments which were performed in flow cells of 0.1 mm optical pathlength using sample concentrations of ≈3–10 mM. The two-dimensional (2D) transient spectro-temporal absorption maps were recorded for delay times up to $\Delta t = 1.3$ ns after excitation. The short optical pathlength and higher sample concentrations resulted in smaller than usual coherent artifacts from cross-phase modulation (XPM) and stimulated Raman scattering (SRS) and very weak signal contributions from solvated electrons generated by multi-photon absorption of the pump pulses in water which were readily subtracted as described.[29,52] GSR time profiles were followed at $\lambda = 244$ nm up to a delay time of $\Delta t = 600$ ps. To ensure reproducibility, all measurements were repeated several times. Error limits given below correspond to the 2σ standard deviation.

UV absorption spectra were taken on a Shimadzu UV-2401 desktop spectrometer. Fluorescence spectra of the mononucleotides dAMP, dGMP and dinucleotide d(pApG) (Midland Certified Reagents) were measured with a Jobin-Yvon FluoroMax-4 spectrometer. The additional phosphate group at the 5′-end of d(pApG) compared to d(ApG) is not expected to have any significant effect here.[1,40–46]

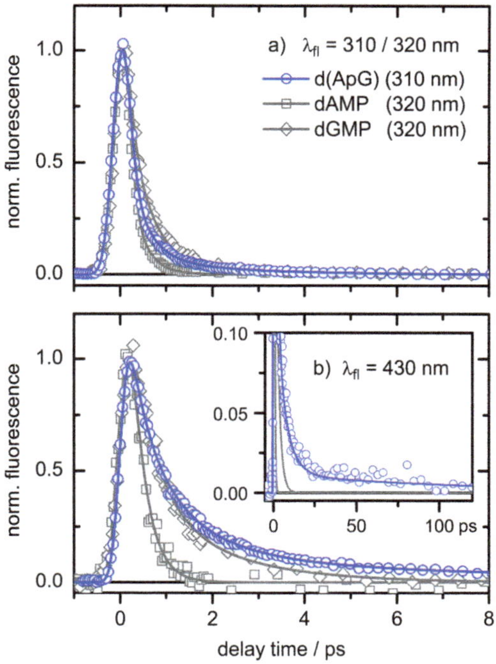

Fig. 3 Fluorescence–time profiles of d(ApG), dAMP and dGMP at a) $\lambda_{fl} = 310$ nm (320 nm for dAMP and dGMP) and at b) $\lambda_{fl} = 430$ nm after excitation at $\lambda = 260$ nm. Open symbols are data points, solid lines represent the best-fit decay curves. The inset in b) shows the data for d(ApG) on an expanded time scale.

3 Results

3.1 Steady-state spectra

The static absorption and fluorescence spectra of the dinucleotide and the two mononucleotides are given in Fig. 2. The absorption of d(ApG) matches that of an equimolar mixture of dAMP and dGMP in the red wing, but is slightly stronger towards the blue (Fig. 2a). The absorption maximum of the dinucleotide is blue-shifted by ≈2.5 nm. The fluorescence spectra, on the other hand, differ strongly both in intensity and in shape (Fig. 2b). As can be seen, the dinucleotide exhibits the aforementioned 'dual fluorescence' with a roughly 0.55 : 0.45 intensity ratio between the 'monomer-like' emission in the UV and the 'excimer-like' emission in the VIS region.

3.2 Transient fluorescence measurements

Fig. 3 displays the observed fluorescence-time profiles for d(ApG) at two wavelengths, $\lambda_{FL} = 310$ nm and $\lambda_{FL} = 430$ nm. The data could be nicely modeled using a sum of up to four exponential decay functions. The time constants increased gradually with the emission wavelength. The fluorescence decay curves for dAMP and dGMP under the same experimental conditions required at most two exponentials for fitting.

In the UV ($\lambda_{FL} = 310$ nm), the emissions by the dinucleotide and the two monomers differ little (Fig. 3a). The fluorescence decays in the sub-picosecond range in all cases. The obtained best-fit lifetimes for d(ApG) are (amplitudes in parentheses)

$$\tau_{FL1} \leq 0.10 \text{ ps},$$

$$\tau_{FL2} = 0.41 \pm 0.05 \text{ ps } (90\%),$$

$$\tau_{FL3} = 2.10 \pm 0.50 \text{ ps } (10\%).$$

The monomer decay curves (measured at $\lambda_{FL} = 320$ nm) are bi-exponential and described by lifetimes of $\tau_{FL1} \leq 0.10$ ps, $\tau_{FL2} = 0.34 \pm 0.11$ ps for dAMP and $\tau_{FL1} = 0.12 \pm 0.07$ ps, $\tau_{FL2} = 0.68 \pm 0.11$ ps for dGMP.

The time profiles for d(ApG) at $\lambda_{FL} = 350$ were similar to those at the shorter wavelength except for the virtual absence of the ≤0.10 ps component. At $\lambda_{FL} = 430$ nm, in contrast, the dinucleotide was observed to emit for much longer times than the nucleotide monomers (Fig. 3b, inset). The fitted d(ApG) fluorescence lifetimes in this 'excimer' emission region[40–46] are

$$\tau'_{FL2} = 0.46 \pm 0.01 \text{ ps } (46\%),$$

$$\tau'_{FL3} = 1.20 \pm 0.30 \text{ ps } (43\%),$$

$$\tau'_{FL4} = 5.90 \pm 1.40 \text{ ps } (10\%),$$

$$\tau'_{FL5} = 118 \pm 60 \text{ ps } (1\%),$$

without ultrashort (≤0.10 ps) component. Under the same conditions, the monomers feature $\tau'_{FL2} = 0.32 \pm 0.07$ ps (dAMP) and $\tau'_{FL2} = 0.53 \pm 0.16$ ps, $\tau'_{FL3} = 1.8 \pm 0.8$ ps (dGMP). The data for GMP could also be approximated by a single

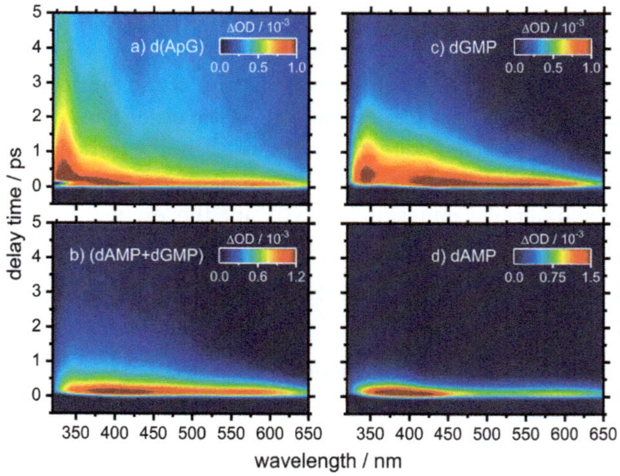

Fig. 4 Two-dimensional spectro-temporal transient absorption maps after 260 nm photoexcitation for a) d(ApG) ($c = 3.2$ mM), b) an equimolar mixture of dAMP and dGMP ($c = 4.8$ mM for each monomer), c) dGMP ($c = 9.1$ mM) and d) dAMP ($c = 9.7$ mM) for delay times up to $\Delta t = 5$ ps. Note the different concentrations of the dinucleotide and the mononucleotides and the different intensity scales (colour codes) in the respective panels.

exponential with $\tau'_{FL2} = 0.86 \pm 0.10$ ps. These lifetimes for the monomers are in good agreement with available literature data.[6,54,55] For d(ApG), on the other hand, the slow (though not very precisely determined) component with $\tau'_{FL5} = 118$ ps hints at the existence of a long-lived excited state, but this state appears to be only barely observable in fluorescence.

3.3 Transient absorption measurements

The measured two-dimensional spectro-temporal transient absorption maps after excitation of d(ApG), dAMP, dGMP and an equimolar mixture of the two monomer nucleotides are displayed for delay times up to $\Delta t = 5$ ps in Fig. 4.

Considering first the two monomers (Fig. 4c–d), dAMP has an intense ESA band with a maximum around $\approx$380 nm for $\Delta t \lesssim 0.5$ ps,[3,5,56] whereas dGMP features two ESA bands with maxima at $\approx$340 nm and $\approx$450 nm that decay within $\Delta t \approx 2$ ps after excitation,[57] slightly more slowly than for dAMP. The sum spectrum for both monomers therefore shows broad ESA over the entire wavelength range with no clear maximum at all and an effective lifetime of 1–2 ps. The measured map for the equimolar dAMP/dGMP mixture in Fig. 4b indeed equals the weighted sum of the two monomer maps.

The 2D transient absorption map for the dinucleotide in Fig. 4a differs strongly from the other three. First, it is the only one with a negative feature near time zero at $\lambda \approx 325$ nm from stimulated emission (SE). Second, the broad positive ESA extending over the entire wavelength range decays mainly within the first 4 ps after excitation, but residues of this ESA are observable well up to $\Delta t \approx 300$ ps

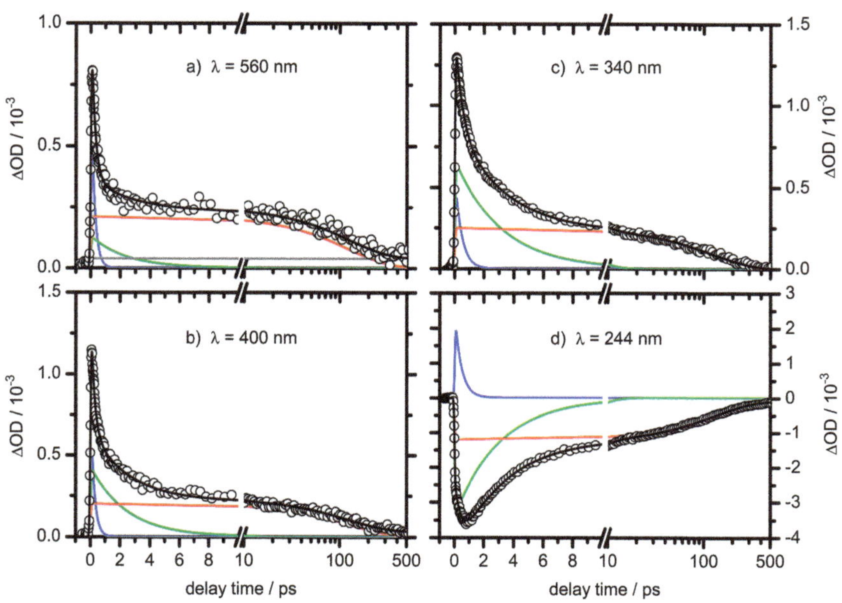

Fig. 5 Transient absorption–time profiles of d(ApG) after 260 nm excitation at a) $\lambda = 560$ nm (ESA), b) $\lambda = 400$ nm (ESA), c) $\lambda = 340$ nm (ESA) and d) $\lambda = 244$ nm (GSR). Open circles show data, solid lines the overall fits (black) and their components (coloured). The small constant offset in the fit at $\lambda = 560$ nm is attributed to the solvated electron absorption. Note the linear time scale from −1 to 10 ps and the logarithmic scale after 10 ps.

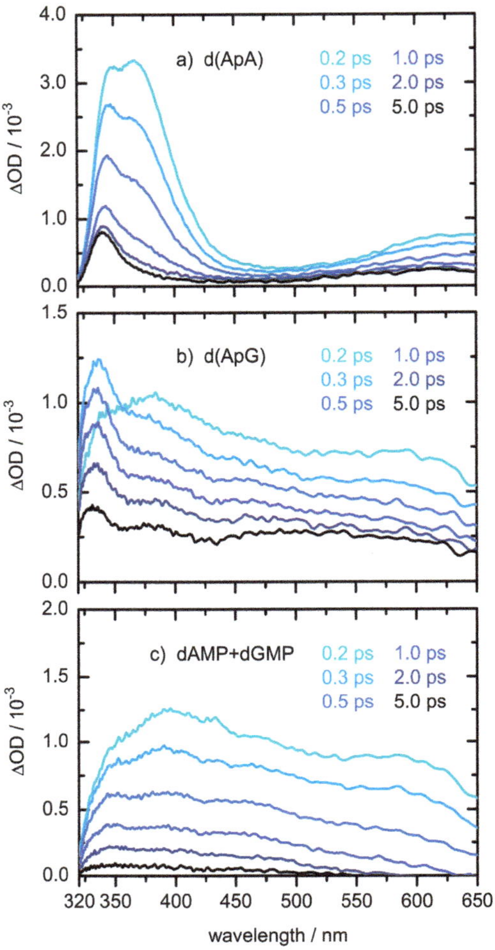

Fig. 6 Transient excited-state absorption spectra at selected delay times after excitation for a) d(ApA),[29] b) d(ApG) and c) an equimolar mixture of dAMP and dGMP.

after excitation (see Fig. 5 below). Moreover, the ESA appears to be weakly structured. A remarkable feature is the distinctive narrow absorption band with a maximum at $\lambda \approx$ 330–340 nm, which remains distinguishable from the broad background for $\approx$300 ps (Fig. 6 below).

To determine the excited-state lifetimes for d(ApG), time profiles were extracted from the 2D maps at 15 wavelengths distributed across the spectrum from λ = 330–680 nm and analysed by global fitting with up to three decaying exponentials. Three representative data curves and their fits are given in Fig. 5a–c. From the analysis, the d(ApG) decay times (and amplitudes depending on wavelength in parentheses) were found to be

$$\tau_{\text{ESA1}} = 0.24 \pm 0.01 \text{ ps } (40\text{–}80\%),$$

$$\tau_{\text{ESA2}} = 2.44 \pm 0.14 \text{ ps } (45\text{–}10\%),$$

$$\tau_{\text{ESA3}} = 124 \pm 4 \text{ ps } (\approx 15\%).$$

Also shown in the ESA decay diagram is the time profile of the ground state recovery (GSR) followed at $\lambda = 244$ nm (Fig. 5d). A similar fit using three exponentials yielded a lifetime constant of $\tau_{GSR1} = 0.45 \pm 0.02$ ps describing the fast initial evolution of the experimental signal likely due to dynamics of the excited state and two time constants of

$$\tau_{GSR2} = 3.10 \pm 0.10 \text{ ps } (74\%),$$

$$\tau_{GSR3} = 124 \pm 4 \text{ ps } (26\%)$$

representing the recovery of the electronic ground state (GSR) as evident by the refilling of the ground state bleach (GSB) signal appearing immediately after excitation. Judged from the relative amplitudes of τ_{GSR2} and τ_{GSR3}, the fraction of long-lived excited states is at least about 26%, but may be as high as $\approx$35% if the amplitude of the fast positive component (τ_{GSR1}) is taken as a measure for longer-lived ESA obscured by the strong GSB at the detection wavelength.

Within experimental errors, the electronic relaxation of d(ApG) seems to be practically complete within $\leq$500 ps after excitation.

4 Discussion

The comprehensive results above demonstrate the advantages of approaching the complex excited-state dynamics in DNA building blocks strategically by several complementary experimental techniques at the same time. Namely, the time-resolved fluorescence measurements reflect mainly the properties of the photoexcited state(s) in the Franck–Condon (FC) region, the departure of the wavepacket from the FC window and the fate of surviving optically bright states at later stages in time. The transient spectro-temporal UV/VIS absorption maps probe the excited-state dynamics as a function of wavelength for longer times *via* observable $S_1 \rightarrow S_n$ ESA even when the $S_1 \rightarrow S_0$ emission turns optically dark. Last but not least, the GSR measurements in the deep-UV check for unknown long-lived optically dark states and showcase the eventual recovery of the electronic ground state. The experiments revealed a hierarchy of processes happening on time scales ranging from $\tau < 100$ fs *via* $\tau = 0.2/0.4$–2.5 ps *via* $\tau \approx 6$–10 ps to $\tau > 100$ ps. The complementary data thus allow for an in-depth assessment of the distinctive radiationless electronic decay mechanisms of the d(ApG) dinucleotide as opposed to the mononucleotides dAMP and dGMP and in comparison to the dynamics of the adenine dinucleotide d(ApA).

4.1 dAMP and dGMP

As already noted, the measured excited-state lifetimes for dAMP and dGMP agree well with available literature data.[3,5,6,54–57] Regarding the ensuing mechanisms, we can discuss only a few selected references next to some overview articles.[4,58–60]

For dAMP,[61–66] consensus appears to have emerged that the ultrashort lifetime of $\tau_1 \leq 0.1$ ps seen in fluorescence is associated with the rapid departure of the excited wavepacket from the FC region of the photoexcited $\pi\pi^*(L_a)$ state, which actually corresponds to the second (S_2) or third excited state (S_3) at the ground-state equilibrium configuration in solution and carries the bulk of the oscillator strength from S_0. Moving through a region, where the $S_3(\pi\pi^*)$, $S_2(\pi\pi^*)$ and $S_1(n\pi^*)$ states are strongly mixed, the wavepacket can reach the S_0 state *via* conical intersections (CIs) of the excited-state and ground-state potential energy hypersurfaces (PEHSs) along two main pathways, which are still intensely debated.[67] One route involves ring puckering at the C^2 atom, the other a distortion of the pyrimidine ring at the C^6 position and NH_2 out-of-plane motion.

Three pathways have been proposed for the non-adiabatic dynamics of dGMP, ring distortion, out-of-plane displacement of the NH_2 group and out-of-plane motion of the CO group.[68–74] dGMP has two spectrally resolved bright $\pi\pi^*$ transitions. The observed biphasic decay dynamics have been assigned to ultrafast $\pi\pi^*(L_b \rightarrow L_a)$ relaxation and subsequent slower motion to the CIs with S_0 across a plateau-like region on L_a.

4.2 Franck–Condon excited-states of d(ApG)

Compared to a single A or G, the complexity of the excited state structure in a homo- or heterodimer such as d(ApA) or d(ApG) increases exponentially. Each purine base has two accessible $\pi\pi^*$ states (L_a and L_b). For a perfectly stacked homodimer with face-to-face orientation of the two chromophores, an exciton model predicts two allowed and two forbidden $\pi\pi^*$-type transitions. Helical distortion in the B-DNA like conformation and static and dynamic disorder lift this simple picture. Nevertheless, the coupling of the transition dipole moments explains the blue-shift of the UV absorption maximum and the hypochromicity of di- and oligonucleotides compared to their monomers. However, the picture is further complicated by the presence of at least one $n\pi^*$ state per monomer in the vicinity of the first $\pi\pi^*$ states and by the possibility of CT transitions. The latter connect, *e.g.*, the highest occupied molecular orbital (HOMO) of one monomer to the lowest unoccupied molecular orbital (LUMO) of the other.

In d(ApA) and longer dA_n strands, the first CT states most likely lie at significantly higher energies than the optically bright $\pi\pi^*$-type states in the FC range.[19,22,23,27] An ultrashort (<50 fs) pump pulse tuned to the first UV absorption band is expected to prepare a neutral exciton delocalised over both bases of the stacked dimer (or delocalised over a few bases in a longer strand). The subsequent temporal evolution of the photoexcited exciton towards the relaxed, energetically stabilised and therefore long-lived excimer involves fast intraband scattering followed by large-amplitude motions of the two adenines forming a neutral excimer (see below).

In heterodimers like d(ApG) and mixed-sequence strands in the FC range, the exciton concept is less useful unless the coupling strength reaches and exceeds the splitting between the excited states of the two chromophores. Using time-dependent density functional theory (TD-DFT) including solvent effects by the polarisable continuum model (PCM), Santoro *et al.* predicted that four of the first ten excited states of d(ApG) in the FC region carry most of the oscillator strength.[75] The lowest CT states were found to be strongly mixed and close in energy with the bright dimer excited states. With the 6-31G(d) basis set, the S_1 and S_7 transitions were identified as $\pi\pi^*(G_{L_a})$ and $\pi\pi^*(G_{L_b})$, respectively, while S_2 and S_4 were of mixed $\pi\pi^*(A_{L_a}/G \rightarrow A_{CT})$ character. S_3, S_5 and S_8 were of $n\pi^*$ character, S_6 was the $\pi\pi^*(A_{L_b})$ transition with low oscillator strength, and S_9 and S_{10} led to energetically higher CT states.

At higher levels of theory, the calculated four $\pi\pi^*$-related transitions were strongly mixed, with significant CT character in all four, and more evenly distributed oscillator strengths.[75] Thus, a pronounced exciplex/CT character of the excited states of d(ApG) in the FC range is expected. Moreover, it appears quite likely that the flexibility of d(ApG) allows for a structural transformation to an energetically trapped exciplex with correspondingly prolonged lifetime at significantly lower energy than the bright $\pi\pi^*$-type excited dimer states.

4.3 Monomer-like excited dinucleotides

The 0.2 to $\approx$1 ps excited-state lifetimes observed by the fluorescence up-conversion and transient absorption experiments ($\tau_{ESA1} = 0.24$ ps, τ_{FL2}, $\tau'_{FL2} = 0.4$–0.5 ps, $\tau'_{FL3} = 1.2$ ps) are very similar to the corresponding values for the mononucleotides and are therefore attributed to fast electronic relaxation of unstacked, 'monomer-like' excited dimers [$d(ApG)^*_m$]. The measured $\tau_{ESA2} = 2.44$ ps, however, already appears slightly long for simple 'monomer-like' decay dynamics. It may belong to slower, sterically hindered dimers or to dimers just entering the slow lane to the excimer/exciplex (see below).

By comparison with earlier work,[52,56] the $\tau_{GSR2} = 3.10$ ps value is attributed to the electronic relaxation of monomer-like excited dinucleotide molecules and identified as cooling time of 'vibrationally hot' $d(ApG)^{\#}_{(S_0)}$ molecules after their return to S_0. Remarkably, this time is about 50% longer than observed elsewhere[52,56] despite the lower vibrational temperature from one-photon excitation of a dinucleotide as compared to a monomer. An explanation by slower electronic deactivation of hindered dinucleotides seems likely from this point of view.

4.4 Relaxation to excimer/exciplex states

Theory predicted the existence of a relaxed, energetically stabilised neutral excimer after photoexcitation of d(ApA) and longer dA_n strands with two chemically 'bound' adenines at a considerably shorter distance than in B-DNA ($\approx$2.9 Å *vs.* 3.4 Å) with face-to-face (F2F) orientation in the minimum energy configuration.[24,27,76] In this neutral excimer state, denoted as $d(ApA)^*_{X(F2F)}$, the two adenine subunits lie on top of each other with maximal orbital overlap and significant π-electron density between the two purine rings. A similar configuration is likely for the relaxed exciplex of d(ApG) in its minimum energy state $d(ApG)^*_{X(min)}$ as well,[76] although a perfect face-to-face orientation is not possible in this case with two chemically different subunits. Obviously, the transformations of the photoexcited dimers from their FC states [$d(ApA)^*_{(FC)}$ and $d(ApG)^*_{(FC)}$] to the excimer/exciplex [$d(ApA)^*_{X(F2F)}$ and $d(ApG)^*_{X(min)}$] in local potential energy minima well below the bright $\pi\pi^*$-type dimer states require a great deal of large-amplitude motion of the two subunits relative to each other besides solvent displacement and (re)solvation. Judged from the isomerisation times of photochromic switches, *e.g.* azobenzenes,[77] where the internal forces steering the molecules are far stronger than in the weakly attractive case of two nucleobases, we expect time scales for relaxation to the excimer/exciplex of several picoseconds. Our experimental data give some insight here:

First, the ultrafast disappearance of the stimulated emission (SE) at $\lambda \approx$ 330 nm close to time-zero in the transient 2D absorption map (Fig. 4a) on the same time scale as the fastest fluorescence decay component ($\tau_{FL1} \leq 0.1$ ps) at about the same wavelength shows that the earliest transitions in the photoexcited dinucleotide happen in ≤0.1 ps. These transitions were first observed[29,50] for d(ApA)* and now also for d(ApG)*. It is unclear whether the SE and fastest fluorescence arise more from 'monomer-like' or from 'stacked' dinucleotides, but it is striking that no significant SE was found in the 2D maps for the monomers (Fig. 4b–d). This suggests (partial) relaxation within the photoexcited $\pi\pi^*$ stacks to 'relaxed' excited dimers [$d(ApA)^*_{(FCrelax)}$ and $d(ApG)^*_{(FCrelax)}$] as mechanism. To some extent, this is related to the supposed $S_3 \rightarrow S_2 \rightarrow S_1$ transitions in the monomers in ≤0.1 ps. That time appears too short for large-amplitude motions of the entire purine rings. Further, the 'relaxed' excited states remain observable in fluorescence (lifetime τ_{FL2}, see above). As has been noted, however, the FC and relaxed FC states of d(ApG)* are expected to have pronounced CT character, in contrast to the exciton situation for d(ApA)*. In longer strands, the intra-band exciton relaxation leads to localisation of the excitation on two bases within 1 to at most a few ps.[12]

Next, as briefly introduced above, the relaxation from the $d(ApA)^*_{(FCrelax)}$ resp. $d(ApG)^*_{(FCrelax)}$ states to the final excimer/exciplex states involves torsional motion from the B-DNA to an F2F conformation and relative motion of the two chromophore units towards each other next to solvent displacement and (re)solvation. The time window, within which these processes occur, must range from between $\Delta t \approx 0.5$ ps (τ'_{FL2}) or ≈ 2.5 ps (τ_{ESA2}) after excitation to $\Delta t \approx 6$ ps (τ'_{FL3}) after excitation in the d(ApG) case and to $\Delta t \approx 8$–10 ps[29] in the d(ApA) case. The intermediate states in this $\Delta t \approx 10$ ps window after excitation are unlikely to correspond to local energy minima, as the downhill motion on the PEHS should be more or less continuous. During this process, on the other hand, they likely gain in excimer/exciplex character, which is why they will be referred to as $d(ApA)^*_{X(BDNA)}$ and $d(ApG)^*_{X(BDNA)}$ in the following. Since CT states are nominally optically dark in emission, the very weak residual fluorescence of d(ApG) at long delay times (*cf.* Fig. 3b) despite the fact that ≈25–35% (see results) of the photoexcited molecules have not yet returned to the ground state indicates that the CT/exciplex minimum is reached after $\Delta t \approx 6$ ps (τ_{FL3}) at the earliest. We consider the long-lived $d(ApG)^*_{X(min)}$ CT/exciplex as optically 'dark grey' here, and not plain black. Judged from the more intense fluorescence with >100 ps lifetime in the d(ApA) case,[50] the neutral $d(ApA)^*_{X(F2F)}$ excimer is labeled as 'bright grey' in this context. The long lifetimes naturally account for the higher fluorescence quantum yields of π-stacked dinucleotides as compared with the monomers.

Eventually, Fig. 6 depicts transient spectra extracted from the 2D maps at selected delay times for d(ApG) in the center as compared to d(ApA) above and the equimolar mixture of dAMP and dGMP below. Showing distinctive, rather narrow, long-lived absorption bands around $\lambda = 330$ nm in the d(ApG) case and between $\lambda \approx 330$ and 375 nm in the d(ApA) case, these transient spectra reveal features that appear to be characteristic and unique for the excimer/exciplex states of the stacked dinucleotides.

For d(ApA), two partially overlapping UV ESA bands are observed which peak at $\lambda \approx 348$ and at $\lambda \approx 368$ nm in addition to the weaker broad ESA in the VIS region with a maximum at $\lambda = 630$ nm (Fig. 6a). The intense 368 nm ESA band disappears in only ≈1 ps and may therefore be attributed to 'monomer-like' excited states or to a short-lived higher $\pi\pi^*$ excited state of the stacked dimer, while the spectral narrowing and blue-shift of the 348 nm ESA band within ≈5–10 ps hints at the abovementioned intermediate $d(ApA)^{*}_{X(BDNA)}$-type states en route to the long-lived $d(ApA)^{*}_{X(F2F)}$ excimer. Being trapped in a potential energy minimum on the excited PEHS, the latter has a lifetime of $\tau = 380$ ps according to the observed ESA decay time, the longest fluorescence lifetime and the ground state recovery time.[29] Su *et al.* interpreted their similar results for the GSR after 267 nm photoexcitation of d(ApA) in terms of $\approx$46% of stacked and $\approx$64% of unstacked dimers in thermal equilibrium.[49] We did, however, notice differences between the transient spectra of d(ApA) and dAMP at short delay times, which indicate that the excited-state absorption properties of 'unstacked' dimers and dAMP are not quite the same.[29]

In the case of d(ApG) (Fig. 6b), the $\lambda \approx 330$ nm 'exciplex' band peaks ≈ 0.5 eV to the blue compared to the maximum of the broad ESA after excitation of the equimolar dAMP/dGMP mixture. Although this $d(ApG)^{*}_{X}$ band appears less pronounced than for $d(ApA)^{*}_{X}$, the shape of the spectrum in the $\lambda = 320$–400 nm region is very different from that of the dAMP/dGMP mixture. Regrettably, the SE observed immediately after UV excitation may obscure a possible delayed rise of the exciplex feature. Compared to d(ApA), however, the lack of spectral evolution in the d(ApG) spectrum in the first few ps after excitation is striking. This may be related to the pronounced CT character of the bright excitation transitions for d(ApG) explained above. The lifetime of $\tau = 124$ ps of the eventually formed long-lived CT/exciplex state $d(ApG)^{*}_{X(min)}$ and the time for its slow return to the ground state is shorter by a factor of about 3 as compared to d(ApA) and in good agreement with a measurement for the ribose nucleotide ApG ($\tau = 105$ ps) by Takaya *et al.*[48] Regarding the mechanism, Olaso-Gonzaléz *et al.* proposed thermal activation followed by nonadiabatic decay along the known monomer-like pathways.[24]

4.5 Implications for other di- and oligonucleotides

The extent of π-stacking between neighbouring bases in DNA oligonucleotides is highly dependent on strand length, exact base sequence, and experimental conditions such as pH, temperature and ion concentration. Their effect on the overall excited-state dynamics is therefore hard to quantify. Disturbed π-stacking likely accounts for many of the current problems in the interpretation and comparison of time-resolved measurements of different oligonucleotides.[4,10–18,47–49]

Here, we pinpointed common characteristics, but also differences between the dinucleotides d(ApG) and d(ApA). Both are supposed to adopt B-DNA conformations in thermal equilibrium in neutral aqueous solution, but are flexible enough to relax to non-B-DNA like (*e.g.*, 'face-to-face') excimer structures with lower energy relative to the ππ*-type bright states upon photoexcitation. Remarkably, even the relatively minor exchange of one purine base with another led to a pronounced decrease in the excited-state lifetime. Naturally, the partial 'locking' of the bases in a longer B-DNA oligonucleotide strand should lead to less well stabilised excimers and thereby a decrease in the excited-state lifetime. This effect is countered by the higher stacking fractions in longer strands,[49] but enhanced by a varying base sequence.[12] Correlations of the exciplex decay rates with the thermodynamic driving force estimated from the ionisation energy of the donor minus the electron affinity of the acceptor have met with some success,[48] while the stacking fractions in comparison with a series of $d(A)_n$ oligonucleotides provided rather good guidance.[49]

Systematic studies of additional model systems, preferably other dinucleotides, followed by tri- and tetranucleotides and by short self-complementary hairpins, will presumably shed more light onto this complex matter. Since theoretical calculations suggested significant differences between the AG and GA sequences regarding the extent of stacking, nature of the FC excited states and their coupling with the exciplex/CT states,[75] it will be interesting to compare the dynamics of d(ApG) and d(GpA) and additional dinucleotides. The so far missing d(GpG) and the differences between RNA and DNA dinucleotides should be checked next.

5 Conclusions

In summary, we have studied the complex electronic relaxation dynamics of the DNA dinucleotide d(ApG) by means of femtosecond time-resolved fluorescence and absorption spectroscopy. The most pronounced difference between the dinucleotide and its monomer nucleotides that was found is the strikingly longer excited-state lifetime of the former. The observed multi-exponential decay dynamics hint at step-wise transitions along complex pathways. We found evidence in the spectra for the formation of an excimer state on the stacked dinucleotide d(ApG) with lifetime $\tau = 124 \pm 4$ ps. This is shorter compared with d(ApA), whose excimer state presumably adopts a face-to-face conformation. We attribute this difference to a less orbital overlap in d(ApG) and thereby weaker stabilisation. Nevertheless, both dinucleotides exhibit a common spectral 'excimer feature', the narrow excited-state absorption band around $\lambda = 330$ nm. The $\approx$0.5 eV blue-shift as compared to the broad ESA maximum of a mixture of the mononucleotides shows considerable energetic stabilisation. This feature may be of great help in future studies of the photo-initiated quantum molecular dynamics of other di- and oligonucleotides. We attributed our observation to the formation of the excimer in its minimum energy configuration below the bright $\pi\pi^*$-type excited states in time $\Delta t \approx 5$–10 ps after excitation. The subsequent slow return to the ground state in several hundred picoseconds refills the initial ground state bleach signal. Little to no room appears left for additional other, longer-lived dark states in the dynamics. The nucleotide monomers dAMP and dGMP exhibit only weak transient absorption around $\lambda = 330$ nm, and both decay to their electronic ground states within <1–2 ps. Time-resolved emission measurements confirmed the picture derived from the transient absorption experiments and revealed that the excimer state of d(ApG) is nearly optically dark regarding fluorescence.

Acknowledgements

This work has been supported by the Deutsche Forschungsgemeinschaft. The authors thank J. Kleber for taking the emission spectra of dAMP, dGMP and d(pApG).

Notes and references

1 J. Eisinger and R. G. Shulman, *Science*, 1968, **161**, 1311–1319.

2 P. R. Callis, *Annu. Rev. Phys. Chem.*, 1983, **34**, 329–352.
3 C. E. Crespo-Hernández, B. Cohen, P. M. Hare and B. Kohler, *Chem. Rev.*, 2004, **104**, 1977–2019.
4 C. T. Middleton, K. de La Harpe, C. Su, Y. K. Law, C. E. Crespo-Hernández and B. Kohler, *Annu. Rev. Phys. Chem.*, 2009, **60**, 217–239.
5 J.-M. L. Pecourt, J. Peon and B. Kohler, *J. Am. Chem. Soc.*, 2000, **122**, 9348–9349.
6 J. Peon and A. H. Zewail, *Chem. Phys. Lett.*, 2001, **348**, 255–262.
7 T. Gustavsson, A. Sharonov and D. Markovitsi, *Chem. Phys. Lett.*, 2002, **351**, 195–200.
8 C. E. Crespo-Hernández, B. Cohen and B. Kohler, *Nature*, 2005, **436**, 1141–1144.
9 As is customary in the literature on DNA, the terms 'excimer' and 'exciplex' will be used essentially synonymously in this text to describe the formation of a (comparatively long-lived) bound complex of two nucleobases upon excitation with shared electron density between them regardless of whether the two nucleobases are identical or not. Strictly speaking, the definition of an excimer or exciplex (as 'excited homodimer' or 'excited heterodimer', respectively) describes the formation of a bound dimer in the excited electronic state from two units that are not connected in their ground state, while the nucleobases as the chromophores of interest in a di- or oligonucleotide are linked *via* the sugar–phosphate backbone. But the important point in the present case is the formation of a bound complex with shared electron density between the two bases on excitation. Accordingly, the two nucleobase chromophores can be regarded as electronically unbound in the ground state, the bound state arises on electronic excitation of a subunit only. An exciplex with two different chromophores usually implies at least some charge transfer (CT) character of the excited state due to the different ionization energies and electron affinities of the subunits, an excimer may be neutral regarding the charges on two identical subunits.
10 C. E. Crespo-Hernández, B. Cohen and B. Kohler, *Nature*, 2006, **441**, E8.
11 D. Markovitsi, F. Talbot, T. Gustavsson, D. Onidas, E. Lazzarotto and S. Marguet, *Nature*, 2006, **441**, E7.
12 N. K. Schwalb and F. Temps, *Science*, 2008, **322**, 243–245.
13 W.-M. Kwok, C. Ma and D. L. Phillips, *J. Phys. Chem. B*, 2009, **113**, 11527–11534.
14 S. O. Konorov, H. G. Schulze, C. J. Addison, C. A. Haynes, M. W. Blades and R. F. B. Turner, *Open Spectrosc. J.*, 2009, **3**, 9–20.
15 B. Kohler, *J. Phys. Chem. Lett.*, 2010, **1**, 2047–2053.
16 I. Vayá, T. Gustavsson, F.-A. Miannay, T. Douki and D. Markovitsi, *J. Am. Chem. Soc.*, 2010, **132**, 11834–11835.
17 I. Vayá, J. Brazard, T. Gustavsson and D. Markovitsi, *Photochem. Photobiol. Sci.*, 2012, **11**, 1767–1773.
18 D. Markovitsi, T. Gustavsson and I. Vayá, *J. Phys. Chem. Lett.*, 2010, **1**, 3271–3276.
19 F. Santoro, V. V. Barone and R. Improta, *Proc. Natl. Acad. Sci. U. S. A.*, 2007, **104**, 9931–9936.
20 D. Nachtigallovà, P. Hobza and H.-H. Ritze, *Phys. Chem. Chem. Phys.*, 2008, **10**, 5689–5697.
21 E. B. Starikov, G. Cuniberti and S. Tanaka, *J. Phys. Chem. B*, 2009, **113**, 10428–10435.
22 A. W. Lange, M. A. Rohrdanz and J. M. Herbert, *J. Phys. Chem. B*, 2008, **112**, 6304–6306.
23 A. W. Lange and J. M. Herbert, *J. Am. Chem. Soc.*, 2009, **131**, 3913–3922.
24 G. Olaso-González, M. Merchán and L. Serrano-Andrés, *J. Am. Chem. Soc.*, 2009, **131**, 4368–4377.
25 I. Conti, P. Altoè, M. Stenta, M. Garavelli and G. Orlandi, *Phys. Chem. Chem. Phys.*, 2010, **12**, 5016–5023.
26 Y. Lu, Z. Lan and W. Thiel, *Angew. Chem., Int. Ed.*, 2011, **50**, 6864–6867.
27 R. Improta and V. Barone, *Angew. Chem., Int. Ed.*, 2011, **50**, 12016–12019.
28 F. Plasser, A. J. A. Aquino, W. L. Hase and H. Lischka, *J. Phys. Chem. A*, 2012, **116**, 11151–11160.
29 M. C. Stuhldreier, M. Malicki, K. Röttger, C. Schüler, J. Kleber, T. Muskat and F. Temps, to be submitted, 2013.
30 M. Y. Warshaw and I. Tinoco, *J. Mol. Biol.*, 1966, **20**, 29–38.
31 C. R. Cantor, M. M. Warshaw and H. Shapiro, *Biopolymers*, 1970, **9**, 1059–1077.
32 J. L. Sussman, I. Barzilav, M. Keren-Zur and Y. Lapidot, *Biochim. Biophys. Acta, Nucleic Acids Protein Synth.*, 1973, **308**, 189–197.
33 N. P. Johnson and T. Schleich, *Biochemistry*, 1974, **13**, 981–987.
34 C.-H. Lee, F. S. Ezra, N. S. Kondo, R. H. Sarma and S. S. Danyluk, *Biochemistry*, 1976, **15**, 3627–3639.
35 C. S. M. Olsthoorn, C. A. G. Haasnoot and C. Altona, *Eur. J. Biochem.*, 2005, **106**, 85–95.

36 C. S. M. Olsthoorn, L. J. Bostelaar, J. J. van Boom and C. Altona, *Eur. J. Biochem.*, 1980, **112**, 95–110.
37 C. S. M. Olsthoorn, L. J. Bostelaar, J. F. M. De Rooij, J. H. van Boom and C. Altona, *Eur. J. Biochem.*, 1981, **115**, 309–321.
38 C. S. M. Olsthoorn, J. Doornbos, H. P. M. de Leeuw and C. Altona, *Eur. J. Biochem.*, 1982, **125**, 367–382.
39 Z. Vokčov, M. Budăšnsk, I. Rosenberg, B. Schneider, J. Šponer and V. Sychrovsk, *J. Phys. Chem. B*, 2009, **113**, 1182–1191.
40 J. Eisinger, M. Gueron, R. G. Shulman and T. Yamane, *Proc. Natl. Acad. Sci. U. S. A.*, 1966, **55**, 1015–1020.
41 M. Daniels and J. P. Morgan, *Chem. Phys. Lett.*, 1978, **58**, 283–286.
42 J. P. Morgan and P. R. Callis, *Photochem. Photobiol.*, 1979, **29**, 1107–1113.
43 J. P. Morgan and M. Daniels, *Photochem. Photobiol.*, 1980, **31**, 101–113.
44 J. P. Morgan and M. Daniels, *Photochem. Photobiol.*, 1980, **31**, 207–213.
45 C. S. Shaar, J. P. Morgan and M. Daniels, *Photochem. Photobiol.*, 1984, **39**, 747–753.
46 M. Daniels, C. S. Shaar and J. P. Morgan, *Biophys. Chem.*, 1988, **32**, 229–237.
47 I. Buchvarov, Q. Wang, M. Raytchev, A. Trifonov and T. Fiebig, *Proc. Natl. Acad. Sci. U. S. A.*, 2007, **104**, 4794–4797.
48 T. Takaya, C. Su, K. de La Harpe, C. E. Crespo-Hernández and B. Kohler, *Proc. Natl. Acad. Sci. U. S. A.*, 2008, **105**, 10285–10290.
49 C. Su, C. T. Middleton and B. Kohler, *J. Phys. Chem. B*, 2012, **116**, 10266–10274.
50 M. C. Stuhldreier, C. Schüler, J. Kleber and F. Temps, *Ultrafast Phenomena XVII*, Oxford, 2011, pp. 553–555.
51 N. K. Schwalb and F. Temps, *J. Phys. Chem. A*, 2009, **113**, 13113–13123.
52 K. Röttger, R. Siewertsen and F. Temps, *Chem. Phys. Lett.*, 2012, **536**, 140–146.
53 J. Kleber, Diploma Thesis, Kiel, 2009.
54 D. Onidas, D. Markovitsi, S. Marguet and T. Gustavsson, *J. Phys. Chem. B*, 2002, **106**, 11367–11374.
55 T. Pancur, N. K. Schwalb, F. Renth and F. Temps, *Chem. Phys.*, 2005, **313**, 199–212.
56 J.-M. L. Pecourt, J. Peon and B. Kohler, *J. Am. Chem. Soc.*, 2001, **123**, 10370–10378.
57 V. Karunakaran, K. Kleinermanns, R. Improta and S. A. Kovalenko, *J. Am. Chem. Soc.*, 2009, **131**, 5839–5850.
58 L. Serrano-Andrés and M. Merchán, *J. Photochem. Photobiol., C*, 2009, **10**, 21–32.
59 T. Gustavsson, R. Improta and D. Markovitsi, *J. Phys. Chem. Lett.*, 2010, **1**, 2025–2030.
60 M. Barbatti, A. J. A. Aquino, J. J. Szymczak, D. Nachtigallovà, P. Hobza and H. Lischka, *Proc. Natl. Acad. Sci. U. S. A.*, 2010, **107**, 21453–21458.
61 C. M. Marian, *J. Chem. Phys.*, 2005, **122**, 104314.
62 S. Perun, A. L. Sobolewski and W. Domcke, *J. Am. Chem. Soc.*, 2005, **127**, 6257–6265.
63 L. Blancafort, *J. Am. Chem. Soc.*, 2006, **128**, 210–219.
64 L. Serrano-Andrés, M. Merchán and A. C. Borin, *Chem.–Eur. J.*, 2006, **12**, 6559.
65 M. Barbatti and H. Lischka, *J. Am. Chem. Soc.*, 2008, **130**, 6831–6839.
66 I. Conti, M. Garavelli and G. Orlandi, *J. Am. Chem. Soc.*, 2009, **131**, 16108–16118.
67 M. Barbatti, Z. Lan, R. Crespo-Otero, J. J. Szymczak, H. Lischka and W. Thiel, *J. Chem. Phys.*, 2012, **137**, 22A503.
68 H. Chen and S. Li, *J. Chem. Phys.*, 2006, **124**, 154315.
69 C. M. Marian, *J. Phys. Chem. A*, 2007, **111**, 1545–1553.
70 S. Yamazaki and W. Domcke, *J. Phys. Chem. A*, 2008, **112**, 7090–7097.
71 S. Yamazaki, W. Domcke and A. L. Sobolewski, *J. Phys. Chem. A*, 2008, **112**, 11965–11968.
72 L. Serrano-Andrés, M. Merchán and A. C. Borin, *J. Am. Chem. Soc.*, 2008, **130**, 2473–2484.
73 M. Barbatti, J. J. Szymczak, A. J. A. Aquino, D. Nachtigallovà and H. Lischka, *J. Chem. Phys.*, 2011, **134**, 014304.
74 B. Heggen, Z. Lan and W. Thiel, *Phys. Chem. Chem. Phys.*, 2012, **14**, 8137–8146.
75 F. Santoro, V. Barone, A. Lami and R. Improta, *Phys. Chem. Chem. Phys.*, 2010, **12**, 4934–4948.
76 N. Öksüz and F. Temps, unpublished results.
77 T. Pancur, F. Renth, F. Temps, B. Harbaum, A. Krüger, R. Herges and C. Näther, *Phys. Chem. Chem. Phys.*, 2005, **7**, 1985–1989.

Faraday Discussions RSC Publishing

PAPER

Photodynamics of Lys^+-Trp protein motifs: Hydrogen bonds ensure photostability

Matteo Guglielmi, Manuel Doemer, Ivano Tavernelli and Ursula Rothlisberger*

Received 11th March 2013, Accepted 22nd March 2013
DOI: 10.1039/c3fd00037k

Cation-π interactions such as Lys^+-Trp, are highly abundant structural motifs in proteins. Both, experimental and theoretical studies of small prototypical gas phase systems, H^+Trp, $H^+Trp\cdot(H_2O)_n$ and H^+Gly-Trp, indicate such an arrangement as potential hot spot for photodamage and photoinstability. Here, we study the photodynamical properties of a Lys^+-Trp pair in the protein human serum albumin (HSA) using nonadiabatic mixed time-dependent density functional theory / molecular mechanics simulations (TDDFT/MM). These simulations show that the findings for small protonated Trp complexes are largely transferable to a more complex protein environment. Under partially hydrated ("dry" conditions), when the $-NH_3^+$ head group is not fully solvated, photoexcitation of the tryptophan leads indeed to rapid photodissociation of the proximal charged amino group. In contrast, photostability is well maintained under fully solvated conditions when the lysine head group is fully hydrogen-bonded. In this case, photodynamics takes place in a π–π* state without interference of fast dissociative σ^*_{C-N} or σ^*_{N-H} channels. These results highlight the crucial role of hydrogen bonds in ensuring the photostability of essential biological building blocks.

1 Introduction

Solar light is the basic energy source for life on Earth, initializing the photosynthetic process in plants, which is at the basis of all food chains that ensure survival of all species. However, solar radiation is at the same time a source of energy and a source of potential molecular damage. Interaction of visible light or near-UV radiation with matter can induce electronic excitations, which in turn can lead to phenomena like photoionization, photodissociation, or formation of radical species. Therefore, biological macromolecules had to be selected by evolutionary pressure to be stable towards interaction with light. The photostability of nucleobases in DNA or RNA that are integral parts of the genetic code and its faithful transcription, is particularly critical in this respect.[1,2] In fact, aromatic rings can absorb UV-light and easily undergo photoinduced chemical

Laboratory of Computational Chemistry and Biochemistry, Ecole Polytechnique Fédérale de Lausanne, Lausanne, Switzerland. E-mail: Ursula.roethlisberger@epfl.ch; Fax: +41 21 6930320; Tel: +41 21 6930325

modifications[3,4] and nature has developed an arsenal of strategies to prevent photodamage[5,6] or repair photolesions in efficient ways.[7–9] Many experimental and theoretical investigations have been devoted to the elucidation of the underlying mechanisms of the intrinsic photostability of individual nucleic acids and base pairs.[10–17] The general consensus that has been reached in these studies is that DNA and RNA are protected from photodamage through either excited state proton transfer or ring puckering structural changes[18,19] that rapidly lead the system to conical intersections with the ground state and thus efficient radiationless deactivation.

Proteins are intrinsically more stable towards light-damage than DNA due to the fact that their electronic structure is mostly transparent to visible-near UV radiation. However, the presence of aromatic side chains (tryptophan, in particular) exposes them to light-damage as well. At difference to DNA/RNA, very little is known about the photoprotection mechanisms that nature has put in place to prevent light-induced damage to protein building blocks. Theoretical studies on single neutral or zwitterionic amino acid – water clusters (Gly-$(H_2O)_2$,[20] Tyr-$(H_2O)_2$ and Trp-$(H_2O)_2$)[21] have proposed that photoexciation of the zwitterionic forms to the first optically allowed state can lead to NH_3 detachment, in agreement with the high quantum yields for ammonia in UV photolysis experiments. Calculations of neutral di- and tripeptides in the gas phase have also identified intramolecular proton transfer processes, reminiscent of those found for nucleic acids, as efficient means to ensure the photostability of specific conformers.[22–24] Recent cold-ion spectroscopy experiments[25] have shown that the excited state properties of small amino acid complexes can be highly influenced by the environment and that the addition of as few as two water molecules can drastically alter the excited state properties. It is therefore not evident to what extent the conclusions drawn from the study of small gas phase complexes are also relevant for the situation in real proteins.

Here, we study the photodynamical properties of a commonly observed protein motif, the Lys^+-Trp cation-π interaction, in a realistic protein environment. Such interactions are present in about one third of all known proteins with Trp residues in their sequence. Tryptophan that acts as the UV-sensitive chromophore in this arrangement, is the most frequently used optical probe for proteins[26–29] and both its ground state and excited state properties have been studied intensively.[30–32] Recent conformer-selective, vibrationally resolved experiments of prototypical protonated tryptophan complexes at low temperature ($\sim$10 K) suggest that positively-charged groups in the vicinity of Trp can lead to charge-transfer induced fast photodissociation phenomena and that these photoinstabilities might be deactivated by microsolvation of the positively charged site.[25]

We have performed a characterization of the static and dynamic ground and excited state properties of bare and microsolvated protonated tryptophan $[H^+Trp]\cdot(H_2O)_n$ ($n = 0,2$) and the H^+-Gly-Trp dipeptide.[25,33] For all of these systems, experimental data is available to validate our theoretical approach.[25] We have then extended these studies to investigate the *in situ* photodynamical properties of the cation-π motif, Lys^+-Trp, in the protein human serum albumin (HSA) using nonadiabatic mixed quantum mechanical/molecular mechanical TDDFT/MM simulations of the entire protein in aqueous solution.

HSA (PDB code: 1AO6[34]) shown in Fig. 1, is the most common protein in blood (30–50 g l^{-1}) and contains a single Lys^+-Trp motif, where the positive charge of Lys

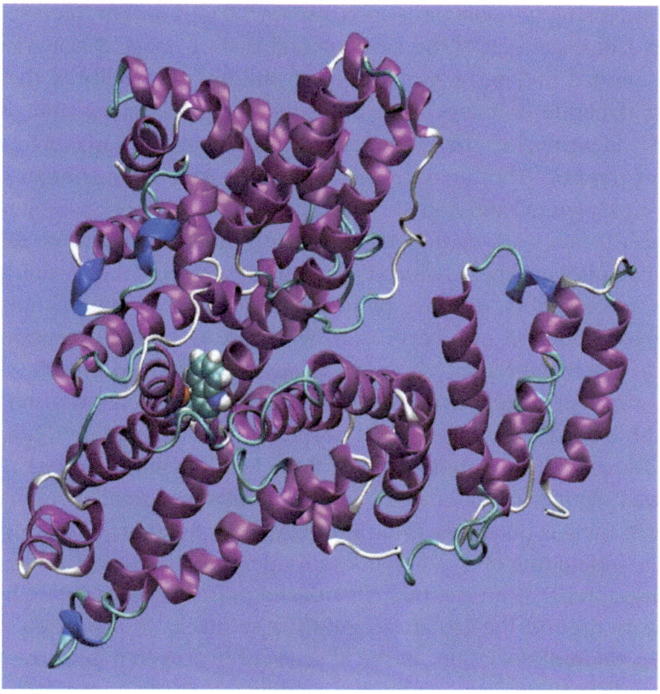

Fig. 1 Human Serum Albumin (PDB code: 1AO6). The location of Trp214 is highlighted in a van der Waals representation.

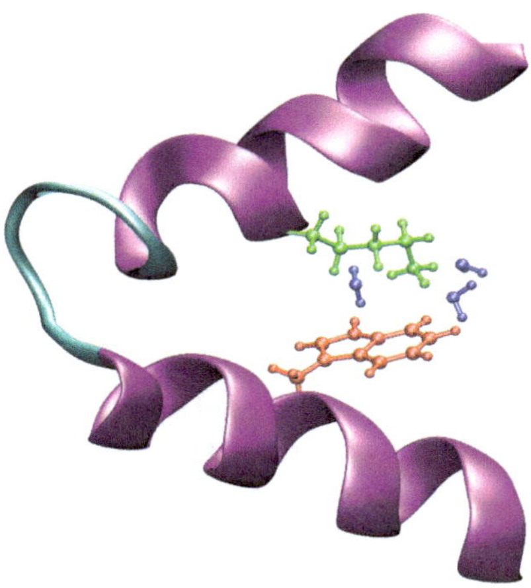

Fig. 2 Zoom in to the crevice of human serum albumin containing the buried Trp-Lys motif. Trp214, Lys199, and three solvating water molecules are colored in red, green and blue, respectively. This configuration represents the starting geometry for the photoexcited dynamics of HSA.

is at a distance of 5.5 Å from the center of mass of the Trp side chain (Fig. 2). HSA is a monomeric globular protein constituted by three structurally similar alpha helical domains. The Trp residue (Trp214) is located in a deep pocket surrounded by water molecules. The solvation dynamics of HSA was the subject of extensive recent studies using ultrafast laser spectroscopy[35] based on Trp fluorescence and site directed mutagenesis.[36] The cavity contains on average 8 water molecules directly H-bonded to Trp214, which (due to the high flexibility of the protein) are in fast dynamic exchange with the bulk water molecules. Bhagavan and co-workers[36] identified a series of mutations, which significantly affect the fluorescence of Trp214. They mainly involve nearby positively charged residues (Arg218, Lys195, Lys199 and Arg257) and cause a significant quenching of the Trp fluorescence. This is a strong indication for the participation of these residues in the photodynamics of Trp214, which is the central subject of the present study. To probe the solvent effect, we have performed photoexcited state simulations of both fully solvated HSA and "dry" HSA, where in a computer experiment, we removed the water molecules solvating the positively charged Lys199.

Anticipating our results from the TDDFT/MM simulations, we find that the presence of buried water molecules solvating the charged Lys head group increases dramatically the photostability of the Lys^+-Trp motif, protecting it from light-induced damages. In analogy to the results on small protonated Trp complexes in the gas phase, we find that the vicinity of a bare positively charged group to the indole ring of tryptophan can render the system potentially prone to photodamage and that on the other hand, the presence of an intact hydrogen-bond network around the Lys^+ head group is sufficient to stabilize the system against NH_3 photodissociation. In the absence of hydrogen-bonds, the presence of low-lying dissociative $\pi\sigma^*_{C-N}$ and $\pi\sigma^*_{N-H}$ channels leads to rapid photodissociation. In contrast, if the Lys head group is fully solvated, these dissociative channels are destabilized through the formation of hydrogen bonds and the only accessible state reachable under UV irradiation is a non-dissociative $\pi\pi^*$ state from where radiationless de-excitation might occur *via* a conical intersection to the ground state. Our conclusions are based on the study of a prototypical protein system, human serum albumin, for which a number of theoretical and experimental studies exist. The proposed mechanism is however applicable to any polypeptide chain containing aromatic residues (in particular Trp) in the vicinity of positively charged groups.

Computational details

Small molecule studies

The reference values for the relative energies of different conformers of H^+Trp and $H^+Trp(H_2O)_2$ were determined with the CBS-C method (following the same extrapolation procedure as for the CBS-Q approach,[37] but employing coupled cluster singles and doubles with perturbative triples CCSD(T)[38,39] in place of quadratic configuration interaction QCISD(T).[40] For the sake of comparison, we also evaluated the energetics at the DFT[41,42] level with a selection of exchange–correlation functionals BLYP,[43,44] B3LYP,[44,45] M05,[46] M05-2X,[47] M06,[48] and M06-2X[48] in conjunction with the 6-31+G(d,p),[49,50] aug-cc-pVDZ[51] basis set. All calculations were performed with the Gaussian G09 package[52] using an ultra fine integration grid. Geometry optimizations and the calculations of the harmonic vibrational frequencies for the zero-point energy (ZPE) corrections have been performed using tight convergence criteria (RMS force set to 10^{-5} a.u.). In addition, BLYP-D3 values employing Grimme's DFT-D3 dispersion correction[53,54] were obtained with the Turbomole 6.4 software package[55] employing the aug-cc-pVDZ basis set, the m5 integration grid, convergence criteria of 10^{-5} a.u. for the Cartesian nuclear forces and 10^{-7} a.u. for the electronic degrees of freedom. More details about the search for low-energy conformers and an extensive discussion of the performance of different theoretical methods in predicting the most stable structures are given in ref. 33.

For the lowest-energy conformers of H^+Trp, and $H^+Trp(H_2O)_2$, we have calculated vertical excitations with both TDDFT/B3LYP and the singles and doubles second-order approximate coupled-cluster model (CC2).[56] B3LYP calculations were performed using Gaussian03,[57] and CC2 calculations were done with TURBOMOLE 5.8[58] within the resolution-of-the-identity (RI) approximation for the evaluation of the electron repulsion integrals (RI-CC2).[59] For both TDDFT/B3LYP and RI-CC2 calculations we used an aug-cc-pVDZ basis set.[60] We have also performed nonadiabatic excited state TDDFT dynamics[61] in the Tamm-Dancoff approximation (TDA).[62] All dynamical ground state (DFT) and excited state (TDDFT) MD simulations were carried out with the CPMD code[63] using soft norm-conserving nonlocal Troullier–Martins pseudopotentials[64] and a 70 Ry energy cutoff for the plane wave expansion of the wave functions. The inherent periodicity in the plane wave calculations is circumvented by solving Poisson's equation for nonperiodic boundary conditions.[65] We use a cubic cell of edge 17 Å for the bare protonated tryptophan (H^+Trp) and 23 Å for the doubly solvated species ($H^+Trp\cdot(H_2O)_2$). Since the use of hybrid functionals is prohibitive within a plane wave based scheme, excited state dynamics were run with a GGA (Perdew-Burke Ernzerhof PBE[66]) functional. The main difference between B3LYP and PBE descriptions consists in a constant shift of all calculated transitions to lower excitation energies by *ca.* 0.6–0.7 eV (when using PBE). The results of DFT/PBE TDDFT dynamics were validated by RI-CC2 calculations at different geometries along the MD trajectory. Surface crossing probabilities were evaluated by the Landau–Zener semiclassical approach.[67,68] All energies and gradients were converged to a maximum deviation of 10^{-7} au. A time integration step of 15 au (0.36 fs) was used for the ground state simulations, while for the TDDFT MD runs it was reduced to 4 au (0.10 fs). Oscillator strengths were computed according to Bernasconi *et al.*[69] Further details can be found in ref. 25.

Simulations of human serum albumin

Our computational setup is based on the 1AO6 pdb structure of HSA,[34] which was solvated by 45702 water molecules and placed in a 116 × 103 × 121 $Å^3$ orthorhombic super cell. Titrable side chains were protonated according to their standard pK_a values in aqueous solution. The system was relaxed for 30 ns of classical MD simulation at room temperature using the AMBER/parm03 force field.[70]

HSA contains a single Lys^+-Trp motif, the one between Lys199 and Trp214. During the classical MD equilibration, the Lys199 side chain adopts two main equally populated conformations: in the first, the NH_3^+ moiety is fully solvated by water molecules, while in the second, one NH bond forms a hydrogen-bond with His242. In both cases, the NH_3^+ group remains close to the indole ring of Trp214. In fact, the nitrogen atom stays within 5.5 Å from the ring for more than 85% of the entire MD simulation. As a starting configuration for the photoexcited state dynamics of HSA, we have selected a geometry with a fully-solvated NH_3^+ moiety, in which the Trp214 and Lys199 side chains lay in close proximity. The photoexcitation dynamics of HSA was investigated by LR-TDDFT coupled to hybrid QM/MM MD simulations. In our setup, the side chains of Lys199 and Trp214, and the three water molecules solvating Lys199 were described at the DFT level with the PBE exchange–correlation functional.[66] The rest of the system was treated at the classical mechanics level. For the QM part, we use a similar computational scheme as for the excited state dynamics on the small protonated Trp complexes in the gas phase, *i.e.* core-valence interactions were integrated out *via* norm-conserving Martins-Troullier pseudopotentials,[64] the electronic wavefunctions were expanded in plane waves up to a cutoff of 90 Ry and the integration of the non-local parts of the pseudopotential was obtained *via* the Kleinman-Bylander scheme[71] for all atoms. The interaction between the classical and the quantum parts was described by a fully Hamiltonian hierarchical electrostatic coupling scheme.[72,73] All excited state QM/MM simulations were carried out with the software package CPMD[63] with a time integration step of 4 au (0.1 fs). Electronic vertical transitions were calculated in the Tamm-Dancoff approximation,[62] and non-adiabatic crossings in the excited dynamics were evaluated by a Landau–Zener semiclassical approach[67,68]. Due to the high computational costs, we could only afford to run one single trajectory for HSA under wet, respectively dry conditions. Furthermore, in the case of wet HSA, three close lying excited states that cross frequently necessitate a large number of independent trajectories to gain a fully representative picture. Since such a sampling is clearly not feasible, we decided in this case to follow the excited state dynamics in one of the three and use surface hops to remain in this state all throughout the trajectory.

Results

Ground and excited state properties of $H^+Trp(H_2O)_n (n = 0, 2)$, and H^+Gly-Trp

To validate our computational approach, we first performed calculations on protonated tryptophan clusters for which quantitative experimental data are available about the number of low-energy conformers at low temperatures, their vibrational fingerprints and their optical properties and excited state behavior.[25]

The conformational space of bare protonated tryptophan is characterized by the three main dihedral angles $\varphi_1 = C_\alpha–C_\beta–C_\gamma–C_\delta$; $\varphi_2 = N–C_\alpha–C_\beta–C_\gamma$; and $\varphi_3 = N–C_\alpha–C–O$. Systematic enumeration of all possible conformers,[33] and determination of their relative energetics at the CBS-C level, leads to the identification of the two lowest-energy conformers depicted in Fig. 3.

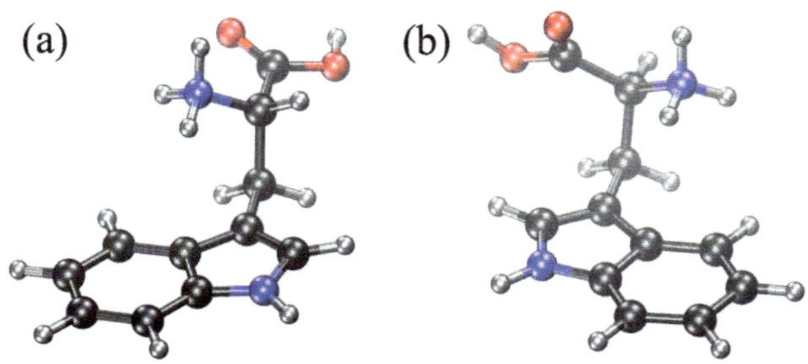

Fig. 3 Lowest energy conformers of H^+Trp.

Table 1 Five lowest vertical excitations energies for H^{+}Trp with different methods using an aug-cc-pVDZ basis set

Excited-state character	RI-CC2 [eV]	TDDFT/B3LYP [eV]	Exp. [eV]
$\pi\sigma^*$	4.46	4.28	4.35
$\pi\pi^*$	4.80	4.71	
$\pi\pi^*$	5.01	4.73	
$\pi\pi^*$CO (dark)	>5.0	3.43	
$\pi\pi^*$CO (dark)	>5.0	3.89	

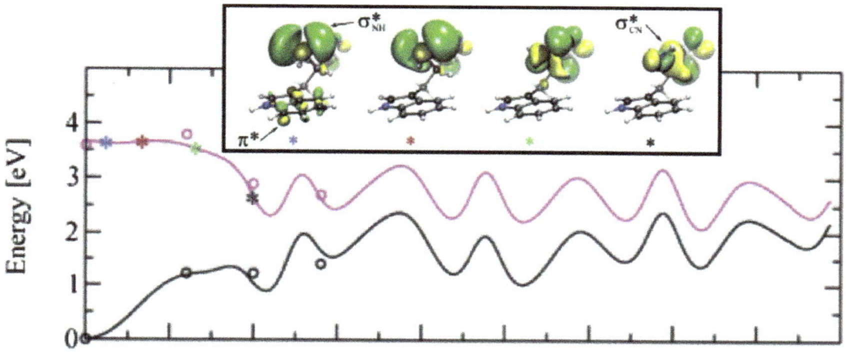

Fig. 4 TDDFT/PBE ground (black) and excited running state (purple) energies of photoexcited H^{+}Trp during photodissociation of NH_3. Open circles give the corresponding RI-CC2 values for selected configurations along the trajectory. Dominant Kohn–Sham orbital contributions to the electronic excited state during TDDFT dynamics at different times are shown in the inset. To facilitate comparison, RI-CC2 excited state energies are shifted down by a constant amount in order to best match the TDDFT curve.

At the CBS-C level including zero-point energy corrections, conformer (b) is only 0.5 kcal mol^{-1} higher in energy than conformer (a) while all other structures are in an energy range ≥1–5 kcal mol^{-1}. These results are in excellent agreement with experiment, which identifies the presence of a major and a minor conformer at ultracold temperatures. Furthermore, comparison of the conformer-selective high-resolution vibrational spectra with the harmonic vibrational frequencies of (a) and (b) calculated with MO5-2X, confirms these two structures unequivocally as the two conformers observed in the experiment.[33]

Impressively, the energetic ordering of all conformers is fully reproduced using the density functionals M05-2X, MO6, and MO6-2X with average energetic root mean square deviations (RMSD) with respect to the CBS-C results of 0.1–0.3 kcal mol^{-1}. In particular, the lowest two conformers (a) and (b) are all predicted correctly with relative energy separations between 0.3–0.6 kcal mol^{-1}. The results from BLYP and B3LYP calculations are less remarkable (with RMSD of 1.0 and 0.8 kcal mol^{-1}, respectively) but both methods are able to identify the two lowest energy conformers within an energy threshold of 1 kcal mol^{-1}. The inclusion of explicit dispersion corrections (BLYP-D3) does not alter the picture (RMSD 0.9 kcal mol^{-1}).

We have calculated the vertical excitations of the lowest energy conformers with RI-CC2, and TDDFT/B3LYP both in combination with an aug-cc-pVDZ basis set on all atoms (Table 1). The first optically allowed transition of $\pi\sigma^*$ character is

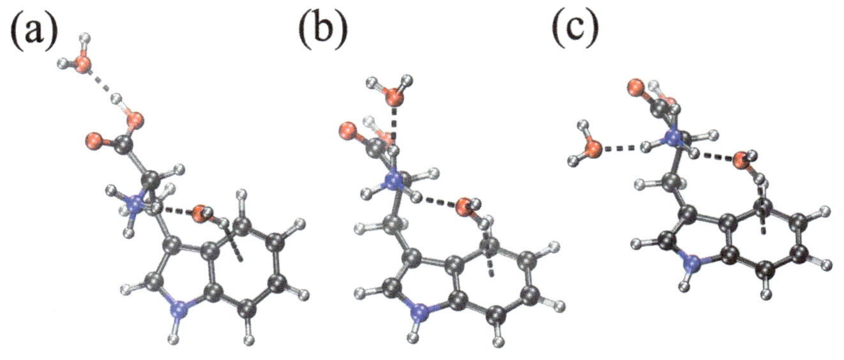

Fig. 5 Lowest energy conformers of $H^+Trp(H_2O)_2$.

predicted at an energy of 4.42 eV (RI-CC2), respectively at 4.28 eV (TDDFT/B3LYP) in good agreement with the experimental value of 4.35 eV.[25] Due to the well-known failure for charge transfer excitations,[74,75] TDDT/B3LYP predicts two dark states of charge-transfer character at erroneously low energies. However, as previously pointed out[76] and further confirmed by our TDDFT dynamics simulations, these states are pure "spectators" and do not affect the photophysics of this system. The same holds for TDDFT/PBE calculations that result in the same ordering of excited states as TDDFT/B3LYP with a constant downshift of 0.6–0.7 eV.

The first optically allowed excitation involves the promotion of one electron from the π system of the indole ring to a mixture of σ^*_{N-H} (57%), σ^*_{C-N} (6%), indole π^* (19%), and carboxyl π^*_{CO} (8%) orbitals. The precise relative contributions of these three components are strongly geometry dependent and change during the excited state dynamics. At the vertical excitation from the equilibrium geometry, the σ^*_{N-H} contribution is the most significant one. Fig. 4 displays the time evolution of the H^+Trp system upon vertical excitation to the first photoaccessible state obtained from TDDFT/PBE dynamics. During the initial fast relaxation phase (10 fs), the σ^*_{N-H} and π^* contributions decrease progressively, while the σ^*_{C-N} antibonding character increases leading to rapid photodissociation of NH_3. This is the result of a photoinduced electron transfer from the π system of the indole to the positively charged NH_3^+ group. For selected points along the

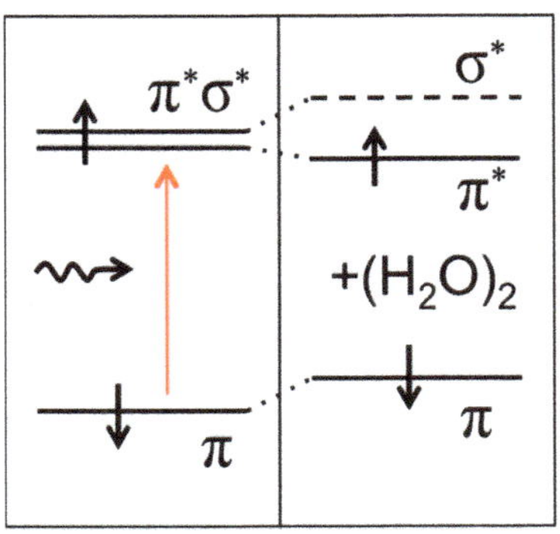

Fig. 6 Schematic molecular orbital diagram for solvated and bare protonated tryptophan.

trajectory, we have performed RI-CC2 calculations (shown with open circles in Fig. 4). The shape of both the potential energy surfaces in the ground as well as in the first excited state are in very good agreement with the TDDFT data, reconfirming that the TDDFT/PBE dynamics produces qualitatively correct results and that the two π^*_{CO} charge transfer states do not interfere in the dynamics.

In the case of doubly solvated protonated tryptophan $H^+Trp(H_2O)_2$, we identified the three conformations shown in Fig. 5 as the lowest-energy minima. All of them have an identical backbone conformation of tryptophan and agree in the placement of one of the two water molecules, differing only in the binding site of the second one. According to CBS-C, the lowest-energy isomer (a) is followed by two almost degenerate conformers (b) and (c), at only ~0.5 kcal mol^{-1} higher energy. All feature one water molecule interacting with the indole ring and the NH_3-group. Conformers (b) and (c) have both water molecules attached to the NH_3-group, in contrast to the lowest energy conformation (a), which is characterized by a second water molecule hydrogen bonded to the COOH group.

Again, the results are in excellent agreement with experiment, in which one major and two minor conformers are observed at low temperature.[25] A calculation of the vibrational spectrum confirms the assignment to the three conformers shown in Fig. 5 with root mean square deviations between experimentally observed and calculated vibrational frequencies as low as ≈5 cm^{-1}. Like in the case of bare protonated tryptophan, the Minnesota functionals are all able to predict the correct lowest energy conformations and perform with energetic RMSDs with respect to CBS-C values of ~0.7 kcal mol^{-1}. Also B3LYP identifies the correct lowest energy conformation, but yields an RMSD of above 1.5 kcal mol^{-1}. BLYP-D3 on the other hand is not able to identify the correct lowest energy conformation.

The presence of two water molecules drastically alters the electronic structure of the low-energy excited states of protonated tryptophan. The transitions to σ^*_{N-H} orbitals are shifted up by ~1.3 eV due to the presence of hydrogen bonds. As a result, the first photoaccessible state of $H^+Trp\cdot(H_2O)_2$ populates what is primarily a π^* orbital in agreement with RI-CC2 calculations. Because of the solvent-induced separation of the π^* state from the dissociative σ^*_{C-N} and σ^*_{N-H} manifold, TDDFT-MD performed on this state does not lead to any appreciable changes in the electronic structure, and unlike the case of bare H^+Trp, we do not observe any fast covalent fragmentation channel. Instead, the initial release of energy after vertical excitation eventually leads to water dissociation, as observed experimentally.

The qualitative change in electronic properties between bare and solvated protonated tryptophan is summarized schematically in Fig. 6.

The photoinstability observed in the case of protonated tryptophan suggests the presence of positive charges close to the indole ring as a general motif for potential photodamage. In fact, both theory and experiment show that rapid photodissociation also occurs for H^+Gly-Trp, in which the positive charge is further removed from tryptophan. Furthermore, if we remove the covalent connection between the N-terminal charge and Trp by constructing a computer model with a "cut backbone", the same effect persists, suggesting that the photoinduced charge transfer from the indole ring to the dissociative σ^* state on the charged ammonium group is not a "through-bond" but rather a "through-space" effect. This supports that tryptophan residues in the proximity of positively charged residues could indeed be vulnerable regions for potential photodamage.

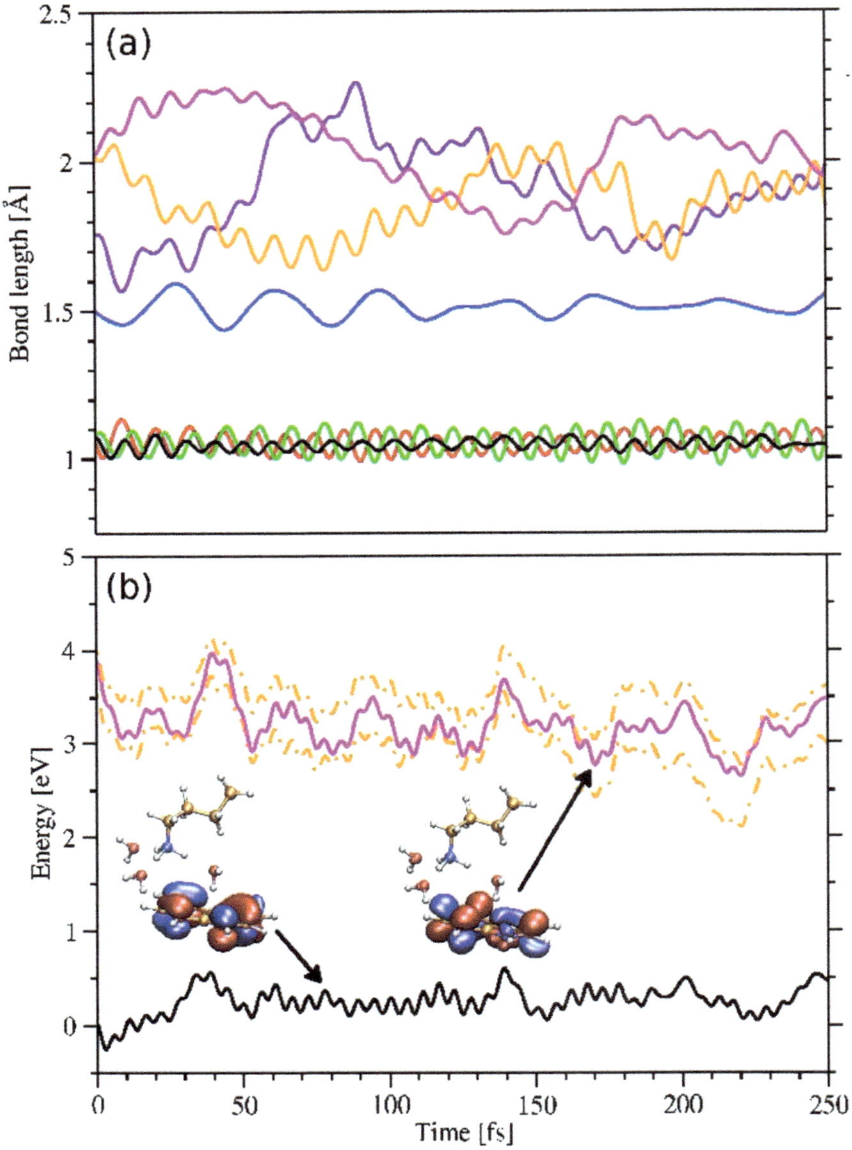

Fig. 7 Properties of photoexcited Lys199^{+}-Trp214 motif in the region described at DFT/TDDFT level of theory. (a) Time evolution of relevant bond lengths: N–H (black, red, and green lines), C–N (blue line) and H_2O–H (magenta, violet and orange lines). (b) Time evolution of ground (black line), running excited-state (magenta line) and close lying excited state (orange) energies. The dominant Kohn–Sham orbital contribution to the electronic excited state dynamics is also shown. Since a representative sampling of many trajectories is computationally not feasible, we have chosen to follow the dynamics in one of the three characteristic states.

Interactions of positively charged arginines or lysines with tryptophan residues are frequently observed cation-π interactions in proteins. An analysis of 45′865 protein structures from the protein database shows in fact that from the over 80% of all proteins containing at least one tryptophan residue roughly 21% show Lys^{+}-Trp and 29% Arg^{+}-Trp motifs, *i.e.* every second protein contains at least one Trp

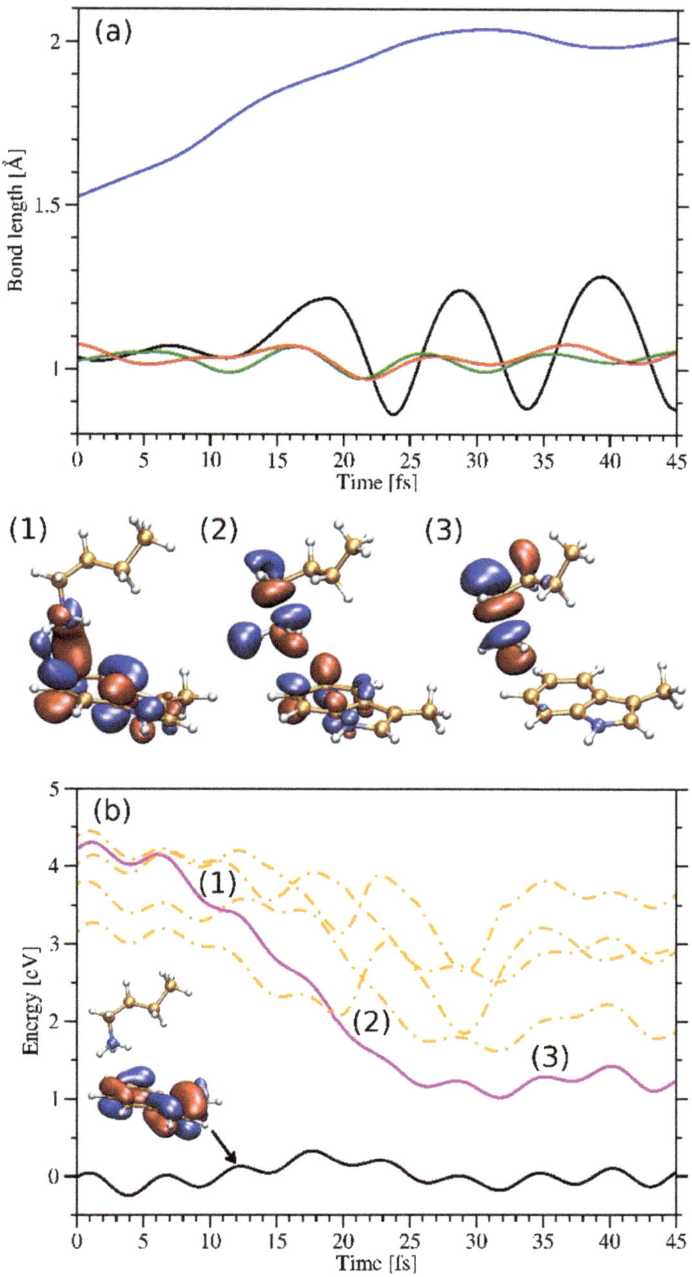

Fig. 8 Properties of photoexcited desolvated Lys199^{+}-Trp214 motif in the region described at DFT/TDDFT level of theory. (a) Time evolution of relevant bond lengths: N–H (black, red, and green lines) and C–N (blue line). (b) Time evolution of ground (black line), running excited-state (magenta line) and close lying excited state (orange) energies. The dominant Kohn–Sham orbital contributions to the electronic excited state dynamics, at different times, are also shown.

residue close to a positively charged group. To investigate to what extent our findings for small tryptophan containing clusters are transferable to a realistic protein environment, we performed a study of the photodynamical properties of a

Lys^+-Trp motif ($Lys199^+$-Trp214) in the protein human serum albumin both under "wet" (*i.e.* fully solvated) and under "dry" conditions (with a reduced number of solvating water molecules).

Photodynamics of HSA under fully hydrated conditions

In fully hydrated conditions, the first optically accessible state (~4.0 eV) has mainly $\pi\pi^*$ character, with a minor admixture of unoccupied water orbitals. Starting from this excited state, we have performed a QM/MM TDDFT molecular dynamics trajectory for a total simulation time of 250 fs. Fig. 7 displays the time evolution of the system in the region described at DFT/TDDFT level of theory. After a short period of relaxation (~60 fs), the dynamics leads to purification of the $\pi\pi^*$ character associated with the state that drives the dynamics.

The system evolves therefore into a long-lived non-dissociative electronic state that can eventually relax through a radiative or nonradiative mechanism back to the ground state. The presence of long-lived structurally stable excited states is directly related to the presence of a dense protein-water H-bond network in the Trp214 cavity. In fact, as shown below, the electronic structure of the Trp214/Lys199 moiety changes dramatically upon desolvation of the pocket region.

Photodynamics of HSA under dry conditions

In order to probe the photostabilizing effect of the solvent, we have performed the same simulation as for fully solvated $Lys199^+$-Trp214, removing the three water molecules hydrogen-bonded to Lys199. In this case, the charged residue Lys199 does not form any H-bonds with neighboring groups. The first optically allowed vertical transition at ~4.2 eV consists of an excitation from the π system of Trp214 to a state of mixed π^*, σ^*_{C-N}, and σ^*_{N-H} characters. During the initial fast relaxation phase (~20 fs), the σ^*_{N-H} and π^* contributions decrease progressively, while the σ^*_{C-N} antibonding character increases, as depicted in Fig. 8. These rapid changes in the electronic state induce major structural alterations. In particular, the initial stretching of one N–H bond ($t < 20$ fs) is followed by elongation of the C–N bond that leads to rapid dissociation of the NH_3 group (Fig. 8a) in analogy to the cases of unsolvated protonated tryptophan complexes in the gas phase.

Conclusions

Experimental and theoretical studies of small prototypical gas phase systems, H^+Trp, $H^+Trp\cdot(H_2O)_n$ and H^+Gly-Trp, suggest that UV-sensitive tryptophan residues in proximity of a positive charge can act as potential sources for photoinduced damage. Our results from TDDFT/MM simulations of a Lys^+-Trp motif in the protein human serum albumin show that the molecular mechanisms found for small gas phase clusters are also at work in a fully realistic protein environment. We find that the presence of buried water molecules solvating the charged Lys^+ head group increases dramatically the photostability of the Lys^+-Trp motif, protecting it from light-induced damages whereas under "dry" conditions with an incompletely solvated Lys^+, photoexcitation populates low-lying dissociative $\pi\sigma^*_{C-N}$ and $\pi\sigma^*_{N-H}$ channels that lead to rapid destruction of the protein. In contrast, if the Lys head group is fully solvated, these dissociative channels are destabilized through the formation of hydrogen bonds and the only accessible state reachable under UV irradiation is a non-dissociative ππ* state from where radiationless de-excitation might occur *via* a conical intersection to the ground state. Our results suggest an additional role of water as the "solvent of life" in protecting essential biological components from immediate photodamage.

References

1 A. A. Vink, V. Shreedhar, L. Roza, J. Krutmann and M. L. Kripke, *J. Photochem. Photobiol., B*, 2001, **65**, 101–104.
2 *From DNA Photolesions to Mutations, Skin Cancer and Cell Death*, ed. E. Sage, R. Drouin and M. Rouabhia, The Royal Society of Chemistry, 2005.
3 J. S. Taylor, *Acc. Chem. Res.*, 1994, **27**, 76–82.
4 M. Münzel, C. Szeibert, A. F. Glas, D. Globisch and T. Carell, *J. Am. Chem. Soc.*, 2011, **133**, 5186–5189.
5 J.-M. L. Pecourt, J. Peon and B. Kohler, *J. Am. Chem. Soc.*, 2000, **122**, 9348–9349.
6 W. J. Schreier, T. E. Schrader, F. O. Koller, P. Gilch, C. E. Crespo-Hernández, V. N. Swaminathan, T. Carell, W. Zinth and B. Kohler, *Science*, 2007, **315**, 625–629.
7 R. C. S. Rupert, S. H. Goodgal and R. M. Herriott, *J. Gen. Physiol.*, 1958, **41**, 451–471.
8 T. Todo, H. Takemori, H. Ryo, M. Lhara, T. Matsunaga, O. Nikaido, K. Sato and T. Nomura, *Nature*, 1993, **361**, 371–374.
9 D. B. Yarosh and M. L. Kripke, *Mutat. Res., Fundam. Mol. Mech. Mutagen.*, 1996, **350**, 255–260.
10 Z. G. Lan, E. Fabiano and W. Thiel, *J. Phys. Chem. B*, 2009, **113**, 3548–3555.
11 M. Barbatti and H. Lischka, *J. Am. Chem. Soc.*, 2005, **130**, 6831–6839.
12 G. Groenhof, L. V. Schäfer, M. Boggio-Pasqua, M. Goette, H. Grubmüller and M. A. Robb, *J. Am. Chem. Soc.*, 2007, **129**, 6812–6819.
13 S. Ullrich, T. Schulz, M. Z. Zgierski and A. Stolow, *Phys. Chem. Chem. Phys.*, 2004, **6**, 2796–2801.
14 A. Abo-Riziq, L. Grace, E. Nir, M. Kabelac, P. Hobza and M. S. de Vries, *Proc. Natl. Acad. Sci. U. S. A.*, 2005, **102**, 20–23.
15 T. Schultz, E. Samoylova, W. Radloff, I. V. Hertel, A. L. Sobolewski and W. Domcke, *Science*, 2004, **306**, 1765–1768.
16 C. E. Crespo-Hernández, B. Cohen and B. Kohler, *Nature*, 2005, **436**, 1141–1144.
17 W.-M. Kwok, C. Ma and D. L. Phillips, *J. Am. Chem. Soc.*, 2008, **130**, 5131–5139.
18 C. M. Mariam, *J. Chem. Phys.*, 2005, **122**, 104314-1–104314-13.
19 L. Serrano-Andres, M. Merchán and A. C. Borin, *J. Am. Chem. Soc.*, 2008, **130**, 2473–2484.
20 A. L. Sobolewski and W. Domcke, *Chem. Phys. Lett.*, 2008, **457**, 404–407.
21 A. L. Sobolewski, D. Shemesh and W. Domcke, *J. Phys. Chem. A*, 2009, **113**, 542–550.
22 D. Shemesh, A. L. Sobolewski and W. Domcke, *J. Am. Chem. Soc.*, 2009, **131**, 1374–1375.
23 D. Shemesh, C. Hattig and W. Domcke, *Chem. Phys. Lett.*, 2009, **482**, 38–43.
24 M. Marazzi, U. Sancho, O. Castano, W. Domcke and L. M. Frutos, *J. Phys. Chem. Lett.*, 2010, **1**, 425–428.
25 S. R. Mercier, O. V. Boyarkin, A. Kamariotis, M. Guglielmi, I. Tavernelli, M. Cascella, U. Rothlisberger and T. R. Rizzo, *J. Am. Chem. Soc.*, 2006, **128**, 16938–16943.
26 A. G. Szabo and D. M. Rayner, *J. Am. Chem. Soc.*, 1980, **102**, 554–563.
27 J. W. Petrich, M. C. Chang, D. B. McDonald and G. R. Fleming, *J. Am. Chem. Soc.*, 1983, **105**, 3824–3832.
28 V. Subramaniam, D. G. Steel, A. Gafni, In *Topics in fluorescence spectroscopy*; Geddes, C. D., Lakowicz, J. R., ed.; Kluwer Academic Publishers: Boston, 2002, pp. 43–65.
29 T. Cellmer, M. Buscaglia, E. R. Henry, J. Hofrichter and W. A. Eaton, *Proc. Natl. Acad. Sci. U. S. A.*, 2011, **108**, 6103–6108.
30 D. Nolting, C. Marian and R. Weinkauf, *Phys. Chem. Chem. Phys.*, 2004, **6**, 2633–2640.
31 Y. Iketaki, T. Watanabe, S. Ishiuchi, M. Sakai, T. Omatsu, K. Yamamoto, M. Fujii and T. Watanabe, *Chem. Phys. Lett.*, 2003, **372**, 773–778.
32 J. Léonard, D. Sharma, B. Szafarowicz, K. Torgasin and S. Haacke, *Phys. Chem. Chem. Phys.*, 2010, **12**, 15744–15750.
33 M. Doemer, M. Guglielmi, P. Athri, I. Tavernelli, and U. Rothlisberger (to be published).
34 S. Sugio, A. Kashima, S. Mochizuki, M. Noda and K. Kobayashi, *Protein Eng., Des. Sel.*, 1999, **12**, 439–446.
35 W. H. Qiu, L. Y. Zhang, O. Okobiah, Y. Yang, L. J. Wang, D. P. Zhong and A. Zewail, *J. Phys. Chem. B*, 2006, **110**, 10540–10549.

36 A. Siemiarczuk, C. E. Peterson, C.-E. Ha, J. Yang and N. V. Bhagavan, *Cell Biochem. Biophys.*, 2004, **40**, 115–122.
37 J. W. Ochterski, G. A. Petersson and J. A. Montgomery Jr, *J. Chem. Phys.*, 1996, **104**, 2598–2619.
38 C. Hampel, K. A. Peterson and H.-J. Werner, *J. Chem. Phys.*, 1992, **190**, 1–12.
39 J. D. Watts, J. Gauss and R. J. Bartlett, *J. Chem. Phys.*, 1993, **98**, 8718–8733.
40 P. M. Mayer, C. J. Parkinson, D. M. Smith and L. Radom, *J. Chem. Phys.*, 1998, **108**, 604–615.
41 P. Hohenberg and W. Kohn, *Phys. Rev.*, 1964, **136**, B864–870.
42 W. Kohn and L. Sham, *Phys. Rev. A*, 1965, **140**, 1133–1138.
43 A. D. Becke, *Phys. Rev. A: At., Mol., Opt. Phys.*, 1988, **38**, 3098–3100.
44 C. Lee, W. Yang and R. Parr, *Phys. Rev. B*, 1988, **37**, 785–789.
45 A. Becke, *J. Chem. Phys.*, 1993, **98**, 5648–5652.
46 Y. Zhao, N. E. Schultz and D. G. Truhlar, *J. Chem. Phys.*, 2005, **123**, 161103.
47 Y. Zhao, N. E. Schultz and D. G. Truhlar, *J. Chem. Theory Comput.*, 2006, **2**, 364–382.
48 Y. Zhao and D. G. Truhlar, *Theor. Chem. Acc.*, 2008, **120**, 215–241.
49 M. M. Francl, W. J. Pietro, W. J. Hehre, J. S. Binkley, M. S. Gordon, D. J. DeFrees and J. A. Pople, *J. Chem. Phys.*, 1982, **77**, 3654–3665.
50 M. J. Frisch, J. A. Pople and J. S. Binkley, *J. Chem. Phys.*, 1984, **80**, 3265–3269.
51 D. Woon and T. Dunning Jr, *J. Chem. Phys.*, 1993, **98**, 1358–1371.
52 M. J. Frisch, G. W. Trucks, H. B. Schlegel, G. E. Scuseria, M. A. Robb, J. R. Cheeseman, G. Scalmani, V. Barone, B. Mennucci, G. A. Petersson, H. Nakatsuji, M. Caricato, X. Li, H. P. Hratchian, A. F. Izmaylov, J. Bloino, G. Zheng, J. L. Sonnenberg, M. Hada, M. Ehara, K. Toyota, R. Fukuda, J. Hasegawa, M. Ishida, T. Nakajima, Y. Honda, O. Kitao, H. Nakai, T. Vreven, J. A. Montgomery, Jr., J. E. Peralta, F. Ogliaro, M. Bearpark, J. J. Heyd, E. Brothers, K. N. Kudin, V. N. Staroverov, R. Kobayashi, J. Normand, K. Raghavachari, A. Rendell, J. C. Burant, S. S. Iyengar, J. Tomasi, M. Cossi, N. Rega, J. M. Millam, M. Klene, J. E. Knox, J. B. Cross, V. Bakken, C. Adamo, J. Jaramillo, R. Gomperts, R. E. Stratmann, O. Yazyev, A. J. Austin, R. Cammi, C. Pomelli, J. W. Ochterski, R. L. Martin, K. Morokuma, V. G. Zakrzewski, G. A. Voth, P. Salvador, J. J. Dannenberg, S. Dapprich, A. D. Daniels, Ö. Farkas, J. B. Foresman, J. V. Ortiz, J. Cioslowski, D. J. Fox, *Gaussian 09, Revision A.1*, Gaussian, Inc., Wallingford CT, 2009.
53 S. Grimme, J. Antony, S. Ehrlich and H. Krieg, *J. Chem. Phys.*, 2010, **132**, 154104.
54 S. Grimme, S. Ehrlich and L. Goerigk, *J. Comput. Chem.*, 2011, **32**, 1456–1465.
55 TURBOMOLE V6.4 2012, 1989–2007 University of Karlsruhe and Fz. Karlsruhe GmbH, since 2007 TURBOMOLE GmbH; available from http://www.turbomole.com.
56 O. Christiansen, H. Koch and P. Jørgensen, *Chem. Phys. Lett.*, 1995, **243**, 409–418.
57 G. T. M. Frisch, H. Schlegel, G. Scuseria, J. C. M. Robb, J. Montgomery, T. V. Jr., K. Kudin, J. Burant, J. Millam, J. T. S. Iyengar, V. Barone, B. Mennucci, G. S. M. Cossi, N. Rega, G. Petersson, M. H. H. Nakatsuji, M. Ehara, K. Toyota, J. H. R. Fukuda, M. Ishida, T. Nakajima, O. K. Y. Honda, H. Nakai, M. Klene, J. K. X. Li, H. Hratchian, J. Cross, C. A. V. Bakken, J. Jaramillo, R. Gomperts, O. Y. R. Stratmann, A. Austin, C. P. R. Cammi, J. Ochterski, P. Ayala, G. V. K. Morokuma, P. Salvador, V. Z. J. Dannenberg, S. Dapprich, M. S. A. Daniels, O. Farkas, D. Malick, K. R. A. Rabuck, J. Foresman, Q. C. J. Ortiz, A. Baboul, S. Clifford, B. S. J. Cioslowski, G. Liu, P. P. A. Liashenko, I. Komaromi, R. Martin, T. K. D. Fox, M. Al-Laham, C. Peng, M. C. A. Nanayakkara, P. Gill, W. C. B. Johnson, M. Wong, C. Gonzalez, and a. J. Pople, *Gaussian 03*, Gaussian, Inc., Wallingford, CT, 2004.
58 R. Ahlrichs, M. Bär, M. Häser, H. Horn and C. Kölmel, *Chem. Phys. Lett.*, 1989, **162**, 165–169.
59 F. Weigend, M. Häser, H. Patzelt and R. Ahlrichs, *Chem. Phys. Lett.*, 1998, **294**, 143–152.
60 D. E. Woon and T. H. Dunning Jr, *J. Chem. Phys.*, 1993, **98**, 1358–1371.
61 J. Hutter, *J. Chem. Phys.*, 2003, **118**, 3928–3934.
62 S. Hirate and M. Head-Gordon, *Chem. Phys. Lett.*, 1999, **314**, 291–299.
63 Computer Code CPMD, Copyright IBM Corp 1990–2008, Copyright MPI für Festkörperforschung Stuttgart 1997–2001.
64 N. Troullier and J. Martins, *Phys. Rev. B: Condens. Matter*, 1991, **43**, 1993–2006.
65 G. J. Martyna and M. Tuckerman, *J. Chem. Phys.*, 1999, **110**, 2810–2821.
66 J. P. Perdew, K. Burke and M. Ernzerhof, *Phys. Rev. Lett.*, 1996, **77**, 3865–3868.
67 G. A. Jones, B. K. Carpenter and M. N. Paddon-Row, *J. Am. Chem. Soc.*, 1998, **120**, 5499–5508.
68 E. Tapavicza, I. Tavernelli and U. Rothlisberger, *J. Phys. Chem. A*, 2009, **113**, 9595–9602.
69 L. Bernasconi, M. Sprik and J. Hutter, *J. Chem. Phys.*, 2003, **119**, 12417–12431.

70 Y. Duan, C. Wu, S. Chowdhury, Mathew C. Lee, G. Xiong, W. Zhang, R. Yang, P. Cieplak, R. Luo, T. Lee, J. Caldwell, J. Wang and P. Kollman, *J. Comput. Chem.*, 2003, **24**, 1999–2012.
71 L. Kleinman and D. M. Bylander, *Phys. Rev. Lett.*, 1982, **48**, 1425–1428.
72 A. Laio, J. VandeVondele and U. Rothlisberger, *J. Chem. Phys.*, 2002, **116**, 6941–6947.
73 A. Laio, J. VandeVondele and U. Rothlisberger, *J. Phys. Chem. B*, 2002, **106**, 7300–7307.
74 A. Dreuw, J. L. Weisman and M. Head-Gordon, *J. Chem. Phys.*, 2003, **119**, 2943–2946.
75 A. Dreuw and M. Head-Gordon, *J. Am. Chem. Soc.*, 2004, **126**, 4007–4016.
76 G. Gregoire, C. Jouvet, C. Dedonder and A. Sobolewski, *Chem. Phys.*, 2006, **324**, 398–404.

Faraday Discussions RSC Publishing

PAPER

Coherent exciton transport driven by torsional dynamics: a quantum dynamical study of phenylene-vinylene type conjugated systems

R. Binder, J. Wahl, S. Römer and I. Burghardt*

Received 12th December 2012, Accepted 23rd January 2013
DOI: 10.1039/c3fd20148a

We present a quantum dynamical study of exciton transfer across a torsional defect that locally breaks the π-conjugation in an oligo-(*p*-phenylene vinylene) (OPV) fragment. A site-based vibronic coupling Hamiltonian is used which is formulated in a comparative fashion (i) for a Frenkel exciton basis, assuming localized electron-hole pairs whose superposition yields a delocalized exciton, and (ii) more accurately, for a Merrifield type exciton basis including spatially separated electron-hole pairs. Starting from a partially delocalized ("spectroscopic unit") initial condition, the observed transfer dynamics is found to involve two characteristic time scales: (i) a very rapid, coherent transient on a 10–100 femtosecond scale, largely determined by Rabi type oscillations modulated by bond-length-alternation modes, and (ii) a slower time scale involving the planarization of the torsional coordinates that determines the onset of a quasi-stationary exciton-polaron state, and in the process leads to a "healing" of the torsional defect within $\sim$ 500 femtoseconds. The dynamics obtained from the full electron-hole basis *vs.* Frenkel basis are in good agreement. In the full electron-hole dynamics, the transients are found to involve a rapid expansion and subsequent contraction of the electron-hole coherence size. Quantum dynamical simulations for a minimal six-site model involving 36 states and 22 vibrational modes, were carried out using the multiconfiguration time dependent Hartree (MCTDH) method.

1 Introduction

Exciton dynamics, or excitation energy transfer (EET), play a prominent role in the photoresponse of extended molecular aggregates, like light harvesting systems and semiconducting polymers. While EET is conventionally described in terms of Förster theory,[1,2] *i.e.*, within a perturbative framework that disregards the role of coherent effects, a series of recent experiments[3–6] have provided compelling evidence that the transfer of excitation on ultrafast time scales can be guided by quantum coherence, despite the presence of an environment, *e.g.*, in a solvent, protein, or polymer. Conjectures have been made regarding the possible role of correlated environmental

Goethe University Frankfurt, Institute for Physical and Theoretical Chemistry, Max-von-Laue-Str. 7, 60438 Frankfurt am Main, Germany. E-mail: burghardt@theochem.uni-frankfurt.de

fluctuations in generating these long-lived coherences.[4,5,7,8] Furthermore, coherent excited-state phonon dynamics have been observed experimentally for semiconducting polymers[9] and related systems, like carbon nanotubes,[10] on a time scale of tens to hundreds of femtoseconds (fs). Even torsional conformational changes can contribute to coherent effects on ultrafast time scales.[11]

Photogenerated excitons in semiconducting polymer species are typically delocalized over 2 to 12 monomer units,[4] as can be inferred from experiments measuring transient absorption anisotropy decays.[12] The intrinsic spatial extension of the excitons is often described in terms of "chromophores" or "spectroscopic units"[13,14] whose spatial boundaries can be defined by conformational defects, *e.g.*, kinks and torsional deformations.[15–18] The role of conformation in defining the exciton extension is, however, controversial[14,19] as are the details of the elementary exciton transport and relaxation events on ultrafast time scales.[20,21]

Against this background, we report here a full quantum dynamical study of ultrafast exciton transfer across a defect produced by a torsional displacement that locally breaks the π-conjugation. Starting from an excitonic wavefunction localized on either side of the conjugation break, coherent ultrafast transfer dynamics is observed, accompanied by a trend towards planarization. The exciton spreads rapidly across the defect, and a quasi-stationary state is reached on a time scale of several hundred femtoseconds. These results relate to the experimental observations reported in ref. 22 where it was conjectured that torsional conformational change leads to a dynamical expansion of the exciton by several monomer units, thus allowing for efficient downhill EET.

We employ a model Hamiltonian tailored to EET in poly-(*p*-phenylene vinylene) (PPV) type semiconducting polymers, parametrized as a function of high-frequency bond-length-alternation (BLA) and low-frequency ring-torsional modes. We focus on intra-chain transfer, which is known to provide the principal exciton transport mechanism in PPV type systems in solution,[4,23] and is most relevant for the discussion of torsional relaxation. By contrast, inter-chain transfer becomes increasingly important in films.[23–25] The model parametrization is in part based upon *ab initio* calculations for oligo-(*p*-phenylene vinylene) (OPV) species.[26,27] The model is formulated in a comparative fashion (i) for a Frenkel exciton[2,28] basis, assuming localized electron-hole pairs whose superposition yields a delocalized exciton, and (ii) more accurately, for a generalized Merrifield type[29] exciton basis involving spatially separated electron-hole pairs. The latter description accounts for the fact that the exciton coherence length is typically of the order of several monomer units in PPV and related systems[30–37] such that a Frenkel picture is not strictly applicable. (However, the system is still in a small-radius exciton regime, rather than a large-radius Wannier-Mott regime.[29,38,39]) As a result, a picture of the transfer process in terms of ultrafast electron-hole dynamics is obtained.

A minimal version of the model, involving a six-site system representing a phenylene-vinylene hexamer will be studied using the multiconfiguration time dependent Hartree (MCTDH) method.[40] This method provides an accurate quantum dynamical treatment which allows one to examine, not only the population transfer across the defect, but also the role of excitonic coherence in the transfer dynamics. This approach should be considered as complementary to classical and semiclassical approaches,[17,20,41–44] as well as kinetic descriptions[45–47] for the dynamics of PPV type systems. The present study significantly extends a preliminary investigation reported in ref. 48.

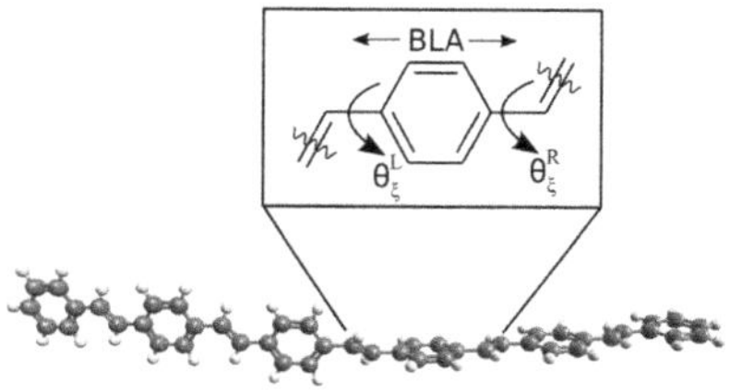

Fig. 1 Schematic illustration of the oligo-(*p*-phenylene vinylene) type hexamer studied in this work and definition of the basic monomer unit with "left" (L) and "right" (R) ring torsional coordinates and a bond-length alternation coordinate. The defect considered in the present simulations corresponds to a ±45° twist leading to an orthogonal arrangement of the left and right segments.

The remainder of the paper is organized as follows. Section 2 presents the general form of the model Hamiltonian in the basis of single-excited exciton states, Section 3 describes the quantum dynamical approach, Section 4 analyzes the observed EET dynamics, and Section 5 offers the discussion and conclusion.

2 Model Hamiltonian

We employ a site-based model Hamiltonian that is set up on the basis of general electron-hole (e–h) states $|\nu_e\mu_h\rangle \equiv |\nu\mu\rangle$ where the conduction-band electron is located at site $\nu_e = \nu$ while the valence-band hole is located at site $\mu_h = \mu$. All other sites remain in their ground states. Each site is defined in terms of a phenylene-vinylene monomer unit as depicted in Fig. 1, and carries "left" (L) and "right" (R) ring torsional modes, as well as a bond length alternation (BLA) mode. The discussion will be restricted to singlet states, such that spin indices do not appear explicitly. We have chosen the most general e–h basis as a starting point since electronic structure studies indicate that the lowest-energy excitonic manifold of PPV type systems is not of pure Frenkel type, but involves spatially separated e–h pairs extending over several monomer units.[32–37] The general, Merrifield type[29] e–h basis reduces to a Frenkel basis if the electrons and holes are always localized at the same site, $\nu = \mu$ and $\nu' = \mu'$. The Frenkel basis can alternatively be interpreted as referring to the center-of-mass motion of the e–h pair.[21] In the dynamical study reported below, a comparative analysis will be given of the EET process represented in the full e–h basis *vs.* the Frenkel basis. We will refer to states $|\nu\nu\rangle$ as Frenkel excitons, while states $|\nu\mu\rangle$, $\nu \neq \mu$, will be denoted charge transfer (CT) excitons.

2.1 Hamiltonian in electron-hole basis

The general form of the Hamiltonian reads as follows,

$$\hat{H} = \sum_{\nu\mu} \sum_{\nu'\mu'} H_{\nu\mu,\nu'\mu'} |\nu\mu\rangle\langle\nu'\mu'| \tag{1}$$

with

$$H_{\nu\mu,\nu'\mu'} = \delta_{\nu\nu'}\delta_{\mu\mu'}\left(H^{\rm intr}_{\nu\mu,\nu'\mu'} + H^{\rm tors}_{\nu\mu,\nu'\mu'} + H^{\rm BLA}_{\nu\mu,\nu'\mu'} + H^{\rm bath}_{\nu\mu,\nu'\mu'}\right) + \delta_{\nu\mu}\delta_{\nu'\mu'}H^{\rm Frenkel}_{\nu\mu,\nu'\mu'} + H^{\rm CT}_{\nu\mu,\nu'\mu'} \tag{2}$$

where all terms can carry a dependence on vibrational coordinates as detailed in the following section. The first four terms on the r.h.s. of eqn (2) are diagonal in the e–h basis, and correspond to the intrinsic e–h pair interaction (H^{intr}), along with vibronic contributions relating to ring-torsional modes (H^{tors}) and BLA modes (H^{BLA}) as well as bath modes (H^{bath}) acting as a source of dissipation. The last two terms involve off-diagonal couplings, which are either of Frenkel type (H^{Frenkel}), or else of charge transfer type (H^{CT}), *i.e.*, involving Frenkel excitonic couplings or electron/hole transfer integrals.

Besides the seminal work by Merrifield[29] and its generalization to include exciton–phonon coupling,[39,49] the Hamiltonian of eqn (1)–(2) is closely related to an e–h Hamiltonian employed by Mukamel and collaborators[50,51] in a Wannier function basis, based upon a Pariser–Parr–Pople (PPP) Hamiltonian. A related formulation is given by Karabunarliev and Bittner,[35] see also ref. 52.

Parallel to the Hamiltonian of eqn (1)–(2), a reduced Frenkel model will be employed, where all terms involving CT excitons have been discarded,

$$\hat{H}_{\text{Frenkel}} = \sum_{\nu}\sum_{\nu'} H_{\nu\nu,\nu'\nu'} |\nu\nu\rangle\langle\nu'\nu'| \tag{3}$$

with

$$H_{\nu\nu,\nu'\nu'} = \delta_{\nu\nu'}\left(H^{\text{intr}}_{\nu\nu,\nu'\nu'} + H^{\text{tors}}_{\nu\nu,\nu'\nu'} + H^{\text{BLA}}_{\nu\nu,\nu'\nu'} + H^{\text{bath}}_{\nu\nu,\nu'\nu'}\right) + H^{\text{Frenkel}}_{\nu\nu,\nu'\nu'} \tag{4}$$

In eqn (4) all individual terms are defined in the same fashion as in eqn (2).

In the following, the various contributions to the Hamiltonian eqn (1) and (2) – and, hence, the reduced Hamiltonian eqn (3) and (4) – are specified. All parameters are compiled in Tables 1 to 4.

2.1.1 Electron-hole on-site contributions. The first diagonal contribution to the Hamiltonian eqn (2) is the intrinsic e–h interaction,

Table 1 Electron-hole interaction parameters, transfer integrals, and Frenkel couplings according to eqn (5) and eqn (10)–(13), respectively. All values given in atomic units

e_0	ε_r	r_{ru}	r_0
7.90×10^{-2}	1	12.7	12.7
t	j_4	j_2	j_0
9.78×10^{-2}	3.23×10^{-2}	1.20×10^{-2}	5.03×10^{-3}

Table 2 Torsional potential parameters according to eqn (7), for $S = G,E$. In addition, force constants k_S and frequencies ω_S are indicated, from the expansion of the G/E potentials around their respective equilibrium geometries. The moment of inertia takes the value $I_{\text{tors}} = 5.91 \times 10^5$. All values are given in atomic units

a_G	b_G	c_G	$k_G(\theta_{0,G} = 24°)$	$\omega_G(\theta_{0,G} = 24°)$
-7.07×10^{-4}	-2.44×10^{-3}	9.54×10^{-4}	8.78×10^{-3}	1.22×10^{-4}
a_E	b_E	c_E	$k_E(\theta_{0,E} = 0°)$	$\omega_E(\theta_{0,E} = 0°)$
-8.72×10^{-4}	-2.59×10^{-2}	-2.85×10^{-3}	1.50×10^{-1}	-5.04×10^{-4}

Table 3 BLA potential parameters pertaining to eqn (8). The BLA mass takes the value $m_{BLA} = 5.49 \times 10^3$. All values are given in atomic units

k_G	ω_G	$x_{0,G}$	k_E	ω_E	$x_{0,E}$
2.92×10^{-1}	7.29×10^{-3}	1.91×10^{-1}	2.38×10^{-1}	6.59×10^{-3}	7.44×10^{-2}

Table 4 Bath parameters specifiying the system-bath couplings $c_{B,i} = (4\omega\Delta\omega\gamma\omega/\pi)^{1/2}$ according to eqn (9), as further described in the text. All values are given in atomic units unless otherwise noted

T_P[fs]	$\Delta\omega$	γ
1500	1.01×10^{-4}	8.06×10^{-5}

$$H^{\text{intr}}_{\nu\mu,\nu'\mu'} = \delta_{\nu\nu'}\delta_{\mu\mu'}\left(e_0 - \frac{1}{4\pi\varepsilon_0\varepsilon_r r_{\nu\mu}}\right) \tag{5}$$

where $r_{\nu\mu}$ is the e–h distance as a function of the sites (ν,μ), $r_{\nu\mu} = |\nu - \mu| r_{\text{ru}} + r_0$, with r_{ru} the repeat unit length, and r_0 the intrinsic e–h distance for a Frenkel exciton. Further, e_0 is the exciton binding energy[53,54] for a localized e–h pair such that $H^{\text{intr}}_{\nu\nu,\nu\nu}=0$.[55] The relative permittivity was set to $\varepsilon_r = 1$.[56]

Next, vibronic contributions $\{H^{\text{tors}},H^{\text{BLA}}\}$ to the diagonal portion of eqn (2) correspond to excitonic potentials involving combinations of monomer ground state (G) and excited-state (E) potentials (see also ref. 48),

$$H^{\text{vibronic}}_{\nu\mu,\nu'\mu'}(\{x_{\xi i}\}) = \delta_{\nu\nu'}\delta_{\mu\mu'}\sum_{\xi=1}^{N_{\text{site}}}\left(-\frac{1}{2}\frac{\partial^2}{\partial x_{\xi i}^2} + \text{v}_i^G(x_{\xi i})(1-\delta_{\nu\xi}) + \text{v}_i^E(x_{\xi i})\delta_{\nu\xi}\right) \tag{6}$$

where $\{x_{\xi i}\}$ represents a set of site-local modes, N_{site} is the number of monomer units, and v_i^G and v_i^E denote ground and excited-state monomer potentials. For a given e–h state (ν,μ), the location of the electron ($\nu_e = \nu$) determines the site participating with an excited-state monomer potential (E). In the present model, each site contributes two torsional angles and a BLA coordinate, $x_{\xi i}\rightarrow\{\theta_\xi^L,\theta_\xi^R,x_\xi^{\text{BLA}}\}$ as depicted in Fig. 1. Mass-weighted coordinates are used throughout. Hence, $H^{\text{vibronic}}_{\nu\mu,\nu'\mu'}(\{x_{\xi i}\}) = H^{\text{tors}}_{\nu\mu,\nu'\mu'}\left(\left\{\theta_\xi^L\right\}\right) + H^{\text{tors}}_{\nu\mu,\nu'\mu'}\left(\left\{\theta_\xi^R\right\}\right) + H^{\text{BLA}}_{\nu\mu,\nu'\mu'}\left(\left\{x_\xi^{\text{BLA}}\right\}\right)$.

For the torsional part of the Hamiltonian, the ground and excited-state potentials are given in terms of a truncated Fourier series,[27]

$$\text{v}^S_{\text{tors}}(\theta)=a_S\cos(\theta) + b_S\cos(2\theta) + c_S\cos(4\theta) - a_S - b_S - c_S \tag{7}$$

where $S = \{G,E\}$. Importantly, these potentials account for the fact that the equilibrium geometry of the electronic ground state (G) is non-planar, with $\theta_{0,G} \sim 24°$, while the excited-state potentials (E) induce planarization and have a larger curvature than the ground state potentials,[6,26,57,58] see Table 2.

For the BLA part of the Hamiltonian, harmonic state-dependent potentials are used,

$$\text{v}^S_{\text{BLA}}(x) = \frac{1}{2}\omega_S^2(x - x_{0,S})^2 \tag{8}$$

obtained again by a fit to electronic structure data.[26] The BLA displacements x_ξ^{BLA} were defined in terms of differences of double and single bond lengths on the vinylene and phenylene moieties.[26,42]

Finally, the bath Hamiltonian H^{bath} is taken to act on a subset of "active" torsional coordinates θ that are located at the junction exhibiting a conjugation break. H^{bath} takes the standard form,[2,59]

$$H^{\text{bath}}_{\nu\mu,\nu'\mu'} = \delta_{\nu\nu'}\delta_{\mu\mu'}\sum_{i=1}^{N_B}\left(-\frac{1}{2}\frac{\partial^2}{\partial x_{B,i}^2}+\frac{1}{2}\omega_{B,i}^2\left(x_{B,i}-\frac{c_{B,i}}{\omega_{B,i}^2}\theta\right)^2\right) \tag{9}$$

The system-bath couplings $\{c_{B,i}\}$ are sampled from an Ohmic spectral density, $c_{B,i} = (2\omega\Delta\omega J_{\text{ohm}}(\omega)/\pi)^{1/2}$, where $\Delta\omega$ is a discretized frequency interval and $J_{\text{ohm}}(\omega) = 2\gamma\omega\exp(-\omega/\Lambda)$ with the friction coefficient γ and the cutoff frequency Λ, providing a phenomenological damping effect.[59,60] In the present treatment, dissipation is restricted to two active torsions (θ^L, θ^R) whereas the BLA modes are only indirectly subject to energy loss due to vibronic coupling effects.

2.1.2 Frenkel couplings and transfer integrals. Two types of e–h couplings are introduced in eqn (2), *i.e.*, Frenkel type couplings between localized e–h pairs ($\nu = \mu$, $\nu' = \mu'$) and electron/hole transfer integrals between delocalized e–h pairs ($\nu \neq \mu$ and/or $\nu' \neq \mu'$). Both couplings are restricted to nearest-neighbor interactions. The Frenkel type couplings correspond to two-particle e–h exchange interactions according to ref. 50,51, while the transfer integrals correspond to electron and hole single-particle terms which are here assumed to be symmetric for electrons *vs.* holes.

The Frenkel-type couplings are taken to depend on the difference torsional angle $\theta^{\text{diff}}_{\nu'\nu} = 1/2(\theta^R_{\nu'} - \theta^L_{\nu})$ for a pair of neighboring sites,

$$H^{\text{Frenkel}}_{\nu\mu,\nu'\mu'} = \delta_{\nu\mu}\delta_{\nu'\mu'}\left(\delta_{\nu',\nu+1}J_{\text{Frenkel}}\left(\frac{1}{2}\left(\theta^R_{\nu}-\theta^L_{\nu'}\right)\right)+\delta_{\nu',\nu-1}J_{\text{Frenkel}}\left(\frac{1}{2}\left(\theta^R_{\nu'}-\theta^L_{\nu}\right)\right)\right) \tag{10}$$

where[27]

$$J_{\text{Frenkel}}(\theta) = (j_4 - j_0)\cos^4(\theta) + (j_2 - j_0)\cos^2(\theta) + 2j_0 \tag{11}$$

The relevant parameters are again taken from the fit of electronic structure data for OPV oligomer species[26] to a Hückel type Hamiltonian.[27]

The electron and hole transfer integrals also act as nearest-neighbor couplings, and are in turn taken to depend upon the difference torsional angle $\theta^{\text{diff}}_{\nu'\nu} = 1/2(\theta^R_{\nu'} - \theta^L_{\nu})$,

$$\begin{aligned} H^{\text{CT}}_{\nu\mu,\nu'\mu'} = {} & \delta_{\nu\nu'}\left(\delta_{\mu',\mu+1}t_{\text{CT}}\left(\frac{1}{2}\left(\theta^R_{\mu}-\theta^L_{\mu'}\right)\right)+\delta_{\mu',\mu-1}t_{\text{CT}}\left(\frac{1}{2}\left(\theta^R_{\mu'}-\theta^L_{\mu}\right)\right)\right) \\ & + \delta_{\mu\mu'}\left(\delta_{\nu',\nu+1}t_{\text{CT}}\left(\frac{1}{2}\left(\theta^R_{\nu}-\theta^L_{\nu'}\right)\right)+\delta_{\nu',\nu-1}t_{\text{CT}}\left(\frac{1}{2}\left(\theta^R_{\nu'}-\theta^L_{\nu}\right)\right)\right) \end{aligned} \tag{12}$$

where a standard functional form of the coupling is employed[61–63]

$$t_{\text{CT}}(\theta) = t\cos(\theta) \tag{13}$$

The Frenkel couplings and transfer integrals are of similar magnitude (see Table 1), and take large values – of the order of 1–2 eV – in the monomer-based representation chosen here.

2.2 Potential surfaces

The Hamiltonian of eqn (1)–(2), or its simplified variant eqn (3)–(4), generate a series of high-dimensional coupled potential energy surfaces (PESs) pertaining to the e–h states. These PESs, which correspond to a diabatic representation,[64] determine the transfer events in terms of transitions induced by the off-diagonal coupling terms described in the preceding section. In the general e–h basis, the number of PESs is N_{site}^2, as compared with N_{site} surfaces in the Frenkel case.

Fig. 2 illustrates the geometry dependence of several representative potential terms, *i.e.*, (i) the torsional dependence of the localized e–h configurations $|33\rangle$ ($\nu = \mu = 3$) and $|44\rangle$ ($\nu' = \mu' = 4$) and the Frenkel coupling connecting these (panel (a)), (ii) the torsional dependence of the Frenkel exciton $|33\rangle$ ($\nu = \mu = 3$) and the CT exciton $|43\rangle$ ($\nu' = 4, \mu' = 3$), and the transfer integral which couples these (panel (b)), and (iii) the BLA dependence of the Frenkel configurations $|33\rangle$ and $|44\rangle$ (panel (c)).

The Frenkel coupling is seen to lift the degeneracy of the $|\nu\nu\rangle$ and $|\mu\mu\rangle$ configurations. This coupling is maximal at the planar geometry, entailing a maximal gap between the adiabatic states obtained by diagonalizing the Hamiltonian. At $\theta^{\mathrm{diff}} = \pi/2$, the coupling is minimal such that a near-degeneracy occurs between the adiabatic states. This is in agreement with the observed near-degeneracy, at the twisted geometry, of the ^1B_u and ^1A_g adiabatic states of the lowest singlet manifold in the relevant oligomer species.[26] The transfer integral acting between Frenkel and CT excitons supports this trend, with the cosine coupling of eqn (13), which is again maximal at the planar geometry.

The BLA modes act diagonally creating high-frequency level fluctuations modulating the effect of the torsion-dependent couplings.

For PPV type oligomers, a quinoid structure is favored in the excited state, leading to a planarization trend upon photoexcitation,[65] as mentioned above. Hence, torsional dynamics is a characteristic feature of the photoexcited species, starting from an initial condition which is generally non-planar, due to the slightly twisted ground-state equilibrium geometry, $\theta_{0,G} \sim 24°$,[26] and a statistical distribution of initial geometries around $\theta_{0,G}$. In the present study, we assume a

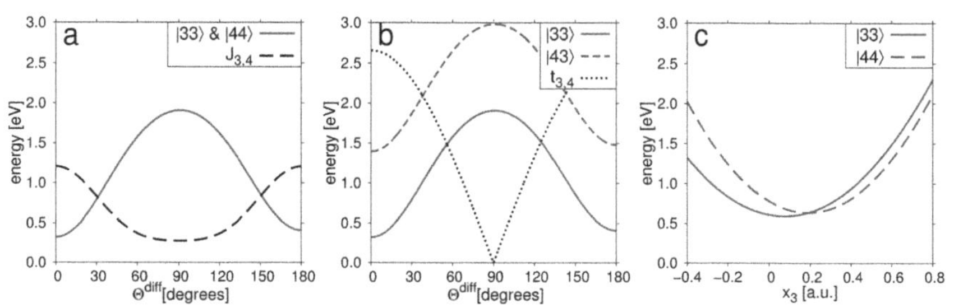

Fig. 2 Geometry dependence of the diabatic potentials used in this study: (a) torsional potentials pertaining to the Frenkel type configurations $|33\rangle$ and $|44\rangle$ whose degeneracy is lifted by the coupling $J_{3,4}$; (b) torsional potentials of the Frenkel exciton $|33\rangle$ and the CT exciton $|43\rangle$ which are coupled by the transfer integral $t_{3,4}$; (c) BLA dependence of the Frenkel configurations $|33\rangle$ and $|44\rangle$.

more pronounced initial torsional displacement, due to a random defect that is subsequently "healed" by the torsional dynamics.

3 Quantum dynamical methods

Quantum dynamical calculations are carried out for both the e–h basis and the Frenkel basis using the MCTDH method.[66–69] In the MCTDH approach, the wavefunction is expanded in terms of Hartree products, *i.e.*, products of time-dependent so-called single-particle functions (spfs) $\varphi_j^{(\kappa)}(\mathbf{x}_\kappa,t)$,

$$|\psi(\mathbf{x}_1,\ldots,\mathbf{x}_f,t)\rangle = \sum_{\mu,\nu}^{N_{\text{site}}}\sum_{j_1}^{n_1}\cdots\sum_{j_f}^{n_f} A_{j_1,\ldots,j_f,\nu\mu}(t)\prod_{\kappa=1}^{f}\varphi_{j_\kappa}^{(\kappa)}(\mathbf{x}_\kappa,t)|\nu\mu\rangle \tag{14}$$

with the time-dependent expansion coefficients $A_{j_1,\ldots,j_f,\mu\nu}(t)$ and time-dependent spf configurations, whose evolution is given in terms of nonlinear variational dynamical equations. The spfs are further expanded in a primitive grid basis, typically a discrete-variable representation (DVR).[66] Eqn (14) corresponds to the so-called *single-set* form of the wavefunction,[66] where electronic states are treated on equal footing with other degrees of freedom; this form is best adapted to the large number of states (here, up to N_{site}^2=36) that are included in the dynamics.

In the hexamer system, 22 vibrational modes were propagated on $N_{\text{site}} = 6$ or N_{site}^2=36 non-adiabatically coupled potential surfaces. The modes include 6 BLA modes with 30 DVR grid points per mode, two torsional modes with 256 grid points each, and 14 bath modes with 20 to 50 grid points per mode. Since the torsional degrees of freedom necessitate large grids and 6 to 8 spfs, only the torsions $\{\theta_3^R,\theta_4^L\}$ at the central junction were included explicitly. In the ansatz eqn (14), the spfs can be either one-dimensional or multi-dimensional (so-called combined-mode spfs). In the latter case, the number of configurations is significantly reduced, which is of special importance for calculations with a large number of degrees of freedom. In the present example, the BLA modes and bath modes were combined into 2-mode or 3-mode particles, with the number n_κ of basis functions in a given subspace between 2 and 3. The overall number of configurations was $N_{\text{config}} = 1\,492\,992$ in the calculations reported below. With these conditions, the calculations can be considered reasonably accurate but not fully converged.

The calculations were restricted to zero temperature, but include a collection of explicit bath modes, see eqn (9), whose frequencies and couplings are sampled from an Ohmic spectral density, as described in Section 2.1.1. The bath parametrization is chosen in such a way that the bath efficiently absorbs the energy contained in the central torsional modes and thus induces a "cooling" of the excitonic subsystem as well. To this end, the inverse friction coefficient is chosen of the order of the torsional period,[70] $\gamma^{-1} = 300$ fs. In the present simulations, independent baths with a minimal number of modes (here, 7 modes) were taken to act on the individual torsional modes. The Poincaré recurrence time[59] was $T_P = 1500$ fs. This system-bath interaction model can be straightforwardly extended to include more active coordinates. Further, a bath can be constructed, possibly based upon a realistic spectral density, that acts in a correlated fashion[7,43] on several system modes.

Prior to real-time calculations, imaginary-time (relaxation) propagation was carried out in order to create the initial, partially delocalized exciton state to the "left" of the torsional defect. In line with a vertical excitation picture, the initial state corresponds to the lowest adiabatic exciton state of the selected fragment, taken at the ground-state equilibrium geometry. In the present model, the ground state equilibrium geometry was artificially modified by introducing a constraint potential which creates an equilibrium for the torsional modes at the (3,4) junction near the ±45° degree twisted geometry. This leads to orthogonally arranged fragments as depicted in Fig. 1.

Finally, for the purpose of analysis, time-evolving density matrix elements will be calculated,

$$\langle \rho_{\nu\mu,\nu'\mu'} \rangle(t) = \mathrm{Tr}\{|\nu'\mu'\rangle\langle\nu\mu|\hat{\rho}(t)\} \tag{15}$$

where the trace (Tr) is taken over the electronic and phonon degrees of freedom of the density operator $\hat{\rho}(t) = |\psi(t)\rangle\langle\psi(t)|$. The diagonal elements

$$P_{\nu\mu}(t) = \delta_{\nu\nu'}\delta_{\mu\mu'}\langle \rho_{\nu\mu,\nu'\mu'} \rangle(t) \tag{16}$$

correspond to e–h state populations; in the Frenkel case, these are restricted to populations of the form $P_{\nu\nu}(t)$. The off-diagonal elements of eqn (15) with $\nu \neq \nu'$ and/or $\mu \neq \mu'$ represent coherences.

4 EET dynamics

As explained above, excitation energy transfer is monitored starting from an initial condition which corresponds to a delocalized exciton on the "left" fragment. Despite the minimal size of the trimer sub-fragment, this state can be taken to correspond to the excitonic ground state, *i.e.*, the first state (^1B_u) of the lowest-lying singlet exciton manifold; this state carries the predominant oscillator strength in PPV type systems.[6,26,71] We will denote this initial exciton ground state on the left fragment as $(^1B_u)_L$ for clarity. If we analyze the same state from the perspective of the combined (L + R) system, the initial state is asymmetrically localized and essentially represents a coherent superposition of the first two hexamer eigenstates, $(^1B_u)_{L+R}$ and $(^1A_g)_{L+R}$.[26,27] As borne out by electronic structure analysis,[26,33,37,72] the excitonic eigenstates can be well approximated as particle-in-a-box states with an increasing number of nodes. The initial nonstationary superposition of the two lowest (L + R) states results in a rapid delocalization that is a determining factor for the observed dynamics.

As mentioned above, the initial state is not fully vibrationally relaxed since a vertical excitation from the ground state at fixed nuclear geometry is assumed. Further, the torsional angles at the central junction are displaced so as to form a defect as illustrated in Fig. 1. Due to the mismatch between the initial condition $\theta^{R/L}(t=0) = \pm 45°$ and the planar equilibrium geometry of the excitonic states (see above), a vibrationally "hot" torsional wavepacket is produced. This enhances a trend that is already present if photoexcitation occurs from the slightly twisted ground state minimum ($\theta_{0,G} = 24°$).

In the following, we report on a comparative study for the full e–h model of eqn (1)–(2) and the reduced Frenkel model of eqn (3)–(4), respectively.

4.1 Frenkel exciton dynamics

Fig. 3 to 5 illustrate the transfer dynamics observed using the Frenkel model of eqn (3)–(4). Starting from the initial condition described above, localized on the

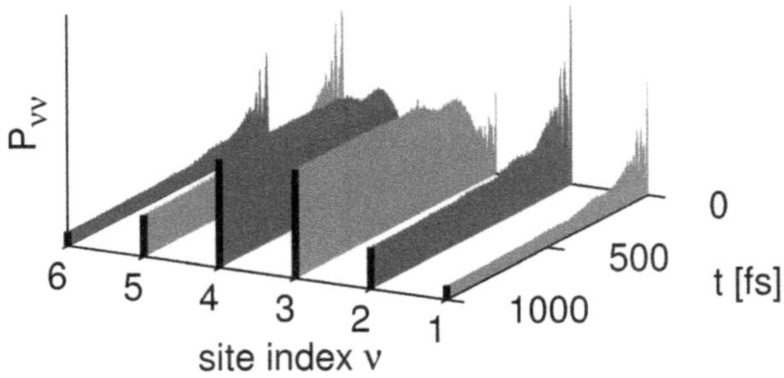

Fig. 3 Time-dependent site populations $P_{\nu\nu}$ for the Frenkel exciton dynamics studied in Section 4.1. An initial, highly oscillatory dynamics within the first ~ 100 fs is followed by an evolution towards a stationary state corresponding to the "trapped" exciton ground state.

left trimer fragment, coherent transfer dynamics is initiated on a very rapid time scale of several femtoseconds. Oscillatory transients extend up to ~ 100 fs, followed by a transition to a stationary state on a longer time scale of ~ 500 fs (see Fig. 3). On the latter time scale, one also observes the planarization of the active torsional coordinates under the effect of dissipation, as illustrated in Fig. 5. Within this time scale, the exciton spreads across the torsional defect, thus healing the defect.

The initial, very rapid transfer dynamics is induced by large nearest-neighbor couplings that act even while the wavepacket is still centered on the twisted geometry. The Rabi type transfer is modulated by energy level fluctuations induced by the bond-length alternation dynamics, giving rise to the rapid oscillations observed within a ~ 100 fs time window. Following the initial spread of the chromophore across the defect, a quasi-stationary state is subsequently reached due to the interplay of BLA and torsional modes on the characteristic time scale of planarization. This quasi-stationary state can be characterized as a trapped ground-state exciton,[26,42,73,74] with a distribution of site occupancies that peaks at the center of the hexamer while the BLA modes are displaced in a symmetric fashion, as shown in Fig. 4. The observed BLA displacements (of the order of 0.06 Å) are in good agreement with the OPV electronic structure calculations of ref. 26 However, the trapped state is still not fully vibrationally relaxed since BLA fluctuations persist for several hundreds of femtoseconds; the BLA displacements illustrated in Fig. 4 have therefore been obtained by averaging over several BLA periods.

The quasi-stationary state represents the singlet exciton ground state $(^1B_u)_{L+R}$ of the enlarged ("left-plus-right") fragment which is reached due to efficient energy dissipation *via* the torsional degrees of freedom. In the absence of dissipation, a strong admixture of the first excited state $(^1A_g)_{L+R}$, with a node at the center of the lattice, is found.[75] Analysis of the adiabatic state populations as a function of time confirms that only the first two adiabatic states are involved in the dynamics.[76]

Fig. 5 illustrates the relation between the torsional dynamics and the time-evolving excitonic coherence $\langle\rho_{33,44}\rangle(t)$ at the central junction. The difference torsional coordinate $1/2(\theta_3^R-\theta_4^L)$ is shown, which defines the initial condition $(\theta^R(t=0)=-\theta^L(t=0))$ and essentially remains the active coordinate while the sum coordinate does not play a significant role.[75] Following the rapid initial

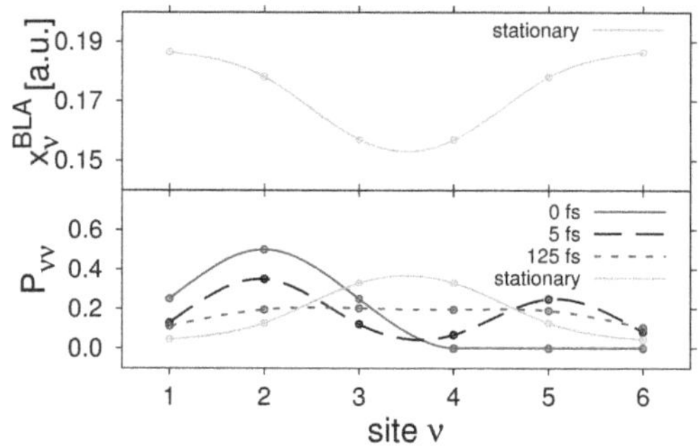

Fig. 4 Frenkel exciton dynamics: BLA displacements $\langle x_\nu^{BLA}\rangle$ (upper panel) characteristic of the trapped stationary exciton state, juxtaposed with site populations $P_{\nu\nu}$ at selected times (lower panel). Here, the site displacements and populations have been connected to form continuous distributions. The initial state, centered to the left of the conjugation break, is seen to immediately evolve towards a symmetric distribution spanning the complete lattice. Subsequently, the exciton contracts to yield the stationary state that is mirrored by the BLA lattice displacements.

transients described above, the coherence evolution towards a stationary state is guided by the torsional coordinate.

The imaginary part of the coherence, $\mathrm{Im}\langle\rho_{33,44}\rangle(t)$, determines the state-to-state population flux[77] whereas the real part, $\mathrm{Re}\langle\rho_{33,44}\rangle(t)$, indicates that a stationary coherent superposition state is reached. The transient dynamics determined by $\mathrm{Im}\langle\rho_{33,44}\rangle(t)$ is seen to persist for $\sim$ 100 fs; this time interval agrees with the strongly oscillatory initial dynamics observed in Fig. 3. Following this, the stationary state described above is reached by 500–600 fs.

4.2 Full e–h dynamics

A similar, but not identical picture is obtained from the more complete model of eqn (1)–(2) using a full electron-hole basis. This model yields a time-dependent picture of the process in terms of electron-hole dynamics. To analyze the dynamics, we introduce e–h center-of-mass and e–h distance ("coherence length") coordinates and the associated occupancies, as further explained in Appendix A.

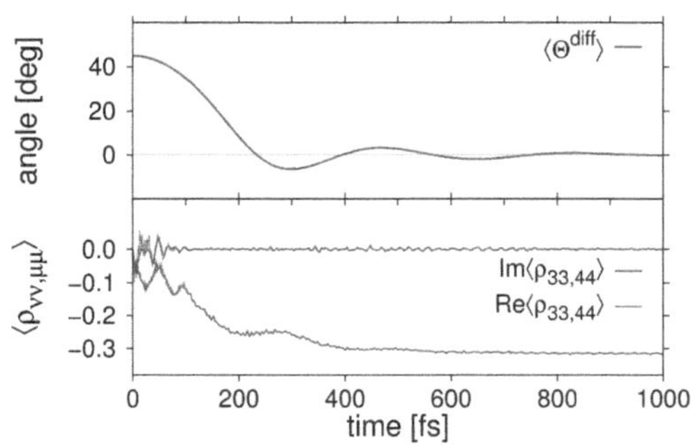

Fig. 5 Frenkel exciton dynamics: Time evolution of the difference torsional angle $\langle\theta_{3,4}^{diff}\rangle = 1/2(\langle\theta_3^R\rangle - \langle\theta_4^L\rangle)$ juxtaposed with the evolution of the excitonic coherence matrix element $\langle\rho_{33,44}\rangle$ (real and imaginary parts). Following initial transients characterized by a non-zero imaginary part of the excitonic coherence, the planarization of the difference torsional coordinate is seen to guide the evolution towards a stationary coherent superposition.

We first compare several snapshots of the time-evolving populations, where we now use the center-of-mass exciton populations P_ξ^{CM} as introduced in eqn (A2), see Fig. 6. The figure shows that the shape of the excitonic distribution starts to resemble the stationary population distribution at ~50 fs, but a quasi-stationary exciton-polaron state is only reached by ~500 fs, similarly to the Frenkel case. However, coherence decay takes place more rapidly, due to the significantly larger number of available pathways. Indeed, coherence matrix elements $\text{Im}\langle \rho_{\nu\mu,\nu'\mu'} \rangle$ decay on a very short time scale of ~ 50 fs.[76] The stationary state as such is similar to the Frenkel state, even though it apparently contains an admixture of the second (and possibly higher) eigenstate(s) of the enlarged fragment, and is therefore less strongly peaked at the center of the hexamer fragment.

The full e–h dynamics are illustrated in Fig. 7 where the occupation probabilities of the e–h states, $P_{\mu\nu} = \langle \rho_{\nu\mu,\nu\mu} \rangle$, see eqn (16) and eqn (A.1), are shown as a function of time. These are in one-to-one correspondence with the squared transition densities obtained from electronic structure analysis,[33] see also a recent discussion of time-dependent transition density matrices.[78] As shown in Fig. 7, the spatial extension of the exciton increases rapidly and tends towards a superposition of the first two excitonic eigenstates of the enlarged fragment. Note that no perfect e–h symmetry is observed, which is due to the fact that the diagonal elements of the Hamiltonian eqn (1)–(2) are not symmetric with respect to e–h exchange. Similar observations have been made for the transition densities computed for OPV oligomer species.[26]

Fig. 8 shows the time evolution of the coherence size distribution P_Δ^{CS} as defined in eqn (A.3). This distribution reflects the combined weight of all (ν,μ) combinations with a given *e–h* distance or coherence length $\Delta = \nu - \mu$, see Appendix A. A transient evolution is observed up to about 300 fs – slightly less than the interval of 500 fs needed to obtain a stationary distribution of populations. Electron-hole separations of one or two monomer units are shown to contribute significantly, as expected from electronic structure analysis.[30–37] The standard deviation σ,

$$\sigma = \left(\sum_{\Delta=-(N_{site}-1)}^{(N_{site}-1)} \Delta^2 P_\Delta^{CS} - \overline{\Delta}^2 \right)^{1/2} ; \overline{\Delta} = \sum_{\Delta=-(N_{site}-1)}^{(N_{site}-1)} \Delta P_\Delta^{CS} \tag{17}$$

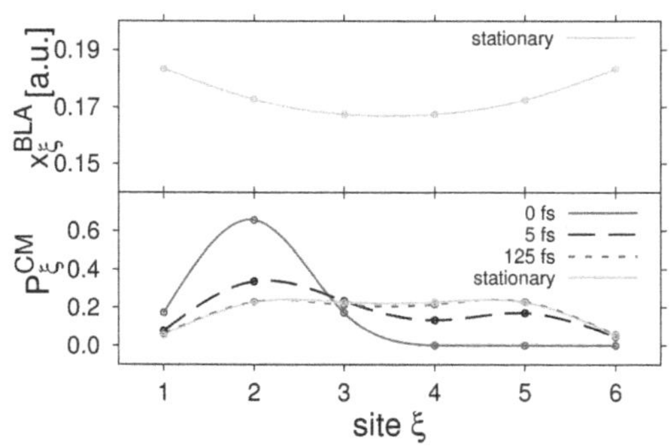

Fig. 6 Full e–h dynamics: Analogously to Fig. 4, BLA displacements $\langle x_\xi^{BLA} \rangle$ (upper panel) characteristic of the trapped stationary exciton state, juxtaposed with site (center-of-mass) populations P_ξ^{CM} at selected times (lower panel). Even though the shape of the excitonic distribution resembles the stationary state early on (see trace at ~ 50 fs), the latter is only reached by ~ 500 fs as in the Frenkel case.

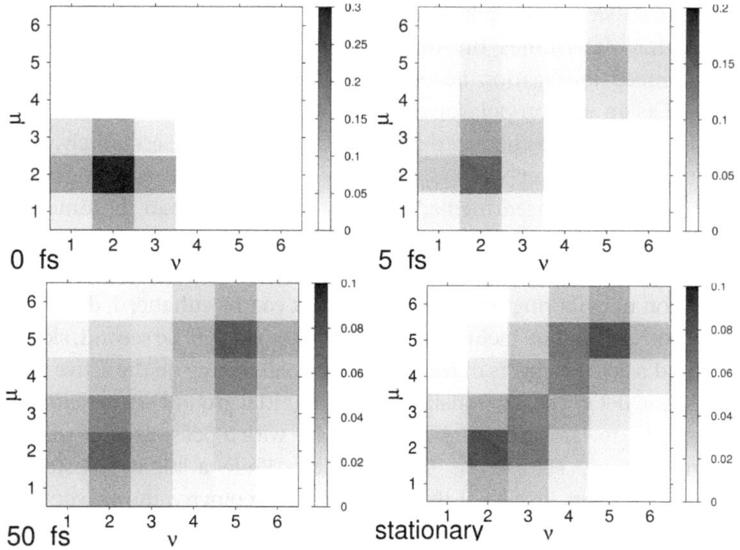

Fig. 7 Full e–h dynamics: Occupation probabilities of the e–h states, $P_{\nu\mu} = \langle\rho_{\nu\mu,\nu\mu}\rangle$ shown for selected times. The axes (ν,μ) refer to the electron and hole positions defining the state $|\nu\mu\rangle$.

shown in the upper panel of Fig. 8,[79] provides a measure of the time-evolving "size" of the exciton. Interestingly, the exciton is seen to expand initially up to $\sigma \sim 1.85$ until the size settles down to a stationary value of $\sigma \sim 1.4$. While the present model parametrization cannot be expected to yield quantitative accuracy, these values are reasonable when compared with the abovementioned electronic structure analyses.

5 Discussion and conclusions

To summarize, the dynamics observed in the Frenkel case as compared with the full e–h picture yield similar results, exhibiting two characteristic time scales: (i) a very short time scale characterized by transients in a 10–100 fs window, largely determined by Rabi oscillations that are modulated by high-frequency BLA

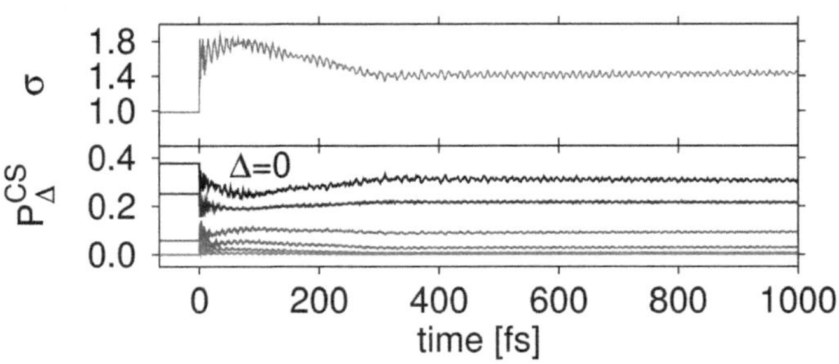

Fig. 8 Full e–h dynamics: Time evolution of the standard deviation σ (see eqn (17)) of the coherence size distribution (upper panel) and individual coherence size components P_{Δ}^{CS} (lower panel). In the lower panel, the components $\Delta = \pm n$, $n = 1,\ldots,5$ are pairwise, very similar, but differ slightly at short times. (The value of $\Delta = 0$, which is dominant, is explicitly labeled in the figure.) Following a discontinuity at time $t = 0$, due to the sudden onset of the inter-fragment coupling, a significant initial expansion of the exciton size occurs, followed by a subsequent contraction.

modes, and (ii) a slower time scale involving the planarization of the torsional coordinates that determines the onset of a quasi-stationary state. This state exhibits enhanced localization accompanied by lattice distortions and can be characterized as an exciton-polaron state.

The first, rapid phase is characterized by electronic (or more accurately, vibronic) coherence that persists over ~100 fs, *i.e.*, longer than decoherence times of ~ 20–30 fs expected for condensed media,[80] but slightly shorter than the time scale of ~250 fs observed in room temperature experiments.[4–6] The comparatively long coherence time is likely due to dynamical correlations between the torsional and BLA motions on neighboring fragments; this effect can be enhanced, depending on how explicit correlations are included in the Hamiltonian.[8,43] The second, slow phase is sensitive to dissipative effects due to an external bath acting on the active torsional coordinates. The net effect is a transition from an initial ground state exciton on the left sub-fragment, towards a ground state exciton – with a possible admixture of the next higher excitonic state – on the hexamer. While a rapid transfer across the torsional defect sets in even at the initial twisted geometry, the slower torsional dynamics is essential in reaching a quasi-stationary state. (Indeed, no stationary state is reached in the absence of the torsional modes.[76]) Eventually, a "healing" of the torsional defect is observed on a time scale of ~ 500 fs, which coincides with the onset of a quasi-stationary state. This time scale is in general agreement with the one recently predicted for dynamical relaxation and localization in PPV type systems using an Ehrenfest simulation method.[81]

The BLA modes have not been explicitly subjected to damping, and indeed high-frequency vibrational coherence is seen to persist for hundreds of femtoseconds. In this sense, the exciton-phonon state that is reached by about 500–600 fs is quasi-stationary. While damping mechanisms for the BLA modes should be invoked in a more complete model, time-resolved experiments have shown that high-frequency oscillations indeed tend to persist for many hundred femtoseconds,[9] even in solution phase (contrasting with the interpretation that BLA relaxation occurs on a time scale < 50 fs[21]). Spectroscopically observable states may therefore correspond to quasi-stationary exciton states, rather than fully vibrationally relaxed species.

Despite the finite dimensionality of the system, the observed transfer dynamics is found to be effectively irreversible. This is directly connected to the interplay of high-frequency BLA and low-frequency torsional modes: while the former induce high-frequency level fluctuations, the latter modulate the site-to-site coupling and provide a quasi-continuum of vibronic states. Both aspects are essential in generating femtosecond-scale dephasing (decoherence) and efficient energy redistribution. Similar observations have been made in a recent analysis[82] where it was shown that exciton-phonon coupling on finite-size lattices, in the absence of an external bath, leads to relaxation effects on intermediate time scales, following coherent short-time evolution. In addition, the external bath modes that have been added in the present model lead to torsional damping and, hence, an effective energy loss of the excitonic subsystem. As a consequence, there is a tendency to reach the exciton ground state of the enlarged hexamer fragment.

While the full e–h dynamics leads to a similar picture as the Frenkel dynamics, the former is more complex since it necessitates an interpretation in terms of both the delocalization length, *i.e.*, the distribution in the center-of-mass coordinate, P_{ξ}^{CM}, and the coherence size, P_{Δ}^{CS}. In particular, the dynamical evolution of the coherence size may have specific effects that cannot be captured by an effective Frenkel picture. Thus, the observed expansion of the exciton size during the transient phase is an interesting effect that distinguishes the full e–h description. In larger assemblies, CT excitons with a significant e–h separation can act as precursors to current formation.

The present system has been chosen to be symmetric, with left and right fragments of equal size, such that localized fragment states can be straightforwardly interpreted as linear combinations, with equal weights, of the system's eigenstates. However, essentially all observations regarding the dynamics carry over to less symmetric situations; for example, we carried out a parallel study of a pentamer system.[75] In general, a spatially confined initial condition with a local spectrum of exciton states will interact (near-)resonantly with neighboring conjugated fragments, thereby causing delocalization (see also the discussion in ref. 18). Due to the interplay of bond-length-alternation and torsional modes, the transfer dynamics and relaxation towards the ground-state exciton become efficient and irreversible.

Our analysis was restricted to a minimal six-site system allowing for an explicit quantum dynamical treatment involving up to 36 states and more than 20 vibronically coupled modes. While exciton trapping in the present finite-size system is accurately captured by the dynamical calculations, a larger lattice size is necessary to study exciton migration and localization in a realistic setting.[21,47,81,83] However, we believe that the present analysis gives some clues as to the mechanism and time scale of formation of quasi-stationary local exciton ground states (also denoted LEGS[20,81,84]), which have been recognized to play a central role in exction localization and migration.

To complement the present analysis, the inclusion of temperature effects and of conformational disorder will be of key importance. These will mainly modify the slower time scale mentioned above, determined by the torsional coordinates. A statistical analysis either in the monomer representation, or else in a "spectroscopic unit" picture, could build upon the parametrization employed in the present analysis.

A Electron-hole center-of-mass and distance distributions

The full e–h probability distribution $\{P_{\nu\mu}\}$ is given as (see eqn (16)),

$$P_{\nu\mu} = \langle \rho_{\nu\mu,\nu\mu} \rangle = \langle \psi | \hat{\mathscr{P}}_{\nu\mu} | \psi \rangle = |\langle \nu\mu | \psi \rangle|^2 \tag{A.1}$$

where the projector $\hat{\mathscr{P}}_{\nu\mu} = |\nu\mu\rangle\langle\nu\mu|$ was used. From this discretized probability distribution, two marginal distributions are derived which correspond to the center of mass of the e–h pair, $\xi = (\nu + \mu)/2$, and the e–h distance ("coherence size"), $\Delta = \nu - \mu$, respectively,

$$P_{\xi}^{\rm CM} = \sum_{\substack{\nu,\mu=1 \\ \nu+\mu=2\xi}}^{N_{\rm site}} \langle \psi | \hat{\mathscr{P}}_{\nu\mu} | \psi \rangle \tag{A.2}$$

and

$$P_{\Delta}^{\rm CS} = \sum_{\substack{\nu,\mu=1 \\ \nu-\mu=\Delta}}^{N_{\rm site}} \langle \psi | \hat{\mathscr{P}}_{\nu\mu} | \psi \rangle \tag{A.3}$$

As the center-of-mass coordinate ξ can take half-integer values, a binning approach has been used to evaluate eqn (A.2). Here, the occupancies of half-integer ξs are redistributed over the nearest neighbors,

$$\bar{P}_{\xi}^{\mathrm{CM}} = \frac{1}{2} P_{\xi-\frac{1}{2}}^{\mathrm{CM}}(1-\delta_{\xi,1}) + P_{\xi}^{\mathrm{CM}} + \frac{1}{2} P_{\xi+\frac{1}{2}}^{\mathrm{CM}}(1-\delta_{\xi,N_{\mathrm{site}}}) \tag{A.4}$$

In the main text, P_{ξ}^{CM} and $\bar{P}_{\xi}^{\mathrm{CM}}$ will be employed interchangeably, with the understanding that $\bar{P}_{\xi}^{\mathrm{CM}}$ has been used in the actual computation.

Acknowledgements

We thank Fabio Sterpone, Hans Lischka, Felix Plasser, Adelia Aquino, and Aditya Panda for discussions, and Eric Bittner for exchange at an early stage of this project.

References

1 T. Förster, *Ann. Phys.*, 1948, **2**, 55.
2 V. May and O. Kühn, *Charge and Energy Transfer Dynamics in Molecular Systems*, 3rd ed., VCH-Wiley, 2011.
3 H. Lee, Y.-C. Cheng and G. Fleming, *Science*, 2007, **316**, 1462.
4 E. Collini and G. D. Scholes, *Science*, 2009, **323**, 369.
5 E. Collini and G. D. Scholes, *J. Phys. Chem. A*, 2009, **113**, 4223.
6 I. Hwang and G. D. Scholes, *Chem. Mater.*, 2011, **23**, 610.
7 J.-L. Brédas and R. Silbey, *Science*, 2009, **323**, 348.
8 X. Chen and R. J. Silbey, *J. Chem. Phys.*, 2010, **132**, 204503.
9 G. Lanzani, G. Cerullo, C. Brabec and N. S. Sariciftci, *Phys. Rev. Lett.*, 2003, **90**, 047402.
10 C. Manzoni, A. Gambetta, E. Menna, M. Meneghetti, G. Lanzani and G. Cerullo, *Phys. Rev. Lett.*, 2005, **94**, 207401.
11 J. Clark, T. Nelson, S. Tretiak, G. Cirmi and G. Lanzani, *Nat. Phys.*, 2012, **8**, 225.
12 M. M. Grage, P. W. Wood, A. Ruseckas, T. Pullerits, W. Mitchell, P. L. Burn, I. Samuel and V. Sundström, *J. Chem. Phys.*, 2003, **118**, 7644.
13 S. N. Yaliraki and R. J. Silbey, *J. Chem. Phys.*, 1996, **104**, 1245.
14 W. Beenken, *Phys. Status Solidi A*, 2009, **206**, 2750.
15 D. Hu, J. Yu, K. F. Wong, B. Bagchi, P. J. Rossky and P. F. Barbara, *Nature*, 2000, **405**, 1030.
16 K. F. Wong, M. S. Skaf, C.-Y. Yang, P. J. Rossky, B. Bagchi, D. Hu, D. Hu and P. F. Barbara, *J. Phys. Chem. B*, 2001, **105**, 6103.
17 T. E. Dykstra, E. Hennebicq, D. Beljonne, J. Gierschner, G. Claudio, E. R. Bittner, J. Knoester and G. D. Scholes, *J. Phys. Chem. B*, 2009, **113**, 656.
18 W. Barford, D. G. Lidzey, D. V. Makhov and A. J. H. Meijer, *J. Chem. Phys.*, 2010, **133**, 044504.
19 W. Beenken and T. Pullerits, *J. Phys. Chem. B*, 2004, **108**, 6164.
20 W. Barford, I. Boczarow and T. Wharram, *J. Phys. Chem. A*, 2011, **115**, 9111.
21 W. Barford, E. R. Bittner and A. Ward, *J. Phys. Chem. A*, 2012, **116**, 10319.
22 S. Westenhoff, W. Beenken, R. H. Friend, N. C. Greenham, A. Yartsev and V. Sundström, *Phys. Rev. Lett.*, 2006, **97**, 166804.
23 D. Beljonne, G. Pourtois, C. Silva, E. Hennebicq, L. M. Herz, R. H. Friend, G. D. Scholes, S. Setayesh, K. Müllen and J. L. Brédas, *Proc. Natl. Acad. Sci. U. S. A.*, 2002, **99**, 10982.
24 T. E. Dykstra, V. Kovalevskij, X. Yang and G. D. Scholes, *Chem. Phys.*, 2005, **318**, 21.
25 H. Yamagata and F. C. Spano, *J. Chem. Phys.*, 2012, **136**, 184901.
26 A. N. Panda, F. Plasser, A. Aquino, I. Burghardt and H. Lischka, *J. Phys. Chem. A.*, 2013, **117**, 2181.
27 R. Binder, S. Römer, F. Plasser, J. Wahl, A. N. Panda, A. Aquino, H. Lischka and I. Burghardt, *An Ab Initio Based Effective Frenkel Hamiltonian for Phenylene Vinylene Type Conjugated Systems*, 2013, to be submitted.
28 J. Frenkel, *Phys. Rev.*, 1931, **37**, 1276.
29 R. E. Merrifield, *J. Chem. Phys.*, 1961, **34**, 1835.
30 S. Mukamel, S. Tretiak, T. Wagersreiter and V. Chernyak, *Science*, 1997, **277**, 781.
31 M. Rohlfing and S. G. Louie, *Phys. Rev. Lett.*, 1999, **82**, 1959.

32 S. Tretiak, A. Saxena, R. L. Martin and A. R. Bishop, *Chem. Phys. Lett.*, 2000, **331**, 561.
33 S. Tretiak and S. Mukamel, *Chem. Rev.*, 2002, **102**, 3171.
34 W. Barford, R. J. Bursill and R. W. Smith, *Phys. Rev. B: Condens. Matter*, 2002, **66**, 115205.
35 S. Karabunarliev and E. R. Bittner, *J. Chem. Phys.*, 2003, **118**, 4291.
36 N. Kirova, *Polym. Int.*, 2008, **57**, 678.
37 F. Plasser and H. Lischka, *J. Chem. Theory Comput.*, 2012, **8**, 2777.
38 M. Hoffmann, *Chem. Phys.*, 2000, **258**, 73.
39 M. Hoffmann, *Thin Films and Nanostructures*, 2003, **31**, 221.
40 M. H. Beck, A. Jäckle, G. A. Worth and H. D. Meyer, *Phys. Rep.*, 2000, **324**, 1.
41 S. Tretiak, A. Saxena, R. L. Martin and A. R. Bishop, *Phys. Rev. Lett.*, 2002, **89**, 097402.
42 F. Sterpone and P. J. Rossky, *J. Phys. Chem. B*, 2008, **112**, 4983.
43 M. J. Bedard-Hearn, F. Sterpone and P. J. Rossky, *J. Phys. Chem. A*, 2010, **114**, 7661.
44 I. Franco and P. Brumer, *J. Chem. Phys.*, 2012, **136**, 144501.
45 B. Van Averbeke and D. Beljonne, *J. Phys. Chem. A*, 2009, **113**, 2677.
46 T. Liu and A. Troisi, *J. Phys. Chem. C*, 2011, **115**, 2406.
47 J. Singh, E. R. Bittner, D. Beljonne and G. D. Scholes, *J. Chem. Phys.*, 2009, **131**, 194905.
48 F. Sterpone, R. Martinazzo, A. N. Panda and I. Burghardt, *Z. Phys. Chem.*, 2011, **225**, 541.
49 A. Stradomska, W. Kulig, M. Slawik and P. Petelenz, *J. Chem. Phys.*, 2011, **134**, 224505.
50 H. X. Wang and S. Mukamel, *Chem. Phys. Lett.*, 1992, **192**, 417.
51 H. X. Wang and S. Mukamel, *J. Chem. Phys.*, 1992, **97**, 8019.
52 I. Burghardt, E. R. Bittner, H. Tamura, A. Pereverzev and J. Ramon, Ultrafast photophysics of organic semiconductor junctions, in: *Energy transfer dynamics in biomaterial systems*, Springer Series in Chemical Physics, Vol. 93, ed. I. Burghardt, V. May, D. A. Micha, E. R. Bittner, Heidelberg, 2009.
53 G. D. Scholes, *ACS Nano*, 2008, **2**, 523.
54 E. Moore, B. Gherman and D. Yaron, *J. Chem. Phys.*, 1997, **106**, 4216.
55 The exciton binding energy e_0 is thus taken to correspond to the Coulombic energy difference between an e–h pair at large separations and a localized e–h pair ($\nu = \mu$), $e_0 = -(4\pi\varepsilon_0\varepsilon_r)^{-1}(1/r_{\mu\nu}-1/r_0), r_{\mu\nu} \gg r_{\mathrm{ru}}$.
56 Even though a value of $\varepsilon_r = 2.3$ would be better adapted to PPV, we use a simplified parametrization in our model. Also, we have chosen a simple Coulomb form here, while alternative formulations may involve the analytical expression $E_{eh} = e_0/\varepsilon_0\varepsilon_r(1+\eta r_{\nu\mu}^2)^{1/2}$
57 J. Gierschner, H.-G. Mack, L. Lüer and D. Oelkrug, *J. Chem. Phys.*, 2002, **116**, 8596.
58 S. Karabunarliev, E. R. Bittner and M. Baumgarten, *J. Chem. Phys.*, 2001, **114**, 5863.
59 U. Weiss, *Quantum Dissipative Systems*, 3rd ed., World Scientific Singapore, 2008.
60 M. Nest and H.-D. Meyer, *J. Chem. Phys.*, 2003, **119**, 24.
61 M. Hultell and S. Stafström, *Phys. Rev. B: Condens. Matter Mater. Phys.*, 2007, **75**, 104304.
62 J. Kirkpatrick, *Int. J. Quantum Chem.*, 2008, **108**, 51.
63 P. Prins, F. C. Grozema and L. Siebbeles, *J. Phys. Chem. B*, 2006, **110**, 14659.
64 H. Köppel, W. Domcke and L. S. Cederbaum, *Adv. Chem. Phys.*, 1984, **57**, 59.
65 Note that the torsional excited-state potential $v_{\mathrm{tors}}^E(\theta)$ exhibits a minimum at the planar geometry ($\theta_{0,E} = 0°$). Likewise, the excitonic potential eqn (7) exhibits a minimum at the planar geometry, due to the dominance of the stiffer potential $v_{\mathrm{tors}}^E(\theta)$ in the linear combination of monomer potentials.
66 M. H. Beck, A. Jäckle, G. A. Worth and H.-D. Meyer, *Phys. Rep.*, 2000, **324**, 1.
67 H.-D. Meyer, U. Manthe and L. S. Cederbaum, *Chem. Phys. Lett.*, 1990, **165**, 73.
68 U. Manthe, H.-D. Meyer and L. S. Cederbaum, *J. Chem. Phys.*, 1992, **97**, 3199.
69 G. A. Worth, M. H. Beck, A. Jäckle and H. Meyer, 2006, The MCTDH Package, Version 9 (2007); H.-D. Meyer, Version 8.3 (2002). See http://www.pci.uni-heidelberg.de/tc/usr/mctdh/.
70 From the harmonic approximation to the ground and excited state torsional potentials, see Table 2, the respective periods are $T_G = 1250$ fs and $T_E = 300$ fs.
71 N. T. Harrison, G. R. Hayes, R. T. Phillips and R. H. Friend, *Phys. Rev. Lett.*, 1996, **77**, 1881.
72 W. Barford and N. Paiboonvorachat, *J. Chem. Phys.*, 2008, **129**, 164716.
73 E. Artacho, M. Rohlfing, M. Côté, P. D. Hayes, R. J. Needs and C. Molteni, *Phys. Rev. Lett.*, 2004, **93**, 116401.
74 I. H. Nayyar, E. R. Batista, S. Tretiak, A. Saxena, D. L. Smith and R. L. Martin, *J. Phys. Chem. Lett.*, 2011, **2**, 566.
75 *Master thesis "Quantum dynamics of excitons in semiconducting polymers"*, R. Binder, Goethe University Frankfurt, 2012.
76 R. Binder, J. Wahl, I. Burghardt, unpublished material.
77 S. Mukamel, *Principles of Nonlinear Optical Spectroscopy*, Oxford University Press, New York/Oxford, 1995.
78 Y. Li and C. A. Ullrich, *Chem. Phys.*, 2011, **391**, 157.

79 In eqn (17), note that $\sum_{\Delta=-(N_{\rm site}-1)}^{N_{\rm site}-1} P_{\Delta}^{\rm CS} = 1$.
80 O. V. Prezhdo and P. J. Rossky, *Phys. Rev. Lett.*, 1998, **81**, 5294.
81 O. R. Tozer and W. Barford, *J. Phys. Chem. A*, 2012, **116**, 10310.
82 V. Pouthier, *J. Chem. Phys.*, 2012, **137**, 114702.
83 S. Saini and B. Bagchi, *Phys. Chem. Chem. Phys.*, 2010, **12**, 7427.
84 D. V. Makhov and W. Barford, *Phys. Rev. B: Condens. Matter Mater. Phys.*, 2010, **81**, 165201.

Faraday Discussions RSC Publishing

PAPER

Quantum-classical effective-modes dynamics of the $\pi\pi^* \rightarrow n\pi^*$ decay in 9H-adenine. A quadratic vibronic coupling model†

David Picconi,[ab] Francisco José Avila Ferrer,[cd] Roberto Improta,[e] Alessandro Lami[c] and Fabrizio Santoro*[c]

Received 11th December 2012, Accepted 10th January 2013
DOI: 10.1039/c3fd20147c

We present mixed quantum-classical simulation of the internal conversion between the lowest energy $\pi\pi^*$ (S_{L_a}) and $n\pi^*$ (S_n) excited electronic states in adenine in the gas phase, adopting a quadratic vibronic model (QVC), parametrized with the help of PBE0 density functional calculations. Our approach is based on a hierarchical representation of the QVC Hamiltonian and a subsequent treatment of the most relevant coordinates at accurate time-dependent quantum level and of the other 'bath' modes at classical level. We predict an ultrafast transfer (~30 fs) of ≈75% of the initial population excited on S_{L_a} to S_n. Within an adiabatic picture, on the same timescale the wave packet concentrates almost completely on the lowest S_1 state, where however it shows a very broad distribution with different characteristics (due to the different 'diabatic' character). It is shown that the proposed methodology offers a practicable route to describe the quantum dynamics of internal conversion processes in large semi-rigid systems.

I Introduction

The interest in the photochemistry of isolated nucleobases is motivated by the likely presence of fast and effective 'monomer-like' deactivation channels within single and double strands DNA, which might play an important role to decrease photodamage.[1–4] Because of this, all the most refined experimental and

[a]*Scuola Normale Superiore di Pisa, Piazza dei Cavalieri 7, I-56126 Pisa, Italy*
[b]*Technische Universität München, Lehrstuhl für Theoretische Chemie, Lichtenbergstraße 4, D-85747 Garching, Germany*
[c]*CNR–Consiglio Nazionale delle Ricerche, Istituto di Chimica dei Composti Organo Metallici (ICCOM-CNR), UOS di Pisa, Area della Ricerca, via G. Moruzzi 1, I-56124 Pisa, Italy. E-mail: fabrizio.santoro@iccom.cnr.it*
[d]*University of Málaga, Physical Chemistry, Faculty of Science, Málaga, 29071, Spain*
[e]*CNR–Consiglio Nazionale delle Ricerche, Istituto di Biostrutture e Bioimmagini (IBB-CNR), Via Mezzocannone 16, I-80134, Napoli, Italy*

† Electronic supplementary information (ESI) available: Derivation of the parameters of the QVC Hamiltonian from electronic computations according to an "adiabatic" approach. See DOI: 10.1039/c3fd20147c

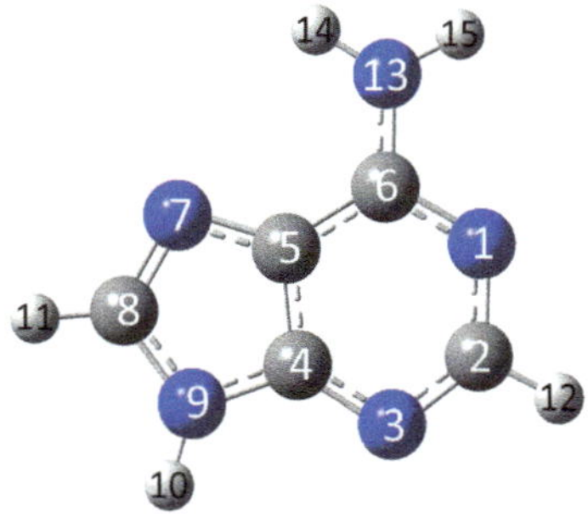

Fig. 1 The structure of 9H-adenine (Ade) with atom numbering.

computational techniques for the study of ultrafast photophysical and photochemical processes have been applied to the study of nucleobases, making these molecules fundamental test-cases for our understanding of the basic chemical-physical effects modulating the dynamics of the excited electronic states.[4] Adenine (see Fig. 1) is an excellent example of the complexity of the microscopic processes involved in the deactivation of a simple organic molecule, also when occurring on a sub-ps time scale and in the gas phase,[5–20] as witnessed by the large number of experimental and computational studies (see ref. 5 for a more complete review) related to its photoexcited-state behavior.

Molecular beams experiments show five resolved vibronic bands within 0.15 eV only (4.40–4.55 eV).[17] Furthermore, the excited state non-radiative dynamics strongly depend on the excitation wavelength and is multi-exponential, with an ultrafast component ($\tau_1 \approx 0.1$ ps) and a slower, though sub-ps ($\tau_2 \approx 0.75$–1 ps),

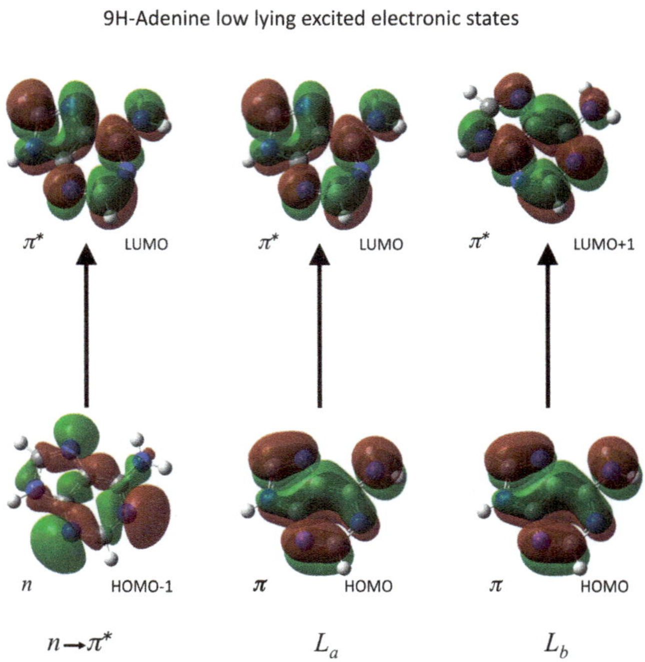

Fig. 2 Molecular orbitals involved in the $n\pi^*$ (S_n), L_a (S_{L_a}) and L_b (S_{L_b}) excitations.

one.[18–20] Calculations[5–15] agree that three different singlet excited states lie close in energy in the Franck–Condon (FC) region, contributing to the lowest energy absorption band (see Fig. 2 for a sketch of the involved molecular orbitals): (i) a bright $\pi\pi^*$ excited state, mainly corresponding to a HOMO → LUMO excitation, usually labeled as L_a (we name it S_{L_a}), and carrying most of the oscillator strength; (ii) another bright $\pi\pi^*$ state, though less intense than L_a, that can be described as resulting from HOMO → LUMO + 1 (dominant) and HOMO-2 → LUMO (not sketched in Fig. 2), and it is usually labeled as L_b (we name it S_{L_b}); (iii) a dark state with $n\pi^*$ character (HOMO-1 → LUMO), that in the following will be labeled as S_n. There is a general consensus that S_n is the lowest energy excited state in the gas phase, whereas the energy ordering between S_{L_a} and S_{L_b} (which however are within 0.2 eV) is still a matter of debate. Concerning the non-radiative decay mechanisms, τ_1 is ascribed to decay of S_{L_a} directly to S_0[14] or to an underlying state ($S_{L_a} \rightarrow S_n$ or $S_{L_a} \rightarrow S_{L_b}$),[5,7] while τ_2 is assigned to S_0 recovery from these latter states (with special emphasis on the $S_n \rightarrow S_0$ decay).[5,7] It has also been proposed that both τ_1 and τ_2 are associated with the $S_{L_a} \rightarrow S_0$ decay.[15] Two critical conical intersections (CoI) have been located between the lowest energy adiabatic states and S_0, one involving the out-of-plane motion of the C_2-H group (see atom labeling in Fig. 1), usually labeled ^2E, and another involving the out-of-plane motion of the C_6-NH_2 group, labeled $^6S_1$.[5] Although for non-planar geometries the three diabatic states are strongly mixed, ^2E CoI can be associated with the path on S_{L_a}, and $^6S_1$ with that on S_n.[15] Finally, for higher excitation energies a deactivation channel involving the breaking of the N_9-H bond could be active, accessed through a Rydberg $\pi\sigma^*$ state ($S_{L_a} \rightarrow \pi\sigma^* \rightarrow S_0$).[12,19]

A number of semiclassical (trajectory surface hopping, TSH)[21] dynamical simulations of the deactivation of photoexcited adenine in the gas phase have

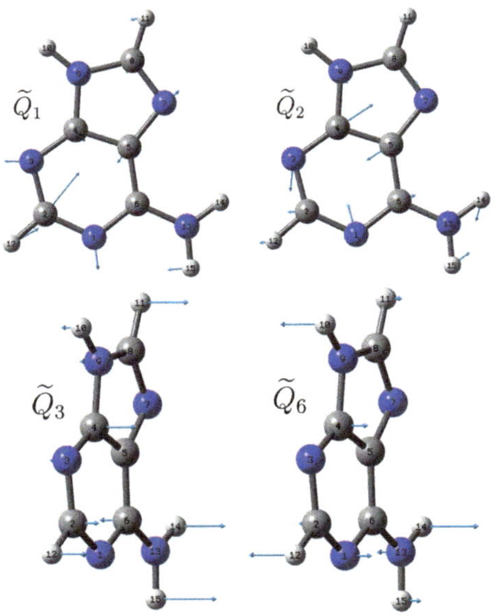

Fig. 3 Effective modes of the first QVC block ($\tilde{Q}_1$, $\tilde{Q}_2$ and $\tilde{Q}_3$) and first non-total-symmetric effective mode of the second block ($\tilde{Q}_6$), directly coupled to the inter-state coupling mode $\tilde{Q}_3$. The coordinate $\tilde{Q}_6$ involves an out-of-plane displacement of the hydrogen atom in position 9.

been reported in the literature,[5–11] and they have been critically compared in a recent contribution.[5] These approaches are very powerful and straightforwardly applicable but, for their nature, treat the nonadiabatic transfer, a pure quantum phenomenon, in an approximate way. Nonetheless, to the best of our knowledge no quantum dynamical (QD) simulation has been attempted up to now for photoexcited adenine. Such a study is challenging, due to the the high dimensionality of the system (39 nuclear degrees of freedom, DoF), and to the highly distorted structures of the CoIs requiring the use of internal coordinate representation (making more complex the form of the associated kinetic operator).

In this contribution we describe an effective computational strategy to face the challenge of performing QD simulations in systems with a large number of DoF, focusing on semirigid systems where harmonic approximation provides a suitable framework for the description of the PES of the involved diabatic states. Starting from an original idea by Cederbaum and coll.,[22–24] our approach is based on the adoption of a hierarchical representation of the nonadiabatic Hamiltonian in blocks, rigorously designed so that the first blocks define a reduced-dimensionality Hamiltonian that describes accurately the short-time dynamics of the full-dimensionality system, and the more blocks are considered the longer is the time for which the simulation is accurate. The protocol proposed in ref. 22–24 for the definition of the hierarchy was limited to linear vibronic coupling (LVC) Hamiltonians, accounting only for displacements of the equilibrium positions of the diabatic states examined. Quadratic vibronic coupling (QVC) model,[25,26] on the contrary, accounts also for quadratic differences of the PES thus introducing Duschinsky mixings and frequency changes. These effects have been shown to significantly affect the lineshape of absorption spectra (see ref. 27 and ref. 28 for a recent review) and we have therefore recently proposed a general iterative protocol to build up a hierarchical representation of the QVC Hamiltonian, already successfully used to simulate the $\pi\pi^* \rightarrow n\pi^*$ transfer in thymine.[29] The DoF belonging to the first blocks of the hierarchy are those most relevant for the QD, in as much as they drive the initial-state dynamics and rule the inter-state coupling, while the other DoF enter at later time and can be essentially considered as bath modes. A number of strategies have been proposed to deal with these bath modes within the LVC model, as for example their description in terms of Gaussian wavepackets[30] or the usage of spectral densities.[31,32] Here, in analogy to what was proposed in the literature by Miller *et al.*[33] we adopt an effective hybrid quantum-classical approach for the dynamics driven by the QVC Hamiltonian, where bath modes are treated classically and the coupled equations for the quantum and classical evolutions are solved simultaneously.

This mixed quantum-classical QVC approach is applied to the nonadiabatic dynamics of photoexcited 9H-adenine (hereafter Ade), focusing on the $S_{L_a} \rightarrow S_n$ population transfer for planar geometries, without accounting for ground state S_0 recovery. The validity of our analysis is therefore limited to the ultrafast dynamics in the first ≈ 50–100 fs (where distortions out-of-plane are not significant), *i.e.* the timescale of the fastest decay (τ_1) detected in experiments. The simultaneous analysis of the evolution of adiabatic and diabatic populations, naturally available from a wavepacket (WP) propagation, will be exploited to bridge the gap with TSH simulations, which usually provide information on the adiabatic populations, and is beneficial to properly compare with experimental evidences since, what is actually monitored in case of emission to S_0, for example, is the diabatic S_{L_a} population.[34,35]

II Theory

A Diabatic states

Quantum dynamics is usually performed in a diabatic representation, where the kinetic nonadiabatic couplings, neglected in Born–Oppenheimer approximation, are replaced by potential couplings, which are smooth functions of nuclear coordinates.

Labeling ($|S_1\rangle$, $|S_2\rangle$) the two adiabatic states involved in the process under study, the corresponding diabatic states ($|S_n\rangle$, $|S_{L_a}\rangle$) are obtained by a geometry ($\boldsymbol{Q}$) dependent transformation,

$$\begin{pmatrix} |S_n\rangle \\ |S_{L_a}\rangle \end{pmatrix} = \begin{pmatrix} \cos\theta(\boldsymbol{Q}) & \sin\theta(\boldsymbol{Q}) \\ -\sin\theta(\boldsymbol{Q}) & \cos\theta(\boldsymbol{Q}) \end{pmatrix} \begin{pmatrix} |S_1\rangle \\ |S_2\rangle \end{pmatrix}. \tag{1}$$

A dipole-based diabatization scheme, already adopted for uracil[36,37] and thymine,[29,38] that defines a dark S_n and a bright S_{L_a} state with respect to S_0, is suitable to define the angle $\theta(\boldsymbol{Q})$ also for Ade. In fact at C_S geometries, for symmetry reasons the state $|S_1\rangle$ is dark (apart from a very weak component perpendicular to the molecular plane that will not be considered hereafter) and $|S_2\rangle$ is bright, so they coincide with the diabatic states.

At difference, at C_1 non-planar geometries we define the transformation (1) in such a way that the 'diabatic transition' $|S_n\rangle \leftarrow |S_0\rangle$ is dipole-forbidden at all geometries:

$$\begin{aligned} \mu_{n0} &= \langle S_n|\mu|S_0\rangle = \cos\theta(\boldsymbol{Q})\langle S_1|\mu|S_0\rangle + \sin\theta(\boldsymbol{Q})\langle S_2|\mu|S_0\rangle \\ &= \mu_{10}(\boldsymbol{Q})\cos\theta(\boldsymbol{Q}) + \mu_{20}\sin\theta(\boldsymbol{Q}) = 0. \end{aligned} \tag{2}$$

At all the geometries where we performed the diabatization (see section II.B) the directions of the vectors μ_{10} and μ_{20} are almost identical, so we have dropped the vector notation in the above formula.

Upon adiabatic-to-diabatic transformation the potential, in the space of the electronic states, is no longer diagonal and the diabatic Hamiltonian looks as

$$\mathscr{H} = \begin{pmatrix} \mathcal{H}_1 & \mathcal{H}_{12} \\ \mathcal{H}_{12} & \mathcal{H}_2 \end{pmatrix} = \begin{pmatrix} \mathcal{T} & 0 \\ 0 & \mathcal{T} \end{pmatrix} + \begin{pmatrix} W_1(\boldsymbol{Q}) & W_{12}(\boldsymbol{Q}) \\ W_{12}(\boldsymbol{Q}) & W_2(\boldsymbol{Q}) \end{pmatrix}, \tag{3}$$

where $\mathcal{T}$ is the kinetic operator and the diabatic surfaces (W_1, W_2, W_{12}) are related to the adiabatic PES (V_1,V_2) by the relations:

$$\begin{aligned} W_1 &= V_1\cos^2\theta + V_2\sin^2\theta; \\ W_2 &= V_1\sin^2\theta + V_2\cos^2\theta; \\ W_{12} &= (V_2 - V_1)\sin\theta\cos\theta. \end{aligned} \tag{4}$$

For all the geometries, eqn (2) and (4) allow to uniquely determine the diabatic potential energies given the adiabatic ones.

B Quadratic vibronic coupling Hamiltonian for wave packet propagation

In order to define the Hamiltonian for the quantum WP propagations, we adopt a fully QVC model for the intra-state diabatic potentials (W_1 and W_2), and a linear approximation for the diabatic coupling (W_{12}).[29] The diabatic coupling is

$$\mathcal{H}_{12} = \Lambda_{12}^T\boldsymbol{Q}, \tag{5}$$

and its gradient Λ_{12} is evaluated by numerical differentiation of W_{12} at the Franck–Condon point ($\boldsymbol{Q} = \mathbf{0}$), following the diabatization procedure of section IIA.

In this work, for the definition of W_1 and W_2 diagonal PES we adopted an approach denoted 'adiabatic' in the recent literature,[27] meaning that the point of expansion is the minimum of the PES. Details are given in the Electronic Supporting Information (ESI).† The matrices of the ground- (ω) and excited-state (ω_α, for state α) frequencies are obtained by calculations at the minima of the ground and the excited states. This allows to compute also the Duschinsky matrices $\mathbb{J}_\alpha$[39] needed, together with ω_α, to obtain the force constants of the adiabatic excited-states along the ground-state normal modes. Diabatic force constants can then be derived from the adiabatic ones with a straightforward procedure that exploits the coincidence of diabatic and adiabatic equilibrium geometries (see Appendix).

In the following the Hessian matrices of the diabatic surfaces are named $\mathbb{B}_\alpha$. Starting from the QVC Hamiltonian, the simplified LVC model can be obtained by assuming that the excited-states normal modes and frequencies are identical to those of the ground state, $\mathbb{J}_\alpha$, $\omega_\alpha = \omega$. This is equivalent to assuming a so-called adiabatic shift (AS) model for the the diagonal diabatic PES.[27]

C Hierarchical reduction of the dimensionality of the problem

We here briefly discuss, without entering details of the derivation, the automatic iterative procedure that generalizes the LVC hierarchical transformation[22–24] to QVC Hamiltonians.[29] Our approach is based on the construction of an orthogonal transformation between the dimensionless normal modes, so to define a new system of effective coordinates partitioned into blocks. By truncating the transformed Hamiltonian to the terms defined by the coordinates up to the first, second, *etc.* block, we obtain the first, second, *etc.* member of the hierarchy.

We start by building up the first block of the hierarchy by the minimal number of effective modes that concentrate all the linear terms of the Hamiltonian (three modes for two coupled surfaces),

$$\begin{cases} \boldsymbol{Q}_1^{(1)} = \mathcal{N}[\Lambda_1] \\ \boldsymbol{Q}_1^{(2)} = \mathcal{N}[\Lambda_2 - \boldsymbol{Q}_1^{(1)}(\boldsymbol{Q}_1^{(1)})^T\Lambda_2] \\ \boldsymbol{Q}_1^{(3)} = \mathcal{N}[\Lambda_{12} - \boldsymbol{Q}_1^{(1)}(\boldsymbol{Q}_1^{(1)})^T\Lambda_{12} - \boldsymbol{Q}_1^{(2)}(\boldsymbol{Q}_1^{(2)})^T\Lambda_{12}] \end{cases}, \tag{6}$$

where the effective coordinate $\boldsymbol{Q}_n^{(j)}$ is the j-th coordinate of the n-th block and the operator $\mathcal{N}[\,]$ simply normalizes. Then we define the matrices $\mathbb{F}_\alpha = \mathbb{B}_\alpha + \hbar\omega$. The second block contains the coordinates coupled to those of the first block by the matrices $\mathbb{F}_1$ and $\mathbb{F}_2$,

$$\begin{cases} \text{for } j = 1,3 \\ \boldsymbol{Q}_2^{(j)} = \mathcal{N}\left[\mathbb{F}_1\boldsymbol{Q}_1^{(j)} - \mathbb{P}_1\mathbb{F}_1\boldsymbol{Q}_1^{(j)} - \sum_{k<j}\boldsymbol{Q}_2^{(k)}(\boldsymbol{Q}_2^{(k)})^T\mathbb{F}_1\boldsymbol{Q}_1^{(j)}\right] \\ \boldsymbol{Q}_2^{(j+3)} = \mathcal{N}\left[\mathbb{F}_2\boldsymbol{Q}_1^{(j)} - \mathbb{P}_1\mathbb{F}_2\boldsymbol{Q}_1^{(j)} - \sum_{k<j+3}\boldsymbol{Q}_2^{(k)}(\boldsymbol{Q}_2^{(k)})^T\mathbb{F}_2\boldsymbol{Q}_1^{(j)}\right] \end{cases}, \tag{7}$$

where $\mathbb{P}_1 = \sum_{j=1}^{3}\mathbf{Q}_1^{(j)}(\mathbf{Q}_1^{(j)})^T$ is the projector of the modes of the first block.

In general, the $(n + 1)$-th block can be built up on the ground of the n-th one,

$$\mathbb{P}_n = \sum_{j=1}^{3\cdot 2^{n-1}} \boldsymbol{Q}_n^{(j)} (\boldsymbol{Q}_n^{(j)})^T,$$

$$\begin{cases} \text{for } j = 1, 3\cdot 2^{n-1} \\ \boldsymbol{Q}_{n+1}^{(j)} = \mathcal{N}\left[\mathbb{F}_1 \boldsymbol{Q}_n^{(j)} - (\mathbb{P}_n + \mathbb{P}_{n-1})\mathbb{F}_1 \boldsymbol{Q}_n^{(j)} - \sum_{k<j} \boldsymbol{Q}_{n+1}^{(k)} (\boldsymbol{Q}_{n+1}^{(k)})^T \mathbb{F}_1 \boldsymbol{Q}_n^{(j)}\right] \\ \boldsymbol{Q}_{n+1}^{(j+3\cdot 2^{n-1})} = \mathcal{N}\left[\mathbb{F}_2 \boldsymbol{Q}_n^{(j)} - (\mathbb{P}_n + \mathbb{P}_{n-1})\mathbb{F}_2 \boldsymbol{Q}_n^{(j)} - \sum_{k<j+3\cdot 2^{n-1}} \boldsymbol{Q}_{n+1}^{(k)} (\boldsymbol{Q}_{n+1}^{(k)})^T \mathbb{F}_2 \boldsymbol{Q}_n^{(j)}\right] \end{cases}. \tag{8}$$

The first two blocks define the first and the second member of the hierarchy, which are a 3- and a 9-dimensional Hamiltonian, respectively. For LVC Hamiltonian $\mathbb{F}_1 = \mathbb{F}_2 = 2\hbar\omega$, eqn (8) simplifies, and each block of the hierarchy has dimension 3. Our protocol can be regarded as a generalization of Lanczos tridiagonalization[40,41] and the underlying idea is to block-tridiagonalize the Hamiltonian terms that induce an exchange of quanta between the initially excited modes (those displaced by the electronic excitation or connected to the diabatic coupling that are included in the first block) and the remaining ones.[29] For an LVC model these couplings are the only quadratic terms in the Hamiltonian, and it can be shown that the more blocks are included in the dynamics, the larger the number of moments of the absorption spectrum that can be exactly reproduced.[22,24,29] The n-th order moment $\mathcal{M}_n$ is connected to the n-th order time derivative (at $t = 0$) of the autocorrelation function of the doorway state promoted on the excited state by light absorption. In this sense, the larger number of moments are correctly reproduced, the longer time the dynamics is accurate.[22,24] In QVC Hamiltonian there are additional quadratic terms and, putting them in a block-tridiagonal form, does not prevent the possibility that, already in the early steps of the dynamics, they create a couple of quanta along any pair of modes (if coupled), independently of their position in the hierarchy.[29] For those terms block-tridiagonalization is avoided. Nonetheless, by application to the $\pi\pi^* \rightarrow n\pi^*$ transfer in thymine we showed that our transformed block-tridiagonal Hamiltonian can reproduce exactly the first-moment and accurately the successive moments, and that the first and the second reduced-dimensionality Hamiltonians suffice to reproduce the short- and intermediate-time dynamics (up to $\approx$ 100 fs) of the full-coordinate system in a reasonably accurate manner.[29]

D Mixed quantum-classical dynamics

The rigorous definition of the hierarchical system of effective modes leads to identification of a certain number of 'relevant' modes, those of the first members of the hierarchy. The remaining modes, let us say those beyond the second block, activate only at later times and are dynamically less relevant, behaving, in some sense, as 'bath' modes. Since the quantum description of such coordinates is computationally expensive, it is difficult to include them in a full-quantum simulation.

In this work we investigate the possibility and the benefits of a description of the 'bath' modes in classical terms, an idea already proposed in the literature[33] but here coupled with the hierarchical representation of the Hamiltonian that provides rigorous partition of the coordinates set, so that only the less relevant for the dynamics are treated approximately. Formally, we consider the complete

Hamiltonian of the system $\mathscr{H}(\boldsymbol{Q}_c, \boldsymbol{P}_c)$, where a number of coordinates and momenta, denoted as $(\boldsymbol{Q}_c, \boldsymbol{P}_c)$, are treated as time-dependent parameters, and not as quantum operators.

Thus, the wavefunction $|\psi, t\rangle$ depends only on the quantum degrees of freedom, and its time evolution is given by the Schrödinger equation with a time-dependent Hamiltonian,

$$i\hbar\partial_t|\psi, t\rangle = \mathcal{H}(\boldsymbol{Q}_c(t), \boldsymbol{P}_c(t))|\psi, t\rangle. \tag{9}$$

On the other hand, the quantities $(\boldsymbol{Q}_c, \boldsymbol{P}_c)$ evolve as classical variables, subject to forces represented by the mean Hamiltonian

$$\langle\mathscr{H}\rangle(\boldsymbol{Q}_c, \boldsymbol{P}_c) = \frac{\langle\Psi, t|\mathscr{H}(\boldsymbol{Q}_c, \boldsymbol{P}_c)|\Psi, t\rangle}{\langle\Psi, t|\Psi, t\rangle}, \tag{10}$$

hence they evolve according to Hamilton equations,

$$\dot{\boldsymbol{Q}}_c = \frac{\partial\langle\mathscr{H}\rangle}{\partial\boldsymbol{P}_c} = \left\langle\frac{\partial\mathscr{H}}{\partial\boldsymbol{P}_c}\right\rangle; \tag{11a}$$

$$\dot{\boldsymbol{P}}_c = -\frac{\partial\langle\mathscr{H}\rangle}{\partial\boldsymbol{Q}_c} = -\left\langle\frac{\partial\mathscr{H}}{\partial\boldsymbol{Q}_c}\right\rangle. \tag{11b}$$

In summary, the simultaneous evolution of the quantum and classical degrees of freedom is described by eqn (10) and (11a)–(11b), implying that the classical variables follow the mean field of the quantum part of the mixed quantum-classical system, and *vice versa*.

In order to recover the quantum effects arising from the coordinates approximated as classical, the phase space points at time zero $(\boldsymbol{Q}_c(0), \boldsymbol{P}_c(0))$ must be sampled according to a statistical distribution to simulate an initial WP, and the results over a number of trajectories must be averaged.

III Computational details

The minima and the frequencies of ground and excited states have been determined by using PBE0/6-31G(d) and time dependent (TD)-PBE0/6-31G(d) calculations,[42] checking their accuracy by test calculations employing larger basis sets. All the electronic calculations were performed by the Gaussian09 program.[43]

These results have been exploited to define the parameters of the diabatic Hamiltonians following the indications in section IIB, in the ESI† and in the Appendix. The hierarchy of reduced-dimensionality Hamiltonians was built using the procedure of section IIC and each new coordinate was re-orthogonalized to all the previous ones (even if it is not formally necessary, see eqn (8)), to cure errors due to finite-precision arithmetics.

Quantum and mixed quantum-classical dynamical simulations have been performed for Ade using, as quantum modes, the coordinates of the first and first + second block, 3 and 9 effective modes respectively for QVC model and of the first, first + second, and first + second + third blocks, 3, 6 and 9 modes respectively, for LVC model. We adopted a short-time iterative Lanczos integrator[40,41,44] for the WP propagation, expanding the wave packet on a tensor product basis $\psi_{\alpha,n} = |\alpha; n_1, n_2,\ldots\rangle$ of the electronic diabatic states (α) and harmonic oscillators wavefunctions $|n_1, n_2, \ldots\rangle$. For each coordinate k, a maximum number of wavefunctions w_k was chosen, according to the symmetry and the block of the hierarchy the coordinate belongs to. In fact, the displacements leading from the FC point to the minima of the PESs are almost exclusively concentrated in the first-block modes, and the a$''$ modes have no displacement, so the effective modes beyond the first block and the non total-symmetric ones require less basis functions. In particular, for $k = 1, 2$ (first block a$'$ coordinates) it was set $w_k = 12$, for $k = 3$ $w_k = 8$, and for $k > 3$ it was set $w_k = 5$. The basis set was also cut: among all the possible basis states, we selected those for which $\langle \psi_{\alpha,n}|\mathscr{H}|\psi_{\alpha,n}\rangle = E_0 \pm \Delta E$ where $E_0 = \langle \psi, t|\mathscr{H}|\psi, t\rangle$ is the WP average energy and ΔE is a convenient threshold, fixed in our case to 17 500 cm^{-1}. The mixed quantum-classical simulations with 3 quantum modes and 36 classical modes were repeated increasing both w_k and the threshold on energy, verifying that the dynamics is not significantly altered, at least in the time scale relevant for the discussion below.

The initial WP was defined assuming a FC transition from the ground S_0 vibrational state to the S_{L_a} state, so that the initial vibrational wavefunction is Gaussian. In the quantum-classical cases the initial classical coordinates and momenta were sampled using a Gaussian probability distribution function and the results were averaged over 20 trajectories which, in the present case, allows to achieve a satisfactory convergence of the results. In this case the Hamilton equations were integrated using a Bulirsch–Stoer extrapolation method.[45]

IV Results

A Potential energy surfaces and their stationary points

At the PBE0/6-31G(d) level, the C_S S_0 stationary point of Ade exhibits an imaginary frequency, corresponding to a pyramidalization of the nitrogen (N_{13}) of the amino group, and it is a saddle-point ~4 cm^{-1} less stable than the absolute C_1 minimum (the sum of the angles around N is 356.7°). With the 6-31+G(d,p) basis set, the C_S structure becomes the absolute S_0 minimum, and in following it will be the S_0 reference geometry (min-S_0). The C_S stationary points located on the excited-states

Table 1 Relative energies (eV) of the low lying electronic excited states of 9H-adenine computed with different basis sets in ground and excited states C_S stationary-points located at PBE0/6-31G(d) level

	6-31G(d)	6-31G(d,p)	6-31+G(d,p)	6-311+G(2d,2p)
	min-S_0			
S_0 (ref E)	0.000	0.000	0.000	0.000
S_n	5.181	5.164	5.152	5.132
S_{L_a}	5.313	5.303	5.174	5.115
S_3 (S_{L_b})	5.521	5.518	5.420	5.357
	min-S_n			
S_0	0.686	0.681	0.647	0.665
S_n	4.569	4.546	4.534	4.534
S_{L_a}	5.479	5.464	5.318	5.281
S_3 (S_{L_b})	6.127	6.116	5.984	5.936
	min-S_{L_a}			
S_0	0.317	0.316	0.308	0.335
S_n	4.995	4.976	4.971	4.986
S_{L_a}	5.009	4.999	4.880	4.855
S_3 (S_{L_b})	5.593	5.589	5.477	5.436

Table 2 Energies (in eV) for the excited (diabatic) states of adenine at the minima of the ground and the excited states, relative to the minimum of the ground state. A comparison is made between the values computed at the TD-PBE0/6-31G(d) level, and those obtained by the harmonic model PES described in the paper. The minimum energy of the Conical Intersection (min-CoI) $S_{La}/S_{n\pi}$ predicted by the model is also reported

	$W_n(Q)$		$W_{L_a}(Q)$		$W_{L_a}(Q) - W_n(Q)$	
Q	TD-PBE0	Model	TD-PBE0	Model	TD-PBE0	Model
FC	5.181	5.176	5.313	5.303	0.132	0.127
min-S_n	4.569	4.569	5.479	5.475	0.910	0.906
min-S_{L_a}	4.995	5.000	5.009	5.009	0.014	0.009
min-CoI	—	5.010	—	5.010	—	0.000

will be labeled according to their diabatic character min-S_n and min-S_{L_a}; however, in C_S symmetry the diabatic states S_n and S_{L_a} always coincide with S_1 or S_2 adiabatic states. In Table I we report the energies of S_0 and the lowest-excited states (namely S_n, S_{L_a} and S_{L_b}) at min-S_0, min-S_n and min-S_{L_a} stationary points obtained in C_S symmetry at PBE0/6-31G(d) level of theory, and in Fig. 2 we sketch the molecular orbitals involved in the transitions. Independently of the minimum considered, S_{L_a} and S_{L_b} are stabilized when increasing the dimension of the basis sets, (~0.15–0.2 eV extending the basis set from 6-31G(d) to 6-311+G(2d,2p)). The minimum min-S_n lies on S_1 both according to 6-31G(d) and 6-31+G(d,p) calculations but it is a saddle point, less stable by 240 cm^{-1} (6-31+G(d,p) data) than the absolute S_1 minimum, exhibiting pyramidalization at N_{13}. The minimum min-S_{L_a} lies on S_2 and shows all positive frequencies according to 6-31G(d) calculations (see Table 1) while, according to 6-31+G(d,p) calculations, it lies on S_1 and exhibits a remarkable (i950 cm^{-1}) imaginary frequency, corresponding to a distortion of the fused rings where N_1 and N_3, and C_2 and C_4 are alternatively displaced out-of-plane. It was not possible to locate a C_1 minimum on the S_1 PES with a predominant S_{L_a} character, since optimizations always end in a minimum corresponding to S_n.

We defined the harmonic PES of the S_n and S_{L_a} diabatic states according to the recipe described in Section IIA and in the Appendix. To that end we decided to neglect N_{13} pyramidalization, which should not significantly affect $S_{L_a} \rightarrow S_n$ transfer on the ultrafast timescale ($\approx$ 50–100 fs), just taking the absolute value of the S_1 imaginary frequency. This choice is also validated *a posteriori* by the analysis of the dynamics showing that such mode mainly enters the modes of the second and successive blocks of the hierarchy (more precisely, only 9% in $\tilde{Q}_3$), while the extent of the $S_{L_a} \rightarrow S_n$ transfer is dominated by the dynamics along the first block of the hierarchy. Hence, it can be expected that introduction of a double-well profile along the N_{13} pyramidalization would only marginally affect the conclusions of our study (since it can modulate the population transfer only at times >40–50 fs).

Table 2 compares the diabatic energies predicted by the QVC model Hamiltonian at some relevant geometries to those obtained directly at PBE0/6-31G(d) level of theory. On balance, our electronic results are consistent with previous analysis.[5,15] PBE0 predicts a minimum for S_n[5] where the rings remain planar (but exhibits a pyramidalization at N_{13}) and a planar minimum for S_{L_a} diabatic PES with all positive frequencies. However, when S_{L_a} mostly corresponds to S_1 (*i.e.* according to 6-31+G(d,p) results, or when N_9 is methylated, results not shown), the S_1 adiabatic PES shows a driving force (*i.e.* an imaginary frequency) towards C_2 pyramidalization, as in the CASPT2 path leading to the CoI with S_0.[15] In our analysis such tendency is partly due to the S_{L_a}/S_n coupling (S_{L_a} Hessian has all positive frequencies) and partly to S_{L_a} intrinsic characteristics (when S_1 mainly corresponds to S_n, it does not exhibit any imaginary-frequency mode with C_2 pyramidalization contribution).

B Population dynamics

Various reduced-dimensionality models have been constructed by truncation of the hierarchy described in section IIC to the first (3q) and the first + second (9q) blocks of modes for QVC and to the first (3q), the first + second (6q) and the first + second + third (9q) blocks of modes for LVC; in the label 3,6,9 indicate the number of coordinates adopted, and "q" their treatment at the quantum time-dependent level. The QVC effective modes belonging to the first block and the a'' mode of the second block are sketched in Fig. 3.

Fig. 4 compares the LVC and QVC predictions for the time-evolution of the population of the diabatic bright S_{L_a} state. For QVC models (3q) and (9q) simulations predict a very similar population decay up to 40 fs, with a fast and effective transfer of ≈ 75% to S_n. At 40 fs the (9q) curve has a bump and then becomes almost stationary around ≈ 0.2 (≈ 80% population transfer). On the contrary, the (3q) curve exhibits population beatings. Similar behavior is seen for the LVC model where after ~40 fs the (3q) simulation predicts the occurrence of beats that are however damped when the effects of additional coordinates are taken into account. The close similarity of (6q) and (9q) LVC results up to 200 fs indicates our calculation to reproduce satisfactorily the dynamics of the full-dimensionality system (39 DoF) on this timescale. The right panel of Fig. 4 also directly compares the most accurate (9q) LVC and QVC calculations: their predictions start differing already at 30 fs but always remain similar, showing that inclusion of quadratic couplings modifies only slightly the dynamics and the S_n yield in the long-term

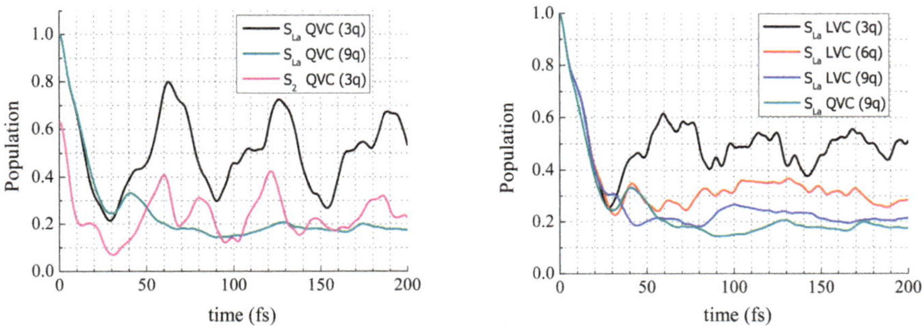

Fig. 4 Diabatic population of the S_{L_a} state with respect to time, according to QVC (left panel) and LVC (right panel) models. For the QVC model results with three (3q, first block of the hierarchy) and nine (9q, first + second block) effective modes are presented, while for LVC models with three (3q, first block of the hierarchy), six (6q, first + second block) and nine (9q, first + second + third block) effective modes were adopted. For QVC model also the time-evolution of the S_2 adiabatic state population is reported.

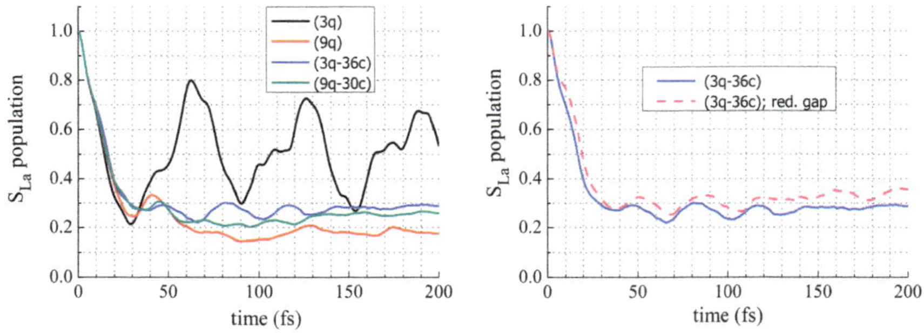

Fig. 5 Diabatic population of the S_{L_a} state with respect to time, according to the QVC model. In the left panel the results obtained with three (3q, first block of the hierarchy) and nine (9q, first + second block) effective modes, treated quantum-mechanically, are compared with those of mixed quantum/classical calculations: (3q-36c) three effective modes + 36 classical oscillators, (9q-30c) nine effective modes + 30 classical oscillators. In the right panel the (3q-36c) results are reported for the original QVC model, extracted from PBE0/6-31G(d) data, and for a modified model where the S_{L_a} PES is stabilized by 0.15 eV to mimic the effect of a larger electronic basis set (red. gap).

limit. Probably this is, at least partially, due to the fact that, as discussed in section IIB, the LVC and QVC models were obtained in an "adiabatic" framework, *i.e.* the LVC and QVC diabatic PES have the same minima and the same energies at the minima, which are the points where the WPs on S_n and S_{L_a} place themselves on a long-term scale. A more extensive analysis of the effect on population transfer of different possible choices to build up the LVC model was given in ref. 29 for the case of thymine, where QVC predictions differentiate from the LVC one already in the ultrafast regime ($\approx$ 20 fs), and on the long-term scale their predictions for the yield of the internal conversion vary from 0.85 to 0.7.

In the following, we focus on QVC results, analyzing the effect of the remaining blocks of the hierarchy on the dynamics. Fig. 5 compares the predictions of the (3q) and (9q) fully-quantum models with those exploiting mixed quantum-classical Hamiltonians. In the spirit of the hierarchy, the latter are built from the pure quantum models by adding the remaining coordinates, and treating their time evolution classically. In particular, since Ade has 39 normal modes, we have a model with 3 quantum and 36 classical effective coordinates (3q-36c), and a model with 9 quantum and 30 classical modes (9q-30c). It is very interesting to notice that (3q-36c) and the (9q-30c) population curves almost coincide up to 50 fs, but then they remain very close even at 200 fs, and rather similar to the (9q) curve, though the inclusion of all the modes slightly decreases the long-term limit of the S_n yield. On the grounds of these results we can safely state that the amount of $S_{L_a} \rightarrow S_n$ population transfer in Ade is 75–80%. The S_n state is populated to a large extent ($\approx$65%) in the first 20 fs, then the transfer continues slowly on and the population reaches the final pseudo-stationary value (small fluctuations) 70 fs after the photoexcitation.

Table 1 indicates that, at the increase of the basis set, S_{L_a} is stabilized with respect to S_n by a similar amount of energy ($\sim$0.15 eV) in the three stationary points min-S_0, min-S_{L_a}, and min-S_n. To investigate the basis-set effect, we thus computed the (3q-36c) QVC dynamics after shifting by −0.15 eV the S_{L_a} PES. As reported in the right panel of Fig. 5, such stabilization has a small effect on the dynamics, slightly reducing (by less than 5% at all times) the S_n yield.

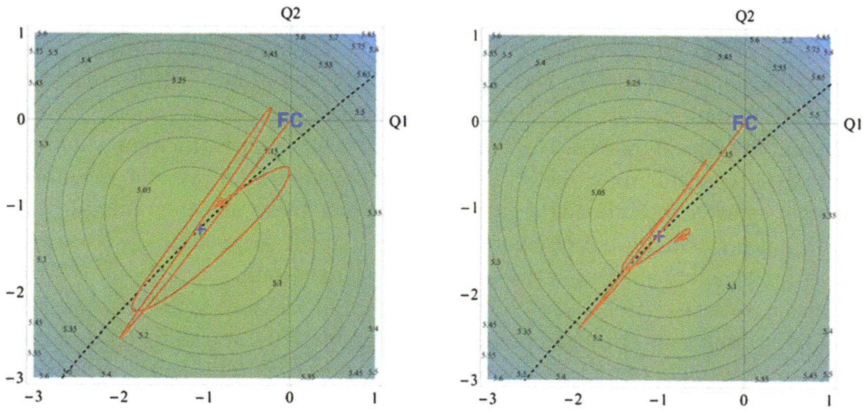

Fig. 6 Mean trajectories of the WP for the (3q) and (9q) QVC models (left and right panels, respectively), over the S_{L_a} PES, for the first 45 fs of dynamics. For the contour lines of the PES, the coordinates different from $\tilde{Q}_1$ and $\tilde{Q}_2$ are set to zero. The dashed line and the cross indicate the crossing seam and its minimum, respectively.

C Different roles of the first- and the next-blocks modes

Fig. 4 shows that the LVC and QVC population curves for the (3q) model and those for the other models have a quite different behavior. The major effect of the coordinates beyond the first three ones (first block of the hierarchy) is to damp the population beatings. Of course, this is a consequence of the effect of the latter coordinates on the WP motion, which mainly occurs along the first block modes, leading, at a first approximation, to the two minima of the PESs.

As pointed out in several works,[46,47] the population beatings are often indicative of an intense dynamics across the intersection seam. The mean trajectories of the S_{L_a} WP for the QVC (3q) and the (9q) models, reported in Fig. 6 in the plane of the two total-symmetric first block modes, show that in the (3q) case the amplitude of the motion transverse to the intersection seam is much larger than in the (9q) case. In the latter case the WP moves essentially along the seam, and the higher number of coordinates leads to faster relaxation towards the minimum.

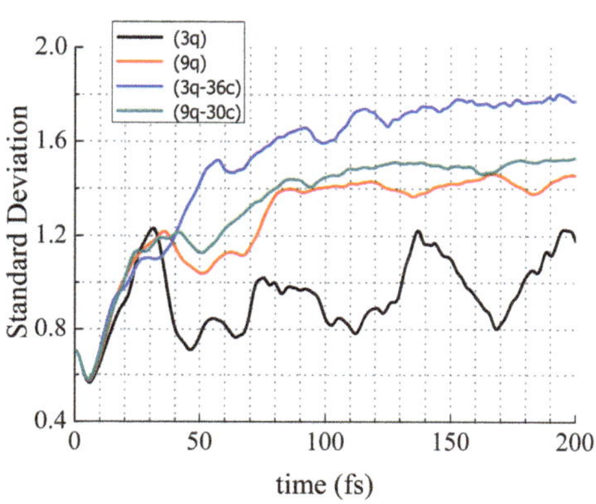

Fig. 7 Extension of the WP along the coupling coordinate $\tilde{Q}_3$ for the different QVC models used for the propagation.

Another important effect of the bath modes on the dynamics is due to the widening of the WP along the coupling mode, *i.e.* the non total-symmetric coordinate of the first block, $\tilde{Q}_3$. This effect is documented in Fig. 7 that reports the standard deviation, $\sqrt{\langle\tilde{Q}_3^2\rangle - \langle\tilde{Q}_3\rangle^2}$, of the S_{L_a} WP along the coupling mode, for the different dynamical models. The minimal (3q) model predicts the right widening up to ~35 fs. At this time, the relevant part of the internal conversion dynamics (70% of transfer) has already occurred, so one can argue that the remaining a″ coordinates modulate the fluctuations of the diabatic populations, but do not affect the absolute amount of the transfer to a large extent.

Also in this case, the effect of the blocks of coordinates next to the first one is to damp the amplitude of the fluctuations of the width of the WP along $\tilde{Q}_3$, and this leads to attainment of pseudo-stationarity in a short time. This has already been found with the (3q-36c) model, but the curves in Fig. 7 show that this model overestimates the long-term limit of the standard deviation, indicating that a more accurate description of the dynamics along the coupling mode requires a quantum description of the first + second block modes, (9q) and (9q-30c) models.

To conclude, the picture of the dynamics can be summarized as follows. In the first 25–35 fs a large amount of population (~65%) is transferred from the S_{L_a} to the S_n diabatic state, and the motion occurs almost exclusively on the first block modes. Then the next coordinates of the hierarchy come into play, causing damping of the motion along the first two block total-symmetric modes, of the widening of the WP along the coupling mode and of the diabatic population fluctuations. At later times the populations behave in a pseudo-stationary manner, and the extent of the S_n yield is ~75%.

V Discussion

The involvement of the S_n state in the ultrafast photoactivated dynamics of Ade has been lively debated in the literature; some authors[14,15] have indeed suggested that the motion of the system on the S_{L_a} PES can give account of all the sub-ps features. In particular, according to Conti *et al.*[15] τ_1 is associated with the motion from the FC point towards a plateau on S_{L_a}, involving in-plane vibrational relaxation plus some ring puckering, while τ_2 to the motion from this plateau towards the ^{2}E CI. Our calculations clearly indicate instead that a substantial amount (~75%) of S_{L_a} is transferred to S_n within 35–50 fs after the excitation. Fig. 4 also compares the time evolution of the S_2 adiabatic and S_{L_a} diabatic populations, according to (3q) QVC model. The first thing to notice is that, already at $t = 0$, 40% of the photoexcited population is in S_1 state, even if it is almost dark at the ground-state geometry. This is a pure quantum phenomenon due to the delocalization of the vibrational wavefunction and to the coupling of the two diabatic states. At later times the S_2 population follows the same beatings of the (3q) S_{L_a} one but it is sensibly smaller. Therefore, on average more than 80% of the excited population is on S_1 at $t > 30$ fs, and it is likely that this percentage would increase taking into account the effect of the other coordinates (this is unfortunately not feasible for the computational cost of the required multidimensional integration). This finding is in substantial agreement with what is reported in ref. 5 where however the transition S_2/S_1 is not discussed in detail.

Our prediction that in a very short time (30 fs) most of the population decays on S_n state nicely agrees with the experimental time-resolved photoelectron

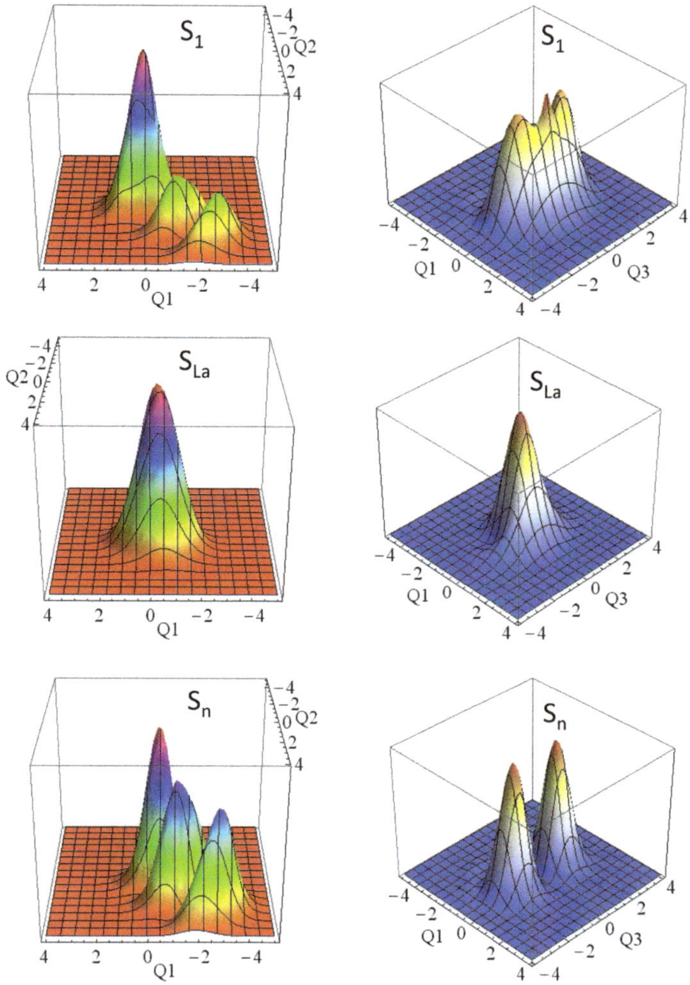

Fig. 8 Probability density of the WP at $t = 20$ fs in the $\{\tilde{Q}_1, \tilde{Q}_2, 0.5\}$ plane (left), and in the $\{\tilde{Q}_1, -0.2, \tilde{Q}_3\}$ plane (right) on S_1 (top), S_{L_a} (middle), and S_n (bottom) PES.

spectra of Ade showing that ionization to lowest energy cation state (IP = 8.48 eV),[18] ascribed to S_{L_a},[48] has a lifetime of 35–70 fs (depending on the excitation wavelength).[19]

As shown in Fig. 8, already at 20 fs the S_1 WP assumes a multimodal distribution with peaks reminiscent of the wave packets running on the diabatic PES. Notice, for example, that the WP in the $\{\tilde{Q}_1, \tilde{Q}_2\}$ plane is centered close to the origin of the axes on S_{L_a} while elongating toward the min-S_n structure on S_n. In the $\{\tilde{Q}_1, \tilde{Q}_3\}$ plane instead, the S_{L_a} WP is maximum at $\tilde{Q}_3 = 0$, where it has a nodal plane on S_n PES. The corresponding WPs on S_1 exhibit characteristics of both the S_{L_a} and S_n components. On a long-term limit, the adiabatic and diabatic WPs become very broad but, clearly, the average positions of the diabatic ones reach pseudo-stationary values in the vicinity of the corresponding minima of the diabatic PES (data not shown), *i.e.* min-S_{L_a} ~25% of the population) and min-S_n ($\approx$ 75% of the population), so that it is reasonable to assume that they feel different forces and can follow different decay paths. It is also interesting to notice that the effect of the coupling along a″ modes is to spread along these modes the S_n WP, and later on also the S_{L_a} one, thus promoting the motion toward non-planar structures.

We can only speculate on the time evolution at later times. According to the nice analysis of Barbatti *et al.*[5] the two populations S_{L_a} and S_n might be thought to preferentially travel toward the ^2E and $^6S_1$ CoIs with S_0, respectively. However both paths toward ^2E and $^6S_1$ CoIs require strong distortion of the planarity and, during this motion, remarkable exchange of populations between the two channels might be operative, most of all if the latter are not separated by substantial energy barriers, or the barriers are lower than those to reach the CoIs. As a matter of fact, as we reported in section IVA, at PBE0 level it was not possible to locate a S_1 minimum with substantial S_{L_a} character, since optimization always falls into the non-planar basin where S_1 exhibits a S_n character. However, it has been suggested[5] that PBE0 underestimates the S_{L_a}/S_n coupling so that it is indeed possible that an increase of the S_{L_a}/S_n diabatic coupling (acting along non-planar coordinates) would lead also in our simulations to the appearance (and population) of a single S_1 non-planar minimum with mixed S_{L_a}/S_n character.

From a more general point of view, the possibility of switching between adiabatic and diabatic description is very informative. Ultrafast dynamical processes are investigated through time-resolved pump–probe spectroscopies where the laser pump excites the system and the laser probe interrogates the evolving system, measuring some specific observable, typically delivering decay times, and sometimes amplitudes in principle connected to the amount of the population transfer. It is clear that any theoretical prediction should be compared to these observations but a general question might be raised: does the probe monitor the diabatic or the adiabatic populations?

As far as fluorescence upconversion (FLUP) is considered, theoretical models show that in a situation like the one of Ade, *i.e.* two coupled excited states, one bright and the other dark with respect to the ground state, the emission to the ground state monitors the diabatic bright population (S_{L_a}) and not the adiabatic populations.[34,35] To exemplify, this means that if the S_1 state has a different diabatic character (S_{L_a} or S_n) in a different region of its PES, like it happens in Ade (min-S_n and min-S_{L_a}) according to 6-31+G(d,p) calculations, two different dynamical processes bringing the WP to concentrate alternatively in one of these basins would give rise to very different FLUP signals, showing a complete decay in the first case and no decay at all in the second case.

The situation is less evident for a probe inducing a transition to a cationic state, since even the latter might have different diabatic characters in different regions of the PES. However, simplifying the discussion to a transition to the ground-state of the cation D_0 with a very well-defined electronic configuration, according to Koopmans' approximation we can predict that the probe monitors the population of the diabatic states, the configuration of which differ by only one electron with respect to D_0. In the specific case of Ade such an experiment should therefore monitor S_{L_a} population. According to this discussion, the 35–70 fs decay observed for the time-resolved photoelectron spectrum toward D_0 in Ade,[19] should be interpreted as a decay to S_n diabatic state and not to S_1 adiabatic one.

VI Conclusions

We have presented a feasible methodology to simulate the quantum dynamics of internal conversions in large semi-rigid systems where the diabatic PES can be described within harmonic approximation. This is based on a recent generalization of the hierarchical representation of LVC Hamiltonian[22–24] to QVC models,[29] that gives the possibility to account for the effect of Duschinsky mixings and frequency changes in the different PES. In this contribution we show that such hierarchical representation, allowing to determine which modes drive the dynamics and which mainly act as a bath, is particularly suited for including the effect of the bath modes in a classical way, performing mixed quantum-classical dynamical simulations.[33]

The proposed computational protocol has been applied to the study of the $S_{L_a} \rightarrow S_n$ internal conversion in Ade in the gas phase, based on PBE0 electronic PES, documenting an ultrafast but non-complete transfer to S_n state and an almost complete decay on the S_1 PES. Here, however, the WP shows a broad distribution where different parts tend to behave like those of the two diabatic states. Our results indicate that the dynamics is ruled by the modes of the first two blocks of the hierarchy while the effect of the others is mainly to damp the beats in the diabatic population, thus increasing "on average" the extent of population transfer, and can be adequately described in a classical way.

In the specific case of Ade, the differences between LVC and QVC models are small, while they were more sensible for thymine.[29] However, it is worthy of note that QVC model allowing each diabatic PES to have a different Hessian, together with the proposed protocol to actually extract these Hessians from the adiabatic ones, provides a nice additional analysis tool for the process under study. In fact, in the particular case of Ade this allows to document that at PBE0/6-31G(d) level the diabatic S_{L_a} state is characterized by a planar minimum with all-positive frequencies and that the removal of planarity of the ring on the adiabatic S_1 PES is due to the S_{L_a}/S_n coupling, information that cannot be accessed by analysing the properties of the adiabatic states only.

Appendix A: Constructing the diabatic force constant matrices from the adiabatic ones

Here we explain our procedure for determining the Hessian of the PESs of the diabatic S_n and S_{L_a} states from our *ab initio* calculations. The starting point is the fact that at C_S geometries the adiabatic and diabatic states coincide, and the gradient of the intrastate coupling is non-totalsymmetric (a″) and can be computed by numerical differentiation of the $S_0 \rightarrow S_n$ transition dipole moment.

Assuming a quadratic form for the diabatic PESs and a linear diabatic coupling we have, for $\nu = 3n_{\text{atoms}} - 6$,

$$W_n(Q_1, \ldots, Q_\nu) = E_n + \sum_j^\nu \lambda_j^{(n)} Q_j + \frac{1}{2} \sum_{i,j}^\nu B_{ij}^{(n)} Q_i Q_j \tag{A1}$$

$$W_{L_a}(Q_1, \ldots, Q_\nu) = E_{L_a} + \sum_j^\nu \lambda_j^{(L_a)} Q_j + \frac{1}{2} \sum_{i,j}^\nu B_{ij}^{(L_a)} Q_i Q_j \tag{A2}$$

$$W_{nL_a}(Q_1,\dots,Q_\nu)=\sum_j^{\nu}\lambda_j^{(nL_a)}Q_j. \tag{A3}$$

Accordingly, the adiabatic PESs are given by

$$V_{1,2}(\boldsymbol{Q})=\frac{W_n(\boldsymbol{Q})+W_{L_a}(\boldsymbol{Q})}{2}\pm\sqrt{\left(\frac{W_n(\boldsymbol{Q})-W_{L_a}(\boldsymbol{Q})}{2}\right)^2+(W_{nL_a}(\boldsymbol{Q}))^2}. \tag{A4}$$

The first and second derivatives of the adiabatic PESs with respect to the normal modes can be straightforwardly expressed in terms of the diabatic ones. If we define

$$\Gamma(\boldsymbol{Q})=\frac{W_n(\boldsymbol{Q})+W_{L_a}(\boldsymbol{Q})}{2}, \tag{A5}$$

$$\Delta(\boldsymbol{Q})=\left(\frac{W_n(\boldsymbol{Q})-W_{L_a}(\boldsymbol{Q})}{2}\right)^2+(W_{nL_a}(\boldsymbol{Q}))^2, \tag{A6}$$

we have

$$\frac{\partial V_{1,2}(\boldsymbol{Q})}{\partial \boldsymbol{Q}_j}=\frac{\partial\Gamma}{\partial \boldsymbol{Q}_j}\pm\frac{\dfrac{\partial\Delta}{\partial \boldsymbol{Q}_j}}{2\sqrt{\Delta(\boldsymbol{Q})}} \tag{A7}$$

and

$$\frac{\partial^2 V_{1,2}(\boldsymbol{Q})}{\partial Q_j\partial Q_k}=\frac{\partial^2\Gamma}{\partial Q_j\partial Q_k}\pm\frac{\dfrac{\partial^2\Delta}{\partial Q_j\partial Q_k}}{2\sqrt{\Delta(\boldsymbol{Q})}}\mp\frac{\dfrac{\partial\Delta}{\partial Q_j}\dfrac{\partial\Delta}{\partial Q_k}}{4\sqrt{\Delta^3(\boldsymbol{Q})}}. \tag{A8}$$

Now we evaluate the required quantities corresponding to a generic planar geometry, $\boldsymbol{Q}_0\in C_S$, where the diabatic coupling is zero (for example the minima of the S_{L_a} and S_n PESs):

$$\frac{\partial^2\Gamma}{\partial Q_j\partial Q_k}=\frac{B_{ij}^{(n)}+B_{ij}^{(L_a)}}{2}, \tag{A9}$$

$$\sqrt{\Delta}=\frac{|\Delta E|}{2}, \tag{A10}$$

$$\frac{\partial\Delta}{\partial Q_j}\frac{\partial\Delta}{\partial Q_k}=\frac{\Delta E^2}{4}\left(\frac{\partial W_n}{\partial Q_j}-\frac{\partial W_{L_a}}{\partial Q_k}\right)\left(\frac{\partial W_n}{\partial Q_k}-\frac{\partial W_{L_a}}{\partial Q_k}\right), \tag{A11}$$

$$\frac{\partial^2\Delta}{\partial Q_j\partial Q_k}=\frac{1}{2}\left(\frac{\partial W_n}{\partial Q_j}-\frac{\partial W_{L_a}}{\partial Q_k}\right)\left(\frac{\partial W_n}{\partial Q_k}-\frac{\partial W_{L_a}}{\partial Q_k}\right)+\frac{\Delta E}{2}\left(\frac{\partial^2 W_{L_a}}{\partial Q_j\partial Q_k}-\frac{\partial^2 W_{L_a}}{\partial Q_j\partial Q_k}\right)+2\frac{\partial W_{nL_a}}{\partial Q_j}\frac{\partial W_{nL_a}}{\partial Q_k}. \tag{A12}$$

where $\Delta E = W_{L_a} - W_n$ computed at the adiabatic minimum of interest (note that $\Delta E < 0$ at S_2-min).

Substituting eqn (A9)–(A12) into eqn (A8) we finally get

$$\left(\frac{\partial^2 V_{1,2}(\boldsymbol{Q})}{\partial Q_j \partial Q_k}\right)_{\mathbf{Q}_0\in C_s} = \frac{B_{ij}^{(n)} + B_{ij}^{(L_a)}}{2} \mp \mathrm{sign}(\Delta E)\frac{B_{ij}^{(n)} - B_{ij}^{(L_a)}}{2} \pm \frac{2}{|\Delta E|}\lambda_j^{(nL_a)}\lambda_k^{(nL_a)}. \quad \text{(A13)}$$

which gives the adiabatic force constants in terms of the diabatic ones. Since the former are computed *ab initio*, the latter are obtained by the reverse formula.

Acknowledgements

The authors acknowledge support from MIUR (FIRB 'Futuro in Ricerca' RBFR10Y5VW and RBFR08DUX6-003) and the Italian Institute of Technology (IIT-Seed HELYOS). F.A. acknowledges support from the EU People Program, Marie Curie Actions (G.A. no. 246550).

References

1 C. E. Crespo-Hernandez, B. Cohen, P. M. Hare and B. Kohler, *Chem. Rev.*, 2004, **104**, 1977.
2 C. T. Middleton, K. de La Harpe, C. Su, U. K. Law, C. E. Crespo-Hernández and B. Kohler, *Annu. Rev. Phys. Chem.*, 2009, **60**, 217.
3 D. Markovitsi, T. Gustavsson and I. Vayá, *J. Phys. Chem. Lett.*, 2010, **1**, 3271.
4 T. Gustavsson, R. Improta and D. Markovitsi, *J. Phys. Chem. Lett.*, 2010, **1**, 2025.
5 M. Barbatti, Z. Lan, R. Crespo-Otero, J. J. Szymczak, H. Lischka and W. Thiel, *J. Chem. Phys.*, 2012, **137**, 22A503.
6 M. Barbatti and H. Lischka, *J. Am. Chem. Soc.*, 2008, **130**, 6831.
7 E. Fabiano and W. Thiel, *J. Phys. Chem. A*, 2008, **112**, 6859.
8 R. Mitric, U. Werner, M. Wohlgemuth, G. Seifert and V. Bonacic-Koutecky, *J. Phys. Chem. A*, 2009, **113**, 12700.
9 M. Barbatti, A. J. A. Aquino, J. J. Szymczak, D. Nachtigallová, P. Hobza and H. Lischka, *Proc. Natl. Acad. Sci. U. S. A.*, 2010, **107**, 21453.
10 Y. Lei, S. Yuan, Y. Dou, Y. Wang and Z. Wen, *J. Phys. Chem. A*, 2008, **112**, 8497.
11 A. N. Alexandrova, J. C. Tully and G. Granucci, *J. Phys. Chem. B*, 2010, **114**, 12116.
12 S. Perun, A. L. Sobolewski and W. Domcke, *J. Am. Chem. Soc.*, 2005, **127**, 6257.
13 L. Serrano-Andres, M. Merchan and A. C. Borin, *Proc. Natl. Acad. Sci. U. S. A.*, 2006, **103**, 8691.
14 L. Serrano-Andres, M. Merchan and A. C. Borin, *Chem.–Eur. J.*, 2006, **12**, 6559.
15 I. Conti, M. Garavelli and G. Orlandi, *J. Am. Chem. Soc.*, 2009, **131**, 16108.
16 H. Satzger, D. Townsend, M. Z. Zgierski, S. Patchkovskii, S. Ullrich and A. Stolow, *Proc. Natl. Acad. Sci. U. S. A.*, 2006, **103**, 10196.
17 (*a*) C. Plützer and K. Kleinermanns, *Phys. Chem. Chem. Phys.*, 2002, **4**, 4877; (*b*) N. J. Kim, H. Kang, Y. D. Park and S. K. Kim, *Phys. Chem. Chem. Phys.*, 2004, **6**, 2802; (*c*) E. Nir, K. Kleinermanns, L. Grace and M. S. de Vries, *J. Phys. Chem. A*, 2001, **105**, 5106.
18 S. Ullrich, T. Schults, M. Z. Zgierski and A. Stolow, *J. Am. Chem. Soc.*, 2004, **126**, 2262.
19 N. L. Evans and S. Ullrich, *J. Phys. Chem. A*, 2010, **114**, 11225.
20 C. Z. Bisgaard, H. Stazger, S. Ullrich and A. Stolow, *ChemPhysChem*, 2009, **10**, 101.
21 J. C. Tully, *Faraday Discuss.*, 1998, **110**, 407.
22 L. S. Cederbaum, E. Gindensperger and I. Burghardt, *Phys. Rev. Lett.*, 2005, **94**, 113003.
23 E. Gindensperger, I. Burghardt and L. S. Cederbaum, *J. Chem. Phys.*, 2006, **124**, 144103.
24 E. Gindensperger, H. Köppel and L. S. Cederbaum, *J. Chem. Phys.*, 2007, **126**, 034106.
25 H. Köppel, W. Domcke and L. S. Cederbaum, *Adv. Chem. Phys.*, 1984, **57**, 59.
26 A. Raab, G. A. Worth, H.-D. Meyer and L. S. Cederbaum, *J. Chem. Phys.*, 1999, **110**, 2.
27 M. Biczysko, J. Bloino, F. Santoro and V. Barone, in *Computational Strategies for Spectroscopy: from Small Molecules to Nanosystems*, ed. V. Barone, Wiley, Chichester, UK, 2011, p. 361.
28 A. Lami and F. Santoro, in *Computational Strategies for Spectroscopy: from Small Molecules to Nano Systems*, ed. V. Barone, Wiley, Chichester, UK, 2011, p. 475.
29 D. Picconi, A. Lami and F. Santoro, *J. Chem. Phys.*, 2012, **136**, 244104.
30 I. Burghardt, K. Giri and G. A. Worth, *J. Chem. Phys.*, 2008, **129**, 174104.
31 R. Martinazzo, K. H. Hughes, F. Martelli and I. Burghardt, *Chem. Phys.*, 2010, **377**, 21.
32 R. Martinazzo, B. Vacchini, K. H. Hughes and I. Burghardt, *J. Chem. Phys.*, 2011, **134**, 011101.

33 H. Wang, M. Thoss and W. H. Miller, *J. Chem. Phys.*, 2001, **115**, 2979.
34 H. D. Meyer and H. Köppel, *J. Chem. Phys.*, 1984, **81**, 2605.
35 F. Santoro, C. Petrongolo and A. Lami, *J. Chem. Phys.*, 2000, **113**, 4073.
36 F. Santoro, R. Improta and V. Barone, *Theor. Chem. Acc.*, 2009, **123**, 273.
37 R. Improta, V. Barone, A. Lami and F. Santoro, *J. Phys. Chem. B*, 2009, **113**, 14491.
38 D. Picconi, V. Barone, A. Lami, F. Santoro and R. Improta, *ChemPhysChem*, 2011, **12**, 1957.
39 F. Duschinsky, *Acta Physicochim. URSS*, 1937, **18**, 132.
40 C. Lanczos, *Res. Nat. Bur. Stand.*, 1950, **45**, 225.
41 T. J. Park and J. C. Light, *J. Chem. Phys.*, 1986, **85**, 5870.
42 (*a*) C. Adamo and V. Barone, *J. Chem. Phys.*, 1999, **110**, 6158; (*b*) M. Enzerhof and G. E. Scuseria, *J. Chem. Phys.*, 1999, **110**, 5029.
43 *Gaussian 09, Revision A.2*, M. J. Frisch, G. W. Trucks, H. B. Schlegel, *et al.*, Gaussian, Inc., Wallingford CT, 2009.
44 C. Leforestier, R. H. Bisseling, C. Cerjan, M. D. Feit, R. Friesner, A. Guldberg, A. Hammerich, G. Jolicard, W. Karrlein, H.-D. Meyer, N. Lipkin, O. Roncero and R. Kosloff, *J. Comput. Phys.*, 1991, **94**, 59.
45 W. H. Press, S. A. Teukolsky, W. T. Vetterling and B. P. Flannery, in *Numerical Recipes*, Cambridge University Press, p. 1192.
46 H. Köppel, *Chem. Phys.*, 1983, **77**, 359.
47 M. Döscher and Köppel, *Chem. Phys.*, 1997, **225**, 93.
48 M. Barbatti and S. Ullrich, *Phys. Chem. Chem. Phys.*, 2011, **13**, 15492.

DISCUSSIONS

General discussion

DOI: 10.1039/c3fd90018e

Professor Scholes opened the discussion of the paper by Professor Eberhard Riedle: I have a question about the malachite green experiments. The idea of the S_x state is very interesting. It is curious to me that it carries no oscillator strength, given that it is assigned to a charge transfer in a relatively small molecule. Could you elaborate a little on the nature of the state? I wonder whether you might see delayed luminescence from this state, or if there is some reason it decays rapidly to the ground state?

Professor Dr Riedle answered: The S_x state indeed carries some oscillator strength to a higher state. This is seen as an extremely broad—about 2 eV—transient absorption from 600 nm downward in Fig. 4 of the paper. This very width of the spectrum could well be explained by a transition from a CT state to a state with more LE character and largely differing geometry. As even the ground state bleach is completely recovered within less than 10 ps, we would not expect any delayed luminescence. Just as we can imagine the ultrafast population of this state by torsion of the phenyl rings and an associated conical intersection, would we expect the flexibility of the system to lead to an ultrafast return to the ground state with charge recombination.

Professor Buma said: In your experiments on TMCPO you come to the conclusion that ultimately the triplet is populated. You reach this conclusion on the basis of the "long" lifetime of the transient at 235 nm. However, in your paper you report this transient only up to 100 ps, have you recorded this transient also for longer delay times, and what lifetime do you find then? From a lifetime point of view this transient could as well be associated with the $S_1(\pi\pi^*)$ state. Is there any reason to exclude this possibility ? One way of checking this might be to see what the influence of O_2 is on the signal.

Professor Dr Riedle responded: From high level quantum chemistry calculations we know that only the $S_2(^1\pi\pi^*)$ and $S_1(^1\pi\pi^*)$ states and the manifold of triplet states are located within the relevant range of the excitation energy. We can see a clear evidence from a global fit that there are two relaxation steps with distinctive spectra. Therefore the first transition occurring within about 250 fs has to be assigned to the $S_2 \rightarrow S_1$ relaxation and for the 4 ps transition only the ISC is left. The lifetime of the triplet state or states is roughly 100 ns (see answer to the question below). So far we have not varied the O_2 content in the solution, but this would hopefully add even more certainty to our assignment.

Professor Orr-Ewing said: The question posed by Professor Buma raises an important issue about interpretation of recovery (or otherwise) of the ground state bleach. In this case, the photo-excited TMCPO is deduced to relax to a triplet state with high quantum yield because the ground state bleach does not recover on the timescale of the experiments. However, alternative photochemical pathways such as dissociation or isomerization might also account for this observation. Are you confident that such competing mechanisms are not active in TMCPO?

Professor Dr Riedle replied: In addition to the fs and ps measurements reported, we also performed measurements with a tunable ns excitation source and the synchronized fs continuum for detection. We find the same spectral signature at low ns delay times as we find in the ultrafast recordings at high ps times. They can be identified as triplet absorption and ground state bleach at a short wavelength. Both components decay to zero with a time constant of about 100 ns in an aerated solution. Any photochemical pathways such as dissociation or isomerization would lead to a persistent absorption. As this is not observed within experimental precision, the photochemical pathways can be ruled out as a major contribution.

Dr Roberts addressed Professor Dr Riedle and Professor Scholes: Earlier, Prof. Scholes commented on the nature of the 'S_x' state in malachite green (MG), which relaxation is proposed to occur to in Prof. Riedle's discussion paper, after initial excitation to S_1, and specifically questioned why no fluorescence is observed from S_x. This comment/question extends upon this point.

It strikes me that, at least structurally, MG shares some similarities with 4-substituted dimethyl-aminobenzenes, which have long been regarded as benchmark systems for studying intramolecular charge transfer (ICT) processes.[1] Arguably, the most prolific of these molecules is 4-(dimethylamino)benzonitrile (DMABN), which is notorious for its dual-colour fluorescence.[2] In the vertical Franck–Condon region, *ab initio* calculations by Robb and co-workers determine the S_1 state of DMABN to be a locally excited $^1\pi\pi^*$ state (analogous to the anti-Kekulé structure of the S_1 state in benzene[3]), however, the higher lying S_2 state (also a $\pi^* \leftarrow \pi$ transition) exhibits strong ICT character, transferring electron density from the lone pair on the N-atom towards the CN group in the 4-position;[4] interestingly, recent calculations by Stavros, Paterson and co-workers also indicate similar behavior in the more 'fundamental' chromophore unit aniline (aminobenzene).[5] Upon out-of-plane deformation of the $N(CH_3)_2$ moiety with respect to the phenyl ring, the S_1 and S_2 states cross in DMABN, yielding an S_1/S_2 conical intersection, and subsequent further twisting/torsion of the $N(CH_3)_2$ group stabilizes the ICT character such that it (adiabatically) becomes the lowest lying excited state at these twisted geometries.[4,6]

Firstly, although MG is a more complex system, would it be correct to suggest that the S_x state discussed for MG is akin to the stabilized ICT electronic character at twisted geometries in DMABN?—if this were the case, one would indeed predict both solvent dependent viscosity and polarity effects upon the dynamics in MG, as alluded to in Prof. Riedle's paper. Secondly, from a chemical perspective, the ICT character may provide one explanation for the absence of fluorescence from S_x in MG (in contrast to DMABN). Given the nature of the ICT process described above, S_x in DMABN is stabilized somewhat by the electron withdrawing CN group in the

4-position, however, by this reasoning I would anticipate the carbocation centre in the '4-position' of MG to even further stabilise S_x, potentially opening up an energetically feasible pathway to an S_x/S_0 conical intersection in MG,[7] which it would appear is not energetically open in DMABN. Finally, this leads me to more broadly wonder, to what extent may previous studies on simpler 'model' systems, such as DMABN, usefully inform an understanding of the photochemistry occurring in more complex species such as MG?

1. Z. R. Grabowski, K. Rotkiewicz and W. Rettig, *Chem. Rev.*, 2003, **103**, 3889–4031.
2. E. Lippert, W. Lüder and H. Boos, in *Advances in Molecular Spectroscopy*, ed. A. Mangini, Oxford, 1962, pp. 443–457.
3. L. Blancafort, P. Celani, M. J. Bearpark and M. A. Robb, *Theor. Chem. Acc.*, 2003, **110**, 92–99.
4. I. Gomez, M. Reguero, M. Boggio-Pasqua and M. A. Robb, *J. Am. Chem. Soc.*, 2005, **127**, 7119–7129.
5. G. M. Roberts, C. A. Williams, J. D. Young, S. Ullrich, M. J. Paterson and V. G. Stavros, *J. Am. Chem. Soc.*, 2012, **134**, 12578–12589.
6. W. Fuß, K. K. Pushpa, W. Rettig, W. E. Schmid and S. A. Trushin, *Photochem. Photobiol. Sci.*, 2002, **1**, 225–262.
7. A. Nakayama and T. Taketsugu, *J. Phys. Chem. A*, 2011, **115**, 8808–8815.

Professor Dr Riedle replied: The presented arguments make a lot of sense to me and I'm happy to hear that the proposed CT character of the intermediate state is in line with other considerations. Certainly I hope that neither MG nor DMABN are particularly singular but contribute both to a more general understanding. From an experimental point of view, a rerecording of the transient spectra of these model compounds with the largely improved capabilities would be no major effort—it would, however, need a matched theoretical support.

Professor Buma asked: In your summary section you come to the conclusion that transient electronic spectroscopy has finaly caught up with transient IR spectroscopy. I think that there will intrinsically always remain a difference between the two, and that in principle transient IR will always provide more information. This is quite logical since in a UV/Vis transient absorption spectrum you can see a very limited number of transitions, normally something like two or three or so. In IR spectroscopy, on the other hand, there is a plethora of bands. Additionally, one should bear in mind that electronic spectroscopy looks at the molecule as a whole, and it is hard to zoom in on specific structural details. IR spectroscopy, on the other hand, provides a direct structural fingerprint, and—when appropriate local modes are considered—allows one to really zoom in on a local structure. What is your opinion on this?

Professor Dr Riedle answered: I do not fully agree with your opinion. I think it is largely based on the very well established spectroscopy—both vibrational and electronic—of stable molecules, commonly termed dyes. In these systems indeed the transient electronic spectroscopy is rather boring as one only observes the rattling down the ladder of electronic states. These photophysical processes do not have much beyond the exact time constants to reveal. For truly chemical processes the situation is quite different. We often do not know which sequence of transformations is run through, chemical wisdom many times can only narrow it down to two or three alternatives. This is the situation where I think the new possibilities of electronic transient spectroscopy will prove most useful. We can identify short lived intermediates by their largely changed signatures. These intermediates will often be radicals or ions and they differ from their neutrals and precursors much more in their electronic structure than in their vibrational frequencies. In addition, such a reaction chain will inevitably lead to hot intermediates, which makes the vibrational assignments quite difficult. I grant you that vibrational spectroscopy comes back into the game quite importantly once the overall picture is established and one can look at local structures. The dominant electronic structures is, however, well tested in transient electronic spectroscopy—with a complete measurement performed in just a few hours at most. Let's take this opportunity.

Professor Neumark asked: You interpret your experiments on the photolysis of diphenylmethylbromide in terms of ion-pair formation on a 10 ps timescale. Using your deep UV probe capability, can you directly observe the charge-transfer-to-solvent transition of the bromide anion (in water it starts around 225 nm and peaks at 200 nm)? This observation would strongly support your interpretation.

Professor Dr Riedle replied: I very much appreciate the importance of looking for the "other side", *i.e.* the chloride or bromide anion. There are, however, some major obstacles to this. First, our continuum so far reaches only down to 225 nm and this would be barely enough to see any clear signature. Secondly, the precursor solution absorbs very strongly in the deep UV, about a factor of 100 stronger than at our 270 nm pump wavelength and the deep UV probe light barely reaches the detector. Last but not least, acetonitrile starts to absorb at about 220 nm—a curse of the solution phase investigations. As a result, we are still searching. Possibly a different leaving group with a longer wavelength absorbing anion will help.

Professor Orr-Ewing asked: A further advantage of time-resolved infrared spectroscopy methods is that the frequencies of vibrational bands of intermediate and product species can generally be computed with greater confidence than the electronic bands of the same species. The comparison with computed values then provides support for the assignment of experimentally observed transient features.

Professor Dr Riedle responded: Having a senior colleague next door with decades of experience in transient IR spectroscopy and another one on the other side who does the mentioned calculations, I'm quite aware of the possibilities and difficulties in assigning the vibrational spectra. In the case of relevant OH-, CO- or amine bands I can see good possibilities. I do, however, also see the high degree of congestion and the problem of overlapping solvent bands. For the electronic spectroscopy, the experimentalist has straight forward means of obtaining the needed reference spectra by chemical stabilisation of ions or radicals, by spectroelectrochemistry and by simple measurements on reference compounds. This might not be so useful or easy for biomolecules with complex synthesis, but it sure helps for chemical processes, particularly in photocatalysis.

Dr Santoro asked: If I remember properly, you say that transient absorption is a very powerful tool because in practice there are no dark-states for this spectroscopy. The other face of the medal, in my opinion, is the difficulty, at least from a computational point of view, to have a reliable description of the final states after absorption of the laser probe. In my understanding this also means that it is difficult to use arguments like those based on selection rules to extract information on the "nature" of the evolving state of our system. This kind of information is actually important in the clarification of the mechanism under investigation (*e.g.* does the decay pass through a $\pi\pi^*$ state or a $n\pi^*$ state?). I would like to ask your opinion on this point

Professor Dr Riedle answered: I appreciate the problem of calculating the higher excited states for large molecules. In practice we have followed an empirical approach and mainly rely on comparison to known spectra. One particularly interesting observation we made is that, *e.g.*, for perylene diimides the lowest transitions from the locally excited S_1 state are nearly perfectly identical to the ground state absorption of the anion. This teaches me that these systems seem to be already large enough that the spectroscopy becomes similar to solid state pictures like biexcitons. On the other hand, I think we as experimentalists and just as well the theoreticians have to expand our knowledge on these larger systems and their transient spectra. With time, I'm sure we will learn how to read off the character of a $\pi\pi^*$ state or a $n\pi^*$ state.

Dr Santoro commented: I have a comment and a question that are a follow up of my previous question. From a computational point of view, it is very difficult to identify the final states reached in a transient absorption (TA) experiment. There are many possible final states and they generally form a very dense manifold that is not easy to characterize. For example, in this discussion, the potentialities of TD-DFT for the study of the excited states of rather large systems have been mentioned. I would like to highlight that in its standard implementation it is not sufficient to interpret TA results, because it cannot provide transition probabilities among excited states, so that it is not easy to understand which states are actually accessible from the states populated by the pump laser pulse. To retrieve this information, quadratic-response implementations are needed, like those available, for example, in the Dalton program. In my opinion, at the state of the art, having computational methodologies that help the interpretation of TA experiments is still more challenging than, for example, in the case of fluorescence up-conversion spectra where the final state is the ground-state. Which kind of strategy do you use to deduce the "nature" of the evolving state, when you don't have a detailed knowledge of the "nature" of the states populated by the probe?

Professor Dr Riedle replied: I appreciate your worries and the difficulties theory still has and possibly will continue to have in identifying the highly excited states. My present view is that TA certainly sees changes in the spectrum if the system under investigation relaxes from the originally excited state and thereby we can quantify the kinetics. The excited state will very often be a $\pi\pi^*$ state since these carry the largest transition strength. The next state the relaxation leads to can be a $n\pi^*$ state with a long lifetime. If we deduce the species associated spectra (SAS) from the decay associated difference spectra (DADS) which are the direct result of a global fit, with some luck we find a region of stimulated emission for the $\pi\pi^*$ state, but not or only extremely weakly for a $n\pi^*$ state. In other words, the TA measurement includes the detection of transitions to the electronic ground state, just as fluorescence up-conversion—granted in a more crowded fashion. If the relaxation is directly or later to a nπ* or ππ* state, we can expect a bond cleavage within an extremely short time of around or even below 100 fs. There are two consequences: first, the transient population of these states will be very small, so any spectroscopy will have difficulties to detect them. The signatures will evolve with the large scale geometric changes and a rather broad spectrum can be expected. The main alternative to electronic TA—transient IR spectroscopy—will hardly have sufficient temporal resolution to detect this situation with sharp bands. Secondly and quite helpfully, the dissociative state will lead to a long lived ground state product with a well predictable absorption spectrum that can be readily observed by electronic TA. If there are two possible dissociation channels that we want to discriminate, they often asymptotically lead to differing products that we can distinguish. The described case of diphenylmethylbromide is a good example of this situation and clearly lets us identify the relaxation path. We even know with which branching ratio the two possible channels are transversed. Last but not least, I want to emphasize that any subsequent chemical processes proceed in the electronic ground state and therefore the problem of predicting the excited state spectroscopy is not anymore relevant.

Professor Orr-Ewing asked: Is the structure evident in the UV continua shown in Fig. 1 of your paper a consequence of specific elements of your optical system, or is it a general result of this approach to continuum generation in this spectral region? Does the structure vary from pulse to pulse and therefore affect the sensitivity of the experiment?

Professor Dr Riedle replied: The structure in the continuum spectra is a combination of the instrinsic whitelight generation, the filter used to block the residual pump light, the transmission of the optical elements and the spectral sensitivity of the detector. For a good alignment of the continuum generation the structure is quite stable. We do, however, know that not all students using the spectrometer are equally successful at aligning for a stable continuum. The main tool we developed is an online analysis and a display of the spectrally resolved fluctuations that is used to reduce them to 1% or below over most of the spectrum. Once this level is reached, very high sensitivity is ensured. What is left is the issue of the high correlation between subsequent pulses and neighbouring colours. On the one hand, this allows us to reference and divide out the spectral structure by chopping the pump beam. On the other hand, residual slow variations result in a critical issue as to the fidelity of weak spectral features in the transient spectra. We have recently followed the work of Ernsting and coworkers and implemented single shot referencing. This eliminates the fidelity issue and leads to an order of magnitude increase of sensitivity or shortening of the measurement time.

Professor Bearpark commented: It's still a considerable challenge to calculate a balanced representation of multiple electronic excited states for larger molecules, particularly as their geometries change away from vertical excitation;[1] even before dynamical and environmental effects are also taken into account.

1. M. J. Bearpark, F. Ogliaro, T. Vreven, M. Boggio-Pasqua, M. J. Frisch, S. M. Larkin, M. Morrison, M. A. Robb, "CASSCF calculations for photoinduced processes in large molecules: Choosing when to use the RASSCF, ONIOM and MMVB approximations", *J. Photochem. Photobiol. A*, 2007, **190**, 207–227. DOI: 10.1016/j.jphotochem.2007.05.008

Professor Scholes enquired: Your femtosecond pump–probe experiments with a very broad band probe window are very revealing. Probing a broad profile of electronic transitions will be an important approach for the future. On this topic of future experiments, it seems to me that multidimensional spectroscopy will also play an important role. Advantages of correlating the photo-excited electronic transitions with those radiating the signal after the pump–probe time delay include the ability to more incisively decide kinetic pathways. Population flow from one electronic state to another is indicated by amplitude changes in cross-peaks. Do you have comments on future breakthroughs in femtosecond experiments for studying chemical dynamics?

Professor Dr Riedle responded: I agree with your hope that multidimensional electronic spectroscopy will help us in the future to unravel the complex dynamics encountered in larger molecules. We have indeed been able to assemble a 2D-UV spectrometer with tunable 15 fs pump pulses that we think is second to none. The first experiments, however, teach us a number of important messages. First, even this very large pump bandwidth is not yet able to really cover multiple chromophores or electronic states. This is due to the large bandwidth of UV transitions. We have therefore set out to increase the pump bandwidth to 200 THz or more, corresponding to sub-5 fs pulses. This would even surpass the best values in the visible. Secondly, we compared the classical probing with a replica of the pump pulse to continuum probing as used in our TA spectrometers. We clearly find that the observed cross peaks are only meaningful for the latter, for the pump replica they basically just monitor the pulses and not the molecule. Typically the most informative probe regions are far away from the pump region with its multiple signal contributions. Continuum probing is quite readily done, but it comes with the problem of a significant chirp. A reconstruction from multiple 2D-spectrograms as pioneered by Ogilvie is prohibitive for the more than an octave bandwidth and many hundreds of fs chirp. Compression of the full continuum still seems out of reach. To solve this predicament, I envision to record 2D overview spectrograms to locate the spectral regions of largest information content and then do full time scans with selected excitation and a compressed part of the continuum. In this way I hope to combine the advantages of 2D with those of TA.

Dr Murdock opened the discussion of the paper by Professor Roseanne Sension: The paper by Sension and coworkers[1] highlights some of the shortcomings of transient UV pump/UV-visible probe spectroscopy for obtaining the excited state lifetime of α-phellandrene. The authors suggest that the absence of any discernible excited state absorption in this molecule might be attributable to rapid relaxation into a dark state or simply reflect the lack of any suitable transition within the visible probe region. Changing from a white-light continuum to a broadband IR probe should provide a powerful complementary source of information, and circumvent problems associated with probe transitions with low electronic oscillator strengths (*i.e.* dark states) or that occur at inconvenient absorption wavelengths.

Our recent work on the solution phase photodissociation dynamics of *para*-methylthiophenol (*p*-MePhSH) following excitation at 267 nm illustrates some of the advantages an IR probe can provide.[2–4] When using a broadband UV-visible probe, the overlap between features attributable to the *p*-MePhS($\tilde{X}$) radical and an adduct formed *via* geminate recombination severely hampers the extraction of their

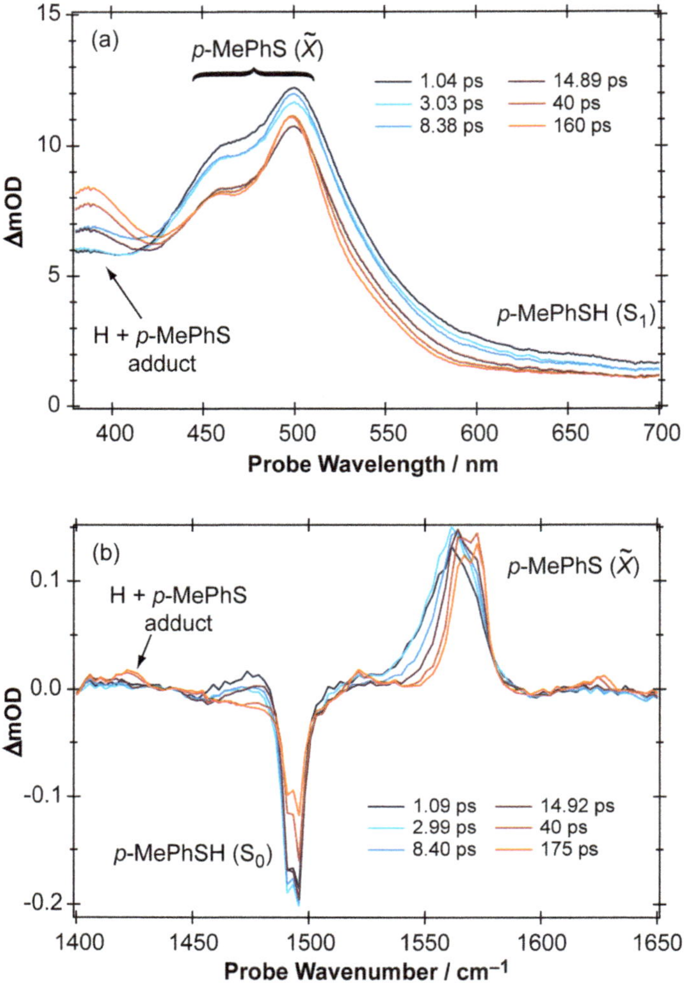

Fig. 1 (a) UV pump/UV-visible probe and (b) UV pump/IR probe spectra measured at different Δt following 267 nm photoexcitation of a 45 mM solution of *p*-MePhSH in acetonitrile flowed through a 100 μm Harrick cell.

attendant kinetics, Fig. 1(a). In contrast, the reduced width of IR transitions (*cf.* their electronic counterparts) leads to much reduced spectral overlap. Further, IR probing (Fig. 1(b)) also allows access to features associated with the parent molecule in its ground electronic state—no transitions of which lie within the range probed by our white light continuum—greatly aiding the resultant kinetic analysis.

1. B. C. Arruda, B. Smith, K. G. Spears and R. J. Sension, *Faraday Discuss.*, DOI:10.1039/c3fd20149j
2. T. A. A. Oliver,Y. Zhang, M. N. R. Ashfold and S. E. Bradforth, *Faraday Discuss.*, 2011, **150**, 439–458.
3. D. Murdock, S. J. Harris, T. N. V. Karsili, G. M. Greetham, I. P. Clark, M. Towrie, A. J. Orr-Ewing and M. N. R. Ashfold, *J. Phys. Chem. Lett.*, 2012, **3**, 3715–3720.
4. S. J. Harris, D. Murdock, Y. Zhang, T. A. A. Oliver, M. P. Grubb, A. J. Orr-Ewing, G. M. Greetham, I. P. Clark, M. Towrie, S. E. Bradforth and M. N. R. Ashfold, *Phys. Chem. Chem. Phys.*, 2013, **15**, 6567–6582.

Professor Sension answered: Ultrafast broadband IR is a powerful complementary technique. The work you cite on *p*-MePhSH is an excellent example. It remains to be seen how useful the technique will be for the ring-opening reactions described in our paper. At this time we think that the lifetime of the α-phellandrene excited state in solution is <500 fs, although it may be longer. A recent study by Weinacht and coworkers[1] reports a very short excited state lifetime for α-phellandrene in the gas phase. The lifetime for α-terpinene in solution is *ca.* 120–160 fs, consistent with the gas phase measurement on this compound.[1,2] Broadband IR is less useful for very short-lived excited states. There is also a problem of overlapping transitions in these polyenes.

1. O. Njoya, S. Matsika and T. Weinacht, *ChemPhysChem*, 2013, **14**, 1451–1455.
2. B. C. Arruda, J. Peng, B. Smith, K. G. Spears and R. J. Sension, *J. Phys. Chem. B*, 2013, **117**, 4696–4704.

Professor Buma commented: You see a striking difference between the transient absorption spectra of DHC and α-terpinene on the one hand, and α-phellandrene on the other hand. In the first case you see a strong excited state absorption in the visible, in the second case not. You have also performed TD-DFT calculations. Do these calculations provide you with indications on how the excited states that are reached in the probe step might be different or shifted in the two cases?

Professor Sension responded: In our calculations we only included the lowest 10 excited states, and this is not sufficient to reach the electronically excited states responsible for the excited state absorption (ESA) in our measurements. An explanation for the difference between the four compounds considered here, DHC and α-terpinene with a strong visible ESA and CHD and α-phellandrene with no visible ESA, will require more extensive calculations. It is useful to note, however, that the one-photon absorption spectra of both CHD and α-phellandrene demonstrate the presence of many states in the range that would be probed by our visible probe.[1,2] Of course, the linear absorption spectrum does not shed light on the selection rules for absorption from the excited state populated by our pump pulse.

1. K. P. Gross and O. Schnepp, *J. Chem. Phys.*, 1978, **68**, 2647–2657
2. N.G. Minnaard and E. Havinga, *Recl. Trav. Chim. Pays-Bas*, 1973, **92**, 1179–1188.

Professor Scholes asked: Being able to assign transient absorption features is an important challenge because it provides the link between the molecular processes we wish to elucidate and the signatures of the participating electronic states in femtosecond measurements. Could you comment on the applicability of DFT methods in this regard? It seems to me that there is an immense challenge for calculating excited state absorption bands because the higher electronic states resonant with the probe are highly correlated as compared to the lowest electronic states. Multiconfiguration reference determinants are likely required.

Professor Sension replied: Accurate calculation of the electronic states of these polyenes are complicated by the need for multiconfiguration methods. The DFT calculations appear to calculate the bright states accessed from the ground state reasonably well, but do not perform as well for the dark states or for geometries that move away from the Franck–Condon region.

Professor Neumark commented: In your experiment, the ground state isomerization dynamics subsequent to ring-opening are probed by their electronic absorption spectra, which are very broad and unstructured. Can you comment on whether it would be preferable to use time-resolved infrared absorption spectroscopy to probe this aspect of the dynamics?

Professor Sension responded: It would be good to use both vibrational and electronic spectroscopy to probe the ground state isomerization dynamics. Because the electronic spectra of the trienes are strong, shift with conformation, and contain some resolved vibrational structure they provide a useful probe of the conformational relaxation. Vibrational spectroscopy could provide additional specific information, of course. Time-resolved IR experiments are complicated by the strength of the vibrational transitions, the relatively modest frequency changes between the different conformations, and the need to look at regions separated from the solvent vibrations. Resonance Raman spectroscopy provides one way to bypass some of these complications. Rich Mathies and coworkers used picosecond time-resolved resonance Raman spectroscopy (TR^3) to look at the ring-opening reactions of cyclohexadiene and α-phellandrene in the early 90's (see ref. 17 of our paper). These TR^3 measurements appear to miss the initial formation of the helical triene product, but observe the formation of the later products with an s-*trans* conformation around one or both single bonds.

Professor Dr Riedle commented: We have recently recorded transient spectra in the range from 245 to 335 nm for 1,3- cyclohexadiene in ethanol after 250 nm pumping with 30 fs pulses (Fig. 2). In the range down to 275 nm shown by Prof. Sension they agree perfectly. In accord with older reports by your group and also Lochbrunner, Fuß and coworkers we observe much more structured spectra below 275 nm after some delay. This is assigned to the evolution of the system to the *all-trans* 1,3,5- hexatriene. The manifold of transient spectra does not display pronounced isosbestic points. Instead, a rather smooth evolution of spectral features and transient absorption strength over the first 100 ps is found. This is in accord with the multiple rotamers developing on the same timescale as vibrational cooling and solvation. In view of this extended data basis, how significant are the biexponential fits reported in your work?

Professor Sension replied: This is an impressive spectrum that you show. Your usable continuum extends about 15 to 20 nm further into the UV than the continuum used in our measurements. Our usable continuum extended to 270, although we truncated our spectral plots at 277 nm because the relative intensities were not reliable on the short wavelength edge. The same appears to be true in the spectrum you plot. The second vibronic peak of the tZt-hexatriene product spectrum should be stronger than the first at long times.

Vibrational relaxation clearly influences the observed transient spectra of all these species. The observed dynamics exhibit the influence of both thermalization and conformational relaxation. The biexponential fit is simply a way to get a handle on the trends observed in the data and should be viewed in this light. The

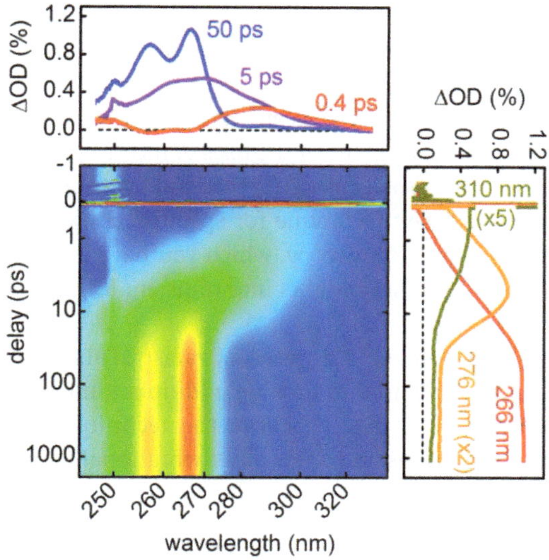

Fig. 2 Transient absorption after excitation of a solution of 1,3-cyclohexadiene in ethanol after pumping with 30 fs pulses at 250 nm. Spectra at selected delay times and time traces at selected wavelengths show the complex spectral evolution in the first 100 ps.

time constants in Table 2 should be viewed as average values as the caption indicates. In fact, the longer time constant represents the cZt → tZt conversion of hot cZt-hexatriene. The room temperature cZt product relaxes much more slowly (see ref. 19 and 22 in our paper).

I'd also like to point out, however, that the absence of pronounced isosbestic points need not be an indication of thermalization or solvation. If three or more species overlap spectrally and convert on similar timescales (within a factor 3 or 4 of each other) pronounced isosbestic points will not be observed because more than two species are generally present during the relaxation process.

Professor Buma asked: In relation to my question before and to the comments made afterwards, I would like to remark that TD-DFT is of course not very well suited for doubly-excited electronic states, and polyenes are of course notorious in ths sense. However, as long as the probe transition in your experiments would only involve single-electron excitations, the TD-DFT calculations might give you an indication of how the final state is effected by substitution, and this in turn might help in understanding the different obervations for CHD and α-phellandrene.

Professor Sension replied: We certainly agree that more extensive calculations, including TD-DFT calculations should give an indication of the nature of the excited states. We intend to pursue such calculations in the future.

Professor Orr-Ewing asked: Do you observe any recovery of the cyclic dienes in these photochemical experiments? Is the pathway to recovery of the ring, as compared to the pathway to the ring-opened trienes, established?

Professor Sension answered: The recovery of the cyclic diene is invisible to our transient absorption experiments. Ground state recovery results in a very weak signal that rapidly approached zero as the molecule cools. As this signal overlaps with the much stronger photoproduct absorption it is not detectable.

Professor Duxbury enquired: In a recent review in *Chem. Soc. Rev.*, 2009, **38**, 1837–1848, Neil Hunt described some of the uses of ultrafast "2D-IR" spectroscopy for "ultrafast insights into biomolecular structure and function". Do you think that a combination of this ultrafast technique for spectroscopy with the ultrafast uv-visible transient absorption spectroscopy, which you have described so clearly in your paper, might give additional insights into transient absorption spectroscopy?

Professor Sension responded: Triggered ultrafast 2D-IR is a powerful technique. Clearly it has the potential to provide additional important insights into photochemical processes. To date I believe the technique has been applied primarily to carbonyl vibrations and a few other vibrational bands. This limits its usefulness for the time being.

Dr Amirjalayer asked: For CHD and α-phellandrene you observe no influence of the solvent properties on their photochemical reactions. For the initial step of the ring-opening reaction leading to the 'hot' products this can be explained assuming a volume-conserving step. However, for the subsequent cooling process the interaction and packing of the solvent molecules around the 'hot' product should play a role. Could you please clarify why—especially for the larger α-phellandrene—the cooling process is independent of the solvent?

Professor Sension responded: There is a small influence of solvent on the subsequent cooling and conformational relaxation of the hexatriene photoproducts formed in the ring-opening reactions of CHD and α-phellandrene, but the solvent effects are quite small. Any clarification of the reason for minor solvent dependence in the cooling and conformational relaxation of DOT formed following the ring-opening reaction of α-phellandrene will require both theoretical simulation and additional experiments.

Surprisingly, there is not yet an adequate understanding of the influence of the solvent on the conformational isomerization of these simple polyenes. In earlier work we found little influence of solvent viscosity on the cZt → tZt reaction of 1,3-HT.[1,2] However, we did see a significant difference between alcohols and alkanes, with a higher effective barrier in alkanes than alcohols. This may reflect the influence of packing around the HT molecule. The effective barriers determined in those experiments are substantially larger than the ground state barriers calculated using either DFT or *ab initio* quantum chemical methods. Including a polarizable continuum model (PCM) for the solvent does not result in any significant difference in the calculated barriers and provides no insight into the experimental results. The barriers calculated for the gZt-DOT and tZg-DOT products formed from α-phellandrene are similar to those calculated for cZt-hexatriene. But there is no experimental evidence that they are accurate either. More work is needed, both experimental and theoretical.

1. N. A. Anderson, S. H. Pullen, L. A. Walker, J. J. Shiang and R. J. Sension, *J. Phys. Chem. A*, 1998, **102**, 10588–10598.
2. D. A. Harris, M. B. Orozco and R. J. Sension, *J. Phys. Chem. A*, 2006, **110**, 9325–9333.

Mr Picconi opened the discussion of the paper by Professor Dr Friedrich Temps: At the end of your paper you mention that further studies of other model systems would help shed light on the interplay between excimer/exciplex and 'monomer-like' states. Another possibility to decipher the nature, energetics and role of the different states in photophysics, could be to study the same system, but in solvents with different polarity. Is this a practicable route?

Professor Dr Temps answered: In order to study regular DNA or RNA dinucleotides with phosphate-sugar backbone in other solvents besides water, *e.g.* in less polar and aprotic solvents, one would have to exchange the counterions Na^+ or K^+ required by the negatively charged phosphate by an organic cation. This is a possible, but somewhat tedious task, requiring special HPLC equipment. Unfortunately, only very small dinucleotide sample amounts are available at reasonable costs. However, we have just succeeded to synthesize chemically bound nucleobase homodimers with a neutral linker chain, which we can also tune in length. Recorded static spectra do indicate the formation of excimers/exciplexes for those systems as well. In addition to water, the prepared dimers are soluble at least in a series of alcohols and in acetonitrile. Femtosecond time-resolved experiments are underway.

Professor Neumark said: When you refer to "monomer-like" excited dinucleotides, do you mean dinucleotides in an unstacked conformation that exhibit monomer-like dynamics? More generally, do the multiple time constants result from distinct ensembles of dinucleotides, some of which are stacked and others unstacked, or do they reflect sequential dynamics in which initially unstacked dinucleotides evolve into a stacked conformation subsequent to electronic excitation?

Professor Dr Temps answered: In the literature, many authors indeed refer to "stacked, excimer/exciplex-like" and "unstacked, monomer-like" dinucleotides, as if these were two different defined ensembles of dinucleotides. Evidence of such a classification stems from literature interpretations of UV absorption and circular dichroism (CD) spectra. However, in the light of our results for both d(ApG) and d(ApA), such an interpretation should be considered with caution. In fact, the observation of "monomer-like" lifetimes for dinucleotides is by no means sufficient for the conclusion that the respective electronic states are monomer-like. The similarity of the lifetimes may simply be fortuitous. From our experience, it does not appear possible to decompose the observed transient spectra into an "excimer/exciplex" and a "monomer" part. Instead, the spectra left after subtraction of the supposed "excimer" contribution differ substantially from those of the mononucleotides. We therefore have to conclude that the alleged "monomer-like" states are in fact substantially perturbed by the second chromophore, even if this second chromophore is not in a stacked configuration. A strong interaction between unstacked bases would indeed be no surprise at all for dipole-coupled excitons. Further analysis of the transient excited-state absorption is underway. In the meantime, we assume an inhomogeneous distribution of dinucleotide configurations and excited states. Even the order of the excited $\pi\pi^*$ and $n\pi^*$ states might change.

Professor Orr-Ewing asked: Why does the fluorescence attributed to the charge transfer or exciplex state appear so apparently strongly in the static spectrum (Fig. 2 of your paper) when the exciplex state of d(ApG) is concluded to be nearly optically dark and so, by implication, non-fluorescing?

Professor Dr Temps answered: Prof. Orr-Ewing raises an interesting question, because the static fluorescence spectra are indeed quite perplexing at first glance. In fact, our time-resolved measurements show a drop in fluorescence intensity at $\Delta t \approx 10$ ps after UV photoexcitation to $\approx 1\%$ of the initial amplitude, even though the ground-state bleach recovery measurements at $\lambda = 244$ nm indicate that some 25–35% of the initially excited dinucleotide molecules have transformed to the long-lived exciplex state. This ratio of $\approx 1/30$ led us to conclude that the exciplex state is "nearly optically dark" (*i.e.*, "nearly non-fluorescent"). In the static fluorescence spectrum, however, the intensity of the red-shifted "exciplex fluorescence" appears strongly enhanced due to the long (≈ 120 ps) lifetime of the state. It is the long observation time of the emission that leads to relatively high intensity. In contrast, the "monomer-like normal fluorescence" in the UV decays with a lifetime of a fraction of a picosecond to a few ps ($\tau \approx 0.4$–2 ps). Without ultrafast electronic deactivation, the optically bright excited state would live for much longer and its intensity in a static fluorescence spectrum would be far higher than that of the exciplex. We emphasize that it is only through the combination of static fluorescence, time-resolved fluorescence, transient excited-state absorption and transient ground-state bleach recovery data that we are able to elucidate the dynamics in its entireness.

Dr Santoro asked: One of the most debated topics in the decay of stacked dimers and oligomers of DNA is whether the long-living excited state has "neutral" or "charge-transfer" character. Do your experiments deliver information of the nature of such state? Would it be possible to extract this kind of information by changing experimental conditions, as for example the solvent (*e.g.* like D. Picconi suggested in a previous question)?

Professor Dr Temps replied: The experiments *per se* do not give information on whether the excimer/exciplex is neutral or of charge-transfer character. What we did note is that d(ApA) shows much stronger fluorescence around 450 nm (in the region of the "excimer" band) in our up-conversion experiments[1] than d(ApG) in the present work. This may hint at a neutral excimer in the d(ApA) case *versus* charge-transfer character in the d(ApG) case, where the difference in redox potentials suggests charge-transfer to be likely. It should indeed be helpful to investigate dinucleotides in different solvents and look for a dependence on solvent polarity, as suggested by Dr Picconi. This is one of the reasons why we started the syntheses of nucleobase dimers with neutral linkers (see above).

1. M. C. Stuhldreier, C. Schüler, J. Kleber, F. Temps, in *Ultrafast Phenomena XVII*, ed. M. Chergui, D. M. Jonas, E. Riedle, R. W. Schoenlein and A. J. Taylor, Oxford University Press, Oxford, 2011, pp. 553–555.

Professor Hirst said: Are you able to comment further on the role played by n-π^* transitions in the experimental absorption spectra of dinucleotides?

Professor Dr Temps replied: In general, nπ* transitions are far too weak to contribute to the experimental absorption spectra of mono-, di- or oligonucleotides from the S_0 ground states. This may change, in principle, in excited electronic states, where the nucleobases in their minimum energy configurations may become nonplanar. In that case, $\pi\pi$* and nπ* states can mix and some oscillator strengths in excited-state absorption and/or in fluorescence may be shared. For nucleobase monomers, however, Kohler and co-workers report that the excited nπ* states cannot be observed in transient absorption at least at the wavelengths used so far. The theoretical studies cited in our paper also found only very low oscillator strengths for the nπ* states.

Professor Buma asked: Fig. 6b in your paper shows that there is a very fast evolution of a band around 390 nm and a band around 600 nm for the dinucleotide d(ApG). These two bands are also visible in the mixture (Fig. 6c), and seem to show a similar evolution. What is the origin of these two bands and how can one understand that they are present in both experiments?

Professor Dr Temps answered: The spectrum of the mixture in Fig. 6c is broad and unstructured because of the addition of the contributions from A and G. The broadness results mostly from the G, which has a nearly unstructured transient spectrum. The fast component between 320–350 nm in Fig. 6b is that of the AG exciplex. The lack of recognizable spectral features in the d(ApG) spectrum at longer wavelengths makes this dinucleotide a worse model system compared to d(ApA) in Fig. 6a, which shows two distinctive bands in the 330–375 nm region with distinct temporal behaviors plus a band in the visible.

Miss Harris addressed Professor Dr Riedle, Professor Sension and Professor Dr Temps: A global fitting procedure is often used in the analysis of transient absorption spectra. These analysis techniques fix the spectral shape of absorption bands corresponding to different species and simply allow the amplitude of the features to evolve with time. The change in amplitude is then assumed to be indicative of population change and kinetic models are subsequently developed to explain the dynamics of the studied process. Could the speakers comment on the validity of these analysis techniques for transient absorption spectra when absorption bands also often evolve spectrally with time, for example when species are formed with high initial vibrational excitation and subsequently cool?

Professor Sension replied: Thermalization of the hot product is always a contributing factor in these experiments and this has to be taken into account. But there are a few important points.

First, in global analysis we do not fix the spectral shape of the absorption bands and allow the amplitudes to vary with time. Rather we assume that there will be a limited number of decay components and couple the time constants across the range of wavelengths probed in the experiment. The amplitudes attached to each decay component can then be used to determine decay associated spectra (DAS) and, given a model for the dynamics, species associated spectra (SAS). Second, when there is a significant contribution from vibrational cooling it is generally not possible to fit the data with a reasonable number of exponential components. An example of this is seen in the transient absorption signal obtained following excitation of α-terpinene reported in ref. 31 (see Fig. 5). Here we find a wavelength dependent decay time on the red edge of the photoproduct spectrum. In dodecane solvent the decay constant ranges from 2.5 ps at 320 nm to 0.2 ps at 360 nm. A global fitting procedure does not model this wavelength dependence well and we did not use a global fitting procedure to model these data. Finally, for the ring-opening reactions compared in our paper, the influence of conformational change on the observed spectra is larger, usually much larger, than the influence of temperature for any given conformation. Thus the observed spectral evolution is dominated by conformational relaxation. The conformational relaxation is modeled reasonably well using a global fitting procedure except in the wings of the spectrum.

Professor Dr Temps answered: The analysis of transient absorption spectra may indeed be very cumbersome and may often be hard to perform without bias. In addition to the issues raised in the question it should be kept in mind that time constant and amplitude parameters from multi-exponential nonlinear least squares fitting are often strongly correlated. The different parameters are better determined if the lifetimes differ by large factors (≥10) and *vice versa*. The cross-correlation problem increases exponentially when two values are of similar magnitude. In all of our fitting analyses, we therefore carefully observe the parameter cross-correlation matrix. Most importantly, however, we emphasize that in our case quite similar (partially even almost identical) lifetime parameters were obtained from time-resolved fluorescence up-conversion, transient excited-state absorption and transient ground-state bleach recovery measurements. Common results obtained by several quite different, but complementary experiments are of greatly enhanced physical reliability compared to data from a single type of experiment.

Considering the analysis of transient absorption spectra with spectrally evolving absorption bands, which are indeed frequently observed, we have found it useful in some cases to analyze the temporal evolutions of the spectrally integrated transient absorption bands to determine decay or rise time values. Even more detailed information can be gained from a combined spectro-temporal band analysis. We have used both these approaches in our analysis of the transient absorption maps of d(ApA). Presumably "unbiased" analysis methods such as singular value decomposition are less helpful in the case of spectrally evolving transient absorption bands, unfortunately. Another approach would be to directly compare experimentally measured and theoretically calculated two-dimensional spectro-temporal absorption maps. We have successfully explored this route in our collaboration with the group of Prof. B. Hartke at Kiel University for the ultrafast ring-opening and ring closing of furyl fulgides.[1,2] The calculation of not only the photo-excited (S_1) but also the higher (S_n) electronic states reached by a transition absorption experiment with sufficient accuracy for direct comparison with experiment is an indeed very formidable challenge for theory, however.

In conclusion, there is no single ideal and unique method analyzing transient absorption spectra. Model-free and model-based approaches should be used in conjunction where possible and different results should be carefully compared and judged.

1. R. Siewertsen, F. Strube, J. Mattay, F. Renth and F. Temps, *Phys. Chem. Chem. Phys.*, 2011, **13**, 3800–3808.
2. J. B. Schönborn, A. Koslowski, W. Thiel and B. Hartke, *Phys. Chem. Chem. Phys.*, 2012, **14**, 12193–12201.

Professor Dr Riedle responded: We typically apply single wavelength fits at a variety of wavelengths first to analyze whether there are some common time constants and how many constants are needed. If this looks promising a global fit is used to obtain the overall picture. A novel tool that proofs quite helpful is a "maximum entropy-based inverse Laplace transform" method developed by Dick and coworkers.[1] This allows to identify all relevant dynamics without *a priori* assuming a model and it also provides uncertainties to the time constants. Even with all necessary care, the mentioned issue of shifting bands and reshaping is the most important challenge at present in the analysis of high quality transient spectra. We have not yet developed a full concept of how to account for the spectral evolutions in a physically meaningful scheme. What we do use so far are partial solutions like the determination of the band maximum shift, band integral fits to extract the electronic dynamics and singular value decomposition approaches. Most importantly, a high degree of scepticism and double checking is called for on the part of the advisor that is presented with an interpretation.

1. R.-J. Kutta, T. Langenbacher, U. Kensy and B. Dick, *Appl. Phys. B*, 2013, **111**, 203.

Dr Oliver opened the discussion of the paper by Professor Ursula Rothlisberger: In your paper, your calculations show that the energy separation between the $\pi\pi^*$ and $\pi\sigma^*$ states increases upon more solvation around the tryptophan residue (Fig. 6). You do not, however, give the relative energies of the $\pi\pi^*$ state upon sequential addition of water molecules. Would you not expect the relative energy of the $\pi\pi^*$ state to exhibit a strong wavelength shift?

Professor Roethlisberger responded: The lowest π–π^* state in protonated Trp is the (covalent) L_b state, which is largely insensitive to changes in the environment. Our TD-DFT calculations with MO6-2X and CAM-B3LYP functionals indicate that the energy of this state is changing by less than 0.03 eV in going from the bare to the solvated Trp^+.

Professor Domcke asked: Spectacular spectroscopic effects have been observed by Rizzo and coworkers in clusters of protonated tryptophan with one or two water molecules. Can your calculations predict spectroscopically observable effects for larger systems such as proteins?

Professor Roethlisberger answered: Experimental observables that we can calculate for large protein systems include absorption and emission spectra and for the case that ultrafast photofragmentation channels are present also predictions of the possible photofragments.

Professor Neumark said: Your results on the effects of solvation by a small number of water molecules are reminiscent of the experiments by Lineberger and co-workers[1] on ion clusters that explored the effects of solvation on caging and recombination dynamics of a photoexcited ionic chromophore. This work showed that a relatively small number of solvent molecules could significantly affect the cluster dynamics through shifting of excited state potential energy surfaces. Does this work have any bearing on your results?

1. A. Sanov and W. C. Lineberger, *Phys. Chem. Chem. Phys.*, 2004, **6**, 2018.

Professor Roethlisberger replied: The results for solvated anionic clusters are indeed similar to our findings in the sense that in both systems the addition of few solvent molecules can lead to dramatic alterations of the photoexcitation dynamics albeit the underlying physical mechanisms are different: in the case of solvated protonated tryptophan, the addition of the hydrogen-bonding water solvent has a direct influence on the electronic structure of the chromophore whereas in the case of anionic solvated clusters like I_2–$(CO_2)_n$, the solvent molecules act predominantly as confinement (caging effect) and energy bath.

Professor Duxbury remarked: Fig. 7 on page 11 shows the "photodynamics of HAS under fully hydrated conditions". In Fig. 7(b) the time evolution of the excited states seems to consist of the evolution of the ground state spectrum plus an additional evolution which resembles a time-shifted evolution of the bond length of C–N. The three peaks in the excited state between 50 and 100 fs appear to resemble the first three oscillations of the C–N behaviour, with the inverted peak at *ca* 175 fs resembling the 4th inverted peak of the C–N time evolution.

In the description of the photodynamics the final sentence on page 10 reads: "After a short period of relaxation (~60 fs) the dynamics leads to a purification of the ππ* character of the state that drives the dynamics." Does this mean that in the evolution of the ground state the position of the Kohn–Sham orbital denotes the successful purification of the ππ* character?

Professor Roethlisberger answered: To give a more intuitive picture of the electronic structure changes during excited state dynamics, we analyze the electronic transition in terms of ground state Kohn–Sham orbital contributions. This analysis shows that at the start of the dynamics the system is in a mixed state with π–π* character and other contributions. The changes in the geometry during the excited state relaxation from the Frank-Condon region induce corresponding changes in the electronic structure that make the π–π* contributions dominant, this is what we call 'purification'.

Dr Worth asked: The protein systems you are studying are charged. How did you treat the long range forces in the direct dynamics simulations? Also, how do you know which tryptophan residues are protonated?

Professor Roethlisberger responded: For the cases, in which the QM region is charged, we use isolating boundary conditions as introduced by Hockney (*Methods Compt. Phys.*, 1970, **9**, 136) or Martyna and Tuckerman (*J. Chem. Phys.*, 1999, **110**, 2810. The small molecule simulations were performed on protonated tryptophan where the additional proton is situated at the NH_3^+ group. This is the most stable proton position predicted theoretically and confirmed experimentally *via* mass spectrometry. For the protein simulations: the tryptophan residues are not protonated, the positive charge is located at a nearby lysine residue. Since the site is highly solvent exposed, and the pK_a of lysine in aqueous solutions is ~10.5 this can be considered as the most likely protonation state.

Professor Buma remarked: Prof. Neumark previously suggested to look for the $\pi\sigma^*$ state in the hydrated complex of tryptophan at higher excitation energies. However, I think that in that case excitation of the $\pi\pi^*$ state will not lead to dissociation. The reason is that along the dissociation coordinate the $\pi\sigma^*$ state will go down in energy but the lower lying $\pi\pi^*$ state will increase in energy. There will thus be a conical intersection between the two states, leading to a fast internal conversion from the $\pi\sigma^*$ state to the $\pi\pi^*$ state. One could of course get an idea of whether the $\pi\sigma^*$ state is leading to dissociation by looking to what extent the Human Serum Albumin is susceptible to UV radiation damage.

Dr Blancafort asked: Zwitterionic tryptophan (Trp) undergoes hydrogen transfer in the (π,π^*) excited state from the NH_3^+ to the indole moiety, as proved experimentally in solution (I. Saito, H. Sugiyama, A. Yamamoto, S. Muramatsu, T. J. Matsuura, *Am. Chem. Soc.*, 1984, **106**, 4286–4287). CASSCF calculations confirm this process (L. Blancafort, D. González, M. Olivucci and M. A. Robb, *J. Am. Chem. Soc.*, 2002, **124**, 6398–6406). In contrast to this, no excited state hydrogen transfer is observed for protonated tryptophan (Trp^+) in experiment or calculations. This is surprising in view of the fact that Trp^+ and zwitterionic Trp only differ in the protonation of the carboxylate group. One would also expect Trp^+ to be more acidic. Do you have an explanation for this difference?

Professor Roethlisberger replied: Our investigation focuses on photodynamics on the <100 fs timescale. For this time window, we find that for bare protonated tryptophan, an ultrafast photodissociation channel is present. Our results do not exclude the existence of additional nonradiative relaxation channels on longer timescales involving, *e.g.*, proton transfer reactions. Your theoretical results for zwitterionic tryptophan indicate that there is an initial barrier for proton transfer from the NH_3^+ group to the indole ring of 3.2 kcal mol^{-1} and the experimental data gives a 22% yield in 30 min for the corresponding photoproduct. Both of these results thus indicate that proton transfer is a relatively slow process that we cannot observe directly in our simulations. The occurrence/preference for different channels can be highly dependent on the geometry, *e.g.* special conformers may favor intramolecular proton transfer over other pathways.

Mr Malis enquired: Have you perhaps included the neighbouring peptide bonds of the Trp residue in the QM part of your HSA model? Peptide bonds can participate in the quenching of the indole π–π^* excited state by an excitation transfer to a n-π^* excited state localized on the peptide bonds.

Professor Roethlisberger replied: The QM part of our HSA simulations only contains the side chain (cut at the Cβ) any transitions involving to the peptide backbone. However, we did full QM (DFT) calculations of a protonated GLY-TRP dipeptide with the backbone and we could not find any low lying electronic excitations that involve the peptide backbone.

Professor Scholes opened the discussion of the paper by Professor Irene Burghardt: You presented very nice calculations showing the role of coherence in the evolution of conjugated polymer states after photo-excitation. In some cases of conjugated polymer films in the presence of an electron acceptor the charge transfer is also happening on a comparable timescale. For instance, we have data showing charge separation between pBTTT and intercalated PCBM occurs in about 30 fs. Do you expect coherence to play a role also in the dynamics of charge transfer?

Professor Burghardt answered: We certainly expect coherence to play a role in the dynamics of charge transfer as well. In fact, we found in recent simulations of oligothiophene–fullerene complexes (*i.e.*, model systems for P3HT-PCBM heterojunctions)[1] that vibronic coherence should guide the transfer process on a characteristic timescale of about 50 fs. This coincides with the timescale of exciton dissociation in this system, which is close to the timescale that you observed for the related pBTTT-PCBM junction (for further details, see my comment on Professor Troisi's paper in General Discussion DOI: 10.1039/c3fd90019c).

1. H. Tamura, R. Martinazzo, M. Ruckenbauer, and I. Burghardt, *J. Chem. Phys.*, 2012, **137**, 22A540; H. Tamura, I. Burghardt, and M. Tsukada, *J. Phys. Chem. C*, 2011, **115**, 9237.

Professor Troisi enquired: Do you expect the phenomenology to change as the system considered becomes larger and therefore more degenerate excitonic states become available?

Professor Burghardt responded: I assume that the phenomenology—in particular regarding the coherent nature of the transfer and the two-timescale process leading to stationary exciton-polaron states—remains essentially unchanged with system size. This is a consequence of the partial localization of the excitonic states on spatially confined sub-fragments ("chromophores") whose size is inferior to the overall lattice size. These sub-fragments give rise to a local spectrum of exciton states that evolve as a function of geometry and energy fluctuations.

The above picture is closely related to the "spectroscopic unit" paradigm that has been extensively discussed in the literature. Here, it is often assumed that the effective confinement is a direct consequence of conjugation breaks or chemical defects. However, as has been pointed out in various recent works,[1] localization is not entirely determined by geometric constraints but should rather be understood as dynamical localization on a disordered lattice structure. Transposing the present study to a larger lattice would allow one to fully explore this aspect—*i.e.*, the dynamics of exciton confinement and exciton migration—along with removing certain limitations as far as the construction of the initial state is concerned.

1. O. R. Tozer and W. Barford, *J. Phys. Chem. A*, 2012, **116**, 10310; W. Barford, *J. Phys. Chem. A*, 2013, **117**, 2665.

Professor Dr Meier said: What is the role of the bath coupled to the central torsional mode? How do the results depend on the chosen bath parameters, like coupling strength and cut-off frequency?

Professor Burghardt replied: The bath coupled to the central torsional mode leads to an efficient damping of this mode which is initially strongly displaced and, hence, vibrationally "hot" at the outset of the dynamics. Here, the bath modes can be taken to subsume the effect of all remaining vibrational modes of the oligomer along with the intermolecular modes connecting neighboring chains. While we are currently working on a realistic representation of the bath, using spectral densities that are constructed from MD simulations, the present contribution has employed a phenomenological, Ohmic bath whose frequency distribution is chosen to be resonant with the torsional frequency and whose coupling strength corresponds to a friction coefficient that ensures a rapid decay on a ~300 fs timescale (see *sec.* 3 and Table 5). The bath thus plays an essential role in two respects: (i) The planarization in the excited state, *i.e.* the "healing" of the torsional defect, depends on an efficient damping by the bath. (ii) At the same time, due to the energy relaxation of the central torsional mode, the energy of the excitonic subsystem is transmitted to the bath, leading to de-excitation of the excitonic system and eventually generating a vibrationally relaxed local exciton ground state (also denoted "LEGS"[1]) on the enlarged fragment, as illustrated in Fig. 3 (here, depicting snapshots of the electron–hole dynamics for an oligothiophene system, with a stationary LEGS state, see our poster contribution P29 and, for comparison, Fig. 7 of our paper). One should perhaps mention that our interpretation differs from the one given in ref. 1 where it is assumed that ultrafast damping of the BLA modes, on a timescale of ~40 fs, leads to a vibrationally relaxed LEGS. In the present analysis, the relaxation of the BLA modes is much less rapid, drawing upon experimental information that points towards coherent BLA oscillations on a timescale of tens to hundred of femtoseconds (see ref. 9 and 10).

1. O. R. Tozer and W. Barford, *J. Phys. Chem. A*, 2012, **116**, 10310; W. Barford, *J. Phys. Chem. A*, 2013, **117**, 2665.

Dr Worth commented: The model Hamiltonian you are using appears to have constant couplings between the electronic states? Is this realistic? What effect would coordinate dependent couplings have?

Professor Burghardt responded: In fact, both the excitonic and charge transfer couplings of our model Hamiltonian are not constant but depend on the torsional coordinates according to eqns (10)–(13), see also Fig. 2a/b. The diabatic site-to-site

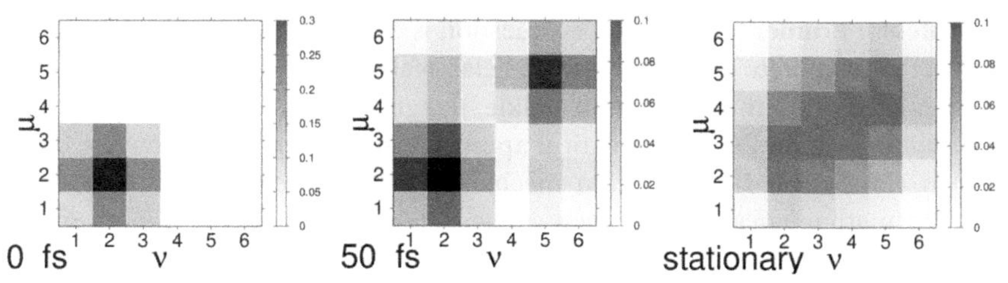

Fig. 3 Complementary to Fig. 7 of paper C3FD20148A, simulations of electron-hole dynamics in an oligothiophene hexamer are shown. Here, the transition to a ground state exciton polaron (LEGS) on the enlarged fragment is very conspicuous (see right panel).The relevant time scale is again of the order of 500 fs.

couplings are maximal at the planar geometry and minimal at the 90° twisted geometry, in line with intuitive overlap arguments. The model, as it stands, does not feature an S_1/S_2 conical intersection at the 90° twisted geometry, in accordance with the electronic structure results reported in ref. 26, where a residual coupling (j_0 of eqn (11)) persists at the 90° twisted geometry. An extension of the model would take into account that the diabatic couplings also carry a dependence on the BLA modes (see also ref. 48 and 61).

Dr Santoro asked: I would like to ask some additional question on the role of the high-frequency BLA modes. Do they also play some role in the dephasing of the EET dynamics, or dephasing processes are only introduced by the low-frequency torsional modes?

Professor Burghardt responded: The BLA modes play a crucial role in the dephasing of the EET dynamics since they cause a rapid modulation of the electronic site-to-site transfer. The correlations between the electronic and BLA degrees of freedom eventually lead to the trapped exciton–polaron state that is observed as the quasi-stationary state in the dynamics simulations. The trapping effect illustrated in Fig. 4 is not observed in the absence of the BLA modes. Conversely, the stationary, trapped state is not reached in the absence of the torsional modes since the latter are essential in providing a quasi-continuum of states that induces an irreversible dynamics on the finite-dimensional lattice.

Professor Scholes asked: I have a question that at first may sound strange. How should we think about coherence in photo-initiated dynamics of complex systems in chemistry or biology? Professor Burghardt examined the off-diagonal density matrix elements, which is an obvious approach (if we can measure, probe, witness, or calculate them). In quantum optics the density matrix would be studied in more detail using measures for entanglement. The point of view that seems to prevail in quantum-classical approaches is that quantum coherence involves superpositions of trajectories (therefore capturing quantum interference of dynamics pathways in phase space). If you examine these approaches carefully I think you will find they each decide differently whether a process is quantum coherent or classically coherent. An even more general measure for coherence is used in statistics: simply that there is correlation. I believe we need to think about what aspects of coherence concern us in chemical dynamics of complex systems (condensed phase, biology) and, on that basis, decide on appropriate measures of coherence.

Professor Brumer responded: The question is excellent, and surely requires an approach common to both quantum and classical mechanics to progress. To this end I call attention to work by Josh Wilkie, Jiangbin Gong, and me that develops and utilizes a common mathematical approach (based upon the Liouville operator in classical and in quantum mechanics) which allows clear distinctions between mathematical issues (such as eigenstates, eigenvalues, Hilbert Space, superposition of states, coherence in time evolution) that are common to classical and quantum mechanics, and those which are not. Sample references are Wilkie and Brumer, *Phys. Rev. A*, 1997, **55**, 27; Wilkie and Brumer, *Phys. Rev.*, 1997, **55**, 43; Gong and Brumer, *Phys. Rev A*, 2006, **73**, 052109. Such an analysis can lead to unexpected conclusions. For example, the latter paper shows that the Born rule has a direct analogue in classical mechanics.

Professor Burghardt replied: This question is indeed of key importance, since coherence, entanglement, and correlations all have similar but not necessarily identical implications. Indeed, the multi-chromophoric systems that typically appear in chemistry and biology involve all these various facets of coherence. For example, the trapped exciton states that are addressed in our contribution are entangled exciton-phonon states that exhibit excitonic coherence in the site representation. Coherent exciton transfer dynamics is guided by the imaginary part of excitonic coherence that gives rise to non-stationary transition flux densities. Site-to-site correlations can be of electronic nature, *i.e.*, involving charge transfer excitons, or involve phonon modes if correlated fluctuations play a role. When defining measures of coherence or correlations, *e.g.*, time-dependent participation ratios, coherence lifetimes, and anisotropy decay times, the chosen representation (adiabatic or diabatic, site-based or fragment-based) will play a crucial role. Further, as suggested by Professor Scholes and by Professor Brumer in his comment, some of the coherence measures relating to wavepacket motion and excitonic transport can be interpreted purely classically, suggesting a broader view on quantum *vs.* classical nonstationary states (see also ref. 1 and 2 here).

In view of the obvious complexity of the dynamical effects, it seems that the experimentally accessible information even from highly specific experiments like two-dimensional electronic spectroscopies, often cannot disentangle, *e.g.*, purely electronic/excitonic and vibronic coherences. Contrary to the measurement of spin coherence in NMR and entangled states in quantum information experiments, the typical case in molecular systems is indeed the parallel occurrence of the various phenomena sketched above. Theoretical models are therefore necessary to identify the main factors at work. To cite a recent example, the analysis of ref. 3 that re-interprets the long-lived quantum coherences observed in photosynthetic systems in terms of local *vibronic* coherences (rather than purely excitonic coherence as previously thought) might well contribute to resolving some of the open issues in this domain.

1. W. H. Miller, *J. Chem. Phys.*, 2012, **136**, 210901.
2. J. S. Briggs and A. Eisfeld, *Phys. Rev. E*, 2011, **83**, 051911.
3. W. Christensson, H. F. Kauffmann, T. Pullerits and T. Mancal, *J. Phys. Chem. B*, 2012, **116**, 7449.

Dr Worth asked: How is the coherent nature of the excitation energy transfer manifested in the simulations? And how does this connect to the experimental observation? Would a density matrix approach provide extra relevant information on these?

Professor Burghardt responded: Since the simulations are carried out in a site-based (monomer) representation, the primary information that is available concerns the site-to-site excitonic coherences (see eqn (15)) which generally do not decay instantaneously but reflect an oscillatory, Rabi type transfer dynamics. This clearly indicates that the transfer cannot be described as a Förster type hopping process. If the experimentally available information relates to a "chromophore" or "spectroscopic unit" representation rather than the site representation (as is the case for the systems considered here), the relevant excitonic coherences can be constructed once the information in the site basis is available. In additon, excitonic (electronic) coherence can be clearly distinguished from vibrational (or vibronic) coherence at the level of the simulations—which may not be straightforwardly possible in the experiment. In principle, simulation thus yields quantitative measures of coherence that can be compared directly with experiment—*e.g.*, excitonic coherences monitored in photon echo experiments—if the representation (site-based, fragment-based, or adiabatic) that corresponds to the spectroscopic measurements is known.

A density matrix representation would involve the direct inclusion of ensemble effects, *e.g.*, relating to a thermal distribution. While the ultrafast processes in question are not thermally activated, the initial conditions could be influenced by such ensemble effects, especially as far as the low-frequency torsional modes are concerned. Furthermore, the coherent character of the dynamics could be reduced due to ensemble-induced decoherence. Some of our simulations for similar systems have shown that (i) the short-time coherence tends to be largely unaffected by temperature effects,[1] and (ii) thermal effects tend to play a minor role in the short-time dynamics.[2] The present simulations have thus been carried out at $T = 0$; however, we are currently working on additional simulations at non-zero temperature.

1. K. H. Hughes, C. D. Christ, and I. Burghardt, *J. Chem. Phys.*, 2009, **131**, 124108.
2. H. Tamura, I. Burghardt, and M. Tsukada, *J. Phys. Chem. C*, 2011, **115**, 9237.

Miss Perveaux opened the discussion of the paper by Dr Fabrizio Santoro: Do you know if your analytical model is capable of describing more complex reaction paths where usually we need to use fitting procedures?[1–3] In benzopyran for example, there is a transition state between the Franck–Condon point and the conical intersection, which means that one needs to go beyond the quadratic vibronic coupling model developed by Cederbaum and coworkers[4] and consider more diabatic states than adiabatic ones.[5,6]

1. Mamadou Ndong, Loic Joubert-Doriol, Hans-Dieter Meyer, André Nauts and Fabien Gatti *et al. J. Chem. Phys.*, 2012, **136**, 034107.
2. E.V. Gromov, A. B. Trofimov, F. Gatti and H. Köppel. *J. Chem. Phys.*, 2010, **133**, 164309.
3. loic Joubert-Doriol, Benjamin Lasorne, Fabien Gatti, Markus Schröder, Oriol Vendrel and Hans-Dieter Meyer, *Comput. Theor. Chem.*, 2012, **990**, 75–89.
4. "Multimode Vibronic Coupling Effects in Molecules", *Int. J. Quantum Chem.*, 1981, **20**, 1981, 251–267.
5. Joaquim Jornet-Somoza, Benjamin Lasorne, Michael A. Robb, Hans-Dieter Meyer, David Lauvergnat, and Fabien Gatti, *J. Chem. Phys.*, 2012, **137**, 084304.
6. B. Lasorne, Michael A. Robb, Hans-Dieter Meyer and F. Gatti, *Chem. Phys.*, 2010, **377**, 30–45.

Dr Santoro answered: Our calculations are based on a hierarchical representation of the quadratic vibronic coupling (QVC) Hamiltonian,[1] *i.e.* the diabatic states have quadratic potential energy surfaces (PES) and are coupled by linear (and quadratic) functions of the nuclear coordinates. Hierarchical representations were first proposed for the linear vibronic coupling model (LVC)[2] and divide the coordinates in blocks, so that only few blocks are necessary to accurately simulate the photoexcited dynamics of the full-dimensionality system on a short timescale. As was shown in ref. 3 for LVC model, they can be straightforwardly generalized to deal with several coupled diabatic states. The price to be paid is the increase of the dimensions of the blocks of the hierarchy. In the most general case, for a model of N coupled states, the first block of the hierarchy already comprises $N^*(N+1)/2$ coordinates. Therefore the reduction of the dimension of the system might not be as remarkable as in the case of two of few coupled PES. Hierarchical representations can be applied also after a system/bath partition of the system. This means that it is possible to account for a general anharmonic PES along some selected coordinates, and to apply the hierarchical representation only to the harmonic bath of modes. Also in this case the dimension of the smallest physically meaningful model increases.

1. D. Picconi, A. Lami and F. Santoro, *J. Chem. Phys.*, 2012, **136**, 244104.
2. E. Gindensperger, I. Burghardt and L. S. Cederbaum, *J. Chem. Phys.*, 2006, **124**, 144103.
3. E. Gindensperger, I. Burghardt and L. S. Cederbaum, *J. Chem. Phys*, 2007, **127**, 124107.

Professor Burghardt responded: It is true that standard vibronic coupling Hamiltonians at the linear or quadratic order are not necessarily suitable to account for reactive dynamics and complex excited-state topologies—even though Dr Santoro's contribution nicely illustrates the usefulness of these models for semi-rigid polyatomics. The concept of "regularized diabatic states",[1] along with various types of fitting procedures, to some extent overcome these limitations. Apart from this, it is often useful, as mentioned by Dr Santoro, to divide the system into a subset of generalized "reaction coordinates" that can involve pronounced anharmonicities, and a larger number of bath modes that are amenable to more approximate treatments, *e.g.*, by reduced-dimensional effective-mode models[2] and their generalization to quadratic order as discussed in Dr Santoro's paper.

1. H. Köppel, *Faraday Discuss.*, 2004, **127**, 35.
2. L. S. Cederbaum, E. Gindensperger, and I. Burghardt, *Phys. Rev. Lett.*, 2005, **94**, 113003; R. Martinazzo, B. Vacchini, K. H. Hughes and Burghardt, *J. Chem. Phys.*, 2011, **134**, 011101.

Dr Worth answered: We have used an extended vibronic coupling model to describe benzene where the reaction path along S_1 to the S_1/S_0 intersection also has a barrier.[1] To do this, it was necessary to include 3rd and 4th order terms to reproduce the topology of the adiabatic surfaces well. This was done using either a 2-state model, in which the barrier on the upper state has to be explicitly described, or a 5-state model in which the barrier is described as a diabatic crossing. The paper on pyrolle by Wolfgang Domcke in this Discussion is another example of where the extended vibronic coupling model is used to describe complex surfaces, again using diabatic functions that account for much of the complex topology. We do, however, find that for systems with multiple crossings and strong coupling between modes the model requires even higher order expansion terms and the fitting becomes difficult.

1. T. J. Penfold, R. Spesyvtsev, O. Kirkby, R. S. Minns, D. Parker, H. H. F. Fielding, and G. A. Worth, *J. Chem. Phys.*, 2012, **137**, 204310–204312.

Professor Domcke said: It can be seen in your paper that the oscillations in the electronic populations are effectively quenched when about six vibrational modes are included, while the population remains around 20% when 30 or more additional modes are included. These additional modes seem to have no effect on the population dynamics. Why does the population not relax to zero when many modes are included?

Dr Santoro responded: The two coupled electronic states are rather similar in energy. What we reported in most of our figures is the decay of the S_{La} diabatic population toward the Sn one. As is shown in Table 1, there are regions of the coordinate space where S_{La} corresponds (or is practically degenerate, depending on the basis set adopted) to the lower-adiabatic excited state S_1. Therefore, in diabatic terms, some population can remain trapped in S_{La}. Fig. 4 (left panel) reports also the decay of the population of the adiabatic S_2 state toward the S_1 one. Calculations are expensive and were performed only for the smallest 3D (3q) model, but they show that, on average, the population of the S_2 state goes sensibly closer to zero than the S_{La} diabatic population does. I believe that increasing the number of coordinates, the S_2 adiabatic population would approximately relax to zero. This might be tested with the hybrid model (3q-36c) *i.e.* treating 3 modes at quantum level and all the other modes at classical level.

Professor Dr Temps addressed Dr Santoro, Dr Worth and Professor Burghardt: The calculation of multiple excited states of large organic chromophores like adenine still poses many challenges. Unfortunately, it appears as if different theoretical methods may systematically predict different orders of excited states, for example for the two $\pi\pi^*$ and the $n\pi^*$ state of adenine. Further, the order could depend on the fluctuating solvent environment. In this situation, it may be a good idea to perform all quantum chemical and dynamics calculations for different inner-shell solvent configurations to estimate ensemble effects and obtain a handle on the size of the possible effects. I'd be very interested in Dr Santoro's view on this.

Dr Santoro responded: While some methods should in principle be considered more accurate than others, I think that as a matter of fact, even in gas phase, none of the computational methods nowadays available is 100% trustworthy, as far as the relative position of closely lying excited states is concerned. Therefore their performance still needs to be validated on the specific system of interest checking whether and to which extent their predictions match experimental observations. In this sense, the availability of several spectroscopic techniques is very important since some states, "dark" with respect to standard one-photon absorption, can be detected by multiphoton absorption or, for example, magnetic circular dichroism. At the state of the art, I think it is always useful to check the robustness of the theoretical predictions with respect to moderate changes of the relative energies; this can be done straightforwardly once a model Hamiltonian has been built.[1] The situation is even more complicated in a solvent. Implicit continuum models like PCM allowed big progress in this field,[2] but they cannot describe specific interactions (this limitation can be cured considering a super-system made up by the solute and some explicit solvent molecules) and give a "mean" representation, *i.e.* do not account for fluctuations. The latter are most properly described through explicit solvent models. A super-system made up by the solute and a suitable cage of solvent molecules can be described within the family of QM/MM methods, which can be rather straightforwardly adopted also for dynamic simulations of photochemical processes, *e.g.* by trajectory surface hopping methods. These systems are however too computationally demanding for quantum-dynamic (QD) simulations. The most natural framework for their study at QD level is probably the on-the-fly direct quantum dynamics;[3,4] G. Worth and I. Burghardt are among the major developers of this methodology and can much better than me foresee when such simulations will be actually routinely performed. Studying the QD of uracil derivatives in acetonitrile and water we faced with a much more limited task,[1] using a hybrid explicit/PCM model of the solvent (it is "static" with respect to the nuclear degrees of freedom of the solvent molecules that have been considered fixed) and focusing on the ultrafast (dozens of fs) timescale. This approach is still a "mean" approach, in the sense that fluctuations are not taken into account, but they might be introduced constructing a number of model Hamiltonians for different positions of the explicit solvent molecules.

On a more general ground I think that the idea you mentioned in the discussion, *i.e.* to organize a network of research groups aiming to apply different experimental and computational techniques to the study of a properly-selected prototype system, on the grounds of a common research plan, might allow a breakthrough in this field.

1. R. Improta, V. Barone, A. Lami and F. Santoro, *J. Phys. Chem. B*, 2009, **113**, 14491.
2. J. Tomasi, B. Mennucci and R. Cammi, *Chem. Rev.*, 2005, **105**, 2999.
3. C. S. M. Allan, B. Lasorne, G. A. Worth and M. A. Robb, *J. Phys. Chem. A*, 2010, **114**, 8713.
4. I. Burghardt, K. Giri and G. A. Worth, *J. Chem. Phys.*, 2008, **129**, 174104.

Professor Burghardt replied: Along the lines of Dr Santoro's comments, the explicit inclusion of the solvent at a dynamic level remains a formidable task to date. On-the-fly QM/MM methods using Gaussian wavepackets, like the *ab initio* multiple spawning method[1] and the variational multiconfigurational Gaussian wavepacket method,[2] still have to overcome insufficient phase-space sampling and lack of convergence with regard to the full quantum dynamics, despite the improvement over surface hopping trajectory schemes. Apart from these shortcomings, these methods provide a suitable platform for assessing solvent effects for individual explicit solvent sonfigurations, as suggested by Professor Temps. Importantly, nonequilibrium solvation effects can be accounted for at this level, which can significantly change the energetics and ultrafast dynamics, especially if charge transfer processes are involved.

1. A. M. Virshup, C. Punwong, T. V. Pogorelov, B. A. Lindquist, C. Ko and T. J. Martinez, *J. Phys. Chem. B*, 2009, **113**, 3280.
2. G. A. Worth, M. A. Robb and I. Burghardt, *Faraday Discuss.*, 2004, **127**, 307; G. A. Worth, M. A. Robb and B. Lasorne, *Mol. Phys.*, 2008, **106**, 2077.

Dr Worth answered: The treatment of solvent effects in excited-state dynamics is indeed the present frontier in quantum dynamics simulations, with the problem of performing reliable, accurate quantum chemistry on large systems with a huge configuration space added to the general problem of high-dimensional quantum dynamics. I agree that QM/MM combined with direct dynamics is probably the way ahead for specific molecules, as grid-based quantum dynamics are simply not flexible enough to treat anything past small models, but there is a long way to go. Prof. Temps mentions the problem of ordering states in excited-state calculations. This will be exacerbated in solution as I am sure that solvent reordering will lead to differential stabilisation of different states and so there will be a need for multiple surfaces to be correctly described in a large region of configuration space. This is not possible at present with the usual CASSCF or TDDFT calculations typically used for larger molecules. EOM-CCSD and MRCI / DFT methods may be better for this, able to provide a range of states in a balanced way, but are not yet combined in QM/MM codes to test further. Prof. Temps also mentions another problem for solvated systems—temperature. Many calculations would be required to sample the ensemble of inner shell configurations, again increasing the effort. Perhaps a density matrix approach might be preferable here.

Dr Malhado asked: I would like to ask two distinct questions concerning your treatment of the dynamics in this system. First, given the similarities in your results labelled 9q and 3q-36c for the quadratic vibronic coupling model one could be tempted to believe that in this case it is not too important how accurate the treatment of the dynamics is as long as the sufficient number of degrees of freedom are included to describe the phenomena. How do your results compare to the trajectory surface hopping calculations performed for 9*H*-adenine cited in your article? Or alternatively, what kind of effect which is relevant for this system do you capture with your quantum dynamical treatment which is absent on trajectory surface hopping calculations?

Second, you use a Ehrenfest mean-field method to describe the classical degrees of freedom in your model, which is known to have a problematic behaviour when reaching a region of low non-adiabatic coupling after crossing a region of strong coupling such as after crossing a conical intersection (see for example J. C. Tully, in *Modern Methods for Multidimensional Dynamics Computations in Chemistry*, ed. D. L. Thompson World Scientific, Singapore, 1998, p. 34). Do you expect this type of problem to affect your results? Furthermore, can you foresee any problem in the description of intramolecular vibrational energy redistribution between your quantum and classical modes?

Dr Santoro replied: Answer to the first question:

The way I look at our results is that the hierarchical representation makes a good job, in the sense that the significant dynamics is properly described by the coordinates belonging to the first and second blocks of the hierarchy. The other coordinates that we included at classical level play a very modest role. Clearly this is true within the quadratic vibronic coupling model we adopted. Specifically, the latter does not account for the relaxation to the ground electronic state which can take place at two different conical intersections (CoIs), both characterized by strongly non-planar and distorted structures. The region of the potential energy surfaces (PES) that connect the FC geometry and these two CoIs is not amenable of a description in terms of harmonic functions. The trajectory surface hopping calculations (TSH) of the decay of the photoexcited 9*H*-Adenine[1] were mostly focused on the decay of the excited-population on the ground state, while they were putting less attention to the initial S_2/S_1 dynamics, which is the main interest of our quantum-dynamical (QD) analysis. In more general terms I think that when the wavepacket travels on two almost-degenerate PES and therefore it can in principle run several times across the conical intersection and has time to spread in the coordinate space (*e.g.* like the S_2/S_1 system in adenine if we forget the channels dealing to S_0), pure quantum effects should be expected to be much more important (and therefore QD results more different from TSH ones) than in a case where the CoI is very peaked and, after a first cross, the population never goes back in the crossing region.

1. M. Barbatti, Z. Lan, R. Crespo-Otero, J. J. Szymczak, H. Lischka and W. Thiel, *J. Chem. Phys.*, 2012, **137**, 22A503.

Answer to the second question:

You are right, the Ehrenfest model has some drawbacks. As already discussed, in our specific case the results show that the motion along the classical coordinates affects only very slightly the evolution of the population of the diabatic states, so that our results are quite robust. I can add that, while working at the implementation of our hybrid classical/quantum propagation, we pragmatically tested both the possible approaches (*i.e.* (i) trajectories driven by the average potential, and (ii) trajectories moving on each of the two diabatic potentials), and the results according to the first (Ehrenfest) option provided results converging faster (increasing the number of coordinates treated at quantum level) to the fully-quantum prediction. However the model is not perfect at all. Concerning your question on the energy redistribution among different modes, consider our specific application, and focus on non-total symmetric modes. The classical modes are coupled to the quantum modes by bilinear couplings xy and, according to our equations, they " feel" the potential mediated on the quantum wavepacket. This is zero by definition for an even wavefunction (*i.e.* the one traveling on the S_{La} PES). This explains why, as shown in Fig. 7 of our paper, to properly describe the spreading of the wavepacket along the coupling mode (*i.e.* the non-total symmetric mode of the first block) it is necessary to treat also the second block of the hierarchy at quantum level.

Professor Domcke commented: This is a general comment concerning the pros and cons of quantum wave-packet calculations and quasiclassical trajectory calculations. One reason why quantum wave-packet calculations are needed, although classical trajectory calculations may be much easier to perform, is spectroscopy. Electronic or vibrational spectra involve coherences among electronic or vibrational states. Such coherences cannot be described with classical trajectory calculations, since the phase information is lost. If only electronic state populations or the energy content of vibrational modes are addressed, then classical trajectory calculations are presumably the most efficient way to get these. If, on the other hand, detailed spectroscopic information, such as four-wave-mixing signals in the time or frequency domain are of interest, then there is no substitute for quantum dynamics calculations or phase-preserving semiclassical calculations.

Dr Graham Worth commented: Prof. Domcke raises an important point. The ease of direct-dynamics simulations using trajectory surface hopping and calculating the potential on-the-fly makes them an attractive proposition for studying photochemistry. They should, however, be used with caution. For validation of models and simulations, we need to be able to compare to experiment, *i.e.* a signal. In the field of ultrafast dynamics this means a spectroscopic signal, for which, as Prof. Domcke writes, there is no substitute for quantum dynamics. However, given the limitations on grid-based quantum dynamics for treating large systems, the question then is how to proceed to enable accurate calculations on molecular systems, possibly in the solution phase. Here there are a few complementary routes. One is to do full quantum dynamics calculations on reduced models to provide a signal. Cheaper classical trajectory calculations can then be used to help validate the model, showing that it includes all the relevant modes and regions of configuration space. A second is to use the MCTDH method,[1] which in the multi-layer version is able to treat the quantum wavepacket of many degrees of freedom.[2] Converging such calculations on a realistic many-dimensional molecular potential function, however, would be a serious challenge. Finally, Gaussian wavepacket based methods such as vMCG,[1] Spawning[3] and Coupled Coherent States,[4] have better scaling properties, like trajectory methods, but provide a wavefunction from which a signal can be calculated. As an example, the Spawning method has been used to provide a time-resolved photo-electron spectrum of DNA bases that can be compared to experiment.[3]

1. G. A. Worth, H.-D. Meyer, H. Koeppel, L. S. Cederbaum and I. Burghardt. *Int. Rev. Phys. Chem.*, 2008, **27**, 569–606.
2. H. B. Wang, I. Pshenichnyuk, R. Hartle and M. Thoss, *J. Chem. Phys.*, 2011, **135**, 244506.
3. H. R. Hudock, B. G. Levine, A. L. Thompson, H. Satzger, D. Townsend, N. Gador, S. Ullrich, A. Stolow and T. J Martinez, *J. Phys. Chem. A*, 2007, **111**, 8500–8508.
4. D. Shalashilin, *J. Chem. Phys.*, 2009, **130**, 244101.

Professor Dr Riedle said: In response to your general comment it should be noted that the electronic coherence between the ground state and the excited state prepared by the pump pulse in an experiment dephases within tens of femtoseconds in solution. This has repeatedly been shown for dye molecules and also for more complex systems. The timescale might only be somewhat longer in biological samples where the chromophore is shielded by proteins. Only a possible vibrational coherence survives either on the electronically excited or the ground state potential energy surface for about a picosecond. Would this not decrease the importance of the quantum dynamics calculations over classical trajectory calculations for many investigations?

Professor Domcke replied: This argument is certainly valid for the "optical coherence" between the electronic ground state and the excited-state electronic manifold. This coherence indeed decays on femtosecond timescales in condensed phases at ambient temperatures. The nonadiabatic quantum dynamics involves, however, also coherences among the excited electronic states, which may persist on somewhat longer timescales. In general, however, it is true that quantum coherences are less relevant in condensed-phase photochemistry than in gas-phase photochemistry.

Professor Dr Riedle addressed Professor Roethlisberger, Professor Burghardt and Dr Santoro: I notice a very strong focus on the concept of vibronic eigenstates even for the investigation and discussion of ultrafast dynamic processes. The fundamental importance of the vibronic eigenstates is certainly justified for stable dye molecules that have many vibrational period times to fully develop the specific character after the optical excitation. For ultrafast processes, however, that proceed on the timescale or even faster than a single vibrational period, the eigenstate is not any longer well defined in my opinion. This is particularly so for chemical processes where bonds are broken or newly formed. Would it therefore not be more appropriate to focus on the concept of wavepackets that evolve on the complex excited states potential energy surfaces?

Dr Santoro replied: I agree with your observation and, indeed, in the analysis of our quantum-dynamical results we did not focus on the discussion of vibronic states but on the behavior of the evolving wave-packets. On the other hand, taking in mind the occurrence of decoherence and dissipative processes, even the picture of a coherent wavepacket might reveal inadequacies at the increase of time. For example, elaborating on your question, it is noteworthy to highlight that, thinking of the wavepacket in terms of a coherent superposition of vibronic eigenstates, interference is essential to explain its localization in space. From the computational point of view, this suggests that the modeling of decoherence processes is rather delicate, since different dephasing models (acting for example among the eigenstates and/or on the density matrix in the coordinate representation) may lead to different representations of the evolving distribution, with different predictions in terms of its localization/ delocalization. Surely, many detailed theoretical analyses can be found in literature, but the problem of how to effectively model these effects in the analysis of the photochemistry and photophysics of complex systems is still a matter of debate. I think that methods like ML-MCTDH, allowing a microscopic description of systems with many degrees of freedom, offer at the state-of-the-art one of the best routes to analyze this kind of problems.

Professor Burghardt responded: I agree with Professor Riedle and Dr Santoro in that the wavepacket picture is most appropriate to capture ultrafast processes (especially if significant changes of the nuclear geometry are involved), and I also share Dr Santoro's view that decoherence and localization can modify the conventional picture of coherent wavepacket motion.

For example, the exciton dynamics described in our contribution represents a case where a transition from coherent wavepacket motion to a stationary exciton–phonon state on a lattice can be observed. Here, we discussed the dynamics in terms of a local excitonic eigenstate picture which offers a more transparent interpretation than the site (monomer) based representation that was used in the actual calculations. The dynamics is taken to depart from the lowest excitonic eigenstate of one of the subfragments, which essentially represents a coherent superposition of the S_1 (^1B_u) and S_2(^1A_g) adiabatic states of the oligomer. This initial condition thus corresponds to a minimal excitonic wavepacket which rapidly evolves in time due to the large electronic site-to-site couplings. Further, we did not refer to vibronic eigenstates, since a wavepacket picture is indeed more appropriate for the vibrational motion (for both torsional and bond-length-alternation modes). In particular, the time-evolving coordinate expectation value of Fig. 5a reflects the motion of the torsional wavepacket at the central junction. Eventually, the system reaches a stationary state that can be interpreted as a "dressed" exciton eigenstate, or incoherent superposition of such dressed eigenstates.

Professor Roethlisberger responded: We do not use the concept of vibronic eigenstates. Our surface hopping dynamics is based on swarms of classical trajectories as a mimic of the quantum probability distribution. Explicit nuclear quantum effects such as the quantization of vibrational states, nuclear tunneling and quantum coherence are not captured within this semiclassical approach. To describe these effects one has indeed to use wavepacket dynamics, which is however unaffordable for very extended and high-dimensional systems.

Dr Worth asked: We have been discussing whether it is correct to model the excitation process using diabatic states. Do experimentalists think their experiments access adiabatic or diabatic states?

Professor Whitaker responded: I suppose the answer depends on the nature of the experiment. But in the case of a time resolved photoelectron experiment, the obvious way to interpret the electron's kinetic energy spectrum (KES) is as a measure of the adiabatic binding energy of the ionizable electrons at any given moment in time following excitation to the excited state surface. The question as to how one joins the "dots" as time goes on is answered by the dynamics; if one sees a sudden change in the KES one says that a non-adiabatic transition has occurred and the experimentalist neatly side-steps the original question!

Dr Nix, University of Leeds, communicated: This is an important point. There has always been confusion in the use of the terms 'diabatic, adiabatic and non-adiabatic' with reference to electronic states and molecular dynamics. The 'diabatic' states show conservation of electronic configuration, typically crossing in configuration space without mixing. Mixing terms are then introduced to recover the observed topologies and energies. Conversely, adiabatic states make no attempt to conserve a given electronic configuration and hence contain the coupling terms *a priori*.

Professor Dr Riedle addressed Professor Roethlisberger, Professor Burghardt and Dr Santoro: I appreciate the importance of the distinction between adiabatic and diabatic descriptions. In all the discussed situation the main emphasis is still on the energetic position of the involved molecular levels. These are to be compared to the experimentally observed transition energies. I would like to argue that in this well established method of approach the importance of the transition strengths is heavily neglected. They can directly report on the electronic character or the change of them during the transition. As such they likely render even more decisive information on the system evolution. I therefore would like to encourage the community to pay increased attention to the transition strengths and utilize them in the analysis.

Dr Santoro responded: I completely agree. In my opinion the analysis of the transition strengths is fundamental for understanding the details of a photochemical mechanism. This is also the reason why, during the discussion of Professor Riedle's paper, I was pointing out that, from a computational point of view, it is challenging to obtain a reliable description of the final excited states of a transient absorption experiment. Such difficulty clearly concerns also the transition strengths among the excited states populated by the pump pulse and those populated by the probe pulse. In more general terms a mechanism is explained in terms of a model made up by "reference" states with well-distinguished and recognizable identities and their mutual interactions/transformations. I think that, in most of the cases, being weakly dependent on the nuclear structure, diabatic states are more suited as "reference states" than adiabatic states are.

Professor Burghardt answered: I entirely agree that transition strengths provide key information, besides the energetics. In our discussion of exciton transfer, we implicitly used information on transition strengths, in that the initial condition was chosen to correspond to the bright state—*i.e.*, the local exciton ground state—on one of the sub-fragments. In this sense, the dynamics simulations we presented match the experimental situation.

Professor Roethlisberger answered: I agree with you that the oscillator strength is an important property. It is however, even more difficult to get accurate theoretical information for the intensities than for the state energies and some approximations like the frequently used Tamm–Dancoff approximation within TDDFT or the approximate CC2 method do not yield reliable oscillator strengths.

Dr Santoro remarked: I also would like to comment on the rhetorical question I raised: "do pump–probe experiments monitor diabatic or adiabatic populations". I asked it with the intent of stimulating some discussion on this subject. I have the feeling that, in some cases, the disagreement among different theses on a photochemical pathway is more ostensible than real, and is partly due to the fact that different representations are implicitly considered. Of course my question was too general and the answer depends on the specific system and the specific probe. Focusing, for example, on fluorescence up-conversion (FLUPS) it has been shown already in the eighties[1] that in a situation like adenine, *i.e.* two coupled electronic states, one bright and the other dark, with respect to the ground state, FLUPS monitors the diabatic "bright" population, while in the case of a conical intersection between the ground state and the lowest-exited adiabatic state, like it occurs in NO_2,[2] FLUPS mainly monitors the adiabatic S_1 population.

1. H. D. Meyer and H. Koeppel, *J. Chem. Phys.*, 1984, **81**, 2605.
2. F. Santoro, C. Petrongolo and A. Lami, *J. Chem. Phys.*, 2000, **113**, 4073.

Faraday Discussions RSC Publishing

PAPER

Ultrafast excited state dynamics of the green fluorescent protein chromophore and its kindling fluorescent protein analogue

Kiri Addison, Ismael A. Heisler, Jamie Conyard, Tara Dixon, Philip C. Bulman Page and Stephen R. Meech*

Received 19th February 2013, Accepted 1st March 2013
DOI: 10.1039/c3fd00019b

Fluorescent proteins exhibit a very diverse range of photochemical behaviour, from efficient fluorescence through photochromism to photochemical reactivity. Remarkably this diverse behaviour arises from chromophores which have very similar structures. Here we describe measurements and modelling of the excited state dynamics in the chromophores of GFP (HBDI) and the kindling fluorescent protein, KFP (FHBMI). The methods are ultrafast fluorescence spectroscopy with sub 50 fs time resolution and the modelling is based on the Smoluchowski equation. The excited state decays of both chromophores are very fast, longer for their anions than for the neutral form and independent of wavelength. Detailed studies show the mean fluorescence wavelength to be independent of time. The excited state decay times are also observed to be a very weak function of solvent polarity and viscosity. These results are modelled utilising recently calculated potential energy surfaces for the ground and excited states as a function of the twist coordinates about the two bridging bonds of the chromophore. For FHBMI and the scarce data on the neutral HBDI the calculations are not successful suggesting the need for refinement of these potential energy surfaces. For HBDI in methanol the simulation is successful provided a strong dependence of the radiationless decay rate on the coordinate is assumed. Such dependence should be included in future calculations of excited state dynamics. When the simulations are extended to more viscous solvents they fail to reproduce the observed weak viscosity dependence. The implications of these results for the nature of the coordinate leading to radiationless decay in the chromophore and for the photodynamics of fluorescent proteins are discussed.

Introduction

Soon after it was first isolated and expressed, the green fluorescent protein (GFP) became established as a key tool in cell biology.[1–4] GFP, and its more photostable

School of Chemistry, University of East Anglia, Norwich Research Park, Norwich NR4 7TJ. E-mail: s.meech@uea.ac.uk

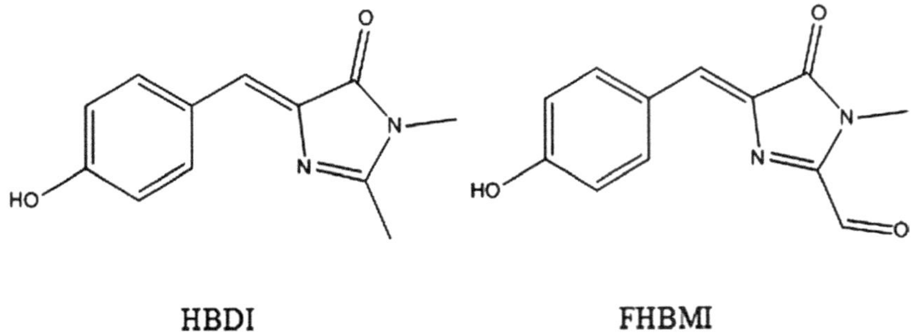

Fig. 1 The chemical structures of HBDI and FBHMI in their neutral states. The anion states are deprotonated at the phenol OH.

and fluorescent mutants,[5] are readily cloned and expressed attached to a target protein, making the expression and subsequent location of that protein in the living cell visible to fluorescence imaging. This new technology sparked a revolution in bioimaging which continues to the present time. The chromophore responsible for the green fluorescence, 4-hydroxybenzylidene-2,3-dimethylimidazolinone (HBDI, Fig. 1) was characterized and isolated by Niwa and co-workers.[6,7] The chromophore is formed *in vivo* in an intramolecular reaction between three adjacent amino acids.[8] Mutations in GFP can modify either the chromophore or its immediate environment in a way that tunes the spectrum across the entire visible range, facilitating multicolour imaging and the imaging of protein-protein interactions through fluorescence resonance energy transfer.[5] Recent advances in protein engineering mean that the chromophore can be modified with unnatural amino acids, expanding the range of possible colours.[9] It has been shown that some relatively minor changes to the chromophore (such as methylation in the phenyl group, the location of the OH group) can have a dramatic effect on the electronic spectrum.[10,11]

This remarkable advance in bioimaging stimulated a search for other fluorescent proteins (FPs) and a broad family has now been identified.[5,12–14] Intriguingly, HBDI (or chromophores very similar to it) demonstrate an extraordinarily diverse range of photophysical behaviour in these different protein environments. While the anionic state of the chromophore in wild type GFP is strongly fluorescent, the same state in other FPs is only weakly fluorescent (such proteins being better labelled as chromoproteins). In other FPs the chromophore undergoes a photochemical reaction with an adjacent amino acid to generate a new chromophore with a different emission spectrum.[15,16] Such proteins (*e.g.* kaede) have been used as 'optical highlighters' to label specific populations of proteins *in situ*. Other proteins, such as dronpa and its relatives, exhibit photochromism, where the protein can be reversibly and repeatedly photochemically switched between two optical states.[17–19] These photochromic FPs have proven extremely valuable in the development of novel ultra-resolution fluorescence microscopies with a sub-Rayleigh limit spatial resolution of a few tens of nanometres.[20–22]

This unexpectedly diverse behaviour has fuelled interest in the photophysics of HBDI itself. It was established that the strong fluorescence of GFP is quenched when the protein is denatured, and that HBDI in fluid solution is very weakly fluorescent (quantum yield of *ca* 0.2×10^{-4}). However, fluorescence is recovered by freezing the solution to the glass transition temperature.[6,7,23,24] It was demonstrated through ultrafast spectroscopy that the mechanism of rapid radiationless decay was internal conversion on the picosecond timescale, and that this is a property exhibited by all three readily accessible charge states (neutral, anion and cation) of the chromophore.[25–27] A number of derivatives of HBDI have been prepared that modify, in some cases quite dramatically, the electronic spectra of the chromophore. However, in most cases radiationless decay remains ultrafast.[11,28] One feature these derivatives have in common is the methine bridge linking the two aromatic rings, so it appears likely that the radiationless decay coordinate involves reorganization about this group. Consistent with this, illumination of HBDI in its stable *cis* state was found to lead to formation of the metastable *trans* form.[29] The spectra of the two isomers are similar and their excited state decays are indistinguishable.[30] Further evidence for the importance of structural reorganization in the methine bonds comes from the observed suppression of radiationless decay when motion about these bonds is restricted,[31,32] and from the observation that a sterically crowded derivative of HBDI, with a twisted ground state, exhibits enhanced radiationless relaxation.[11]

Detailed experimental studies of the excited state decay of HBDI showed that the decay rate is only moderately sensitive to solvent viscosity and solvent polarity, although fluorescence is recovered as the glass transition temperature is approached, *i.e.* in very viscous media.[26,27,33] The observed temperature dependence and weak dependence on viscosity suggested that a volume-conserving coordinate couples the excited state to the ground state over a small or zero energy barrier. The weak dependence on solvent polarity suggests that the radiationless decay does not require stabilization of a charge transfer state.

Early theoretical calculations suggested that the most likely volume-conserving coordinate associated with the methine bridge, the "hula twist", is energetically up-hill, and so incompatible with the observed near barrierless ultrafast relaxation.[34,35] Thus, the details of the radiationless decay mechanism and the mechanism by which the protein matrix renders the chromophore strongly fluorescent are incompletely understood. There have been a number of high quality theoretical calculations concerning the excited state dynamics of HBDI.[36–46] In a number of cases these calculations (which mainly refer to the gas phase) have investigated excited state dynamics, and two possible pathways leading to radiationless decay to the ground state have been suggested. These involve rotations about either of the two methine bridge bonds. In both cases the coordinates are barrierless or nearly so, consistent with the observed ultrafast decay in fluid solution. These calculations are also in good agreement with recent experimental observation of the ultrafast dynamics of HBDI in the gas phase, where the initially excited Franck–Condon state relaxed to a twisted intermediate in a few hundred femtoseconds, and from the twisted state to the ground state in about 1 ps.[47] The success of these calculations in the gas phase is important, but does not address the mechanism of fluorescence enhancement in the protein. There have been a number of attempts to incorporate some representation of the medium into the calculations, either as a continuum or using the actual structure of the protein.[44,48] However, none of these include dynamics associated with the environment.

In this work we extend our time-resolved studies of HBDI fluorescence to include a detailed analysis of the time-dependent emission spectrum. We also compare data on HBDI with measurements on a newly synthesised derivative of HBDI related to the chromophore of the kindling fluorescent protein (KFP), specifically 2-formyl-4-hydroxybenzylidene-3-methylimidazolinone, FHBMI, Fig. 1. KFP is one of the photoactive proteins in which a fluorescent state can be photochemically generated;[49,50] the excited state potential energy surface of the KFP chromophore (2-acetyl-4-hydroxybenzylidene-3-methylimidazolinone, FHBMI) was investigated by Olsen.[37] The time-dependent spectra observed here are modelled using a generalized Smoluchowski equation. Similar models have previously been applied to probe excited state dynamics, initially for electron transfer reactions by Barbara and co-workers and extended to excited state structural dynamics by Glasbeek and co-workers.[51–54] This approach has since found wide application in studies of excited state reactions in a variety of solutes and environments, but usually relying on a model potential energy surface generated from coupling neutral reactant and charge separated (product) harmonic surfaces.[55–61] Here we replace the model PES with one determined from theoretical calculations, specifically the calculations of Olsen for both HBDI and AHBMI in their anion states.[36,37] We then incorporate realistic models of the solvent friction experienced by the assumed reaction coordinate for the FP chromophore. We further extend these calculations using the model PESs generated from the much more limited calculations on the neutral chromophore,[45] and thus discuss the neutral chromophore dynamics. In this way we endeavour to combine realistic calculated potential energy surfaces with a realistic model of the medium.

Experimental

The experimental arrangement for fluorescence upconversion has been described in detail elsewhere, and is based on a 1 W 100 MHz titanium sapphire laser generating sub 20 fs pulses centered at 800 nm.[55] The excitation wavelength was usually the second harmonic at 400 nm with power up to 9 mW. Fluorescence was detected by a reflective microscope objective and focused into a thin nonlinear crystal where it was mixed with the residual 800 nm pulses, which were routed *via* a delay stage that controls the timing with femtosecond accuracy. The resulting sum frequency intensity samples the instantaneous fluorescence intensity at the arrival time of the 800 nm pulse. Thus the temporal resolution is limited only by the pulsewidths of the 400 and 800 nm pulses. Careful control of the dispersion in each beam yields a time resolution of 45 fs, determined by measuring the upconverted Raman scattering of heptane. An example of one time-resolved measurement on neutral HBDI in methanol is shown in Fig. 2. As has been noted previously the decay is ultrafast and departs somewhat from a simple single exponential.[27]

HBDI was prepared by standard methods, and the synthesis of FHBMI will be detailed elsewhere.[6] All samples were made up to have an OD of roughly 0.5 at the excitation wavelength in a 1 mm pathlength cell (typically a concentration of 0.5 mM).

To construct the time-resolved emission spectra, fluorescence decay profiles were recorded across the emission spectrum by selecting the phase matching condition in the up-converting nonlinear crystal and tuning the wavelength of the

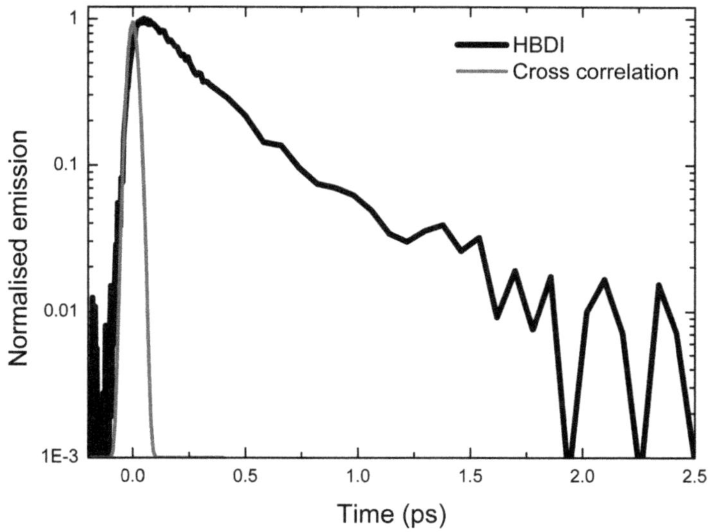

Fig. 2 Example of the measured fluorescence decay of neutral HBDI (black). The up-conversion of the solvent Raman signal scattered from the pump pulse is shown (red). This determines the time resolution of the experiment.

SFG detection monochromator. These individual wavelength-resolved temporal profiles are area normalized to the corresponding intensity in the steady state emission spectra to create a three dimensional time–intensity–wavelength surface. In a number of cases the range of detection wavelengths was limited by scattered pump radiation (a particular problem with short pulse broadband radiation). In these cases two-photon excitation was used, as has been described by Joo *et al.*[62] For these experiments a separate optical pathway for the pump beam was constructed, again with separate dispersion compensation. In this case the time resolution was measured as 60 fs.

Results and discussion

Steady state absorption and emission data for FHBMI are compared with those for HBDI in Fig. 3. In Table 1 the decay parameters for fluorescence decays measured at the peak wavelength of the emission spectra are also compared. Both absorption and emission spectra for FHBMI are slightly red-shifted from that of HBDI for both the neutral and anionic forms. The small shift in the anion spectrum on replacement of a weakly donating methyl group with the electron-withdrawing CHO group is in marked contrast to the large effect of exchanging a hydrogen for methyl on the phenolic ring reported previously.[11] The relatively benign nature of the methyl for formyl substitution is confirmed by the fluorescence decay times which are very similar for HBDI and FHBMI in both neutral and anion states. The similarity of HBDI and FHBMI in both charge states contrasts with the very different spectroscopic and photochemical properties of GFP and KFP.[63,64] This is further evidence of the critical role played by protein structure and by specific interactions between the chromophore and adjacent amino acid residues in determining the photophysical properties of FPs.

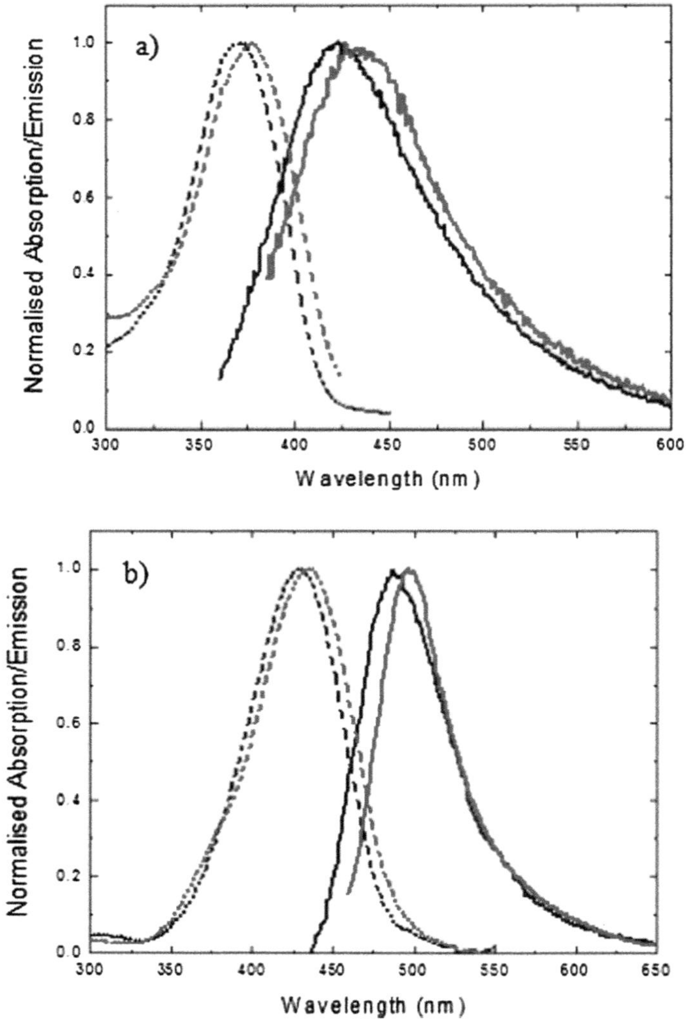

Fig. 3 Steady state absorption (dashed line) and emission (solid line) spectra of HBDI (black) and FHBMI (blue) in methanol: (a) neutral form (b) anionic form.

Table 1 Results of fits of the fluorescence decay to a sum of two exponentials. The weights are a_i, the decay times τ_i and the weighted mean decay time $<\tau>$

	Neutral HBDI	Anionic HBDI	Neutral FHBMI	Anionic FHBMI
a_1	1.30	0.64	0.60	0.82
τ_1/ps	0.28	0.57	0.31	0.35
a_2	—	0.54	0.62	0.29
τ_2/ps	—	0.69	0.27	1.12
$<\tau>$ /ps	0.28	0.62	0.29	0.55

The most complete representation of excited state dynamics comes not from the single wavelength measurements reported in Table 1, but from observations of the temporal evolution of the entire emission spectrum. Such time-resolved

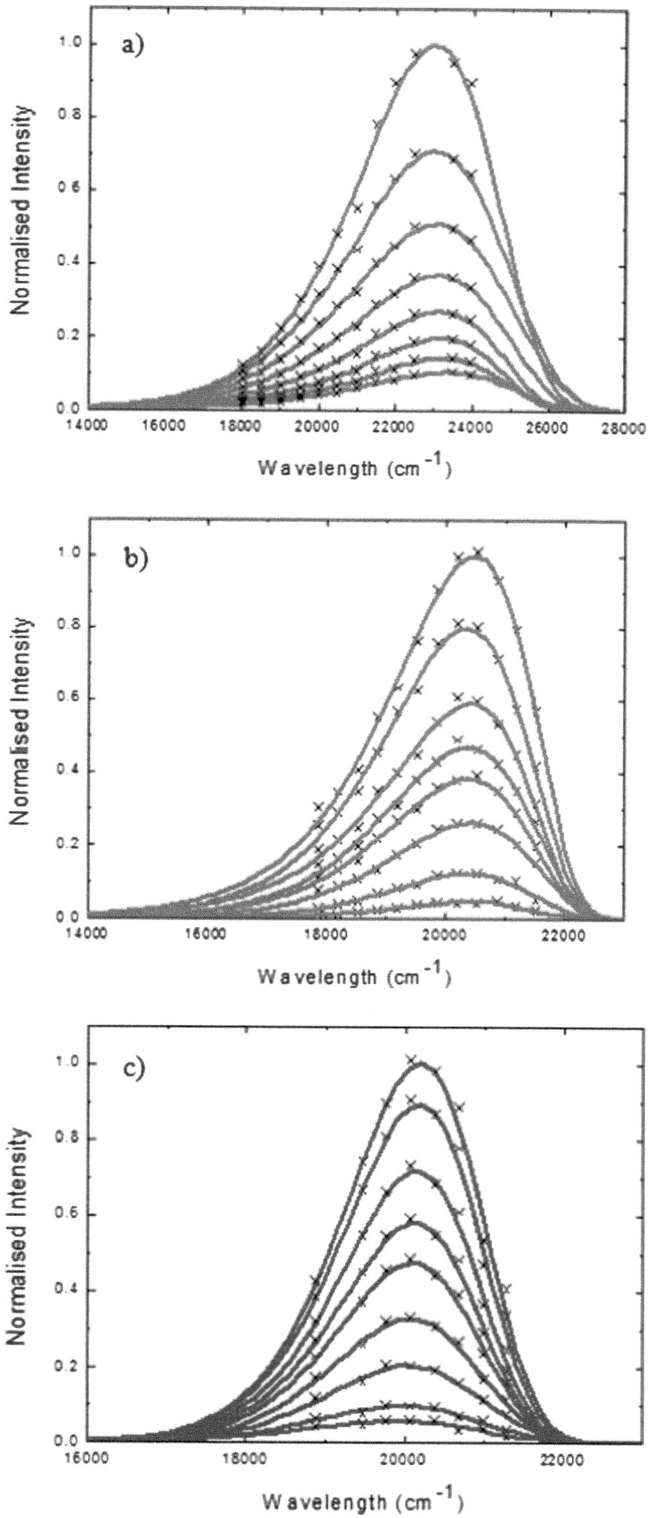
a)
Normalised Intensity
0.0
0.2
0.4
0.6
0.8
1.0
14000
16000
18000
20000
22000
24000
26000
28000
Wavelength (cm -1)
b)
Normalised Intensity
0.0
0.2
0.4
0.6
0.8
1.0
14000
16000
18000
20000
22000
Wavelength (cm -1)
c)
Normalised Intensity
0.0
0.2
0.4
0.6
0.8
1.0
16000
18000
20000
22000
Wavelength (cm -1)

spectra are readily obtained from wavelength-intensity slices of the intensity - time - energy surface described above. Examples are shown for HBDI and FHBMI in methanol in Fig. 4. The data points are well represented by a fit to the log-normal function:

$$F(\bar{\nu}) = h\begin{Bmatrix} \exp[-\ln(2)\{\ln(1+\alpha)/\gamma\}^2] & \alpha > -1 \\ 0 & \alpha \leq -1 \end{Bmatrix} \quad (1)$$

where $\alpha = 2\gamma[\bar{\nu} - \bar{\nu}_p]\Gamma$. The amplitude is h, γ the asymmetry parameter, Γ the width parameter, and $\bar{\nu}_p$ the peak wavenumber. These log-normal parameters form the basis for further analysis of the time-resolved spectra. In Fig. 5 the time dependence of the mean frequency (first moment) and integrated intensity of the log-normal fits are displayed.

The integrated intensity represents the time-dependent excited state population (assuming there is no temporal evolution in the transition moment) and can be seen to be non-single exponential for both chromophores. The anionic states are longer lived than the neutral for both FHBMI and HBDI, consistent with what was observed in the single wavelength study (Table 1). Equally significant is the observation that the mean frequency of the spectrum is essentially independent of time in all cases (Fig. 5a). A time-dependent shift in the spectrum is expected in (at least) two eventualities. First, if the excited electronic state has a significantly different dipole moment to the ground state, then solvation dynamics should stabilise the new excited state, resulting in a time-resolved red shift in the emission spectrum occurring on the timescale of solvent reorganization.[65] Second, if the excited state relaxes along a coordinate which causes closer approach of ground and excited states, and the transition moment remains finite, the emission is predicted to shift to the red with time. Thus, the fact that no time-dependent shift is observed places important constraints on the nature of the excited state dynamics.

The starting point for our simulation of the time-dependent spectra are the potential energy surfaces calculated by Olsen for the ground and excited states of HBDI and AHBMI, a molecule closely related to FHBMI, in their anionic states (which are typically the fluorescent states in FPs). Calculations were made using two different sets of surfaces for the anion, one based on a state-averaged CASSCF model and the second on a perturbation theory-corrected CASSCF model (MR).[37] The calculation was made as a function of the reaction coordinate which includes both double and single bond twist coordinates of the methine bridge (Fig. 1), varying between 0 and 90°, and these together create a single PES. We restrict our attention to chromophores which are initially in their *cis* ground state; we have shown elsewhere that *cis* and *trans* HBDI have indistinguishable excited state decay times, which accords with the very similar shapes of their calculated PES.[66]

These PES are input to the generalized Smoluchowski eqn (2), which describes the diffusive evolution of a population density $\rho(z,t)$ under a diffusion coefficient D (calculated from the Stokes–Einstein–Debye equation) along the calculated PES $[S_1(z)]$ where z is a generalised twist coordinate.[54]

Fig. 4 Time-resolved emission spectra (points) fit to the log–normal function (lines) for (a) neutral HBDI 50 fs to 800 fs (b) anionic HBDI 50 fs to 2ps (c) anionic FHBMI 50 fs to 2 ps. Solvent methanol, excitation wavelength 400 nm.

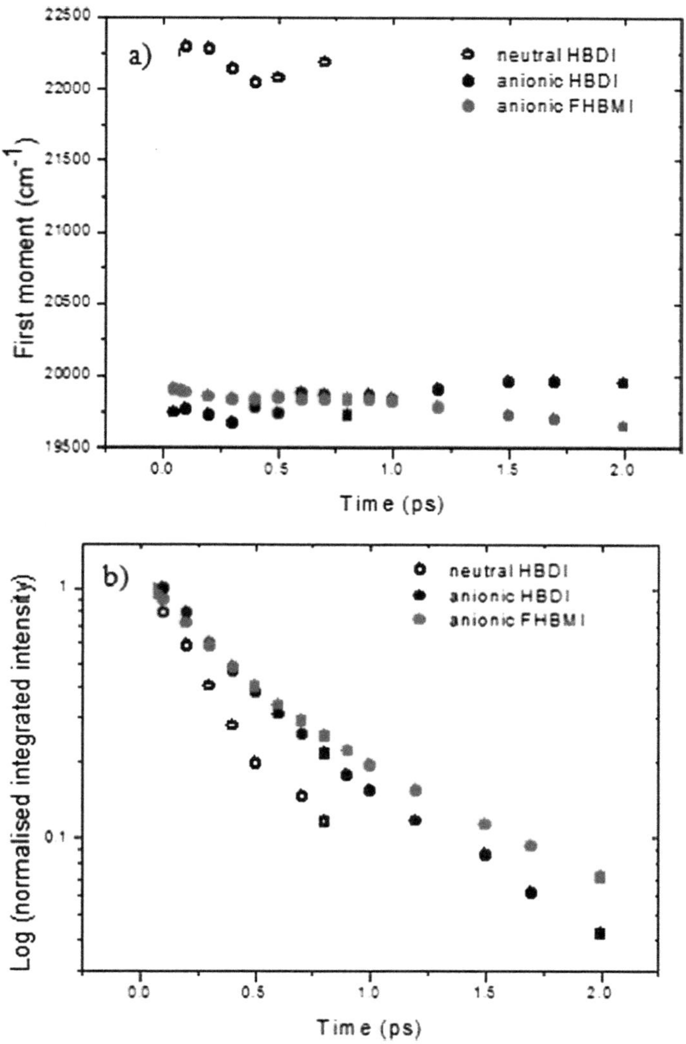

Fig. 5 Time dependence of the (a) first moment and (b) integrated intensity for HBDI (black) and FBHMI (blue) recovered from the log normal plots. Open symbols neutral, closed anion.

$$\frac{\partial}{\partial t}\rho(z,t) = D\frac{\partial}{\partial z}\left(\frac{\partial}{\partial z} + \frac{1}{k_B\mathrm{T}}\frac{\partial}{\partial z}S_1(z)\right)\rho(z,t) - \Gamma(z)\rho(z,t) \qquad (2)$$

In previous approaches using (2) to simulate time-dependent spectra the generally unknown coordinate dependence of the excited state PES was calculated assuming coupling between harmonic bright and dark states.[52,54] In the present case we digitized published PESs for HBDI and FHBMI anions and fit them to polynomials for the calculation of $\rho(z,t)$ using (2). The last term in eqn (2) describes nonradiative decay from the excited state to the ground state, where $\Gamma(z)$ is a coordinate-dependent probability for decay back to the ground state (a 'sink' function). Several forms can be chosen for $\Gamma(z)$ but in this work a pair of Gaussians function was chosen, described by:

$$\Gamma(z) = k_{G1}\frac{1}{\sqrt{\pi}\sigma_1}e^{-\left[\frac{(z-z_{01})}{\sigma_1}\right]^2} + k_{G2}\frac{1}{\sqrt{\pi}\sigma_2}e^{-\left[\frac{(z-z_{02})}{\sigma_2}\right]^2} \tag{3}$$

where k_{G1} and k_{G2} are the rate constants describing coupling of the ground and excited state, σ_1 and σ_2 are the widths and z_{01} and z_{02} are the centres, and we have set $k_{G1} = k_{G2}$, $\sigma_1 = \sigma_2$ and $z_{01} = 1$ and $z_{02} = -1$; the latter choice accords with the calculation of a conical intersection between or close approach of S_0 and S_1 at 90° twist. Consequently the solution to (2) for a particular choice of the diffusion coefficient, D, gives the population dynamics on the upper surface, $\rho(z,t)$. The rate of population decay back to the ground state, however, depends on the overlap of the population with the sink functions (3). Finally, the excited state populations are converted to spectra for comparison with the measurements (Fig. 4 and 5), which requires the choice of a lineshape function $g(\nu_0(z),\nu(z) - \nu_0(z))$, where $\nu_0(z)$ is the coordinate-dependent spacing between the S_0 and S_1 surfaces. The lineshape function is here taken to be the log-normal lineshape. The final requirement is a model for a coordinate-dependent transition moment, $M^2(z)$. These two combine with the calculated population density to give the fluorescence and absorption spectra respectively:[52,54]

$$I_{fl}(t) \propto \int dz\, g(\nu_0(z),\nu(z) - \nu_0(z))M^2(z)\rho(z,t)\nu^3 \tag{4}$$

$$I_A \propto \int dz\, g(\nu_0(z),\nu(z) - \nu_0(z))M^2(z)\rho(z,0). \tag{5}$$

The $M(z)$ in (4) and (5) accounts for the possibility that the transition moment may be a function of the twist coordinate. This has already been shown to be the case for a methylated HBDI derivative in which the substituents twist the methine bond away from the most stable planar state through steric crowding, giving rise to a weaker transition moment.[11] To model the transition moment in (4) and (5) we use an arbitrary function which has the property of going to zero at the 90° twist angles.

$$M^2(z) = \tan^{-1}[a(z + b)] - \tan^{-1}[a(z - b)] \tag{6}$$

The relevant coordinate-dependent parameters for the simulation are plotted in Fig. 6.

Fig. 6 shows that electronic excitation of the planar ground state generates a Franck–Condon excited state, in which the molecule is on a nearly flat region of the PES and able to undergo barrierless or near barrierless twisting about one or both bridging bonds. These features are common to a number of other calculated PESs for HBDI. Excited state dynamics are typically obtained from these calculations by following a minimum energy pathway for the excited state, as it relaxes down the twisting and stretching coordinates. A consequence of this twisting is an increase in the energy of the ground state until at some point the two PESs approach one another or merge at a conical intersection. Once the PESs become sufficiently close it is assumed that internal conversion from excited to ground state occurs, consistent with the known ultrafast relaxation of HBDI in solution.

In line with this scenario we have calculated the excited state dynamics for the HBDI anion in methanol according to eqn (2) (the choice of D being discussed further below). Since the theoretical calculations described above typically

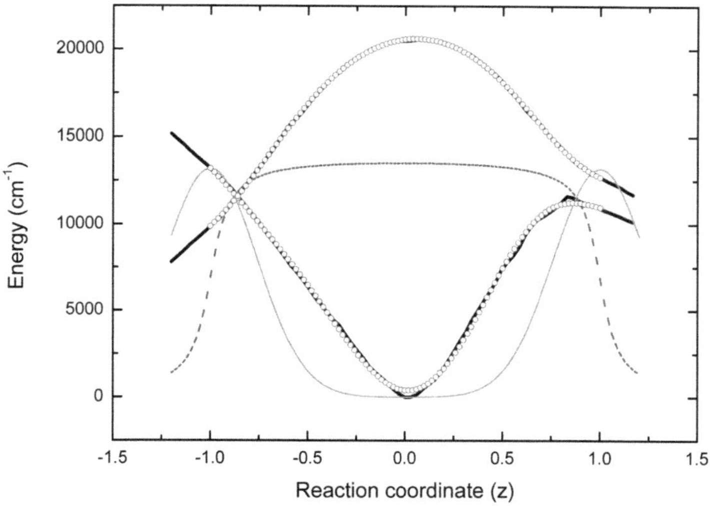

Fig. 6 Parameters used in the simulations. Solid and dashed black lines are the ground and excited state PESs calculated by Olsen. The red circles are polynomial fits to the PES used in the simulation. The green line is the Gaussian decay functions and the dashed blue line is the coordinate dependent transition moment.

include neither a coordinate dependence of the transition moment nor a rate for the radiationless decay, we initially set the second term in (2) to zero to focus on the dynamics of the population density. The resulting $\rho(z,t)$, calculated time-resolved emission spectra and mean emission frequency are plotted in Fig. 7, using the $M^2(z)$ in Fig. 6. This procedure predicts a time-dependent spectral shift in the emission spectrum after excitation, arising from the approach of ground and excited state surfaces with increasing z. This does not agree with the experimentally observed time independent spectrum (Fig. 5). It is shown below that the experimental result can only be reproduced when the simulation incorporates a coordinate dependence of both the transition moment and rate of decay to the ground state.

To simulate the spectra we assume a form for $M^2(z)$ (Fig. 6) and calculate $\rho(z,t)$ adjusting the parameters of $\Gamma(z)$ to obtain a best fit to the calculated absorption and time-dependent emission spectra using (4) and (5). The results for HBDI are shown in Fig. 8. In this case a good fit to the spectral shape and amplitude is obtained, but it does require a strong dependence of the rate of internal conversion on the twist coordinate (the actual example recovered from the fit is shown in Fig. 6). Within this simple one dimensional representation such a coordinate dependence is essential to reproduce the time-independent mean frequency observed (Fig. 5): a narrower Gaussian (*i.e.* a 'pinhole sink') and a smaller k lead to a red shift in the spectrum with time. The results in Fig. 8 were obtained with a fixed $M^2(z)$ and varying $\Gamma(z)$. Clearly these two parameters can have compensating effects, and fixing $\Gamma(z)$ and varying $M^2(z)$ achieves similar (but less good) fits. These two cases do, however, correspond to physically different phenomena, and should be distinguished. In Fig. 8 decay is directly to the ground state, as a consequence of the second term in (2). The alternative corresponds to the build up of a population in the twisted state (analogous to the results seen in

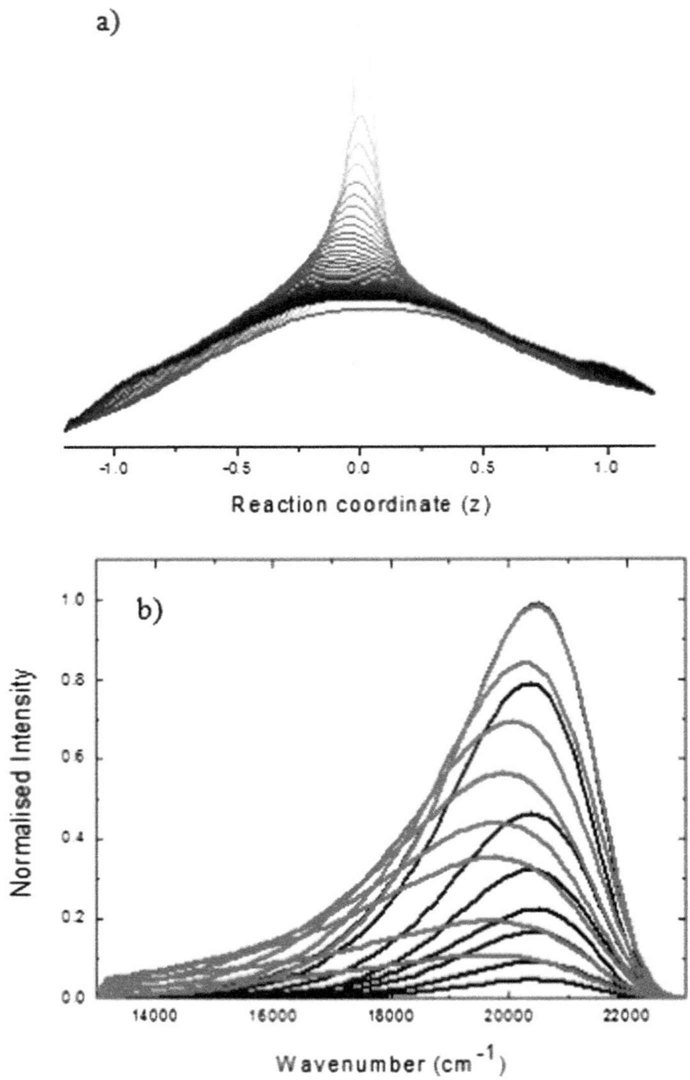

Fig. 7 (a) the $\rho(z,t)$ calculated from (2) with the final term omitted (no decay function) The population decays at Franck-Condon position $z = 0$ and builds up at $z = \pm 1$. (b) The spectra calculated for the data in (a) using (4) and (5).

Fig. 7) which does not contribute to the emission due to a $M^2(z)$ which falls to zero rapidly with increasing $|z|$. These two cases are distinguishable by studying transient absorption spectra, to which a twisted dark state may be expected to contribute. We have measured transient infra-red spectroscopy of HBDI (neutral) in solution and find that the excited state decays on a sub-picosecond timescale to yield a vibrationally hot ground state which cools in picoseconds. This is consistent with efficient internal conversion, *i.e.* a dominant role for $k\Gamma(z)$ in the decay.

Fig. 8a shows the fits obtained for anionic HBDI using the MR surfaces with the hydrodynamic volume used for D (the choices for D being either a measured

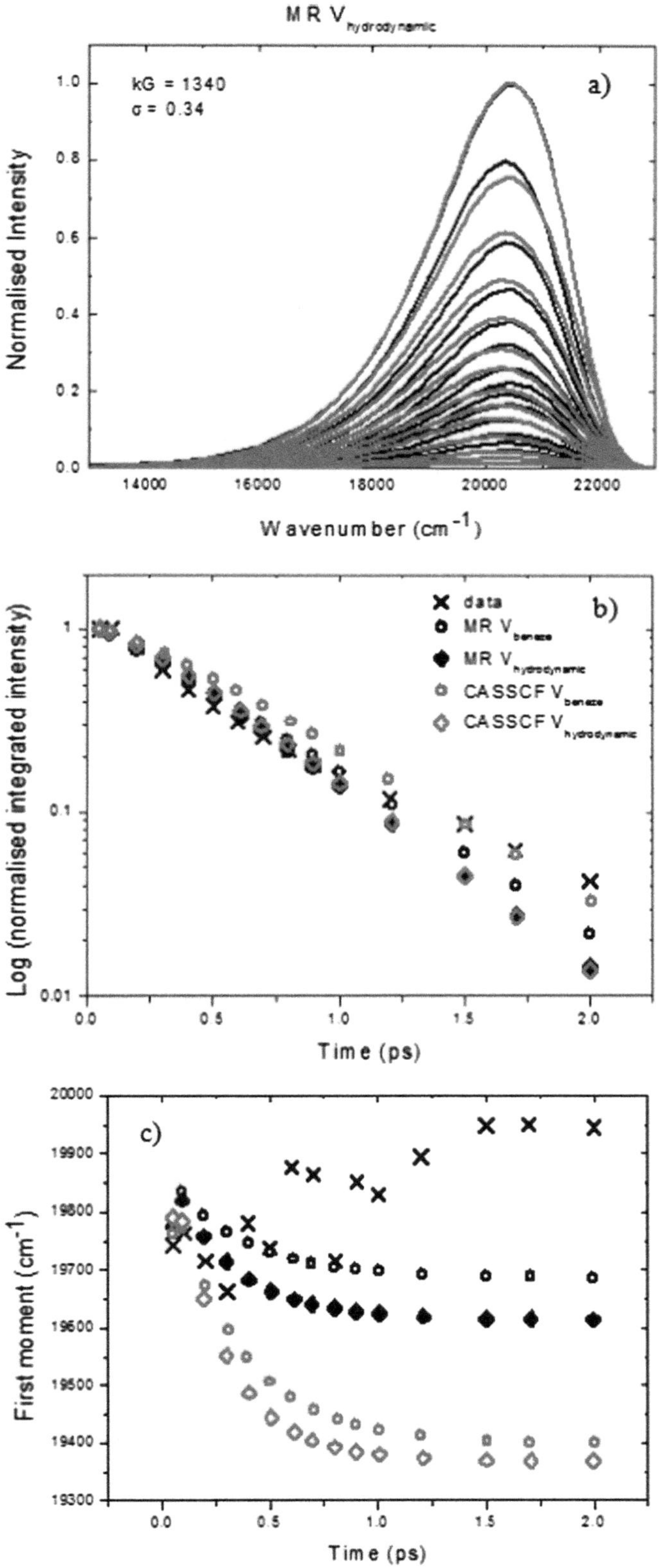

MR V hydrodynamic
kG = 1340
σ = 0.34
a)
Normalised Intensity
Wavenumber (cm-1)
b)
data
MR V benzene
MR V hydrodynamic
CASSCF V benzene
CASSCF V hydrodynamic
Log (normalised integrated intensity)
Time (ps)
c)
First moment (cm-1)
Time (ps)

hydrodynamic volume or a calculated benzene ring volume, see below).[37,67] In S1 the equivalent fits obtained using the volume of benzene for D and the CASSCF surfaces are shown. The time dependence of the integrated intensity and first moment for all fits are shown in 8b and c. As taken from Olsen's paper, the S_0 CASSCF and MR surfaces differ by approximately 1300 cm^{-1} at 0°, while on the S_1 surface, the CASSCF surface is approximately 10700 cm^{-1} higher in energy. Once adjusted vertically to match the experimental absorption and $t = 50$ fs resolved fluorescence spectrum, there is no energy difference between the S_1 states at 0°, where the absorption takes place, but at 90° the flatter MR surface is roughly 7000 cm^{-1} higher in energy. The surfaces provide reasonable fits to the experimental data at early times. Using the benzene volume provides a better fit to the integrated intensity after 1 ps, with the CASSCF surface giving the better fit overall. The hydrodynamic volume is, however, better at earlier times, with little difference between the MR and CASSCF surfaces. None of the fits reproduce the small blue shift seen experimentally, but these variations are on the same scale as the precision of the calculated spectra. All fits required the broad (large σ) and deep (large k) sinks shown in Fig. 6.

In contrast to these results for HBDI, the same calculation failed for FHBMI. Since the experimental spectra do not appear to be significantly different it is likely that the failure reflects a less suitable PES for this molecule. This may arise from one of two causes. First, that there was some shortcoming in calculation for the more complex chromophore. Alternatively, the chromophore we have studied differs from that calculated by the exchange of a methyl for hydrogen (or a formyl for an acetyl substitution); this may have a significant effect on the PES.

In Fig. 9 we compare the calculated potential surface for the HBDI anion with a recent calculation for the energy of the neutral form at 0° and 90° torsion about the double bond.[37] The neutral form of HBDI presents a much more challenging case for accurate calculation than the anion,[45] explaining the relative paucity of theoretical results with which to model the data. To obtain a PES for simulation of the neutral HBDI (Fig. 9a) the anionic surface has been scaled to match the two neutral points at 0° and 90°. A gradient was calculated given by the energy difference between the anionic surface and the two points for the neutral energy. This gradient was applied to the anionic surface for the single bond torsion resulting in a calculated coordinate dependent neutral surface connecting the two calculated points. It can be seen that the calculation apparently predicts a flatter PES for the neutral HBDI than for the anion. The constructed PES were vertically shifted to simulate the observed absorption spectrum and then used to calculate the time-dependent fluorescence spectra, assuming the form of $M(z)$ shown in Fig. 6. In Fig. 9b the calculated time dependence of the integrated area predicts a decay for the neutral which is slower than that of the anion, which is opposite to the experimental observations. This underlines the difficulty of accurately calculating the excited state surface for neutral HBDI. This difficulty sets a significant new challenge for theoretical calculations, since the neutral state of

Fig. 8 (a) Comparison of the spectra calculated using eqn (5) (red) and the log normal fits to the data (black), eqn (1), for the HBDI anion in methanol using the hydrodynamic volume. (b) and (c) Calculated integrated intensity and first moment of the spectra using the volume of benzene (circles) and the hydrodynamic volume (diamonds) for D in (2) (see text for details) for the MR surfaces (black) and the CASSCF surfaces (red). Data are for HBDI (black crosses).

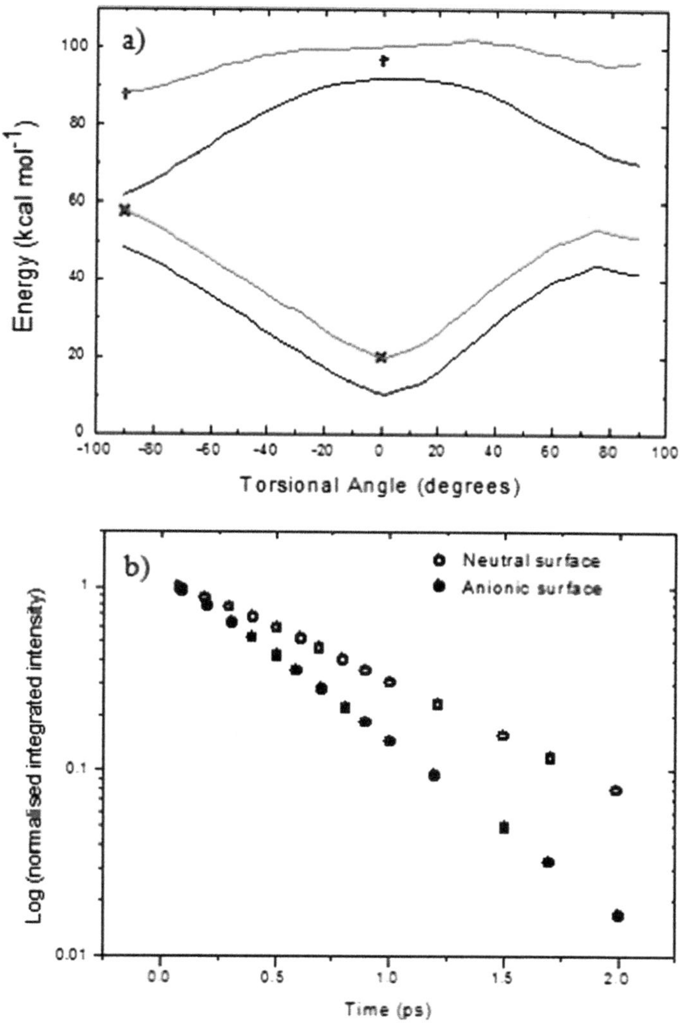

Fig. 9 (a) Anionic HBDI S_0 and S_1 surfaces calculated by Olsen (black). Calculated S_0 and S_1 neutral HBDI points are shown (black crosses). Neutral surfaces were created by scaling the anionic surfaces to the neutral points (red) (see text for details). (b) Time dependence of the integrated intensity of HBDI predicted by the neutral surface (open circles) and Olsen's anionic surface (filled circles).

HBDI is the one most often responsible for the excited state reactions which are critical in second generation fluorescent proteins.[68,69]

Finally we turn to the solvent friction effects on the excited state decay. All of the data above have been presented in methanol solution and the diffusion coefficient has been calculated from the Stokes–Einstein-Debye relation:

$$D = \frac{kT}{6\eta V} \tag{7}$$

where η is the solvent viscosity and V is the volume, where the volume is calculated from the estimated volume of the isomerizing unit. This was estimated in two ways, first from the tabulated volume of a benzene ring and second from the hydrodynamic volume of benzonitrile, measured elsewhere.[70] The results for both volumes are in qualitative agreement with the experimental data (Fig. 8a, b and c) in methanol. This suggests that eqn (7) is an appropriate model for the friction in methanol. However, it is worth noting that the recovery times in methanol are quite close to those observed in the gas phase,[47] suggesting that the rate in this very fluid solvent may be determined by the shape of the PES rather than friction of the medium.

To test this friction model further, the excited state decay times of the HBDI in a series of solvent of varying viscosity were measured (Fig. 10 and Table 2). These measurements were made at two or three wavelengths, which was sufficient to check that the more viscous solvents do not introduce a wavelength dependence (which would in turn indicate a time-dependent spectrum). The excited state

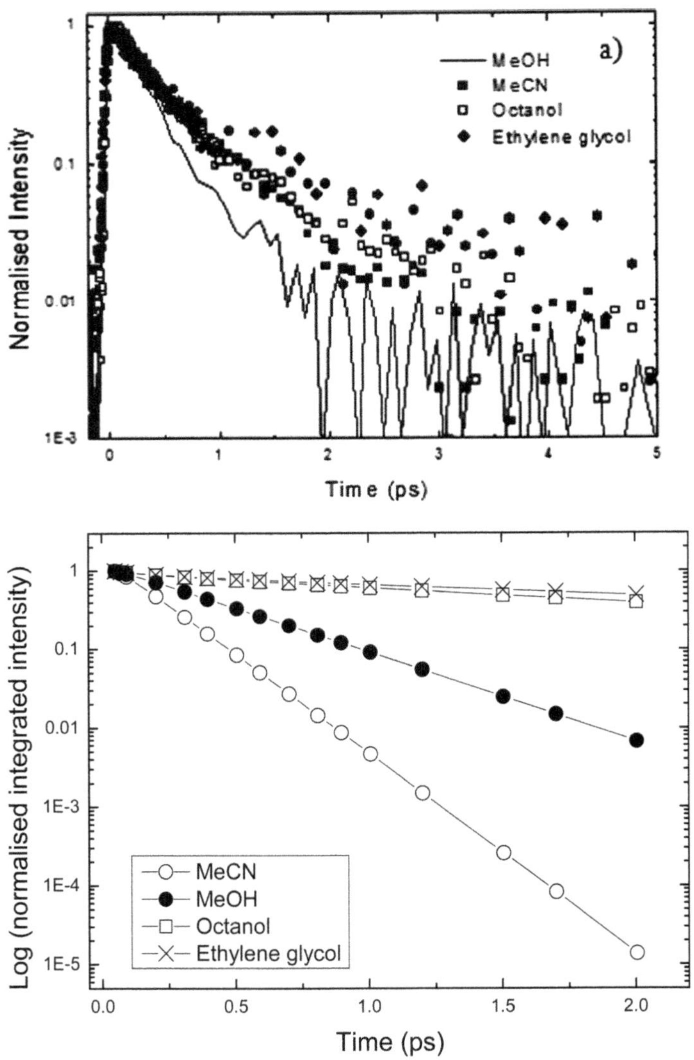

Fig. 10 (a) Time-resolved fluorescence decay for HBDI neutral in a range of solvent. (b) Calculated integrated intensity from (2) where *D* for the appropriate solvent was used and other parameters fixed as for methanol.

Table 2 Solvent dependence of fits of the fluorescence decay to a sum of two exponentials. The weights are a_i, the decay times τ_i and the weighted mean decay time $<\tau>$

Solvent	a_1	τ_1 /ps	a_2	τ_2 /ps	$<\tau>$/ps
Methanol	1.30	0.28	—	—	0.28
Acetonitrile	1.13	0.48	—	—	0.48
Octanol	0.27	0.84	0.92	0.30	0.42
Ethylene glycol	0.30	1.12	0.81	0.30	0.52

decays remained independent of wavelength. The measured ultrafast decay is seen to be only slightly dependent on the solvent viscosity (Fig. 10a) in line with previous observations suggesting a volume-conserving (thus friction independent) reaction coordinate.[11,26,27] In contrast, when we apply the corresponding diffusion coefficient (7) in the calculation a very strong viscosity dependence is predicted, dramatically stronger than that observed experimentally (Fig. 10b). This suggests a significant shortcoming in the model for the diffusion in eqn (2), which probably arises because of the significant friction experienced for motion along the coordinates recovered in the calculations.

Conclusions

Ultrafast fluorescence spectroscopy of the chromophore of GFP (HBDI) and a molecule closely related to the chromophore of KFP (FHBMI) have been measured in their neutral and anionic states. The spectra and dynamics of the two chromophores in solution are similar, in contrast to the behaviour of the two proteins. This highlights the importance of including the protein matrix in determining the photophysical properties of FPs. The time-resolved fluorescence spectra were measured with a 50 fs time resolution. For both molecules and for neutral and anion states the spectral mean frequency is independent of time.

The excited state dynamics of the anionic form of HBDI have been simulated by solving the Smoluchowski equation for diffusion on a potential energy surface. The inputs for the simulations were potential energy surfaces calculated for the chromophore.[37,67] The simulation for HBDI accurately reproduced the experimentally observed dynamics in methanol provided a strong coordinate dependence of the radiationless decay rate and transition moment were taken into account. This suggests that the simplest models for excited state dynamics in FP chromophores, relaxation on an excited state PES to a twisted geometry followed by internal conversion to the ground state at that geometry (*i.e.* at a 'pinhole sink' or conical intersection), are inadequate. To reproduce the experimental behaviour theoretical models need to include the possibility of coordinate-dependent rates of radiationless decay, with decay occurring rather early on along the coordinate. An attempt to simulate the faster dynamics observed in the neutral form of the chromophore on the basis of the very limited data for its excited state PES were not successful, suggesting the need for more detailed calculation on the neutral chromophore, which is of particular importance in photochemically active FPs.

Although the model calculations using a realistic PES were successful for the HBDI anion in methanol they completely fail to correctly predict the observed very weak effect of solvent viscosity on lifetime. As suggested elsewhere the observed weak dependence of the HBDI decay time on solvent viscosity is consistent with a volume-conserving isomerization coordinate, *i.e.* one which displaces a much smaller volume of solvent than either of the single bond rotations which generated the PESs employed here.[26,27] The possibility of a volume-conserving 'hula twist' coordinate promoting radiationless decay has been considered, but calculations show that this is energetically up-hill, which is inconsistent with the observed ultrafast barrierless nature of the internal conversion.[34] An alternative volume conserving radiationless decay coordinate, which was reported to be important in at least some calculations of the excited state dynamics of HBDI, is pyramidalization at the central bridging carbon atom.[35,38,40] Such a shape change appears to be sufficiently small to be consistent with the observed weak viscosity dependence. Relaxation along a pyramidalization coordinate has been invoked to account for the ultrafast excited state double bond rotation observed in a family of synthetic molecular motors.[71,72]

These considerations of the very efficient radiationless decay in HBDI and FHBMI still leave open the question of how the protein tunes the quantum yield of the chromophore between >0.8 and <0.01 depending on the environment. As has been discussed extensively, the electronic spectra of HBDI are shifted significantly on incorporation into GFP.[68] This suggests the possibility that the interaction between HBDI and the protein leads to changes in electronic structure, such that the surfaces calculated for the gas phase may not be an accurate representation of the chromophore in the protein. Alternatively protein - chromophore interactions may provide the friction required to resist relaxation along the barrierless PES. It is difficult to see how simple steric crowding in the protein could resist a single bond rotation coordinate, given that there is sufficient room in the cavity of the β-barrel to allow at least one bond to rotate.[73] This is especially true as full rotation is evidently not required to promote radiationless decay (Fig. 6). It is yet more difficult to see how such steric crowding could prevent isomerization along a volume-conserving coordinate, which is the nature of the coordinate most consistent with the observed viscosity dependence. We speculate that a motion which may be restricted in the protein, but not in free solution, is the length of the chromophore. Both twisting and pyramidalization will lead to some modulation of the length of the chromophore. At least in the highly fluorescent anionic state of wild type GFP the chromophore is strongly H-bonded at both the phenolic and imidazole rings. The stretching of these H-bonds as a consequence of the excited state dynamics may be sufficient to generate the friction required (or equivalently generate a sufficiently high energy barrier) to suppress the excited state structural relaxation leading to rapid internal conversion. This further highlights the importance of including protein – chromophore interactions when simulating the excited state dynamics of FPs.

Acknowledgements

We are grateful to EPSRC for financial support (EP/H025715/1) and K.A. thanks UEA for the provision of a studentship.

References

1 R. Y. Tsien, *Annu. Rev. Biochem.*, 1998, **67**, 509.
2 W. B. Frommer, M. W. Davidson and R. E. Campbell, *Chem. Soc. Rev.*, 2009, **38**, 2833.

3 R. H. Newman, M. D. Fosbrink and J. Zhang, *Chem. Rev.*, 2011, **111**, 3614.
4 V. Sample, R. H. Newman and J. Zhang, *Chem. Soc. Rev.*, 2009, **38**, 2852.
5 R. N. Day and M. W. Davidson, *Chem. Soc. Rev.*, 2009, **38**, 2887.
6 S. Kojima, H. Ohkawa, T. Hirano, S. Maki, H. Niwa, M. Ohashi, S. Inouye and F. I. Tsuji, *Tetrahedron Lett.*, 1998, **39**, 5239.
7 H. Niwa, S. Inouye, T. Hirano, T. Matsuno, S. Kojima, M. Kubota, M. Ohashi and F. I. Tsuji, *Proc. Natl. Acad. Sci. U. S. A.*, 1996, **93**, 13617.
8 R. M. Wachter, J. L. Watkins and H. Kim, *Biochemistry*, 2010, **49**, 7417.
9 K. P. Kent, L. M. Oltrogge and S. G. Boxer, *J. Am. Chem. Soc.*, 2009, **131**, 15988.
10 J. Dong, K. M. Solntsev and L. M. Tolbert, *J. Am. Chem. Soc.*, 2009, **131**, 662.
11 J. Conyard, M. Kondo, I. A. Heisler, G. Jones, A. Baldridge, L. M. Tolbert, K. M. Solntsev and S. R. Meech, *J. Phys. Chem. B*, 2011, **115**, 1571.
12 N. O. Alieva, K. A. Konzen, S. F. Field, E. A. Meleshkevitch, M. E. Hunt, V. Beltran-Ramirez, D. J. Miller, J. Wiedenmann, A. Salih and M. V. Matz, *PLoS One*, 2008, **3**.
13 M. V. Matz, K. A. Lukyanov and S. A. Lukyanov, *BioEssays*, 2002, **24**, 953.
14 R. M. Wachter, *Photochem. Photobiol.*, 2006, **82**, 339.
15 R. Ando, H. Hama, M. Yamamoto-Hino, H. Mizuno and A. Miyawaki, *Proc. Natl. Acad. Sci. U. S. A.*, 2002, **99**, 12651.
16 H. Mizuno, T. K. Mal, K. I. Tong, R. Ando, T. Furuta, M. Ikura and A. Miyawakil, *Mol. Cell*, 2003, **12**, 1051.
17 R. Ando, C. Flors, H. Mizuno, J. Hofkens and A. Miyawaki, *Biophys. J.*, 2007, **92**, L97.
18 S. Habuchi, R. Ando, P. Dedecker, W. Verheijen, H. Mizuno, A. Miyawaki and J. Hofkens, *Proc. Natl. Acad. Sci. U. S. A.*, 2005, **102**, 9511.
19 H. Mizuno, T. K. Mal, M. Walchli, A. Kikuchi, T. Fukano, R. Ando, J. Jeyakanthan, J. Taka, Y. Shiro, M. Ikura and A. Miyawaki, *Proc. Natl. Acad. Sci. U. S. A.*, 2008, **105**, 9227.
20 H. Shroff, C. G. Galbraith, J. A. Galbraith and E. Betzig, *Nat. Methods*, 2008, **5**, 417.
21 A. Vaziri, J. Y. Tang, H. Shroff and C. V. Shank, *Proc. Natl. Acad. Sci. U. S. A.*, 2008, **105**, 20221.
22 M. Bates, B. Huang, G. T. Dempsey and X. W. Zhuang, *Science*, 2007, **317**, 1749.
23 K. L. Litvinenko, N. M. Webber and S. R. Meech, *Bull. Chem. Soc. Jpn.*, 2002, **75**, 1065.
24 N. M. Webber, K. L. Litvinenko and S. R. Meech, *J. Phys. Chem. B*, 2001, **105**, 8036.
25 K. L. Litvinenko, N. M. Webber and S. R. Meech, *Chem. Phys. Lett.*, 2001, **346**, 47.
26 K. L. Litvinenko, N. M. Webber and S. R. Meech, *J. Phys. Chem. A*, 2003, **107**, 2616.
27 D. Mandal, T. Tahara and S. R. Meech, *J. Phys. Chem. B*, 2004, **108**, 1102.
28 K. M. Solntsev, O. Poizat, J. Dong, J. Rehault, Y. B. Lou, C. Burda and L. M. Tolbert, *J. Phys. Chem. B*, 2008, **112**, 2700.
29 V. Voliani, R. Bizzarri, R. Nifosi, S. Abbruzzetti, E. Grandi, C. Viappiani and F. Beltram, *J. Phys. Chem. B*, 2008, **112**, 10714.
30 K. Addison, J. Conyard, T. Dixon, P. C. B. Page, K. M. Solntsev and S. R. Meech, *J. Phys. Chem. Lett.*, 2012, **3**, 2298.
31 L. M. Tolbert, A. Baldridge, J. Kowalik and K. M. Solntsev, *Acc. Chem. Res.*, 2012, **45**, 171.
32 C. C. Hsieh, P. T. Chou, C. W. Shih, W. T. Chuang, M. W. Chung, J. Lee and T. Joo, *J. Am. Chem. Soc.*, 2011, **133**, 2932.
33 K. L. Litvinenko and S. R. Meech, *Phys. Chem. Chem. Phys.*, 2004, **6**, 2012.
34 W. Weber, V. Helms, J. A. McCammon and P. W. Langhoff, *Proc. Natl. Acad. Sci. U. S. A.*, 1999, **96**, 6177.
35 M. E. Martin, F. Negri and M. Olivucci, *J. Am. Chem. Soc.*, 2004, **126**, 5452.
36 S. Olsen and S. C. Smith, *J. Am. Chem. Soc.*, 2007, **129**, 2054.
37 S. Olsen and S. C. Smith, *J. Am. Chem. Soc.*, 2008, **130**, 8677.
38 A. Toniolo, S. Olsen, L. Manohar and T. J. Martinez, *Faraday Discuss.*, 2004, **127**, 149.
39 B. G. Levine and T. J. Martinez, *Annu. Rev. Phys. Chem.*, 2007, **58**, 613.
40 A. Toniolo, G. Granucci and T. J. Martinez, *J. Phys. Chem. A*, 2003, **107**, 3822.
41 I. Polyakov, E. Epifanovsky, B. Grigorenko, A. I. Krylov and A. Nemukhin, *J. Chem. Theory Comput.*, 2009, **5**, 1907.
42 K. B. Bravaya, M. G. Khrenova, B. L. Grigorenko, A. V. Nemukhin and A. I. Krylov, *J. Phys. Chem. B*, 2011, **115**, 8296.
43 I. V. Polyakov, B. L. Grigorenko, E. M. Epifanovsky, A. I. Krylov and A. V. Nemukhin, *J. Chem. Theory Comput.*, 2010, **6**, 2377.
44 P. Altoe, F. Bernardi, M. Garavelli, G. Orlandi and F. Negri, *J. Am. Chem. Soc.*, 2005, **127**, 3952.
45 C. Filippi, M. Zicchedd u and F. Buda, *J. Chem. Theory Comput.*, 2009, **5**, 2074.
46 J. Y. Hasegawa, K. Fujimoto, B. Swerts, T. Miyahara and H. Nakatsuji, *J. Comput. Chem.*, 2007, **28**, 2443.
47 C. R. S. Mooney, D. A. Horke, A. S. Chatterley, A. Simperler, H. H. Fielding and J. R. R. Verlet, *Chemical Science*, 2013.

48 A. Sinicropi, T. Andruniow, N. Ferre, R. Basosi and M. Olivucci, *J. Am. Chem. Soc.*, 2005, **127**, 11534.
49 D. M. Chudakov, V. V. Belousov, A. G. Zaraisky, V. V. Novoselov, D. B. Staroverov, D. B. Zorov, S. Lukyanov and K. A. Lukyanov, *Nat. Biotechnol.*, 2003, **21**, 191.
50 D. M. Chudakov, A. V. Feofanov, N. N. Mudriku, S. Lukyanov and K. A. Lukyanov, *J. Biol. Chem.*, 2003, **278**, 7215.
51 K. Tominaga, G. C. Walker, W. Jarzeba and P. F. Barbara, *J. Phys. Chem.*, 1991, **95**, 10475.
52 K. Tominaga, G. C. Walker, T. J. Kang, P. F. Barbara and T. Fonseca, *J. Phys. Chem.*, 1991, **95**, 10485.
53 M. Glasbeek, H. Zhang and M. J. van der Meer, *J. Mol. Liq.*, 2000, **86**, 123.
54 M. J. van der Meer, H. Zhang and M. Glasbeek, *J. Chem. Phys.*, 2000, **112**, 2878.
55 I. A. Heisler, M. Kondo and S. R. Meech, *J. Phys. Chem. B*, 2009, **113**, 1623.
56 M. Kondo, I. A. Heisler, J. Conyard, J. P. H. Rivett and S. R. Meech, *J. Phys. Chem. B*, 2009, **113**, 1632.
57 M. Kondo, I. A. Heisler and S. R. Meech, *Faraday Discuss.*, 2010, **145**, 185.
58 M. Kondo, I. A. Heisler and S. R. Meech, *J. Phys. Chem. B*, 2010, **114**, 12859.
59 N. Amdursky, Y. Erez and D. Huppert, *Acc. Chem. Res.*, 2012, **45**, 1548.
60 Y. Erez, N. Amdursky, R. Gepshtein and D. Huppert, *J. Phys. Chem. A*, 2012, **116**, 12056.
61 Y. Erez, Y. H. Liu, N. Amdursky and D. Huppert, *J. Phys. Chem. A*, 2011, **115**, 8479.
62 C. H. Kim and T. Joo, *Opt. Express*, 2008, **16**, 20742.
63 Y. A. Tretyakova, A. A. Pakhomov and V. I. Martynov, *J. Am. Chem. Soc.*, 2007, **129**, 7748.
64 M. L. Quillin, D. A. Anstrom, X. K. Shu, S. O'Leary, K. Kallio, D. A. Chudakov and S. J. Remington, *Biochemistry*, 2005, **44**, 5774.
65 M. L. Horng, J. A. Gardecki, A. Papazyan and M. Maroncelli, *J. Phys. Chem.*, 1995, **99**, 17311.
66 K. Addison, J. Conyard, T. Dixon, P. C. Bulman Page, K. M. Solntsev and S. R. Meech, *J. Phys. Chem. Lett.*, 2012, **3**, 2298.
67 S. Olsen and R. H. McKenzie, *J. Chem. Phys.*, 2009, **130**, 184302.
68 S. R. Meech, *Chem. Soc. Rev.*, 2009, **38**, 2922.
69 J. J. van Thor, *Chem. Soc. Rev.*, 2009, **38**, 2935.
70 N. A. Smith and S. R. Meech, *J. Phys. Chem. A*, 2000, **104**, 4223.
71 J. Conyard, K. Addison, I. A. Heisler, A. Cnossen, W. R. Browne, B. L. Feringa and S. R. Meech, *Nat. Chem.*, 2012, **4**, 547.
72 A. Kazaryan, J. C. M. Kistemaker, L. V. Schafer, W. R. Browne, B. L. Feringa and M. Filatov, *J. Phys. Chem. A*, 2010, **114**, 5058.
73 S. L. Maddalo and M. Zimmer, *Photochem. Photobiol.*, 2006, **82**, 367.

Faraday Discussions RSC Publishing

PAPER

Ultrafast dual photoresponse of isolated biological chromophores: link to the photoinduced mode-specific non-adiabatic dynamics in proteins

Anastasia V. Bochenkova*[ab] and Lars H. Andersen[a]

Received 13th December 2012, Accepted 9th January 2013
DOI: 10.1039/c3fd20150c

The anionic wild-type Green Fluorescent Protein (GFP) chromophore defines an entire class of naturally occurring chromophores, which are based on the oxydized tyrosine side chain. The GFP chromophore exhibits an enriched photoinduced non-adiabatic dynamics in the multiple excited-state decay channels. Deactivation includes vibrational resonant photodetachment and internal conversion. Here, we provide detailed insight into the efficiency of different vibrational modes in promoting a selective photoresponse in the bare GFP chromophore anion. We introduce a general theoretical model that is capable of accounting for the alternative non-equivalent pathways in internal conversion, and we outline the factors, by which the photo-initiated response may be altered in this channel. The topography around the planar minimum in S_1 and the two distinct types of the S_1/S_0 conical intersections obtained through high-level *ab initio* calculations provide direct support to the proposed model. There are mode-selective ways to control the photoresponse and to direct it towards a single excited-state decay channel. By tuning the excitation wavelength, the photoresponse may be directed towards the ultrafast non-statistical electron emission coupled with vibrational (de)coherence, whereas a vibrational pre-excitation in the ground state may lead to the ultrafast non-statistical internal conversion through a conical intersection. We also discuss the implication of our results to the photo-initiated non-adiabatic dynamics in the proteins.

1 Introduction

Light-induced processes are ubiquitous in nature. Interaction of molecules with light is central to the vital activities of living organisms and human beings. Photosynthesis, vision in vertebrates, solar energy harvesting and conversion, and light sensing are remarkably efficient processes, and much focus has been placed

[a]*Department of Physics and Astronomy, Aarhus University, DK-8000, Aarhus C, Denmark. E-mail: bochenkova@phys.au.dk; Tel: +45 8715 6353*
[b]*Department of Chemistry, M.V. Lomonosov Moscow State University, 119991 Moscow, Russia*

on elucidating the role played by the protein environment in their primary events which occur on a timescale down to sub-picoseconds. Photoactive proteins attract immense attention not only because of their crucial role for the functioning of living organisms, but they also provide a stage for studying some very elementary processes in nature, such as photoinduced isomerization, light-driven electron transfer, and charge separation. These fundamental events, initiated as a primary response to light absorption, are ultrafast and, at the same time, are rate-determining inside the proteins. In this respect, the intrinsic photoresponse of light-absorbing molecular units in these proteins, and the characteristic timescales of their photo-initiated dynamics are of utmost importance as they provide a well-defined reference to the functioning of the proteins.

Recent advances in modern physics, including ultrafast pulsed laser techniques, new light sources, and ion-storage facilities, combined with a detailed theoretical insight gained by using state-of-the-art methods of quantum chemistry enable the disclosure of new fundamental concepts that lie at the heart of the functioning of the photoactive proteins.[1–5] Remarkably, an ever increasing number of experimental evidence for an important role of quantum effects, such as electronic coherence in photosynthetic proteins[4,5] and non-adiabatic dynamics through conical intersections in visual[3] and signaling[1] photoreceptors, is emerging. Our most recent findings reveal the existence of competing and remarkably fast dual energy transfers between the nuclei and the electrons in the photoinduced channels of the isolated anionic chromophore from the Green Fluorescent Protein (GFP).[6] Importantly, the energy flow may be directed towards electronic as well as nuclear degrees of freedom mediated by specific vibrational modes. This results in vibrationally mediated photodetachment (PD) out of the first electronically bound excited state of the anion or in internal conversion (IC) back to the 'hot' ground state of the anion (see Fig. 1). In the case of the anionic GFP chromophore, the inherent timescales of the photo-initiated reaction dynamics may be down to the tens of femtoseconds when coupled to vibrational coherence of the excited active modes.[6] We consider these non-adiabatic

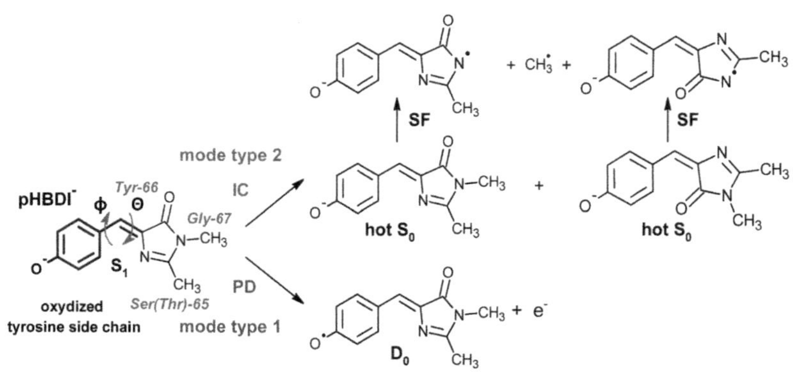

Fig. 1 Intrinsic de-excitation channels mediated by the two types of vibrational modes of the deprotonated GFP chromophore terminated with the two methyl groups: PD directly from S_1 and IC into a vibrationally hot anion in the electronic ground state S_0. From the hot S_0 state statistical fragmentation (SF) may occur. Shown are the two exocyclic angles, ϕ and θ, that induce a non-adiabatic coupling between the S_1 and S_0 surfaces.

couplings in the excited states dynamics of biological chromophores as a key point for various mechanisms by which the proteins efficiently tune their response to the absorption of light. Here, we explore the ways by which the GFP proteins may use this electron-to-nuclei coupling to guide the photochemistry and photophysics upon which their functioning is based.

The wild-type GFP is a 27 kDa protein composed of 238 amino acids. The protein has an 11-stranded β-barrel structure with a single α-helix running through the center.[7] The heart of the protein is the chromophore which is formed at the central helix, well shielded by the hydrophobic core of the protein. The GFP chromophore is a 4-(*p*-hydroxybenzylidene)-5-imidazolinone (pHBI) derivative formed by autocatalytic cyclization of the Ser65-Tyr66-Gly67 tripeptide.[8,9] One important role of the protein environment is to dictate the balance between a neutral chromophore and a deprotonated phenolate ion. Numerous wild-type and mutant green fluorescent proteins contain solely the anionic chromophore and are therefore characterized by a single-peak excitation, as seen in one of the most popular GFPs, enhanced eGFP.[10] In wild-type GFP, the ratio between the two forms is six to one in the electronic ground state,[7] whereas the neutral chromophore exhibits increased acidity in the excited state.[11] An excited-state proton transfer occurring on the sub-picosecond timescale is well known to result in a remarkable blue-to-green light conversion in absorption/emission of the wild-type GFP protein.[12,13] Out of the 238 amino acids, Gly67 is essential for the formation of the chromophore, but others form by substitution the basic principles behind the development of many GFP color variants that are so widely used as *in vivo* fluorescent tags.[10] For this purpose numerous studies have been aimed at fine-tuning the photophysical properties of fluorescent proteins, in particular to work in the far-red spectral regions to enable a mammalian deep tissue imaging.[14,15]

The GFP is strongly fluorescent because the conjugated system of the chromophore with the two rings exists in a nearly co-planar orientation in the *cis*-configuration in the protein. The planarity is maintained by a network of hydrogen bonds which prevents the rings from twisting after the chromophore has been photo-excited. In the gas-phase, as well as in solutions, nothing, however, prevents the chromophore in the excited state from returning to the ground state by internal conversion without emission of fluorescent light, and indeed the search for gas-phase fluorescence has been unsuccessful.[16] The twisting in the central methine bridge along the θ and/or ϕ angles (see Fig. 1) is thought to induce a non-adiabatic coupling between the two surfaces. This results in a very fast excited-state decay *via* a conical intersection.[17–19]

The (sub)picosecond non-radiative deactivation, which leads to efficient fluorescence quenching outside the protein environment, has been extensively studied in various solvents by time-resolved spectroscopy.[20–22] The effect of conformational freedom on the rapid internal conversion has also been studied through structural modifications of the chromophore and through adjustments of supra-molecular systems.[11] All these studies strongly support the idea that internal conversion occurs through the conical intersection(s) induced by twisting in the methine bridge. However, there is still no agreement on the details of the deactivation mechanism. Even in the case of the bare chromophore, there are controversial theoretical predictions[18,19,23,24] for the most favorable IC pathway that includes either a single of the two possible twisting coordinates or both of them (the so called space-conserving 'hula-twist' mechanism[25]).

The experimental solution data clearly show the non-exponential decay kinetics with a dominant sub-picosecond component,[20–22,26] dependence of the fluorescence decay on temperature[20,22] and, in some cases, on viscosity.[20] Iso-viscosity analysis shows that internal conversion is effectively a nearly barrierless process.[27] There is, however, a single activation parameter which has been found to depend linearly on inverse temperature in the markedly multi-exponential decay of the excited electronic state.[22] This has been explained by invoking a model of the two-dimensional dynamics that couples the phenyl-ring rotation to a weakly activated libration of the imidazolinone ring.[22] The latter has been considered as a fast active mode that leads to a conical intersection, while the former has been assigned to a slow coordinate inhomogeneously distributed in S_1. This model directly implies a different character of the two bridge bonds and, hence, the different topographies (shallow *versus* steep) associated with the twisting along the corresponding angles, so that the topographies become reverse upon excitation. However, a simple consideration in the molecular orbital framework suggests that there should be a nearly perfect resonance between the two mesomeric structures in the bare chromophore anion, which completely scrambles the bond orders in the central bridge moiety and leads to a charge delocalization throughout the entire conjugated system.[28] This apparently contradicts the above mentioned model, although solvent effects may play a role in decoupling the two resonance forms.

It can be seen from the above consideration, that although certain progress has been made in describing the mechanism of the deactivation channel, there are still many open questions and contradictions that have to be resolved. It seems that rigorous characterization of the possible types of conical intersections and the corresponding structural motifs is missing even in the case of the isolated anion. In this respect, a simple reasonable model that is capable of giving the entire picture of the excited-state radiationless decay pathways and that is free of a particular computational scheme is highly in demand. Knowledge of the intrinsic properties, such as barrier heights for escaping from the planar minimum (if any) associated with the fluorescent state and pathways to various types of the conical intersections characterized by their relative energies will enable us to list the factors which may be altered in different media and, in particular, in various proteins. For example, the inconsistency in particular IC mechanisms in different media suggests that there are several types of conical intersections that may be altered in energy, depending on the environment. The quantitative picture, then, comes from high-level *ab initio* calculations. Since the theoretical level that is affordable for computational studies is ever increasing, it is of utmost importance to have a qualitative picture that can be further quantitatively refined in the future.

Since many GFP-like proteins are fluorescent, much focus has been put on elucidating the role of the protein in suppressing the competing non-radiative channel. However, there are certain GFP homologues which are non-fluorescent, but they may be reversibly photo-activated.[29,30] Photoinduced isomerization is thought to be responsible for the photo-switching of these kindling fluorescent proteins. Most recent results suggest that diverse GFPs, containing anionic tyrosine-based chromophores, may also be involved in biological photoinduced electron transfer reactions.[31] Therefore, GFP-like proteins show remarkable diversity in terms of their excited-state decay channels, thus enabling a wide range of their possible applications as fluorescent markers in bioimaging, as photoswitches in super-resolution optical microscopy and ultimately as redox sensors in living cells. Such diversity has to be closely related to the intrinsic excited-stated decay channels of the deprotonated GFP chromophore, where specific vibrational modes have recently been shown to facilitate selective photoresponse.[6]

About ten years ago, the first gas-phase absorption measurement of a deprotonated GFP-model chromophore (a dimethyl derivative of pHBI, called pHBDI) was performed.[32] It was found that it absorbs at about the same wavelength in vacuum and in the protein, implying that the sum of perturbations in the wild-type GFP environment has very little effect on the absorption wavelength, which to a large extent is determined by the intrinsic properties of the chromophore. Today it is well established that the spectral diversity (color tuning) of fluorescent proteins is first and foremost a consequence of chemical modifications of their chromophore itself,[10,14,33] and fine tunings are primarily achieved by non-covalent interactions with the surrounding amino acids through mutagenesis. Since the first measurements, action spectroscopy has been extensively used to study the absorption of biologically relevant chromophores and, in particular, the GFP chromophore anion in the gas phase. Various ion storage techniques, including electrostatic storage rings,[34,35] quadrupole ion traps[16,36] and ion cyclotron resonance cells,[37] along with detection of both ionic[16,36] and neutral fragments,[34,35] as well as parent ions loss[16,36,37] after photoexcitation have been used (see Fig. 1). However, several precautions should apply when using a photoresponse of these chromophores for retrieving their absorption profiles. In particular, a sequential multiple photon absorption and competing decay channels have to be taken into account.

Recent photoelectron spectroscopy studies have directly confirmed that electron emission is indeed one of the deactivation channels in the bare anion.[38–40] Moreover, these results strongly suggest that S_1 is electronically bound in the Franck–Condon region, and electron emission occurs resonantly through vibrational autodetachment inside the S_0–S_1 absorption band.[40]

The aim of this discussion is two-fold. In light of our recent findings that excited-state dynamics is governed by the non-adiabatic nature that couples the electronic and nuclear motion in both competing decay channels of the GFP chromophore anion,[6] we first aim at getting a deep insight into the efficiency of different vibrational modes in promoting a selective photoresponse in the bare anion. We introduce a general theoretical model that is capable of accounting for different IC pathways, and we outline the factors by which the photo-initiated response may be altered in this channel. The topography around the planar minimum in S_1 obtained through high-level *ab initio* calculations within the extended multiconfigurational quasi-degenerate perturbation theory[41] provides direct support to the proposed qualitative model. We then explore the wavelength-dependence of the PD/IC branching ratio, where both competing decay channels are mediated by internal vibrational energy redistribution (IVR) in S_1 at low excitation energies. Their lifetimes are determined through the corresponding activation energies. Secondly, we discuss the implication of our results to the photo-initiated non-adiabatic dynamics in the proteins. We show remarkable similarity of the photoinduced nuclear dynamics in the gas phase and in the protein and emphasize the close interrelation between excited-state dynamics and the corresponding spectral shapes. Finally, we discuss the ways, by which the GFP-like proteins may use the intrinsic dual electron-to-nuclei coupling to promote a selective photoresponse.

2 Methods

2.1 Electronic structure calculations and simulation of absorption profiles

Equilibrium geometry parameters and vibrational frequencies in the ground electronic states of the anion and the neutral radical were found using the density functional theory with the hybrid PBE0/(aug)-cc-pVDZ functional. Geometry parameters in the first excited state of the anion were optimized at the XMCQDPT2[41]/CASSCF(14,13)/(aug)-cc-pVDZ level of theory. The XMCQDPT2 excited-state gradient was calculated numerically as a two-sided derivative.

The vertical excitation energy was obtained using the XMCQDPT2/SA(7)-CASSCF(16,14)/(aug)-cc-pVDZ approach within the active space which comprised all valence orbitals of the π-type. The vertical detachment energies were calculated at the XMCQDPT2/SA(2)-CASSCF(14,14) level of theory within the (p-type d-aug)-cc-pVTZ basis set. Oxygen atoms were treated with the entire aug-spdf diffuse shells. The basis set was also augmented with an additional diffuse function of p-type with a particularly small exponent (-10) centered at the middle carbon atom. The π^* orbital of this highly diffuse shell was included in the active space and was used to mimic an electron-detachment process. The ground and ionized states were included in the state-averaging procedure. The XMCQDPT2 effective Hamiltonians were constructed in the frame of the reference spaces spanned by 9 zeroth-order wavefunctions. For all electronic structure calculations the Firefly version 8.0 package,[42] partially based on the GAMESS (US) source code,[43] was used.

The $S_0 \rightarrow S_1$ vibrational profiles at 0 K and 300 K were obtained through a time-domain formalism[44] based on Fourier transforms of the Lax's autocorrelation function[45] within the double harmonic parallel-mode approximation.[46]

2.2 Protein simulations

The model of the Ser 65$\rightarrow$ Thr GFP mutant that contained a permanently deprotonated chromophore[47] was constructed based on the crystal structure 1EMA[48] from the pdb databank.[49] The initial structure was obtained by adding missing hydrogen atoms to the heavy nuclei, assuming a protonated Glu 222 aminoacid residue and treating all His residues as neutral and δ-protonated. The overall net charge of the protein was set to zero by performing a set of point mutations of charged Glu and Asp residues exposed to solvent to their neutral Gln and Asn analogues. Five surface residues (Asp 19, Glu 32, Glu 34, Asp 117, Asp 173) lacking charge-stabilization through the corresponding salt bridges were mutated along with the C-end that was kept neutral. The structure was then extensively solvated by adding ~6500 water molecules and exposed to the preliminary geometry optimization (10 000 steps) followed by constant temperature NVT molecular dynamics (MD) simulations with CHARMM force field parameters.[50] The protein backbone and the chromophore were held fixed during the simulations. Periodic boundary conditions were applied. The equilibration temperature was set to 300 K using a Langevin thermostat. Long-range electrostatic interactions were treated with the particle mesh Ewald method. The simulations were carried out with a 1 fs integration step for 2 ns. After that, the temperature of the system was gradually reduced down to 50 K in steps of 50 K during 1.2 ns. The final MD geometry was then obtained by performing energy minimization during another 10 000 steps. The atomistic model of the S65T-GFP protein was constructed using the VMD program,[51] while all MD simulations were run by using the NAMD 2.7b2 version.[52]

The final annealed MD atomistic model was reduced for the subsequent quantum mechanical/molecular mechanical (QM/MM) treatment by removing outer water molecules located farther than 2.9 Å from the protein, *i.e.*, out of the first solvation shell. The reduced system contained 909 water molecules, while the total number of atoms, including the entire protein, was equal to 5410. The chromophore anion based on the entire Thr 65, Tyr 66 and Gly 67 residues as well as the positively charged Arg 96 and protonated Glu 222 side chains together with all residues and water molecules of the chromophore's binding pocket that formed an extensive hydrogen bonding network were included in the QM part. The chromophore was hence represented by the extended model of 70 atoms based on the backbone and side chains of the residues which were covalently bound to the conjugated system. In total, 154 atoms, including 15 capping hydrogen atoms, were assigned to this subsystem treated using 1550 basis functions at the PBE0/(aug)-cc-pVDZ level of theory, while performing ground state geometry optimization within a mechanical embedding QM/MM scheme. This method implied geometrical constraints to be imposed on the QM part, where the QM and MM interactions were treated by the classical MM force field. The current choice of the QM part accounted correctly for all interactions of the chromophore with the key residues, such as charge transfer, hydrogen bonding, short-range repulsion, treating the entire binding pocket at a sufficiently high level of theory. Additional constraints were imposed on the outer part of the protein, located more than 7 Å from the chromophore, that was held fixed during the QM/MM optimization procedure.

Vibrational analysis was carried out in the frame of the same QM/MM scheme, where the second derivatives of the total QM and MM energy with respect to nuclear coordinates of the extended chromophore were calculated. The partially evaluated Hessian matrix (210×210) was projected to eliminate rotational and translational contaminants and diagonalized, and its all 204 real eigenvalues ensured that the located QM/MM stationary point was a true minimum. Since the size of the extended chromophore was approximately two times larger than that of the conjugated system itself, the calculated medium and high frequencies associated with rather localized normal modes of the conjugated part, such as breathing or stretching, were considered to be close to the 'true' values. These 'true' values may be obtained through diagonalization of the entire Hessian matrix, where the corresponding mixed second derivatives of the energy with respect to the Cartesian coordinates of the chromophore and its protein surrounding are expected to be close to zero.

The equilibrium geometry parameters of the neutral radical chromophore were found in the frame of the same QM/MM scheme, where the partial charges of the radical chromophore used in calculating long-range electrostatic interactions of the QM and MM parts were recalculated at the PBE0/ROHF/(aug)-cc-pvdz level of theory, applying the natural orbital analysis (NBO).[53] The differences in the calculated NBO charges between the isolated chromophore anion and its radical were used to redefine the corresponding force field parameters.

The excited state gradient calculations at the XMCQDPT2/CASSCF(14,13) level of theory were performed using the extended model of the chromophore and all nearby residues that directly formed hydrogen bonds with the conjugated system. The QM part included 110 atoms and was described by 1130 basis functions. The rest of the protein (MM part) was represented by a set of MM point charges and this electrostatic field was incorporated in the one-electron integrals of the electronic Hamiltonian by means of the effective fragment potential (EFP) method,[54] thus allowing the QM electron density to be polarized in the field of the protein. The final excited-state gradient was corrected by subtracting from it the S_0 gradient, calculated at the same level of theory.[6]

The modified PC GAMESS/Firefly version 6.4. and Firefly version 8.0[42] were used for the QM/MM modeling using the mechanical embedding scheme and the QM/EFP electrostatic approach, respectively.

3 Results and discussion

3.1 Intrinsic excited-state decay channels

Action absorption spectroscopy studies have previously revealed two excited state decay channels in pHBDI anion upon excitation to S_1 (see Fig. 1). One of these channels is attributed to the delayed statistical fragmentation in the vibrationally hot ground electronic state of the anion, following fast internal conversion. This channel has been directly confirmed by measuring neutral[34,35] as well as charged[16,36] daughter fragments. The second photodetachment channel has first been identified by indirect measurements of parent ions loss in action spectroscopy experiments and has recently been confirmed by direct registration of outgoing electrons.[38–40] The interplay between the two channels has recently been studied experimentally as well as theoretically,[6] where the emphasis has been on explaining a relatively long excited-state lifetime that enables IC to be the predominant channel in deactivation of the pHBDI anion at the excitation wavelength of the maximum fragmentation yield (482 nm).

Here, we provide a detailed mechanistic insight into the IC channel and also show that the totally symmetric stretching coordinates, which are Franck–Condon active modes (see below), are those that assist photodetachment, as opposed to out-of-plane twisting coordinates which efficiently lead out of the region of very close proximity between the surfaces of D_0 and S_1. The latter is observed even at very small fluctuations of the twisting angles on the flat potential energy surface (PES) of S_1 due to energy increase of D_0.

3.1.1 Vibrational photodetachment. We have shown that an electronically stable anion in the Franck–Condon (FC) region with a very low electron affinity can fully support an experimental finding of a substantial lifetime in the S_1 state.[6] Our calculated vertical detachment energy (VDE) is 2.62 eV, and that places the D_0 state vertically above the S_1 state by ~0.1 eV. The calculated vertical detachment and excitation energies as well as origin shifts for the S_0 and S_1 electronic states of the anion and for the D_0 ground state of the neutral radical may be used to illustrate the picture of embedded vibrational resonances for the anion in S_1 (see Fig. 2).

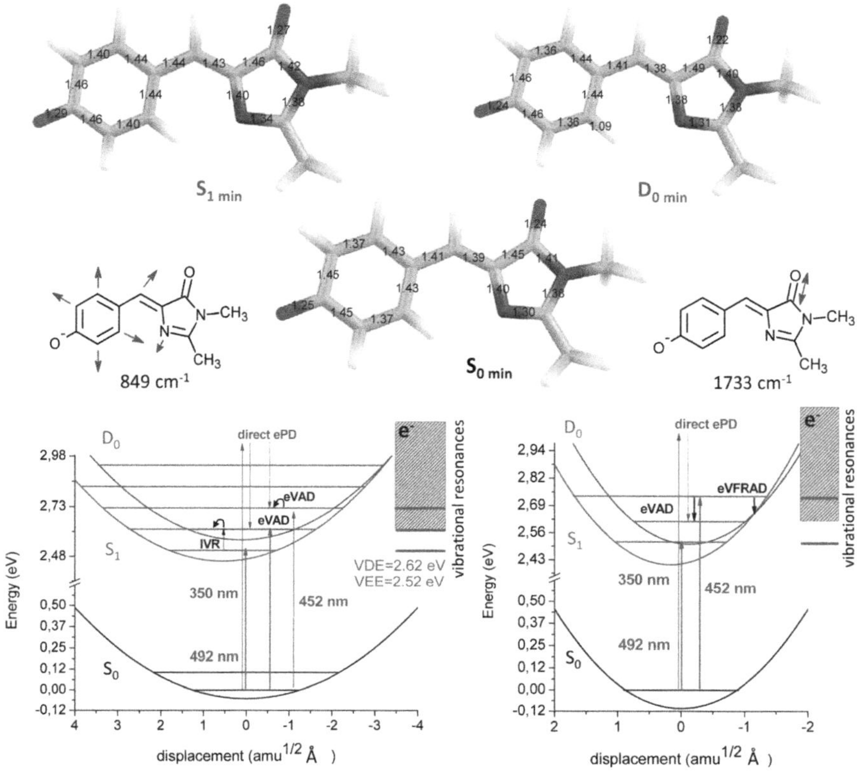

Fig. 2 Upper panel: equilibrium geometry parameters for the planar structures of the S_1, D_0 and S_0 minima. All distances are shown in Å. Lower panel: three one-dimensional quadratic surfaces of the S_0, S_1 and D_0 electronic states shifted according to the calculated minimum displacements along the totally symmetric stretching coordinates and according to the calculated vertical detachment and excitation energies. Note that PD occurs through vibrational resonances out of S_1 within the S_0–S_1 absorption band down to ~350 nm.

Fig. 2 shows quadratically approximated PESs for the three states as a function of the two emission-active mode coordinates. The harmonic approximation is valid here, since these modes refer to the medium and high-frequency stretching coordinates. If not being coupled to the nuclear motion, electron detachment would never occur, since the S_1 state is electronically bound in the FC region. These are vibrational resonances that enable electron autodetachment (VAD). The crossing point of the S_1 and D_0 surfaces is importantly located outside the FC region, thus even at energies where this crossing occurs (~2.7 eV), vibrations are required to reach this point (autodetachment through vibrational Feshbach resonances, VFRAD). At the crossing point, the states are electronically degenerate, and this can be distinguished from VAD that occurs even below the crossing point (and even when a photon energy is less than adiabatic detachment energy). In VAD, the energy has to be transferred from the nuclei to the electrons through their (non-adiabatic) coupling. This means that this coupling may occur not only when two surfaces are close in energy to each other, but also in cases where the states do not cross at all. This is, in fact, a well-known phenomenon.[55]

The mode-specificity in VAD comes from the S_1/D_0 origin shifts. In the case of VAD, the non-adiabatic couplings should depend on the vibrational overlap between the final vibrational wavefunction and the derivative of the initial vibrational wavefunction with respect to the corresponding nuclear coordinates. In the harmonic approximation, such overlap would simply be zero if the two surfaces along the corresponding normal mode are not displaced. Remarkably, these PD-active modes are those that are excited upon the S_0–S_1 transition. This means that the S_1 surface is shifted in the same direction with respect to both the S_0 and D_0 surfaces. In the case of the high-frequency stretching modes, the S_1 and D_0 minima are displaced in the opposite directions with respect to the one of S_0. This is easily recognized by comparing the corresponding equilibrium geometry parameters in S_0, S_1 and D_0 (see Fig. 2).

3.1.2 Internal conversion through conical intersections. As shown above, there are quite controversial findings regarding the mechanism of internal conversion and the possible pathways from the Franck–Condon region in S_1 to the S_1/S_0 conical intersection seam. We summarize the experimental findings of time-resolved experiments in various solvents that are most intriguing for the following discussion:

- a non-exponential decay kinetics of the first excited-state of the anion with a sub-picosecond predominant component;
- a nearly barrierless radiationless conversion;
- a weak dependence on solvent viscosity.

Before discussing our theoretical results, we introduce a simple qualitative picture of the distinct electronic structures that may lead to different types of the conical intersections in the case of the deprotonated GFP chromophore. These types taken separately were previously discussed, although the main emphasis was on the optimized geometry parameters of the points which corresponded to the minimum-energy conical intersections (MECI) located using different levels of theory.[18,19,23,24] It has previously been pointed out that the bright character of the S_0–S_1 transition in pHBDI anion may be easily interpreted in the frame of a simplified molecular orbital picture based on the Hückel model.[28] According to this model illustrated in Fig. 3, the central bridge moiety can be described by invoking a three orbital diagram of the symmetric allyl anion. However, it is easily recognized that this model can not entirely reproduce changes in the character of the electronic structure of the S_0 and S_1 states upon twisting in the methine bridge (the model was not originally applied to the twisted conformations). The two equivalent structures, which are 90° twisted along one of the two bonds, correspond to the cases where one of the three orbitals is completely decoupled from the others. This refers to the ethylene molecule and a single non-interacting orbital. It is then easily recognized that twisting along a single bond would never lead to a conical intersection and further decoupling in the ethylene molecule should occur, resulting in piramidalization of the central carbon atom.

The above-mentioned model suffers from its perfect symmetry, whereas the two parts in the GFP chromophore anion, the phenol (Ph^-) and imidazolinone (Im^-) rings, may hardly be thought of as perfectly symmetric. However, it is appealing to find a simple model that can be applied to a qualitative prediction of possible IC pathways in the pHBDI anion. Fig. 4 introduces such a model based on the asymmetric allyl anion. This model has predictive power and results in two alternative non-equivalent pathways for IC based on:

- two non-equivalent electronic structures of the twisted conformations with a distinct negative charge localized on one of the two parts;

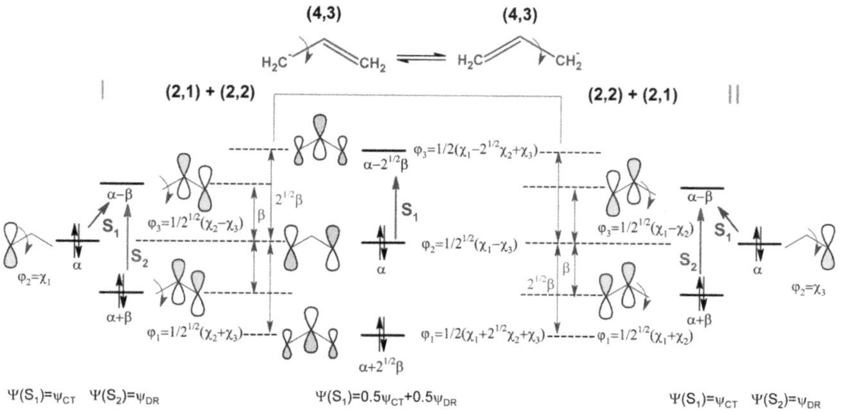

Fig. 3 Hückel model of the symmetric allyl anion with four electrons distributed in three orbitals. The perfect resonance interaction between the two mesomeric structures in the planar configuration is shown. The first excited state may then be described as a linear combination of the charge transfer (CT) and diradical (DR) valence electronic structures with equal weights. Upon twisting, the interaction breaks down and this decouples the system into two non-interacting parts: the non-bonded orbital of the carbon atom and the ethylene molecule. Due to decoupling, the energy gap of the frontier orbitals tends to decrease. The conical intersection associated with the equal contributions of the corresponding valence structures of the ground state and of the first excited state of the CT character is, however, never reached upon twisting about one of the two alternative C–C bonds. Simultaneous partial decoupling of all three orbitals is required to reach quasi-degeneracy of the two electronic states.

• two distinct types of the S_1/S_0 conical intersections based on the different electronic structure;

• similar small barriers in S_1 for initial twisting out of the planar minimum, where there is a strong coupling between the two mesomeric structures;

• two structurally different motifs of the MECIs: a hula-twist (double twist, I) and a (primarily) single twist about only one of the two bonds (II). The actual twist about the two bonds drastically depends on the relative electron affinities of the two subsystems (see Fig. 4) or, in other words, on the degree of asymmetry. The hula-twist structure exhibits a markedly higher degree of pyramidalization of the central carbon atom as induced by an additional decoupling in the neutral counterpart.

• The transfer between the two types of the electronic structures may occur in S_1 in the vicinity of the S_2/S_1 conical intersection.

In the case of the bare GFP chromophore anion, this model predicts that the conical intersection of type II is more favorable and formally corresponds to the isomerization pathway. This results from the fact that the structural motif of the Ph=CH_2 neutral counterpart is known to be an efficient electron acceptor, thus it is readily involved in the excited state of the charge transfer character. The type I hula-twist conical intersection presumably lies higher in energy, as has indeed been previously shown.[18] However, the barrier heights for twisting along the two alternative coordinates are expected to be very small (the two bridge bonds are simultaneously elongated upon excitation, as follows even from the simple Hückel model), even though they may have been expected to be slightly different due to asymmetry. This inevitably leads to splitting of the excited-state population into two alternative branches in S_1, leading out of the fluorescent state. The

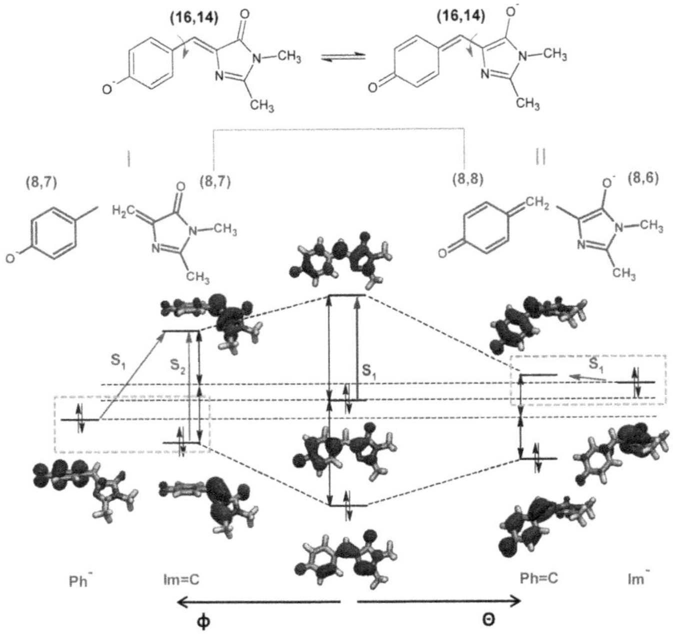

Fig. 4 Qualitative molecular orbital framework for the asymmetric allyl anion, which mimics the central bridge moiety of the deprotonated GFP chromophore. The two alternative pathways along the twisting coordinates are non-equivalent and bring the system to two distinct cases, where only one of them leads to close proximity of the two surfaces of the ground and first excited states (II). The corresponding natural CASSCF(16,14) orbitals are shown for the planar and twisted structures. The latter refers to the transition states along the corresponding coordinates. The decoupling of the orbitals as well as the bonding and anti-bonding character of the C=C bridge bond of the corresponding neutral fragments are easily recognized. The asymmetry is defined by different electron affinities of the phenol and imidazolinone rings. The conical intersection associated with the ϕ coordinate requires an additional twisting along the θ coordinate that further decouples the second C=C bridge bond and also results in piramidalization of the central carbon atom. The geometries of the S_1/S_0 and S_2/S_1 conical intersections simultaneously twisted along the two angles are envisaged for the electronic structure associated with the system I, where the latter presumably lies lower in energy. The lowest energy conical intersection corresponds to the electronic structure of system II, where the geometry is primarily twisted along the θ coordinate.

subpopulation II goes directly to the lowest-energy S_1/S_0 CI, whereas the evolution of the subpopulation I needs special consideration. Remarkably, the system is not trapped along the pathway I, since there should be an alternative route to the ground state rather than the corresponding high-energy hula-twist CI. The subpopulation I is expected to evolve in S_1 towards the same conical intersection of type II, however it requires to pass the region in S_1, where the electronic character of the wavefunction changes. In other words, the system should evolve in the vicinity of the S_2/S_1 conical intersection. A presumably small fraction of the subpopulation I may be reflected back to the fluorescent state. This may cause a non-exponential fluorescence decay with the second component characterized by an increased excited-state lifetime. However, this pathway in S_1 has to be studied in more details.

The planar minimum of the anion does exist in S_1, as follows from the unconstrained geometry optimization procedure. However, as expected, the S_1

potential energy surface is very flat in the vicinity of the fluorescent state. By performing a series of constrained geometry searches along the two alternative twisting coordinates, the corresponding transition states (TS) have been located (see Fig. 5). The barrier heights are similar and equal to ~1.2 kcal mol^{-1} (~0.05 eV).

Remarkably, the change in energy as a function of the corresponding twisting angles is markedly different in both cases. Therefore, the high level *ab initio* calculations strongly support the conclusions drawn above. As a consequence, the type II CI is easily reached by the system, whereas the topography around the type I CI is sloped, as follows from the calculated S_1 and S_0 gradients, and the MECI point (see Fig. 6) lies higher in energy by 4 kcal mol^{-1} with respect to the corresponding transition state. Furthermore, the S_1 potential energy surface is found to

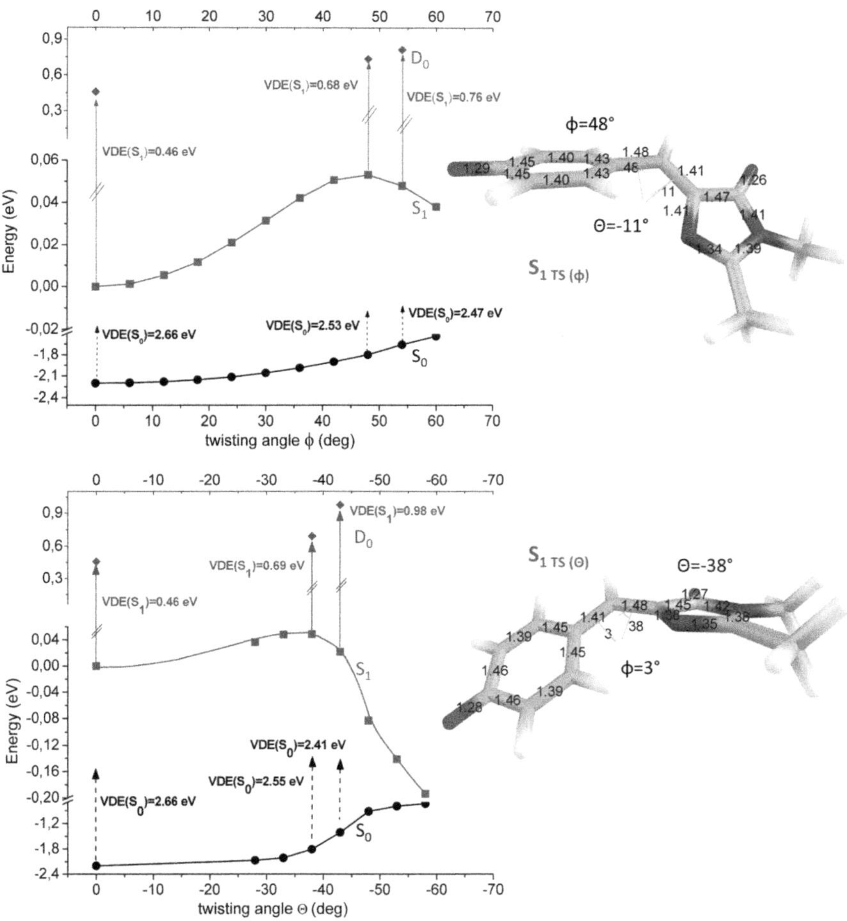

Fig. 5 S_0, S_1 and D_0 energies as a function of the twisting angles, ϕ and θ. Each point is obtained through constrained geometry optimization in S_1 at the XMCQDPT2/SA(2)-CASSCF(14,13)/(aug)-cc-pVDZ level of theory, where the corresponding angle is held fixed, while all the other coordinates are allowed to relax. The first point, which corresponds to the planar configuration, is obtained through entirely unconstrained geometry optimization. The energies of the D_0 and S_0 states are calculated in the S_1 optimized structures. To the right, the equilibrium geometry parameters of the corresponding transition states in S_1 are shown. All distances and angles are depicted in Å and degrees, respectively.

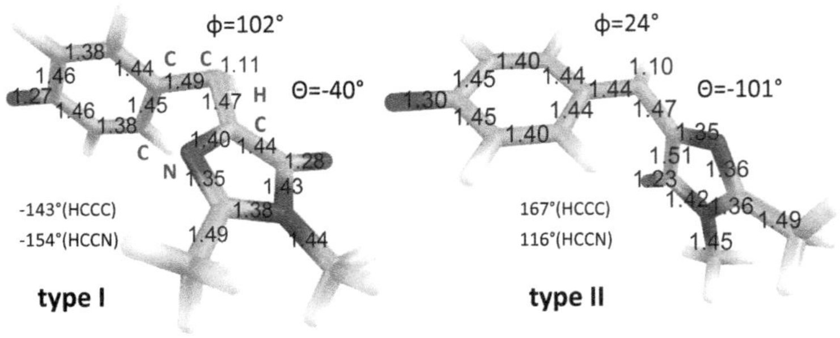

Fig. 6 Geometry parameters of the type I S_1/S_0 MECI (left) and the type II S_1/S_0 MECI (right). The type I exhibits a hula-twist geometry for the bare anion. Note, this type of the S_1/S_0 conical intersections lies higher in energy than the transition state, which leads out of the fluorescent state in S_1. The type II is the lowest-energy MECI, which lies 0.2 eV below the planar minimum in S_1. For this type, small variations of the angle ϕ belong to the conical intersection seam. Due to symmetry consideration, two MECI structures of each type exist. All distances and angles are depicted in Å and degrees, respectively.

be very steep with respect to the out-of-plane twisting coordinate θ, which leads to the type II CI (see Fig. 5). This appears to be consistent with a weak dependence on medium viscosity,[30] since the substantial energy gain enables to overcome solvent friction upon twisting.

3.2 PD/IC branching ratio

The PD and IC rate constants, their branching ratio and the excited-state lifetime may be estimated using the RRKM/QET quasi-equilibrium theory. The average excited-state lifetime equals the inverse of the effective rate constant, which is a sum of the PD and IC microcanonical unimolecular rate constants. The RRKM/QET rate constant is given by the following expression:

$$k = \frac{1}{h}\frac{W^{\ddagger}\left(E - E_0^{\ddagger}\right)}{\rho(E)}, \tag{1}$$

where $W^{\ddagger}$ is the sum of energy levels at the transition state, ρ is the density of states for the minimum-energy structure, $E_0^{\ddagger}$—the activation energy, including zero point vibrational energy correction and h is the Planck constant. The total internal energy E in S_1 available to the system after one-photon absorption is estimated as a sum of vibrational energy in the ground state at 300 K, 0.31 eV, and the difference in the photon energy and the vertical excitation energy (2.52 eV). The exact sums and densities of harmonic oscillator eigenstates are calculated using the direct count algorithm of Beyer and Swinehart,[56] assuming the same set of frequencies in S_1 as in the ground state. The sum of energy levels at the TS is calculated in the reduced dimension, where the reaction mode, out-of-plane twisting at 64.8 cm^{-1}, is excluded. This mode corresponds to twisting about the C–C bond adjacent to the phenol ring. Given that there are two alternative pathways to the S_1/S_0 conical intersection that refer to twisting about one or the other C–C bridge bonds, the IC rate constant is multiplied by a factor of two. The estimated IC lifetimes as a function of the excitation wavelength are shown in Fig. 7.

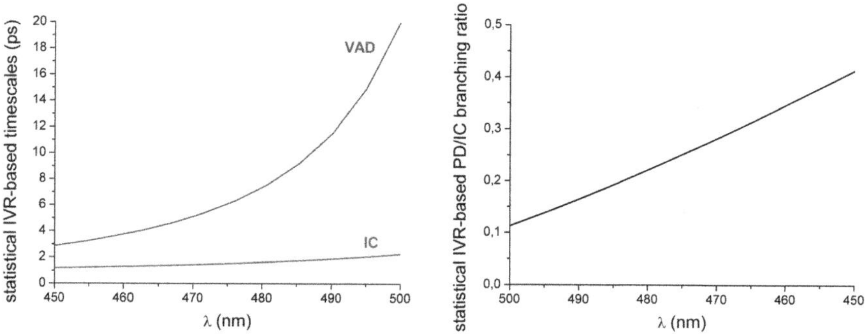

Fig. 7 Calculated statistical VAD and IC lifetimes (left) and their branching ratio (right) as a function of the excitation wavelength.

The VAD rate is determined by the energy transfer to the lowest-frequency emission-active mode of 849 cm^{-1} (0.105 eV). The amount of energy required for VAD corresponds to the 0–1 excitation energy for this mode. This enables a nearly perfect resonance between the S_1 ($\nu = 1$) and D_0 ($\nu = 0$) states (see Fig. 2). The rate constant for the energy transfer may then be statistically estimated using eqn (1). It is proportional to the total number of vibrational states, where the emission-active mode is excited at least to the first level. And it is reciprocally proportional to the density of all states at energy E. Note that the VAD activation energy is two times higher than that of IC. The VAD lifetimes and the PD/IC branching ratio as a function of the excitation wavelength are shown in Fig. 7.

What should specifically be stressed here is that the timescales are very sensitive to the excitation wavelength, since the PD mechanism changes upon increase in photon energy.[6] The timescales discussed above are relevant only at low energies, when both processes are IVR-mediated in the vicinity of the planar minimum in S_1 (governed by the transition and pseudo-transition states crossing). At higher energies, the emission active modes are directly excited, and electron-emission dynamics may occur within a timescale of vibrational decoherence. Support to this comes from the experiment,[6,16] where the IC channel essentially vanishes at energies higher than 2.75 eV (450 nm). To completely surpass IC, PD should be ultrafast and non-statistical. The photoresponse of the GFP chromophore may also be directed towards IC by a vibrational pre-excitation of the IC-active twisting modes. This regime can be explored by sequential pumping of two low-energy photons. In this case, the low-frequency IC-active modes 'statistically' bear enough energy in the ground electronic state after one-photon absorption to overcome the barrier in S_1 after the second photon absorption.

3.3 Tuning by proteins

Two characteristic timescales may be outlined in the photo-initiated dynamics of the GFP chromophore anion at low excitation energies: a sub-picosecond time-scale, when the early-time adiabatic nuclear dynamics in S_1 defines the spectral shape of photoabsorption and a picosecond timescale of the non-adiabatic dynamics in the excited-state decay channels. We discuss below a remarkable similarity between the early-time dynamics in the gas phase and in the protein and outline the implications of our gas-phase findings to the photoresponse of the proteins.

3.3.1 Photoinduced excited-state nuclear dynamics and spectral shapes. The active FC modes of the deprotonated chromophore in the gas phase and in the protein are shown in Fig. 8. To the left, the graphs represent the so-called Huang-Rhys factors S_i. To the right, the graphs show the contributions from each normal mode to the overall band width estimated based on the short-time approximation of the Lax's autocorrelation function[45] that produces a nearly Gaussian absorption band.[46] The Gaussian variance parameter accounts for temperature, the corresponding origin shifts, and frequencies ν_i:

$$(\delta\nu)^2 = \sum_i (\delta\nu_i)^2 = \sum_i S_i\nu_i^2(2\bar{n}_i + 1), \tag{2}$$

where $\bar{n}_i$ is the thermal average occupation number of the i-th mode. It is worth noting that the major contribution to the width comes from the high-frequency modes, while those with low frequencies promote spectral blurring. All most active FC modes refer to in-plane vibrations of the conjugated part of the chromophore both in the gas phase and in the protein. In the latter case, there are also certain modes with non-zero ground-to-excited state minimum displacements attributed to changes in the protein as a response to perturbation within the chromophore. Two groups of modes contribute to the environmental rearrangements: low-frequency modes and deformation vibrations with frequencies below 1600 cm^{-1} of the –OH and –NH groups that are involved in the hydrogen bonding network in the binding pocket of the chromophore.

Remarkably, the high-frequency stretching modes play the same role in the early-time excited-state nuclear dynamics that determines the corresponding spectral shapes both in the gas phase and in the protein (see Fig. 8). These modes

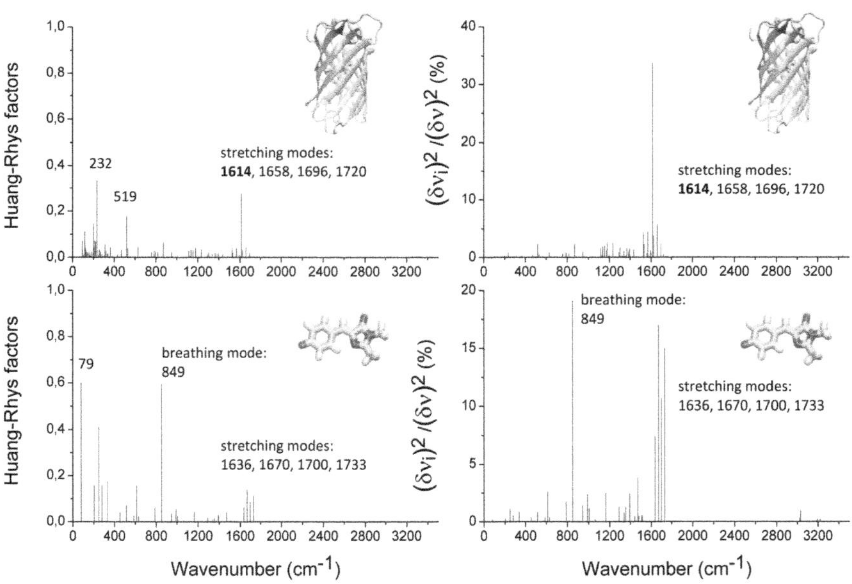

Fig. 8 Active Franck–Condon modes for the $S_0 \rightarrow S_1$ transition in the bare deprotonated GFP chromophore (lower panel) and in the S65T-GFP protein (upper panel). The Huang-Rhys factors, which are defined as squares of origin shifts in the dimensionless normal coordinates divided by a factor of two, and normal mode contributions to the spectral width at 0 K are shown.

define the spectral width, while the large-amplitude lower-frequency bending and breathing modes are, to a large extent, arrested in the protein due to the steric constraints imposed by the environment. The calculated vertical excitation energy is red-shifted by only 0.1 eV in the S65T GFP protein compared to the gas phase, and this is fully consistent with the experimental findings and earlier theoretical predictions.[57] Note that there are small deviations in the absorption maxima of various GFP variants. In particular, the maximum of the wild-type GFP (471 nm at 77 K and 477 nm at 300 K)[58] is blue-shifted as compared to those of the S65T structure (489 nm)[47,59] and of the enhanced S65T/F64L GFP mutant (492 nm)[60] at 300 K.

The time-domain approach used here for simulating photoabsorption profiles also allows one to get valuable insights into the early-time excited state dynamics, such as quantum coherence loss and media response timescales,[61] thus providing a link between the protein dynamics and decoherence rate in the photoexcited chromophore inside the protein. Here, we focus on the excited-state evolution of the chromophore in the gas phase and in the protein and on the associated dynamics. By analyzing the decay of the absolute value of the Lax's autocorrelation function (AF) in time shown in Fig. 9, we identify three characteristic timescales that refer to initial downhill movement of the vibrational wave packet along

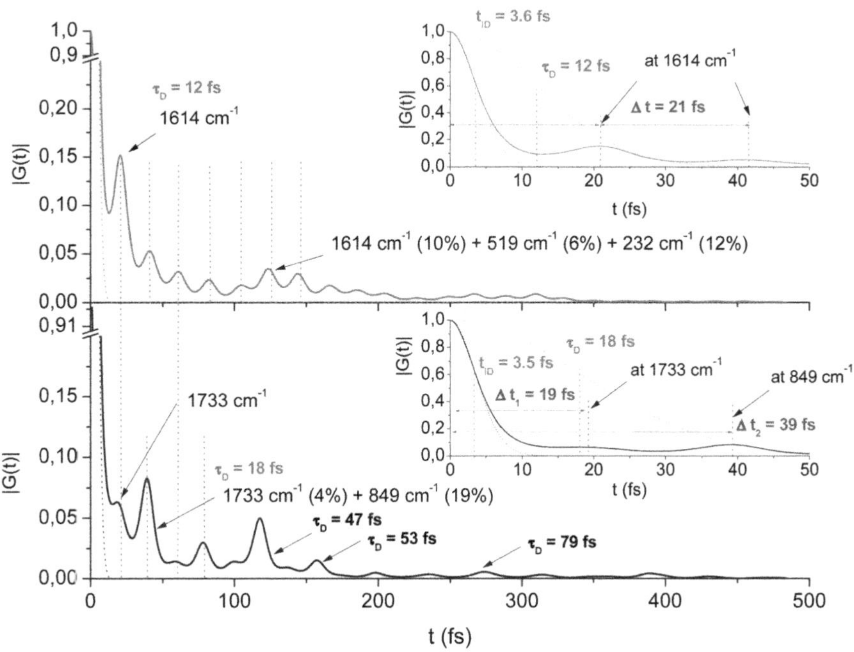

Fig. 9 Decay of the absolute values of Lax's autocorrelation functions in time in the gas phase (lower panel) and in the protein (upper panel) at 0 K. The total propagation time is 0.5 ps. The initial Gaussian AF decays are shown in the insets. Three characteristic timescales can be indicated in the early-time excited-state nuclear dynamics: the initial AF decay time, t_{ID}, the decoherence times, τ_D, and the times Δt, at which recurrence in the time domain occurs. These characteristic timescales may straightforwardly be correlated with the vibrational structure of the photoabsorption band. The decoherence times estimated assuming a Gaussian damping of the subsequent overlap peaks are used as labels for the corresponding vibrational modes in the AF decay.

the path of steepest descent (t_{ID}), the coherence decay while the system wanders in the vicinity of the excited-state minimum (τ_D), and the entire vibrational relaxation associated here with almost complete decay of AF at early times used to define a time cutoff.

The initial AF falloff is described by a Gaussian function in the short-time approximation, $|G(t)| \sim \exp(-t^2/2t_{ID}^2)$ (see Fig. 9), and its Fourier transform produces a nearly Gaussian vibrationally unresolved shape of the photo-absorption band[46] (see Fig. 10). The characteristic time of the Gaussian decay is straightforwardly connected with the Gaussian variance parameter in the frequency domain $t_{ID}^2 \sim 1/(\delta\nu)^2$ (see also eqn (2).) The initial AF decay occurs on a timescale of 3.5 fs, while the spectrum can be obtained within a time of ~500 fs at 0 K.

The coherence loss rates τ_D that refer to the Franck–Condon active modes important in the excited-state decay/reaction channels are determined by the sensitivity of the 'spectator' bath modes evolution to the states of these active modes in the chromophore. These bath modes may well be attributed to the low-frequency modes of the chromophore itself, but also include the modes of the protein. The larger the 'bath', the faster is the decoherence. Even the bare GFP chromophore has quite a number of low-frequency spectator modes (Fig. 8) that are sensitive to the excitation of the chromophore and respond to such a

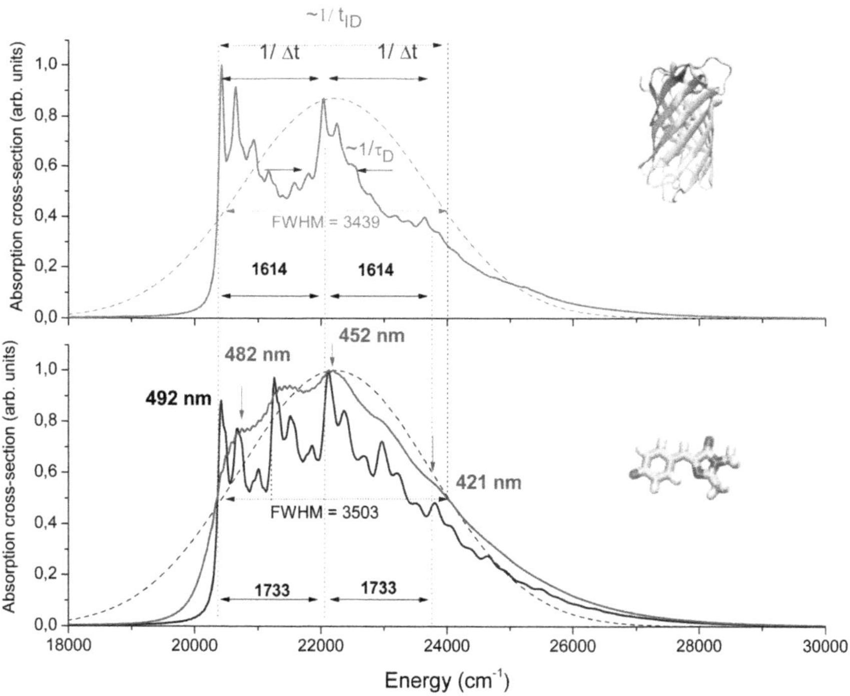

Fig. 10 The calculated S_0–S_1 absorption profiles of the deprotonated GFP chromophore (lower panel) and the S65T-GFP protein (upper panel) at 0 K. The nearly structureless profile is also shown for the bare chromophore at 300 K. The Gaussian-like envelopes, which refer to the initial AF decay time, are depicted as the dashed lines. The structure in the frequency domain is correlated with the characteristic timescales in the early-time excited-state nuclear dynamics. Note the similar full widths at half maximum (FWHM) in the gas phase and in the protein.

perturbation through the corresponding nuclear rearrangements. Their excited-state evolution considerably increases decoherence rates and also results in an inherently poor resolution of the absorption band of the GFP chromophore even at 0 K (Fig. 10). The characteristic time of the coherence loss, which is estimated assuming a Gaussian damping of the subsequent overlap peaks in the AF decay, occurs within ~18 fs at this temperature, while it proceeds faster (~9.5 fs) at room temperature, where the low-frequency bath modes also bear considerable amount of energy, or, in other words, more bath states "interact" with the quantum states of interest.

Remarkably, the vibrational decoherence rate defines a characteristic time for the non-statistical electron emission that occurs above ~2.75 eV, where the emission active modes are directly excited upon photoabsorption and the competing IC channel essentially vanishes.[6] The electron emission dynamics thereby becomes coupled with the vibrational coherence. PD from the emission-active modes should then proceed within a decoherence timescale, otherwise the energy deposited into the active modes is most likely dissipated through couplings to the bath modes upon the excited-state evolution. At room temperature, the coherence is lost on a timescale of ~10 fs, thus determining the timescale of the ultrafast electron dynamics.

The absorption maximum in the protein appears to be red-shifted as compared to the one in the gas phase (see Fig. 10), while the vertical excitation energies are strikingly similar in both media, as discussed above. However, precautions should apply, when trying to interpret the shift in the absorption maximum. It must be stressed that the absorption profile does not necessarily peak at the wavelength of the main transition with the largest relative FC strength that defines vertical excitation energy. This is a well-known fact, but is often overlooked, since such shifts are usually small (300–500 cm^{-1}) and originate from a set of transitions to the states where low-frequency active modes are excited in addition to the main transition. The relative strengths of such transitions are smaller, but their density is obviously higher. This has previously been discussed.[62] In contrast to this, the shift disclosed here is much larger. The maximum rather refers to the transitions where the high-frequency modes are excited. This is not usually expected, since each subsequent higher-energy transition, also accompanied by the same low-frequency satellites as the main transition, has a smaller relative strength according to the corresponding FC progression. The non-trivial aspect here is that there is an accidental degeneracy of the two sets of the active modes, breathing at 849 cm^{-1} and stretches at 1600–1700 cm^{-1} in the case of the bare chromophore. The former breathing mode has a long FC progression, such that its 0–2 excitation coincides in energy and in the corresponding strengths with those of the 0–1 excitation for the high-frequency modes. This causes a peak in the density of states of the active modes, and the maximum in the absorption is blue-shifted down to 452 nm. The breathing mode is arrested in the protein (see Fig. 8) and, hence, absorption peaks closely to the main vibronic transition.

3.4 Photoresponse of the GFP proteins

The photoabsorption profiles and the early-time excited-state nuclear dynamics of the anionic GFP chromophore are strikingly similar in the gas phase and in the protein. Here, the high-frequency stretching vibrational modes play a crucial role. The mode-specificity of the intrinsic non-adiabatic photoresponse of the chromophore[6] should then have a direct implication to the functioning of the GFP proteins.

The well-established paradigm in the functioning of the fluorescent proteins is the IC suppression by the protein environment. The intrinsic IC mechanism introduced here highlights the existence of the two alternative twisted intramolecular charge transfer[30] pathways, which importantly are non-equivalent. This suggests that the following factors may be altered in order to block (or enhance) the radiationless channel:

- The electron affinity of the phenol and imidazolinone rings, which may be selectively changed by the nearby charged aminoacid residues, such as Arg 96. This positively charged residue stabilizes the negative charge on the imidazolinone ring, thus resulting in the destabilization of the lowest-energy type II CI (see Fig. 4).
- The energy gap between the frontier orbitals of the corresponding neutral counterparts, which may be altered by increasing the conjugation in the imidazolinone part, as, for example, seen in the red variants of the fluorescent proteins, including the photoswitchable kindling fluorescent proteins.
- The barrier heights for twisting, which are apparently changed in wild-type GFP through an adjustment of the hydrogen bonding network around the chromophore.

A new emerging scenario[31] is based on the active role of the GFP proteins, containing tyrosine-based chromophores, in the photochemical reactions. Remarkably, the modes that are excited upon photoabsorption should also be most active in the possible electron transfer reactions involving the GFP oxidation, since the S_1/D_0 minimum displacements are large along these high-frequency stretching modes, similar to the gas phase.[6] The electron transfer through resonant tunneling, involving GFPs as electron donors, should then be accompanied by nuclear rearrangements along these particular modes.

The biological function is closely connected to the ways by which the protein environment tunes the intrinsic photoresponse of their chromophores and guides their excited-state evolution. Many GFP proteins prohibit internal conversion of their chromophores, thus possibly directing the decay towards the fast electron transfer governed by the most Franck–Condon active stretching modes. Remarkably, the long timescale fluorescence may then be regarded as a side channel[31,63] only enabled in the absence of relevant electron acceptors.

4 Conclusions

Here, we have provided insight into the efficiency of different vibrational modes in promoting a selective photoresponse in the bare GFP chromophore anion. We have introduced a general theoretical model that is capable of accounting for the alternative non-equivalent IC pathways, and we have outlined factors by which the photo-initiated response may be altered in this channel. The topography around the planar minimum in S_1 obtained through the high-level *ab initio* calculations provides direct support to the proposed qualitative model. Our results are consistent with the experimental data.

We have explored the interplay between the intrinsic ultrafast competing excited-state decay channels. At low excitation energies, statistical IVR-mediated reaction channels may compete with each other on a picosecond timescale in the case of the isolated GFP chromophore anion. IC is a predominant channel at low excitation energies, which is consistent with the most recent experimental findings.[6] There are alternative mode-selective ways to control the photoresponse and to direct it towards a single excited-state decay channel. By tuning the excitation wavelength, the photoresponse may be directed towards the ultrafast non-statistical electron emission coupled with vibrational (de)coherence, whereas a vibrational pre-excitation in the ground state may lead to the ultrafast non-statistical internal conversion through a conical intersection. Effective vibrational pre-excitation of the twisting modes in the ground electronic states may be achieved by sequential photon pumping at low excitation energies above 480 nm. We have then discussed the implication of our results to the photo-initiated non-adiabatic dynamics in the proteins and the ways, by which the GFP-like proteins may use the intrinsic dual electron-to-nuclei coupling to promote a selective photoresponse.

The anionic wild-type GFP chromophore defines an entire class of naturally occurring chromophores which are based on the oxydized tyrosine side chain. Remarkably, the tyrosine-based chromophores are found in various fluorescent and photoactive proteins, like GFPs of diverse origin as well as Photoactive Yellow Protein. Since these proteins have very diverse biological functions, their structurally similar chromophores represent a fascinating class for getting a deep insight into the intrinsic properties and photo-initiated quantum dynamics of their light-absorbing molecules. This will enable to disclose their similarities, which are anticipated based on the common structural motif of the oxidized tyrosine side chain, as well as to outline possible distinctions in their intrinsic photo-behavior and the ways by which proteins may use this for tuning their photoresponse.

Acknowledgements

The work is supported by a Marie Curie European Career Integration Grant within the 7th European Community Framework Programme. A.V.B. also acknowledges support from the Russian Foundation for Basic Research (grant No.11-03-01214), as well as the Supercomputing Center of the M.V. Lomonosov Moscow State University for providing high-performance computing resources. This work was also granted access to the HPC resources of the High Performance Computing Center Stuttgart, Germany made available within the Distributed European Computing Initiative by the PRACE-2IP, receiving funding from the European Community's Seventh Framework Programme (FP7/2007–2013) under grant agreement RI-283493.

References

1 I.-R. Lee, W. Lee and A. H. Zewail, *Proc. Natl. Acad. Sci. U. S. A.*, 2006, **103**, 258–262.
2 C. Fang, R. R. Frontiera, R. Tran and R. A. Mathies, *Nature*, 2009, **462**, 200–204.
3 D. Polli, P. Altoe, O. Weingart, K. M. Spillane, C. Manzoni, D. Brida, G. Tomasello, G. Orlandi, P. Kukura, R. A. Mathies, M. Garavelli and G. Cerullo, *Nature*, 2010, **467**, 440–443.

4 E. Collini, C. Y. Wong, K. E. Wilk, P. M. G. Curmi, P. Brumer and G. D. Scholes, *Nature*, 2010, **463**, 644–647.
5 G. D. Scholes, G. R. Fleming, A. Olaya-Castro and R. van Grondelle, *Nat. Chem.*, 2011, **3**, 763–774.
6 A. V. Bochenkova, J. Rajput, D. B. Rahbek, B. Klærke and L. H. Andersen, unpublished work.
7 R. Y. Tsien, *Annu. Rev. Biochem.*, 1998, **67**, 509–544.
8 M. Ormö, A. B. Cubitt, K. Kallio, L. A. Gross, R. Y. Tsien and S. J. Remington, *Science*, 1996, **273**, 1392–1395.
9 F. Yang, L. G. Moss and G. N. Phillips, *Nat. Biotechnol.*, 1996, **14**, 1246–1251.
10 D. Chudakov, M. Matz, S. Lukyanov and K. Lukyanov, *Physiol. Rev.*, 2010, **90**, 1103–1163.
11 L. M. Tolbert, A. Baldridge, J. Kowalik and K. M. Solntsev, *Acc. Chem. Res.*, 2012, **45**, 171–181.
12 M. Chattoraj, B. King, G. Bublitz and S. Boxer, *Proc. Natl. Acad. Sci. U. S. A.*, 1996, **93**, 8362–8367.
13 C. Fang, R. R. Frontiera, R. Tran and R. A. Mathies, *Nature*, 2009, **462**, 200–204.
14 V. Sample, R. H. Newman and J. Zhang, *Chem. Soc. Rev.*, 2009, **38**, 2852–2864.
15 D. Shcherbo, I. Shemiakina, A. Ryabova, K. Luker, B. Schmidt, E. Souslova, T. Gorodnicheva, L. Strukova, K. Shidlovskiy, O. Britanova, A. Zaraisky, K. Lukyanov, V. Loschenov, G. Luker and D. Chudakov, *Nat. Methods*, 2010, **7**, 827–829.
16 M. W. Forbes and R. A. Jockusch, *J. Am. Chem. Soc.*, 2009, **131**, 17038–17039.
17 A. Toniolo, S. Olsen, L. Manoha and T. J. Martinez, *Faraday Discuss.*, 2004, **127**, 149–163.
18 M. E. Martin, F. Negri and M. Olivucci, *J. Am. Chem. Soc.*, 2004, **126**, 5452–5464.
19 I. V. Polyakov, B. L. Grigorenko, E. M. Epifanovsky, A. I. Krylov and A. V. Nemukhin, *J. Chem. Theory Comput.*, 2010, **6**, 2377–2387.
20 A. Kummer, C. Kompa, H. Niwa, T. Hirano, S. Kojima and M. Michel-Beyerle, *J. Phys. Chem. B*, 2002, **106**, 7554–7559.
21 D. Mandal, T. Tahara and S. Meech, *J. Phys. Chem. B*, 2004, **108**, 1102–1108.
22 R. Gepshtein, D. Huppert and N. Agmon, *J. Phys. Chem. B*, 2006, **110**, 4434–4442.
23 W. Weber, V. Helm, J. Mccammon and P. Langhoff, *Proc. Natl. Acad. Sci. U. S. A.*, 1999, **96**, 6177–6182.
24 S. Olsen and R. McKenzie, *J. Chem. Phys.*, 2009, **130**, 184302.
25 R. S. H. Liu and G. S. Hammond, *Chem.-Eur. J.*, 2001, **7**, 4536–4545.
26 K. Solntsev, O. Poizat, J. Dong, J. Renault, Y. Lou, C. Burda and L. Tolbert, *J. Phys. Chem. B*, 2008, **112**, 2700–2711.
27 K. L. Litvinenko, N. M. Webber and S. R. Meech, *J. Phys. Chem. A*, 2003, **107**, 2616–2623.
28 K. B. Bravaya, B. L. Grigorenko, A. V. Nemukhin and A. I. Krylov, *Acc. Chem. Res.*, 2012, **45**, 265–275.
29 K. A. Lukyanov, D. M. Chudakov, S. Lukyanov and V. V. Verkhusha, *Nat. Rev. Mol. Cell Biol.*, 2005, **6**, 885–890.
30 S. R. Meech, *Chem. Soc. Rev.*, 2009, **38**, 2922–2934.
31 A. M. Bogdanov, A. S. Mishin, I. V. Yampolsky, V. V. Belousov, D. M. Chudakov, F. V. Subach, V. V. Verkhusha, S. Lukyanov and K. A. Lukyanov, *Nat. Chem. Biol.*, 2009, **5**, 459–461.
32 S. B. Nielsen, A. Lapierre, J. U. Andersen, U. V. Pedersen, S. Tomita and L. H. Andersen, *Phys. Rev. Lett.*, 2001, **87**, 228102.
33 A. A. Pakhomov and V. I. Martynov, *Chem. Biol.*, 2008, **15**, 755–764.
34 L. H. Andersen, H. Bluhme, S. Boye, T. J. D. Jørgensen, H. Krogh, I. B. Nielsen, S. B. Nielsen and A. Svendsen, *Phys. Chem. Chem. Phys.*, 2004, **6**, 2617–2627.
35 T. Tanabe, M. Saito and K. Noda, *Eur. Phys. J. D*, 2011, **62**, 191–195.
36 M. W. Forbes, A. M. Nagy and R. A. Jockusch, *Int. J. Mass Spectrom.*, 2011, **308**, 155–166.
37 K. Chingin, R. M. Balabin, V. Frankevich, K. Barylyuk, R. Nieckarz, P. Sagulenko and R. Zenobi, *Int. J. Mass Spectrom.*, 2011, **306**, 241–245.
38 D. A. Horke and J. R. R. Verlet, *Phys. Chem. Chem. Phys.*, 2012, **14**, 8511–8515.
39 C. R. S. Mooney, M. E. Sanz, A. R. McKay, R. J. Fitzmaurice, A. E. Aliev, S. Caddick and H. H. Fielding, *J. Phys. Chem. A*, 2012, **116**, 7943–7949.
40 Y. Toker, D. B. Rahbek, B. Klærke, A. V. Bochenkova and L. H. Andersen, *Phys. Rev. Lett.*, 2012, **109**, 128101.
41 A. A. Granovsky, *J. Chem. Phys.*, 2011, **134**, 214113.
42 A. A. Granovsky, Firefly, http://classic.chem.msu.su/gran/firefly/index.html.
43 M. W. Schmidt, K. K. Baldridge, J. A. Boatz, S. T. Elbert, M. S. Gordon, J. H. Jensen, S. Koseki, N. Matsunaga, K. A. Nguyen, S. Su, T. L. Windus, M. Dupuis and J. A. Montgomery, *J. Comput. Chem.*, 1993, **14**, 1347–1363.
44 E. J. Heller, *Acc. Chem. Res.*, 1981, **14**, 368–375.
45 M. Lax, *J. Chem. Phys.*, 1952, **20**, 1752–1760.

46 P. V. Yurenev, M. K. Kretov, A. V. Scherbinin and N. F. Stepanov, *J. Phys. Chem. A*, 2010, **114**, 12804–12812.
47 R. Heim, A. Cubitt and R. Tsien, *Nature*, 1995, **373**, 663–664.
48 M. Ormö, A. Cubitt, K. Kallio, L. Gross, R. Tsien and S. Remington, *Science*, 1996, **273**, 1392–1395.
49 http://www.rcsb.org/.
50 A. MacKerell Jr, D. Bashford, M. Bellott, R. Dunbrack Jr, J. Evanseck, M. Field, S. Fischer, J. Gao, H. Guo, S. Ha, D. Joseph-McCarthy, L. Kuchnir, K. Kuczera, F. Lau, C. Mattos, S. Michnick, T. Ngo, D. Nguyen, B. Prodhom, W. Reiher III, B. Roux, M. Schlenkrich, J. Smith, R. Stote, J. Straub, M. Watanabe, J. Wiorkiewicz-Kuczera, D. Yin and M. Karplus, *J. Phys. Chem. B*, 1998, **102**, 3586–3616.
51 W. Humphrey, A. Dalke and K. Schulten, *J. Mol. Graphics*, 1996, **14**, 33–38.
52 J. Phillips, R. Braun, W. Wang, J. Gumbart, E. Tajkhorshid, E. Villa, C. Chipot, R. Skeel, L. Kalé and K. Schulten, *J. Comput. Chem.*, 2005, **26**, 1781–1802.
53 E. D. Glendening, J. K. Badenhoop, A. E. Reed, J. E. Carpenter, J. A. Bohmann, C. M. Morales and F. Weinhold, *NBO 5.G., Theoretical Chemistry Institute*, University of Wisconsin, Madison, WI, http://www.chem.wisc.edu/nbo5.
54 P. Day, J. Jensen, M. Gordon, S. Webb, W. Stevens, M. Krauss, D. Garmer, H. Basch and D. Cohen, *J. Chem. Phys.*, 1996, **105**, 1968–1986.
55 J. Simons, *J. Phys. Chem. A*, 2008, **112**, 6401–6511.
56 S. E. Stein and B. S. Rabinovitch, *J. Chem. Phys.*, 1973, **58**, 2438–2445.
57 K. B. Bravaya, M. G. Khrenova, B. L. Grigorenko, A. V. Nemukhin and A. I. Krylov, *J. Phys. Chem. B*, 2011, **115**, 8296–8303.
58 H. Lossau, A. Kummer, R. Heinecke, F. Pöllinger-Dammer, C. Kompa, G. Bieser, T. Jonsson, C. Silva, M. Yang, D. Youvan and M. Michel-Beyerle, *Chem. Phys.*, 1996, **213**, 1–16.
59 K. Brejc, T. Sixma, P. Kitts, S. Kain, R. Tsien, M. Ormö and S. Remington, *Proc. Natl. Acad. Sci. U. S. A.*, 1997, **94**, 2306.
60 H. Hosoi, S. Yamaguchi, H. Mizuno, A. Miyawaki and T. Tahara, *J. Phys. Chem. B*, 2008, **112**, 2761–2763.
61 O. V. Prezhdo and P. J. Rossky, *Phys. Rev. Lett.*, 1998, **81**, 5294–5297.
62 E. Kamarchik and A. I. Krylov, *J. Phys. Chem. Lett.*, 2011, **2**, 488–492.
63 K. A. Lukyanov, E. O. Serebrovskaya, S. Lukyanov and D. M. Chudakov, *Photochem. Photobiol. Sci.*, 2010, **9**, 1301–1306.

Faraday Discussions RSCPublishing

PAPER

Excited state dynamics of Photoactive Yellow Protein chromophores elucidated by high-resolution spectroscopy and *ab initio* calculations

Eric M. M. Tan, Saeed Amirjalayer, Bert H. Bakker and Wybren J. Buma*

Received 5th December 2012, Accepted 21st December 2012
DOI: 10.1039/c2fd20139a

We report on experimental high-resolution spectroscopic studies in combination with advanced theoretical calculations that focus on the excited-state dynamics of various forms of the chromophore of the Photoactive Yellow Protein (PYP), and the dependence of these dynamics on conformational and isosteric structure, as well as the biological environment. Three-colour nanosecond multiphoton ionization pump–probe studies confirm and extend previous conclusions that the dominant decay channel of the lowest excited $\pi\pi^*$ state (the so-called V′ state) of methyl-4-hydroxycinnamate is picosecond internal conversion to the adiabatically lower $n\pi^*$ state, and enable us to resolve apparent contradictions with picosecond pump–probe studies. Comparison of multiphoton ionization and laser induced fluorescence excitation spectra leads to the assignment of the hitherto elusive excitation spectrum of the $V(\pi\pi^*)$ state. Complexation of methyl-4-hydroxycinnamate with water radically changes the excited-state dynamics; internal conversion to the $n\pi^*$ state is absent, and bond isomerization channels instead play a prominent role. Excited states of the thio-ester compound, the form in which the chromophore is present in PYP, have till the present study remained out of reach of gas-phase studies. The excitation spectra obtained here show a broad, almost structureless band system, giving evidence for enhanced nonradiative decay channels. The gas-phase results will be discussed in the context of results from ultrafast studies on these two chromophores in solution.

Introduction

Light is used ubiquitously in nature to drive important processes such as photosynthesis, but also as a means to perceive and respond to environmental changes as occurs in vision, phototropism and phototaxis.[1] Common to all these processes is that absorption of light triggers conformational changes in a light-sensitive signalling protein that ultimately lead to a response at the cellular level. Nature uses to this

Van't Hoff Institute for Molecular Sciences, Faculty of Science, University of Amsterdam, Science Park 904, 1098 XH Amsterdam, The Netherlands. E-mail: W.J.Buma@uva.nl

purpose small chromophores embedded in the protein whose photochemical response leads to large-scale structural changes of the entire protein.

Photoisomerization is one of the most widely used methods to translate light absorption into biological function.[1] Ideally, excitation of the chromophore leads exclusively to the desired photoresponse. However, in practice, the electronic states that are accessed upon photon absorption are also subject to a large number of competing processes such as internal conversion to other electronically excited states, vibrational relaxation, and solvation relaxation. It is therefore important to determine how the geometrical and electronic structure of such chromophores changes upon light absorption, how these changes are influenced by specific groups and how these groups steer the chromophore towards an optimal biological functioning, and how coupling of the chromophore to the protein and a solvent influences its photochemical and photophysical properties.

The Photoactive Yellow Protein (PYP) is one of the proteins used in nature for light-induced signalling. PYP was originally discovered in the *Ectothiorhodospira halophila* bacterium.[2,3,4] It utilizes isomerisation of a chromophore derived from *trans*-4-hydroxycinnamic acid (*para*-coumaric acid (*p*CA)) to initiate a photocycle that ultimately leads to a negative phototaxis towards blue light,[5] thereby protecting the bacterium from its potential harmful effects. PYP is in many aspects an ideal system for the study of biological photoreceptors. It is relatively small (125 residues, 14 kDa), robust towards chemical changes, and can easily be crystallized. As a result, its *modus operandi* has been the subject of many experimental and theoretical studies.[6–9] Although nowadays a comprehensive description of the structural changes occurring during the photocycle of PYP has been acquired,[10] many aspects of the initial processes in the photoexcited states are still matter of debate. There is therefore considerable interest in the photochemical and photophysical properties of the PYP chromophore, and, in particular, on how to relate the properties of the isolated chromophore to its properties when it is part of the protein.

The most simple representation of the PYP chromophore is *p*CA. Gas-phase studies of *p*CA have for a long time not been possible.[11–13] Recently, we succeeded in recording the first excitation and absorption spectra under supersonic beam conditions. These studies enabled the identification of the lowest electronically excited state and the geometry changes associated with electronic excitation.[14] *Ab initio* calculations on *p*CA[15–17] indicate that three states are important for the description of the lower excited singlet manifold: the V′($\pi\pi^*$) state, in a one-configuration picture described by the HOMO → LUMO + 1 excitation, the V($\pi\pi^*$) state described by the HOMO → LUMO excitation, and the nπ^* state described by the n → LUMO excitation. For vertical excitation from the ground state these states are respectively S_1, S_2, and S_3. However, calculation of the adiabatic excitation energies shows that in that case the nπ^* state has the lowest excitation energy, although the order of the two $\pi\pi^*$ states remains unchanged.[15]

In the protein *p*CA is (i) present in its anionic form, (ii) covalently linked to the Cys69 residue *via* a thioester bond, and (iii) forms hydrogen bonds on the phenolate side with the Glu46 and Tyr42 residues. Theoretical studies have therefore focused on how the properties of the lower excited singlet states depend on the anionic nature of the chromophore in the protein, the role of the sulphur atom, and the influence of hydrogen bonding interactions.[18–23] From these studies it has become clear that each of these aspects has its own consequences for the excited state manifold. Replacing the oxygen atom by a sulphur atom makes the nπ* state both vertically as well as adiabatically the lowest excited singlet state, and reverses the order of the V′(ππ*) and V(ππ*) states. In the anion of the thio-compounds the V(ππ*) state is the lowest excited singlet state, the energy differences with the other two excited states becoming much larger. Hydrogen bonding gives rise to pronounced energetic shifts.

In recent years a number of gas-phase spectroscopic studies have been performed aimed at elucidating the properties of *p*CA-related compounds more similar to the form of the chromophore in the protein, including gas-phase studies on the anion.[24–26] The ester functionality influences the structural and excited-state properties of the chromophore; studies of molecular systems like methyl-4-hydroxycinnamate (historically indicated as OM*p*CA, see Scheme 1) in which this functionality is incorporated are therefore important to assess the influence of the protein environment. LIF[27] and REMPI[28] studies on OM*p*CA show vibrationally resolved excitation spectra that indicate that the geometry changes of the molecule upon excitation are relatively small, in good agreement with theoretical predictions for the V′(ππ*) state. The LIF and REMPI studies show a striking influence of hydrogen bonding. In the OM*p*CA–H_2O cluster the origin transition to the V′(ππ*) state is shifted by 641 cm^{-1}, and its intensity is very much enhanced with respect to the bare molecule. REMPI studies on the V′(ππ*) state of *p*CA concluded that internal conversion to the nπ* state is an important decay channel.[14] The same conclusion was reached for the V′(ππ*) state of OM*p*CA.[28] The observed intensity enhancement of transitions in the water cluster has therefore been interpreted in

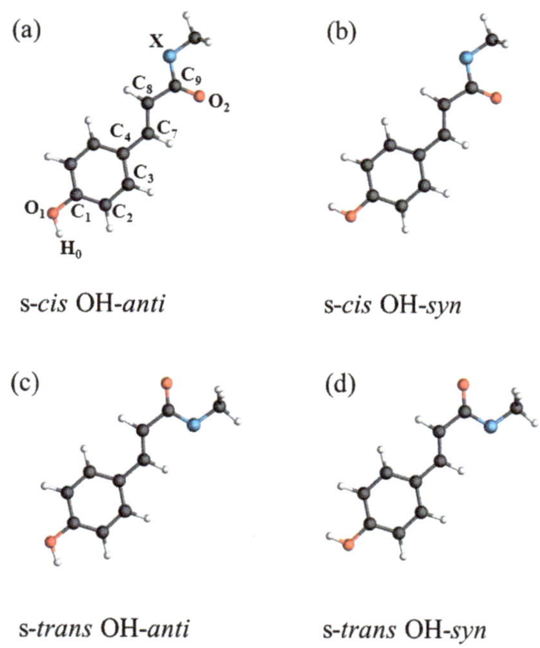

Scheme 1 Molecular structure of compounds studied and employed atom labelling. Methyl-4-hydroxycinnamate corresponds to X=O and is designated in the text as OM*p*CA, the thio ester corresponds to X=S and is referred to in the text as TM*p*CA. Drawn are the four low-energy conformations that have been identified in previous molecular beam experiments on OM*p*CA: (a) s-*cis* OH-*anti*, (b) s-*cis* OH-*syn*, (c) s-*trans* OH-*anti* and (d) s-*trans* OH-*syn*.

terms of a reduced internal conversion rate to the $n\pi^*$ state. Recent ps time-resolved REMPI studies on OM*p*CA and OM*p*CA–H_2O suggest on the other hand an explanation in terms of excited-state isomerisation.[29]

A principal goal of the present work is to determine the importance of the various possible decay channels of the $V'(\pi\pi^*)$ state of OM*p*CA. We furthermore want to determine how complexation with water influences the electronic properties of the chromophore, and how this in turn is reflected in the excited-state dynamics. In the protein the $V(\pi\pi^*)$ state is the lowest excited singlet state. Studies of its properties are therefore important. Although in the LIF study on OM*p*CA some bands were tentatively assigned as transitions to the $V(\pi\pi^*)$ state, its structural and dynamic properties have so far remained unclear. Here, we will identify its excitation spectrum and study its decay channels. Finally, we will study the methyl-4-hydroxycinnamate thio ester (TM*p*CA) and report for the first time its excitation spectrum under molecular beam conditions. Theory predicts that in TM*p*CA the $V(\pi\pi^*)$ state is the first excited $\pi\pi^*$ state. Yet, under solution conditions OM*p*CA and TM*p*CA show rather similar excited-state dynamics.[30,31] Comparison of the gas-phase excitation spectra of OM*p*CA and TM*p*CA will enable us to assess how the electronic manifold changes by sulphur substitution, and how solute–solvent interactions influence the excited-state dynamics.

Experimental details

In our studies two- and three-colour Resonance Enhanced MultiPhoton Ionization (REMPI) spectroscopy has been used to record excitation and depletion spectra of molecules cooled in a supersonic free jet expansion. The molecular beam setup employed for these measurements has been described in detail before.[32,33] Briefly, the setup makes use of a pulsed nozzle construction in which the sample is placed in an oven that is heated to the temperature required to generate a high enough vapour pressure of the sample. A molecular beam is then produced by mixing the vapour with a rare gas, in the present experiments Ne at a pressure of 2 bar, and expanding this mixture into a differentially pumped vacuum with a pulsed nozzle (General Valve Iota One system). The nozzle has an orifice of 0.5 mm and is kept 5 °C higher in temperature than the oven to avoid clogging of the nozzle. After expansion, the molecular beam is skimmed with a conical skimmer (2 mm) and enters the ionization chamber where mass-resolved ion detection is performed using a reflectron time-of-flight mass spectrometer (R.M. Jordan Co.). Clusters of chromophores with water are generated by mixing the carrier gas with small amounts of water vapour.

A two-colour (1 + 1′) Resonance Enhanced Two-Photon Ionization (RE2PI) scheme has been employed to record excitation spectra. In these experiments a frequency-doubled dye laser (Sirah Precision Scan) pumped by a 30 Hz Nd:YAG laser (Spectra Physics Lab 190) was used for electronic excitation. A series of dyes including DCM, Rhodamine 640, Rhodamine 610, Rhodamine 590, Pyromethen 597 and Fluorescein 548, in combination with frequency-doubling by a KDP crystal, enabled us to cover the frequency region from 30000 to 36500 cm^{-1}. Typical excitation energies were 2–3 mJ per pulse. The second colour was used to ionize electronically excited states and was produced by an ArF excimer laser (Neweks PSX-501) with typical pulse energies of 5 mJ per pulse. Both laser beams were introduced in a counter-propagating fashion into the spectrometer without further focusing. In the two-colour (1 + 1′) RE2PI experiments these two laser systems were synchronized in time using a delay generator (Stanford Research Systems DG535). The same delay generator was used to study the decay dynamics of excited states by recording the decay of the RE2PI signal as a function of the time delay between excitation and ionization.

To obtain conformation-specific spectra we have utilized UV-UV depletion spectroscopy.[34] UV-UV depletion spectroscopy monitors the signal from a probe laser that is fixed at the frequency of one particular band in the excitation spectrum, and is associated exclusively with a single conformation. Scanning the pump laser then leads to a depletion of this signal as a result of the depopulation of the ground state each time the pump laser is in resonance with a transition of the selected conformer. In the present study we used either one- or two-colour RE2PI to generate a probe signal. To generate the probe signal in the one-colour RE2PI scheme the Nd:YAG pumped dye laser system described above was used. In the two-colour RE2PI scheme electronic excitation occurred with a dye laser system consisting of a dye laser (Lumonics HD500) pumped by an excimer laser (Lambda Physik CompexPro 205) operating on XeCl. The output of this dye laser was frequency-doubled with a KDP or BBO crystal using an autotracking system (INRAD AutoTracker II), and was introduced co-linearly with the 193 nm ionization beam into the spectrometer. Typical pulse energies employed for $S_1 \leftarrow S_0$ excitation were 0.5–1.0 mJ. Depletion of the ground state took place in these three-colour experiments with the Nd:YAG pumped dye laser system. Depletion and probe beams were introduced in a counter-propagating fashion into the spectrometer with a typical time delay of 160 ns. For the absolute determination of the wavelength of both dye lasers a Burleigh WA-4500 wavemeter was used. Since in previous studies[27,28] the wavelength had been calibrated in an indirect manner, the excitation frequencies reported here differ slightly from those reported previously.

Methyl-4-hydroxycinnamate was obtained from ABCR with a purity of 95% and used without further purification. The thiomethyl ester derivative of *para*-coumaric acid was synthesised according to the procedure described previously.[35] In our experiments OM*p*CA and TM*p*CA the samples were heated up to 165 °C and 185 °C, respectively, in a glass bottle placed in the oven assembly.

Theoretical details

Using the spin-component scaled version of the approximate coupled-cluster singles-and-doubles model (SCS-CC2),[36,37] the vertical and adiabatic excitation energies for the excited states of interest of the investigated molecules systems were calculated. The geometry of each structure was optimized in C_s symmetry in the respective electronic state (S_0–S_3) using the RICC2 module[38,39] as implemented in the TURBOMOLE package.[40] For all calculations the cc-pVTZ basis set,[41] together with the standard auxiliary basis sets,[42] was employed.

Results and discussion

Excited state dynamics of OM*p*CA

The (1 + 1′) RE2PI excitation spectrum of OM*p*CA in the 32700–35700 cm^{-1} excitation region is shown in Fig. 1a. This spectrum shows a considerably better

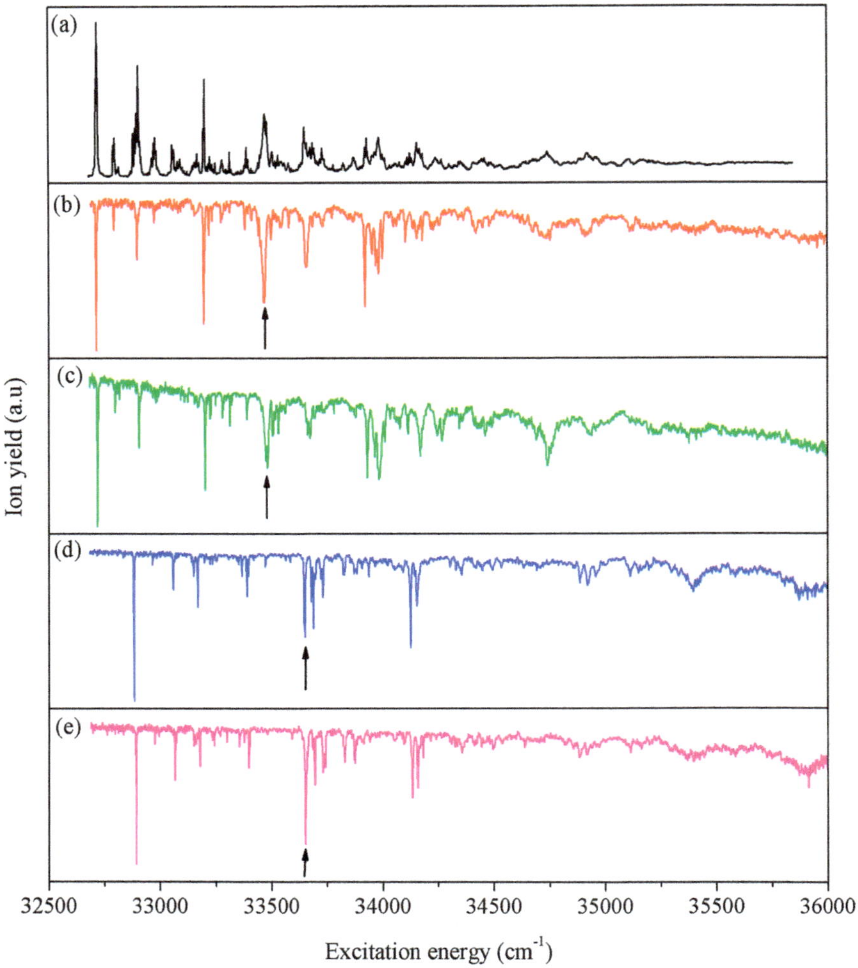

Fig. 1 (a) (1 + 1′) RE2PI spectrum of OM*p*CA and UV-UV depletion spectrum of the (b) s-*cis* OH-*syn* (c) s-*cis* OH-*anti* (d) s-*trans* OH-*anti* (e) s-*trans* OH-*syn* conformers. The bands indicated with a black arrow in the depletion spectra are assigned to the $S_2 \leftarrow S_0$ origin transitions (see text).

resolution than the previously reported one-colour RE2PI excitation spectrum.[28] Moreover, well-resolved vibrational activity is observed up to higher excitation energies. The increase in resolution can be attributed to the fact that the (1 + 1′) RE2PI signal is significantly larger than the (1 + 1) RE2PI signal under the same excitation intensity conditions. As a result, lower laser intensities can be used for the excitation step in the two-colour scheme as compared to the one-colour scheme. Indeed, comparison of the line widths in the LIF spectrum with the line widths in the one-colour RE2PI spectrum gave in the past reason to believe that it is hard to avoid partial saturation in one-colour RE2PI.[28] As will be argued below, the observation that the two-colour signal is much stronger than the one-colour signal provides direct evidence for the conclusion that the electronic state that is excited is not the same state from which ionization takes place.

OM*p*CA can adopt four different conformations under our molecular beam conditions (see Scheme 1). To investigate to what extent the excited-state dynamics depend on the finer details of the molecular structure, it is necessary to obtain the excitation spectrum of each of the conformers separately. Such conformation-specific excitation spectra obtained by UV-UV depletion spectroscopy are reported in Fig. 2b–2e for the s-*cis* OH-*syn*, s-*cis* OH-*anti*, s-*trans* OH-*anti* and s-*trans* OH-*syn* conformer, respectively, with $S_1(V'(\pi\pi^*)) \leftarrow S_0$ origin transitions located at 32711.0 ± 0.2 cm^{-1}, 32715.9 ± 0.2 cm^{-1}, 32879.5 ± 0.2 cm^{-1}, and 32887.7 ± 0.2 cm^{-1}. Interestingly, we find that the line widths of the transitions associated with the s-*cis* conformers are systematically larger than those of the s-*trans* conformers. As a result, the excitation spectra of the s-*trans* species show well-resolved bands up to higher excitation energies than the s-*cis* species.

The difference in line widths might indicate that the s-*cis* and s-*trans* conformers have different excited-state lifetimes. The spectral resolution of our experiments, however, does not enable us to observe single rotational transitions. The widths of the vibronic bands are thus in principle determined by a combination of homogeneous lifetime broadening and inhomogeneous rotational broadening, and do not allow us to determine the excited-state lifetime directly from the band widths. Comparison of the widths in the (1 + 1′) RE2PI spectrum with those in the UV-UV depletion spectra show that the bands in the former spectrum are systematically broader. For example, when the origin bands are fitted with a Lorentzian line shape we find in the UV-UV depletion spectra widths of 3.2 ± 0.1, 3.3 ± 0.2, 2.0 ± 0.1, and 2.3 ± 0.1 cm^{-1}, while in the (1 + 1′) RE2PI spectrum they have widths of 3.3 ± 0.2, 4.8 ± 0.1, 2.5 ± 0.3, and 2.7 ± 0.3 cm^{-1}.

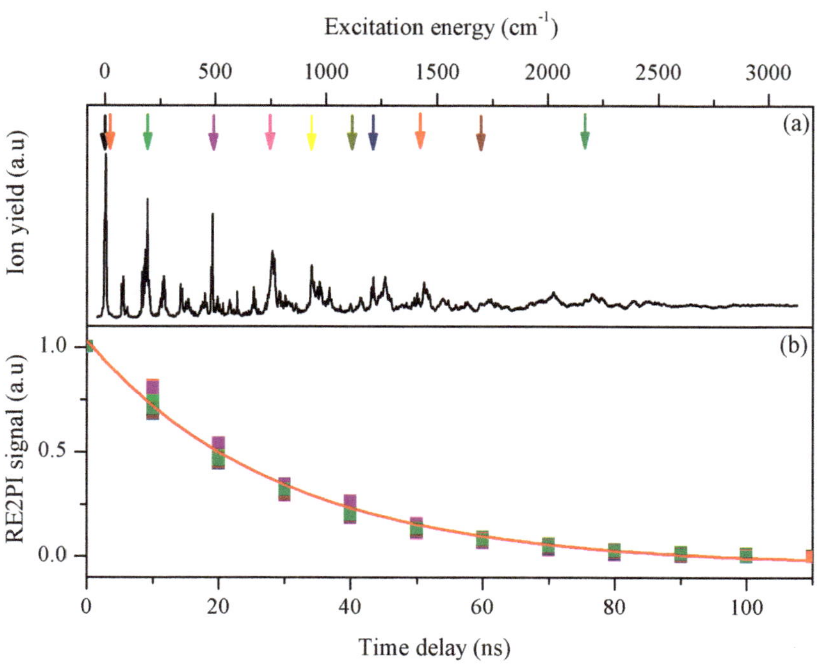

Fig. 2 (a) Excitation spectrum of OM*p*CA. The energy scale has been shifted to the excitation energy of the $S_1 \leftarrow S_0$ origin transition of the s-*cis* OH-*syn* conformer at 32711.0 cm^{-1}. (b) Normalized decays of the (1 + 1′) RE2PI signal at various excitation energies above the origin of the s-*cis* OH-*syn* conformer: 0.0, 4.3, 191.1, 486.1, 753.9, 936.4, 1111.7, 1204.6, 1417.6, 1703.1, and 2169.0 cm^{-1}.

This increase suggests that inhomogeneous rotational broadening plays a non-negligible role in determining the overall width. The spectral width of our laser is smaller than the line width. If the band is inhomogeneously broadened, the probe step in the UV-UV depletion experiments monitors selectively only part of the unresolved rotational transitions. This, in turn, leads to narrower lines when the pump frequency is scanned in depletion spectra than in the $(1 + 1')$ RE2PI spectrum where this selection is absent. We thus conclude that the lifetimes of 1.5 and 2.5 ps for the s-*cis* and s-*trans* conformations, respectively, as determined from the Lorentzian fits of the UV-UV depletion spectra should be considered as lower limits for the lifetime.

The conclusion that the lifetime of the $V'(\pi\pi^*)$ state is in the low picosecond regime is in agreement with time-resolved studies by picosecond UV-UV pump–probe spectroscopy.[29] In these studies it was reported that the lifetime of the s-*cis* and s-*trans* conformations is 8 and 9 ps, respectively, with an uncertainty of 3 ps.[43] Both the line width measurements and the time-resolved studies lead us to expect that studies of the decay of the $(1 + 1')$ RE2PI signal as a function of the time delay between excitation and ionization lasers will merely show the convolution of their ns time profiles. Fig. 2 shows that the opposite is actually true. In this Figure the normalized $(1 + 1')$ RE2PI signal decay is plotted for several excitation energies up to $\sim$2200 cm^{-1} above the $S_1(V'(\pi\pi^*)) \leftarrow S_0$ origin transition of the s-*cis* OH-*syn* conformer. In all cases the decay can be fitted very well with a mono-exponential decay with a time constant of 29.3 ± 1.6 ns. The only way to reconcile these apparently contradictory observations is to conclude that excitation from the ground state to the optically bright state is followed by a fast internal conversion to another state with a ns lifetime. Importantly, our experiments demonstrate that this state can still be ionized. It therefore must be an electronically excited state that does not show up directly in the RE2PI excitation spectrum. In the unlikely scenario that internal conversion would occur to a hot ground state species -which in that case would need to decay to some other species on the observed time scale of ns- ionization would energetically still be possible, but would be precluded because of the highly unfavourable vibrational overlap between the initial neutral and the final ionic states (*vide infra*).

Ab initio calculations on *p*CA predict that for vertical excitation the $n\pi^*$ state is S_3, but that adiabatically it is the lowest electronically excited state.[15,16] CC2

Table 1 Vertical and adiabatic excitation energies (eV) of the lower electronically excited states of the s-*cis* OH-*anti* conformer of OM*p*CA and the OM*p*CA-H_2O cluster calculated at the SCS-CC2 /cc-pVTZ level. Oscillator strengths for vertical excitation are given in parentheses

	OM*p*CA		OM*p*CA + H_2O	
	Vertical	Adiabatic	Vertical	Adiabatic
$V'(\pi\pi^*)$	4.599 (0.221)	4.348	4.523 (0.437)[a]	4.216
$V(\pi\pi^*)$	4.819 (0.627)	4.585	4.718 (0.437)[a]	4.522
$n\pi^*$	5.203 (9×10^{-5})	4.202	5.240 (9×10^{-5})	4.236

[a] Upon complexation with water the excited-state characters of S_1 and S_2 are no longer exclusively associated with the HOMO → LUMO + 1 and HOMO → LUMO excitations as they are in the $V'(\pi\pi^*)$ and $V(\pi\pi^*)$ states of the bare chromophore, but become strongly mixed.

calculations reported in Table 1 show that in OM*p*CA the V′(ππ*), V(ππ*), and nπ* states have the same state ordering.[44] We therefore conclude that the excitation spectrum displayed in Fig. 1 shows transitions to the V′(ππ*) and possibly the V(ππ*) states. Transitions to the nπ* state should be present at lower excitation energies but have so far escaped detection because of the small oscillator strength of the nπ* ← S_0 transition. The calculations thus strongly suggest that internal conversion to the nπ* state is responsible for the ps decay of the excited bright state. The nπ* state, in turn, decays with a time constant of 30 ns to the ground state.

In the present experiments with ns lasers we observe decay of the pump–probe REMPI signal on a ns time scale. This would seem to be at odds with the results of the ps pump–probe studies in which a decay of the ionization signal is observed on a ps time scale.[29] Consideration of the experimental conditions under which both experiments have been performed reveals, however, that the ps studies actually provide further evidence that internal conversion takes place to the nπ* state. In the ps experiments the ionization laser frequency was fixed at 32590 cm^{-1} (4.041 eV). For excitation of the s-*cis* conformers at 32710 cm^{-1} the total energy is therefore barely above (150 cm^{-1}) the adiabatic ionization energy (65154 cm^{-1}).[28] After internal conversion from the V′(ππ*) state the nπ* state is populated. For this state we calculate a vertical ionization energy to D_0 of 5.751 eV, which is much higher than the employed 4.041 eV photons. Franck–Condon arguments thus tell us that in the ps experiments ionization from the nπ* state will be highly inefficient.[45] A second aspect that should be taken into account is the ionic state to which ionization takes place.[46,47] In a one-configuration picture the ground state D_0 of the ion can be described as a π^{-1} state. One thus expects efficient ionization to D_0 from the V′(ππ*) and V(ππ*) states, but not from the nπ* state. The latter state, instead, will preferentially ionize to the n^{-1} state, one of the electronically excited states of the ion, but this state is not accessible in the ps experiments. Both arguments thus imply that the ps experiments are "blind" to the hot nπ* states created after internal conversion. What is therefore observed in the ps experiments is the decay of the electronically excited state that can be ionized (in this case the V′(ππ*) and/or V(ππ*) state), but not the electronic state to which it decays.

In the ns experiments, on the other hand, the ionization laser frequency is fixed at 51813 cm^{-1} (6.424 eV), which according to our calculations is enough for efficient ionization of hot nπ* states. In fact, the situation and numbers are rather similar to the previously reported situation for *p*CA for which ionization with 193 nm was necessary to observe any REMPI signal at all.[14] In the ns pump–probe experiments on OM*p*CA we therefore observe a similar fast decay of the initially excited V′(ππ*) state as seen in the ps experiments, although this time not in the time domain but in the frequency domain. However, since there is enough energy for efficient ionization in the ns experiments, we are not "blind" to the hot nπ* states and can observe their time evolution.

On the basis of *ab initio* calculations on the potential energy surfaces of ground and excited states along the $C_7{=}C_8$ double bond isomerization coordinate (see Scheme 1 for atom numbering), it has been suggested that a barrierless *trans–cis* isomerization process of the double bond *via* a conical intersection of the S_1 and S_0 state might contribute to the nonradiative decay dynamics of S_1.[29] The present results indicate that it is unlikely that such a process determines dominantly the decay of S_1, even though we recognize that in the ns experiments we are "blind" to the hot ground-state species that would be created in such a process.[48] On the other hand, our calculations indicate that the V′(ππ*) state has a stable minimum on its potential energy surface, which would preclude a barrierless isomerization process.

Excited state dynamics of OM*p*CA-H_2O

The (1 + 1′) RE2PI excitation spectrum detected at the mass of the OM*p*CA-H_2O cluster is displayed in Fig. 3a. Overall, the spectrum reproduces the spectrum detected with LIF,[27] although the relative intensity of the origin band of the s-*trans* conformers appears somewhat low. Plots of the ratio of band intensities in the LIF and RE2PI spectra (not shown) do not show any systematic trends, and therefore do not give reason to believe that there is a fundamental difference in the non-radiative decay channels of the s-*cis* and s-*trans* conformers. Fig. 3b and 3c show

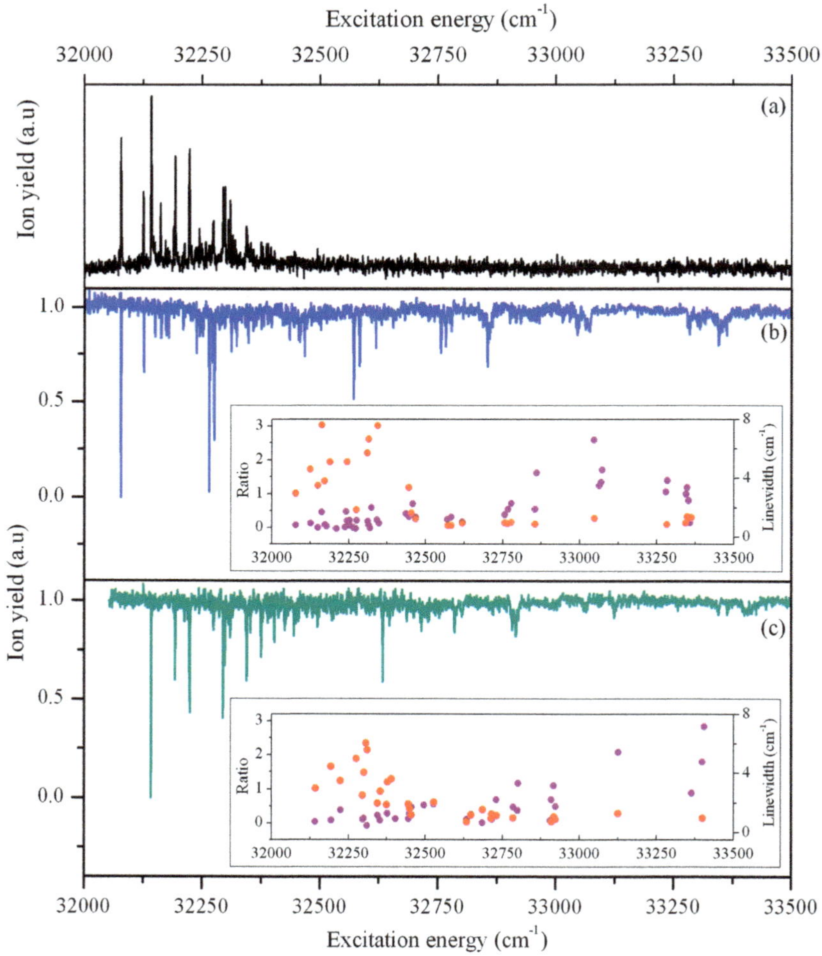

Fig. 3 (a) RE2PI excitation spectrum of OM*p*CA-H_2O and UV-UV depletion spectra of (b) s-*cis* of OM*p*CA-H_2O and (c) s-*trans* OM*p*CA-H_2O. The insets of panels (b) and (c) show the ratio of band intensities in the RE2PI and depletion spectra (red) as well as their line width (purple).

three-colour UV-UV depletion spectra obtained at probe excitation energies of 32078.0 and 32142.0 cm^{-1}, respectively, associated with the $S_1 \leftarrow S_0$ origin transitions of the s-*cis* and s-*trans* conformers.[49,50] Careful analysis of these origin transitions show that they are actually composed of two close-lying transitions that we associate with the OH-*syn* and OH-*anti* conformations, thereby confirming the conclusions drawn in the LIF hole burning experiments.[28] Compared to the spectra of the bare molecule, the band shape of the origin transitions in the water cluster changes significantly. Fits with a Gaussian line shape now give better results than fits with a Lorentzian line shape, and the bands have a much smaller FWHM, which appears to be determined by inhomogeneous rotational broadening.

Interestingly, the intensities of the one- and two-colour RE2PI signals are rather similar for the OM*p*CA-H_2O cluster, in contrast to the situation for the bare molecule where the two-colour signal is considerably enhanced. In the first instance one might suspect that this is due to the decrease of the ionization energy of the cluster compared to the bare molecule (0.442 eV),[28] which could lead to a more efficient ionization of the $n\pi^*$ state in the cluster. However, experiments in which the decay of the REMPI signal is monitored as a function of the delay between excitation and ionization show a signal that follows the convolution of the time profiles of the two lasers and does not give any indication of a long-living state. This conclusion is in agreement with the ps pump–probe studies in which lifetimes of 930 ± 200 and 700 ± 200 ps were reported at the s-*cis* and s-*trans* origin transitions.

We thus conclude that in the cluster the internal conversion pathway to the $n\pi^*$ state is blocked near the vibrationless level, and that the dominant decay channel of the $V'(\pi\pi^*)$ state is internal conversion to the ground state. Further support for these conclusions is found in the results of *ab initio* calculations on the vertical and adiabatic excitation energies of the lower electronically excited states in the OM*p*CA-H_2O cluster (Table 1). We find that the two $\pi\pi^*$ states are red-shifted -the red shift of the V′ state being twice as large as that of the V state- whereas the $n\pi^*$ state undergoes a blue shift. Such shifts are consistent with the general rule that solvation by proton-donating solvents causes a red shift of $\pi\pi^*$ transitions, while $n\pi^*$ states are blue-shifted by the increased stabilisation of the ground state compared to the excited state.[51] As a result of the opposing shifts of the $\pi\pi^*$ and $n\pi^*$ states, the adiabatic excitation energy of the $n\pi^*$ state in the cluster becomes *higher* than that of the $V'(\pi\pi^*)$ state, thereby explaining the experimentally observed absence of the $V'(\pi\pi^*) \rightarrow n\pi^*$ internal conversion pathway.

Fig. 2 and 3 show that complexation of OM*p*CA with water changes the general features of the RE2PI excitation spectrum of the chromophore considerably. For the bare molecule extensive vibrational activity is observed but in the water cluster such activity vanishes ~400 cm^{-1} above the origin. The UV-UV depletion spectra reported in Fig. 3b and 3c, on the other hand, show vibrational activity up to much higher excitation energies. The loss in signal intensity in the RE2PI spectrum is therefore not due to a loss in absorption intensity, but should be attributed to the onset of a non-radiative decay channel to a state with a significantly reduced ionization cross section.[52,53] In agreement with this conclusion we find that the excitation spectrum that is obtained when the two UV-UV depletion spectra are added correlates well with the RE2PI spectrum reported for ps laser excitation.[29]

Importantly, however, the higher spectral resolution of the present experiments gives us access to conformation-specific excitation spectra and decay dynamics. In order to obtain further information on the threshold of the decay channel and on its conformational dependence we have plotted in Fig. 3b and 3c the ratio of band intensities in the RE2PI and depletion spectra, normalizing this ratio to 1.0 at the electronic origin transitions of the s-*cis* and s-*trans* conformers. The same Figures show the line widths of transitions in the depletion spectra. The dependence of both the intensity ratio as well as the line width on the excitation energy clearly indicates that at higher excitation energies an additional decay channel opens up. For the s-*cis* conformers this occurs ~400–500 cm^{-1} above the origin, for the s-*trans* conformers at slightly higher excess energies (~600 cm^{-1}).

A priori various pathways come to mind as possible candidates for such a decay channel. What is important to bear in mind when considering these channels is the experimental observation that the state to which decay takes place cannot be ionized. As will be argued below, this condition excludes any mechanism that produces OM*p*CA in electronically excited states such as photodissociation, internal conversion to the $n\pi^*$ state, and conformational changes in electronically excited states. Photodissociation of the complex into water and OM*p*CA leads to electronically excited OM*p*CA, which subsequently can be ionized by the ionization laser. One would thus still be able to observe OM*p*CA-H_2O resonances, albeit not in the mass channel of the complex but in that of the bare molecule. Mass-resolved wavelength scans of the excitation region of interest monitoring the OM*p*CA mass do not show such resonances, and we therefore have to conclude that photodissociation is not responsible for the increased decay rate.

In the bare molecule the excited-state dynamics are dominated by internal conversion to the $n\pi^*$ state. *Ab initio* calculations and RE2PI decay measurements show that in the complex this channel is blocked near the vibrationless level (*vide supra*). Table 1 indicates that in the cluster the adiabatic excitation energy of the $n\pi^*$ state is only slightly higher than that of the $V'(\pi\pi^*)$ state and one might therefore assume that internal conversion comes into play at higher excitation energies. In such a case the decay rate of $V'(\pi\pi^*)$ levels is enhanced, leading to broader bands, but one is not able to explain the loss of signal in the RE2PI spectrum since the $n\pi^*$ state populated after internal conversion can still be ionized. Using a similar reasoning one can exclude mechanisms based on bond isomerization in the electronically excited state without further decay to the electronic ground state.

Radiationless decay mechanisms that involve internal conversion to the electronic ground state, on the other hand, are in excellent agreement with our experimental observations as they produce hot ground state molecules that cannot be ionized by single-photon absorption, and thus lead to a loss of signal in the RE2PI spectrum. One of these mechanisms is *trans–cis* isomerization of the $C_7{=}C_8$ double bond *via* a conical intersection between S_1 and S_0 as suggested previously in the ps pump–probe experiments.[29] In recent years it has become clear from theoretical studies on the primary isomerization pathway of the chromophore that the *trans–cis* isomerization process not only involves torsion around the $C_7{=}C_8$ double bond, but also other vibrational coordinates, most notably torsion around the C_4–C_7 single bond.[19,22,54,55] Single-bond isomerization in the excited state might thus as well accelerate internal conversion to the ground state. In fact, femtosecond upconversion studies on the viscosity dependence of the excited-state dynamics of neutral OM*p*CA have been interpreted as indicating that in OM*p*CA isomerization occurs *via* a concerted motion of double and single bonds.[31]

Our experiments show that in the bare molecule *trans–cis* isomerization is of minor importance while it dominates the excited-state dynamics of the OM*p*CA-H_2O cluster, implying that *trans–cis* isomerization is facilitated considerably by complexation with water. Such a conclusion finds further support in previous experimental results and in the results of our calculations. Molecular beam IR absorption experiments have shown that water complexation decreases the phenolic OH stretch frequency by 156 cm^{-1} and leads to a large increase of its transition dipole moment.[28] Because of the hydrogen bonding, we anticipate that in the cluster a quinone-like electronic configuration, in which the double bond character of the $C_7{=}C_8$ bond is reduced, gains importance. Inspection of the equilibrium geometry of the V′(ππ*) state in the absence and presence of water indeed reveals that upon complexation the $C_7{=}C_8$ double bond length increases by 0.006 Å, and that in the benzene ring the C_2–C_3 and C_5–C_6 bond lengths decrease while the other bond lengths in the ring increase slightly. Interestingly, the C_4–C_7 bond length remains unaffected. Furthermore, in agreement with the EOM-CCSD data,[27] a comparison of the computed oscillator strengths on the CC2 level for the bare and complexated OM*p*CA indicate a strong mixing of the V and V′ state for the OM*p*CA-H_2O cluster. The change of excited state character upon complexation also involves an increase of the antibonding character with respect to the $C_7{=}C_8$ bond, which promotes the *trans–cis* isomerization pathway. In a zeroth-order view on the isomerization process one might thus say that water complexation leads to loss of double bond character of the $C_7{=}C_8$ double bond, which in turn lowers the barrier associated with torsion around this bond and thereby facilitates access to geometries where internal conversion to the electronic ground state is enhanced.

The S_2 (V(ππ*)) state of OM*p*CA

Up to this point, we have only considered the S_1 (V′(ππ*)) state as the carrier of the RE2PI spectra. Yet, *ab initio* calculations[15,19,27] indicate that in the excitation energy range explored in Fig. 1, transitions to the S_2 (V(ππ*)) state are expected to be present as well, certainly when taking into account that the V(ππ*) ← S_0 oscillator strength is predicted to be larger than the oscillator strength of the V′(ππ*) ← S_0 transition. In the following we will argue that the RE2PI spectrum indeed contains contributions from transitions to both states. To this purpose we will consider the line intensities in the RE2PI and LIF excitation spectra, and the Franck–Condon factors predicted for the V′(ππ*) ← S_0 and V(ππ*) ← S_0 transitions. Fig. 4 displays both the (1 + 1′) RE2PI excitation spectrum measured in the present study and the previously reported LIF spectrum.[27] Comparison of the two spectra leads to the conclusion that some 700 cm^{-1} above the lowest origin transition the intensities in the two spectra start to become quite different. The UV-UV depletion spectra displayed in Fig. 1 show that concurrently the bands become broader.

Further input is provided by the Franck–Condon factors reported in ref. 27 for the V′(ππ*) ← S_0 transition. Here it was found that relatively small changes in geometry occur upon excitation to the V′(ππ*) state. Accordingly, the vibronic

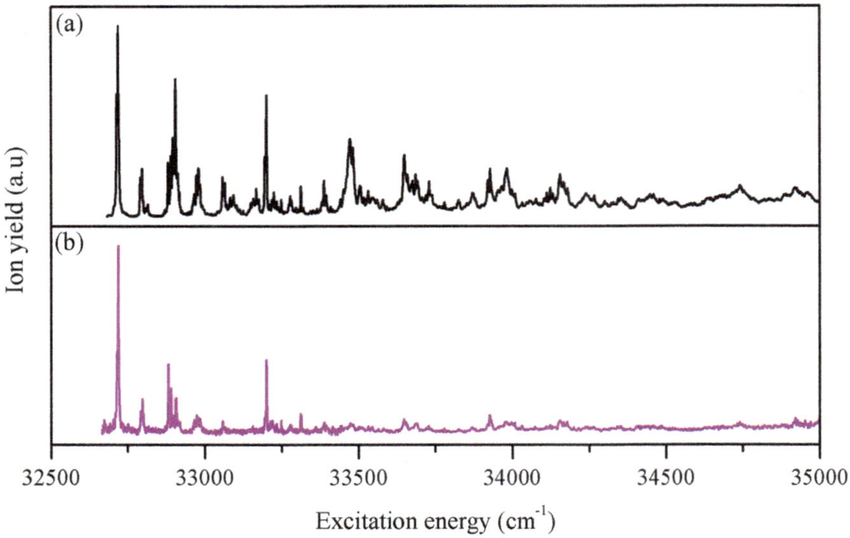

Fig. 4 (a) (1 + 1′) RE2PI and (b) LIF excitation spectrum of OM*p*CA.

activity for the V′(ππ*) ← S_0 transition is rather limited and actually follows quite nicely the activity observed in the LIF spectrum. The differences between LIF and RE2PI spectra therefore are not caused by a loss in fluorescence intensity as would occur, for example, if an additional non-radiative channel would open up 700 cm^{-1} above the vibrationless level of the V′(ππ*) state. Instead, the appropriate interpretation should be that in the RE2PI spectra transitions to another electronic state are visible which are much weaker in the LIF spectrum. Based on the predicted excitation energy of the V(ππ*) state, we associate these transitions with the V(ππ*) state. On the basis of line widths and deviations from expected Franck–Condon intensities we tentatively assign the V(ππ*) ← S_0 origin transition of the s-*cis* OH-*syn*, s-*cis* OH-*anti*, s-*trans* OH-*anti*, and s-*trans* OH-*syn* conformers to the 33466.7 ± 0.2, 33473.9 ± 0.2, 33645.3 ± 0.2, and 33650.8 ± 0.2 cm^{-1} bands in their UV-UV depletion spectra (see Fig. 1). A coarse comparison of the observed vibronic activity in the S_2 ← S_0 part of the single-conformation excitation spectra (Fig. 1) with the activity predicted in ref. 27 shows a reasonable agreement. A more detailed inspection reveals that they contain a mixture of broad and more narrow transitions. The narrow transitions most likely are associated with S_1 vibronic transitions that obtain transition intensity by vibronic coupling to the S_2 state.

In particular in the UV-UV depletion spectra of the s-*cis* OH-*syn* and s-*cis* OH-*anti* conformers, a reasonable strong transition is observed at about 484 cm^{-1} from the origin which does not fit the predicted Franck–Condon intensity.[56] However, dispersed emission spectra recorded at these excitation energies[27] have shown that in the emission spectra unpredicted activity is present at similar vibrational energies as well. The transition should thus be assigned to a S_1 ← S_0 vibronic transition for which the intensity apparently is not correctly predicted by the Franck–Condon calculations.

Although the vibronic activity in the RE2PI spectra can now be well understood, the intensity differences between the RE2PI and LIF spectra require some further attention. Kasha's rule tells us that, in the absence of S_2 decay channels other than internal conversion to S_1, line intensities in the two spectra should be the same. The observation that this is not the case implies that the S_2 state has an additional – in this case dominating – decay channel that brings the molecule into a state from which it can be ionized, but from which radiative decay to the ground state is less efficient. The logical candidate for this pathway is direct internal conversion to the $n\pi^*$ state. This conclusion is further supported by the observation that for transitions assigned to S_2 the RE2PI signal decays with a similar time constant as for transitions assigned to S_1 (see Fig. 2).

Now that the $S_2 \leftarrow S_1$ spectrum has been identified in the bare chromophore, one might wonder where the $S_2 \leftarrow S_1$ spectrum is located for the OM*p*CA-H_2O cluster. Table 1 indicates that the energy difference between the origin transitions of S_1 and S_2 should be roughly 600 cm^{-1} larger in the cluster than in the bare molecule. Although a number of broader transitions are present in this part of the spectrum (see Fig. 3), there are no other indications in terms of intensity or other criteria that would warrant to assign them as transitions to S_2.

Excited state dynamics of TM*p*CA

The (1 + 1′) RE2PI jet-cooled excitation spectrum of TM*p*CA is shown in Fig. 5. In contrast to OM*p*CA for which the excitation spectrum shows well-resolved vibrational structure, the spectrum of TM*p*CA only displays a broad band system with little structure. The band system has an onset at about 29 700 cm^{-1} and a first clearly discernible inflection point at about 30 400 cm^{-1}. The red shift of ~3000 cm^{-1} with respect to the excitation spectrum of OM*p*CA is in the range expected from theoretical calculations[19] which predict that the vertical excitation energies of

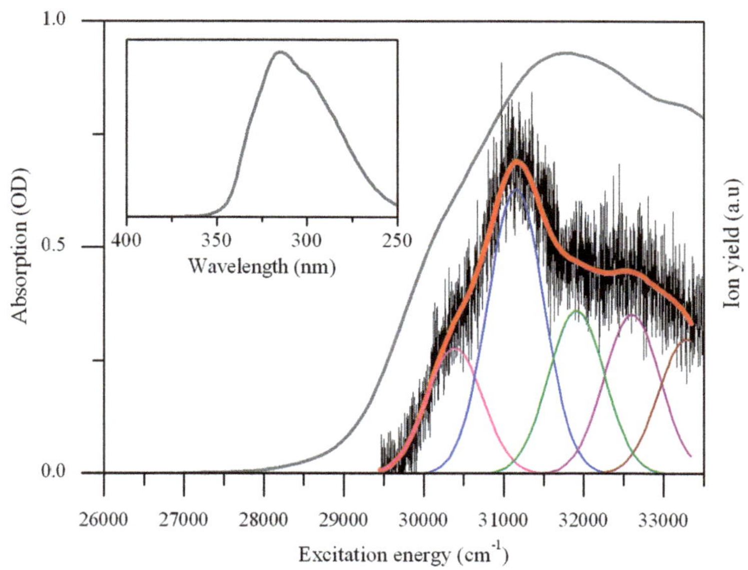

Fig. 5 (1 + 1′) RE2PI excitation spectrum of TM*p*CA (black). The red spectrum is the spectrum obtained by fitting the experimental spectrum with five Gaussian bands. The grey spectrum is the absorption spectrum recorded for TM*p*CA dissolved in n-hexane at room temperature. A more extended absorption spectrum under such conditions is displayed in the inset of the figure.

the V′(ππ*) state in OM*p*CA and the V(ππ*) state in TM*p*CA differ by 0.1 to 0.3 eV. To exclude that imperfect cooling in the molecular beam is responsible for the observed broadening, we have performed various experiments in which molecular beam conditions such as the carrier gas (He, Ne, or Ar), backing pressure, and heating temperature of the sample have been varied. However, none of these experiments gave rise to better resolved excitation spectra. The broadening of the bands is therefore an intrinsic property of TM*p*CA and not due to experimental conditions. Band broadening is indicative of a short-lived excited state, a conclusion that finds further support in the RE2PI signal intensities which are considerably lower for TM*p*CA than for OM*p*CA. In this respect our experimental observations parallel what is observed in REMPI spectroscopic studies on nucleobases and base pairs where efficient coupling of excited states to the ground state *via* conical intersections lead to broad, unresolved REMPI excitation spectra.[57]

Although the spectrum shows little structure, it is clear that it cannot be represented by a single Gaussian or Lorentzian band. A coarse inspection of the inflections in the spectrum suggests that it is composed of at least four bands. To obtain further information on possible values of band widths and energy spacings we have fitted the spectrum with a sum of Gaussian bands. In view of the limited resolution we have taken in the first instance a minimal approach and assumed that all of these bands have the same width and are equally spaced from each other. We then find that the spectrum can be well fitted (see red spectrum in Fig. 5) with the sum of five Gaussian bands with a width of 352.5 ± 5.6 cm^{-1} and with a spacing of 725 ± 40 cm^{-1}. In Fig. 5 the jet-cooled excitation spectrum of TM*p*CA is compared with the absorption spectrum recorded in a solution of n-hexane at room temperature. The solution spectrum is in good agreement with spectrum reported before for TM*p*CA in an aqueous buffer solution[30] albeit that the n-hexane spectrum shows somewhat more resolution and is displaced to slightly higher excitation energies. Comparison of the jet-cooled and solution-phase spectra shows that the bands in the jet-cooled spectrum are considerably more narrow and confirms our conclusion that the band widths in the former spectrum are not determined by the cooling conditions in the molecular beam.

Calculations predict that upon going from OM*p*CA to TM*p*CA significant changes occur in the ordering of the lower electronically excited states.[15,19] The V′(ππ*) and V(ππ*) state are reversed, the V(ππ*) state becoming the lowest excited ππ* state, while the nπ* state is now both vertically as well as adiabatically the lowest excited singlet state. It is thus reasonable to conclude that the large differences between the (1 + 1′) RE2PI excitation spectra of OM*p*CA and TM*p*CA are associated with differences in the structural and dynamic properties of the V′(ππ*) and V(ππ*) states. A further interesting aspect that appears from the calculations is that they predict that the oscillator strengths for the two ππ* states are quite different for the two compounds. At the EOM-CCSD level the oscillator strength of the V(ππ*) $\leftarrow S_0$ transition in OM*p*CA is about seven times as large as that of the V′(ππ*) $\leftarrow S_0$ transition, in TM*p*CA both transitions have roughly the same oscillator strength. The (1 + 1′) RE2PI spectrum depicted in Fig. 5 therefore most likely contains contributions from transitions to both states, and one should be cautious to give too much weight to the Gaussian fit discussed above. Nevertheless, the low-energy part of the spectrum definitely indicates that a ~725 cm^{-1} vibrational mode plays a prominent role in the spectrum and that much smaller band widths than 350 cm^{-1} cannot be accommodated by the spectrum (however, *vide infra*).

Although we cannot rigorously exclude it, it seems unlikely that the band width of 350 cm^{-1} can be attributed entirely to lifetime broadening since it would imply an excited-state lifetime on the order of a few tens of femtoseconds. More likely is an explanation that combines extensive Franck–Condon activity of low-frequency modes in the electronically excited state with lifetime broadening. For OM*p*CA we have observed that internal conversion to the $n\pi^*$ state dominates the excited-state dynamics. One might therefore speculate that in TM*p*CA internal conversion to the $n\pi^*$ state is enhanced, leading to larger line widths. Such a scenario in which bands with a width on the order of tens of wavenumbers are observed in jet-cooled excitation spectra due to internal conversion to another long-living excited state are not without precedence.[58] However, measurements of the decay of the (1 + 1′) RE2PI signal in TM*p*CA do not give any evidence for a possible decay to a longer-living state.[59]

The considerations discussed above, in combination with the photochemical properties of the V($\pi\pi^*$) state appearing from theoretical and experimental studies, suggest that photoisomerization is the most likely candidate for the observed broadening. In this respect it is worth noticing that calculations we performed on the ground and excited states of TM*p*CA readily identify the 725 cm^{-1} mode active in the spectrum as a mode that has a large contribution of double bond torsion and to a minor extent of single bond torsions. The same calculations enable us to make a rough estimate of the minimum lifetime of the excited state compatible with the observed 350 cm^{-1} width. To this purpose we assume that the width is caused by unresolved Franck–Condon activity of a low-frequency mode, and that the transitions to each of these vibronic levels are lifetime-broadened. The calculations indicate that the minimal spacing in such a Franck–Condon progression is at least 20 cm^{-1}, implying that the transitions should have a similar width to become indistinguishable.[60] Transient absorption studies of the excited-state dynamics of the neutral form of TM*p*CA in aqueous solutions report an excited-state lifetime of 3.7 ps,[30] which corresponds to a homogeneous line width of 1.4 cm^{-1}. Such a line width clearly is incompatible with the gas-phase excitation spectrum. The present results thus suggest that the radiationless decay rate to the ground state is substantially larger in the gas phase than in solution. Assuming that this decay rate is closely related to bond isomerization processes, this implies that solvation impedes these isomerization processes.

Conclusions

Understanding the photophysics and photochemistry of photoactive proteins requires a profound understanding of the structural and dynamical properties of the lower excited states of the chromophore that initiates the photocycle, and the influence exerted on these primary steps by external factors like substituents and solvent molecules. The present work has provided a detailed view on the properties of electronically excited states of chromophores of the Photoactive Yellow Protein. Nanosecond RE2PI excitation studies in combination with pump–probe studies of the first optically bright state in OM*p*CA (the V′($\pi\pi^*$) state) provide unambiguous evidence that internal conversion to the adiabatically lower-lying $n\pi^*$ state is an important decay channel in the ps decay of this state. Although there are formally no dark states in multiphoton ionization, the present study has emphasized the importance of the energy of the photon used for ionization of the molecule from an excited state, in particular when electronically excited states are studied that are not the lowest excited state, and/or when ionization to the ground ionic state has a low cross section on account of the electronic character of the excited and ionic states. Comparison of RE2PI and LIF excitation spectra and consideration of the Franck–Condon activity visible in conformation-specific excitation spectra has led us to propose that the excitation spectrum of the S_2 state of OM*p*CA (the V($\pi\pi^*$) state) starts some 760 cm^{-1} higher in energy than the $S_1 \leftarrow S_0$ excitation spectrum. The differences between the RE2PI and LIF excitation spectra indicate that internal conversion to the nπ^* state is also for S_2 the dominant decay channel.

Complexation of OM*p*CA with water leads to large differences in the excited-state dynamics of the V′($\pi\pi^*$) state, and has a large impact on isomerization pathways. The V′($\pi\pi^*$) → nπ^* internal conversion pathway is completely shut down near the bottom of the V′($\pi\pi^*$) well, and appears to remain shut down for higher excitation energies as well. This observation has been rationalized by *ab initio* calculations which show that complexation with water lowers the excitation energy of the $\pi\pi^*$ states and increases that of the nπ^* state to such an extent that the V′($\pi\pi^*$) state becomes the lowest excited singlet state in the OMpCA-H_2O cluster. In agreement with previous ps pump–probe experiments[29] we find that at excess vibrational energies of some 400–600 cm^{-1} an additional radiationless decay channel to the ground state, which we associate with isomerization, opens up. In the present experiments we cannot determine the final distribution over the various possible single- and double-bond isomers in the electronic ground state. At the same time it is clear that such information would provide key information on salient details of the potential energy surface of the excited state. We are therefore presently setting up experiments similar to the ones reported in ref. 53 which will enable us to probe the ground state after internal conversion from the excited state.

It is interesting to notice that both transient absorption[30] and fluorescence upconversion[31] experiments on neutral OM*p*CA in aqueous solution conclude that the excited state has a lifetime of 2–3 ps, and that both find no evidence for the presence of a long-lived excited state. This demonstrates that under such conditions internal conversion to the nπ^* state is indeed not a competitive process, in agreement with the present results. Moreover, for excitation of excited-state levels above the barrier of 400–500 cm^{-1} in the s-*cis* conformers and 600 cm^{-1} in the s-*trans* conformers, the measured lifetime is in good agreement with the lifetime determined in the ps pump–probe experiments (faster than 8 ps)[29] and with our line width measurements. In view of the conclusion of the present studies that the excited-state dynamics of OM*p*CA are affected by complexation with hydrogen-bonding solvents, it would be of interest to perform similar time-resolved experiments on OM*p*CA in non-hydrogen-bonding solvents. The present studies suggest that under such conditions the excited state might have a longer lifetime.

The present study has reported the first jet-cooled excitation spectrum of TM*p*CA. In striking contrast with the excitation spectrum of OM*p*CA, the excitation spectrum of TM*p*CA is broad and almost featureless, and is shifted by about 3000 cm^{-1} to lower energies. These differences have been rationalized in terms of the excited-state dynamics associated with the V′($\pi\pi^*$) and V($\pi\pi^*$) states. In view of the different electronic nature of the two states, such differences are not unexpected. What is actually more surprising – given that the two chromophores have a different lowest excited ππ* state – is that in solution OM*p*CA and TM*p*CA have quite similar excited-state lifetimes (2.1 and 3.7 ps, respectively). These observations once more highlight the importance of solute–solvent interactions for the excited-state dynamics of these two chromophores, and indicate that further time-resolved studies on OM*p*CA and TM*p*CA exploring a wider range of solvents would be useful.

Acknowledgements

This work was supported by The Netherlands Organization for Scientific Research (NWO). S. A. would like to thank the Deutsche Akademie der Naturforscher Leopoldina – German National Academy of Sciences for a Leopoldina research fellowship (grant number LPDS 2011-18).

References

1 M. A. Van der Horst and K. J. Hellingwerf, *Acc. Chem. Res.*, 2004, **37**, 13.
2 T. E. Meyer, *Biochim. Biophys. Acta*, 1985, **806**, 175.
3 T. E. Meyer, E. Yakali, M. A. Cusanovich and G. Tollin, *Biochemistry*, 1987, **26**, 418.
4 T. E. Meyer, J. C. Fitch, R. G. Bartsch, G. Tollin and M. A. Cusanovich, *Biochim. Biophys. Acta*, 1990, **1016**, 364.
5 W. W. Sprenger, W. D. Hoff, J. P. Armitage and K. J. Hellingwerf, *J. Bacteriol.*, 1993, **175**, 3096.
6 K. J. Hellingwerf, J. Hendriks and T. Gensch, *J. Phys. Chem. A*, 2003, **107**, 1082.
7 M. L. Groot, L. J. G. W. van Wilderen, D. S. Larsen, M. A. van der Horst, I. H. M. van Stokkum, K. J. Hellingwerf and R. van Grondelle, *Biochemistry*, 2003, **42**, 10054.
8 G. Groenhof, M. Bouxin-Cademartory, B. Hess, S. P. de Visser, H. J. C. Berendsen, M. Olivucci, A. E. Mark and M. A. Robb, *J. Am. Chem. Soc.*, 2004, **126**, 4228.
9 N. Mataga, H. Chosrowjan and S. Taniguchi, *J. Photochem. Photobiol., C*, 2004, **5**, 155.
10 F. Schotte, H. S. Cho, V. R. I. Kaila, H. Kamikubo, N. Dashdorj, E. R. Henry, T. J. Graber, R. Henning, M. Wulff, G. Hummer, M. Kataoka and P. A. Anfinrud, *Proc. Natl. Acad. Sci. U. S. A.*, 2012, **109**, 19256.
11 W. L. Ryan, D. J. Gordon and D. H. Levy, *J. Am. Chem. Soc.*, 2002, **124**, 6194.
12 M. de Groot and W. J. Buma, *J. Phys. Chem. A*, 2005, **109**, 6135.
13 S. Smolarek, A. Vdovin, E. M. M. Tan and W. J. Buma, *J. Phys. Chem. B*, 2011, **115**, 1275.
14 S. Smolarek, A. Vdovin, D. L. Perrier, J. P. Smit, M. Drabbels and W. J. Buma, *J. Am. Chem. Soc.*, 2010, **132**, 6315.
15 E. V. Gromov, I. Burghardt, H. Köppel and L. S. Cederbaum, *J. Phys. Chem. A*, 2005, **109**, 4623.
16 Q.-S. Li and W.-H. Fang, *Chem. Phys.*, 2005, **313**, 71.
17 R. Send, V. R. I. Kaila and D. Sunholm, *J. Chem. Theory Comput.*, 2011, **7**, 2473.
18 E. V. Gromov, I. Burghardt, H. Köppel and L. S. Cederbaum, *J. Am. Chem. Soc.*, 2007, **129**, 6798.
19 E. V. Gromov, I. Burghardt, J. T. Hynes, H. Köppel and L. S. Cederbaum, *J. Photochem. Photobiol., A*, 2007, **190**, 241.
20 Y. Wang and H. Li, *J. Chem. Phys.*, 2010, **133**, 034108.
21 D. Zuev, K. B. Bravaya, T. D. Crawford, R. Lindh and A. I. Krylov, *J. Chem. Phys.*, 2011, **134**, 034310.
22 E. V. Gromov, I. Burghardt, H. Köppel and L. S. Cederbaum, *J. Phys. Chem. A*, 2011, **115**, 9237.
23 D. Zuev, K. B. Bravaya, M. V. Makarova and A. I. Krylov, *J. Chem. Phys.*, 2011, **135**, 194304.
24 I.-R. Lee, W. Lee and A. H. Zewail, *Proc. Natl. Acad. Sci. U. S. A.*, 2006, **103**, 258.
25 T. Rocha-Rinza, O. Christiansen, J. Rajput, A. Gopalan, D. B. Rahbek, L. H. Andersen, A. V. Bochenkova, A. A. Granovsky, K. B. Bravaya, A. V. Nemukhin, K. L. Christiansen and M. B. Nielsen, *J. Phys. Chem. A*, 2009, **113**, 9442.
26 M. Almasian, J. Grzetic, J. van Maurik, J. D. Steill, G. Berden, S. Ingemann, W. J. Buma and J. Oomens, *J. Phys. Chem. Lett.*, 2012, **3**, 2259.
27 M. de Groot, E. V. Gromov, H. Köppel and W. J. Buma, *J. Phys. Chem. B*, 2008, **112**, 4427.

28 S. Smolarek, A. Vdovin, E. M. M. Tan, M. de Groot and W. J. Buma, *Phys. Chem. Chem. Phys.*, 2011, **13**, 4393.
29 D. Shimada, R. Kusaka, Y. Inokuchi, M. Ehara and T. Ebata, *Phys. Chem. Chem. Phys.*, 2012, **14**, 8999.
30 M. Vengris, D. S. Larsen, M. A. van der Horst, O. F. A. Larsen, K. J. Hellingwerf and R. van Grondelle, *J. Phys. Chem. B*, 2005, **109**, 4197.
31 A. Espagne, D. H. Paik, P. C. Barret, M. M. Martin and A. H. Zewail, *ChemPhysChem*, 2006, **7**, 1717.
32 S. Smolarek, A. Vdovin, A. M. Rijs, C. A. van Walree, M. Z. Zgierski and W. J. Buma, *J. Phys. Chem. A*, 2011, **115**, 9399.
33 A. M. Rijs, E. R. Kay, D. A. Leigh and W. J. Buma, *J. Phys. Chem. A*, 2011, **115**, 9669.
34 C. Manca, C. Tanner, S. Coussan, A. Bach and S. Leutwyler, *J. Chem. Phys.*, 2004, **121**, 2578.
35 D. S. Larsen, M. Vengris, I. H. M. van Stokkum, M. A. van der Horst, R. A. Cordfunke, K. J. Hellingwerf and R. van Grondelle, *Chem. Phys. Lett.*, 2003, **369**, 563.
36 O. Christiansen, H. Koch and P. Jørgensen, *Chem. Phys. Lett.*, 1995, **243**, 409.
37 A. Hellweg, S. Grün and C. Hättig, *Phys. Chem. Chem. Phys.*, 2008, **10**, 4119.
38 C. Hättig and F. Weigend, *J. Chem. Phys.*, 2000, **113**, 5154.
39 C. Hättig, A. Hellweg and A. Köhn, *Phys. Chem. Chem. Phys.*, 2006, **8**, 1159.
40 TURBOMOLE V6.1, 2009; a development of the University of Karlsruhe and Forschungszentrum Karlsruhe GmbH, 1989–2007, TURBOMOLE GmbH since 2007. http://www.turbomole.com (accessed July 8, 2012).
41 A. K. Wilson, D. E. Woon, K. A. Peterson and T. H. Dunning, *J. Chem. Phys.*, 1999, **110**, 7667.
42 F. Weigend, A. Köhn and C. Hättig, *J. Chem. Phys.*, 2002, **116**, 3175.
43 The spectral resolution of the laser system employed in ref. 29 did not allow the *syn* and *anti* conformations to be distinguished.
44 CC2 and EOM-CCSD calculations on the vertical excitation energies of the OM*p*CA chromophore have been reported before in ref. 19. Our calculations are in excellent agreement with these results.
45 M. de Groot, R. W. Field and W. J. Buma, *Proc. Natl. Acad. Sci. U. S. A.*, 2009, **106**, 2510.
46 V. Blanchet, M. Z. Zgierski and A. Stolow, *J. Chem. Phys.*, 2001, **114**, 1194.
47 M. Schmitt, S. Lochbrunner, J. P. Shaffer, J. J. Larsen, M. Z. Zgierski and A. Stolow, *J. Chem. Phys.*, 2001, **114**, 1206.
48 The hot ground state species resulting from internal conversion of the electronic excited state to the ground state cannot be ionized efficiently with one photon – even at an ionization wavelength of 193 nm – on account of vibrational overlap arguments.
49 In ref. 27, the 32078.0 and 32142.0 cm^{-1} transitions were assigned to the s-*trans* and s-*cis* conformations, respectively. Recent mid-IR absorption studies show, however, that this assignment should be reversed.[50]
50 E. M. M. Tan, S. Amirjalayer, S. Smolarek, A. Vdovin, A. M. Rijs and W. J. Buma, *J. Phys. Chem. B*, 2013, **117**, 4798.
51 G. J. Brealey and M. Kasha, *J. Am. Chem. Soc.*, 1955, **77**, 4462.
52 C. Manca, C. Tanner and S. Leutwyler, *Int. Rev. Phys. Chem.*, 2005, **24**, 457.
53 J. J. Newby, C. W. Muller, C.-P. Liu and T. S. Zwier, *J. Am. Chem. Soc.*, 2010, **132**, 1611.
54 A. Yamada, S. Yamamoto, T. Yamamoto and T. Kakitani, *J. Mol. Struct.: THEOCHEM*, 2001, **536**, 195.
55 M. Boggio-Pasqua, M. A. Robb and G. Groenhof, *J. Am. Chem. Soc.*, 2009, **131**, 13580.
56 In the depletion spectra of the s-*trans* conformers a transition is observed at similar vibrational energies though with a somewhat lower intensity.
57 M. S. de Vries and P. Hobza, *Annu. Rev. Phys. Chem.*, 2007, **58**, 585.
58 J. A. Sebreez and T. S. Zwier, *Phys. Chem. Chem. Phys.*, 2012, **14**, 173.
59 Theoretically it could be possible that the $n\pi^*$ state in TM*p*CA has a shorter lifetime than in OM*p*CA due to a larger spin–orbit coupling on account of the presence of the sulphur atom. In that case one would not observe either a long-living state.
60 The 20 cm^{-1} frequency is associated with the methyl torsional mode. Though the orientation of the methyl group might change upon excitation, it seems unlikely that it would lead to a progression encompassing 350 cm^{-1}. At frequencies of 50 and 80 cm^{-1}, on the other hand, skeleton modes are found that appear to be more likely to be Franck–Condon active.

PAPER

Designs for molecular circuits that use electronic coherence

Francesca Fassioli,†[ab] Daniel G. Oblinsky†[a] and Gregory D. Scholes*[a]

Received 5th February 2013, Accepted 26th February 2013
DOI: 10.1039/c3fd00009e

The mounting evidence of recent years regarding long-lived coherent dynamics of electronic excitations in several light-harvesting antenna proteins suggests the possibility of realizing and exploiting light-initiated quantum dynamics in synthetic molecular devices based on electronic energy transfer. Inspired by the field of molecular logic, we focus this discussion on the prospect of using quantum coherence to control the direction of energy flow in a molecular circuit. As a prototype system we consider a circuit consisting of three chromophores that deliver energy to two trap chromophores. Our aim is to control to which trap the energy is more likely to be delivered. This is achieved by switching one of the circuit chromophores *ON* and *OFF* from the system, such that the direction of energy flow substantially changes from the ON and OFF states of the circuit. We find that quantum coherence can allow a significant ability to direct energy transfer in the circuit. However, when realistic levels of noise are added, quantum coherence only slightly improves the ability to direct electronic energy in comparison to a classical hopping mechanism.

Introduction

We are interested in the question of whether it is feasible to realize photo-initiated quantum dynamical effects in condensed phase supramolecular systems. If so, what would we gain from such effects? Conversely, if photo-induced dynamics can be influenced by non-trivial quantum effects, how will that inspire the design of supramolecular systems? These questions motivate the present discussion. We consider a specific example of a molecular *circuit* that directs excitation to a pair of chromophore traps. This idea of circuits designed to "process" excitation energy was conceived recently when a group of us forecast the next steps after learning piecewise the mechanisms employed by photosynthetic organisms to harvest sunlight.[1]

[a]*Lash Miller Chemical Laboratories, Institute for Optical Sciences and Centre for Quantum Information and Quantum Control, University of Toronto, 80 St. George St., Toronto, Canada. E-mail: gscholes@chem.utoronto.ca; Fax: +416 946 8775; Tel: +416 946 7532*
[b]*University College London, Department of Physics and Astronomy, Gower Street, London WC1E6BT, UK*

† Both of these authors contributed equally to this work.

Chemistry is essentially defined by non-additive effects as atoms combine to form molecules. Elegant examples of quantum interference abound at the molecular scale. One example is the interplay of through space and through bond interactions that determines orbital energies.[2–4] Compelling precedents for inter-chromophore quantum interference phenomena include superexchange-mediated electron transfer reactions.[5–10] More recently quantum interference effects in charge transport have been uncovered.[11–13]

Despite the examples mentioned above, quantitative models for photo-induced dynamics are generally non-additive descriptions. In other words, we use Fermi's golden rule and a diabatic basis for energy and electron reactions to provide a rate constant for the transition from reactant to product. The success of these theories (Marcus theory, Förster theory, *etc*) may be partly ascribed to their relation to how condensed phase photo-induced dynamics are measured by experimentalists, usually by detecting the time evolution of populations. Therefore a key to addressing the questions posed above is to develop experiments that detect mechanistic aspects of dynamics, not solely kinetics.[14] Encouraging steps in that direction have been attained through the development of multidimensional optical spectroscopies.[15–33]

Control over the pathways of energy flow would impact many applications of energy transfer. This kind of control may stimulate the field of *molecular logic*.[34,35] The idea that molecules can manipulate and process information using logic operations has spawned studies of molecule-based systems.[36,37] To implement molecule-based systems photochromes are used. Photochromes are special chromophores envisioned to be photoswitchable, bistable molecules that can act as transistors, implementing binary logic operations.[38] When several of these photochromes are combined the interactions between them allow for development of simple Boolean logic gates to more complex logic devices using all-photonic input and output. This is achieved through selective isomerization of individual photochromes, modifying the wavelengths of light which they absorb and fluoresce at.[39,40]

In the present work we focus on electronic energy transfer in a simple supramolecular model system. It is interesting to ask the question how does electronic energy flow beyond the Förster limit? What do non-additive effects "look like"? One example is when strongly interacting molecules act as effective donors and acceptors, which is described by generalized Förster theory.[41–47] In this case, excitation is delocalized over several chromophores but the transfer is still a classical random hopping of excitation. It has emerged recently that coherent energy transfer might be found in photosynthetic systems. That possibility is suggested by 2-dimensional electronic spectroscopy experiments, which are providing evidence that quantum coherent dynamics of excitations are much longer-lived than previously realized.[16,18,19,48,49] This suggests the possibility that electronic energy transfer in light-harvesting proteins could actually proceed quantum coherently on a timescale comparable to the energy transfer process.[50–59] The prospect that energy can flow aided by coherence in the biological environment, rather solely *via* a Förster mechanism, opens up new possibilities for technological applications that would not be possible otherwise.[60] In particular, it might allow control over the pathways of electronic energy transfer through the exploitation of multi-path interference effects.[61] Our aim is to examine this question by determining how performance of the circuit varies with the level of environmental noise (reorganization energy), and more specifically with coherent/incoherent nature of electronic energy transfer.

Model

As a proof of concept, we propose an energy transfer circuit consisting of five chromophores: three circuit chromophores and two trap chromophores (see Fig. 1a). We focus on energy transfer under a low influx of excitations, such that dynamics are restricted to the zero and single-exciton subspace. In this case, the electronic Hamiltonian is given by the standard Frenkel exciton Hamiltonian,

$$H_{el} = \sum_{m} \varepsilon_m |m\rangle\langle m| + \sum_{n \neq m} V_{mn}(|m\rangle\langle n| + |n\rangle\langle m|) \tag{1}$$

where $|m\rangle$ represents the state with chromophore m excited and all others in the ground state, ε_m is the site energy and V_{mn} is the electronic coupling between chromophores m and n. All chromophores are assumed to have equal site energies so we set $\varepsilon_m = 0$ for all m.

We explore configuration space by varying the electronic couplings between all five chromophores. This is accomplished by rotating the transition dipole moment of each of the three circuit chromphores in the plane of the page in a systematic manner. The transition dipole moments of the trap chromophores are kept fixed. The electronic couplings are then calculated at each circuit configuration using the ideal dipole approximation,

$$V_{mn} = C\,\frac{\hat{d}_m \cdot \hat{d}_n - 3\left(\hat{d}_n \cdot \hat{r}_{mn}\right)\left(\hat{d}_m \cdot \hat{r}_{nm}\right)}{|\hat{r}_{nm}|^3} \tag{2}$$

where $\hat{r}_{nm}$ is the vector connecting the centres of the chromophores, $\hat{d}_{m(n)}$ is the unitary vector describing the direction of the transition dipole moment of chromophore $m(n)$ and C is a scaling factor. The parameters were chosen such that the strongest couplings are on the order of 100 cm^{-1}. As a reference, the strongest electronic coupling in the Fenna-Matthews-Olson complex is $|V|$= 106 cm^{-1}.[62]

The ultimate goal is to direct energy flow within the circuit configuration. To accomplish this (Fig. 1), we *switch* chromphore 3 on and off. In practice this could

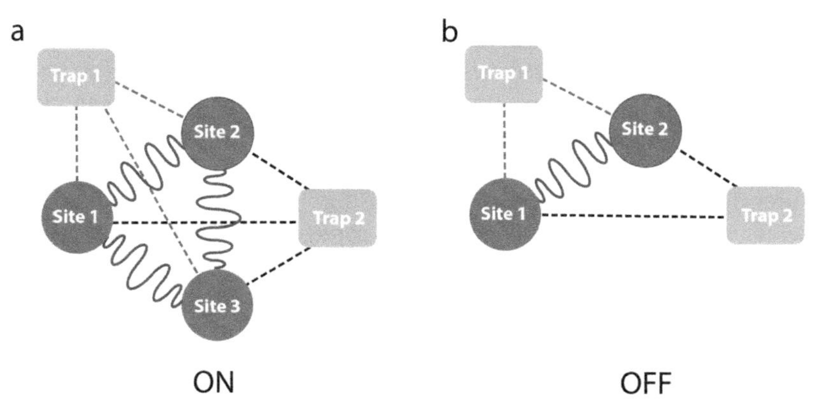

Fig. 1 Energy transfer circuit. (a) ON state, with chromophore 3 electronically coupled to the remaining chromophores. (b) OFF state, with chromophore 3 detached from the system.

be easily accomplished using an isomerization or pH-dependant colour change. With all five chromophores present, the system is in the *ON* state and if the chromophore is removed (or bleached) at site 3 the system is in the *OFF* state. Our aim is for this switching to direct energy to the traps: in the ON state the excitation has to be trapped at trap 2 and in the OFF state the excitation has to be trapped at trap 1. Our metric is to optimize the contrast of the trapping probabilities so that $\eta_2^{ON} > \eta_1^{ON}$ and $\eta_1^{OFF} > \eta_2^{OFF}$, where η_i is the probability of trapping at site i. Trapping is modelled as an incoherent and irreversible process.[63] For simplicity, and also because we are interested in fast energy transfer on the ps timescale, we neglect exciton loss due to fluorescence relaxation (ns timescale). Then, since the excitation can only leave the system through trap 1 or trap 2, placing the restriction that $\eta_1 + \eta_2 = 1$. In a circuit where excitation is not preferentially trapped $\eta_1 = 0.5$ and $\eta_2 = 0.5$.

To simulate energy transfer dynamics two methodologies were implemented, the hierarchical equations of motion (HEOM) and Förster resonance energy transfer leading to a Pauli master equation (PME).[64–66] We use the HEOM to correctly interpolate between the limits of strong- and weak-electronic coupling. Specifically, we use the scaled HEOM approach as it has been shown to converge faster than the original formulation.[67,68] Simulation of electronic energy transfer in the limit of strong to intermediate-electronic coupling using the PME, does not accurately predict the energy transfer dynamics.[69–73] By using the PME we artificially restrict energy transfer to occur by an incoherent hopping mechanism between individual chromophores. We can contrast this with the HEOM model to get an idea of the significance of coherent dynamics. For simplicity we assume that each chromophore couples to a local bath, and that the form of this coupling is characterized by a spectral density $J(\omega)$ of Ohmic form with a Drude cutoff, $J(\omega) = 2\lambda\gamma\omega/(\omega^2 + \gamma^2)$, where λ is the reorganization energy and determines the strength of the coupling to the bath and γ^{-1} is the bath correlation time. In order to make the designed circuits more feasible for practical applications, we focus on room temperature, $T = 300$ K.

Results

Simulations of each circuit configuration were carried out using the HEOM and the PME at reorganization energies of $\lambda = 35$, 100, 200 and 500 cm^{-1}, in addition to the purely coherent evolution ($\lambda = 0$), allowing for exploration across the spectrum of energy transfer regimes.[74] In all cases we fix $\gamma = 50$ cm^{-1} and assume a rate of trapping equal at each trap of $k = 1$ ps^{-1}. Convergence between the two theories starts to occur for reorganization energies greater than $\lambda = 500$ cm^{-1}, with pronounced differences occurring at smaller reorganization energies. Using the previously defined criteria for success ($\eta_2^{ON} > \eta_1^{ON}$ and $\eta_1^{OFF} > \eta_2^{OFF}$), in the purely coherent regime ($\lambda = 0$), 21% of the total configurations analysed exhibit the desired directed trapping. The number of configurations satisfying the trapping criteria at $\lambda = 35$ cm^{-1} is also 21%. This number decreases as more noise is introduced into the system, converging to the PME at $\lambda = 500$ cm^{-1} with only 13% of the configurations satisfying the trapping criteria.

Using box plots in Fig. 2 we depict the performance of the configurations satisfying the trapping criteria ($\eta_2^{ON} > \eta_1^{ON}$ and $\eta_1^{OFF} > \eta_2^{OFF}$). The plots show the distribution of probability of trapping at trap 2 in the ON state (left) and

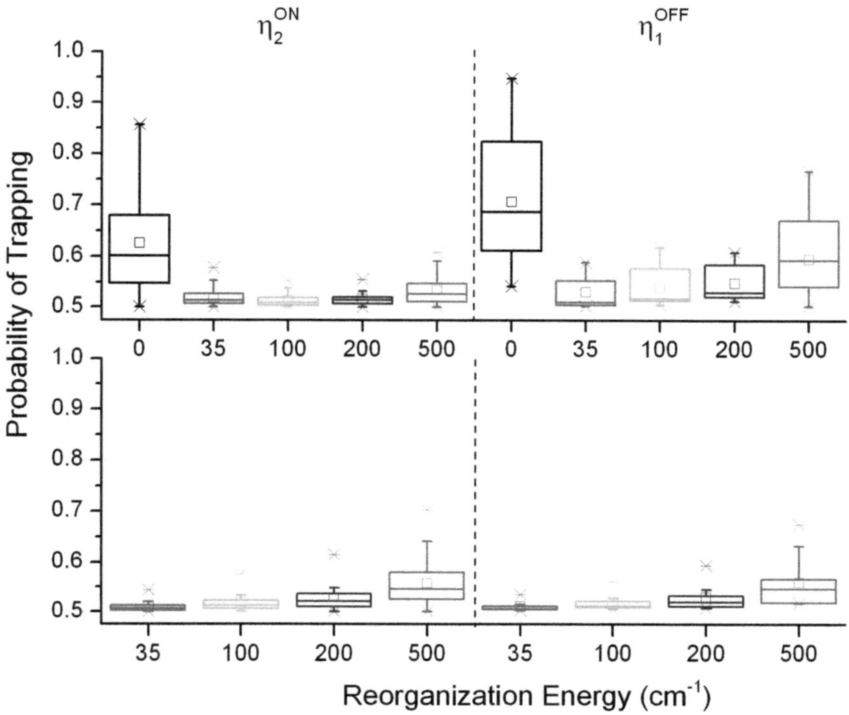

Fig. 2 Trapping probability at trap 2 in the ON state, η_2^{ON} (left of dotted line) and trapping probability at trap 1 in the OFF state, η_1^{OFF}(right of dotted line). Dynamics were modelled using the HEOM (top) and PME (bottom). Initial state is $\rho(0) = |1\rangle\langle 1|$.

probability of trapping at trap 1 in the OFF state (right) when simulated using the HEOM (top) and the PME (bottom) as a function of reorganization energy. The mean is depicted by a □, the median by a horizontal line in the box, the top and bottom edges of the box are the 25th and 75th percentiles, upper and lower whiskers represent the 5th and 95th percentiles, × are the predicted 1st and 99th percentiles and finally – are the maximum and minimum values found for the successful configurations.

Trapping probabilities under purely coherent dynamics are shown together with the HEOM calculations (top). The coherent regime exhibits the highest trapping probabilities and wider distribution. Some configurations are exceptional in their ability to direct energy transfer, *e.g.* $\eta_2^{ON} = 0.81$ and $\eta_1^{OFF} = 0.83$. In such configurations trapping at one or the other trap is almost completely suppressed in the *ON* or *OFF* state of the circuit. When coupling to the environment ($\lambda > 0$) is included, calculations based on the HEOM show that η_2^{ON} exhibits a non-monotonic behaviour. The median, maximum value and width of the probability distribution significantly decrease and varies slowly across all reorganization energies in both ON and OFF states, reaching the largest trapping probabilities at $\lambda = 500\ \text{cm}^{-1}$. In contrast, the classical model shows that the probability distribution broadens and the median increases as reorganization energy is increased, predicting then a higher ability to direct the excitation to a particular trap for large reorganization energies. These results suggest that, at low reorganization energies, non-classical transport may give an advantage over a purely Förster transfer to direct excitation energy.

To determine if there is a trend in the ability to achieve a significant bias in trapping of an excitation between the ON and OFF states, the difference of the two trapping probabilities, $\Delta\eta_{2,2} = \eta_2^{ON} - \eta_2^{OFF}$, is considered in Fig. 3. (top). Also the difference $\Delta\eta_{2,1} = \eta_2^{ON} - \eta_1^{OFF}$ is shown in Fig. 3 (bottom). Again $\Delta\eta_{2,2}$ exhibits a non-monotonic behaviour as a function of reorganization energy for the HEOM calculations. The largest difference in the performance of the circuit between *ON* and *OFF* states is achieved in the coherent regime while including noise tends to result in a more equally distributed trapping in both traps. The PME predicts on the contrary, a monotonic behaviour such that a greater difference between *ON* and *OFF* is achieved for larger reorganization energy. We have performed calculations based on PME for larger reorganization energies that confirm this trend and show that the maximum value for $\Delta\eta_{2,2}$ converges to ~0.37. This value is still much lower than the difference obtained in the purely coherent regime, suggesting again that quantum transport may outperform classical transport in directing energy transfer. The difference in trapping probabilities $\Delta\eta_{2,1}$ shows if a certain bias is associated with trapping depending on whether there are five or four chromophores present in the circuit. The PME calculations show a symmetric distribution, and a median and mean situated close to $\Delta\eta_{2,1} = 0$. When the dynamics are simulated with the HEOM the simulation shows a bias towards a larger η_1^{OFF}compared to η_2^{ON}. This skew in the trapping probabilities is true for all

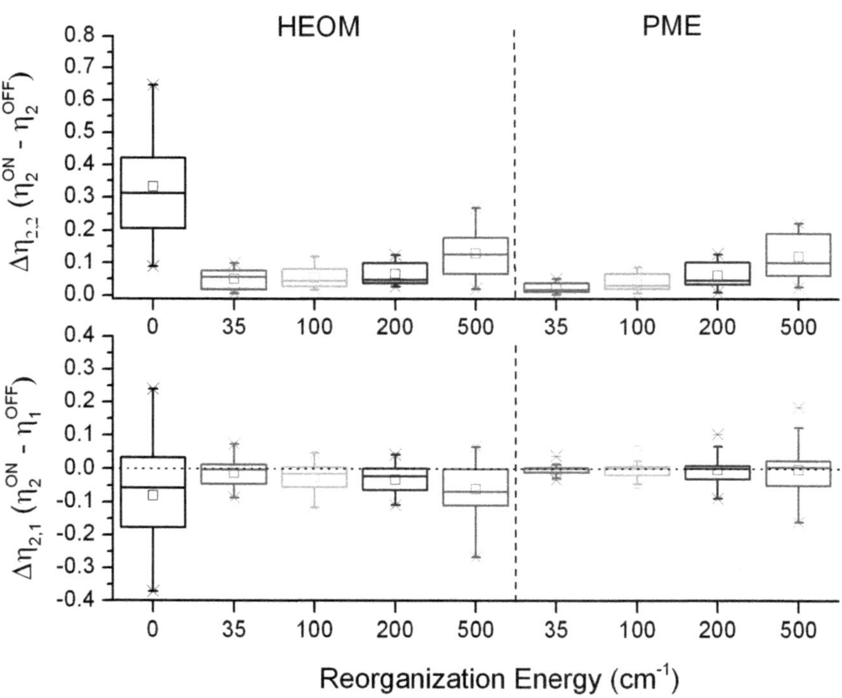

Fig. 3 Difference in trapping probabilities of configurations that satisfy energy directing criteria. $\eta_2^{ON} - \eta_2^{OFF}$ (top) and $\eta_2^{ON} - \eta_1^{OFF}$ (bottom) calculated using HEOM (left of dotted line) and PME (right of dotted line).

reorganization energies, and becomes more pronounced as the reorganization energy is increased. This is perhaps due to the initial state considered, that favours trapping in trap 1, however it is unclear why this effect only appears in the HEOM and not the PME approach.

The probabilities of trapping for the configuration that can "best" direct energy through the circuit, *i.e.* that exhibits the largest difference in trapping between *ON* and *OFF* states $\Delta\eta_{2,2}$, at a given reorganization energy are shown in Table 1. When the circuit is in the *ON* state PME simulations predict a better ability to direct the energy towards a particular trap than the HEOM simulations. This trend is then reversed when the circuit is placed in the OFF state, where dynamics based on HEOM outperforms the classical dynamics predicted by PME. The net effect can be appreciated by considering the difference in trapping between *ON* and *OFF* states, $\Delta\eta_{2,2}$. From Fig. 3 and Table 1 the general conclusion drawn is that at low reorganization energies, quantum transport is able to better direct energy transfer than a classical process. Directing excitation energy at large reorganization energies, when only exciton hopping between sites dominates is more effective than intermediate reorganization energies but, is still not as successful as the purely coherent regime.

It is not possible to *a priori* exactly control the level of environmental noise and thus one should opt for designs that are able to control energy flow in a range of energy transfer scenarios, such as the robust configuration shown in Table 2. We will briefly discuss the underlying features that allow this robust functioning.

Table 1 Trapping probability of the "best" configuration in directing energy transfer at given reorganization energy[a]

	HEOM			PME		
$\lambda[\mathrm{cm}^{-1}]$	η_2^{ON}	η_1^{OFF}	$\Delta\eta_{2,2}$	η_2^{ON}	η_1^{OFF}	$\Delta\eta_{2,2}$
0	0.86	0.79	0.64	—	—	—
35	0.52	0.58	0.09	0.54	0.51	0.05
100	0.52	0.59	0.12	0.58	0.51	0.08
200	0.53	0.60	0.12	0.61	0.51	0.11
500	0.60	0.67	0.26	0.70	0.52	0.22

[a] Note that in all cases $\eta_1 + \eta_2 = 1$.

Table 2 Trapping probability of the robust configuration that will direct energy transfer at any reorganization energy[a]

	HEOM			PME		
$\lambda[\mathrm{cm}^{-1}]$	η_2^{ON}	η_1^{OFF}	$\Delta\eta_{2,2}$	η_2^{ON}	η_1^{OFF}	$\Delta\eta_{2,2}$
0	0.61	0.68	0.29	—	—	—
35	0.52	0.58	0.10	0.51	0.53	0.04
100	0.52	0.59	0.11	0.52	0.55	0.07
200	0.53	0.60	0.12	0.54	0.58	0.11
500	0.58	0.67	0.25	0.58	0.63	0.21

[a] Note that in all cases $\eta_1 + \eta_2 = 1$.

Fig. 4 shows the electronic couplings (left) and the exciton energies (eigenenergies of H_{el}, eqn (1)) of the robust configuration in the *ON* and *OFF* states of the circuit. Site 2 is strongly coupled to sites 1 and 3, while site 3 is also strongly coupled to trap 2. In the classical regime when energy is transferred incoherently between states localized on individual chromophores, the effectiveness of the circuit in achieving directed transfer can be understood simply by recalling that the transfer rate is proportional to the square of the electronic coupling V^2. Then, in the ON state, transfer from site 1 to trap 2 is mediated by chromophore 3, while in the absence of chromophore 3, *i.e.* *OFF* state, chromophore 2 mediates transfer to trap 1. In the opposite limit of strong electronic coupling, both coherent dynamics and environmentally induced relaxation between excitons take place and the delocalized character of excitations becomes important. To understand then how delocalized excited states contribute to the robustness of the circuit in this regime, we focus on the energy landscape of the circuit in the *OFF* state as in the *ON* state trapping is almost equally distributed on both traps. The strong coupling between chromophores 1 and 2 allows for fast delocalization of the excitation initially localized on site 1 to chromophore 2, and the intermediate coupling between sites 2 and trap 1 will allow some delocalization of excitation to trap 1 as well. Simultaneously environmental relaxation will occur. In this case, the highest exciton is delocalized over the strongly coupled sites 1 and 2 and will mostly transfer energy to the second highest exciton, which has a larger participation of trap 1 than trap 2.

All of the results presented consider that initially the excitation is fully localized on site 1. This kind of initial state is usually assumed in scenarios where only one chromophore in the system is connected to some input source of excitation. However, even if this is the case, transfer from the light-collecting antenna to the system may be better described as a transfer between excitons rather than sites, especially when the dynamics are close to coherent. In order to examine the influence of the initial state on our results, we have calculated trapping probabilities when the initial state is a statistical mix of excitons. If only site 1 is coupled

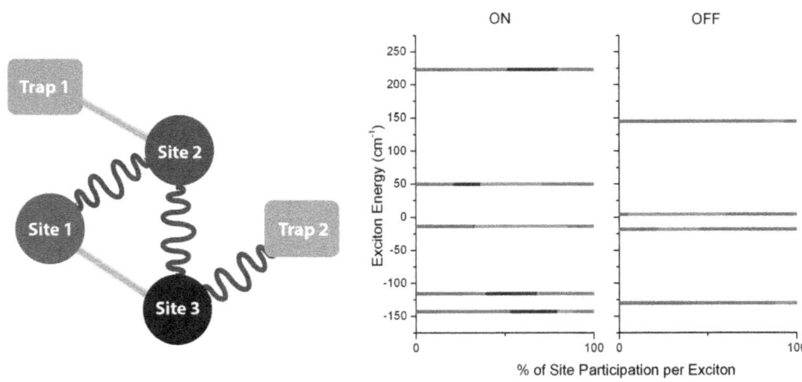

Fig. 4 The "robust" circuit configuration. Left: Electronic couplings: curled red lines represent strong coupling (~100 cm^{-1}) and grey straight lines represent intermediate couplings (~60 cm^{-1}). Right: Energy levels in the *ON* and *OFF* state. Colours indicate the participation of each site in a given exciton $|\alpha\rangle$, quantified by the overlap $|\langle i|\alpha\rangle|^2$, where $|i\rangle$ denotes the electronic state localized on site *i*.

to the source of excitation, then we can approximate the initial state as $\rho(0) = \sum_\alpha |c_1^\alpha|^2 |\alpha\rangle\langle\alpha|$, where $|\alpha\rangle$ denotes an eigenstate of H_{el} and $c_1^\alpha = \langle 1|\alpha\rangle$ is the overlap between exciton α and site 1. We find that in the purely coherent regime, the trapping probabilities are the same with both (localized and statistical mix of excitons) initial states. This can be explained because the trapping probability is a time-integrated quantity and therefore, if coherent dynamics are fast in comparison to the lifetime, trapping will mostly depend on the initial population of each exciton (rather than on coherences). In both our initial states, the exciton populations are the same, explaining then the equal probabilities of trapping.

Conclusions

Control over the pathways of electronic energy flow on the nanoscale is desirable for photo-induced processes initiated by resonant energy transfer. So far, it is only clear how to direct energy through an energy funnel in the classical hopping limit of energy transfer. In this work we have investigated whether quantum coherence can provide new and perhaps better ways for achieving control over excitation energy flow. We considered a model molecular circuit that directs energy to two traps. We analysed a wide range of circuit configurations and found that purely coherent dynamics allows for a better ability in directing energy to a specific trap. However, in the presence of realistic noise, coherence did not help direct excitation remarkably compared to incoherent hopping. Some configurations were found to be more robust than others, performing well in a range of energy transfer regimes. Electronic coherence contributed to the robustness.

Acknowledgements

We thank DARPA for funding under the Quantum Effects in Biological Environments (QuBE) program. The United States Air Force Office of Scientific Research and the Natural Sciences and Engineering Research Council of Canada (NSERC) are also acknowledged for financial support.

Notes and references

1 G. D. Scholes, G. R. Fleming, A. Olaya-Castro and R. van Grondelle, *Nat. Chem.*, 2011, **3**, 763.
2 R. Hoffmann, E. Heilbronner and R. Gleiter, *J. Am. Chem. Soc.*, 1970, **92**, 706.
3 R. Hoffmann, *Acc. Chem. Res.*, 1971, **4**, 1.
4 M. N. Paddon-Row, *Acc. Chem. Res.*, 1982, **15**, 245.
5 C. A. Naleway, L. A. Curtiss and J. R. Miller, *J. Phys. Chem.*, 1991, **95**, 8434.
6 K. D. Jordan and M. N. Paddon-Row, *Chem. Rev.*, 1992, **92**, 395.
7 A. B. Ricks, G. C. Solomon, M. T. Colvin, A. M. Scott, J. Chen, M. A. Ratner and M. R. Wasielewski, *J. Am. Chem. Soc.*, 2010, **132**, 15427.
8 D. N. Beratan, J. N. Onuchic, J. R. Winkler and H. B. Gray, *Science*, 1992, **258**, 1740.
9 H. B. Gray and J. R. Winkler, *Proc. Natl. Acad. Sci. U. S. A.*, 2005, **102**, 3534.
10 T. R. Prytkova, I. V. Kurnikov and D. N. Beratan, *Science*, 2007, **315**, 622.
11 S. Wohlthat, G. C. Solomon, N. S. Hush and J. R. Reimers, *Theor. Chem. Acc.*, 2011, **130**, 815.
12 J. P. Bergfield, G. C. Solomon, C. A. Stafford and M. A. Ratner, *Nano Lett.*, 2011, **11**, 2759.
13 C. M. Huedon, H. Valkenier, T. Markussen, K. S. Thygesen, J. C. Hummelen and S. J. van der Molen, *Nat. Nanotech.*, 2012, **7**, 304.
14 G. D. Scholes, T. Mirkovic, D. B. Turner, F. Fassioli and A. Buchleitner, *Energy Environ. Sci.*, 2012, **5**, 9374.
15 T. Brixner, J. Stenger, H. M. Vaswani, M. Cho, R. E. Blankenship and G. R. Fleming, *Nature*, 205, **434**, p. 625.

16 G. S. Engel, T. R. Calhoun, E. L. Read, T. -K. Ahn, T. Mancal, Y.-C. Cheng, R. E. Blankenship and G. R. Fleming, *Nature*, 2007, **446**, 782.
17 E. Collini and G. D. Scholes, *Science*, 2009, **323**, 369.
18 E. Collini, C. Y. Wong, K. E. Wilk, P. M. G. Curmi, P. Brumer and G. D. Scholes, *Nature*, 2010, **463**, 644.
19 G. Panitchyangkoon, D. Hayes, K. A. Fransted, J. R. Caram, E. Harel, J. Z. Wen, R. E. Blankenship and G. S. Engel, *Proc. Natl. Acad. Sci. U. S. A.*, 2010, **107**, 12766.
20 I. P. Mercer, Y. C. El-Taha, N. Kajumba, J. P. Marangos, J. W. G. Tisch, M. Gabrielsen, R. J. Cogdell, E. Springate and E. Turcu, *Phys. Rev. Lett.*, 2009, **102**, 057402.
21 G. Panitchayangkoon, D. V. Voronine, D. Abramavicius, J. R. Caram, N. H. C. Lewis, S. Mukamel and G. S. Engel, *Proc. Natl. Acad. Sci. U. S. A.*, 2011, **108**, 20908.
22 G. H. Richards, K. E. Wilk, P. M. G. Curmi, H. M. Quiney and J. A. Davis, *J. Phys. Chem. Lett.*, 2012, **3**, 272.
23 G. S. Schlau-Cohen, T. R. Calhoun, N. S. Ginsberg, E. L. Read, M. Ballottari, R. Bassi, R. van Grondelle and G. R. Fleming, *J. Phys. Chem. B*, 2009, **113**, 15352.
24 C. Y. Wong, R. M. Alvey, D. B. Turner, K. E. Wilk, D. A. Bryant, P. M. G. Curmi, R. J. Silbey and G. D. Scholes, *Nat. Chem.*, 2012, **4**, 396.
25 F. Milota, J. Sperling, A. Nemeth, D. Abramavisius, S. Mukamel and H. F. Kauffmann, *J. Chem. Phys.*, 2009, **131**, 054510.
26 D. M. Jonas, *Annu. Rev. Phys. Chem.*, 2003, **54**, 425.
27 S. M. Gallagher, A. W. Albrecht, T. D. Hybl, B. L. Landin, B. Rajaram and D. M. Jonas, *J. Opt. Soc. Am. B*, 1998, **15**, 2338.
28 S. Mukamel, *Annu. Rev. Phys. Chem.*, 2000, **51**, 691.
29 D. Abramavicius, B. Parmieri, D. V. Voronine, F. Sanda and S. Mukamel, *Chem. Rev.*, 2009, **109**, 2350.
30 Y. C. Cheng and G. R. Fleming, *Annu. Rev. Phys. Chem.*, 2009, **60**, 241.
31 M. Cho, *Chem. Rev.*, 2008, **109**, 2350.
32 S. T. Cundiff, T. H. Zhang, A. D. Bristow, D. Karaiskaj and X. C. Dai, *Acc. Chem. Res.*, 2009, **42**, 1423.
33 R. M. Hochstrasser, *Proc. Natl. Acad. Sci. U. S. A.*, 2007, **104**, 14190.
34 F. Remacle and R. D. Levine, *Eur. Phys. J. D*, 2007, **42**, 49.
35 F. Remacle and R. D. Levine, Inter- and intra- molecular level logic devices. *Information Technology IV, Nanotechnology*, ed R. Waser (Wiley, Weinheim), Vol 4, p. 213.
36 A. P. de Silva, H. Q. N. Gunaratne and C. P. McCoy, *Nature*, 1992, **364**, 42.
37 D. Gust, T. A. Moore and A. L. Moore, *Faraday Discuss.*, 2012, **155**, 9.
38 J. Andréasson, S. D. Straight, T. A. Moore, L. A. Moore and D. Gust, *J. Am. Chem. Soc.*, 2008, **130**, 11122.
39 S. D. Straight, P. A. Liddell, Y. Terazono, T. A. Moore, A. L. Moore and D. Gust, *Adv. Funct. Mater.*, 2007, **17**, 777.
40 J. Andréasson, U. Pischel, S. D. Straight, T. A. Moore, A. L. Moore and D. Gust, *J. Am. Chem. Soc.*, 2011, **133**, 11641.
41 D. Beljonne, C. Curutchet, G. D. Scholes and R. J. Silbey, *J. Phys. Chem. B*, 2009, **113**, 6583.
42 S. Jang, M. D. Newton and R. J. Silbey, *Phys. Rev. Lett.*, 2002, **275**, 319.
43 G. D. Scholes and G. R. Fleming, *J. Phys. Chem. B*, 2000, **104**, 1854.
44 S. Jang, Y. J. Jung and R. J. Silbey, *Chem. Phys.*, 2002, **275**, 319.
45 S. Jang, M. D. Newton and R. J. Silbey, *Phys. Rev. Lett.*, 2004, **92**, 218301.
46 G. D. Scholes, X. J. Jordanides and G. R. Fleming, 2001, **105**, p. 1640.
47 H. Sumi, *J. Phys. Chem. B*, 1999, **103**, 252.
48 H. Lee, Y. C. Cheng and G. R. Fleming, *Science*, 2007, **316**, 1462.
49 D. B. Turner, R. Dinshaw, K. K. Lee, M. S. Belsley, K. E. Wilk, P. M. G. Curmi and G. D. Scholes, *Phys. Chem. Chem. Phys.*, 2012, **14**, 4857.
50 P. Rebentrost, M. Mohseni, I. Kassal, S. Lloyd and A. Aspuru-Guzik, *New J. Phys.*, 2009, **11**, 033003.
51 A. Ishizaki and G. R. Fleming, *New J. Phys.*, 2010, **12**, 055004.
52 D. Hayes, G. Panitchyangkoon, K. A. Fransted, J. R. Caram, J. Wen, K. F. Freed and G. S. Engel, *New J. Phys.*, 2010, **12**, 065042.
53 D. Abramavicius and S. Mukamel, *J. Chem. Phys.*, 2011, **134**, 174504.
54 F. Caycedo-Soler, A. W. Chin, J. Almeida, S. F. Huelga and M. B. Plenio, *J. Chem. Phys.*, 2012, **136**, 155102.
55 J. R. Caram, N. H. C. Lewis, A. F. Fidler and G. S. Engel, *J. Chem. Phys.*, 2012, **136**, 104505.
56 F. Fassioli, A. Olaya-Castro and G. D. Scholes, *J. Phys. Chem. Lett.*, 2012, **3**, 3136.
57 E. J. O'Reilly and A. Olaya-Castro, 2012, arXiv1301.6970v1.
58 A. Kolli, E. J. O'Reilly, G. D. Scholes and A. Olaya-Castro, *J. Chem. Phys. Lett.*, 2012, **137**, 174109.
59 A. Ishizaki and G. R. Fleming, *Proc. Natl. Acad. Sci. U. S. A.*, 2009, **106**, 17255.

60 G. R Fleming, G. S. Schlau-Cohen, K. Amarnath and J. Zaks, *Faraday Discuss.*, 2012, **155**, 27.
61 H. Hossein-Nejad, A. Olaya-Castro and G. D. Scholes, *J. Chem. Phys.*, 2012, **136**, 024112.
62 M. Cho, H. M. Vaswani, T. Brixner, J. Stenger and G. R. Fleming, *J. Phys. Chem. B*, 2005, **109**, 10542.
63 A. Olaya-Castro, C. F. Lee, F. Fassioli-Olsen and N. F. Johnson, *Phys. Rev. B: Condens. Matter Mater. Phys.*, 2009, **78**, 085115.
64 R. Alden, R. Islampour, H. Ma, A. A. Villaeys and S. H. Lin, *Density Matirx Method and Femtosecond Processes*, World Scientific: Tempe Arizona, 1991.
65 Y. Tanimura and R. Kubo, *J. Chem. Phys.*, 1989, **58**, 101.
66 A. Ishizaki and G. R. Fleming, *J. Chem. Phys.*, 2009, **130**, 234111.
67 Q. Shi, L. Chen, G. Nan, R. X. Xu and Y. Yan, *J. Chem. Phys.*, 2009, **130**, 084105.
68 J. Zhu, S. Kais, P. Rebentrost and A. Aspuru-Guzik, *J. Phys. Chem. B*, 2011, **115**, 1531.
69 W. T. Simpson and D. L. Peterson, *J. Chem. Phys.*, 1957, **26**, 588.
70 G. W. Robinson and R. P. Frosch, *J. Chem. Phys.*, 1962, **37**, 1962.
71 T. F. Soules, *Phys. Rev. B: Solid State*, 1971, **3**, 262.
72 D. L. Dexter, T. H. Förster and R. H. Knox, *Phys. Status Solidi B*, 1969, **34**, K159.
73 M. Yang and G. R. Fleming, *Chem. Phys.*, 2002, **275**, 355.
74 W. T. Simpson and D. L. Peterson, *J. Chem. Phys.*, 1957, **26**, 588.

Faraday Discussions RSCPublishing

PAPER

Time-resolved photoemission on the attosecond scale: opportunities and challenges

Renate Pazourek,* Stefan Nagele* and Joachim Burgdörfer*

Received 14th January 2013, Accepted 28th January 2013
DOI: 10.1039/c3fd00004d

The interaction of laser pulses of sub-femtosecond duration with matter opened up the opportunity to explore electronic processes on their natural time scale. One central conceptual question posed by the observation of photoemission in real time is whether the ejection of the photoelectron wavepacket occurs instantaneously, or whether the response time to photoabsorption is finite leading to a time delay in photoemission. Recent experimental progress exploring attosecond streaking and RABBIT techniques find relative time delays between the photoemission from different atomic substates to be of the order of ~20 attoseconds. We present *ab initio* simulations for both one- and two-electron systems which allow the determination of both absolute and relative time delays with ~1 attosecond precision. We show that the intrinsic time shift of the photoionization process encoded in the Eisenbud–Wigner–Smith delay time can be unambiguously disentangled from measurement-induced time delays in a pump-probe setting when the photoionized electronic wavepacket is probed by a modestly strong infrared streaking field. We identify distinct contributions due to initial-state polarization, Coulomb-laser coupling in the final continuum state as well as final-state interaction with the entangled residual ionic state. Extensions to multi-electron systems and to the extraction of time information in the presence of decohering processes are discussed.

1 Introduction

Following electronic dynamics in real time, watching the formation or breaking of a chemical bond, the transfer of electrons from one constituent to another, or the ejection of electrons from an atom or molecule has been a dream and challenge of time-resolved quantum physics for a long time. The extent to which such ultrafast processes are accessible to measurements at all has remained a matter of debate. Observing the temporal electronic evolution requires interrogation of the system on ultrafast time scales. Yet, such probes are subject to both time-energy and position-momentum uncertainty. Moreover, they are prone to distortion of the evolution to be observed. To put it simply: the challenge is not just to take

Institute for Theoretical Physics, Vienna University of Technology, 1040 Vienna, Austria, EU. E-mail: renate.pazourek@tuwien.ac.at; stefan.nagele@tuwien.ac.at; joachim.burgdoerfer@tuwien.ac.at

"snapshots" of ultrafast electronic motion but also to identify, to extract, and to resolve the information on the underlying processes such snapshots may contain.

Through recent advances in the creation of (sub) femtosecond laser pulses, the real-time observation of electronic dynamics in atoms, molecules, and solids has come into reach.[1–4] Phase-controlled few-cycle infrared (IR) laser pulses (cycle period $T_{IR} \simeq 2.6$ fs at $\lambda = 800$ nm) and $\sim$100 attosecond XUV pulses, temporally well correlated with each other through to the underlying high-harmonic generation (HHG) process open the opportunity to confine the photoelectric effect to a $\sim$100 attosecond time interval and to interrogate the excited or ionized electronic state by the few-cycle IR field. This so called "attosecond streaking"[5–8] technique is now being utilized to explore a wide range of fundamental photophysical and photochemical processes in the time domain, including Auger[9] and autoionization processes,[10] photoemission from surfaces,[11] and photoemission of atoms.[12] Closely related techniques include RABBIT ("reconstruction of attosecond harmonic beating by interference of two-photon transitions")[13–17] and the attosecond clock for circularly polarized fields.[18,19] Similarly, in attosecond transient absorption (ATA)[20,21] the IR pulse creates the wavepackets while the modulation of the absorption of the attosecond XUV pulse probes the time evolution of the electronically excited system. These variants of attosecond chronoscopy commonly promise to deliver real time information on electronic processes on ultrafast time scales. The ultimate goal is not only to observe, but to actively control electronic motion by light fields, *i.e.*, "lightfield electronics".[22–24] An understanding of the wealth of information to be extracted is still in its infancy and is only now beginning to emerge.

In the present manuscript we address one prototypical example, the "time delay" in photoemission. Time and time delay as a quantum dynamical observable was originally introduced by Eisenbud, Wigner and Smith (EWS)[25–27] for resonant scattering. Photoemission representing a half-scattering process allows the application of this concept. The corresponding delay, t_{EWS}, can be viewed as the finite response time in the formation of the outgoing electronic wavepacket during the photoabsorption event. In this sense, photoionization is not instantaneous, as conventionally thought, but the departure of the outgoing wavepacket is temporally shifted relative to the arrival of the XUV pulse, typically by a few attoseconds. We present an outline of a theoretical framework which unambiguously indentifies t_{EWS} as an observable accessible by attosecond streaking. The key is the determination of measurement-induced phase shifts due to the presence of the IR field which tends to taint the timing information. We discuss both one- and two-electron systems and address the influence of electronic correlation on the time-resolved photoemission processes. We point to future directions aiming at the observation and control of electronic motion in more complex nanoscale structures. Unless otherwise stated, atomic units (a.u.) are used.

2 Time delay in photoemission

Association of time, t, with a Hilbert space operator following standard correspondence identities,

$$t = -i\frac{\partial}{\partial E}, \tag{1}$$

faces fundamental conceptual difficulties[28,29] as the spectrum of the canonically conjugate Hamilton operator, H, is bounded from below. Eisenbud,[25] Wigner[26] and Smith[27] pointed out that, when restricted to the domain of scattering states, an expectation value of the time operator can be constructed,

$$t_{\mathrm{EWS}} = -iS^{\dagger}(E)\frac{\partial}{\partial E}S(E), \tag{2}$$

where $S(E)$ is the scattering operator (matrix). t_{EWS} corresponds to the observable time delay in scattering. A prototypical case is resonant scattering where t_{EWS} describes the "sticking time" of the incoming particle due to transient trapping in a quasi-bound state before leaving the interaction region as an outgoing wavepacket. This sticking time amounts to the time delay of the outgoing wavepacket relative to that of the incoming wavepacket passing by the scattering region in the absence of the interaction potential. In the special case of potential scattering with spherical symmetry, the S matrix is diagonal in the angular momentum representation S_{ℓ},

$$S_{\ell}(E) = e^{2i\delta_{\ell}(E)}. \tag{3}$$

The corresponding time delay for the partial wave, ℓ, follows from Eq. 2 as

$$t_{\mathrm{EWS}}(E,\ell) = 2\frac{\mathrm{d}}{\mathrm{d}E}\delta_{\ell}(E) \tag{4}$$

and is given by the energy variation of the partial wave scattering phase shift $\delta_{\ell}(E)$. Photoemission, the focus of the following discussion, corresponds to a half-scattering process in which only in the exit channel a matter wavepacket resides in the continuum, while in the entrance channel the wave function represents a bound rather than a continuum state. Accordingly, for photoemission by a dipole transition to a specific partial wave ℓ, the EWS time is only one-half of Eq. 4,

$$t_{\mathrm{EWS}}(E,\ell) = \frac{\mathrm{d}}{\mathrm{d}E}\delta_{\ell}(E). \tag{5}$$

More generally, when the final state does not consist of just a single partial wave, the time delay is given by the matrix element of the dipole transition operator, $\vec{d}$, between the initial bound state, ψ_{i}, and the final scattering state, ψ_{f}, describing emission into the solid angle (θ, φ),

$$t_{\mathrm{EWS}}(E,\theta,\varphi) = \frac{\mathrm{d}}{\mathrm{d}E}\arg\left[\langle\psi_{\mathrm{f}}(E,\theta,\varphi)|\vec{d}|\psi_{\mathrm{i}}\rangle\right]. \tag{6}$$

When only one partial wave in the continuum is accessed, Eq. 6 reduces to Eq. 5.

The time delay, t_{EWS}, can be alternatively extracted from the motion of the outgoing wavepacket without explicitly invoking the S matrix.[30] Such a wavepacket is formed by a coherent superposition of a band of continuum states around a central energy, E_0. The time evolution of the radial position of its crest, $r_{\mathrm{c}}(t)$, or its first moment, $\langle r\rangle_t$, eventually follows asymptotically $(t \to \infty)$ the motion of a free particle,

$$r_{\mathrm{c}}(t) = v_{\mathrm{g}}(t - t_{\mathrm{EWS}}), \tag{7}$$

$$\langle r \rangle_t = v_g(t - t_{EWS}), \tag{8}$$

after the wavepacket has left the scattering region, however delayed[31] by the time t_{EWS}. In Eq. 7 and Eq. 8 the group velocity v_g of the wavepacket is given by

$$v_g = \left(\frac{\partial E}{\partial k}\right)_{E=E_0}. \tag{9}$$

One key condition of the applicability of Eqs. 4 to 8 is that the interaction potential is short-ranged such that the wavepacket reaches the free-particle motion at large distances.[26,27]

Both the challenges and the opportunities of time-resolved photoemission are direct consequences of the implications and limitations of Eqs. 4 to 8. Typical intrinsic delay times t_{EWS} are only of the order of ~10 attoseconds. Such time scales have only now become accessible with the advances of attosecond chronoscopy. However, except for the case of photodetachment, the exit-channel interaction between the outgoing electron and the ion core is Coulombic ($\sim 1/r$) and, thus, the intrinsic atomic or molecular EWS time delay is, strictly speaking, not well defined.[26,27,32,33] We show in the following that, with suitable modifications, t_{EWS} can, indeed, be theoretically unambiguously determined and, furthermore, experimentally extracted from attosecond streaking or RABBIT measurements.

The extraction of t_{EWS} from the radial expectation value $\langle r \rangle_t$ (Eq. 8) in the absence of Coulomb interactions in the exit channel is illustrated in Fig. 1 for the simulated photodetachment by an attosecond XUV pulse (pulse duration $\tau_p = 200$ as, energy $\omega = 80$eV, intensity $I = 10^{13}$ W cm^{-2}) of an electron initially bound in an s-like state by an exponentially screened Coulomb (or Yukawa) potential,

$$V_Y(r) = -\frac{Z}{r}\exp(-r/a), \tag{10}$$

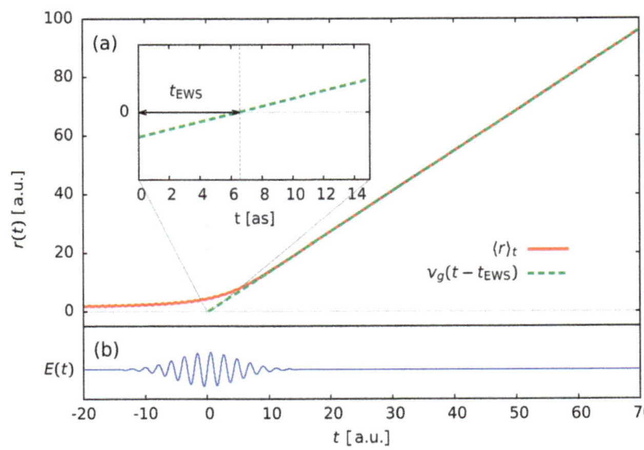

Fig. 1 (a) Extraction of t_{EWS} from the linear extrapolation of the time dependence of $\langle r \rangle_t$ (Eq. 11), for photoionization from an initial 1s state from a Yukawa potential (Eq. 10) with $Z = 3.8166$ and $a = 0.5$, which results in a binding energy of -2 a.u. The intercept with the t axis (inset) gives t_{EWS} in excellent agreement with the direct calculation for the S matrix (Eq. 5). (b) The temporal variation of the XUV pulse.

with the screening length a. For later comparison with photoionization of He^+ we choose the charge parameter Z for a given screening length, a, such that the binding energy of the detached and the ionized $He^+(1s)$ electron coincide ($\varepsilon_i = -2$ a.u.). The radial expectation value rapidly approaches the linear distance *vs.* time relation (Eq. 8) after the conclusion of the XUV pulse (FWHM) at $t \approx 4$ a.u. ($\sim$100 as). Here and in the following, the peak of the XUV-pulse envelope is chosen to be at $t = 0$. Tracing back the linear time dependence $\langle r \rangle_t$ to small times allows to determine t_{EWS} from the intercept with the t axis as $t_{EWS} = 6.6$ attoseconds. This value agrees with the one extracted from the p-wave ($\ell = 1$) phase shift $\delta_{\ell=1}$ at $E = \omega + \varepsilon_i$, (Eq. 5), to within 0.1 attoseconds, illustrating the equivalence of the time delay given by the S matrix (Eq. 5) and of the wavepacket delay (Eq. 8) for photodetachment with sub-attosecond level precision.

For photoionization of H(1s) under otherwise identical conditions a dramatically different picture emerges (Fig. 2). The delay times as extracted from the intercept of the linear extrapolation with the t axis,

$$t_{EWS} = t - \frac{\langle r \rangle_t}{k} \tag{11}$$

strongly depend on the time t (or position $\langle r \rangle$) when the linear backward extrapolation (Eq. 8) is applied. Convergence to a finite value cannot be achieved no matter for how long and how far the wavepacket is propagated into the asymptotic regime. This divergence reflects the fundamental difficulty in applying standard S-matrix scattering theory and the concept of time delay to photoionization where long-range Coulomb interactions are ubiquitously present.[32,33] This difficulty is closely related to the well-known logarithmic phase distortion by the Coulomb field which precludes equispaced wave crests and nodes, even as $r \to \infty$.

In order to extend the concept of the EWS time delay to photoionization, the logarithmic phase distortion should be included from the outset in the determination of the time delay for the Coulombic systems. We therefore start from the energy and position dependent phase of the Coulomb wave,

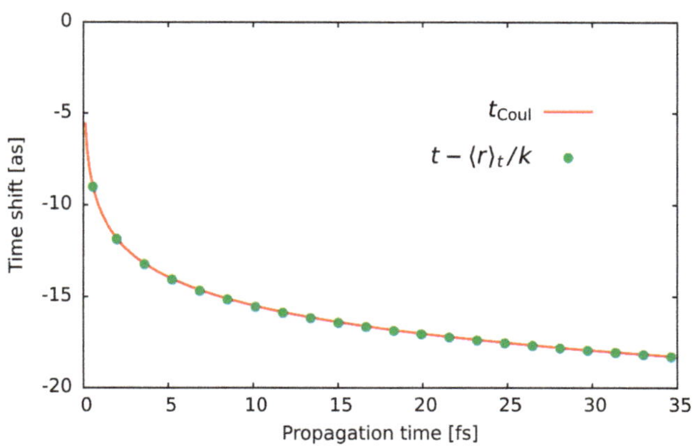

Fig. 2 Comparison between the time delay determined from the linear extrapolation with the slope taken at increasing propagation times (Eq. 11, points) with the Coulomb time $t_{Coul}(E, \ell = 1, r = kt)$ (Eq. 14, line) for ionization from the H(1s) state with an XUV pulse with $\omega = 80$eV.

$$\phi^{\text{Coul}}(E,\ell,r) = \sigma_\ell^C(E) + \frac{Z}{k}\ln(2kr), \tag{12}$$

where $k = \sqrt{2E}$ is the asymptotic momentum (or wavenumber) and

$$\sigma_\ell^C(E) = \arg\Gamma(1 + \ell - i\eta). \tag{13}$$

The r-independent term (Eq. 13) is the Coulomb analogue to the phase shift in standard scattering theory for short-ranged potentials and is often referred to as the "Coulomb phaseshift" with $\eta = Z/k$, the Coulomb–Sommerfeld parameter. Extending now the recipe for determining the time delay (Eq. 5) to the Coulomb phase (Eq. 12) we start from the spectral derivative

$$t_{\text{Coul}}(E,\ell,r) = \frac{\partial}{\partial E}\phi^{\text{Coul}}(E,\ell,r) = t_{\text{EWS}}^C(E,\ell) + \Delta t_{\text{Coul}}(E,r) \tag{14}$$

where we have introduced the Coulomb analogue of the EWS time delay associated with the Coulomb phaseshift as

$$t_{\text{EWS}}^C(E,\ell) = \frac{\partial}{\partial E}\sigma_\ell^C(E) \tag{15}$$

and an additional contribution due to the logarithmic distortion of the wavefront,

$$\Delta t_{\text{Coul}}(E,r) = \frac{Z}{(2E)^{\frac{3}{2}}}\left[1 - \ln\left(2\sqrt{2E}r\right)\right]. \tag{16}$$

It is precisely this additional time shift, Δt_{Coul}, that accounts for the r, or (since $r \approx kt$), t dependence of the time delay when extracted from the linear extrapolation. $t_{\text{Coul}}(E, \ell, r)$ (Eq. 14) agrees to within the graphical resolution with $t - \langle r\rangle_t/k$ for the wavepacket in the Coulomb field (Fig. 2). We therefore arrive at the relation between the effective time shift ("time delay") and the distance-time relation for a Coulomb wavepacket as

$$t_{\text{Coul}}(E,\ell,r = kt) = t - \frac{\langle r\rangle_t}{k}, \tag{17}$$

generalizing the relation Eq. 11 for short-ranged potentials to the case of Coulomb exit-channel interactions. From the Coulomb time shift t_{Coul} (Eq. 14) the Coulomb EWS time delay (Eq. 15) can be determined by subtracting from t_{Coul} the correction

$$\Delta t_{\text{Coul}}(E,r = kt) = \frac{Z}{(2E)^{\frac{3}{2}}}\left[1 - \ln(4Et)\right] \tag{18}$$

evaluated at $r = kt$. The correction given by Eq. 18 is universal as it depends only on the energy $E = k^2/2$ of the electron and the strength of the Coulomb field, Z, but is independent of the partial wave, ℓ, or the initial state to be photoionized. It accounts for the "slowing down" of the outgoing electron by the retarding Coulomb field resulting in an apparent negative timeshift (see Fig. 2).

The Coulomb time shift (Eq. 14) possesses a remarkably close classical counterpart first explored in a little known paper by Clark.[34] Solving the equation of motion for a hyperbolic Kepler trajectory of fixed classical angular momentum, L, we find a classical time shift $t_{\text{Coul}}^{\text{cl.}}$ relative to that of a free electron at large distance as

$$t_{\rm Coul}^{\rm cl.}(E,L,r=kt)=t(r)-\frac{r}{k}\approx\frac{Z}{(2E)^{\frac{3}{2}}}\left[1-\ln\left(\frac{4Et}{\sqrt{\eta^2+L^2}}\right)\right]. \tag{19}$$

Eq. 19 can be rewritten in terms of the logarithmic distortion term of quantum scattering (Eq. 16),

$$t_{\rm Coul}^{\rm cl.}(E,L,r=kt)=\frac{Z}{(2E)^{\frac{3}{2}}}\ln\left(\sqrt{\eta^2+L^2}\right)+\Delta t_{\rm Coul}(E,r=kt). \tag{20}$$

Comparing now Eq. 14 with Eq. 20 suggests to identify the first term in Eq. 20 as the classical analogue to the intrinsic Eisenbud–Wigner–Smith delay,

$$t_{\rm EWS}^{\rm C,\,cl.}(E,L)=\frac{Z}{(2E)^{\frac{3}{2}}}\ln\left(\sqrt{\eta^2+L^2}\right). \tag{21}$$

This identification can be readily justified by an asymptotic expansion of the quantum EWS delay (Eq. 15), expressed in terms of the digamma function $\Psi(x)$,

$$t_{\rm EWS}^{\rm C}(E,\ell)=\frac{Z}{(2E)^{\frac{3}{2}}}\Re\left[\Psi(1+\ell-i\eta)\right]. \tag{22}$$

In the (semi) classical limit of large arguments $|x|$,

$$\Psi(1+\ell-i\eta)\sim\ln(1+\ell-i\eta), \tag{23}$$

reached for either large ℓ or large Sommerfeld parameter, η, we find

$$t_{\rm EWS}^{\rm C}(E,\ell)\overset{|x|\gg 1}{=}\frac{Z}{(2E)^{\frac{3}{2}}}\ln\left(\sqrt{(1+\ell)^2+\eta^2}\right)\simeq\frac{Z}{(2E)^{\frac{3}{2}}}\ln\left(\sqrt{L^2+\eta^2}\right)=t_{\rm EWS}^{\rm C,\,cl.}(E,L), \tag{24}$$

in complete agreement with the intrinsic delay of the classical trajectory (Eq. 21), see Fig. 7 below. Note that in the (semi) classical limit, the proper mapping of the classical angular momentum L onto the ℓ quantum number is given by[35,36] $L=\ell+1/2$. This result has far-reaching implications for time-resolved photoionization studies: distortion of the timing information due to long-range Coulomb interactions in the exit channel can be accounted for both classically and quantum mechanically to a high degree of accuracy, thereby allowing to clearly disentangle intrinsic delay times in complex atoms or molecules from Coulomb induced time shifts.

3 Attosecond streaking of photoemission

Attosecond streaking has developed into one of the most important and versatile tools of attosecond science. It is a variant of a pump-probe technique in which the attosecond XUV pulse (with, typically, τ_p = 100 to 300 as) serves as a pump creating the photoelectron wavepacket, while the carrier-envelope phase (CEP) controlled few-cycle IR pulse streaks, *i.e.*, probes the wavepacket. In direct analogy to classical streaking, attosecond streaking maps time information onto energy thereby allowing to extract time information from the photoelectron spectrum with attosecond precision.

Point of departure for attosecond streaking is that the momentum of the emitted electron receives in the presence of the IR field a ponderomotive shift,[5,7,8,37]

$$\vec{p}_f(\tau) = \vec{p}_0 - \vec{A}_{\mathrm{IR}}(\tau), \tag{25}$$

given by the value of the (rescaled) vector potential $A_{\mathrm{IR}} = \tilde{A}_{\mathrm{IR}}(\tau)/c$ at the instant of the arrival in the continuum, τ. In Eq. 25, $\vec{p}_0$ is the asymptotic momentum associated with the energy of the photoelectron, $E_0 = p_0^2/2$, ($E_0 = \omega_{\mathrm{XUV}} + \varepsilon_i$), in the absence of the streaking field. Consequently, the momentum $\vec{p}_f(\tau)$ and the energy $E_f(\tau) = p_f^2(\tau)/2$ in the presence of the IR field become functions of τ via $\vec{A}_{\mathrm{IR}}$. Since the temporal distribution of the IR laser field $\vec{F}_{\mathrm{IR}}(\tau)$ and of the vector potential $\vec{A}_{\mathrm{IR}}(\tau)$,

$$\vec{A}_{\mathrm{IR}}(\tau) = \int_{\tau}^{\infty} \vec{F}_{\mathrm{IR}}(\mathrm{t})\mathrm{d}t, \tag{26}$$

can be well controlled with subcycle precision, Eq. 25 amounts to a mapping of the arrival time in the continuum, τ, onto a modulation of the linear momentum and of the energy, $E_f(\tau)$, of the photoelectron.

The implicit assumption underlying Eq. 25 is that the emitted electron attains instantaneously the asymptotic momentum $\vec{p}_0$ on a time scale that is resolved by attosecond streaking. This approximation of a sudden transition holds for the streaking of Auger decay with typical lifetimes of femtoseconds, for which this technique was first pioneered.[9] For dynamical processes on the scale of a few attoseconds and, in particular, for photoemission where the long-range Coulomb interaction implies that the local momentum of the outgoing electron $\vec{p}(r)$ approaches the asymptotic momentum only after ~100 attoseconds, application of the mapping Eq. 25 requires modification. Furthermore, the presence of the IR field may distort the timing information to be extracted. Since for easily resolvable energy shifts IR fields with intensities of the order $I_{\mathrm{IR}} \approx 10^{11} - 10^{12}\mathrm{W\ cm^{-2}}$ are needed, distortion effects generally cannot be neglected. As we will discuss in the following, the influence of the IR field on the time shift in photoemission extracted from streaking is significant, yet remarkably insensitive to the strength of the streaking field.

We first probe the extraction of timing information of photoemission for the detachment of a 1s electron bound by the Yukawa potential (Eq. 10). The photodetachment is initiated by an attosecond pump pulse with its peak amplitude centered at $t = 0$, an intensity of $10^{13}\mathrm{W\ cm^{-2}}$, and a pulse duration $\tau_{\mathrm{p}}^{\mathrm{XUV}} = 200$ as (FWHM). We expect for such short-ranged potentials the emission time (arrival in the continuum) to be delayed by the EWS time, *i.e.*, the ponderomotive momentum $\vec{A}_{\mathrm{IR}}(\tau)$ (Eq. 25) to be shifted by t_{EWS}, $\vec{A}_{\mathrm{IR}}(\tau + t_{\mathrm{EWS}})$. In turn, the time shift determined by streaking should agree with this intrinsic atomic time delay (Fig. 1),

$$t_{\mathrm{S}} = t_{\mathrm{EWS}}, \tag{27}$$

and, hence

$$\vec{p}_f(\tau) = \vec{p}_0 - \vec{A}_{\mathrm{IR}}(\tau + t_{\mathrm{S}}) = \vec{p}_0 - \vec{A}_{\mathrm{IR}}(\tau + t_{\mathrm{EWS}}). \tag{28}$$

In the simulation,[38] as well as in the experiment,[12] t_S can be determined by a fit of the first moment of the emission line in the time modulated spectrum (usually for emission parallel to the electric field polarization direction) to the vector potential $\vec{A}_{IR}(\tau)$. An example of such a so called spectrogram will be shown below in section 5 (Fig. 10). Indeed, we find Eq. 27 to be fulfilled for a wide range of final kinetic energies E corresponding to different pump frequencies ω_{XUV} and for different screening lengths (Fig. 3) to within sub-attosecond precision. As discussed in Section 2, the pair of parameters of charge and screening length is adjusted such that the initial-state binding energy remains constant at the value $\varepsilon_{1s} = -2$ a.u. corresponding to $He^+(1s)$. For later reference we note that polarization of the initial state by the IR field is negligible. The excellent agreement of t_S with t_{EWS} validates streaking as a measurement protocol of fundamental intrinsic timing information (Eq. 28). It is worth stressing the remarkable level of time resolution in reach. The arrival of the photoelectron in the continuum is delayed by $t_{EWS} \approx 6$ as (Fig. 1) relative to the peak of the XUV pulse at $t = 0$ when photoemission, if instantaneous, would, on average, occur. Yet, this minute delay, almost two orders of magnitude shorter than τ_p^{XUV} of the pump pulse is unambiguously and accurately accessible.

A drastically different picture emerges for photoionization with Coulombic long-range interactions present (Fig. 4). In that case the streaking time t_S is not directly associated with t_{EWS}^C, but is strongly influenced by the Coulomb-field induced time shift (Eq. 18).[38–40] Remarkably, the time shift t_S calculated by the solution of the TDSE in its full dimensionality[41] agrees quite well with a fully classical calculation of the time shift, $t_S^{cl.}$, using the classical trajectory Monte Carlo (CTMC) method.[42,43] This agreement indicates that the larger share of the difference between t_S and t_{EWS}^C is due to the logarithmic distortion term in Eq. 16 which has a direct classical counterpart (see Eq. 20).

For a deeper insight into the interplay between the Coulomb and laser fields and the resulting time shifts in a streaking setting, we consider a typical trajectory

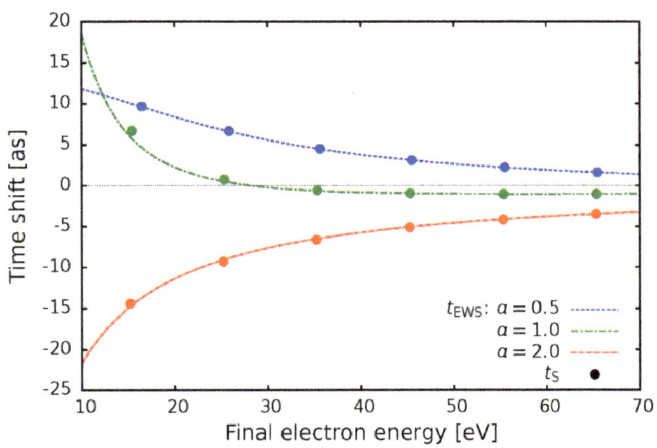

Fig. 3 Streaking time shifts t_S (dots) extracted from quantum mechanical streaking simulations for the short-ranged Yukawa potential V_Y (Eq. 10) with screening lengths $a = 0.5$, $a = 1$ and $a = 2$ a.u. The streaking IR laser field has a wavelength of 800 nm, a duration of 3 fs and an intensity of 10^{12} W cm^{-2}. The Eisenbud–Wigner–Smith delay times t_{EWS} (lines) are determined from the spectral variation of the scattering phase (Eq. 5).

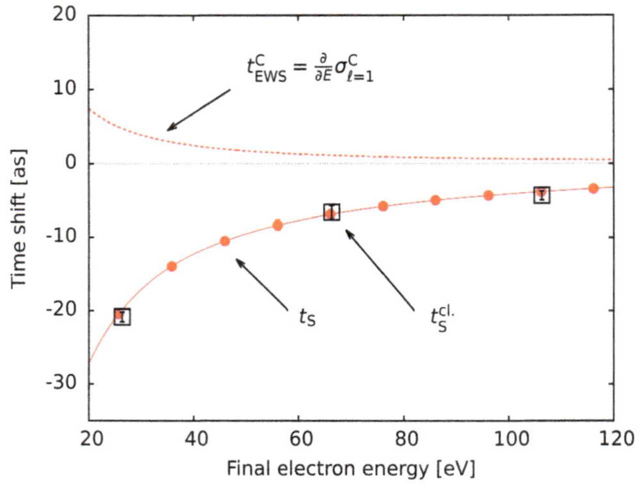

Fig. 4 Comparison between the streaking time shift t_S and the Coulomb EWS time delay t^C_{EWS} for photoionization of hydrogen as a function of the kinetic energy E of the outgoing electron ($E = \omega + \varepsilon_i$). t_S (red dots) agrees well with the classical streaking time shift $t_S^{cl.}$ (black boxes) calculated with the CTMC method.

of outgoing electrons.[38] The final momentum after taking off at $\vec{r}(\tau) \approx 0$ with momentum $\vec{p}_i$ near the ionic core is given by

$$\vec{p}_f(\tau) = \vec{p}_i + \int_\tau^\infty dt\, \vec{a}\left[\vec{F}_C(\vec{r}(t)), \vec{F}_{IR}(t)\right], \tag{29}$$

where $\vec{a}\,[\vec{F}_C,\vec{F}_{IR}]$ is the acceleration in the combined laser and Coulomb fields. In the limit of a vanishing Coulomb field, $\vec{F}_C(t) = 0$, the acceleration $\vec{a}$ reduces to $\vec{a}\,[\vec{F}_{IR}(t)] = -\vec{F}_{IR}(t)$, $\vec{p}_i = \vec{p}_0$, and Eq. 29 becomes equal to Eq. 25. In the simultaneous presence of the Coulomb field, the corrected mapping

$$\vec{p}_f(\tau) = \vec{p}_0 - \vec{A}_{IR}(\tau) + \int_\tau^\infty dt\left(\vec{a}\left[\vec{F}_C(\vec{r}(t)), \vec{F}_{IR}(t)\right] - \vec{a}_C\left[\vec{F}_C(\vec{r}(t)), 0\right] - \vec{a}_{IR}\left[0, \vec{F}_{IR}(t)\right]\right), \tag{30}$$

replaces Eq. 25 as the working equation for streaking in the presence of a Coulomb field where we have used that in absence of the laser field the integral over $\vec{a}_C[\vec{F}_C(\vec{r}(t)),0]$ describes the approach of the initial momentum $\vec{p}_i$ to the asymptotic momentum $\vec{p}_0$ and for vanishing Coulomb field the contribution from $\vec{a}_{IR}[0,\vec{F}_{IR}(t)$ can be recast as $-\vec{A}_{IR}(\tau)$.

Eq. 30 can be viewed as the classical realization of Coulomb-laser coupling (*cf.* 44,45) in the exit channel through the modification of the trajectory probing the Coulomb field by the simultaneous presence of the IR field. It treats the IR field and the Coulomb field non-perturbatively and on equal footing. Note that within a CTMC treatment an average over the microcanonical ensemble of starting positions $\vec{r}(\tau)$ representing the initial state is taken and, therefore, the results are independent of a particular choice of initial conditions. This differs from a recent semi-classical model[46] where $\vec{r}(\tau)$ is introduced as a matching parameter influencing the resulting time shift. One remarkable and, at first glance, counterintuitive consequence of Eq. 30 is that the time shift due to the interplay between the Coulomb and laser fields, or Coulomb-laser coupling, t_{CLC}, is, to leading order, independent of the IR laser intensity. Taylor expansion of the CLC correction term in Eq. 30 using that both F_C and F_{IR} vanish for $t \to \infty$, leads to

$$\int_\tau^\infty \mathrm{d}t\left(\vec{a}\left[\vec{F}_C(\vec{r}(t)),\vec{F}_{\mathrm{IR}}(t)\right]-\vec{a}_C\left[\vec{F}_C(\vec{r}(t)),0\right]-\vec{a}_{\mathrm{IR}}\left[0,\vec{F}_{\mathrm{IR}}(t)\right]\right)\approx c(E,Z,\omega_{\mathrm{IR}})\vec{F}_{\mathrm{IR}}(\tau), \tag{31}$$

with $c(E, Z, \omega_{\mathrm{IR}})$ a constant depending on the kinetic energy of the emitted electron, the strength of the Coulomb field in the exit channel, and the frequency, but not on the field strength F_{IR} of the IR field. Inserting Eq. 31 into Eq. 30 yields, in the absence of any intrinsic delay, the Coulomb modified streaking equation (see Eqs. 28, 35),

$$\vec{p}_f(\tau) = \vec{p}_0 - \alpha\vec{A}_{\mathrm{IR}}(\tau + t_{\mathrm{CLC}}), \tag{32}$$

with

$$t_{\mathrm{CLC}} = \frac{1}{\omega_{\mathrm{IR}}}\tan^{-1}c(E,Z,\omega_{\mathrm{IR}}) \tag{33}$$

independent of F_{IR}, and

$$\alpha = \frac{F_{\mathrm{IR}}}{\omega_{\mathrm{IR}}}\left(1 + c(E,Z,\omega_{\mathrm{IR}})^2\right)^{1/2}. \tag{34}$$

Consequently, the total time shift t_S mapped out by streaking and given by the time (or phase) shift of the vector potential $\vec{A}_{\mathrm{IR}}(\tau + t_S)$ contains now two contributions, the intrinsic atomic EWS delay t_{EWS} (Eq. 27) and the contributions from Coulomb-laser coupling (Eq. 33),

$$t_S = t_{\mathrm{EWS}}^C + t_{\mathrm{CLC}}. \tag{35}$$

The important consequence of the additivity (Eq. 35) is that EWS delays for photoionization in the presence of long-range Coulomb interactions become accessible by attosecond streaking provided that the time shift t_{CLC} can be independently determined. Key for the latter is that t_{CLC} is universal for Coulomb exit channel interactions, *i.e.*, is independent of the atomic species and initial state. Since the Coulomb-laser coupling is of classical origin, we can determine t_{CLC} classically by subtracting from the classical streaking time $t_S^{\mathrm{cl.}}$ determined by the CTMC the classical limit of the EWS time delay $t_{\mathrm{EWS}}^{\mathrm{C,cl.}}$ (Eq. 21),

$$t_{\mathrm{CLC}}(E,\ell) = t_S^{\mathrm{cl.}}(E,L) - t_{\mathrm{EWS}}^{\mathrm{C,cl.}}(E,L). \tag{36}$$

Fig. 5 illustrates the determination of t_{CLC} for the prototypical case of hydrogen 1s photoionization by an XUV pulse ($I_{\mathrm{XUV}} = 10^{13}\,\mathrm{W\,cm^{-2}}$, $\tau_p^{\mathrm{XUV}} = 200$ as) where the angular momentum of the ionized electron is $\ell = 1$ and the classical $L = 1 + 1/2$ (see Eq. 21). The accuracy of t_{CLC} as determined classically can be checked independently by applying Eq. 36 to the full TDSE simulation for t_S and the quantum EWS delay t_{EWS} (Eqs. 6, 35). These two completely independent methods for determining t_{CLC} agree with each other on a sub-attosecond precision level

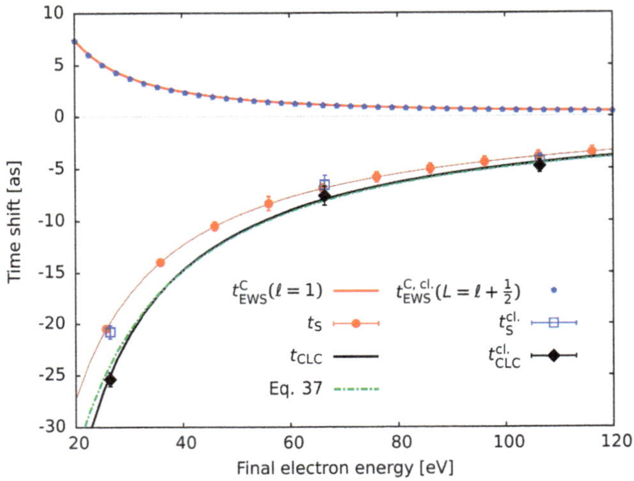

Fig. 5 Photoionization of H(1s) by an XUV pulse ($I_{XUV} = 10^{13}$W cm^{-2}, $\tau_p = 200$ as) as a function of E ($E = \omega_{XUV} + \varepsilon_{1s}$) streaked by an IR field with $I_{IR} = 10^{12}$ W cm^{-2}, $\omega_{IR} = 0.057$a.u. (800 nm). Shown are both the quantum and classical EWS delay time and the streaking time t_S. Classical and quantum t_{CLC} determined by subtraction (Eqs. 35,36) agree with each other to within the graphical resolution when taking into account that the classical angular momentum corresponding to a quantum wave with well defined ℓ is $L = \ell + 1/2$. The analytic estimation Eq. 37 coincides as well.

allowing for unambiguous identification of the Coulomb distortion on the timing information extracted by streaking.

The analysis of Coulomb-specific contributions to emission time delays discussed in section 2 provides further insight into the origin of the time shift t_{CLC} present in Coulomb interacting systems. A starting point is the remarkable similarity between the expression for the Coulomb time shift $t_{Coul}(E, \ell, r = kt)$ (Eq. 14) in the absence of an IR field and t_S (Eq. 35) in the presence of a streaking field. This comparison suggests to associate the additional time shift due to the logarithmic distortion $\Delta t_{Coul}(E, r = kt)$ (Eq. 18) with t_{CLC}. We recall that Δt_{Coul} is, just as t_{CLC} is, independent of the state-specific properties such as the partial wave, ℓ, of the continuum electron involved. It was therefore neglected in early discussions of time delay.[34] In the present case of an explicitly time-dependent streaking field, $\Delta t_{Coul}(E, t)$ becomes dynamically relevant and observable through its explicit dependence on time. We therefore surmise that the time t_0 over which the shift Δt_{Coul} is accumulated is given by the time over which the streaking field significantly varies. We therefore identify

$$t_{CLC}(E) = \frac{Z}{(2E)^{3/2}}\left(1 - \ln\left(2k^2 t_0\right)\right) \simeq \frac{Z}{(2E)^{3/2}}\left(1 - \ln(0.37ET_{IR})\right). \tag{37}$$

In the last step of Eq. 37 we have chosen t_0 to achieve a near perfect fit to the numerically determined t_{CLC}. Because of the logarithmic dependence on t_0, t_{CLC} depends only relatively weakly on the precise value chosen for t_0 within the range $\omega_{IR} t_0 \lesssim 1$. Eq. 37 provides a simple and remarkably accurate analytic estimate for t_{CLC}. We have numerically checked that Eq. 37 (for $t_0 = 0.092T_{IR}$) accounts correctly for the dependence on the strength of the Coulomb field (Z), the period

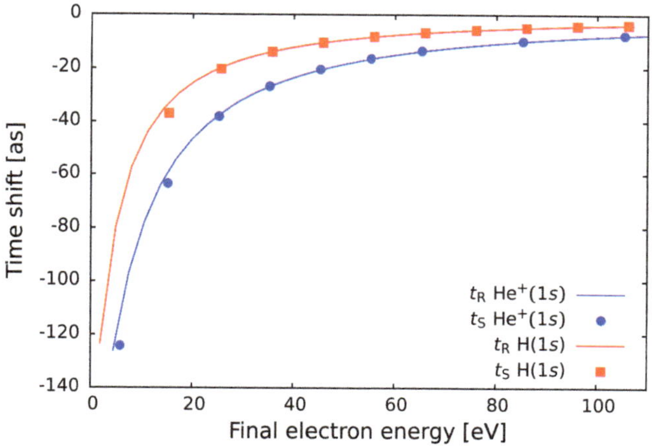

Fig. 6 Comparison between the streaking times t_S and RABBIT time shifts t_R. For streaking single attosecond pulses with τ_p= 200 as and variable ω_{XUV} are used, for RABBIT APTs built of harmonics from $q = 9$ to $q = 81$ of the fundamental ω_{IR} ($\lambda = 800$nm) are employed.

T_{IR} of the streaking field and the kinetic energy E of the ionized electron. Eqs. 35 to 37 provide a self-contained protocol to extract timing information on the photoemission for both short-ranged (Eq. 28) and Coulomb final state interactions on the attosecond scale, provided no other long-range interactions are present.

We conclude this section with a comparison between streaking and an alternative interferometric method for extracting time information, "reconstruction of attosecond harmonic beating by interference of two-photon transitions" (RABBIT).[13–17,40,47,48] The time shift observed by RABBIT, t_R, can also be written as a sum of an intrinsic atomic delay and an IR field induced shift,[16,40,48,60,74] referred to in this context as continuum-continuum coupling, t_{cc}. It is now instructive to directly and quantitatively compare t_R with t_S. In Fig. 6 we compare t_R and t_S for the photoemission of $He^+(1s)$ and $H(1s)$ over a wide range of energies.[49] Since RABBIT employs attosecond pulse trains (APTs) rather than single attosecond pulses, the frequency range in Fig. 6 is covered by APTs encompassing harmonics ranging from $q = 9$ to $q = 81$ of the fundamental ω_{IR} with $\lambda = 800$ nm. The excellent agreement underlines that t_{EWS} can be unambiguously extracted by conceptually entirely different methods provided that the additional contributions t_{CLC} or the equivalent t_{cc} are accounted for.

4 Dipole–laser interactions

Interactions decaying asymptotically as $V \sim r^{-2}$ provide an interesting case at the borderline between short- and long-ranged interactions. A prominent example is the centrifugal potential

$$V(\ell, r) = \frac{\ell(\ell+1)}{2r^2}. \tag{38}$$

Standard scattering theory applies to $V(\ell, r)$ and gives for a free particle and for all short-ranged potentials decaying faster than r^{-2}, a constant phase shift $-\ell\pi/2$

and hence a vanishing contribution to the time delay. In the simultaneous presence of the Coulomb potential, however, the centrifugal potential gives rise to an additional non-trivial ℓ-dependent contribution[50] to t^{C}_{EWS} through the non-separable dependence of the argument of the digamma function on ℓ (Eq. 22), see Fig. 7. Analogously, the classical $t^{\mathrm{C,cl.}}_{\mathrm{EWS}}$ depends on the L dependence of the Kepler hyperbola (Eq. 21). For the Coulomb exit channel interaction, the non-trivial contribution of the centrifugal potential to the Coulomb–EWS time delay can be determined by the difference to the $\ell = 0$ partial wave where the centrifugal potential vanishes,

$$t^{\Delta\ell}_{\mathrm{EWS}}(E,\ell) = t_{\mathrm{EWS}}(E,\ell) - t_{\mathrm{EWS}}(E,0). \tag{39}$$

Despite being the result of the interplay with the long-range Coulomb interaction these contributions are fully accounted for by the Coulomb EWS time delay.

Another important case are (permanent) dipole interactions also decaying as $V_d \sim r^{-2}$. Their influence on streaking was first discussed by Baggesen *et al.*[51,52] A prototypical case is the photoionization of degenerate hydrogenic manifolds where dipolar interactions are present in the entrance channel and Coulomb interactions in the exit channel. We consider in the following ionization of $He^{+}(n = 2)$. While for the states $He^{+}(2s)$ and $He^{+}(2p_1)$ only one partial wave in the continuum is accessed ($\ell = 1$ and $\ell = 2$, respectively), for $He^{+}(2p_0)$ and the parabolic states[53] $He^{+}(n = 2, k = \pm 1)$ the dipole selection rule for photoionization leads to the coherent superposition of partial waves with $\ell = 0, 1, 2$ and, according to Eq. 6, to variations of the Coulomb–EWS times with the angle of photoemission.

The two parabolic states feature a permanent electric dipole moment[53] of $d_{k=\pm1} = \pm 3n/2Z$. The effect of the initial state polarization on the photoionization transition matrix element is fully contained in the Coulomb–EWS time (see Eq. 6). However, in the presence of the IR streaking field, its interaction with the initial-state dipole leads to a time-dependent initial-state energy shift

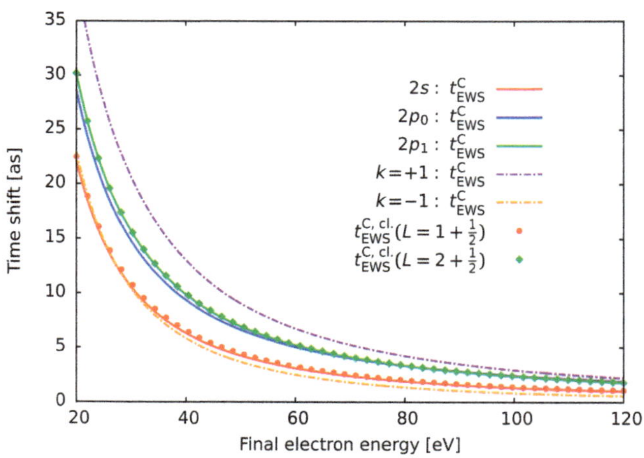

Fig. 7 Coulomb EWS times for the photoionization of $He^{+}(n = 2)$ substrates as a function of the kinetic energy $E = \omega_{\mathrm{XUV}} + \varepsilon_{1s}$. Also shown is $t^{\mathrm{C,cl.}}_{\mathrm{EWS}}$ for different $L = \ell + 1/2$ (Eq. 21).

$$\Delta E(\tau) = -\vec{d}\vec{F}_{\mathrm{IR}}(\tau) \tag{40}$$

and, in turn, to a phase shift

$$\Phi(t) = -\vec{d}\int_{-\infty}^{t} \vec{F}_{\mathrm{IR}}(t')\mathrm{d}t' = \vec{d}\vec{A}_{\mathrm{IR}}(t). \tag{41}$$

This dipole-laser coupling (dLC) induced energy and phase shift gives rise to additional streaking-measurement related time shifts which distort the intrinsic time information to be extracted. Indeed, the streaking time t_S for the $He^+(n = 2)$ initial states significantly differ from those predicted by Eq. 35 (Fig. 8). This dLC contribution can be easily accounted for by including the IR field induced energy shift of the initial state.[51] The additional modulation of the final-state energy (see Eq. 40), neglecting terms to second order in the streaking field, is given by

$$E_f(\tau) = \frac{p_f^2(\tau)}{2} = E_0 - \vec{p}_0\vec{A}(\tau) - \vec{d}\vec{F}_{\mathrm{IR}}(\tau). \tag{42}$$

In direct analogy to Eq. 32 and Eq. 33, the dipole contribution $\sim F_{\mathrm{IR}}(\tau)$, being $\pi/2$ out of phase with the streaking momentum $A_{\mathrm{IR}}(\tau)$, gives rise to an additional time shift[51] of the streaking momentum

$$\alpha_{\mathrm{dl}}A_{\mathrm{IR}}(\tau + t_{\mathrm{dLC}}^{(i)}) = A_{\mathrm{IR}}(\tau) + d_k F_{\mathrm{IR}}(\tau)/p_0, \tag{43}$$

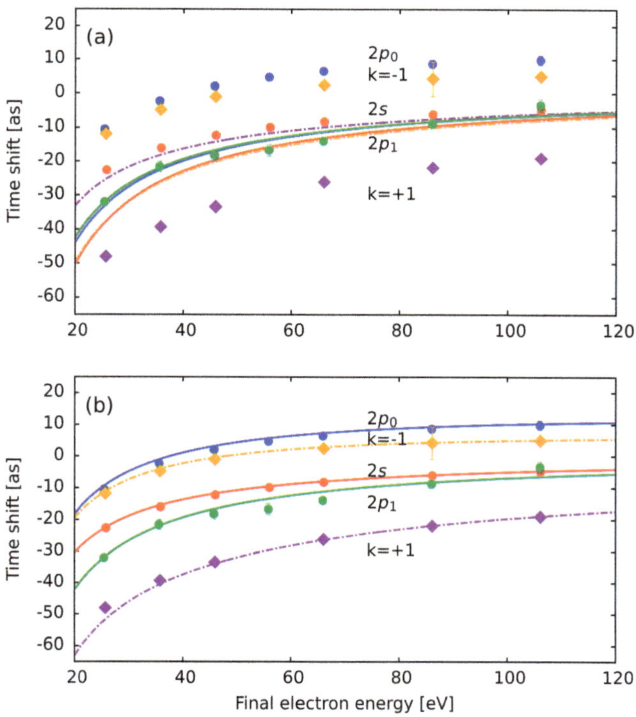

Fig. 8 Streaking time shifts t_S (points) for $He^+(n = 2)$ substates, 2s, $2p_0$, $2p_1$ and the parabolic states ($n = 2$, $k = \pm 1$) compared to (a) the analytic prediction without initial-state polarization (Eq. 35, lines), (b) including the dipole-laser coupling contribution (Eq. 46, lines).

evaluated for emission along the laser polarization axis, with

$$t_{\mathrm{dLC}}^{(i)} = \frac{1}{\omega_{\mathrm{IR}}} \tan^{-1}\left(-\frac{d_k \omega_{\mathrm{IR}}}{p_0}\right) \tag{44}$$

and

$$\alpha_{\mathrm{dLC}} = \left(1 + \left(\frac{d_k \omega_{\mathrm{IR}}}{p_0}\right)^2\right)^{1/2}. \tag{45}$$

Thus, for photoionization of atomic or molecular initial states (i) with a permanent electric dipole the relation Eq. 35 between the observable streaking time shift t_S and the intrinsic EWS time delay t_{EWS} has to be modified to

$$t_S = t_{\mathrm{EWS}} + t_{\mathrm{CLC}} + t_{\mathrm{dLC}}^{(i)}. \tag{46}$$

Eq. 46 can account for all time shifts observed for the degenerate hydrogenic manifolds with (sub-)attosecond precision (Fig. 8b). Most remarkably, Eq. 46 allows us to accurately account for the corrections to initial states that have no permanent dipole (2s, $2p_0$) but are highly polarizable due to the degeneracy of the manifold. Expanding $|2s\rangle$ and $|2p_0\rangle$ in terms of the two parabolic states

$$|n = 2, \ell = 0\ (\ell = 1)\rangle = \frac{1}{\sqrt{2}}\Big(|k = 1\rangle + (-)|k = -1\rangle\Big), \tag{47}$$

the effective induced dipole moment follows as

$$\langle d_{\mathrm{eff}}^{\ell} \rangle = \frac{1}{2|c_\ell|^2} \sum_k d_k |c_k|^2 \tag{48}$$

where $|c_k|^2$ and $|c_\ell|^2$ are the square moduli of the dipole transition matrix elements (Eq. 6) from the initial state k or ℓ to the continuum with final momentum along the polarization axis. We note that the present findings of a dLC shift of polarizable states without a permanent dipole moment is at variance with recent claims.[52]

5 Many-electron systems

The analysis of the preceding section focussed on one (active) electron systems for which both classical and quantum dynamics in a streaking setting can be solved (to within numerical precision) exactly, and, moreover, accurate analytical estimates for the time-shifts involved can be given. Since in the pioneering experiment the relative time delay between photoionization of the 2s and 2p electrons was measured for neon,[12] the influence of many-electron and correlation effects on the intrinsic time delay and on streaking-measurement related time shifts became important issues. We briefly address this topic with the help of two prototypical two (active) electron systems for which numerically exact solutions of the TDSE are still feasible.[54–56]

We first discuss a fully correlated two-active electron (TAE) model[57] for two non-equivalent electrons denoted in the following as (1s, 2p). These two active electrons move in a mean field of $N - 2$ passive electrons. We choose a background potential, V_{N-2}, such that the ionization potential of the 2p electron

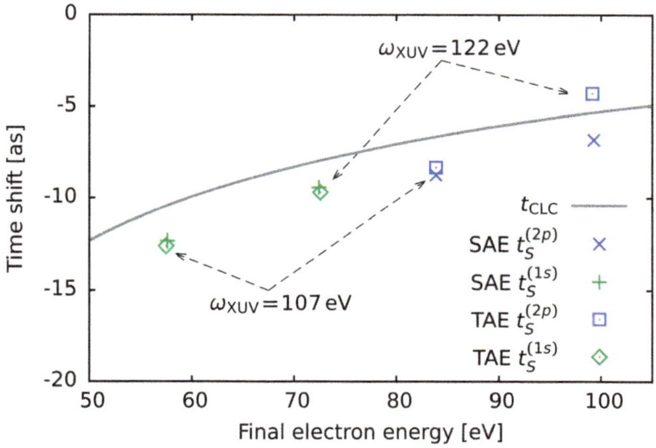

Fig. 9 Streaking times t_S for a fully correlated two-electron "neon model" atom (see text) employing a two-active electron (TAE) description and comparison with a mean-field single-active (SAE) model. Also shown is t_{CLC} (Eq. 33).

matches that of the Ne 2p electron while the ionization potential of the 1s electron corresponds to that of the Ne 2s subshell. This model thus shares several features of the neon atom including identical kinetic energies, identical t_{CLC} (Eq. 33), and identical angular momenta of the two electrons, the difference in emission time of which has been experimentally explored.[12] While the model is not intended to provide a realistic approximation to the 10-electron system it allows to fully account for the effect of e–e interactions of the two non-equivalent electrons on streaking. Moreover, by comparison with a corresponding single-active electron (SAE) model for electrons moving in a corresponding $N-1$ electron mean field potential $V_{N-1}(r)$, e–e effects on both the EWS time delay and on the measurement induced time shifts can be probed. Fig. 9 illustrates similarities and differences of the observable streaking shifts t_S within TAE and SAE approaches for two different XUV energies. Significant differences between the TAE and SAE models of about 2.5 as can be observed for emission of the 2p electron at $\omega_{XUV} = 122\text{eV}$. This difference can be traced to differences in the (negative) EWS delay times, while the CLC contribution (also shown in Fig. 9) is identical within the TAE and SAE models. We have numerically verified the additivity relation Eq. 35 with sub-attosecond precision. Note that the initial-state polarizability for this model system with well separated energy levels is negligible and, hence, $t_{\text{dLC}}^{(i)} = 0$. We also stress that additional contributions to the streaking-related time shifts analogous to t_{CLC} due to short-ranged portions of many-electron potentials V_N, as recently proposed,[46] can be ruled out based on the present numerical evidence. Moreover, simple semi-classical considerations argue strongly against additional CLC-like contributions from short-ranged potentials: for typical XUV frequencies currently accessible by HHG, kinetic energies of the released electrons are $E \lesssim 100\text{eV}$. Their large de-Broglie wavelength $\lambda_{\text{dB}} = 2\pi/\sqrt{2E} \gtrsim 3\text{a.u.}$ precludes the spatial resolution of the short-ranged component of atomic potentials. Consequently, quasi-classical trajectory modifications induced by the interplay of short-ranged atomic potentials and the laser field in analogy to Eq. 30 are unlikely to occur.

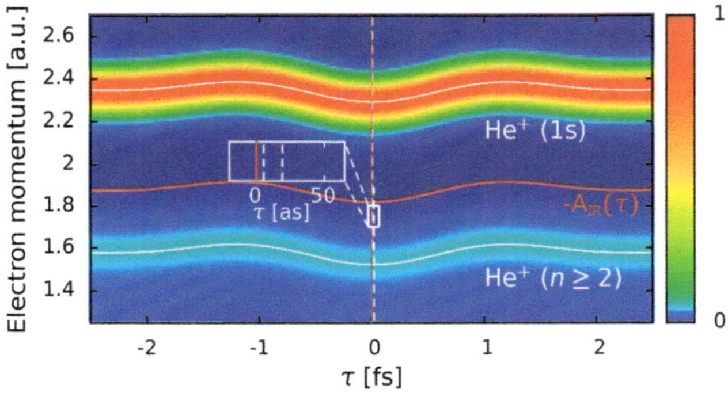

Fig. 10 Numerically simulated streaking spectrogram for helium and emission into a forward cone (opening angle 10°) around the polarization axis. XUV pulse with $\omega_{XUV} = 100$ eV, $\tau_p = 200$ as, $I_{XUV} = 10^{12}$ W cm^{-2}, streaking field with $I_{IR} = 4 \cdot 10^{11}$ W cm^{-2}, $\lambda = 800$nm. Top: main line $He^+(1s)$, bottom: shake-up satellite (predominantly) to $He^+(n = 2)$.

The experimentally observed relative delay of $\Delta t_S \approx 21$ as between the 2p and 2s electron of neon is, so far, theoretically unaccounted for. Several theoretical approaches contained in the original paper,[12] as well as more recent works employing the random phase approximation with exchange for the photoionization matrix element,[58] time-dependent R-matrix theory,[59] and many-body perturbation theory[60] fall short by about a factor two to three. For the present Ne model system (Fig. 9) we find $\Delta t_S \approx 4$ as.

Ionization of helium accompanied by shake-up forming so called spectroscopic "correlation satellite" lines represents the prototypical case of photoionization strongly driven by electron–electron interactions of two equivalent electrons.[61,62] For these processes fully *ab initio* simulations for an attosecond streaking setting are feasible.[63] We focus in the following on the shake-up to $n = 2$ which becomes accessible for XUV energies exceeding the sum of the ionization potential (I_p) of He and the excitation energy $\Delta E(n = 1 \rightarrow 2)$ in He^+. The streaking spectrogram (Fig. 10) of the main line ($He^+(1s)$) and satellite ($He^+(n = 2)$) reflects the weakness of the shake-up channel (< 5% of the main line) and indicates the experimental challenge to perform a chronoscopic measurement with attosecond precision.

While for the main line, the sum of t_{CLC} and t_{EWS} (Eq. 35) accurately accounts for the numerically determined t_S, for the shake-up states to $n = 2$ Eq. 35 fails (Fig. 11). Note that the initial-state polarization of the helium ground state can be safely neglected as confirmed by the agreement for the main line and, thus, $t^{(i)}_{dLC} = 0$. In this case, it is the high polarizability of the degenerate hydrogenic $He^+(n = 2)$ manifold of the residual ion that causes the discrepancy between the extracted streaking time and contributions from Coulomb-laser coupling and the EWS time.

The starting point is the observation that the two parabolic states $He^+(n = 2, k = \pm 1)$ possess a large permanent dipole moment (see section 4). The long-range interaction between the ionized electron and the residual ion in the exit channel contains now both a Coulombic and a dipolar interaction, $V_d = -\vec{d}\vec{r}/r^3$. Their direct contributions to t_{EWS} in the absence of a streaking field is included in the exact calculation of the dipole transition matrix for single ionization of He employing the extraction method based on exterior complex scaling.[64,65,75] In the presence of a streaking field, however, both long-range portions give additional

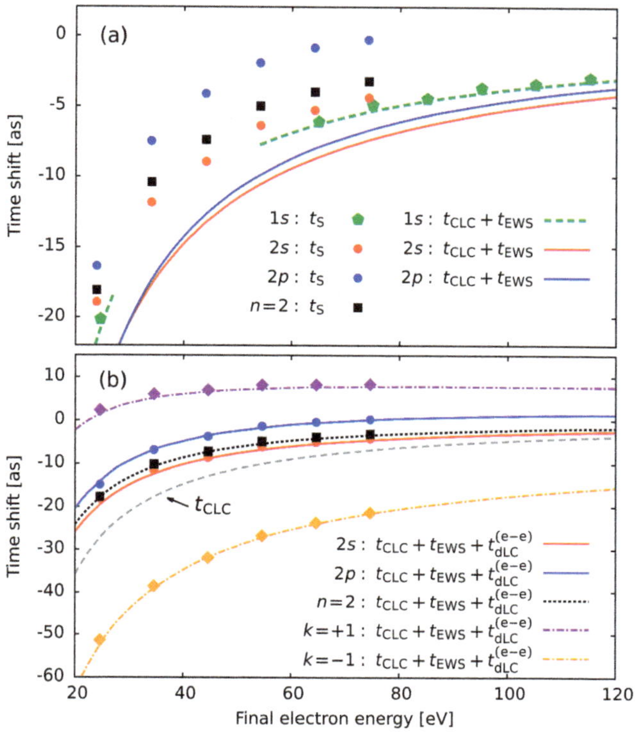

Fig. 11 Streaking of shake-up ionization of helium (laser parameters as in Fig. 10). (a) Comparison between the prediciton Eq. 35 (lines) and the *ab initio* two-electron streaking data (points). (b) Comparison between Eq. 51 (lines) and streaking data for all $n = 2$ substates and the $n = 2$ shell average.

contributions. The Coulomb tail results in the time shift t_{CLC} (Eq. 33) while the dipolar interaction between the bound and ionized electron gives rise to a novel contribution which is a true field-induced electron-electron effect. The asymptotic two-electron state representing one continuum and one bound electron in the presence of the streaking field has the time-dependent energy (or time derivative of its phase),

$$E_{e-e}(\tau) = \varepsilon_{nk} + E - \vec{p}_0\vec{A}_{IR}(\tau) - \vec{d}_k\vec{F}_{IR}(\tau), \tag{49}$$

where ε_{nk} is the binding energy of the residual parabolic state and E the kinetic energy of the ionized electron at zero field. The remaining two terms account for the interaction of the IR field with the continuum electron and with the dipole moment of the residual electronic bound state. The physical picture underlying the IR field induced dipole shift to streaking is illustrated in Fig. 12. The energy modulations of the parabolic states $k = 1$ and $k = -1$ are out of phase by π relative to each other (Fig. 12b). Following the derivation Eqs. 43 to 45 we find for the two-electron dipole laser coupling induced time shift (evaluated for emission along the laser polarization axis)

$$t_{dLC}^{(e-e)} = \frac{1}{\omega_{IR}}\tan^{-1}\left(\frac{d_k\omega_{IR}}{p_0}\right). \tag{50}$$

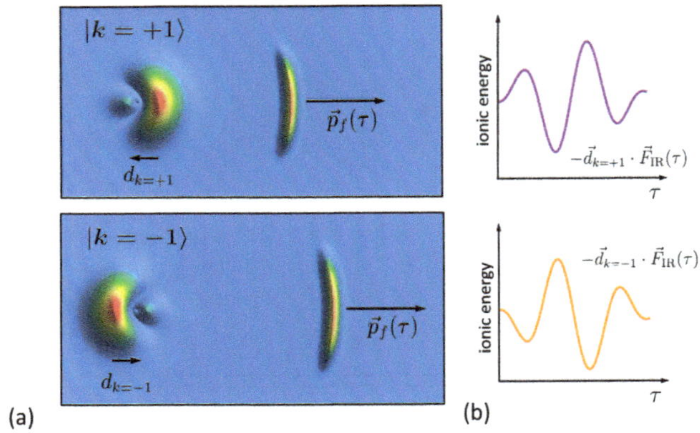

Fig. 12 (a) Two-electron dipole-laser interaction in the exit channel. (b) Resulting modulation of the kinetic energy of the ionized electron emitted at time τ.

Note the opposite sign of the dipole term relative to the corresponding expression Eq. 44 for the entrance channel perturbation of the one-electron problem. Adding this term to Eq. 46,

$$t_{\rm S} = t_{\rm EWS} + t_{\rm CLC} + t_{\rm dLC}^{(i)} + t_{\rm dLC}^{(\rm e-e)}, \tag{51}$$

with $t_{\rm dLC}^{(i)} \approx 0$ in the present case, leads to near-perfect agreement for all $n = 2$ shake-up substates (Fig. 11b) confirming, once more, the additivity rule for laser-modified long-range interactions. As in the one-electron case, not only ionic final states with a permanent dipole moment $|n, k, m\rangle$ but also highly polarizable states in degenerate manifolds with zero static dipole moment $|n, \ell, m\rangle$ suffer a dipole-laser induced time shift (see Eq. 48). A non-zero contribution survives even when an average over all $n = 2$ substates is taken (Fig. 11b). Conceptually, Eq. 50 and Eq. 51 reflect the entanglement in the exit channel. As a result, the streaking time shift of the observed ionized electron depends on the dipole moment of the substate of the residual ion. In turn, in such a setup the final quantum state $|n, \ell, m\rangle$ or $|n, k, m\rangle$ becomes accessible in a non-destructive measurement, *i.e.*, without directly observing it.

6 Outlook: complex systems

We conclude the discussion of time-resolved photoemission by streaking with an outlook toward the extraction of dynamic information of more complex many-electron systems. As the number of degrees of freedom increases, following the dynamic evolution on a state-resolved level becomes impossible. Instead of the fully coherent N-electron evolution, the partially coherent dynamics on the level of one- (or few-) particle reduced density matrices takes center stage. The streaked electron thus carries information on the interactions with the "bath" of the electrons and ionic cores it is immersed in prior to and during the emission process. Instead of an EWS time delay associated with an individual scattering state, dephasing, decoherence, and relaxation times become accessible. Concomitantly, the strict Fourier reciprocity between the time domain and energy domain is broken. While for a fully coherent photoionization process the spectral variation of the scattering phase (see Eq. 6) determining the intrinsic atomic time delay, is, in principle, indirectly accessible in the energy domain by high-resolution angular spectroscopy, time-resolved photoemission in complex open systems can provide novel information on decohering processes such as collisions, energy exchange, and relaxation not easily accessible in the spectral domain. We therefore conjecture that the full potential of time-resolved photoelectron spectroscopy, *i.e.*, chronoscopy, will emerge when applied to complex systems. Exploiting these opportunities, both experimentally and theoretically, remains a challenge.

One case in point is time-resolved photoemission from a tungsten surface.[11] The time delay between conduction band electrons and core levels were associated with transport effects and steep variation of the group velocity near the points in k-space where Bragg reflections occur.[11,66] Within a classical transport theory, multiple scattering processes and energy loss due to plasmon excitation have been shown to effectively contribute to the observed time delay.[67] However, accounting quantitatively for the experimentally observed time delay has remained a widely open question.[68,69]

An intriguing scenario at the borderline between an atomic and condensed-matter system is photoemission from an endohedral C_{60} complex, $A@C_{60}$, where an atom A, for example xenon, with a well-defined core-level emission line is located at the center of the C_{60} cage[70,71] (Fig. 13a). The cage will act as a finite-size bath with which the photoelectron interacts. Point of departure of a simplified SAE model is an effective potential for the core electron bound to the center atom $V_A(r)$ with a long-range Coulomb tail. As the outgoing electron traverses the cage, it experiences a short-ranged potential created by the shell of C_{60} atoms. A simple model potential for the shell is[72]

$$V_{\text{shell}}(r) = \begin{cases} -V_0 & \text{for} \quad r_0 \le r \le r_0 + \Delta \\ 0 & \text{otherwise} \end{cases} \tag{52}$$

with r_0 the inner-radius of the C_{60} shell, Δ its width and V_0 the mean potential on a one-electron level. The short-ranged admixture would lead to an energy dependent modulation δt_{EWS} of the intrinsic atomic time delay, t^A_{EWS},

$$t_{\text{EWS}}(E) = t^A_{\text{EWS}}(E) + \delta t_{\text{EWS}}(E) \tag{53}$$

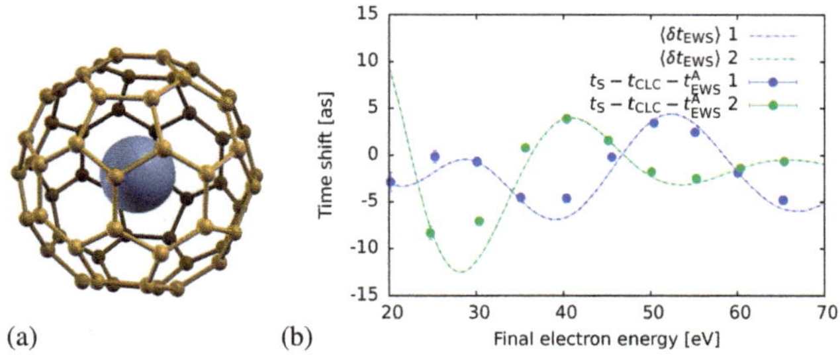

Fig. 13 (a) Endohedral complex, $He^+@C_{60}$, schematically. (b) Variation of t_S (dots) due to the modulation of t_{EWS} (lines, averaged over the XUV spectrum) as a function of the kinetic energy E of the outgoing electron, see Eq. 53. For V_{shell} we have chosen two sets of values previously used in the literature.[72] 1: $V_0 = -0.302$ a.u., $\Delta = 1.9$ a.u., $r_0 = 5.89$ a.u., 2: $V_0 = -0.422$ a.u., $\Delta = 1.25$ a.u., $r_0 = 6.01$ a.u.

on the fully coherent one-electron level. We find, for He^+ enclosed in a C_{60} cage, good agreement for the prediction for the streaking time (Eq. 35) when the modified EWS time Eq. 53 is used (Fig. 13b). Future studies will go beyond such an SAE model and novel features are expected to emerge. The outgoing electron will interact with the 240 valence electrons of the C_{60} shell which represent environmental degrees of freedom. Among the channels to be considered are collective excitations, *i.e.*, surface and bulk plasmons, and collisional excitation and ionization of the shell. Moreover, the highly polarizable C_{60} shell is expected to modify the streaking process itself. It is hoped that experimental data for streaking of such nanoplasmonic systems become accessible in the near future.[73]

Acknowledgements

The authors would like to thank Johannes Feist, Katharina Doblhoff-Dier, Stephan Preisinger, Christoph Lemell, Károly Tőkési, and Georg Wachter for valuable input. This work was supported by the FWF-Austria (SFB NEXTLITE, SFB VICOM and P23359-N16), and in part by the National Science Foundation through XSEDE resources provided by NICS and TACC under Grant TG-PHY090031. The computational results presented have also been achieved in part using the Vienna Scientific Cluster (VSC). RP acknowledges support from the TU Vienna Doctoral Program Functional Matter.

References

1 P. H. Bucksbaum, *Science*, 2007, **317**, 766–769.
2 P. B. Corkum and F. Krausz, *Nat. Phys.*, 2007, **3**, 381–387.
3 M. F. Kling and M. J. J. Vrakking, *Annu. Rev. Phys. Chem.*, 2008, **59**, 463–492.
4 F. Krausz and M. Ivanov, *Rev. Mod. Phys.*, 2009, **81**, 163–234.
5 M. Drescher, M. Hentschel, R. Kienberger, G. Tempea, C. Spielmann, G. A. Reider, P. B. Corkum and F. Krausz, *Science*, 2001, **291**, 1923–1927.
6 M. Hentschel, R. Kienberger, C. Spielmann, G. A. Reider, N. Milosevic, T. Brabec, P. Corkum, U. Heinzmann, M. Drescher and F. Krausz, *Nature*, 2001, **414**, 509–513.
7 J. Itatani, F. Quéré, G. L. Yudin, Ivanov, F. Krausz and P. B. Corkum, *Phys. Rev. Lett.*, 2002, **88**, 173903.
8 M. Kitzler, N. Milosevic, A. Scrinzi, F. Krausz and T. Brabec, *Phys. Rev. Lett.*, 2002, **88**, 173904.
9 M. Drescher, M. Hentschel, R. Kienberger, M. Uiberacker, V. Yakovlev, A. Scrinzi, T. Westerwalbesloh, U. Kleineberg, U. Heinzmann and F. Krausz, *Nature*, 2002, **419**, 803–807.
10 M. Wickenhauser, J. Burgdörfer, F. Krausz and M. Drescher, *Phys. Rev. Lett.*, 2005, **94**, 023002.
11 A. L. Cavalieri, N. Müller, T. Uphues, V. S. Yakovlev, A. Baltuška, B. Horvath, B. Schmidt, L. Blümel, R. Holzwarth, S. Hendel, M. Drescher, U. Kleineberg, P. M. Echenique, R. Kienberger, F. Krausz and U. Heinzmann, *Nature*, 2007, **449**, 1029–1032.
12 M. Schultze, M. Fiess, N. Karpowicz, J. Gagnon, M. Korbman, M. Hofstetter, S. Neppl, A. L. Cavalieri, Y. Komninos, T. Mercouris, C. A. Nicolaides, R. Pazourek, S. Nagele, J. Feist, J. Burgdörfer, A. M. Azzeer, R. Ernstorfer, R. Kienberger, U. Kleineberg, E. Goulielmakis, F. Krausz and V. S. Yakovlev, *Science*, 2010, **328**, 1658–1662.
13 V. Véniard, R. Taïeb and A. Maquet, *Phys. Rev. A: At., Mol., Opt. Phys.*, 1996, **54**, 721–728.
14 P. M. Paul, E. S. Toma, P. Breger, G. Mullot, F. Auge, P. Balcou, H. G. Muller and P. Agostini, *Science*, 2001, **292**, 1689–1692.
15 E. S. Toma and H. G. Muller, *J. Phys. B: At., Mol. Opt. Phys.*, 2002, **35**, 3435–3442.
16 K. Klünder, J. M. Dahlström, M. Gisselbrecht, T. Fordell, M. Swoboda, D. Guénot, P. Johnsson, J. Caillat, J. Mauritsson, A. Maquet, R. Taïeb and A. L'Huillier, *Phys. Rev. Lett.*, 2011, **106**, 143002.
17 D. Guénot, K. Klünder, C. L. Arnold, D. Kroon, J. M. Dahlström, M. Miranda, T. Fordell, M. Gisselbrecht, P. Johnsson, J. Mauritsson, E. Lindroth, A. Maquet, R. Taïeb, A. L'Huillier and A. S. Kheifets, *Phys. Rev. A: At., Mol., Opt. Phys.*, 2012, **85**, 053424.

18 P. Eckle, M. Smolarski, P. Schlup, J. Biegert, A. Staudte, M. Schoffler, H. G. Muller, R. Dorner and U. Keller, *Nat. Phys.*, 2008, **4**, 565–570.
19 A. N. Pfeiffer, C. Cirelli, M. Smolarski, R. Dorner and U. Keller, *Nat. Phys.*, 2011, **7**, 428–433.
20 E. Goulielmakis, Z.-H. Loh, A. Wirth, R. Santra, N. Rohringer, V. S. Yakovlev, S. Zherebtsov, T. Pfeifer, A. M. Azzeer, M. F. Kling, S. R. Leone and F. Krausz, *Nature*, 2010, **466**, 739–743.
21 H. Wang, M. Chini, S. Chen, C. H. Zhang, F. He, Y. Cheng, Y. Wu, U. Thumm and Z. Chang, *Phys. Rev. Lett.*, 2010, **105**, 143002.
22 E. Goulielmakis, V. S. Yakovlev, A. L. Cavalieri, M. Uiberacker, V. Pervak, A. Apolonski, R. Kienberger, U. Kleineberg and F. Krausz, *Science*, 2007, **317**, 769–775.
23 M. Schultze, E. M. Bothschafter, A. Sommer, S. Holzner, W. Schweinberger, M. Fiess, M. Hofstetter, R. Kienberger, V. Apalkov, V. S. Yakovlev, M. I. Stockman and F. Krausz, *Nature*, 2013, **493**, 75–78.
24 A. Schiffrin, T. Paasch-Colberg, N. Karpowicz, V. Apalkov, D. Gerster, S. Muhlbrandt, M. Korbman, J. Reichert, M. Schultze, S. Holzner, J. V. Barth, R. Kienberger, R. Ernstorfer, V. S. Yakovlev, M. I. Stockman and F. Krausz, *Nature*, 2013, **493**, 70–74.
25 L. Eisenbud, PhD thesis, Princeton University, 1948.
26 E. P. Wigner, *Phys. Rev.*, 1955, **98**, 145–147.
27 F. T. Smith, *Phys. Rev.*, 1960, **118**, 349–356.
28 W. Pauli, *Handbuch der Physik*, 1st edn, 1926, vol. 23, pp. 1–278.
29 W. Pauli, *Handbuch der Physik*, 2nd edn, 1933, vol. 24, pp. 83–272.
30 C. A. A. de Carvalho and H. M. Nussenzveig, *Phys. Rep.*, 2002, **364**, 83–174.
31 W. Brenig and R. Haag, *Fortschr. Phys.*, 1959, **7**, 183–242.
32 D. Bollé, F. Gesztesy and H. Grosse, *J. Math. Phys.*, 1983, **24**, 1529–1541.
33 P. A. Martin, *Acta Phys. Austriaca, Suppl.*, 1981, **23**, 159.
34 C. W. Clark, *Am. J. Phys.*, 1979, **47**, 683.
35 R. E. Langer, *Phys. Rev.*, 1937, **51**, 669–676.
36 J. Burgdörfer, C. O. Reinhold, J. Sternberg and J. Wang, *Phys. Rev. A: At., Mol., Opt. Phys.*, 1995, **51**, 1248–1252.
37 E. Constant, V. D. Taranukhin, A. Stolow and P. B. Corkum, *Phys. Rev. A: At., Mol., Opt. Phys.*, 1997, **56**, 3870–3878.
38 S. Nagele, R. Pazourek, J. Feist, K. Doblhoff-Dier, C. Lemell, K. Tőkési and J. Burgdörfer, *J. Phys. B: At., Mol. Opt. Phys.*, 2011, **44**, 081001.
39 C. H. Zhang and U. Thumm, *Phys. Rev. A: At., Mol., Opt. Phys.*, 2010, **82**, 043405.
40 J. M. Dahlström, D. Guénot, K. Klünder, M. Gisselbrecht, J. Mauritsson, A. L'Huillier, A. Maquet and R. Taïeb, *Chemical Physics*, 2013, **414**, 53–64.
41 X. M. Tong and S. I. Chu, *Chem. Phys.*, 1997, **217**, 119–130.
42 R. Abrines and I. C. Percival, *Proc. Phys. Soc.*, 1966, **88**, 861–872.
43 K. I. Dimitriou, D. G. Arbó, S. Yoshida, E. Persson and J. Burgdörfer, *Phys. Rev. A: At., Mol., Opt. Phys.*, 2004, **70**, 061401.
44 O. Smirnova, M. Spanner and M. Y. Ivanov, *J. Phys. B: At., Mol. Opt. Phys.*, 2006, **39**, S323–S339.
45 O. Smirnova, A. S. Mouritzen, S. Patchkovskii and M. Y. Ivanov, *J. Phys. B: At., Mol. Opt. Phys.*, 2007, **40**, F197–F206.
46 M. Ivanov and O. Smirnova, *Phys. Rev. Lett.*, 2011, **107**, 213605.
47 H. G. Muller, *Appl. Phys. B: Lasers Opt.*, 2002, **74**, s17–s21–s21.
48 J. M. Dahlström, A. L'Huillier and A. Maquet, *J. Phys. B: At., Mol. Opt. Phys.*, 2012, **45**, 183001.
49 S. Nagele, S. Preisinger, R. Pazourek and Burgdörfer, *In preparation*, 2013.
50 R. Pazourek, S. Nagele, K. Doblhoff-Dier, J. Feist, C. Lemell, K. Tökési and J. Burgdörfer, *J. Phys. Conf. Ser.*, 2012, **388**, 012029.
51 J. C. Baggesen and L. B. Madsen, *Phys. Rev. Lett.*, 2010, **104**, 043602.
52 J. C. Baggesen and L. B. Madsen, *Phys. Rev. Lett.*, 2010, **104**, 209903.
53 H. A. Bethe and E. E. Salpeter, *Encycl. Phys.*, 1957, **35**, 88–439.
54 E. S. Smyth, J. S. Parker and K. T. Taylor, *Comput. Phys. Commun.*, 1998, **114**, 1–14.
55 B. I. Schneider, J. Feist, S. Nagele, R. Pazourek, S. X. Hu, L. A. Collins and J. Burgdörfer, *Quantum Dynamic Imaging*, Springer, 2011, ch. 10.
56 J. Feist, S. Nagele, R. Pazourek, E. Persson, B. I. Schneider, L. A. Collins and J. Burgdörfer, *Phys. Rev. A: At., Mol., Opt. Phys.*, 2008, **77**, 043420.
57 S. Nagele, R. Pazourek, J. Feist and J. Burgdörfer, *Phys. Rev. A: At., Mol., Opt. Phys.*, 2012, **85**, 033401.
58 A. S. Kheifets and I. A. Ivanov, *Phys. Rev. Lett.*, 2010, **105**, 233002.
59 L. R. Moore, M. A. Lysaght, J. S. Parker, H. W. van der Hart and K. T. Taylor, *Phys. Rev. A: At., Mol., Opt. Phys.*, 2011, **84**, 061404.

60 J. M. Dahlström, T. Carette and E. Lindroth, *Phys. Rev. A: At., Mol., Opt. Phys.*, 2012, **86**, 061402.
61 M. Y. Amusia, *Atomic photoeffect*, Plenum Press, New York, 1990.
62 A. Starace, *Springer Handbook of Atomic, Molecular, and Optical Physics*, Springer, New York, NY, 2006, ch. 24, pp. 379–390.
63 R. Pazourek, J. Feist, S. Nagele and J. Burgdörfer, *Phys. Rev. Lett.*, 2012, **108**, 163001.
64 A. Palacios, C. W. McCurdy and T. N. Rescigno, *Phys. Rev. A: At., Mol., Opt. Phys.*, 2007, **76**, 043420.
65 M. Liertzer, J. Feist, S. Nagele and J. Burgdörfer, *Phys. Rev. Lett.*, 2012, **109**, 013201.
66 C. H. Zhang and U. Thumm, *Phys. Rev. Lett.*, 2009, **102**, 123601.
67 C. Lemell, B. Solleder, K. Tőkési and J. Burgdörfer, *Phys. Rev. A: At., Mol., Opt. Phys.*, 2009, **79**, 062901.
68 A. L. Cavalieri, F. Krausz, R. Ernstorfer, R. Kienberger, P. Feulner, J. V. Barth and D. Menzel, *Dynamics at Solid State Surfaces and Interfaces*, Wiley-VCH, 2010, pp. 537–553.
69 S. Neppl, PhD thesis, Technische Universität München, München, 2012.
70 A. L. D. Kilcoyne, A. Aguilar, A. Müller, S. Schippers, C. Cisneros, G. Alna'Washi, N. B. Aryal, K. K. Baral, D. A. Esteves, C. M. Thomas and R. A. Phaneuf, *Phys. Rev. Lett.*, 2010, **105**, 213001.
71 V. K. Dolmatov and S. T. Manson, *J. Phys. B: At., Mol. Opt. Phys.*, 2008, **41**, 165001.
72 V. K. Dolmatov, J. L. King and J. C. Oglesby, *J. Phys. B: At., Mol. Opt. Phys.*, 2012, **45**, 105102.
73 S. Zherebtsov, T. Fennel, J. Plenge, E. Antonsson, I. Znakovskaya, A. Wirth, O. Herrwerth, F. Suszmann, C. Peltz, I. Ahmad, S. A. Trushin, V. Pervak, S. Karsch, M. J. J. Vrakking, B. Langer, C. Graf, M. I. Stockman, F. Krausz, E. Ruhl and M. F. Kling, *Nat. Phys.*, 2011, **7**, 656–662.
74 T. Carette, J. M. Dählstrom, L. Argenti and E. Lindroth, *Phys. Rev. A: At., Mol., Opt. Phys.*, 2013, **87**, 023420.
75 L. Argenti, R. Pazourek, J. Feist, S. Nagele, M. Liertzer, E. Persson, J. Burgdörfer and E. Lindroth, *Phys. Rev. A: At., Mol., Opt. Phys.*, 2013, **87**, 053405.

Faraday Discussions

RSC Publishing

PAPER

How quasi-free holes and electrons are generated in organic photovoltaic interfaces

Alessandro Troisi*

Received 6th December 2012, Accepted 11th January 2013
DOI: 10.1039/c3fd20142b

We introduced a minimal model of the donor–acceptor interface encountered in organic solar cells to explain the efficient generation of free charges in these systems and we investigated the nature of charge transfer states formed by a delocalized exciton in the donor component. Contrarily to the generally accepted view excitons do not generate strongly bound hole-electron pairs, but relatively delocalized charge transfer states with energy very close to the energy of free holes and electrons. These states are kinetically more accessible from the exciton state and very close in structure to the free hole and electron states. The most relevant molecular parameter that affects the rate of exciton dissociation is the electronic coupling between donor orbitals and acceptor orbitals, *i.e.* the band widths of the donor and acceptor materials. Moreover, and to some surprise, we find that the process of charge separation is mostly a purely electronic process and a very similar physics can be described neglecting the role of nuclear degrees of freedom altogether.

1 Introduction

The great technological interest in the development of organic solar cells has generated a good number of new problems of fundamental nature related to the charge and exciton generation and transport in these devices.[1–5] In organic solar cells an electron donor (D) and an acceptor (A) organic material with suitable energy levels are put in contact. The absorption of solar radiation typically generates excitons in the electron donor. The exciton can dissociate at the interface between the donor and acceptor to generate free holes (in the donor) and electrons (in the acceptor), but the actual mechanism for the formation of free charges is not truly understood. Conventionally it is believed that, before dissociating, the exciton migrates near the interface where a bound charge transfer (CT) state is formed (see Fig. 1a). This state should subsequently generate free holes and electrons, but obviously this process is kinetically unfavourable and a bound CT state seems more likely to be destroyed by hole-electron recombination.[6,7] The idea that a bound CT state is the intermediate between

Department of Chemistry and Centre of Scientific Computing, University of Warwick, U.K. E-mail: a.troisi@warwick.ac.uk

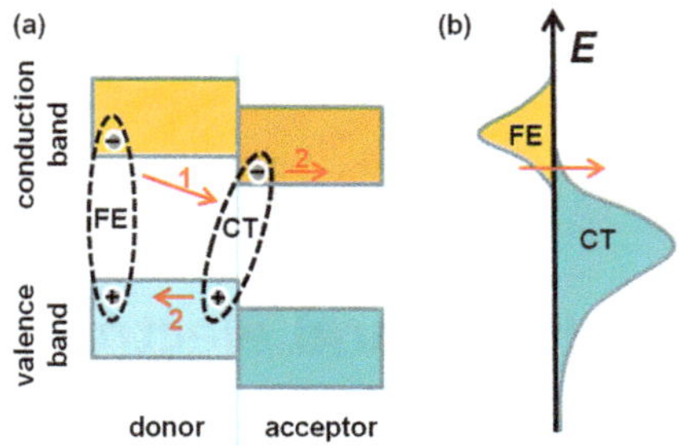

Fig. 1 (a) A conventional schematic of the charge separation in molecular organic solar cells based on a one-electron picture. A Frenkel exciton (FE) migrates from the bulk to the donor–acceptor interface where a charge transfer (CT) bound state is formed (1). This state will dissociate, in a separate event (2), into a free hole and electron. (b) In this paper we consider instead the generation of CT states from the FE state adopting a multi-electron picture, *i.e.* we compute the rate of transfer from each FE state to each CT state and we will determine the nature of the CT state. We will not assume that FE states generate bound CT states at the interface.

exciton and free charges seems not reconcilable with the experiments showing that most excitons generate free charges in good solar cells.[8,9]

A number of proposals have been put forward to explain the efficient generation of free charges. It was proposed, but a quantitative theory is still missing, that the excess energy generated when the bound CT state is formed from the exciton is not immediately dissipated in heat but it is used to promote the dissociation into free charges.[10] Considering that some interfaces can develop an intrinsic interface dipole moment it was suggested that this is exploited to facilitate the charge separation process.[11,12] Many authors have reached the conclusion that the delocalization of the charges in the CT state may reduce the barrier for generating free charges.[13–16] Most importantly, device modelling seems to agree with electrical measurements if one could assume that exciton dissociation leads to relatively delocalized holes and electrons.[17,18]

To make progress with this problem one needs to identify what the nature of the CT state generated by the exciton is and, possibly, one should express the rate for the generation of such a CT state in terms of a few key system parameters. We have recently proposed a model for long range exciton separation in organic solar cell interfaces where we computed the likelihood that an exciton far from the interface can generate directly quasi-free electrons and holes without an intermediate state at the interface.[19] In that work we made a number of assumptions on the nature of the disorder in the system that, while realistic, hide some of the interesting physics of the process. In this paper we study instead the absolute minimal Hamiltonian model that explains the efficient process of charge generation in organic solar cells.

The key elements of our model are the explicit inclusion of the hole-electron interaction and the description in terms of multi-electron states instead of one-electron states (Fig. 1b). To introduce the main idea of this paper through an intuitive example, we can consider a system composed by a donor and *two* acceptor fragments (DAA). The initial state can be thought of as a state where the excitation resides on the donor (D*AA). There are two possible charge transfer states which are two linear combinations of states (D^+A^-A) and (D^+AA^-) where the charge is localized on the two different acceptors. Since the two CT states have rather different energies it is expected that one of the CT state will have a strong (D^+A^-A) character and the other CT state will have a strong (D^+AA^-) character. It is not possible to tell without a model which of the two states will be more likely to be formed because (D^+A^-A) is more strongly coupled to the initial state, but (D^+AA^-) is energetically closer to the initial state. The naïve picture that the exciton at the interface generates a bound hole-electron pair may be very misleading and we will see in this paper that the formation of CT states with *larger* hole-electron separation is often the most kinetically favoured product.

We consider in this paper the highly idealized situation of two perfectly crystalline donor and acceptor materials sharing a defect-less interface. We consider the materials to be molecular materials so that the initial state is a Frenkel exciton in the donor (delocalized because there are no defects). We consider finite size clusters (of different sizes) in the absence of information on the actual delocalization length of the exciton. Apart from some simplifications introduced to limit the number of parameters, the model is representative of junctions realized from gas phase deposition,[20] although it is more common that one of the phases is amorphous. Most of the conclusions of this paper, however, are immediately generalizable to more complicated and common interfaces.

2 Initial and final states

Electronic Hamiltonian

We consider two adjacent and finite two-dimensional lattices, the donor (D) and acceptor (A) lattices as illustrated in Fig. 2. Both lattices have a rectangular shape, a square unit cell of side L, dimensions $n_z \times n_x$ and share a side of length n_x. The system supports two sets of states:

(i) Frenkel exciton (FE) states (initial states), where a Frenkel type exciton is delocalized over all donor sites;[21]

(ii) Charge transfer (CT) states (final states), where a hole and an electron are separated in the donor and acceptor portions of the system respectively.[10]

The total Hamiltonian can be written as

$$\hat{H} = \hat{H}_{FE} + \hat{H}_{CT} + \hat{V} \tag{1}$$

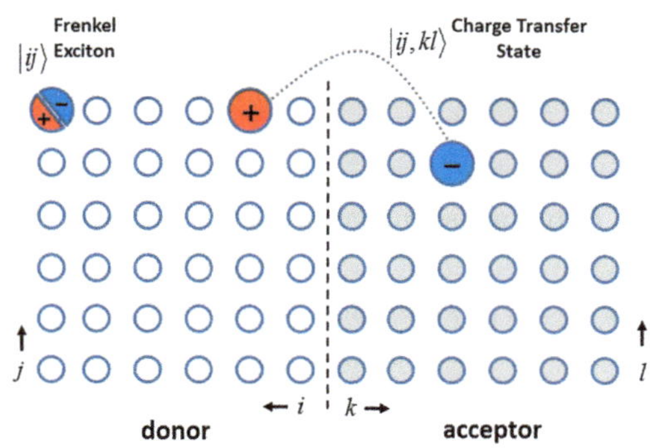

Fig. 2 Schematic of the clusters of donor and acceptor molecules considered in this work. Lattice sites (i,j) and (k,l) are occupied by donor and acceptor molecules respectively. Frenkel exciton states localized on the donor molecule and charge transfer states between the donor and acceptor are indicated as $|ij\rangle$ and $|ij, kl\rangle$ respectively.

where $\hat{H}_{FE}$ is the Hamiltonian for the excited states localized on the donor, $\hat{H}_{CT}$ the Hamiltonian for the charge transfer states and $\hat{V}$ the coupling between these two sets of states.

If the FE states are delocalized over the D sites, the only FE characteristics that turn out to be important for the charge separation process are the width and position of the Frenkel exciton band. For this reason, a very simple form for $\hat{H}_{FE}$ can be chosen:

$$\hat{H}_{FE} = \sum_{ij} E_{ex}|ij\rangle\langle ij| + \sum_{\{ij,i'j'\}} J|ij\rangle\langle i'j'| \tag{2}$$

Here, the set $\{|ij\rangle\}$ represents an electronic configuration with the exciton localized in site (i,j). E_{ex} would be the energy of such a configuration if the molecule in site (i,j) does not interact with the others (or it is isolated). J is the exciton coupling between excited states localized on neighbouring molecules (the symbol $\{ij,i'j'\}$ indicates that the summation is limited to the nearest neighbouring pairs of sites). Neglecting the coupling between non nearest neighbours has no real effect on the physics described here, as it will be clearer from the results section.

The CT states are assumed to be described by a linear combination of states $\{|ij,\ kl\rangle\}$ where an electron has been transferred from a neutral molecule on site (i,j) to the molecule on site (k,l). The energy of state $|ij,\ kl\rangle$ is given by the energy of the cation of the donor E_{D+}, plus the energy of the anion of the acceptor E_{A-}, plus the Coulombic interaction between them. We set the value of $E_{D+} + E_{A-}$ as the zero of our energy scale. The Coulombic interaction is conveniently expressed as

$$-\left(\frac{1}{4\pi\varepsilon_0}\frac{e^2}{\varepsilon_r L}\right)\left(\frac{1}{r_{ij,kl}}\right) = -\frac{C}{r_{ij,kl}} \tag{3}$$

Where the constant C incorporates the effect of the lattice spacing L and the relative dielectric constant ε_r. $r_{ij,kl}$ is the distance between site (i,j) and (k,l) in the units of the lattice spacing L. To include the electronic coupling between the states $\{|ij,\ kl\rangle\}$ we use the theory of configuration interaction assuming that $|ij,\ kl\rangle$ corresponds to an excitation from the HOMO of the molecule in site (i,j) to the LUMO of the molecule in site (k,l). The Hamiltonian matrix elements in this case are:[22]

$$\langle ij,\ kl|\hat{H}_{CT}|i'j',k'l'\rangle = \delta_{ii'}\delta_{jj'}F^{A-LUMO}_{kl,k'l'} - \delta_{kk'}\delta_{ll'}F^{D-HOMO}_{ij,i'j'} \tag{4}$$

where $F^{A-LUMO}_{kl,k'l'}$ is the Fock matrix element (orbital coupling) between the LUMO orbitals localized on site (k,l) and (k',l') and $F^{D-HOMO}_{ij,i'j'}$ is the Fock matrix element between the HOMO orbitals localized on site (i,j) and (i',j'). These matrix elements decrease exponentially with the distance between sites and it is generally acceptable to include only the nearest neighbour's term, which we indicate as V_A and V_D. The final form for the charge transfer states Hamiltonian is

$$\begin{aligned}\hat{H}_{CT} = &\sum_{ij,kl} -\frac{C}{r_{ij,kl}}|ij,kl\rangle\langle ij,kl| + \sum_{ij}\sum_{\{kl,k'l'\}} V_A|ij,kl\rangle\langle ij,k'l'| \\ &- \sum_{kl}\sum_{\{ij,i'j'\}} V_D|ij,kl\rangle\langle i'j',kl|\end{aligned} \tag{5}$$

To derive a plausible form for the coupling between Frenkel and CT excitons we consider again the theory of configuration interaction and assume that the

Frenkel excited states $|ij\rangle$ are described by an excited configuration from the HOMO to the LUMO of the molecule in site (i,j). If we consider that the Fock matrix elements between non-neighbouring sites are negligible, the only non-zero matrix elements between states $|ij\rangle$ and $|ij, kl\rangle$ involve the Fock matrix element of the LUMO of the donor in site $(i,1)$ (*i.e.* at the interface) and the LUMO of the acceptor in site $(k = i,1)$ (*i.e.* also at the interface). Setting this matrix element to the constant V_{DA} the coupling between CT and FE states can be written as:

$$\hat{V} = \sum_i V_{DA}|i1\rangle\langle i1, i1| \quad (6)$$

The process we wish to study is the transition between FE states and CT states in the limit of small coupling between them. We therefore diagonalize $\hat{H}_{FE}$ and $\hat{H}_{CT}$ separately and treat $\hat{V}$ as a perturbation. We indicate with $E_{FE}^{(n)}$ and $E_{CT}^{(m)}$ the eigenvalues of the unperturbed Hamiltonians and we express their eigenfunctions as linear combination of the basis functions as:

$$\Psi_{FE}^{(n)} = \sum d_{ij}^{(n)}|ij\rangle \quad (7)$$

$$\Psi_{CT}^{(m)} = \sum c_{ij,kl}^{(m)}|ij, km\rangle \quad (8)$$

Structure of the CT states

We set $C = 0.96$ eV throughout this paper, corresponding to lattice distance $L = 5$ Å[23] and a relative dielectric constant of $\varepsilon_r = 3$,[24,25] *i.e.* to physically plausible values for the interface between two organic crystals. We set $V_D = V_A$ in the range 0.02–0.10 eV which is typical for molecules with good electronic coupling.[26,27] In the results section we discuss briefly the more general situation where the coupling along the x and z direction is different.

The total number of CT states increases very rapidly with the size of the cluster (it is $(n_z \times n_x)^2$) and, to analyse the nature of the eigenstates more efficiently, we use the density of states, the charge distribution and the charge participation ratio. The density of CT states is defined as usual as:

$$DOS_{CT}(E) = \sum_m \delta(E - E_{CT}^{(m)}) \quad (9)$$

Where $\delta(E)$ is the Dirac delta function, approximated in this paper by a normalized Gaussian of width 0.05 eV. To visualize the charge distribution for each state, we used the matrices $D_{ij}^{(m)} = \sum_{kl} |c_{ij,kl}^{(m)}|^2$ and $A_{kl}^{(m)} = -\sum_{ij} |c_{ij,kl}^{(m)}|^2$ representing the positive and negative charge on the donor site (i,j) and acceptor site (k,l) for state m.

Fig. 3a represents the density of states (DOS), for models with various values of V_D, and Fig. 3c shows the charge distribution for a few selected states for $V_D = 0.08$ eV. At very low energy there are few states corresponding to a tightly bound hole and electron near the interface. The energy of these states is close to the Coulombic interaction $(-C)$ between charges in neighbouring lattice sites. The comparison of Fig. 3a and 3c shows that, as the energy increases between -1.0 and -0.2 eV, the DOS increases and the CT states become more delocalized over the donor and acceptor clusters. The peak of the DOS corresponds to fairly

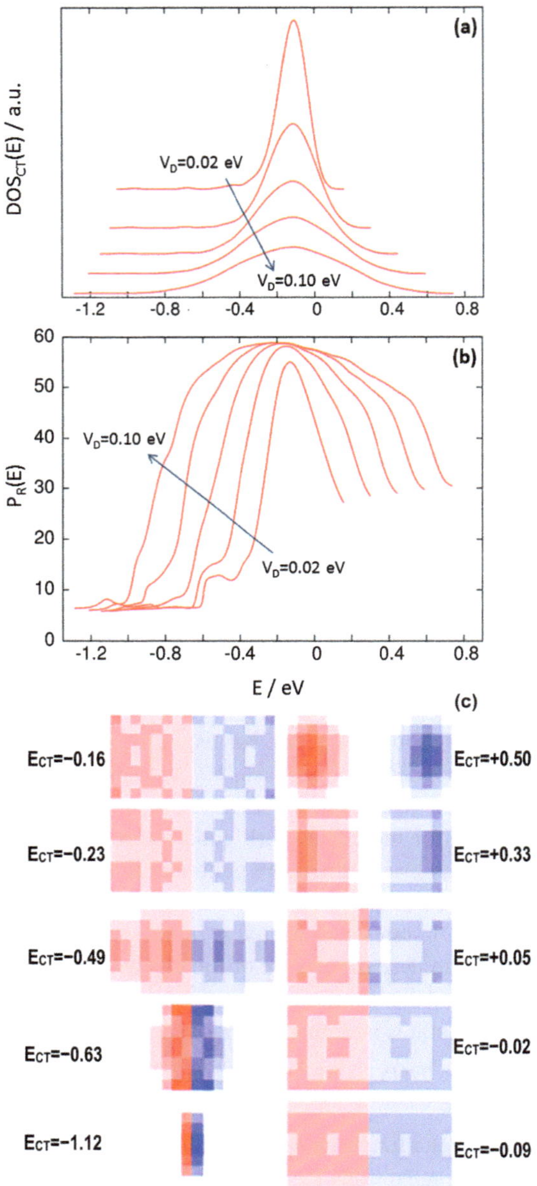

Fig. 3 (a) Density of states of the CT states for different values of the parameter V_D (plots are offset vertically for clarity). (b) Inverse participation ratio as a function of the energy for the same system and parameters. (c) Illustration of the charge distribution for CT states at different energies for $V_D = 0.08$ eV (dark red and blue correspond to highly positive and negative charges, white corresponds to no charge). In all cases a cluster with $n_x = n_y = 8$ was used.

delocalized states. The energy of these very delocalized states is slightly negative because of the relatively small Coulombic attraction between the holes and electrons. In the highest energy CT state, the charges are separated away from the interface. Having chosen $V_D = V_A$, the charge distribution is symmetrical with respect to the interface.

For an even more compact and useful description of the localization of the CT states at different energies we can define the inverse participation ratio[28] as

$$P_R^{(m)} = \left(\sum_{ij} D_{ij}^{(m)2}\right)^{-1} = \left(\sum_{kl} A_{kl}^{(m)2}\right)^{-1} \tag{10}$$

The equality holds because of the symmetry of the problem (and $V_D = V_A$). $P_R^{(m)}$ measures how many molecules share the positive (or negative) charge and can range from 1 to $n_z \times n_x$ in our problem. To illustrate how the average participation ratio changes with the energy of the state and the coupling V_D we have shown in Fig. 3b the function

$$P_R(E) = \frac{\sum_m P_R^{(m)} \delta(E - E_{CT}^{(m)})}{\sum_m \delta(E - E_{CT}^{(m)})} \tag{11}$$

representing the average energy dependent participation ratio. As implied by the more detailed representation in Fig. 3c, the lowest energy states are strongly bound CT pairs at the interface and, as the energy increases, the charge becomes more delocalized. $P_R(E)$ at low energies is close to n_x (*i.e.* to the localization over the first line of molecules at the interface) and the maximum of $P_R(E)$ is not far from the maximum allowed (*i.e.* the total size of the cluster $n_z \times n_x$). The small decrease of $P_R(E)$ for high energies is due to a partial localization of the charge opposite to the interface in the high energy CT states.

Fig. 3b also shows that CT excitons are more bound (more localized) for small coupling V_D. If we consider that these states are originating from the interaction between quasi free hole and electron in the donor and acceptor material, larger coupling is associated with lighter effective masses and less localization. Several papers have successfully described the CT interface exciton in this way[10,29,30] focusing mostly on the lowest energy CT states. Our task however is to describe the generation of any CT state from Frenkel excitons.

3 Rate for the generation of charge transfer states

If we indicate with $k^{(n)\to(m)}$ the rate constant for the process from $\Psi_{FE}^{(n)}$ to $\Psi_{CT}^{(m)}$, the total rate for the process of generating any CT state can be expressed as:

$$k_{TOT} = \sum_n \frac{\exp(-E_{FE}^{(n)}/k_B T)}{Q} \sum_m k^{(n)\to(m)} \tag{12}$$

Where $Q = \sum_n \exp(-E_{FE}^{(n)}/k_B T)$, with $k_B T$ being the thermal energy. The Boltzmann factor implies that we are assuming equilibrium among the FE electronic states. The initial and final states are very weakly coupled (see below for a justification) and the rate constant can be evaluated from the theory of non-adiabatic electron transfer reactions.[31] The electron transfer is accompanied by a deformation in the equilibrium geometry of the molecule which is quantified by the parameter λ, known as the reorganization energy. If the nuclear modes can be treated classically the rate of charge transfer can be expressed by the Marcus equation:[31]

$$k^{(n)\rightarrow(m)} = \frac{2\pi}{\hbar} \frac{|V_{nm}|^2}{(4\pi\lambda_m k_B T)^{1/2}} \exp\left(-\frac{\left(E_{CT}^{(m)} - E_{FE}^{(n)} + \lambda_m\right)^2}{4\lambda_m k_B T}\right) \tag{13}$$

where V_{nm} is the electronic coupling between the initial and final states. In writing eqn (13) we have noted that the reorganization energy λ_m depends on the final state states m (depending on the localization of the charge the reorganization energy may change). The combination of the two factors containing the term λ_m are normally referred to as the Franck–Condon factor. When the reorganization energy in the molecular crystals is decomposed according to the nuclear modes that contribute to it,[32] it turns out that more than half of the reorganization energy is due to high frequency modes that cannot be treated classically. A more general expression taking into account the quantum nature of nuclear modes can be used but, as we will see in the result session, this is not really needed.

The coupling V_{nm} can be obtained from eqn (6)–(8):

$$V_{nm} = \left\langle \Psi_{FE}^{(n)} \middle| \hat{V} \middle| \Psi_{CT}^{(m)} \right\rangle = V_{DA} \sum_{i=1,n_x} d_{i1}^{(n)} c_{i1,i1}^{(m)} \tag{14}$$

Except when we have very small clusters of 1–4 sites, or when we consider very localized states, we have generally $\left| \sum_{i=1,n_x} d_{i1}^{(n)} c_{i1,i1}^{(m)} \right| \ll 1$, *i.e.* the coupling V_{nm} is always sufficiently small enough to allow the use of a perturbative expression, such as eqn (13). It is evident from eqn (13)–(14) that the parameter V_{DA} will only modulate the final rate (through its square) but has otherwise no effect on the physics that we wish to describe. For this reason V_{DA} was set to the physically plausible value of 0.08 eV and left unchanged.

The reorganization energy is smaller for more delocalized states and it is approximately inversely proportional to the size of the system.[33] This has been verified computationally for polymers of different lengths,[34] for polyacenes of different sizes[35] and for a large set of conjugated molecules.[33] To incorporate this effect in the model we assumed that, for the computation of the rate $k^{(n)\rightarrow(m)}$, λ_m decreases with the increase of delocalization of the charge in the CT state as $\lambda_m = \lambda_0/P^{(m)}$, where $P^{(m)}$ is the number of molecules that share a positive or a negative charge as defined in eqn (10) and λ_0 is the reorganization energy for a system with only one donor and one acceptor molecule. λ_0 was set to 0.25 eV, based on calculations of organic donor–acceptor pairs reported in literature.[32] We discuss the (modest) effect of changing λ_0 within a plausible range 0.15–0.40 eV in the results section. This scaling of the reorganization energy with the localization length is valid as long as the system does not form small polarons, a situation that is likely for the most common high mobility materials.[23,27]

Table 1 summarizes the parameters used in the model, or their range when a range was used. The bottom of the excitonic band for a large cluster will be at $E_{ex} - 4|J|$ and this value should be greater than zero (the energy of separated D^+ and A^-) in order to have charge separation thermodynamically favoured for any choice of the other parameters. We have therefore set $J = 0.06$ eV, based on a typical excitonic coupling between organic crystals[36] and $E_{ex} = 0.25$ eV. We have also considered narrower ($E_{ex} = 0.13$ eV, $J = 0.03$ eV) and broader ($E_{ex} = 0.37$ eV, $J =$

Table 1 The parameters used in the model presented in the paper

Symbol	Value or Range (eV)	Description
C	0.96	Nearest neighbour Coulombic attraction
$V_D = V_A$	0.02–0.1	Donor or Acceptor orbital coupling
E_{ex}	0.13–0.37	Isolated exciton energy
J	0.03–0.09	Frenkel exciton coupling
V_{DA}	0.08	Donor–Acceptor orbital coupling
λ_0	0.25	Reorganization energy for isolated DA pair

0.09 eV) excitonic bands above $E = 0$ as reported in the results section. Instead, it is probably not useful to consider excitonic bands at much higher energy (as in real devices that would dissipate too much energy and would probably involve further excited states of the acceptor).

Considering the plausible range of parameters suitable for organic solar cell interfaces, it is already evident that the reaction that brings a Frenkel exciton at the bottom of the excitonic band ($E_{FE}^{(n)} \sim [0.01 - 0.1]$ eV) into a strongly bound charge transfer state ($E_{CT}^{(m)} \sim -1$ eV) is always occurring in the Marcus inverted region, *i.e.* where $E_{CT}^{(m)} - E_{FE}^{(m)} + \lambda_m < 0$. This fact is essentially determined by the size of typical organic molecules (setting boundaries to both the reorganization energy and the Coulombic interaction) and the low dielectric constant of the medium. In the Marcus inverted region the reaction is faster for smaller values of $E_{FE}^{(m)} - E_{CT}^{(m)}$, *i.e.* it is to be expected that the Frenkel exciton will not dissociate to form the most stable CT state, but a CT state closer in energy to the initial Frenkel state.

4 Results and discussion

Injection from the lowest energy FE state

Before studying the charge separation rate for different system parameters it is convenient to consider in detail the process of charge separation from the lowest energy FE state, which we can denote as $\Psi_{FE}^{(1)}$. The total rate of charge separation from this state is $k^{(1)\rightarrow All} = \sum_m k^{(1)\rightarrow(m)}$ and the probability that the system is found in state $\Psi_{FE}^{(m)}$ after the charge separation has taken place is $k^{(1)\rightarrow(m)}/k^{(1)\rightarrow All}$. Therefore, the probability of having a given electronic energy after the charge separation process can be computed as:

$$P(E) = \frac{\sum_m k^{(1)\rightarrow(m)} \delta(E - E_{CT}^{(m)})}{\sum_m k^{(1)\rightarrow(m)}} \tag{15}$$

Fig. 4a compares the function $P(E)$ with $DOS_{CT}(E)$ for the CT states. The key observation from this comparison is that the charge separation process selects states around $E \sim 0$ and is not correlated at all with the DOS of the CT states. The rate (see eqn (13) and below) is determined by two terms: the coupling $|V_{mn}|^2$, which roughly decreases with increasing the energy, and the Franck–Condon factor, that has a maximum for $E_{CT}^{(m)} - E_{FE}^{(1)} + \lambda_m \sim 0$.

To describe more quantitatively the relation between the two terms it is convenient to define the energy averaged squared coupling between FE and CT states as:

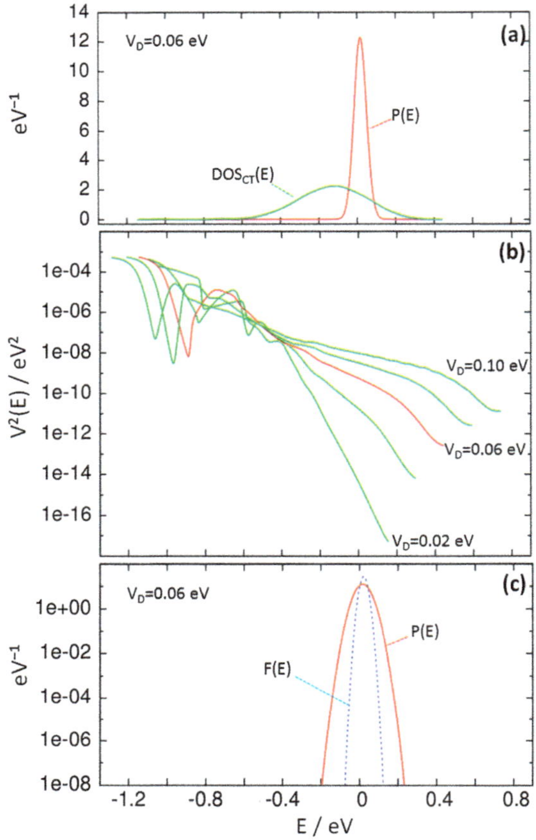

Fig. 4 (a) Comparison between the DOS of the CT state and the probability $P(E)$ of the final state being at energy E. (b) Average squared electronic coupling between initial and final states, defined in eqn (16), for $V_D = 0.02, 0.04, 0.06, 0.08, 0.10$ eV. (c) Comparison between the Franck Condon factor $F(E)$ and $P(E)$, to illustrate that the maxima of the two functions coincide. In all panels $E_{ex} = 0.25$, $J = 0.06$ eV.

$$V^2(E) = \frac{\sum_m |V_{1m}|^2 \delta(E - E_{CT}^{(m)})}{\sum_m \delta(E - E_{CT}^{(m)})} \tag{16}$$

This coupling decreases with increasing energy because low energy states are near the interface and therefore strongly coupled with the CT state (see Fig. 3c and eqn (14)). Fig. 4b shows that this decrease is approximately exponential at sufficiently large energies.

The Franck–Condon factor depends on the specific state m considered. In Fig. 4c we plot for reference the function

$$F(E) = \frac{1}{(4\pi\lambda(E)k_BT)^{1/2}} \exp\left(-\frac{\left(E - E_{FE}^{(1)} + \lambda(E)\right)^2}{4\lambda(E)k_BT}\right) \tag{17}$$

where the energy average reorganization energy was estimated as $\lambda(E) = \lambda_0/P(E)$.

The Franck–Condon energy dependence dominates over the energy dependence of the effective coupling and the majority of the CT states are formed closer to the maximum of the Franck–Condon factor (shifted to lower energies because of the exponential decrease of the coupling). Essentially all the results presented in this section can be understood in terms of the contrasting energy dependence of the two functions $V^2(E)$ and $F(E)$ that contribute to the charge separation rate.

Effect of the Intermolecular coupling V_D

In Fig. 3a we have seen that the effect of the coupling V_D ($=V_A$) was to broaden the DOS_{CT}. However, this effect is not particularly important for the computation of the $k^{(1)\to \text{All}}$ rate of charge separation. More important is that CT states at the energy where the charge transfer takes place are more delocalized for larger V_D, and, therefore, the coupling between these states and the Frenkel exciton is larger. Fig. 4b shows that for $E > -0.2$ eV (where the CT actually takes place) the average square coupling $V^2(E)$ increases with increasing the coupling V_D. Consequently the rate of charge separation increases for increased values of V_D, as reported in Fig. 5. The interesting result is therefore that the donor and acceptor materials with broader bands (smaller effective masses) will facilitate not only the transport of holes and electrons, but also the formation of charge separated states.

Another interesting point is that the charge separation process does not involve low energy bound CT states at the interface. The fact that these bound CT states can be more or less delocalized depending on the value of V_D[10] is not of direct relevance for the problem of charge separation. In realistic situations, if the band widths of the donor or acceptor are too small, the charge separation rate can be too slow and the exciton will more likely relax to the ground state without generating charges.

Fig. 5 presents a comparison of the rates k_{TOT} and $k^{(1)\to \text{All}}$ for different values of V_D to show that the two rates are only marginally different. This happens because the charge separation occurs from states at the bottom of the excitonic

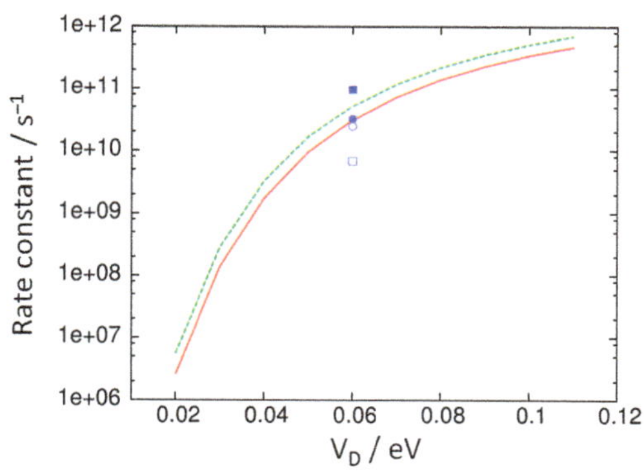

Fig. 5 The solid and dashed line represent the rates k_{TOT} and $k^{(1)\to \text{All}}$ with $E_{ex} = 0.25$ eV and $J = 0.06$ eV. Full and open circles indicate the rate k_{TOT} computed for a broader ($E_{ex} = 0.37$ eV and $J = 0.09$ eV) or narrower ($E_{ex} = 0.13$ eV and $J = 0.03$ eV) excitonic band, but with the same band minimum energy. Full and open squares indicate the rate computed for an excitonic band shifted to lower ($E_{ex} = 0.15$ eV and $J = 0.06$ eV) and to higher ($E_{ex} = 0.35$ eV and J = 0.06 eV) energies.

band. Consequently, the width of the excitonic band (associated to the exciton effective mass and the parameter J) is not expected to be a crucial parameter. To verify this fact we report in the same figure the total rate for $V_D = 0.06$ eV and with excitonic bands 50% broader or narrower: the total rate is essentially unaffected by the exciton band-width.

Slightly more important is the energy position of the lowest excitonic band edge, which determines the initial state energy. Consistent with the charge separation occurring in the inverted Marcus region, the charge separation rate (square symbol in Fig. 5) decreases when the exciton band energy is shifted upwards. However, as we noted above, these parameters cannot be set freely as an excitation energy too low will make the charge separation process thermodynamically unfavourable and an excitation energy too high will make the charge separation process too energy wasteful.

Effect of the cluster shape (n_x and n_y)

It is illustrative to compute the total rate of charge separation k_{TOT} for different shapes of the donor–acceptor cluster (Fig. 6). We considered for example clusters with $n_x = 4$ and n_z varying from 1 to 12 (see inset in the same figure). For thin interfaces ($1 < n_z < 3$) k_{TOT} increases with the increase of the cluster size. This is because the cluster is too thin to develop high energy delocalized CT states and the rate is decreased by an unfavourable Franck–Condon factor. When the clusters are thicker than $n_z = 4$, k_{TOT} does not change much but decreases as the size of the cluster increases. This is because, for larger clusters, the delocalized exciton is less coupled with the CT state (it is more "diluted"). It is remarkable that this decrease is so slow, *i.e.* if the exciton is delocalized, the formation of charge separated states may happen when the "centre" of the exciton is many layers away from the interface. This is in strong contrast with the view that an exciton should be localized at the interface to be split into free charges. Of course, in practice, an exciton cannot be unlimitedly delocalized, but delocalization over tens or hundreds of molecules seems to be the most consistent with the available spectroscopic data.[37,38]

Similar calculations repeated with fixed small n_z and variable n_x give qualitatively similar results (not shown), *i.e.* what promotes efficient charge separation is

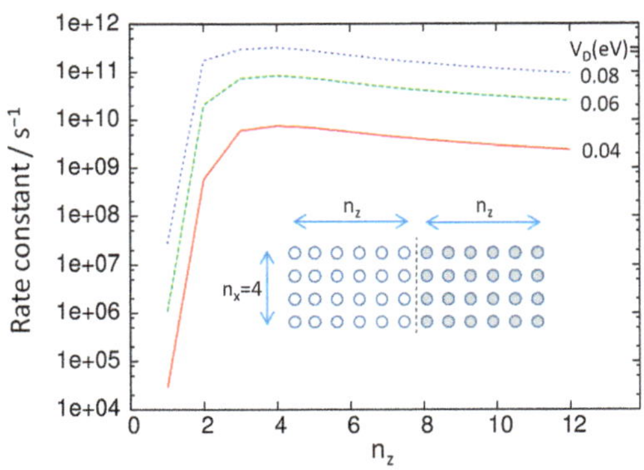

Fig. 6 Total rate costant k_{TOT} for clusters of different shape (same length of the interface n_x and varying thickness n_z).

the availability of CT states at the right energy and it is equivalent if the hole and electrons in these states are separated along the z or the x direction. In more realistic systems the bandwidth along the x and z direction is different and one can think that broader bands in the z directions are the most favourable to prevent charge recombination. Very interesting, these ideas begin to be explored experimentally, for example with time-resolved two-photon photoemission.[39]

The large cluster limit

As the cluster becomes larger the reorganization energy becomes smaller. In the limit of $\lambda_m \to 0$ the expression of the rate changes radically becoming:

$$\lim_{\lambda_m \to 0} k^{(n)\to(m)} = \lim_{\lambda_m \to 0} \frac{2\pi}{\hbar} \frac{|V_{nm}|^2}{(4\pi\lambda_m k_B T)^{1/2}} \exp\left(-\frac{\left(E_{CT}^{(m)} - E_{FE}^{(n)} + \lambda_m\right)^2}{4\lambda_m k_B T} \right)$$
$$= \frac{2\pi}{\hbar} |V_{nm}|^2 \delta(E_{CT}^{(m)} - E_{FE}^{(n)}) \quad (18)$$

The expression on the right hand side is just a Fermi Golden rule expression that would describe the charge separation if this were an entirely electronic process with no contribution from the nuclear degrees of freedom. The $8 \times 8(\times 2)$ cluster that was used to illustrate the results in this paper is already very close to this minimum. Fig. 7 shows the rate computed for a range of values of λ_0, including the limit $\lambda_0 \to 0$ (until numerical errors becomes important). Maybe surprisingly, the reorganization energy that dramatically affects the charge transfer rate in single pairs of donor and acceptor is of modest importance for the system containing a cluster of donor and acceptor molecules: the charge separation rate in the limit $\lambda_0 \to 0$ is only half of that computed with the reference value used in this paper ($\lambda_0 = 0.25$ eV). This fact suggests that, although the process of charge separation in these interfaces has been principally considered a "chemical" process involving nuclear reorganization, it is to a large extent a purely electronic process and can be described to a good degree of accuracy by a purely electronic Hamiltonian. Also for this reason we did not present the results where eqn (13) is improved to take into account the quantum nature of some vibrations.

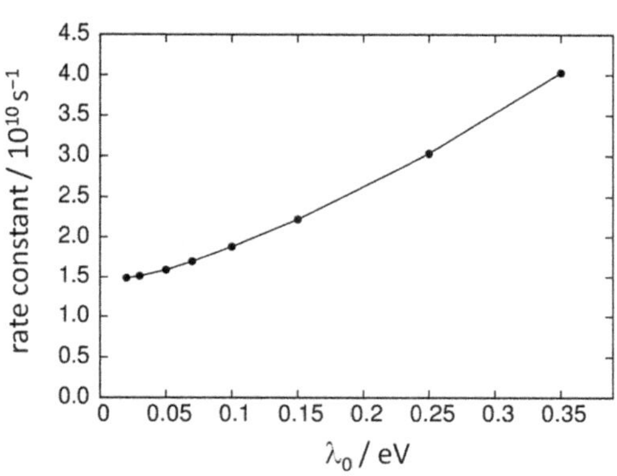

Fig. 7 The total rate constant for different values of λ_0 (and in the limit $\lambda_0 \to 0$).

5 Conclusions

We have studied a minimal model of the donor–acceptor interface encountered in organic solar cells to investigate the nature of the most likely charge transfer state formed from the dissociation of an exciton in the donor. The main findings of this work can be summarized as follows:

(1) Excitons do not generate strongly bound hole-electron pairs, but relatively delocalized charge transfer states with energy very close to the energy of free holes and electrons. These states are kinetically more accessible from the exciton state and very close in structure to the free hole and electron states.

(2) The most relevant molecular parameter that affects the rate of exciton dissociation is the electronic coupling between donor orbitals and acceptor orbitals, *i.e.* the band widths of the donor and acceptor materials. These parameters are important, not because they generate more delocalized CT states at the interface (which do not play any role in the charge separation process), but because they increase the effective coupling between the initial Frenkel exciton state and the final charge transfer (delocalized) state.

(3) The process of charge separation is, to a large extent, an electronic process and a very similar physics can be described neglecting the role of nuclear motions altogether. This happens because the delocalization of hole and electron dramatically decreases the coupling with the nuclear degrees of freedom.

Setting aside the many idealizations present in our model, whose relevance will be investigated in future works, the key difference between the presented description and the commonly accepted picture is that we did not use a single electron approximation to describe the charge separation process. It is customary in solid state physics to consider all charge transfer processes as occurring by electron transfer from one orbital to another, with the energy difference of the process given by the energy difference between the initial and final orbital energy. This single electron picture (*e.g.* in Fig. 1a) is accurate if we can assume that the energy of an electron is not affected by the occupation of the other orbitals, but this is clearly not true for the charge transfer state where the hole and electron wavefunctions generated by a single Frenkel exciton are entangled. The band diagram of holes and electrons is not at all representative of the process of charge separation which needs to be described in term of *states*. Once we have described the process as a mostly electronic process involving two sets of states (Fig. 1b) we made considerable progress and realized that there is no real kinetic barrier for the generation of free changes. Crucially, we have considered *all* charge transfer states, not just the ones that are generally assumed to be important and this is the main departure from other state-based models. The scheme of Fig. 1b remains valid if we move from the idealized situation described in this paper to a more realistic interface, involving disorder and more localized states, a situation that was investigated in ref. 19. The problem of explaining the efficient generation of free carriers in organic solar cells seems now really manageable if the few steps above are taken.

A number of recent experimental evidences seem to support the view of charge separation proposed in this paper. The necessity of a sufficiently large cluster of acceptors to promote charge separation was proposed by the Eindhoven group[40] and the importance of highly crystalline interfaces to promote delocalization has been highlighted a few times for polymer solar cells.[8,41] The temperature independent generation of CT states measured by many authors[15,42,43] is also fully consistent with the picture proposed in this paper, where the role of nuclear motions becomes marginal and the charge transfer happens by population transfer between quasi degenerate Frenkel and CT states.

There are two directions of further investigation suggested by this work. First, one should relax some of the approximations made to reduce the number of parameters of the model (*e.g.* $V_D = V_A$, square lattice arrangement, identical coupling along the *x* or *y* direction) and substitute these with realistic evaluations for systems with known morphology at the interface. Secondly, having realized that there is a strong coupling between exciton delocalization and charge separation dynamics, it becomes even more important to look again at the source of exciton localization in molecular crystals (and other organic semiconductors). While it was traditionally thought that the exciton migration and exciton dissociation process in organic photovoltaics can be considered independently, this and other works[19] suggest that the dynamics of the exciton and its dissociation into hole and electron are not really separable.

Acknowledgements

This work was supported by the Leverhulme Trust and by the European Research Council.

References

1 A. Mishra and P. Bäuerle, *Angew. Chem., Int. Ed.*, 2012, **51**, 2020–2067.
2 S. Uchida, J. G. Xue, B. P. Rand and S. R. Forrest, *Appl. Phys. Lett.*, 2004, **84**, 4218–4220.
3 P. Peumans, A. Yakimov and S. R. Forrest, *J. Appl. Phys.*, 2003, **93**, 3693–3723.
4 C. J. Brabec, S. Gowrisanker, J. J. M. Halls, D. Laird, S. Jia and S. P. Williams, *Adv. Mater.*, 2010, **22**, 3839–3856.
5 P. W. M. Blom, V. D. Mihailetchi, L. J. A. Koster and D. E. Markov, *Adv. Mater.*, 2007, **19**, 1551–1566.
6 P. Peumans and S. R. Forrest, *Chem. Phys. Lett.*, 2004, **398**, 27–31.
7 C. L. Braun, *J. Chem. Phys.*, 1984, **80**, 4157–4162.
8 H. Ohkita, S. Cook, Y. Astuti, W. Duffy, S. Tierney, W. Zhang, M. Heeney, I. McCulloch, J. Nelson, D. D. C. Bradley and J. R. Durrant, *J. Am. Chem. Soc.*, 2008, **130**, 3030–3042.
9 S. H. Park, A. Roy, S. Beaupre, S. Cho, N. Coates, J. S. Moon, D. Moses, M. Leclerc, K. Lee and A. J. Heeger, *Nat. Photonics*, 2009, **3**, 297–U295.
10 X. Y. Zhu, Q. Yang and M. Muntwiler, *Acc. Chem. Res.*, 2009, **42**, 1779–1787.
11 S. Mothy, M. Guillaume, J. Ide, F. Castet, L. Ducasse, J. Cornil and D. Beljonne, *J. Phys. Chem. Lett.*, 2012, **3**, 2374–2378.
12 S. R. Yost, L.-P. Wang and T. Van Voorhis, *J. Phys. Chem. C*, 2011, **115**, 14431–14436.
13 A. A. Bakulin, A. Rao, V. G. Pavelyev, P. H. M. van Loosdrecht, M. S. Pshenichnikov, D. Niedzialek, J. Cornil, D. Beljonne and R. H. Friend, *Science*, 2012, **335**, 1340–1344.
14 A. V. Nenashev, S. D. Baranovskii, M. Wiemer, F. Jansson, R. Osterbacka, A. V. Dvurechenskii and F. Gebhard, *Phys. Rev. B: Condens. Matter Mater. Phys.*, 2011, **84**, 035210.
15 D. H. K. Murthy, M. Gao, M. J. W. Vermeulen, L. D. A. Siebbeles and T. J. Savenije, *J. Phys. Chem. C*, 2012, **116**, 9214–9220.
16 A. Petersen, A. Ojala, T. Kirchartz, T. A. Wagner, F. W. Urthner and U. Rau, *Phys. Rev. B*, 2012, **85**.
17 C. Deibel, T. Strobel and V. Dyakonov, *Phys. Rev. Lett.*, 2009, **103**, 036402.
18 H. van Eersel, R. A. J. Janssen and M. Kemerink, *Adv. Funct. Mater.*, 2012, **22**, 2700–2708.
19 D. Caruso and A. Troisi, *Proc. Natl. Acad. Sci. U. S. A.*, 2012, **109**, 13498.
20 P. Peumans, S. Uchida and S. R. Forrest, *Nature*, 2003, **425**, 158–162.
21 J. Frenkel, *Phys. Rev.*, 1931, **37**, 17.
22 A. Szabo and N. Ostlund, *Modern Quantum Chemistry*, Dover, 1996.
23 A. Troisi, *Org. Electron.*, 2011, **12**, 1988–1991.
24 L. J. A. Koster, V. D. Mihailetchi and P. W. M. Blom, *Appl. Phys. Lett.*, 2006, **88**, 093511.

25 A. J. Moulé and K. Meerholz, *Appl. Phys. Lett.*, 2007, **91**.
26 V. Coropceanu, J. Cornil, D. A. da Silva, Y. Olivier, R. Silbey and J. L. Bredas, *Chem. Rev.*, 2007, **107**, 926–952.
27 A. Troisi, *Chem. Soc. Rev.*, 2011, **50**, 4039–4046.
28 H. Aoki, *J. Phys. C: Solid State Phys.*, 1983, **16**, 205–208.
29 Q. X. Yang, M. Muntwiler and X. Y. Zhu, *Phys. Rev. B: Condens. Matter Mater. Phys.*, 2009, **80**, 115214.
30 M. Wiemer, A. V. Nenashev, F. Jansson and S. D. Baranovskii, *Appl. Phys. Lett.*, 2011, **99**, 013302.
31 P. F. Barbara, T. J. Meyer and M. A. Ratner, *J. Phys. Chem.*, 1996, **100**, 13148–13168.
32 T. Liu and A. Troisi, *J. Phys. Chem. C*, 2011, **115**, 2406–2415.
33 A. Devos and M. Lannoo, *Phys. Rev. B: Condens. Matter*, 1998, **58**, 8236–8239.
34 S. S. Zade, N. Zamoshchik and M. Bendikov, *Acc. Chem. Res.*, 2011, **44**, 14–24.
35 M. Malagoli, V. Coropceanu, D. A. da Silva and J. L. Bredas, *J. Chem. Phys.*, 2004, **120**, 7490–7496.
36 H. Yamagata, J. Norton, E. Hontz, Y. Olivier, D. Beljonne, J. L. Bredas, R. J. Silbey and F. C. Spano, *J. Chem. Phys.*, 2011, **134**.
37 S.-H. Lim, T. G. Bjorklund, F. C. Spano and C. J. Bardeen, *Phys. Rev. Lett.*, 2004, **92**, 107402.
38 D. H. Arias, K. W. Stone, S. M. Vlaming, B. J. Walker, M. G. Bawendi, R. J. Silbey, V. Bulovic and K. A. Nelson, *J. Phys. Chem. B*, 2012, DOI: 10.1021/jp3086717.
39 G. J. Dutton and S. W. Robey, *J. Phys. Chem. C*, 2012, **116**, 19173–19181.
40 D. Veldman, O. Ipek, S. C. J. Meskers, J. Sweelssen, M. M. Koetse, S. C. Veenstra, J. M. Kroon, S. S. van Bavel, J. Loos and R. A. J. Janssen, *J. Am. Chem. Soc.*, 2008, **130**, 7721–7735.
41 D. P. McMahon, D. L. Cheung and A. Troisi, *J. Phys. Chem. Lett.*, 2011, **2**, 2737–2741.
42 J. Piris, T. E. Dykstra, A. A. Bakulin, P. H. M. van Loosdrecht, W. Knulst, M. T. Trinh, J. M. Schins and L. D. A. Siebbeles, *J. Phys. Chem. C*, 2009, **113**, 14500–14506.
43 R. Mauer, I. A. Howard and F. Laquai, *J. Phys. Chem. Lett.*, 2010, **1**, 3500–3505.

DISCUSSIONS

General discussion

DOI: 10.1039/c3fd90019c

Professor Neumark opened the discussion of the paper by Professor Meech: In your paper, you point out that the calculations that appear to reproduce the relaxation dynamics for HBDI fail badly for FHBMI, and you speculate that "the failure reflects a less suitable PES for this molecule." The two molecules look quite similar, so how different can the potential energy surfaces be with respect to the reaction coordinate in Fig. 6?

Professor Meech replied: The spectroscopy (figure 3) dynamics (table 1) and potential energy surfaces (ref. 37) are all similar, so the failure to fit the time resolved data is surprising. We do apply a number of constraints in the fitting procedure including fitting absorption steady state and Stokes loss (ref. 55) and a change in any one of these could prevent a successful analysis; we did not analyse systematically the origin of the poor fit, though different initial conditions were tried with the same result. The key point from these data is the overall similarity of the two chromophores in solution compared with the very different photophysics of the respective proteins.

Professor Sension asked: In your paper you present data for the decay of HBDI neutral that is largely independent of solvent for solvents ranging from acetonitrile to ethylene glycol. Yet your model predicts a very large viscosity dependence. Doesn't this suggest that the potential surface used in the simulations is incorrect? Or would you attribute the failure of the calculations to some other factor?

Professor Meech replied: There are two possibilities. One is indeed that the potential energy surfaces calculated (which generally involve rotation about one or other of the bridging bonds) is not the correct reaction coordinate in solution, and the reaction proceeds by another (volume conserving) coordinate; in that case the observed viscosity independence is not unexpected. We mention the pyramidalization coordinate in the paper, as this coordinate does feature in some calculations, and is of a more volume conserving character. The second possibility is that the surface is correct, but that coupling to the ground state occurs much earlier along the coordinate than a full twist (which is where the conical intersection is usually calculated). This is rather crudely represented in our model by the broad $\Gamma(z)$ function. The small scale motion required in that case may be insensitive to medium friction.

Dr Amirjalayer commented: In your work you show that there is no solvent dependency on the photochemistry of the GFB chromophore both with respect to the viscosity and (also to some degree) to the polarity of the solvent. Based on these results and the possibility of having a volume-conserving decay coordinate, could you please comment on how you expect the protein environment to influence the properties of the chromophore?

Professor Meech replied: The chromophore in the protein has a very different electronic spectrum to any solvent, being significantly red-shifted. It has proven very difficult to reproduce this behaviour outside the protein. Thus one possibility is that the particular H-bonding environment gives rise to a different electronic structure, with different excited state surfaces, conical intersections, *etc.* A second possibility concerns the H-bonding environment itself. We speculate in the paper that the strong H-bonding between the phenyl ring O^- and the protein in the anionic form (the most strongly fluorescent state of GFP) may have to be broken for the reaction to proceed, providing an effective barrier in the protein environment. It is interesting that proteins with different H-bond environments for the chromophore (dronpa for example) have much faster excited state decays.

Professor Temps asked: The proposed coordinate dependence of the radiationless decay rate is not really as surprising as it might look at first glance because in reality the transitions will occur with some probability anywhere in the vicinity of the multi-dimensional conical intersection seam. The usually plotted conical intersection 'point' in a one-dimensional plot of the potential energy surfaces against the reaction coordinate is just the minimum of that seam. Surfaces hopping calculations for a number of systems have indeed shown that the hops may not necessarily happen close to the minimum intersection point. Would it be possible from the simulations to extract information on the energetic distance of the coupled diabatic states or on the dependence of the coupling element on coordinate space?

Professor Meech replied: This is an important question. With the simple 1D model we used requires relatively efficient excited state decay remote from the CI. It is probably a step too far to try to infer from our data details on coupling elements, but the calculations you mention (and those cited by Dr Malhado below) suggest progress in this direction is possible.

Dr Malhado asked: I wish to comment on the excited state dynamical model used in Professor Meech's article to interpret the experimental results. In the article, the gas phase potential energy profile along a one dimensional generalized reaction coordinate is used, with the dynamics being described by the

Smoluchowski equation and a coordinate dependent decay rate function. In such a simple picture much of the complexity of the interaction of the system with the molecular environment (which has a strong influence on the photophysics/photochemistry of these systems as mentioned in the article) is buried on the coordinate dependent decay function $\Gamma(z)$. Interpretation of this function and the different physical effects leading to it is thus highly non-trivial. Nine years ago, at another Faraday Discussion, Burghardt and Hynes [1] presented a model to describe a photoizomerization reaction occurring through a conical intersection in solution. This model includes the effect of the solvent in a simple way and has the virtue of including dynamic solvent effects which are important on the ultra-fast time scale. The model shows how the solvent may have a strong influence on the potential energy surfaces, changing the topography of the conical intersection, and changing the reaction path.[1,2,3] A high degree of care should thus be used in interpreting $\Gamma(z)$, since while non-adiabatic transitions can occur away from conical intersections, and indeed may not occur mainly exactly at the conical intersection seam,[3] it is highly unlikely that they occur too far from the seam where the energy gap between the ground and excited state is large.

The use of the Smoluchowski equation may also be inadequate in this case and explain the difficulties described in the article to reproduce the experimental results. The Smoluchowski equation is valid in the overdamped limit and considers that the velocity correlation function $\langle v_z(0).v_z(t)\rangle$ decays very fast compared to the time scale of the dynamics (that it is a δ function). This is however an unlikely hypothesis in the ultra-fast time scale and in this particular system where the dynamics in solution is on the same time scale as in the gas phase. A first improvement over this treatment would be to use a time dependent diffusion coefficient (which would provide challenges to the physical interpretation of its own) or the use of a more complete Fokker–Planck equation. Recently we have published calculations for an photo-isomerization reaction in solution[3] where we have included solvent dissipative effects using a generalized Langevin equation, which not only considers the finite time scale of the velocity correlation function, but also the finite time scales of the environment/dissipation forces acting on the system.

1. I. Burghardt, L. S. Cerderbaum, J. T. Hynes, *Faraday Discuss.*, 2004, **127**, 395–411
2. J. P. Malhado, J. T. Hynes, *J. Chem. Phys.*, 2012, **137**, 22A543
3. J. P. Malhado, R. Spezia, J. T. Hynes, *J. Phys. Chem. A*, 2011, **115**, 3720–3735

Professor Meech answered: A key conclusion of the present paper is that, if the calculated excited state surface is correct, decay to the ground state must occur rather far from the conical intersection. This is represented rather crudely by $\Gamma(z)$. Calculations of the type you describe seem likely to be important in understanding this phenomenon in more detail. Unfortunately a more sophisticated model of the excited state dynamics for a large molecule in a realistic solvent requires more detailed calculations than we are presently capable of performing.

Concerning the assumptions of the Smoluchowski equation, you are quite right to say that the timescales we are working under are fast, and therefore possibly outside the overdamped limit implicit in our application of equation (2). We have however looked for evidence of coherent dynamics in the HBDI reaction and not found them, while in other excited state reactions in solution we have resolved them more successfully.[1,2] In previous applications of equation (2) we have made use of frequency dependent friction,[3] as described earlier by Hynes.[4] It was not applied here as it complicates further the fitting, and will not change the qualitative conclusions of Figure 10.

1. J. Conyard, K. Addison, I. A. Heisler, A. Cnossen, W. R. Browne, B. L. Feringa and S. R. Meech, *Nature Chem.*, 2012, **4**, 574.
2. M. Kondo, I. A. Heisler and S. R. Meech, *Faraday Discuss.*, 2010, **145**, 185.
3. I. A. Heisler, M. Kondo and S. R. Meech, *J. Phys. Chem. B*, 2009, **113**, 1623.
4. G. van der Zwan and J. T. Hynes, *J. Chem. Phys.*, 1982, **76**, 2993.

Professor Stolow said: The failure of the adiabatic approximation is based on the time scale of a process being fast compared to an inverse level spacing. More specifically for polyatomic molecular dynamics, the time scales of specific vibrational motions will be highly relevant for any discussions of diabatic *versus* adiabatic passage through a conical intersection. I would expect that even innocuous chemical substituents, such as alkyl groups, would vary the relative frequencies of some modes relative to others. Would such 'inertial' effects not be important in affecting dynamics at conical intersections?

Professor Meech answered: Inertial effects can certainly modulate reactivity is isomerzing systems at conical intersections. There are already a few calculations on relevant examples, where the effect of hydrogen/deuterium exchange was studied.[1] This seems the best experimental approach, and we are currently exploring site specific H/D exchange synthetically. Unfortunately methyl substitution turned out to be far from innocuous. At least on the phenyl ring of HBDI large spectral shifts were observed, suggesting modifications of the potential energy surfaces.[2] If the synthesis is successful we expect the observation of inertial effects to depend on the time scale for the reaction relative to the time for solvent friction effects to become significant.

1. T. Montagnese, F. Ogliaro, S. Wilsey and M. J. Bearpark., *J. Phys. Chem. A*, 2008, **112**, 13239–13243.
2. J. Conyard, M. Kondo, I. A. Heisler, G. Jones, A. Baldridge, L. M. Tolbert, K. M. Solntsev, and S. R. Meech, *J. Phys. Chem. B*, 2011, **115**, 1571–1577.

Professor Fielding said: Professor Meech has presented an elegant experimental and theoretical study of the ultrafast excited state dynamics of the green fluorescent protein chromophore in solution. In the gas-phase, electron detachment has been shown to be an important competing channel, close to the Franck–Condon region.[1–5] Would Professor Meech like to comment on whether electron transfer processes might be relevant competing processes in solution?

1. M. W. Forbes and R. A. Jockusch, *J. Am. Chem. Soc.*, 2009, **131**, 17038.
2. D. A. Horke and J. R. R. Verlet, *PCCP*, 2012, **14**, 8511–8515.
3. C. R. S. Mooney, M. E. Sanz, A. R. McKay, R. J. Fitzmaurice, A. E. Aliev, S. Caddick and H. H. Fielding, *J. Phys. Chem. A*, 2012, **116**, 7943–7949.

4. Y. Toker, D. B. Rahbek, B. Klaerke, A. V. Bochenkova and L. H. Andersen, *Phys. Rev. Lett.*, 2012, **109**, 128101.
5. D. Ghosh, A. Acharya, S. C. Tiwari and A. I. Krylov, *J. Phys. Chem. B*, 2012, **116**, 12398–12405.

Professor Meech replied: I am cautious about drawing firm conclusions on electron transfer reactions in proteins on the basis of measurements of the gas phase electron detachment. The rate of an electron transfer reaction will certainly be a function of the energy of the acceptor, as well as the donor. Further the reorganisation energy is generally a property of the medium, at least for electron transfer reactions between large aromatic molecules. Finally, for HBDI, Krylov and coworkers (*J. Chem. Phys.*, 2011, **135**, 194304) suggested that hydration by a single water molecule converts the autoionizing excited state into a bound state. Thus there are at least three ways in which the environment might perturb the rate of any electron transfer reaction. Of course the interplay between gas phase experiments, calculations and measurements in solution and in proteins will continue to be central to future progress. There is currently a need to extend the gas phase experiments to solvated molecules, and for much better characterisation of electron transfer reactions in fluorescent proteins, where there are currently rather few well characterised examples.

Professor Bearpark asked: Can you explain why you chose the calculations of Olsen as a reference point for your model (Fig. 6)?

Professor Meech answered: There are a number of high level calculations on HBDI. However, Olsen and co-workers have extended these to other derivatives relevant to different members of the fluorescent protein family.[1,2] We wished to model our data on FHBMI, and Olsen has calculated on the related AHBMI molecule.

1. S. Olsen and S. D. Smith, *J. Am. Chem. Soc.*, 2007, **129**, 2054.
2. S. Olsen and S. D. Smith, *J. Am. Chem. Soc.*, 2008, **130**, 8677

Dr Bochenkova remarked: Branching in internal conversion is key in understanding complex dynamics of fast radiationless decay in the GFP chromophores. Two branches that refer to the twisting about one or the other bridging bonds are non-equivalent and only one of them directly leads to internal conversion through a conical intersection (see *Faraday Discuss. 163*/13, 2013, DOI: 10.1039/c3fd20150c). However, energy barriers for escaping from a planar minimum (fluorescent state) along these coordinates are similarly small ($\sim$0.05 eV for the HBDI anion in the gas phase). Such kinetic equivalence inevitably leads to a splitting of the excited-state population, where one branch repopulates the fluorescent state in time. Therefore, the scheme that has to be considered includes one irreversible branch (decay of the fluorescent state in internal conversion through a conical intersection) and one reversible branch (decay and gain of the fluorescent state through equilibrium with the excited-state twisted intermediate). Such a scheme directly leads to an analytical solution with a two-exponential decay of the fluorescent state observed experimentally in various solvents. It is also fully consistent with an experimental observation that the mean fluorescence wavelength is independent of time. The observed weak viscosity dependence is traced to a very steep topography of the potential energy surface after crossing a barrier, which corresponds to a rather small value of the twisting angle (38 degrees) along the irreversible branch. The same two-exponential excited-state decay has recently been observed in the gas phase (H. H. Fielding, J. R. R. Verlet *et al.*, *Chem. Science*, 2013, **4**, 921–927), although the two-exponential kinetics have been revealed by tracing the evolution of the excited-state twisted intermediate, rather than of the fluorescent state. Nevertheless, the disclosed timescales are those that equally refer to the fluorescent state decay provided that the kinetic model discussed above is used, rather than a sequential two-step mechanism as originally proposed by the authors of this paper (radiationless decay of the fluorescent state to the ground state through a twisted excited-state intermediate).

A simple consideration in the molecular orbital framework (supported by the calculated electron binding energies of the corresponding counterparts of the anionic chromophore upon twisting) is capable of accounting for different twisted intramolecular charge transfer pathways in internal conversion. The topography around the planar minimum in the excited state and along the two decay branches of the isolated HBDI anion obtained through high-level *ab initio* calculations within the extended multi-configuration quasi-degenerate perturbation theory provides direct support to this scheme. Finally, the conical intersection of the other type associated with twisting about the bond adjacent to the phenol ring (a reaction coordinate of the inefficient branch in internal conversion) requires an additional twisting about the other bridging bond that further decouples molecular orbitals of the bridge moiety and, as a result, leads to pyramidalization at the central carbon atom. This conical intersection is referred to as a hula-twist, but it also includes a pyramidalization coordinate. However, it lies higher in energy than the transition state that directly leads to the lowest lying conical intersection. Therefore, this pathway cannot be considered as rate-determining in the fast radiationless decay of the GFP chromophore.

Professor Meech answered: Your two channel decay model (one populating an intermediate dark state that thermally couples back to the bright state and one going direct to the ground state) would certainly reproduce the biexponential decay. However, for a rigorous test both the decay time and the weight of the two components must be modelled. In addition the long lived dark state should be observed in transient absorption experiments. We are currently conducting such experiments. Thus far transient absorption data in the visible and IR do not show a clear dark intermediate state.

Miss Perveaux opened the discussion of the paper by Dr Bochenkova: A very interesting result is presented in Fig. 10. We can notice in the absorption spectrum that the full widths at half maximum in the gas phase and in the protein are the same. This means that the Franck–Condon gradient is identical in both cases.

This is surprising because I would have expected the protein to have an impact on the potential energy surface of the chromophore throughout different kinds of interactions (geometrical constraint, electronic coupling) and thus alter the width of the spectrum.[1] What is the role played by the protein if it does not modify the potential energy surface of the chromophore?

1. G. A. Worth, H.-D. Meyer, and L. S Cederbaum., *J. Chem. Phys.*, 1998, **109**, 3518.

Dr Bochenkova responded: The spectral widths of the GFP chromophore anion are indeed remarkably similar in the gas phase and in the protein. This is highlighted in our paper and clearly indicates a similar early-time excited-state nuclear dynamics in both environments. The high-frequency stretching modes play a crucial role here, since the major contribution to the width comes from these modes, while those with low frequencies promote a spectral blurring. These Franck–Condon active modes refer to in-plane stretching vibrations that reflect the electron density redistribution upon excitation and are not space-demanding. Therefore, the large origin shifts and the corresponding gradients at the Franck–Condon point along these particular modes are intrinsic to the properties of the chromophore, connected to the nature of the first excited and ground states, which is the same in both environments. However, such similarity in the spectral widths does not allow us to generalize this conclusion to the entire excited-state potential energy surface. In fact, the GFP protein environment is expected to modify the excited-state potential energy surface of the bare chromophore by increasing barrier heights for intramolecular twisting, thus preventing a non-radiative de-excitation of the chromophore through internal conversion. This part of the potential energy surface is not probed at all by the early-time excited-state nuclear dynamics, which determines the spectral shape (the corresponding modes are even not exited upon the S_0–S_1 transition), and requires experimental studies that trace excited-state evolution on longer timescales using time-resolved pump–probe techniques. In fact, the part of the potential energy surface that is probed by absorption spectroscopy is rather small compared to those parts that may play a role in the excited-state decay channels.

Professor Neumark said: Your paper considers how exciting various vibrational modes of the GFP anion affects the branching between autodetachment and fragmentation. My understanding is that the absorption spectrum of the GFP anion shows no vibrational structure regardless of which channel is being monitored, so what experimental evidence is there for vibrational mode selectivity? Also, what evidence do you have that autodetachment occurs on the anion excited state surface as opposed to occuring after internal conversion to the anion ground state?

Dr Bochenkova responded: Our paper considers the mode-specificity in the excited-state decay channels. These are vibrational autodetachment (VAD) out of S_1 and internal conversion through a conical intersection S_1/S_0. The branching between these channels (rather than autodetachment and fragmentation) may directly be accessed by using a time-domain approach to absorption spectroscopy, where action spectra are recorded according to the timescale of action. In the case of the GFP chromophore anion, there are intrinsically different timescales at which the action takes place out of S_1 and out of S_0. Being transferred to the ground state, the anion with high internal energy undergoes fast intramolecular internal energy redistribution (IVR) on a timescale of picoseconds, so that anything that happens after has a statistical nature. The reason behind this is a very high vibrational level density of the anion in the S_0 state in the energy region accesses after one-photon absorption. The direct count algorithm together with calculated vibrational frequencies give a density of $\sim 10^{27}/cm^{-1}$ at an excitation energy of 2.9 eV (corresponding to 480 nm excitation energy and 0.3 eV internal energy at 300 K). The hot ground state may then proceed and decay by statistical fragmentation, as well as through a delayed electron ejection process out of the ground state with a relative strength that depends on the corresponding activation energies. The experimental vertical detachment energy (VDE, which is here very close to adiabatic detachment energy) has been measured to be in the range of 2.7–2.8 eV.[1–3] The activation energy for fragmentation is of the same order, but most likely lower. The barrier height has been estimated to be 2.1 eV.[4] Therefore, both statistical IVR-mediated pathways in S_0, which facilitate either electron emission or bond fragmentation, should appear in the delayed channel. The delayed electron emission out of S_0 has indeed been observed in photoelectron spectroscopy studies,[2] however, together with the clear signatures of VAD out of S_1.[1,2] Importantly, the branching ratio in the S_1 state is not affected by possible multiple pathways in the hot S_0 state and may be directly accessed by recording the prompt (out of S_1) and delayed (out of S_0) action separately. This may directly be compared with the theoretically estimated branching ratio based on the disclosed mechanisms of the excited-state decay channels and topography of the S_1 state in the Franck–Condon region and along reaction modes, as it has been done in our paper.[5] Indeed, our theoretical estimations are very close to those found experimentally.

The prompt signal is due to electron emission out of the S_1 state which is *bound* in the Franck–Condon region according to both our calculations[5] and experimental results.[1–3] In principle, a non-resonant direct channel is also open at energies above the VDE. In the case of a strong coupling of the embedded vibrational resonances to the electron continuum observed for very diffuse ground states of anions,[6] the resonant and non-resonant channels are mixed to a large extent with intensity being diluted through a set of interacting states. However, the coupling is weak in the case of the bound valence-type S_1 state of the GFP chromophore anion. The contribution from the direct photodetachment channel to the prompt action spectrum can be considered negligible within the excitation energies of the S_1 absorption band (2.48–3.26 eV) compared to that of the bright resonant absorption due to much smaller oscillator strength for the direct transition to the continuum. Support for this comes from photoelectron spectroscopy studies, where the distribution of kinetic energies of the outgoing electrons in the prompt channel is monitored as a function of excitation wavelengths.[1] A low kinetic energy together with a wavelength-independent narrow shape of the electron-energy distribution in the wide range of excitation energies of 2.48–2.92 eV gives evidence that the prompt electron emission within the S_1 absorption band occurs resonantly through vibrational autodetachment out of the S_1 state. A tiny contribution from the direct non-resonant channel in addition to the major resonant contribution may only be recognized at 2.92 eV,[1] appeared as a shoulder at 0.2 eV. It gives an electron binding energy of 2.72 eV. However, this contribution is small even at this high excitation energy, which is well above the VDE.

The prompt and delayed action spectra have a maximum at the same wavelength at 482 nm. In addition, the prompt spectrum significantly extends to the blue part and has major adjacent shoulders at 452 nm and 422 nm and a weak one at 472 nm. Similar features were observed at the quadrupole ion trap with almost the same wavelength assignments (470, 450 and 425 nm).[7] The experimental spectral shape is rather broad caused by the large manifold of contributing vibrational levels and the observed features are poorly resolved. Nevertheless, a full theoretical account of the observed broad features may then be given by comparing the S_0–S_1 simulated spectrum shown in Fig. 10 in our paper[5] and the experimental prompt and delayed action spectra. Importantly, the calculated major spectral feature, its total width, is consistent with the experimental data measured both in ELISA and in the quadrupole ion trap.[7] This gives additional evidence that the structures (even though poorly resolved) in the experimental prompt channel are explained as vibrational autodetachment out of the anionic S_1 state.

Direct experimental evidence for vibrational mode selectivity may be gained through future femtosecond time-resolved measurements, which include pre-excitation of specific vibrational modes. However, as explained above, photoelectron and time-domain action spectroscopy studies, as well as the recently published time-resolved photoelectron pump-probe experiments[8] give substantial evidence concerning the nature of the first excited-state of the anion, which is bound, and the mechanism of the prompt electron emission at low excitation energies, which is based on a coupling between the nuclear and electronic dynamics. The detailed insights into the mechanisms of the excited-state decay channels based on theory and presented in our paper[5] reveal the modes that most efficiently induce non-adiabatic couplings between the S_0/S_1 and S_1/D_0 states. If not being coupled to the nuclear motion, electron detachment would never occur out of the electronically bound S_1 state below the S_1/D_0 crossing point located out of the Franck–Condon region. The mode-specificity in VAD comes from the calculated S_1/D_0 origin shifts. The modes characterized by the largest shifts serve as reaction coordinates, since electron ejection has to be accompanied by the nuclear rearrangements along these particular modes. Furthermore, the non-adiabatic couplings in VAD should depend on vibrational overlap between the final vibrational wavefunction and the derivative of the initial vibrational wavefunction with respect to the corresponding nuclear coordinates. At low excitation energies, the final state refers to the ground vibrational level, and such overlap is larger if the two surfaces are displaced further along the corresponding normal modes.

Finally, the emission-active modes are shown to be directly excited upon photoabsorption. Upon their excitation in the blue part of the spectrum, electron ejection becomes a dominating channel. At low excitation energies, the calculated statistical non-radiative lifetimes based on the identified reaction modes agree well with those found experimentally.[8]

1. Y. Toker, D.B. Rahbek, B. Klærke, A.V. Bochenkova, and L. H. Andersen, *Phys. Rev. Lett.*, 2012, **109**, 128101.

2. D. A. Horke and J. R. R. Verlet, *Phys. Chem. Chem. Phys*, 2012, **14**, 8511–8515.
3. C. Mooney, M. E. Sanz, A. R. McKay, R. J. Fitzmaurice, A. E. Aliev, S. Caddick, and H. H. Fielding, *J. Phys. Chem. A*, 2012, **116**, 7943–7949.
4. L. H. Andersen *et al.*, *Phys. Chem. Chem. Phys.*, 2004, **6**, 2617–2627.
5. A. V. Bochenkova and L.H. Andersen, Faraday Disc., 2013, DOI: 10.1039/c3fd20150c.
6. N. I. Hammer *et al.*, *Science*, 2004, **306**, 675–679.
7. M.W. Forbes and R. A. Jockusch, *J. Am. Chem. Soc.*, 2009, **131**, 17038–17039.
8. C. R. S. Mooney, D. A. Horke, A. S. Chatterley, A. Simperler, H. H. Fielding and J. R. R. Verlet, *Chem. Science*, 2013, **4**, 921–927.

Dr Verlet addressed Professor Neumark and Dr Bochenkova: In response to the question by Professor Neumark about the low energy photoelectrons observed in the GFP chromophore photoelectron spectra. This low energy feature in the photoelectron spectra are in fact structured and contain two components. There is a statistical peak at very low kinetic energy, which we believe arises from thermionic emission from the ground state. This can be distiguished from a feature arising from vibrational autodetachment (VAD) discussed by Dr Bochenkova. Interestingly, both features are observed in the photoelectron spectra measured in our group (*PCCP* , 2012, **14**, 8511), but only the VAD feature is seen in the work by the Andersen group (*Phys. Rev. Lett.*, 2012, **109**, 128101).

Professor Neumark answered: Looking at Fig. 4 of your *PCCP* paper, it appears that thermionic emission from the ground state of the anion is the dominant process, with a small amount of structured signal that might be from the S_1 state. The structured signal is what you call the VAD signal, right? But this signal appears to be missing in the Andersen PRL at the same excitation energy (2.58 eV); all they see is the unstructured low energy signal that you attribute (correctly, I believe) to thermionic emission.

Dr Bochenkova answered: In Fig. 4 of the *PCCP* paper from the group of Prof. Verlet, there are indeed two contributions to photoelectron spectra. One of them at nearly zero kinetic energy is highlighted by an exponential decay and attributed to thermionic electron loss from the ground state of the anion following internal conversion. The other feature peaks at ~0.05 eV and is associated with VAD. In the PRL paper of Prof. Andersen, the contribution from the electron emission out of the ground state is missing, *i.e.*, the contribution arising from the presence of electrons with nearly zero kinetic energy, rather than the VAD signal that peaks at higher energies of 0.05 eV. In the latter paper, the resolution is not enough to reveal structures of the VAD signal, however, the maximum of this feature is also located at 0.05 eV (Fig. 5). What has to be underlined is that autodetachment out of S_1 and thermionic emission out of S_0 have intrinsically different timescales (for more details please see the answer to the previous question). The former is attributed to the prompt signal (a timescale of less than 100 ns and, in fact, a ps-timescale at low excitation energies or even faster at higher excitation energies), while the latter occurs on a considerably longer timescale. This is important since the experimental setup for registering outgoing electrons used in the group of Prof. Andersen appears to be blind to the delayed emission of the electrons with low kinetic energy, while the prompt emission coming from VAD is registered and shown in Fig. 5. Finally, the branching ratio addressed in our paper and also estimated by using a time-domain approach to action spectroscopy refers to the branching in the excited-state decay channels: internal conversion and prompt electron emission (VAD). Possible multiple decay channels in the hot ground state following internal conversion neither alter the experimental results, nor change their theoretical account, since the prompt and delayed signals are recorded according to the timescale at which action takes place.

Professor Domcke addressed Dr Bochenkova and Professor Buma: An essential question concerning the relevance of the one-dimensional torsional potential-energy function is the relative time scale of the torsional dynamics and the dynamics of other vibrational modes involved in the dynamics. There are two possibilities: the torsion can be slow or fast compared with other relevant modes. The model of Professor Meech seems to a *priori* assume that the torsion is the slow coordinate. To decide this question, *ab initio* based models including systematically the 3, 4, 5 or 6 most relevant modes should be constructed. Such calculations should be feasible for these medium-size molecules (in the gas phase).

Dr Bochenkova answered: It seems reasonable that the torsional dynamics is the slowest in this case. Based on our theoretical results discussed in paper 13, the excited-state potential energy surface is rather flat in the vicinity of the planar minimum of the GFP chromophore anion. However, there are distinct small barriers associated with intramolecular twisting that induces non-adiabatic couplings between two potential energy surfaces, resulting in internal conversion through a conical intersection. The other coordinate that defines a branching space for the conical intersection seam is represented by a linear combination of in-plane Franck–Condon active modes that reflect the electron density redistribution upon excitation. The dynamics associated with these modes are expected to be much faster. However, it is indeed important to include all slow coordinates to construct a multi-dimensional model for the dynamical treatment. In particular, the two torsional modes and any modes that might be associated with their dynamics, such as a pyramidalization coordinate, have to be considered to access the interplay between them. As a side note, a construction of 3,4,5-dimensional surfaces are feasible, but still very time-consuming and resources-demanding, provided that the high-level multistate multi-reference perturbation theory approaches are used for constrained geometry optimizations in the excited state of the GFP chromophore anion. This level of theory might be important to probe a rather flat excited-state potential energy surface.

Professor Buma asked: In your calculations you have optimised the geometry of deprotonated HBDI in the three relevant electronic states (S_0, S_1, and D_0). However, you have only obtained (harmonic) force fields for S_0 and D_0. In your analysis of the S_0–S_1 absorption profile you have taken for the force field of S_1 the same force field as for S_0. I would first like to remark that you can assess the quality of your force field by comparison with experimental data. Recently, a study has been published on the vibrational absorption spectrum of the S_0 state of deprotonated HBDI.[1] This study shows for example that the frequency of the carbonyl stretch that you calculate at 1733 cm^{-1} is actually found at 1660 cm^{-1}. Secondly, your analysis implicitly assumes that frequencies and normal modes are identical in S_0 and S_1. In view of the electronic character of S_1 and the associated changes in bond order, one can expect significant changes in the frequency of, for example, the stretch modes, and considerable Duschinsky mixing of a number of modes. This might significantly influence your analysis.

1. M. Almasian, J. Grzetic, G. Berden, B. H. Bakker, W. J. Buma, J. Oomens, *Int. J. Mass. Spec.*, 2012, 118–123, 330-332.

Dr Bochenkova replied: In calculating absorption profiles, we indeed use a somewhat simplified picture of multidimensional ground and excited-state harmonic potential energy surfaces displaced relative to each other along each normal mode. Neither frequency change nor Duschinsky rotation is presumed to occur upon electron transition. Although approximations made for treating a nuclear problem may be thought of as an oversimplified approach, it is important to underline that a highly reliable prediction of the ground-to-excited state minimum displacements (origin shifts) play here the most crucial role in simulating major spectral features. We use a highly correlated multistate multi-reference perturbation theory in its invariant version XMCQDPT2 formulated recently.[1] One way to validate an electronic structure method used for calculating an excited-state gradient is to compare the final overall widths in the experimental and simulated spectra. By the time-energy uncertainty principle the broadest feature of the spectrum, its width, is determined by the shortest feature in time. This feature refers to the initial dynamics of vibrational wave packet moving downhill along the path of steepest descent. As a result, an accuracy of the excited-state gradient at the Franck–Condon point essentially determines the major feature of the spectrum, its total width. The simulated vibrational profiles are rather broad and do not give precise spectroscopic information. The broadening is caused by the large manifold of contributing vibrational levels. Importantly, the spectral profiles of the GFP chromophore anion are inherently smooth and mostly structureless in the gas phase and in the protein.

The quality of the force field used here and obtained through a routine DFT approach has extensively been tested before. A scaling factor of 0.96, which also includes the anharmonic corrections, might well be expected for the stretching mode mentioned above. However, the deviations between the calculated and experimental frequencies are expected to be altered for different types of the modes. A search for a 'better' method for the electronic structure calculations, a switch to the methods that go beyond the harmonic approximation, as well as a simple scaling of the frequencies might all be in operation, but drastically depending on the aim of the studies. As it has been outlined before, intrinsically broad spectral profiles leave out the precise spectroscopic information concerning a spacing between the distinct vibronic transitions.

Taking all the approximations together, we do not expect any dramatic changes in the shapes upon going beyond them due to a large manifold of contributing vibrational levels. A set of Franck–Condon active high-frequency stretching modes could indeed be affected to some degree and mixed with each other, but it won't change their role in the dynamics and in the excited-state decay channels. Despite a little precise spectroscopic information, the simulated spectra are of utmost importance, since they allow us to access to the nature of the first excited-state of the isolated anion and to reveal a crucial role of specific vibrational modes in the excited-state decay channels within a certain energy range. These simulated profiles are also required for interpreting the experimental results obtained through action absorption spectroscopy and enable a direct comparison between absorption spectra in different media. Importantly, the spectral widths as well as the major structures in the spectra are fully accounted for in the calculated profiles both as compared to the experimental results, thus justifying the use of the above-mentioned approximations in the gas phase and in the protein.

1. A. A. Granovsky, *J. Chem. Phys.*, 2011, **134**, 214113.

Dr Santoro addressed Professor Buma and Dr Bochenkova: I would like to comment on the necessity/usefulness to take into account frequency changes and Duschinsky mixings in the calculation of absorption spectra. It is true that precision problems can arise and that they may be expected to be more concrete than for ground-state calculations but, on the other side, if the interest is in an accurate reproduction of high-resolution (e.g fluorescence) spectra, anharmonic corrections have to be considered even for the ground-state. What I would like to point out is that frequency changes and Duschinsky mixings (arising from quadratic differences between the potential energy surfaces of the initial and final state) have an impact also on low-resolution spectra[1] so that, depending on the focus of the analysis, it can be worthy to take them into account. This is easily recognized by looking at the first moments of the spectrum lineshape, considered as a distribution, on the ground of results obtained already in the fifties.[2,3] Quadratic couplings shift the barycentre of the spectrum, and redshifts on the order of 0.1 eV are very common.[4] Moreover they contribute a "positive" term to the spectrum standard deviation, *i.e.* they increase the spectral width,[5] and they break the mirror symmetry between the absorption and emission spectra. To mention some specific examples from my own experience, the frequency changes in low-frequency modes is responsible for the strong broadening effect due to temperature in *trans*-stilbene[6,7] and for the remarkable difference of the width of absorption and emission spectra in di-thiophene.[8] In 3methyl-cyclopentanone the vibronic structure associated with the $n\pi^*$ transition is dominated by the CO stretching that exhibits a frequency much lower than in the ground-state. In such a situation, the calculation of the excited-state Hessian allows us to remarkably improve the predicted spacing between the vibronic peaks.

1. F. J. Avila Ferrer and F. Santoro, *Phys. Chem. Chem. Phys.* 2012, **14**, 13549.
2. M. Lax, *J. Chem. Phys.*, 1952, **20**, 1752.
3. J. J. Markham, *Rev. Mod. Phys.*, 1959, **31**, 956.
4. F. J. Avila Ferrer, J. Cerezo, E. Stendardo, R. Improta and F. Santoro, *J. Chem. Theory Comput.*, 2013, **9** , 2072
5. A. Lami and F. Santoro, *Time-Dependent Approaches to Calculation of Steady-State Vibronic Spectra: From Fully Quantum to Classical Approaches* in *Computational Strategies for*

Spectroscopy: From Small Molecules to Nanosystems, ed. V. Barone, John Wiley and Sons, Chichester, U.K., 2011, pp 475–516.
6. J. Tatchen and E. Pollack, *J. Chem. Phys.*, 2008, **128**, 164303
7. F. Santoro, A. Lami, R. Improta and V. Barone, *J. Chem. Phys.*, 2007, **126**, 184102.
8. E. Stendardo, F. J. Avila Ferrer, F. Santoro and R. Improta, *J. Chem. Theory Comput.* 2012, **8**, 4483
9. N. Lin, F. Santoro, X. Zhao, A. Rizzo and V. Barone, *J. Phys. Chem. A*, 2008, **112**, 12401.

Dr Bochenkova replied: Please see the reply to the earlier question from Professor Buma.

Dr Oliver opened the discussion of the paper by Professor Buma: The molecular structure and REMPI excitation spectra of OM*p*CA bear a striking resemblance to phenol.[1] Drawing an analogy between OM*p*CA and phenol may give an alternative explanation for the lifetime broadening discerned in your analysis. In phenol, the $\pi\sigma^*$ state is dissociative with respect to the O–H bond, but not accessible in the vFC region at the electronic $1^1\pi\pi^*$ origin, however, tunnelling under $\pi\pi^*$/ $\pi\sigma^*$ conical intersection can populate this state and lead to O–H bond fission.[2,3,4] Did you look for any H atom photofragments upon UV photoexcitation? Does the fluorescence lifetime of the $1^1\pi\pi^*$ state change upon deuteration of the O–H bond? Phenol-h_6 and phenol-d_6 have fluorescence lifetimes that differ by an order of magnitude[5,6]and is a key signature that H atom tunnelling and subsequent O–H bond fission is a significant excited state decay pathway in phenol and substituted phenols.[1,2,3,7]

1. M. G. D. Nix, A. L. Devine, B. Cronin, R. N. Dixon and M. N. R. Ashfold, *J. Chem. Phys.*, 2006, **125**, 133318.
2. R. N. Dixon, T. A. A. Oliver and M. N. R. Ashfold, *J. Chem. Phys.*, 2011, **134**, 194303.
3. G. A. Pino, A. N. Oldani, E. Marceca, M. Fujii, S. I. Ishiuchi, M. Miyazaki, M. Broquier, C. Dedonder and C. Jouvet, *J. Chem. Phys.*, 2010, **133**, 124313.
4. A. L. Sobolewski, W. Domcke, C. Dedonder-Lardeux, and C. Jouvet, *Phys. Chem. Chem. Phys.*, 2002, **4**, 1093.
5. R. J. Lipert, G. Bermudez and S. D. Colson, *J. Phys. Chem.*, 1988, **92**, 3801.
6. R. J. Lipert and S. D. Colson, *J. Phys. Chem.*, 1989, **93**, 135.
7. T. A. A. Oliver, G. A. King and M. N. R. Ashfold, *Phys. Chem. Chem. Phys.*, 2011, **13**, 14646.

Professor Buma answered: We agree that in many aspects one would expect a very similar behavior of excited states in OM*p*CA and phenol. We have not looked for any H atom photofragments upon UV photoexcitation. However, in a previous paper[1] we have explicitly considered whether tunneling to the $\pi\sigma^*$ state is a dominant decay pathway when exciting the V'($\pi\pi^*$) state by comparing the lifetime broadened band contours of resonances in OM*p*CA and OM*p*CA(d) in which the hydroxyl group is deuterated. We find only small differences that definitely do not parallel the large changes observed for the deuteration of phenol. Moreover, one should also keep in mind that the excited-state lifetimes in OM*p*CA as deduced from the line widths in ref. 1 and as recently reported in ps time-resolved pump–probe REMPI experiments[2] are more than three orders of magnitude shorter than what is observed for phenol and its deuterated form. Other relevant information comes from experiments on the cluster of OM*p*CA and phenol with water. For phenol it has been observed that complexation with water extends the lifetime of the vibrationless level of the $S_1(\pi\pi^*)$ state from about 2 ns to 15–18 ns.[3,4] However, for OM*p*CA much more dramatic effects are observed: the lifetime is increased by more than two orders of magnitude,[1,2] while the present experiments show that there is no longer a REMPI signal that decays on a ns timescale. All these observations strongly suggest that for the presently considered excitation energies the relaxation dynamics of phenol and OM*p*CA are fundamentally different. At the same time, it should be noticed that this does not preclude that at higher excitation energies a relaxation channel involving the $\pi\sigma^*$ state might become important.

1. S. Smolarek, A. Vdovin, E. M. M. Tan, and W. J. Buma, *J. Phys. Chem. B*, 2011, **115**, 1275.
2. D. Shimada, R. Kusaka, Y. Inokuchi, M. Ehara, and T. Ebata, *Phys. Chem. Chem. Phys.* 2012, **14**, 8999.
3. R. J. Lipert, G. Bermudez and S. D. Colson, *J. Phys. Chem.*, 1988, **92**, 3801.
4. R. J. Lipert and S. D. Colson, *J. Phys. Chem.*, 1989, **93**, 135.

Professor Burghardt remarked: Your observation of an unstructured excitation spectrum of the methyl-4-hydroxycinnamate thio ester (TM*p*CA) species is most interesting, all the more so since the associated lifetime seems to be inferior to the lifetime of the same species in aqueous solution. Possibly the isolated species exhibits a direct double-bond torsional pathway leading to rapid S_1–S_0 decay, while water complexation could favor a quinone-like electronic structure and, hence, induce pathways that are similar to those observed for the deprotonated species. A solvent-induced stabilization of the S_1 state might thus be conceivable. From our recent analysis of the S_1 potential surface of the anionic, hydrogen bonded species [Ref. 22 of your paper], S_1 minima at the single-bond twisted geometry are easily accessible in the deprotonated species, and we concluded that the interplay of at least three strongly coupled coordinates needs to be considered, *i.e.*, the double and single bond torsions and a hydrogen out-of-plane (HOOP) mode. Could you possibly carry out experiments and calculations for a TM*p*CA–H_2O cluster - similarly to your study of OM*p*CA–H_2O - to shed light on these pathways in the neutral chromophore–water complex?

Professor Buma responded: The studies we have reported here on TM*p*CA have been a first exploration of the excited-state manifold of this molecule, and definitely need to be followed up by other studies. In my reply to Professor Neumark I already indicated that femtosecond time-resolved photoelectron studies are on this list, but studies of the TM*p*CA–H_2O cluster are certainly on our to-do list as well. We are moreover synthesizing derivatives of TM*p*CA in which bond isomerization is inhibited. We expect that these derivatives will show sharper excitation spectra and thereby provide us with a means to come to a better understanding of the influence of bond isomerization in the excited-state dynamics of TM*p*CA, including the role of the modes that you mentioned.

Professor Neumark remarked: It seems that measuring the photoelectron spectrum resulting from resonant two-photon ionization would provide invaluable information concerning the excited state relaxation pathways in PYP. Do you have plans to carry out such experiments?

Professor Buma responded: In the present experiments we have performed mass-resolved ion detection. I agree with Professor Neumark that resonance enhanced multiphoton ionization in combination with kinetic-energy-resolved photoelectron detection might in principle provide additional information on the excited-state relaxation pathways. The reason for this is on the one hand that in a one-configuration picture the $\pi\pi^*$ and $n\pi^*$ states are expected to ionize to different ionic states. The photoelectron spectrum thus will directly reflect relaxation pathways between these states. On the other hand, when internal conversion takes place from an electronically excited state that is responsible for the resonance enhancement to another electronically excited state, the latter state will be vibrationally excited. The photoelectron spectra before and after internal conversion are thus in general quite different, and offer the possibility to elucidate such relaxation pathways. We therefore indeed plan to perform such experiments, in the first instance using nanosecond lasers, but it is clear that there is much to gain by performing femtosecond time-resolved photoelectron detection.[1] This pertains in particular to the case of TM*p*CA where our excitation spectra strongly suggest that sub-ps excited-state relaxation processes involving bond isomerization take place. The time evolution of the photoelectron spectrum could in this case be able to pin down the details of these structural changes.

1. A. Stolow, A. E. Bragg, and D. M. Neumark, *Chem. Rev.*, 2004, **104**, 1719.

Professor Bearpark addressed Professor Buma, Professor Meech and Dr Bochenkova: Calculations and experiments on isolated chromophores are challenging. Nevertheless we miss the environmental effects – mechanical, electrical, dynamical – which we expect to be considerable. This seems like removing a patient's heart to operate on, yet expecting it to behave as it does in the body?

Professor Buma responded: I agree completely with Dr Bearpark that complex molecular systems can give rise to unique properties at the molecular and material level that often are hard to study by using simple, covalently bound molecular systems as their representation. At the same time, one has to acknowledge that one needs to know the properties of an isolated chromophore before one can assess how these properties are effected by an environment. Modern experimental spectroscopy also recognizes this, and that is exactly the reason that we see a continuous development of novel spectroscopic tools that enable one to go beyond a minimal approach. In the gas phase, examples include the use of laser desorption and helium nanodroplet spectroscopy to study large molecular systems, but also the study of made-to-order noncovalent clusters to study key elements of a molecular environment. The advantage of the gas-phase approach as followed here, is that one can study the influence of these environmental effects one at a time by controllably adding them to the chromophore. In the present study we have seen, for example, that the addition of one single water molecule to the chomophore reverses the order of the V'(ππ*) and nπ* states, and that it lowers the barrier for bond isomerisation. Similarly, we find that attaching the *para*-coumaric acid chromophore to the proteinic backbone by a thioester has dramatic consequences on the order of the electronically excited states. Furthermore, these studies show that the solvent appears to have a much larger influence on the structural dynamics of the chromophore than assumed previously. The mere fact that all of these conclusions had not been reached before in the vast amount of studies on PYP and its chromophores emphasizes in this respect even more the necessity to perform this type of studies.

Professor Meech answered: The analogy is an interesting one. Of course we learn a lot about the heart by removing and dissecting it. However, if we want to understand how it interacts with the host it is better to study it *in situ* (certainly from the host's point of view).

Dr Bochenkova answered: In our strategy we adopt a reductionist approach and first strip away the chromophore natural environment to provide a scientific benchmark. We then study the interactions of the chromophore with the hosting protein medium and finally whole proteins. The isolated protein chromophore may have numerous fast response modes, but only one is usually selected by the protein. This raises the question why is this so? It also makes one ask whether it is possible to control or change the selectivity of the photochemical reaction in the protein in a well-defined and rational way. In other words, can we understand and ultimately change the specificity dictated by protein interactions? These questions have been addressed in our paper [A.V. Bochenkova and L.H. Andersen, *Faraday Disc.*, 2013, DOI: 10.1039/c3fd20150c]. Remarkably, the photoabsorption profiles and the early-time excited-state nuclear dynamics of the GFP chromophore anion are found to be strikingly similar in the gas phase and inside the protein, where the high-frequency stretching modes play a crucial role. This has an interesting implication to the biological function of the GFP proteins and, in particular, to their possible role in light-induced electron transfer reactions mediated by the most active Franck–Condon modes.

Professor Fielding opened further discussion of the paper by Dr Bochenkova: Despite the fact that the isolated GFP chromophore anion *in vacuo* has an absorption spectrum that appears to be similar to that of the anionic form of the chromophore in the protein,[1, 2] the excited state dynamics in vacuo[3] are extraordinarily similar to those of the isolated chromophore in solution.[4, 5] We have discussed a paper presented by Dr Bochenkova describing competition between autodetachment and internal conversion for the chromophore *in vacuo* and a paper presented by Professor Meech describing a coordinate-dependent non-radiative lifetime for the chromophore in solution. Could Professor Meech and Dr Bochenkova comment on how we can rationalise almost identical non-radiative lifetimes *in vacuo* and in solution with the two, different, models of the excited state dynamics?

1. S. B. Nielsen, A. Lapierre, J. U. Andersen, U. V. Pedersen, S. Tomita and L. H. Andersen, *Phys. Rev. Lett.*, 2001, **87**, 228102.
2. M. W. Forbes and R. A. Jockusch, *J. Am. Chem. Soc.*, 2009, **131**, 17038.
3. C. R. S. Mooney, D. A. Horke, A. S. Chatterley, A. Simperler, H. H. Fielding and J. R. R. Verlet, *Chemical Science*, 2013, **4**, 921–927.
4. D. Mandal, T. Tahara and S. R. Meech, *J. Phys. Chem. B*, 2004, **108**, 1102–1108.
5. S. R. Meech, *Chem. Soc. Rev.*, 2009, **38**, 2922–2934.

Professor Meech responded: The similarity is surprising. Even if the effect of medium friction in fluid solvents is negligible the isolated molecule will have some excess energy which would be rapidly lost in solution. Further, as Dr Bochenkova comments, the similarity suggests that the twisted state can contribute to the fluorescence decay. Identifying the twisted state in solution is one objective for the future.

Dr Bochenkova answered: Almost identical non-radiative lifetimes of the GFP chromophore anion observed *in vacuo*[1] and in solution[2,3] results from the fact that the same process governs an excited-state decay. In both cases, internal conversion is the fastest rate-determining channel provided that the isolated anion is studied upon excitation with a low-energy photon (pump pulses at 2.48 eV/500 nm as used in Ref. [1]). All competing processes, such as resonant photodetachment (*in vacuo*) or fluorescence, are characterized by considerably longer intrinsic timescales under these conditions. According to our theoretical estimates, an indirect photodetachment timescale is 20 ps upon excitation at 500 nm.[4] The GFP fluorescence occurs on a nanosecond timescale, whereas the experimental[1–3] and theoretical[4] characteristic lifetimes attributed to internal conversion are of the order of (sub)picoseconds. Therefore, the excited-state decay, as well as the fluorescence decay as monitored in solutions, is almost entirely determined by internal conversion rate. This is illustrated below.

As discussed in our paper,[4] branching in internal conversion is crucial for understanding the complex dynamics observed experimentally both in vacuo and in solution. A two-exponential decay of the excited-state population is readily explained considering the following generalized scheme (Fig. 1):

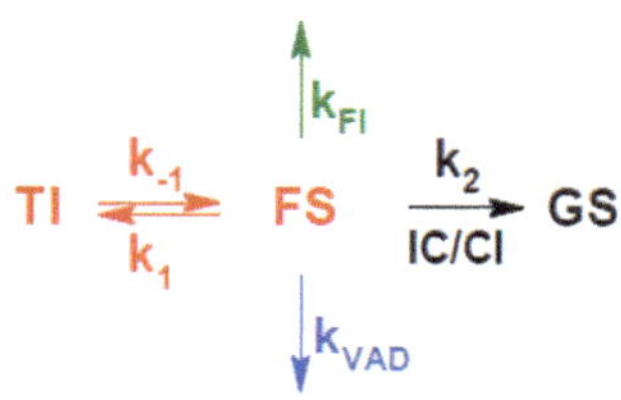

Fig. 1 Scheme for a two-exponential decay of the excited-state population

where FS is the fluorescent state that refers to the planar minimum in the excited state, TI is the twisted transient intermediate, GS is the ground state product and IC/CI denotes internal conversion through the lowest lying conical intersection. Vibrational autodetachment (VAD) out of S_1 and fluorescence (Fl) are shown as competing processes to the decay of the fluorescent state through internal conversion. The transient intermediate, TI, may also undergo internal conversion, however, at a significantly slower rate, since the corresponding branch depicted in red does not directly lead to the associated conical intersection seam.[4] For the sake of simplicity, we omit this stage here. The fluorescence decay (as monitored in solution) can then be written as

$$I_{Fl}(t) = k_{Fl}[FS] = k_{Fl}\left(\frac{k_{-1}-\gamma_1}{\gamma_2-\gamma_1}e^{-\gamma_1 t} + \frac{k_{-1}-\gamma_2}{\gamma_1-\gamma_2}e^{-\gamma_2 t}\right) = a\cdot e^{-\frac{t}{\tau_1}} + b\cdot e^{-\frac{t}{\tau_2}} \quad (1)$$

where τ_1 and τ_2 are the two characteristic excited-state decay timescales defined as inverse of the γ functions, which, in general, depend on the rate constants of all elementary stages. However, the largest rate constants determine the lifetime components:

$$\begin{aligned}\gamma_{1,2} &= \frac{1}{2}(k_1 + k_{-1} + k_2 + k_{VAD} + k_{Fl}) \\ &\mp \frac{1}{2}\sqrt{(k_1 + k_{-1} - k_2 - k_{VAD} - k_{Fl})^2 + 4k_1k_2 + 4k_1k_{VAD} + 4k_1k_{Fl}} \\ &\approx \frac{1}{2}(k_1 + k_{-1} + k_2) \mp \frac{1}{2}\sqrt{(k_1 + k_{-1} - k_2)^2 + 4k_1k_2}\end{aligned} \quad (2)$$

We here assume that k_1, k_2, k_{-1} >> k_{VAD}, k_{Fl} both *in vacuo*, as well as in solution, where one photon detachment doesn't occur since the electron binding energy is higher than that *in vacuo*.

The transient intermediate kinetics is described by the same two lifetime components defined by internal conversion (as monitored *in vacuo*[1]):

$$I(t) \sim [TI] = \frac{k_1}{\gamma_2-\gamma_1}e^{-\gamma_1 t} + \frac{k_1}{\gamma_1-\gamma_2}e^{-\gamma_2 t} = \tilde{a}\cdot e^{-\frac{t}{\tau_1}} + \tilde{b}\cdot e^{-\frac{t}{\tau_2}} \quad (3)$$

Therefore, the excited-state evolution of the fluorescent state and the transient intermediate is both described by the same bi-exponential kinetics. We note here, that a sequential two-step mechanism originally proposed for describing the decay of the excited-state population[1] is not capable of accounting for a two exponential decay of the fluorescent state, as observed in solution.

The similar experimental timescales also allow us to conclude that the corresponding barrier heights that determine the rate constants are almost equally small *in vacuo* (~1 kcal/mol) and in solution. In the GFP proteins, the barriers associated with internal rotation are on the contrary expected to be significantly higher, thus switching the rate-determining process from internal conversion to light emission. Photodetachment of the GFP protein as a whole does not occur upon excitation with a single low-energy photon, and a photo-induced electron transfer, where GFP may serve as an electron donor, only occurs in the presence of proper electron acceptors.[5]

Both absorption spectra and excited-state decay through internal conversion are determined by the photoinitiated nuclear dynamics, occurring, however, on different timescales and influenced by different factors. The spectral profiles are mostly defined through origin shifts between two potential energy surfaces, whereas the excited-state decay is governed by crossing a barrier along reaction modes, which may not have even been excited upon photoabsorption. In the case of the GFP chromophore anion, the corresponding barrier heights in internal conversion are remarkably small both *in vacuo* and in solution, thus it is not surprising that the nonradiative lifetimes are similar in these environments.

It is worth noting that both experimental excited-state lifetime components are of the order of (sub)picoseconds and the transient intermediate lives less than a few picoseconds in the excited state, so that internal conversion may occur faster than vibrational energy dissipation in the excited state through the couplings to solvent molecules, usually occurring on a picosecond timescale. Alternatively, a partial loss of vibrational energy through the couplings to solvent or somewhat higher internal vibrational energy of anions produced in the gas phase may easily be compensated for by even a smaller barrier height for internal conversion in solution as compared to that in the isolated anion. This indeed might be the case in polar solvents, since internal conversion proceeds via twisted intramolecular charge-transfer pathways.

1. C. R. S. Mooney, D. A. Horke, A. S. Chatterley, A. Simperler, H. H. Fielding and J. R. R. Verlet, *Chem. Science*, 2013, **4**, 921–927.
2. D. Mandal, T. Tahara and S. R. Meech, *J. Phys. Chem. B*, 2004, **108**, 1102–1108.
3. S. R. Meech, *Chem. Soc. Rev.*, 2009, **38**, 2922–2934.
4. A.V. Bochenkova and L.H. Andersen, *Faraday Disc.*, 2013, DOI: 10.1039/c3fd20150c.
5. A. M. Bogdanov, A. S. Mishin, I. V. Yampolsky, V. V. Belousov, D. M. Chudakov, F. V. Subach, V. V. Verkhusha, S. Lukyanov and K. A. Lukyanov, *Nat. Chem. Biol.*, 2009, **5**, 459–461.

Dr Verlet said: This is in response to Professor Fielding's question concerning the observed close similarity of the timescales in solution and the gas-phase.

The internal conversion dynamics may be similar, however, in the gas-phase, autodetachment can proceed by vibrational autodetachment, as shown by Dr Bochenkova. The argument that is made is that, at least at 500 nm, the autodetachment is slower than the internal conversion. We have very recently measured the lifetime of the excited state autodetachment following 500 nm excitation, showing that the lifetime is on the order of 30 ps. Hence, the actual lifetime of internal conversion is slightly longer than that measured by the Fielding and Verlet groups (*Chem. Sci.*, 2013, **4**, 921), but not dramatically. At higher photon energies, the observed lifetime may be expected to decrease because of the competing autodetachment channel.

Professor Duxbury opened further discussion of the paper by Professor Buma: On page 12, in Figure 4 the (a) RE2Pi and (b) LIF excitation spectra show similar patterns over most of the range of the excitation energy. However the strong band evident close to 33 500 cm^{-1} in 4(a) appears to be completely absent in Figure 4(b), even though the next three clumps of lines out to 23 250 cm^{-1} are clearly visible. The intensities of the two sub-bands at 34 000 cm^{-1} seem to be reversed on swapping from RE2PI to LIF. On pages 13 to 15, Figure 5 In your description of

TMpCA you note the large contrast between this spectrum and that of OM*p*CA. One possible explanation for this is the large magnitude of the spin–orbit coupling constant of sulphur in comparison with oxygen. In many of the electronic spectra of triatomic molecules such as the molecular ions H_2O^+ and H_2S^+ the effect of the increasing spin–orbit coupling is to produce a strong coupling into other electronic states. The effects of this can range from predissociation, a breaking off the emission spectrum in the excited state as in H_2S^+, to an apparent randomisation of the interaction between electronic states. For example, the singlet triplet coupling in CH_2 is small, but exhibits apparently random fluctutions, whereas that in SiH_2 is very large, leading to the observation of a large variation in the emission lifetime.

Professor Buma answered: Concerning Figure 4 on Page 12: although we agree that in the LIF spectrum the band close to 33 500 cm^{-1} is not as evident as the next three clumps, it certainly is still there. One of the reasons for the differences noticed by Professor Duxbury is that the RE2PI and LIF spectra have been recorded with different laser systems (a Nd:YAG and an excimer pumped dye laser system). As a result, the 33 500 cm^{-1} band in the two spectra are not in the same part of the dye range. Even though we correct for excitation intensities as standard, this might nevertheless still lead to small differences. Another reason for different intensities in the RE2PI and LIF excitation spectra could well be that the two experiments probe different electronic states: in the LIF spectrum we monitor emission which can only originate from V'($\pi\pi^*$) or V($\pi\pi^*$) character and not nπ^* character, while in the RE2PI spectrum both $\pi\pi^*$ and nπ^* characters are monitored. The reversal of the two sub-bands at 34 000 cm^{-1} could thus indicate that the $\pi\pi^*$ levels involved do not have the same internal conversion rate as the nπ^* state.

With respect to the difference in the excitation spectra of OM*p*CA and TM*p*CA, I would like to remark that these differences primarily arise because of the energetic ordering of the V'($\pi\pi^*$) and V($\pi\pi^*$) states. In both molecules the V'($\pi\pi^*$) state is the lowest excited $\pi\pi^*$ state for vertical excitation. However, in TM*p*CA the V($\pi\pi^*$) becomes adiabatically the lowest $\pi\pi^*$ state, and it is this state that has considerable antibonding character in the $C_7{=}C_8$ double bond.

Professor Neumark opened further discussion of the paper by Dr Bochenkova: Is there any evidence that photodetachment of the GFP anion in solution results in autodetachment, producing a solvated electron?

Dr Bochenkova answered: In solution, there is a clear experimental evidence for a two-photon ionization of the GFP chromophore anion, which results in the generation of two species, a hydrated electron and a corresponding radical.[1,2] One-photon ionization does not occur, since the electron binding energy is considerably higher in polar solvents compared to that *in vacuo*. The multiple-photon nature of the ionization process is confirmed by the observed quadratic power dependence of the radical yield as a function of the laser-pulse energy. Furthermore, it has been concluded that ionization proceeds through a sequential two-step mechanism, where the second photon is absorbed by the anion in the S_1 state.[1] In the gas phase, electron emission is induced by the absorption of a single photon.[3]

Vibrational autodetachment out of the S_1 state does not occur in solution, as opposed to the gas phase, since the energy of a single photon is not enough to autoionize the GFP chromophore anion from the first excited state. However, it may well be that a resonant photoinduced electron transfer occurs in the presence of proper electron acceptors. This has not been studied in solution. However, it has been confirmed that GFP proteins, containing tyrosine-based chromophore anions, may serve as electron donors in photochemical reactions with various electron acceptors.[4] We note that there is a substantial difference between the excited-state lifetimes in solution and inside proteins. The photo-induced electron transfer is enabled in the GFP proteins, since many of them prohibit ultrafast de-excitation through internal conversion, whereas it is a dominating decay channel *in vacuo* at low excitation energies,[5,6] as well as in various solvents.[1,2,7] Therefore, the electronic excited-state decay channel is expected to be suppressed in solution, as it has been found *in vacuo* at low excitation energies.[5,6]

1. M. Vengris, I. H. M. van Stokkum, X. He, A. F. Bell, P. J. Tonge, R. van Grondelle, D. S. Larsen, *J. Phys. Chem. A*, 2004, **108**, 4587–4598.
2. K. M. Solntsev, O. Poizat, J. Dong, J. Rehault, Y. B. Lou, C. Burda, L. M. Tolbert, *J. Phys. Chem. B*, 2008, **112**, 2700–2711.
3. M. W. Forbes and R. A. Jockusch, *J. Am. Chem. Soc.*, 2009, **131**, 17038–17039.
4. A. M. Bogdanov, A. S. Mishin, I. V. Yampolsky, V. V. Belousov, D. M. Chudakov, F. V. Subach, V. V. Verkhusha, S. Lukyanov and K. A. Lukyanov, *Nat. Chem. Biol.*, 2009, **5**, 459–461.
5. C. R. S. Mooney, D. A. Horke, A. S. Chatterley, A. Simperler, H. H. Fielding and J. R. R. Verlet, *Chem. Science*, 2013, **4**, 921–927.
6. A. V. Bochenkova and L. H. Andersen, *Faraday Disc.*, 2013, DOI: 10.1039/c3fd20150c.
7. S. R. Meech, *Chem. Soc. Rev.*, 2009, **38**, 2922–2934.

Professor Buma addressed Dr Bochenkova and Professor Meech: The analysis of the excited-state dynamics of HBDI and deprotonated HBDI has so far only considered the ππ* excited states. However, calculations show that there are also a number of low-lying nπ* states.[1] For the chromophore of the Photoactive Yellow Protein we have a similar situation. At the ground state geometry the two lowest ππ* states are S_1 and S_2 and the nπ* state is S_3, but adiabatically the nπ* state becomes the lowest state. One of the consequences is that for this chromophore internal conversion from the ππ* states to the nπ* state is a dominant relaxation channel.[2] Do you think that a similar situation might also occur here, and thus that internal conversion to an adiabatically lower-lying nπ* state could be one of the competing relaxation channels after electronic excitation of the optically bright state ?

1. S. C. Olsen, Dissertation, University of Illinois at Urbana-Champaign, 2004.
2. E. M. Tan, S. Amirjalayer, B. H. Bakker, W. J. Buma, *Faraday Disc.*, 2013, DOI:10.1039/C2FD20139A.

Professor Meech responded: Most authors conclude that the bright state of ππ* character is the S_1 state, including Olsen. It is true that low lying nπ* states have been identified, and so in principle might play a role in the radiationless decay, and this would not require an excited state structure change, consistent with our observation. However, if that were the mechanism then much more marked solved effects might be expected (for example H-bonding solvents will shift nπ* energies, resulting in a variation of the radiationless decay rate; this is not observed). Further the fast radiationless decay is observed for all charge states of HBDI and persists in a number of derivatives (*e.g.* when replacing the OH group with H) which causes the bright state transition energy to be shifted significantly, sometimes by up to 100 nm. It is not clear that the nπ* state would necessarily track these shifts in all cases. On the other hand, constraining motion in the bridge or pre-twisting the ground state has a large effect on radiationless decay. Thus, the weight of experimental evidence for HBDI and its derivatives seems to favour a volume conserving change in excited structure leading to fast internal conversion from the bright excited state to the ground state.

Dr Bochenkova answered: We do not expect that a similar situation might also occur in the case of the deprotonated GFP chromophore. According to the results presented in your paper, the neutral PYP chromophore undergoes internal conversion to a lower-lying nπ* state on a picosecond timescale, resulting in the formation of a long-lived dark intermediate, which decays with a time constant of 30 ns to the ground state. The deprotonated PYP chromophore shows much faster excited-state decay with two lifetime components of 1 ps and 52 ps.[1] In the case of the GFP chromophore anion, femtosecond time-resolved photoelectron spectroscopy has revealed that the excited-state decay has two ultrafast lifetime components of 330 fs and 1.4 ps.[2] The dynamics in vacuo appears to be very similar to that observed in solution, where both upconversion fluorescence and subpicosecond pump–probe measurements give very similar lifetime components.[3,4] The lack of evidence for the presence of a long-lived excited-state intermediate, as it has been found in the case of the neutral PYP chromophore, strongly suggests that the nπ* state does not play a role in the ultrafast excited-state dynamics of the isolated GFP chromophore anion.

1. I.-R. Lee, W. Lee and A. H. Zewail, *Proc. Natl. Acad. Sci. USA*, 2006, **103**, 258.
2. C. R. S. Mooney, D. A. Horke, A. S. Chatterley, A. Simperler, H. H. Fielding and J. R. R. Verlet, *Chem. Science*, 2013, **4**, 921–927.
3. K. M. Solntsev, O. Poizat, J. Dong, J. Rehault, Y. B. Lou, C. Burda, L. M. Tolbert, *J. Phys. Chem. B*, 2008, **112**, 2700–2711.
4. S. R. Meech, *Chem. Soc. Rev.*, 2009, **38**, 2922-2934.

Dr Santoro opened further discussion of the paper by Professor Buma: If I understand properly, considering a system with a bright ππ* state excited by a laser pump and one or more closely lying nπ* (dark) states, Prof. Buma suggests that it is always necessary to locate the nπ* minima since, at those geometries, they might correspond to the lowest-energy excited state and therefore play a role in the decay pathway of the ππ* state. I fully agree that such optimizations are advisable since they provide a more complete picture of the investigated process. However, even if the global minimum of the lowest excited-state has nπ* character, this is not a sufficient condition, in my opinion, to assert that the nπ* state will be involved in the decay process. It seems to me that a key factor is the time that the system spends before returning to the ground-state. If such time is rather short, if the gradient along the ππ* potential energy surface (PES) carries the wavepacket in a region far away from the nπ* minimum, and if the ππ* and nπ* PESs do not cross during the dynamics, the system can directly make a transition to the ground-state (GS) without populating the nπ* state. One can think of it also in terms of a vibronic-states-based, Fermi-Golden-rule description of the possible ππ*->nπ* transfer. The rate constant will be a function of the square of the electronic coupling times a nuclear factor depending on the sum of a number of Franck–Condon factors. If the ππ* and nπ* equilibrium geometries are very different, these FC factors might be very small, giving rise to a ππ*->nπ* decay too slow to compete with the ππ*->GS direct process. A special case occurs when the π* orbital is the same in the ππ* and nπ* transitions, since this may lead to some similarities in the PESs of the two states, and it is not uncommon that their energy gap changes only moderately with the nuclear geometry (see for example the case of uracil[1,2]). In those cases, in my opinion, the recommendation of Prof. Buma is even more stringent.

1. F. Santoro, V. Barone R. Improta, *J. Am. Chem. Soc.*, 2006, **128**, 16312.
2. R. Improta, V. Barone, A. Lami and F. Santoro, *J. Phys. Chem. B*, 2009, **113**, 14491.

Professor Weitzel addressed Dr Bochenkova and Professor Neumark: During the discussion of paper 13 the question was raised by D. Neumark whether part of the signals from A. V. Buchenkova might be attributed to "thermionic emission". The term "thermionic emission" originally refers to the emission of electrons or ions by macroscopic samples operated at specific temperatures as originally applied by Richardson and Dushman (see *e.g.* O. W. Richardson, Thermionic phenomena and the laws which govern them, Nobel Lecture, 1928). It appears to me that the term thermionic emission is strictly applicable only to canonical systems. On the other hand Klots (*Chem.Phys.Lett.*, 1991, **186**, 73–76) has argued that the delayed emission of electrons from hot aggregates carries the signature of thermionic emission. What reasons would justify the usage of the term "thermionic emission" in the current context?

Professor Neumark answered: I and others use the term for emission of electrons from vibrationally hot clusters in their ground electronic state. In that case, one can use the statistical formulation by Klots to model the emission rate and kinetic energy distribution of the ejected electrons. This model can be applied to either canonical or microcanonical ensembles of clusters. Justification for using the term "thermionic emission" derives from the similar physics driving the process in clusters and in macroscopic materials.

Dr Bochenkova answered: Here, the term "thermionic emission" is used to refer to a statistical thermally-activated emission of an electron from isolated anions with a well-defined energy in the hot ground state, following their photo-excitation and subsequent internal conversion. Since a vibrational level density of anions is very high in the ground electronic state upon excitation followed by internal conversion, vibrational relaxation occurs very fast on a timescale of picoseconds. Therefore, any "activated" decay associated with crossing a barrier in the hot ground state has indeed a statistical nature, *i.e.*, all vibrational microstates of the anion with the same energy are equally populated. Under experimental conditions, there is a set of microcanonical ensembles, where the total energy is represented as a sum of a well-defined photon energy and a thermal energy of an ion equilibrated to a given temperature prior to its excitation under vacuum conditions. Therefore, the ground-state electron emission proceeds in a set of microcanonical ensembles with the total energies distributed according to the initial canonical distribution at a given temperature shifted by a well-defined photon energy.

In the microcanonical decay associated with electron emission, activation barrier is associated with an electron binding energy, and the total decay rate per unit time is obtained by integration of all probabilities for emission of an electron with some energy over all kinetic energies of the outgoing electrons. A single probability is then expressed in terms of level densities for the parent anion and for the daughter radical at particular energies. However, it is tempting to derive a simple expression of an Arrhenius type that would be typical for describing the properties of a system of a canonical ensemble. The microcanonical temperature is defined in terms of the derivative of the logarithm of the level density. If we consider a very large microcanonical system (a subsystem of interest plus its heat bath), it is recognized that a small subsystem has a canonical energy distribution. A similar idea may be used for a relatively small isolated anion represented by a set of oscillators to calculate the probability for gathering a particular amount of energy on very few or even on a single degree of freedom associated with a reaction coordinate. The concept of the microcanonical temperature for an isolated finite system is discussed in detail in ref. 1. The Arrhenius type decay may be obtained through a Taylor expansion of the logarithm of the level density in the statistical expression of the microcanonical rate constant. The temperature in the Arrhenius formula is then straightforwardly connected to the initial microcanonical temperature, where the correction depends on the size of the system.[1] The Arrhenius type decay may also be derived in the Klot's formalism of a finite-heat-bath theory, where the system is represented by a canonical ensemble. Therefore, the use of this term is here justified by the similar thermally-activated mechanism for electron emission in micro-canonical decay as compared to that of electron ejection in macroscopic metal systems. Furthermore, the same type of the emission law with a mathematical form similar to the Arrhenius expression may be derived in both cases.

1. J. U. Andersen, E. Bonderup, and K. Hansen, *J. Chem. Phys.*, 2001, **114**, 6518–6525.

Dr Alvaro Galue commented: Generally, charged π-conjugated aromatic-ring molecules exhibit significant electron-nuclei couplings along vibrational modes, and remarkably, along high-frequency in-plane symmetry CC stretching modes (see for instance H Torii, *J. Phys. Chem. A*, 2000, **104**, 413). Therefore, in principle, I would expect that high-frequency in-plane-symmetry CC modes play a role mediating the non-adiabatic dynamics in the deprotonated GFP chromophore anion. My question is how the vibrational mode activity of the GFP chromophore distinguishes from this general fact, and do you expect to find similar Huang-Rhys factors in other tyrosine-based chromophores?

Dr Bochenkova responded: We are primarily interested in the non-adiabatic couplings and minimum energy conical intersections along torsional modes in the GFP chromophore anion. While couplings along in-plane stretching modes are also important for electron emission through vibrational autodetachment (VAD) out of the first excited state, they also depend on vibrational overlap between the final wavefunction and the derivative of the initial wavefunction with respect to the corresponding nuclear coordinates. The mode-specificity in VAD then comes from the calculated S_1/D_0 origin shifts (and the corresponding Huang-Rhys factors). We do expect the similar pattern for other tyrosine-based chromophores based on the nature of the S_0, D_0 and S_1 states in these chromophores. However, the model presented in the paper of H. Torii does not have anything in common with the non-adiabatic couplings between the real ground and excited electronic states. This model is rather designed to describe anomalously large values of the dipole derivatives with respect to nuclear coordinates for the CC stretches in negatively charged conjugated polyenes, thus related to the IR intensities of these vibrational modes. The approach is based on constructing an effective Hamiltonian in the basis of non-existing, purely imaginary resonant states coupled *via* BLA (bond length alternation) type coordinates. It is important that the fictitious resonant states are used to describe the same ground electronic state. Therefore, the couplings between them are not related to the non-adiabatic couplings between different electronic states we are interested in.

Professor Stolow opened the discussion of the paper by Professor Scholes: Is the close packing of dipole oscillators not bound to exhibit coherent phenomena? This would be simply a secondary consequence of proximity and therefore 'coherence' may not be a design criterion used by Nature. How could one decide this?

Professor Scholes responded: It is very likely that the high concentration of chlorophyll together with minimization of concentration quenching are the principle parameters driving the optimization of light harvesting complexes. You are right, it is a good hypothesis that coherence is simply an unavoidable consequence of the closely-packed chromophores. Nevertheless, it is fascinating to note also the exceptionally small reorganization energy of chlorophyll in the light harvesting complexes compared to in solution. That tuning will also enhance coherence.

Mr Picconi communicated: I have two curiosities about the paper.

First, what is the range of time scales which you observe for the trapping in the efficient configurations? How much does this time scale depend on the relative orientation of the dipoles?

Second, as far as I understand, in this work you modify the configuration of the circuit by rotating the site dipoles, but the architecture of the circuit itself stays unvaried. Could it be possible (and interesting) to develop some algorithm in order to find the geometry of the circuit which maximizes the percentage of efficient configurations?

Professor Scholes communicated in reply: In the work we reported on the "molecular circuit" there was not such an interesting breadth of time scales because the complex is small. However, as you suggest, when antenna complexes are larger, there is a huge range of timescales for excitation trapping. I think experiments that can reveal that distribution, rather than focusing just on the average timescale, will be of crucial importance. See *Energy Environ. Sci.*, 2012, **5**, 9374 for further discussion of this point.

In the second part of your question you suggest that a smarter search algorithm is needed. I entirely agree. See my answer to the question of Prof. Orr-Ewing related to this point.

Professor Domcke communicated: I do have slight reservations concerning this model of electronic two-level systems coupled individually to Ohmic baths. This model seems overly simplistic. It is expected that these chromophores possess high-frequency stretching modes as well as slow torsional modes, *etc.* A more realistic modeling of the vibrational dynamics of the individual chromophores may be essential.

Professor Scholes communicated in reply: Absolutely, this is a very important point. I think what is missing from much of the theoretical studies is a detailed treatment of spectroscopy at the molecular level. Recent experimental results from my group are pointing unquestionably to the importance of considering molecular vibronic features in 2D spectra and in coherent analogs of Förster's spectral overlap. I think an issue has been that the 'bath' spectral density is difficult to measure experimentally and the initial work in this direction was concerned with line broadening and solvation (*e.g.* the work of Graham Fleming). In these studies it is the solvent spectral density that matters and the molecular spectroscopy can be ignored. That is not the case in energy transfer work nor in analysis of 2D electronic spectra.

Professor Duxbury communicated: In your presentation you gave a very interesting introduction into designs for molecular circuits that use electronic coherence. In your discussion paper, in references 11–13 you refer to quantum interference effects in charge transport. Reference 14, mechanistic aspects of dynamics, contains a very interesting overview of solar harvesting of light, related to molecular circuits, in which you and your colleagues explore a potential pathway from ecology to coherence. I found this fascinating as you project a possible interdisciplinary approach including quantum engineering in chemistry and biology.

Dr Oliver communicated: It is clear from the model presented in your paper and the current literature, that it is essential to incorporate the protein bath to fully capture the effects of coherent energy transfer. The spectral density will be essential to modeling this properly, but currently there is no way to determine this entirely from experiments. How can this be circumvented in a rigorous manner?

Professor Scholes communicated in reply: This is an important subject: we definitely need to measure, or at least probe, the spectral density of these light harvesting complexes. I do not know how to do this, but I think a very useful first step will be two-dimensional electronic-IR experiments, like those you are performing. It will be important, though, to push the IR probe far to the low frequency region. One outstanding issue is that I expect the protein to be rather heterogeneous as a solvent. How can we probe that? We already know that these light harvesting proteins are highly heterogeneous dielectric media, so perhaps that is a clue as to what experiments will be insightful [Curutchet *et al.*, Photosynthetic light-harvesting is tuned by the heterogeneous polarizable environment of the protein, *J. Am. Chem. Soc.*, 2011, **133**, 3078–3084].

Professor Buma communicated: In your paper and your introductory talk you mention that a number of proteins have been designed in such a way that coherence is preserved. In your model, on the other hand, you come to the conclusion that in the presence of realistic noise coherence only slightly improves the ability to direct electronic energy. Can anything be said on how your model would need to be modified to improve this? Could this then help you in (theoretically) designing better circuits than the ones that you considered in your paper ?

Professor Scholes communicated in reply: I think we were simply naive to imagine we could examine such a limited extent of "phase space" and find interesting coherent "circuits". As indicated in one of the other answers, I think a more sophisticated search/optimization is needed to design the kinds of molecular systems we envisioned when starting the research reported in the paper.

Professor Troisi communicated: Is the comparison of coherence time fair between solution phase and a biological matrix? Would a glass be a better reference for a protein environment?

Professor Scholes communicated in reply: The protein is certainly a unique environment – a lot of direct and indirect evidence points to that. For example, the Stokes shift of chlorophyll is significantly smaller in light-harvesting complexes than in solution. That means the reorganization energy (system-bath) coupling is much smaller. This small reorganization energy may have been selected for as light harvesting complexes evolved. The protein is also extraordinary because of its heterogeneity. It can modify both site energies and electronic couplings of the bound chromophores. The big challenge is to design experiments that more incisively reveal chromophore–protein interactions.

Professor Orr-Ewing asked: The conclusions drawn in your paper are arguably appropriate for the single architecture of chromophores and traps specific to your model, albeit with flexibility in the couplings between these sites. It would be interesting to take the ideas further and test their generality. Is there scope to develop a method to optimize the system with respect to coherent enhancement by introducing a learning algorithm approach to the choice of chromophore, trap and couplings architecture, both in the absence and presence of the decoherence noise?

Professor Scholes communicated in reply: This is exactly what needs to be done. We were rather naive thinking that we could explore such a limited phase space and find an example of how interference effects controlled energy transfer. I think we do need, as you suggest, a learning algorithm in the first instance – it becomes a control problem. What might be even more interesting in the long run is to design supramolecular 'neural networks' that learn to perform.

Professor Herek communicated: Your paper focuses on model systems inspired by nature, where the aim is to engineer molecular circuits to exploit quantum coherence. I think that this bottom approach is very powerful, and will surely yield fundamental insight on intermolecular interactions, at least in clean and simple constructs. Yet I am still very curious to hear your current thinking on the nature of the observed coherence in big and messy systems, like photosynthetic proteins. Is coherence there coincidental, or does it play a role in function?

Professor Scholes communicated in reply: I think coherence, when observed, is mainly a consequence of the high concentration of chromophores in light-harvesting complexes – in that sense it is unavoidable if the key optimization is high chromophore concentration, and hence strong interchromophore electronic coupling. The concentration of chlorophyll in light-harvesting complexes, for example, is 0.25 to 0.6 M and consequently center-to-center distances between chlorophylls are on the order of 10 Å. In terms of transport over long length and timescales, it is unlikely that coherence plays a role – surely it cannot be sustained in complex, disordered networks where there are many pathways connecting donor and acceptor (trap). There is some relevant discussion about this from my point of view in *Energy Environ. Sci.*, 2012, **5**, 9374–9393, but there is a range of opinion in the literature. An interesting recent result is that, when comparing different species of cryptophyte algae, we have found a fascinating comparison between PC645 (*Chroomonas*), which shows strong coherence after photo-excitation, and PC612 from an organism in the *Hemiselmis* genus. Owing to a key structural difference the light-harvesting proteins, identified by Paul Curmi (UNSW, Australia), the PC612 shows no coherence in our experiments. The observation points to the idea that solutions to effective light harvesting both with and without coherence are viable.

Professor Riedle communicated: I find the question, whether electronic coherence also plays a role in natural light photo-excitation intriguing. Recent work has demonstrated a spatial coherence on the order of 100 waves for solar beam radiation.[1] This is far larger than the light harvesting complexes. To my understanding, solar light is not totally incoherent and if we only consider the spectral part overlapping with the rather narrow absorption features, would we not obtain an effective coherence time of tens of femtoseconds? Is it therefore really inconceivable that the coherent behavior observed with ultrafast pulse excitation is still important with solar irradiation?

1. H. Mashaal, A. Goldstein, D. Feuermann and J. M. Gordon, *Opt. Lett*, 2013, **37**, 3516.

Professor Scholes communicated in reply: There is currently discussion in the literature along the lines of whether femtosecond laser experiments are relevant to understanding processes initiated by sunlight. In my opinion, there is a lot of confusion about this topic and also about the 2D electronic spectroscopy (2D ES) experiments that reveal coherent oscillations. You raise an excellent point about the coherence inherent in any light source of finite bandwidth. Two postdocs in my group (Aggie Branczyk and Aurélia Chenu) have been working on this subject for the past few months (paper in preparation!). Indeed sunlight has a coherence time of 10 fs or so, depending on its (finite) spectral bandwidth. Nevertheless, it is an incoherent source so that frequencies absorbed by molecules can be viewed as narrowband 'spikes'; quite different from the coherent, broadband excitation of a femtosecond laser. In addition to how to think about light in the context of absorption and photophysical dynamics, the following points are worth considering:

(a) The 2D ES experiments, that reveal coherent oscillations attributed to electronic coherence (in addition to vibrational coherence), *e.g.* Turner *et al.*, *PCCP*, 2012, **14**, 4857, do *not* claim to reveal coherent energy transfer dynamics. The experiments probe properties of the system Hamiltonian. In other words, they show that coherent effects should be considered in theories for very fast energy transfer dynamics. That conclusion is completely independent of the way the system is photo-excited. In fact, lasers are not even necessary for recording 2D ES data, as we report in a manuscript currently under review: "Coherent Multi-dimensional Optical Spectra Measured Using Incoherent Light" by D. B. Turner *et al.* (b) At least one recent paper claims that dynamics cannot even happen if a system is photo-excited by sunlight. This is the old argument of whether we excite to eigenstates or non-stationary states – debated strongly during the development of radiationless transition theory in the 1960s. The paper by Bixon and Jortner *J. Chem. Phys.*, 1968, **48**, 715 addresses the point well. The review by Karl Freed, *Top. Curr. Chem.*, 1972, **31**, 105 gives a nice discussion of how to think about photo-initiated dynamics that happen regardless of the excitation source. Part of the current misunderstanding, I think, is that it is thought that only steady state experiments like absorption and fluorescence spectroscopy are possible with incoherent light sources like the sun or a (un-pulsed) lamp. This is not the case. For example, fluorescence lifetimes can be measured using phase fluorometry and ultrahigh-time resolution pump–probe type experiments can be performed with incoherent light (*e.g.* Morita and Yajima, *Phys. Rev. A*, 1984, **30**, 2525.

Professor Buma communicated: In your paper and introduction you conclude that for a number of proteins the protein is optimized to reduce decoherence. What would one need to put in your model to reduce decoherence as well? Can we learn from these proteins construction rules for optimal electronic circuits? To what extent do you think it would be possible to actually implement such rules in a real man-made device ?

Professor Scholes communicated in reply: I would say these are among the 'big questions' in this field. It is clear that proteins are remarkable scaffolds for chromophores because they can tweak transition energies locally, they can change orientations of chromophores, and can even modify (significantly and heterogeneously) the dielectric environment, and hence electronic couplings. For example, it has been documented that a single amino acid change in the LH2 light-harvesting complex from purple bacteria can shift the absorption band of bacteriochlorophyll-a by ~7nm. Paul Curmi's structures of cryptophyte antenna proteins, with Beverley Green's sequences, show that addition of a single amino acid at a particular point in the protein completely changes the structure and hence the way that energy transfer happens. We need to learn more about these kinds of phenomena, which requires large-scale comparative studies as I suggested in my brief presentation. The issue of how decoherence can be controlled is a tough one. Correlating fluctuations has been suggested, and this would work. Your question was a topic of debate at the 22nd Solvay Conference in Chemistry (2010) "Quantum Effects in Chemistry and Biology". You might be interested in the discussion, documented in *Procedia Chemistry*, 2011, **3**, issues 1–4, p1–365.

Can we conceive of such 'engineered' environments in chemical analogs? That's a good question and a fascinating idea. I hope it inspires someone with the know-how.

Mr Debnath communicated: I have one specific question:

The magnitude of couplings to bath described in the paper have to be varied quite a lot [up to five times in comparison to the intra-chromophore couplings. The strongest electronic coupling is of the order of 100 cm^{-1}] to get Förster type (incoherent hopping) behavior. Is there some specific rationale for doing that ?

One would normally expect that decoherence would start to show up prominently on a comparable energy scale as far as relative strengths of those couplings are concerned.

Professor Scholes communicated in reply: What we really wanted to do in this study was to ask whether it would be straightforward to control energy transfer dynamics in the weak/intermediate electronic coupling regime using coherence rather than kinetics (incoherent dynamics). Therefore we searched for 'circuits' where a good contrast between on and off was found in the HEOM calculations (that include the account of coherent energy transfer), but where the Pauli master equation calculations (kinetics based on Förster theory) did NOT predict the same contrast between on and off.

Mr Debnath communicated: My second question is of a broader nature – as the paper attempts to examine the role of noise in the performance of coherent excitonic energy transfer, it is not clear if the noise plays any constructive role in general [in the model analysed here it decreases the efficacy, evidently]. Do you think that a correlated bath OR non-ohmic bath could have changed things for the better in terms of performance? What is your comment on this aspect?

Professor Scholes communicated in reply: Subtle differences will be seen as different model baths are considered. In particular the time scale of the bath, not just the amplitude of the fluctuations, matters. Correlations in the bath will look a lot like a weaker coupling to the bath. Real chemical and biological systems in the condensed phase have quite complicated baths, with multiple relaxation time scales. Your question touches on a key point, emphasized in our discussion, that we need to be able to measure the details of the bath using experiment. At present there is some work using fluorescence line narrowing. In terms of theory, there is impressive work on atomistic QM/MM methods to predict the spectral density of light-harvesting proteins by the groups of Kleinekathofer, Aspuru-Guzik and Mennucci.

Mr Debnath communicated: It is mentioned in the paper "We use HOEM to correctly interpolate between the limits of strong and weak electronic coupling".

Given the fact that HEOM scales correctly between weak to strong couplings to bath and not intra-chromophore electronic couplings, I think this remark needs to be clarified.

Professor Scholes communicated in reply: I would put it this way, as described in the paper by Ishizaki and Fleming [A. Ishizaki and G. R. Fleming, *J. Chem. Phys.*, 2009, **130**, 234111], that the HEOM extrapolates sensibly from the Förster limit (localized excitation) to the strong electronic coupling limit (delocalized excitation). Therefore it captures the balance between system bath coupling and the relative magnitude of electronic coupling between the chromophores.

Dr Kirrander opened the discussion of the paper by Professor Burgdoerfer by communicating: Firstly, what is the best way to understand the source of the ionization time-delay in Eq. (5) in your paper? The retardation due to the Coulomb potential is straightforward both in the quantum and classical pictures, as you elegantly illustrate, and it is well known that the phase $\delta(E)$ changes rapidly across resonances giving delayed ionization in the structured continuum of atoms and molecules (see e.g. the discussion below), but what is the physical origin of the delay t_{EWS} in the context of your paper?

As an example of the effect of $\frac{d\delta(E)}{dE}$ on ionization (and dissociation) in the structured continuum, I would like to show a figure taken from our paper[1], which shows the ionization and dissociation flux for an H_2 molecule excited into the dense manifold of pre-dissociated and pre-ionized Rydberg resonances. The non-adiabatic dynamics of the excited molecule are simulated using an accurate time-dependent Multichannel Quantum Defect approach. We show that the ionization and dissociation can be advanced or delayed using appropriate generalized chirps of the optical excitation pulse. Fig. 2 shows the fluxes upon excitation with no (i.e. a flat) optical phase, $\Phi(E)=0$, and with a generalized chirp specifically designed to cancel the material phase, $\Phi(E)=-arg(D_{\bar{ks}}(E))$, where the excitation pulse is written as $\epsilon(E) = \epsilon_0 e^{-\alpha^2(E-E_0)2+i\Phi(E)}$ and $D_{\bar{ks}}(E\ )$ is the complex-valued excitation dipole transition moment, corresponding to the term that appears in Eq. (6) of your paper.

There are a number of points to be made about Fig. 1.

Ionization: For the flat reference pulse (black line) we see the 'natural' flux delayed by changes in the material phase; essentially your Eq. (5). In contrast, the phase-shaped pulse, in which the optical and material phases cancel, shows much less delay since $\frac{d\delta(E)}{dE} = 0$ (in the weak-field domain the material and optical phases are additive). The remaining structure in the phase-cancelled ionization flux stems from the presence of several peaks in the amplitude $|D_{\bar{ks}}\ |$, since the optical pulse spans several resonances and, of course, the time-dynamics correspond to a Fourier transform of *both* phase and amplitude.

Dissociation: The dissociation continuum in this spectral region is essentially flat, meaning that the dissociation moment $D_{\bar{ks}}(E)$ has very little structure. Therefore the flat excitation pulse gives a correspondingly simple-looking peak in the dissociation flux. On the other hand, the chirped pulse designed to cancel the phase in the ionization channel causes significant distortion to the dissociation flux, reflecting the distortion of the optical pulse by the chirp. The optical pulse in

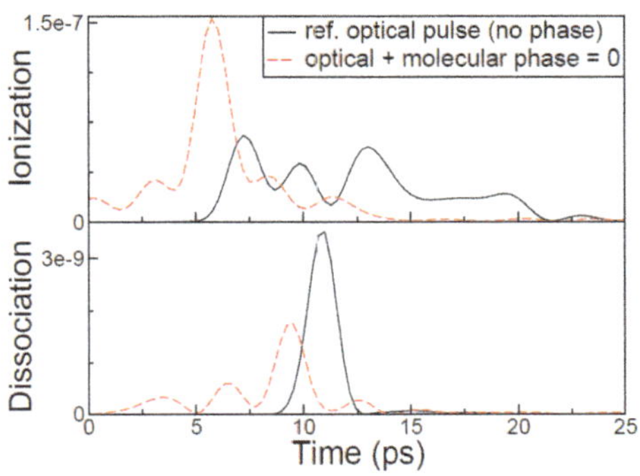

Fig. 2 (Reproduced with minor modifications from Fig. 2 in Ref. 1). Ionization and dissociation flux calculated for excitation with an optical pulse with no chirp (black) and a generalized chirp (red-dashed). The chirp can advance or delay the ionization or dissociation.

this particular example advances dissociation, although the definition of time zero is not unambiguous for complicated pulses.

1. A. Kirrander, H. H. Fielding and Ch. Jungen, Optical phase and the ionization-dissociation dynamics of excited H_2, *J. Chem, Phys.*, 2010, **132**, 024313

Second question. You compare the time delay given by Eq. (5) with Eq. (11) for the Yukawa potential. Fig. 1 in your paper shows very clearly how the two time-delays agree for the screened potential. Could you comment on the effect of the pulse duration on this comparison between Eq. (5) and (11)? Clearly, for a broad bandwidth (i.e. instant) pulse the comparison fails, essentially because it becomes unclear at what energy to evaluate Eq. (5) while $t_{EWS} \rightarrow 0$ in Eq. (11). Equally, for a very narrow bandwidth (i.e. slow) pulse the comparison between Eq. (5) and (11) would be difficult. Is there an optimal pulse duration for a comparison between the two equations?

Professor Burgdoerfer communicated in response to the first questions: As you discuss in your contribution, on a technical level t_{EWS} can be associated with the spectral variation of the transition matrix element. The physical origin and the interpretation is that it takes a finite time to form a wavepacket in the continuum that leaves the atom following photo-absorption by an initially bound electron. It is intuitively plausible that this time depends on the shape of the wavefunction and the binding energy of the initial state as well as on the (mean) energy of the final state in the continuum. Clearly, as the streaking signal involves a folding of the transition matrix element with the XUV field, a chirp of the XUV field would give rise to additional apparent time shifts as your example nicely illustrates. Therefore, extraction of t_{EWS} requires precise control over the phase of the XUV field.

In response to the second question: As t_{EWS} gives the timing of the wavepacket, forming such a well-defined wavepacket is a *conditio sine qua non*. This constraint naturally limits the range of useful pulse durations. For example, a very short pulse and consequently very broad excitation spectrum would lead to overlapping spectra from different initial states and would prevent a state-specific wavepacket formation. Conversely, a very long pulse corresponding to a narrow-band excitation would lead to a rapid dispersion such that its spatial width would exceed the mean distance r. In this context, one may speak of an "optimum pulse duration". In the streaking setting this "optimum" is by default given in terms of pulse duration of the XUV pulse whose duration must be small compared to the optical period of the IR ($\sim$ 1fs) and is for current high-harmonic sources of the order of $\sim$100 as or longer. For even shorter pulses the spectral overlap would prevent resolution of state-specific wavepackets for most atomic, molecular, and condensed matter systems.

Professor Neumark communicated: What is the current thinking on the origin of the 21 as delay reported by the Krausz group for emission from the 2s and 2p orbitals of Ne?

Professor Burgdoerfer communicated in reply: So far, a multitude of theoretical approximations have yielded delays that are a factor 2 (or more) smaller than the experimental data. The origin of this discrepancy is still a widely open question at this moment. Clearly, the experimentally achieved resolution for this remarkably short intrinsic atomic time information poses a considerable challenge to theory and offers novel sensitive tests for the descriptions of time-dependent many-body physics previously inaccessible. Numerically exact descriptions of the 10-electron problem are still out of reach. At the same time, this experiment is the very first of its kind and further improvements and refinements should allow to improve on the accuracy of this number. Clearly, as outlined in section 5 of the paper, the ultimate testing ground will be helium for which numerically exact predictions can be made and experiments should become feasible in the near future as the intensity of HHG light sources can be further improved.

Professor Domcke communicated: You presented exact result for the ionization of the hydrogen atom from the 1s orbital. Can such calculations also be performed for the 2s, 2p, 3d, *etc.* orbitals of the hydrogen atom. Can one learn more about the physics of the time delay in ionzation from such exact calculations?

Professor Burgdoerfer communicated in reply: Yes, calculations of initial states other than the ground state are possible. In our paper we present results for time delays for several *nlm* states of the He^+ (Fig. 8). There we show that the initial-state specific EWS time delay t_{EWS} (nlm) can be extracted with high accuracy. We find strong dependencies on *nlm* and also on the Stark-state quantum number *k*. The reason we did not present the corresponding results for hydrogen you are asking about is that for these more weakly bound excited states even streaking fields of modest intensity (10^{11}–10^{12} W/cm^2) can cause tunnel ionization and the destruction of the states to be investigated thereby blurring the timing information.

Professor Bucksbaum communicated: You express concern in your paper that the strong streaking field exerts an influence on the photoemission process itself, and not only on the emitted electron, because it polarizes the atom. There are, however, two times during the streaking laser cycle when the streaking field does not polarize the atom, namely during its zero crossings. These also correspond to the peak values of the ponderomotive vector potential, which means that if a free electron is born at the instant of a zero in the oscillating field, it will achieve the maximum momentum from the field. Would it not therefore follow that a measurement of the value of phase shift away from the zero crossing required to deliver the maximum ponderomotive momentum to the electron is a direct measurement of the photoemission time delay, without the contamination of the polarization of the atom by the strong field?

Professor Burgdoerfer communicated in reply: As the time delay corresponds to a phase shift, it is, indeed, in principle possible to determine the phase shift locally, for example by determining the delays between zero crossings of the electric field of the IR pulse. This would correspond to the delays between the extremes of the ponderomotive motion shift. While such a probe local in time could possibly reduce the influence of the laser field, it cannot eliminate the underlying Coulomb-laser coupling entirely. One reason is that the XUV pulse duration itself is of the order of a few hundred attoseconds or about a tenth of the optical period T and, hence, the emission time is therefore not confined to the time of zero field. Moreover, the electron propagates in the combined Coulomb- and laser-field effectively for another few hundred attoseconds or another tenth of a period of the optical field. Therefore, the streaking setting inevitably probes the emission process in the presence of the laser field. In practice and in the experiment, the ability to extract time shifts with single-digit attosecond precision from a streaking setting in which the pulse durations of both the XUV field and the IR field are orders of magnitude larger, hinges on the use of the entire temporal and spectral streaking trace that allows to determine the phase shift of a sinusoidal signal with an uncertainty very small compared to π.

Professor Riedle opened the discussion of the paper by Alessandro Troisi by communicating: You emphasized in your introduction that no conical intersections have to be invoked for the generation of quasi-free-holes and electrons in photovoltaic interfaces. We recently reported a 140 fs decay of the optically generated excitons in P3HT interface to Si, an appearance of polarons with the same time constant, and most importantly in this context, a significant recovery of the ground state bleach with the identical 140 fs.[1] Does this not clearly call for a conical intersection? How else would you explain such a correlated behavior and branching?

1. D. Herrmann, S. Niesar, C. Scharsich, A. Köhler, M. Stutzmann, E. Riedle, *J. Am. Chem. Soc.*, 2011, **133**, 18220.

Professor Troisi communicated in reply: A more accurate version of my statement is that conical intersections become intractable objects in the presence of a continuum of electronic states. This situation generates a continuum of conical intersections which suggests that a diabatic representation should be used. The problem of one electronic state coupled with a continuum of electronic states is actually simpler. The dynamics can be ultrafast without involving nuclei. The proposed experiment is neither in agreement or disagreement with the model proposed in the paper which does not include the ground state. However, it is well known that a metal can quench excitation by introducing virtual charge transfer states. For the proposed experiment it is therefore conceivable that the manifold of CT states is a mediator for the exciton relaxation and therefore the formation of a CT state and the re-appearance of ground state population may have the same rate. A quantitative model is however unavailable at the moment.

Professor Burghardt communicated: Your analysis apparently suggests that the role of electron–phonon coupling tends to be negligible in the exciton dissociation process. This seems to contradict our detailed quantum dynamical treatment of related heterojunction systems[1,2] showing that electron–phonon coupling is crucially important for the charge transfer event - similarly to our analysis of exciton transfer reported in this Faraday Discussion.[3] In fact, your description in terms of Marcus type rates may not be able to capture the details of the ultrafast, coherent transfer step at the polymer interface. In view of the proximity of the lattice models used in your work and in our contribution,[3] a full quantum dynamical treatment could be carried out for a minimal site-based heterojunction model, to shed further light on this issue.

For further illustration - and to respond to an earlier question by Professor Scholes - the figure below shows quantum dynamical simulations of the ultrafast exciton break-up in an oligothiophene–fullerene complex,[2] *i.e.*, a model of P3HT-PCBM type heterojunctions. The decay profile features rapid recurrences associated with vibronic coherence, followed by an irreversible transfer due to electron–phonon coupling.

1. H. Tamura, J. Ramon, E. R. Bittner, and I. Burghardt, *Phys. Rev. Lett.*, 2008, **100**, 107402.
2*a*. H. Tamura, R. Martinazzo, M. Ruckenbauer, and I. Burghardt, *J. Chem. Phys.*, 2012, **137**, 22A540, 2*b*. H. Tamura, I. Burghardt, and M. Tsukada, *J. Phys. Chem. C*, 2011, **115**, 10205.
3. R. Binder, J. Wahl, S. Roemer, and I. Burghardt, Faraday Discuss. 163, DOI: 10.1039/C3FD20148A, in press.

Professor Troisi communicated in reply: We shared the very same opinion of Prof. Burghardt when this project was initiated. We have in fact computed the (not-so-small) electron phonon coupling term[1] and we planned to adopt our Ehrenfest-like quantum dynamics model[2] to this problem. However, and this is essentially the key result of our paper, we have noted that very delocalized CT state become accessible at the very same energy of the exciton if one builds a sufficiently large model. These states remains delocalized also when electron phonon coupling is introduced.[3] The reorganization energy for these states is so small that, if one constructs a model neglecting completely the electron–phonon coupling, the results are quantitatively similar. It is not a surprise that the

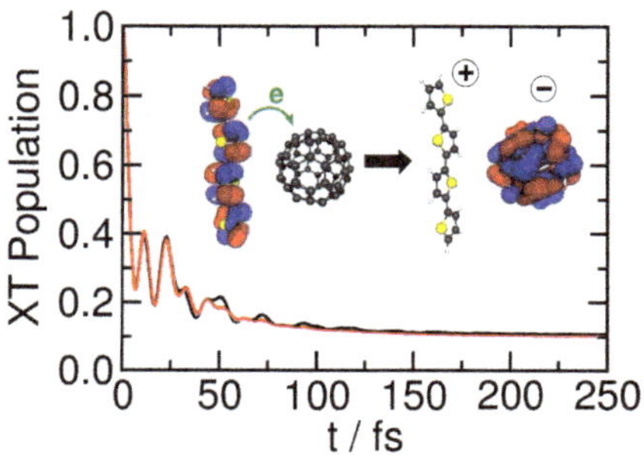

Fig. 3 Quantum dynamical simulations of the ultrafast exciton break-up in an oligothiophene–fullerene complex (adapted from Tamura *et al.*, *J. Chem. Phys.*, 2012, **137**, 22A540).

delocalization of excitons and charge reduces the importance of the electron–phonon coupling (*e.g.* the latter is rightly ignored in inorganic semiconductors), it is a surprise that electron–phonon coupling can start to be ignored also for molecular systems in the presence of a coupling with a quasi-continuum of states. So there is no contradiction: a small model displays important electron-nuclear dynamics, a large model that supports delocalized degenerate states will see an ever decreasing role of the electron-nuclear coupling. I agree that Marcus theory will not capture ultrafast nuclear dynamics for many different reasons. The Marcus *equation*, however, in the limit of vanishing reorganization energy is identical to the Fermi Golden rule and this *will* capture correctly a purely electronic process between electronically degenerate states. Figure 7 and eq. 18 show that we are very close to this limit.

We agree that a full quantum dynamical treatment is desirable and we will certainly pursue it. Knowing what is the closest limiting case would be a good starting point for such an investigation.

1. T. Liu, A. Troisi, *J. Phys. Chem. C*, 2011, **115**, 2406.
2. A. Troisi, G. Orlandi, *Phys. Rev. Lett.* 2006, **96**, 086601.
3. A. Troisi, *Chem. Soc. Rev.*, 2011, **40**, 2347.

Professor Scholes communicated: You have reported a very nice and intuitive model for excitons in an ideal system. I wonder if there is a way to test experimentally such a model? For instance, we published work [*J. Chem. Phys.*, 2009, **131**, 84714] on exactly this kind of model applied to CdSe–CdTe nanocrystal heterostructure spectroscopy. If the type-II band offset is the dominant energy scale, then free carriers are the dominant species populated at low temperature. Therefore we reasoned that temperature dependent photoluminescence (PL) measurements would show the quantum yield of charge-transfer emission (the electron-hole pair bound at the interface) to diminish as the temperature is lowered. In contrast we found the PL quantum yield to increase by two orders of magnitude! Do you have ideas for how to rationalize application of your theory with "real" systems? Given the work your group has done on atomistic simulations of such organic interfaces you are the right person to comment.

Professor Troisi communicated in reply: The PL yield of the electron-hole pair cannot be described by the current model which focuses entirely on the charge separation process. Once the initial long-range charge separated state is formed there are two fates available for the system: (i) separation into free charges and (ii) formation of a bound pair. This bifurcation was not discussed by the model but it seems possible to speculate that any increase of the temperature will favour the process (i) (which is thermodynamically favoured at all temperatures and more so at higher temperatures for entropic reasons). I.e. it is to be expected that lowering the temperature the PL quantum yield from bound CT states should increase. The electrical Hamiltonian of our model is similar to [*J. Chem. Phys.*, 2009, **131**, 84714] but the process studied is different (charge separation in our case, luminescence in the *J. Chem. Phys.* Case). It was not obvious at all that in organic materials one can use the same model used for inorganic materials and the demonstration of this fact is the main novelty of the paper proposed in this discussion. Concerning the application to real systems, all parameters appearing in this model have been computed for real systems by others and by ourselves (charge transfer parameters in *J. Phys. Chem. C*, 2011, **115**, 2406; band and electron phonon coupling in *Adv. Mater.*, 2007, **19**, 2000) and the range used in this paper are realistic. A paper like this one, unlike the computational chemistry works, allows one to explore the range of validity of a proposed mechanism.

Professor Scholes communicated: I think that the way to think about the low energy states you calculate, where the electron and hole are widely separated is to keep in mind the high density of those states. Indeed the states converge to the most widely separated electron-hole configurations, just like a Rydberg series for a molecule. However, to see why the carriers are indeed mobile (and localized) you need to think about energy scales. The transfer integrals, which promote electron and hole hopping between subunits, act to overwhelm the electron–hole attraction in the band of charge transfer exciton states, meaning that the electron and hole act as independent particles. That is, they are mobile (and hence localized) carriers. See *ACS Nano*, 2008, **2**, 523.

Professor Troisi communicated in reply: The penultimate statement of this question is exactly what the paper under discussion states ("The most relevant molecular parameter that affects the rate of exciton dissociation is the [...] the band widths of the donor and acceptor materials", where the band widths are proportional to the transfer integral) and a quantitative description of this fact is in figure 5 (showing the rate *versus* transfer integral). The new element of this paper is that these charge separated states are kinetically more accessible than the charge bound states, *i.e.* it is not true that a bound charge-transfer exciton is formed and then the charges are separated, but the state formed is already almost-separated. Not only the paper considers explicitly the transfer integral but also the fundamental difference between forming a bound hole-electron pair (a process dominated by the Franck–Condon overlap) and the process of forming charge separated states (a process dominated by purely electronic coupling). It is believed by this author that the localization of the charge carrier after dissociation is a property of the bulk (they will more or less localized depending on the level of disorder and electron–phonon or hole-phonon coupling and there is no simple relation between being mobile and being localized - especially in molecular solids). We are currently exploring a version of this model with disordered materials at the interfaces and the differences in the phenomenology are surprisingly small. We believe that the rates of charge separation have not been evaluated before.

Professor Fielding opened further discussion of the paper by Professor Burgdoerfer by communicating: Towards the end of your paper you discuss the possibility of extending time-resolved photoemission to extract dynamical information for more complex systems with many electrons. Are you able to comment further on how realistic this might be, given the discrepancy between experiment and theory for relatively small systems with few electrons.

Professor Burgdoerfer communicated in reply: First of all, it is important to appreciate the remarkably short time scale of 20 attoseconds that has become accessible with this experiment. Both experiment and theory are just at the very beginning to explore and to understand the many-electron dynamics on such time scales and for processes previously inaccessible. I am therefore not surprised that discrepancies on such a small scale of appear at this early stage. The point to be noted is that the characteristic times for electron dynamics in complex systems to be explored are, typically, considerably longer than for this prototypical test case of fundamental quantum-electrodynamics. As discussed in our paper, e.g. the characteristic time for electron transport and inelastic scattering in solids is in the 50 as range, the time-delay in between sequential emission of two electrons may reach ~100 as or longer. Therefore even at the current level of the accuracy, high-resolution attosecond chronoscopy of many dynamical processes is in reach. I am convinced that the level of accuracy will be further improved as this technique matures.

Professor Neumark communicated: Your paper illustrates the complexity of using attosecond streaking as a dynamical probe. The situation is even more complicated for molecules, in which the strong 800 nm streaking pulse can in principle trigger additional dynamics that may be difficult to disentangle from those induced by the attosecond pulse. In our laboratory[1] and elsewhere,[2] attosecond transient absorption looks to be an alternative and possibly very appealing probe of sub (or few) femtosecond dynamics. Here, a broadband attosecond pulse passes through a sample and is spectrally dispersed. The effect of an overlapped few-cycle 800 nm pulse on the resulting spectrum is determined as a function of pulse delay. These experiments have yielded novel light-induced structures that are seen only when the two pulses overlap.

1. S. Chen, M. J. Bell, A. R. Beck. H. Mashiko, M. Wu, A. N. Pfeiffer, M. B. Gaarde, D. M. Neumark, S. R. Leone, and K. J. Schafer, *Phys. Rev. A*, 2012, **86**, 063408.
2. M. Chini, B. Zhao, H. Wang, Y. Cheng, S. X. Hu, and Z. Chang, *Phys. Rev. Lett*, 2012, 073601.

Faraday Discussions

RSC Publishing

PAPER

Active and passive control of zinc phthalocyanine photodynamics

Divya Sharma,*[a] Annemarie Huijser,[a] Janne Savolainen,[b] Gerwin Steen[a] and Jennifer L. Herek[a]

Received 21st December 2012, Accepted 16th January 2013

DOI: 10.1039/c3fd20156b

In this work we report on the ultrafast photodynamics of the photosensitizer zinc phthalocyanine (ZnPc) and manipulation thereof. Two approaches are followed: active control *via* pulse shaping and passive control *via* strategic manipulation in the periphery of the molecular structure. The objective of both of these control experiments is the same: to enhance the yield of the functional pathway and to minimize loss channels. The aim of the active control experiments is to increase the intersystem crossing yield in ZnPc, which is important for application in photodynamic therapy (PDT). Pulse shaping allowed an improvement in triplet to singlet ratio of 15% as compared to a transform-limited pulse. This effect is ascribed to a control mechanism that utilizes multiphoton pathways to higher-lying states from where intersystem crossing is more likely to occur. The passive control experiments are performed on ZnPc derivatives deposited onto TiO_2, serving as a model system of a dye-sensitized solar cell (DSSC). Modification of the anchoring ligand of the molecular structure resulted in an increased rate for electron injection into TiO_2 and slower back electron transfer, improving the DSSC efficiency.

Introduction

Photosensitizers are currently a topic of intense scientific research because of their photophysical properties that hold great promise for application in photodynamic therapy (PDT), blood sterilization, sunlight-activated herbicides and insecticides, biomedical imaging and dye-sensitized solar cells (DSSC).[1] A photosensitizer should have a high absorption cross section at the wavelength suitable for the specific application and excellent photochemical and thermal stabilities. Other important parameters while searching for or designing new photosensitizers involve solubility and tendency to aggregate.

Amongst the most promising photosensitizers are zinc phthalocyanines (ZnPcs), planar aromatic molecules with considerable electron delocalization,

[a]*Optical Sciences group, MESA+ Institute for Nanotechnology, University of Twente, PO Box 217, 7500 AE, Enschede, The Netherlands. E-mail: d.sharma@utwente.nl; Fax: +31 53 489 3511; Tel: +31 53 489 3805*

[b]*Physikalisch-Chemisches Institut, Universität Zürich, Winterthurerstrasse 190, CH-8057 Zürich, Switzerland*

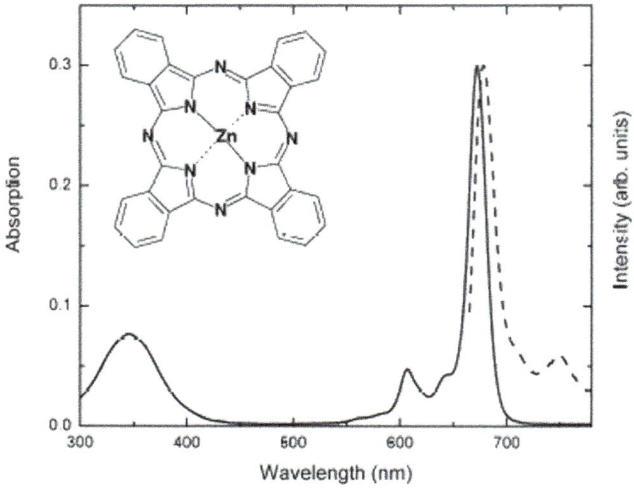

Fig. 1 Absorption and fluorescence spectra of ZnPc in DMSO. The inset shows the molecular structure of ZnPc.

accounting for unique optical properties. The absorption spectrum of ZnPc in dimethyl sulfoxide (DMSO) is shown in Fig. 1. The Q-band ($S_0 \rightarrow S_1$) has a maximum at 672 nm, as well as vibrational progression to the blue side with a shoulder at 645 nm and a weak band at 606 nm. Further in the ultraviolet, the Soret band ($S_0 \rightarrow S_2$) with a maximum at 345 nm can be seen. Also included in Fig. 1 is the fluorescence spectrum. The fluorescence has a maximum at 680 nm, implying a Stokes shift of 8 nm.

PDT is based on three key components: a photosensitizer, a light source and tissue oxygen. The photosensitizer should ideally absorb in the red and near-infrared since human tissue is most transparent in this spectral region. Amongst the most promising photosensitizers for PDT are the phthalocyanines.[1a] The photophysical decay pathways of ZnPc have been studied in detail earlier and are summarized in Fig. 2.[2] Absorption of a photon leads to the formation of a singlet excited state. Decay pathways of the singlet excited state involve fluorescence (F), internal conversion (IC) and intersystem crossing (ISC). Only the latter process leads to the formation of the PDT-required triplet excited state. The reason is that excitation energy transfer (EET) from the photosensitizer triplet state towards

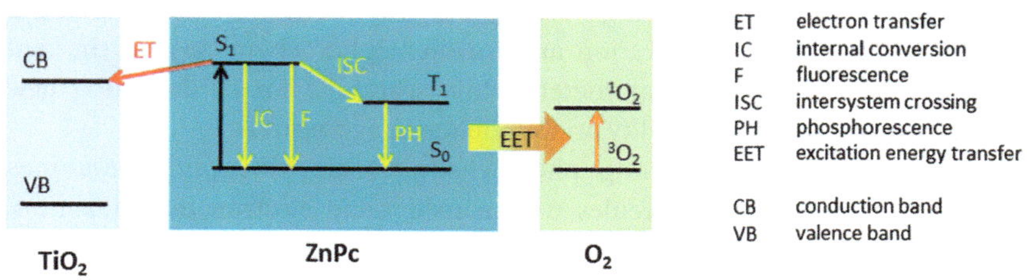

Fig. 2 Jablonski diagram of the photophysical decay pathways in ZnPc (yellow), ZnPc combined with O_2 (yellow and orange, relevant in PDT) and ZnPc combined with an electron acceptor such as TiO_2 (yellow and red, relevant in DSSC).

oxygen leads to the formation of highly reactive singlet oxygen. Singlet oxygen is phototoxic to cancer cells and its presence leads to cell death. A PDT photosensitizer is hence required to have a high ISC yield, which means that the intersystem crossing rate has to be competitive to the unwanted singlet decay channels (F and IC). Manipulation of the photodynamics in favor of ISC is thus essential in the development of efficient PDT photosensitizers.

Since the solar emission photon flux has a maximum between 650 and 750 nm, phthalocyanines are also considered as promising photosensitizers in DSSCs.[1d] DSSCs are a low-cost alternative to conventional silicon-based solar cells and consist of a semiconductor such as TiO_2 coated with a light-absorbing photosensitizing dye. The photophysical pathways of a DSSC are illustrated in Fig. 2 and elaborated in Fig. 3A. Absorption of a photon by the dye leads to formation of a strongly bound electron-hole pair (exciton). Efficient exciton dissociation into charge carriers is only possible at the interface between dye and semiconductor and occurs *via* electron transfer (ET) from the dye into the semiconductor; back electron transfer has to be avoided. Formed charge carriers need subsequently to be transported from the dye–semiconductor interface towards the electrodes. The fact that excitons are usually only able to diffuse over distances of a few nm, while a layer thickness in the order of 100 nm is required for sufficient light-absorption, was the reason for Grätzel and co-workers to introduce the concept of a network of TiO_2 nanoparticles coated with a monolayer of dye molecules and filled with a liquid I^-/I_3^- electrolyte.[3] The use of ruthenium-based photosensitizer dyes has resulted in efficiencies up to 11%,[4] though drawbacks involve the price of ruthenium and lack of absorption in the red/near-infrared region of the solar emission spectrum.

Phthalocyanines not only have the ideal absorption spectrum, but are also highly stable chemically and do not contain precious elements. DSSCs based on phthalocyanines, however, have shown poor device efficiencies that did not exceed 1% for a long time.[5] Only recently significantly higher efficiencies up to

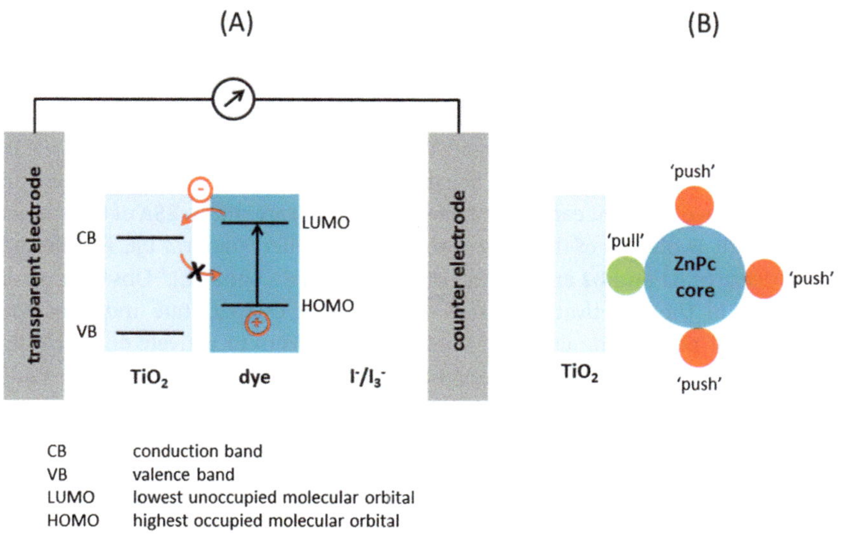

Fig. 3 Photophysical principles of a dye-sensitized solar cell (DSSC) (A) and an asymmetric push–pull ZnPc photosensitizer with electron transfer directed towards the TiO_2 (B).

~5% have been realized.[6] Important tools in the development towards higher efficiencies involve the prevention of aggregation, by the use of a co-solvent or introduction of bulky side groups, and introducing directionality of the electron transfer process towards the TiO_2. The latter has been realized by replacing a symmetric phthalocyanine by an asymmetric equivalent with one electron pulling group and three electron pushing groups, as visualized in Fig. 3B. The electron pushing groups are required to suppress aggregation. Only the electron pulling group should be able to anchor onto the TiO_2 substrate. The observation that adapting the chemical structure from symmetrical into an unsymmetrical push–pull structure leads to substantial improvement in efficiency suggests a significant change in forward and back electron transfer photodynamics. However, ultrafast photodynamics studies have not been reported so far.

In this work we explore the ultrafast photodynamics of ZnPc and manipulation thereof. Two approaches are followed: active control *via* pulse shaping and passive control *via* strategic changes in the periphery of the molecular structure. The aim of these control experiments is to enhance the yield of the functional pathway and minimize loss channels. Actively, this entails illuminating the photosensitizer with an ideally shaped pulse instead of a transform-limited (*i.e.* unshaped) pulse. We will use this approach to manipulate the ISC yield. The objective of the passive control experiment is to manipulate electron transfer from ZnPc into TiO_2 and the undesirable back electron transfer by adapting the chemical structure of the anchoring electron pulling group.

Results

ZnPc photophysics

A detailed description of the pump–probe setup was reported earlier.[2] ZnPc in DMSO was excited at 672 nm (22 fs, 30 nm FWHM). Photophysical processes were monitored in time using a delayed white-light continuum as a probe, which was generated by sending part of the amplified fundamental 775 nm beam through a delay line and into a sapphire window. To avoid any anisotropy effects, the polarization angle between the pump and the probe pulses was set to 54.7°.

Fig. 4 shows the pump–probe data as a function of time and wavelength of ZnPc in DMSO. The 2D-surface is a combination of broad excited state absorption (ESA) bands superimposed with negative signals due to ground state bleach (GSB) and stimulated emission (SE). The negative band (A) reflects a combined GSB and SE. The positive bands at early time scales (B1 and B2) are due to ESA of the singlet excited state. Evolution of this state into a triplet excited state *via* ISC is reflected by the decay of B1 and B2 and the rise of the triplet ESA band (C).[2] Observation of branching in the deactivation of the ZnPc singlet excited state indicates the possibility of performing coherent control experiments to actively enhance the ISC yield *via* pulse shaping. In fact, recent 2D electronic spectroscopy reveals coherences in the excited state dynamics.[7]

Active control

Coherent control is a spectroscopic tool that can drive and control the temporal evolution of a system to a desired product channel by optimally shaping the time-dependent electric field of the laser pulse and has been used to control

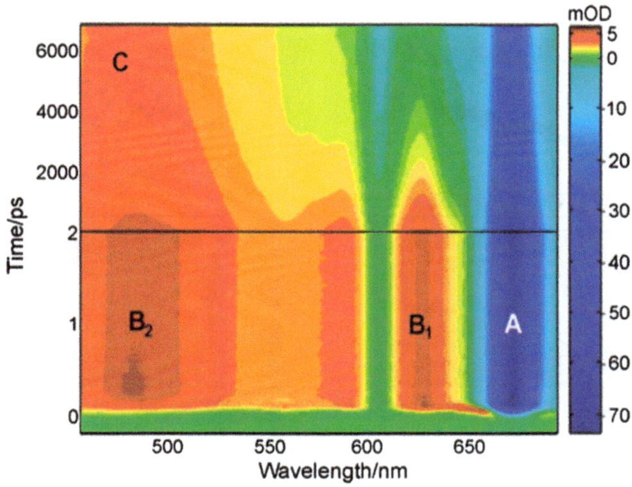

Fig. 4 Pump–probe data as function of time and wavelength of ZnPc in DMSO. The characteristic features of stimulated emission and bleach (A), singlet ESA (B1 and B2) and triplet ESA (C) are indicated. Note the two different time windows and non-linear intensity scaling, used to emphasize the ultrafast dynamics.

isomerization reactions, excitation energy transfer in natural and artificial light-harvesting complexes, selective excitation of vibrational modes and population transfer.[8] The phases and/or amplitudes of different frequency components of the laser pulse are matched to the inherent molecular resonances and dynamics to selectively address and steer them to the target state. Thus, coherent control provides a powerful means for manipulation on the molecular scale, facilitated by active tuning of laser pulse parameters such as frequency, phase, amplitude, and polarization. Here, the aim of the optimization experiment is to find an optimum laser shape that improves the ISC yield and as a consequence the efficiency of ZnPc in generating singlet oxygen.

Since for a polyatomic molecule such as ZnPc the complete Hamiltonian is too complex to determine the driving optical field in advance, an iterative closed-loop feedback optimization[9] was performed. The sample was excited using shaped femtosecond pulses (~672 nm) with a pulse energy of 500 nJ/pulse. The learning loop consists of a pulse shaper, the ZnPc sample and an evolutionary learning algorithm. The evolutionary algorithm was started with a set of random phases; the resulting pulse shapes were tested on the sample. The feedback signal obtained from the transient absorption experiment was evaluated *via* a fitness function, from which a new set of pulse shapes were generated and applied. The feedback signal for the singlet (S) state was obtained by integrating a 10 nm band from the transient spectrum at 20 ps time delay at the maximum of the ESA band (centered at 490 nm). The feedback signal for the triplet (T) state was measured at 25 ns time delay at the maximum of the triplet ESA at 480 nm with a bandwidth of 15 nm. The fitness functions were $f = T/S$ and $f = S/T$, respectively. The evolutionary algorithm iteratively optimized the value of the feedback function (*i.e.* T/S and S/T) relative to that of a transform-limited (TL) pulse. The shape of the optimized pulse was characterized by using a frequency-resolved cross-correlation measurement (X-FROG). Fig. 5 shows the obtained learning curves for T/S and S/T

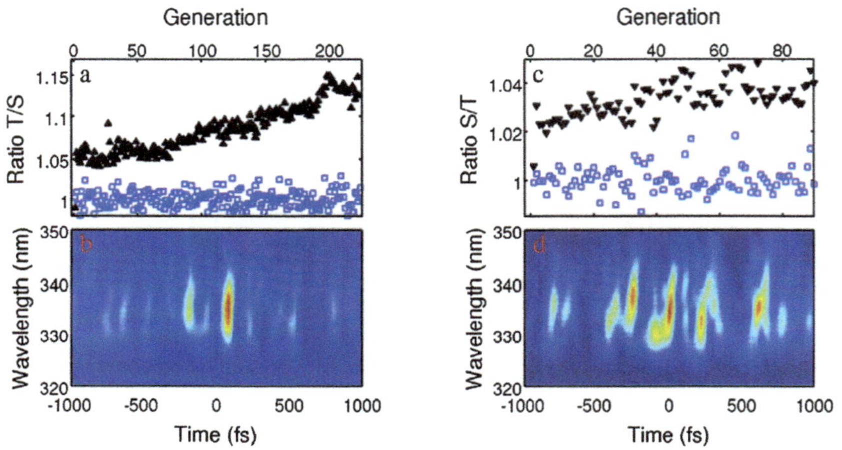

Fig. 5 Optimal control results on the triplet yield of ZnPc. Panel a: The obtained learning curve (black) for the ratio T/S and fitness of the TL pulse measured before each generation (blue). Panel b: The X-FROG surface of the best pulse. Panel c: The obtained learning curve (black) for the ratio S/T and fitness of the TL pulse measured before each generation (blue). Panel d: The X-FROG surface of the best pulse.

and fitness of the TL pulse measured before each generation. Obviously it is possible to increase the T/S ratio by a factor of ~1.15 and the S/T ratio by a factor of ~1.05. The X-FROG traces of the best pulses for both cases show complicated structures that are difficult to interpret directly.

To verify the robustness and to further understand the obtained optimal control results, repeated measurements were done where the fitnesses of the found best pulse shapes were remeasured. Fig. 6a shows the repeated optimization; the T/S ratio is improved by a factor of ~1.10. Also included is the so-called actinic energy of the best individual pulses along the optimization. Actinic energy is simply an integrated sum of the pump spectrum, amplitude mask and the absorption spectrum of the ZnPc. The shown values are normalized to 100% for the flat amplitude mask with no amplitude shaping. The actinic energy first drops

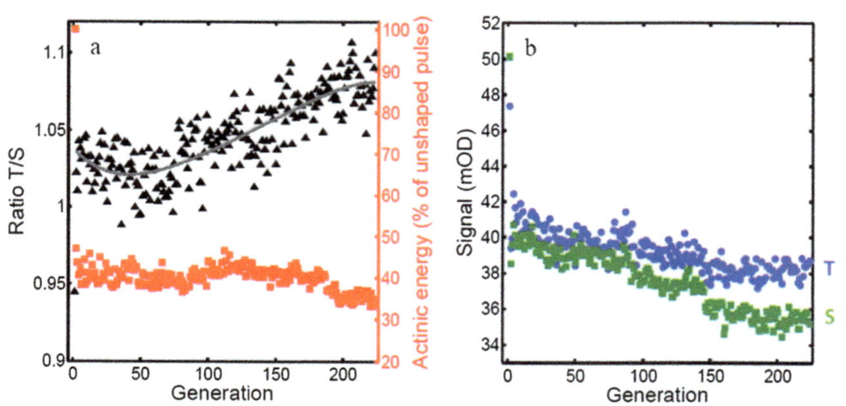

Fig. 6 Repeated measurement of the T/S optimization. Panel a: The ratio T/S (black) and normalized actinic energy (red). The grey line is a polynomial fit to guide the eye. Panel b: The triplet signal (blue circles) and the singlet signal (green squares).

to ~47% of the initial value, which is expected for the random amplitude mask of the first generation. Thereafter, the actinic energy stays fairly constant varying ~13%. It is interesting to compare the relation between the actinic energy and the obtained fitnesses. The initial fall of ~53% in the actinic energy results in an increase of the T/S ratio of ~6%. However, later, along the optimization such correlations cannot be found. For instance, while from generation 50 to 100 the T/S ratio rises with the slightly increasing actinic energy, from generation 150 to 200 the opposite holds: increasing fitness with decreasing actinic energy. Fig. 6b shows the behaviour of the individual signals belonging to T and S during the repeated measurement shown in Fig. 6a. The overall population transfer decreases as both of the channels show decreasing signals. However, it is evident that the difference between the two signals shows a jump with the initial fall of the actinic energy and then slowly expands along the learning curve. In conclusion, the data show that the control involves other features than trivial solutions related to the pulse intensity.

To further investigate the importance of the pulse intensity to the control, we conducted intensity dependence measurements using a TL pulse with varying intensities. Fig. 7 shows S and T signals as a function of pump intensity. A difference in the shape of the saturation curves of S and T is reflected in the fitness value as an increase of the ratio T/S with decreasing pump intensity. This is in good agreement with what was observed in the T/S optimal control experiments: an initial drop in the pump intensity to some 47% of the initial energy of 500 nJ/pulse results in a step-wise increase of the fitness value in the beginning of the optimization.

However, quantitatively the jump from 500 to 235 nJ/pulse with TL pulse causes a larger (~17%) jump in the fitness compared to the one observed in the optimization and the repeated measurement (~6%, see Fig. 6). Further, as is shown in Fig. 7, the fitness variation of the TL pulse in the actinic energy range of the optimization is less than 4% and therefore cannot be behind the learning process. This indicates that the actual shape of the pulse bears importance

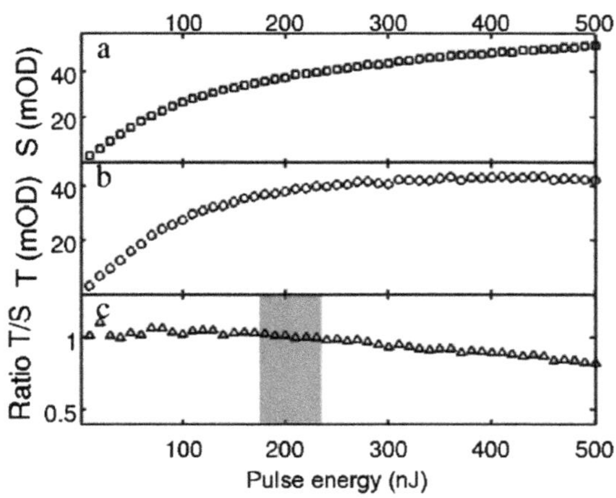

Fig. 7 Intensity dependence of the feedback signals using a TL pulse. Singlet and triplet states show slightly different behaviour resulting in a decreasing ratio T/S with pulse intensity. The grey area indicates the actinic energy regime used by the algorithm in the course of the optimization.

implying the involvement of more complicated mechanisms than signal saturation.

Open-loop phase scans were performed using the amount of linear chirp and the spectral location of π phase steps as variables. Fig. 8a shows the T and S signals and their ratio (T/S) as a function of the applied linear chirp. Both S and T decrease similarly on introducing more negative chirp. Positive chirp leads to a different behaviour: S remains constant whereas T grows gradually. This variation can be seen in the asymmetric shape of the ratio T/S. The overall shape of the chirp scan traces points to a pump–dump mechanism that reduces the population on the excited state demonstrated previously in a study on the fluorescence emission yield of the green fluorescent protein (GFP).[10] As the initially excited wavepacket slides down the potential energy surface of the first excited state, a negatively chirped pulse can dump it back to the ground state whereas a positively chirped pulse is not able to produce such an effect. However, this well established mechanism does not explain the different behaviour of the two channels, *i.e.* the gradual growth that was observed for the T trace. Therefore, chirp scans as a function of the pulse intensity were also done (data not shown). Unlike in the GFP study,[10] overall features remained similar (reduced S and T with negative chirp), but the difference between T and S vanished with reduced intensity. This could be an indication of a multiphoton process involving higher-lying excited states. The evolution of the wavepacket on the potential energy surface of the first excited state can vary the transition probabilities of possible sequential two photon absorption processes. By following this evolution a chirped pulse can enhance the yield of such processes, which then leads to more efficient triplet formation.

To further test the involvement of multiphoton processes, π-phase-jump scanning where the spectral position of a step function with an amplitude of π is scanned over the pump spectrum while recording the feedback signals, was used. The results in Fig. 8b show that the T and S signals are dependent on the location

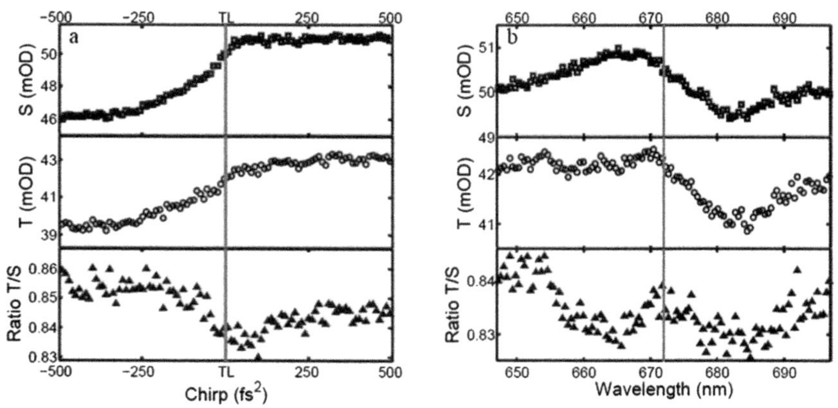

Fig. 8 Open-loop control results. Panel a: Chirp scan, the feedback signals S, T and the T/S ratio when a linear chirp is scanned from negative to positive. The grey line indicates a TL pulse. Panel b: π-phase jump scan where the feedback signals S, T and the T/S ratio are measured when the spectral position of a step function having a π phase jump is scanned over the pump spectrum. The grey line indicates the peak of the resonance.

of the phase jump. The overall shape of the S and T traces resembles the imaginary part of resonance that describes absorption. The phase profile of the pulse modulates the population transfer by compensating the natural phase flip occurring at resonance. Furthermore, the ratio T/S has a different shape having a maximum at the peak of the absorption band at 672 nm, and U-shaped features symmetrically around it. The T/S ratio trace bears a striking resemblance to previous results on multiphoton transitions.[11] The T/S trace is an intermediate example between a narrow atomic transition and a broad molecular transition of a multiphoton control curve. Qualitatively this makes sense as the width of the ZnPc absorption band is between the widths of cesium and Coumarin 6H used in ref. 11. In conclusion, the π-phase jump scan supports the hypothesis that multiphoton processes are involved in the control mechanism. It has previously been shown that unlike in the case of nonresonant two-photon absorption TL pulses are not optimal for two-photon transitions involving an intermediate resonant state.[12] Considering this together with the results from the chirp and π-jump scans, we propose a control mechanism that utilizes multiphoton pathways to higher-lying states from where ISC is more probable, thus leading to enhancement of the triplet yield. A similar mechanism has also been proposed in a previous study on the triplet yield in β-carotene by Buckup *et al.*[13]

Passive control

Two asymmetric push–pull zinc phthalocyanine DSSC photosensitizers were studied, namely TT1 and TT7; the molecular structures are shown in Fig. 9.

The synthesis of TT1 and TT7 was reported in detail earlier.[14] TT1 is an asymmetric zinc phthalocyanine with one electron pulling and anchoring carboxylic acid group, and three *tert*-butyl groups with electron pushing and aggregation suppressing character. TT7 has similar *tert*-butyl groups, however the anchoring group differs substantially. The carboxylic acid group is no longer in direct contact with the phthalocyanine core, but connected *via* a conjugated bridge. The conjugated bridge in TT7 is equipped with an electronegative CN group promoting directionality of the ET process in the direction of the TiO_2. The

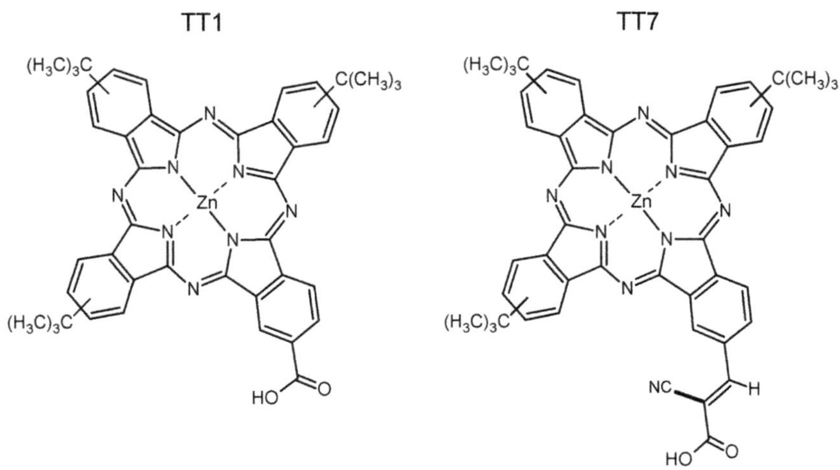

Fig. 9 Molecular structures of zinc phthalocyanine derivatives TT1 and TT7.

modification of the anchoring group can be expected to change both the forward ET from the phthalocyanine into the TiO_2 as well as the unwanted back ET process. The increased distance between the phthalocyanine core and the TiO_2 is expected to slow down both processes. Due to its electronegative character, the CN group might increase the rate of the ET processes. The observation that DSSCs based on the two derivatives show a substantial difference in efficiency (2.55% for TT7 and 3.56% for TT1[14]) suggests a significant effect on the interface dynamics.

Colloidal TiO_2 paste (Ti-Nanoxide HT) was purchased from Solaronix. The particle size is specified to be 8–10 nm and the paste is well-suited for the preparation of transparent electrodes for DSSC. The TiO_2 paste was spread onto a glass sheet using tape as a spacer. A thin film was obtained by raking off the excess of paste with a glass rod. After removing the tape and drying in air, the sample was sintered in air at 450 °C for 30 min. The obtained transparent TiO_2 coated glass substrate was dipped for 4 h into a 0.1 mM solution of the zinc phthalocyanine derivative in ethanol containing 20 mM 3a,7a-dihydroxy-5b-cholanic acid (Cheno, purchased from Sigma Aldrich) and finally rinsed several times with ethanol.

The absorption spectra of the ZnPc derivatives in solution and deposited onto TiO_2 are shown in Fig. 10A and 10B. The Q-band observed for TT1 sensitized TiO_2 at 685 nm is due to monomers and the band at 635 nm is partly due to a vibrational progression of the monomer absorption and due to the presence of aggregates. The corresponding bands for TT7 sensitized TiO_2 are at 645 nm and a shoulder at 690 nm. The absorption spectra of both the TT1 and TT7 sensitized TiO_2 substrates are broadened as compared to those in solutions (Fig. 10A), which can be attributed to an increased inhomogeneity of the molecular environment.

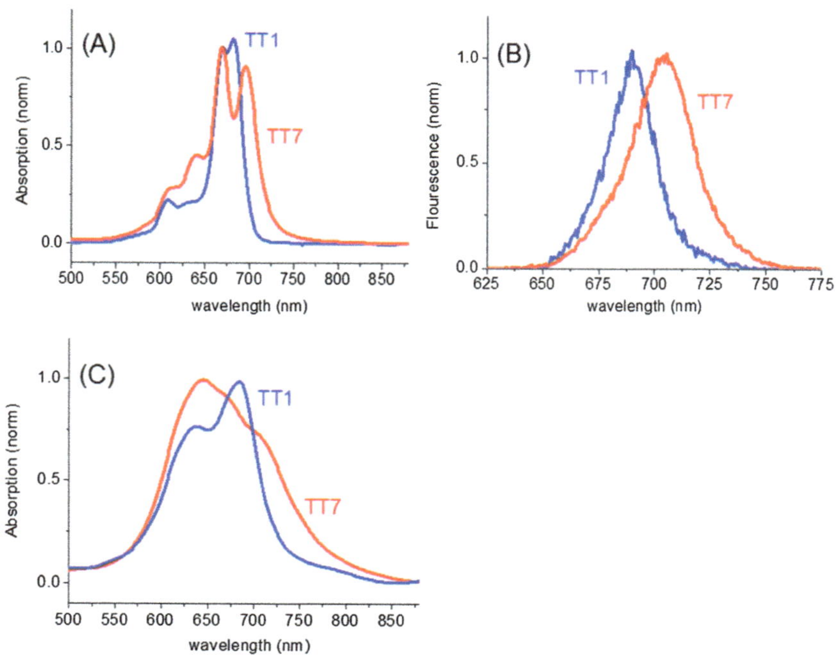

Fig. 10 Absorption (A) and fluorescence spectra (B) of TT1 and TT7 in solution (1×10^{-4} M in ethanol containing 20 mM Cheno) and the absorption spectra (C) of TT1 and TT7 sensitized TiO_2 substrates.

The fluorescence spectra of TT7 in solution exhibits a ~14 nm red shift (peak at 704 nm) as compared to that of TT1 in solution (peak at 690 nm) (see Fig. 10C). The fluorescence of the sensitized TiO_2 substrates was observed to be heavily quenched (data not shown).

Pump–probe data were recorded upon excitation at 680 nm (80 fs, 40 nm FWHM) to minimize the contribution of aggregates to the dynamics. Pump pulses with an energy of 20 nJ/pulse were focused to a spot-size of 130 μm which correspond to a photon flux of 5×10^{14} photons/cm^2 and was verified to be in the linear regime. Transient dynamics were probed by using a white light continuum (470 nm–700 nm) generated by focusing a part of the fundamental (775 nm) into a 2 mm sapphire window. The relative polarization of pump and probe was kept at 54.7°. The photosensitized TiO_2 substrate was mounted on an automated mechanical translational stage to avoid photodegradation.

Electron injection from a photosensitizer into TiO_2 is known to occur with various rate constants, typically an ultrafast component ~100 fs followed by a slower ps component. This is explained in terms of a two-state relaxation model, the fast component is attributed to electron injection from the nonthermalized state whereas the ps component is due to electron injection from the relaxed excited state. This difference in electron injection time arises from a variation of the density of states in the TiO_2 for different energy levels. The biphasic behavior is known to be particularly pronounced for Ru polypyridyl complexes on TiO_2 and applies to a lower extent to organic chromophores.[15] Literature values for zinc phthalocyanine derivatives deposited on TiO_2 are 296 fs[16] and <500 fs.[5b] For a zinc porphyrin derivative a value of 300 fs was reported.[17] Back electron transfer times cover a wide range from the ps up to the ms domain, important factors are the absence/presence of an electrolyte, electrodes and bias voltage.[18] Geminate recombination, *i.e.* back electron transfer to its origin, typically occurs in the ps domain, while back transfer of delocalized electrons is known to be slower.[17]

Fig. 11A shows the transient absorption kinetics of TT1 and TT7 sensitized TiO_2 substrates at 500 nm, where the signal is predominantly due to singlet ESA. Both samples show an initial ultrafast decay ~500 fs, followed by a slower ps decay component, which is much faster than the ESA decay observed for the ZnPc in solution (Fig. 4). The ps component is considerably faster for TT1 (~4 ps) than for TT7 (~15 ps), indicating faster electron injection. The transient spectral

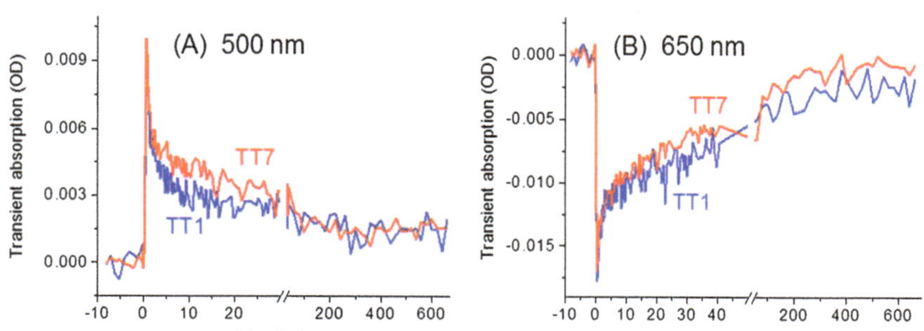

Fig. 11 Transient absorption kinetics of TT1 and TT7 sensitized nanocrystalline TiO_2 at 500 nm (A, representing ESA) and 650 nm (B, representing GSR).

evolution shows the development of a positive band around 560 nm, which can be attributed to the oxidized photosensitizer.[6b,19] This band appears within 1 ps, consistent with the ultrafast decay of the ESA and again reflecting electron injection into the TiO_2.

A signature of back electron transfer from the TiO_2 to the photosensitizer involves the ground state recovery (GSR). GSR is probed at 650 nm, at which wavelength stimulated emission is marginal (see also Fig. 10C). Fig. 11B shows the GSR kinetics of TT1 and TT7 sensitized TiO_2 substrates. Recombination occurs at time scales of 1–2 ps and tens of ps (TT1: 60 ps, TT7: 50 ps) and is slightly faster for TT7 than for TT1.

In summary, strategic manipulation in the periphery of the molecular structure allows passive control of the interface dynamics. The TT1 sensitized TiO_2 substrate shows faster injection and slower recombination as compared to TT7, explaining the lower efficiency observed for a DSSC based on the latter photosensitizer.[14] Apparently the effect of the increased distance between the phthalocyanine core and the TiO_2 for TT7 is greater than the impact of the electronegative CN group resulting in a slower electron injection. The presence of the electron pulling CN group might however promote the back electron transfer from the TiO_2 to the photosensitizer leading to faster recombination.

Conclusions

In this work we report on the manipulation of the ultrafast photodynamics of the photosensitizer zinc phthalocyanine (ZnPc) *via* active and passive control. Pulse shaping allowed an improvement in triplet to singlet ratio of 15% as compared to exposure to a transform-limited pulse. This effect is ascribed to a control mechanism that utilizes multiphoton pathways to higher-lying states from where intersystem crossing is more likely to occur. Manipulation of the periphery of the molecular structure resulted in an increased rate for electron injection into TiO_2 and slower back electron transfer, improving the DSSC efficiency.

Acknowledgements

Prof. Tomás Torres from the University of Madrid is acknowledged for providing the TT1 and TT7 samples and insightful discussions. We are grateful to Jeroen P. Korterik and Frans B. Segerink for technical support. J.L.H. acknowledges the Nederlandse Organisatie voor Wetenschappelijk Onderzoek (NWO) for generous financial support.

References

1 (*a*) M. C. DeRosa and R. J. Crutchley, *Coord. Chem. Rev.*, 2002, **233**, 351; (*b*) M. Ochsner, *J. Photochem. Photobiol., B*, 1997, **39**, 1; (*c*) M. V. Martinez-Diaz, G. de la Torrea and T. Torres, *Chem. Commun.*, 2010, **46**, 7090; (*d*) M. V. Martinez-Diaz, M. Ince and T. Torres, *Monatsh. Chem.*, 2011, **142**, 699.
2 J. Savolainen, D. van der Linden, N. Dijkhuizen and J. L. Herek, *J. Photochem. Photobiol., A*, 2008, **196**, 99.
3 B. Oregan and M. Gratzel, *Nature*, 1991, **353**, 737.
4 M. Gratzel, *J. Photochem. Photobiol., A*, 2004, **164**, 3.
5 (*a*) M. K. Nazeeruddin, R. Humphry-Baker, M. Gratzel, D. Wohrle, G. Schnurpfeil, G. Schneider, A. Hirth and N. Trombach, *J. Porphyrins Phthalocyanines*, 1999, **3**, 230;

(*b*) J. J. He, G. Benko, F. Korodi, T. Polivka, R. Lomoth, B. Akermark, L. C. Sun, A. Hagfeldt and V. Sundstrom, *J. Am. Chem. Soc.*, 2002, **124**, 4922.
6 (*a*) P. Y. Reddy, L. Giribabu, C. Lyness, H. J. Snaith, C. Vijaykumar, M. Chandrasekharam, M. Lakshmikantam, J. H. Yum, K. Kalyanasundaram, M. Graetzel and M. K. Nazeeruddin, *Angew. Chem., Int. Ed.*, 2007, **46**, 373; (*b*) J. J. Cid, M. Garcia-Iglesias, J. H. Yum, A. Forneli, J. Albero, E. Martinez-Ferrero, P. Vazquez, M. Gratzel, M. K. Nazeeruddin, E. Palomares and T. Torres, *Chem.–Eur. J.*, 2009, **15**, 5130; (*c*) M. E. Ragoussi, J. J. Cid, J. H. Yum, G. de la Torre, D. Di Censo, M. Gratzel, M. K. Nazeeruddin and T. Torres, *Angew. Chem., Int. Ed.*, 2012, **51**, 4375; (*d*) S. Mori, M. Nagata, Y. Nakahata, K. Yasuta, R. Goto, M. Kimura and M. Taya, *J. Am. Chem. Soc.*, 2010, **132**, 4054.
7 T. Mančal, N. Christensson, V. Lukeš, F. Milota, O. Bixner, H. F. Kauffmann and J. Hauer, *J. Phys. Chem. Lett.*, 2012, **3**, 1497.
8 Y. Silberberg, *Annu. Rev. Phys. Chem.*, 2009, **60**, 277.
9 J. Savolainen, R. Fanciulli, N. Dijkhuizen, A. L. Moore, J. Hauer, T. Buckup, M. Motzkus and J. L. Herek, *Proc. Natl. Acad. Sci. U. S. A.*, 2008, **105**, 7641.
10 C. J. Bardeen, V. V. Yakovlev, J. A. Squier and K. R. Wilson, *J. Am. Chem. Soc.*, 1998, **120**, 13023.
11 D. Meshulach and Y. Silberberg, *Phys. Rev. A: At., Mol., Opt. Phys.*, 1999, **60**, 1287.
12 N. Dudovich, B. Dayan, S. M. G. Faeder and Y. Silberberg, *Phys. Rev. Lett.*, 2001, **86**, 47.
13 T. Buckup, T. Lebold, A. Weigel, W. Wohlleben and M. Motzkus, *J. Photochem. Photobiol., A*, 2006, **180**, 314.
14 M. Garcia-Iglesias, J. J. Cid, J. H. Yum, A. Forneli, P. Vazquez, M. K. Nazeeruddin, E. Palomares, M. Gratzel and T. Torres, *Energy Environ. Sci.*, 2011, **4**, 189.
15 N. A. Anderson and T. Q. Lian, *Annu. Rev. Phys. Chem.*, 2005, **56**, 491.
16 D. Ino, K. Watanabe, N. Takagi and Y. Matsumoto, *J. Phys. Chem. B*, 2005, **109**, 18018.
17 C. W. Chang, L. Y. Luo, C. K. Chou, C. F. Lo, C. Y. Lin, C. S. Hung, Y. P. Lee and E. W. G. Diau, *J. Phys. Chem. C*, 2009, **113**, 11524.
18 S. A. Haque, Y. Tachibana, R. L. Willis, J. E. Moser, M. Gratzel, D. R. Klug and J. R. Durrant, *J. Phys. Chem. B*, 2000, **104**, 538.
19 P. Charlesworth, T. G. Truscott, R. C. Brooks and B. C. Wilson, *J. Photochem. Photobiol., B*, 1994, **26**, 277.

Faraday Discussions RSC Publishing

PAPER

Strong field control of predissociation dynamics

María E. Corrales,[a] Garikoitz Balerdi,[a] Vincent Loriot,[ab] Rebeca de Nalda[b] and Luis Bañares*[a]

Received 6th December 2012, Accepted 17th December 2012
DOI: 10.1039/c2fd20143g

Strong field control scenarios are investigated in the CH_3I predissociation dynamics at the origin of the second absorption *B*-band, in which state-selective electronic predissociation occurs through the crossing with a valence dissociative state. Dynamic Stark control (DSC) and pump–dump strategies are shown capable of altering both the predissociation lifetime and the product branching ratio.

1 Introduction

Control of photochemical reaction dynamics employing strong laser fields has been a topic of great interest for many years, but only recently demonstrated experimentally.

In general, the subject of laser control (see ref. 1 for an excellent perspective) is understood as the attempt at designing and applying external laser fields that can drive the dynamics of a material system from its initial state to a given desired final state. In a theoretical framework, the "optimal control" algorithms, proposed initially by Rabitz *et al.*[2,3] and Kosloff *et al.*,[4] consisted of trying to find the optimum external field function (as a function of time) that optimized a given final outcome. Several approaches have been identified, and applied, in some cases with extraordinary success, that rely on the control of a single decisive parameter of the laser field: two-pathway quantum interference,[5,6] pump–dump control,[7,8] or stimulated Raman adiabatic passage.[9] Tailored laser pulses, where the phases and amplitudes of different frequency components can be controlled in order to induce constructive and destructive interferences that promote certain mechanisms at the expense of others, can optimally exploit the complexity of the laser–matter interaction to achieve a given goal in molecular dynamics.[10]

The above-mentioned pump–dump strategy for control[7,8] has been demonstrated in a number of experiments for the control of yields or branching ratios in simple reactions (see, for example, ref. 11–13). The pump–dump scheme has been

[a]*Departamento de Química Física, Facultad de Ciencias Químicas (Unidad Asociada CSIC), Universidad Complutense de Madrid, 28040 Madrid, Spain. E-mail: banares@quim.ucm.es; Fax: +34913944135; Tel: +34913944228*

[b]*Instituto de Química Física Rocasolano, CSIC, C/Serrano 116, 28004 Madrid, Spain*

extensively employed not only as a control tool, but also in ultrafast time resolved spectroscopy to explore the molecular transient states; *e.g.*, Zgrablić *et al.*[14] showed strong dependence of the photoisomerization yield of all *trans* retinal with solvent polarity after absorption of a 400 nm pump pulse using ultrafast transient absorption spectroscopy and a pump–dump–probe scheme, and Chen *et al.* investigated the bond length influence over the ionization dependence of the electrons in the valence orbitals of I_2 using a pump–dump–probe technique and Fourier-transform spectroscopy.[15]

A different approach was employed by Sussman *et al.*[16] to modify the Br*/Br branching ratio in IBr photodissociation with a strong 1730 nm control pulse when its delay respect to the pump pulse was changed. This so-called Dynamic Stark Control (DSC) technique is based on the dynamic modification of the Hamiltonian during the evolution of the system. This is done with a strong control pulse by the non-resonant dynamic Stark effect (NRDSE) to enhance or inhibit specific reaction channels. Interestingly, in this demonstration the laser is non resonant with any transition of the molecule, which makes it a universal technique suitable for a large number of quantum systems.

Understandably, in most of the molecular dynamics control studies performed up to now, the target to be optimized has been the branching ratio between the products of a reaction. However, some mechanisms to control the lifetime of excited states have also been proposed. Sussman *et al.* calculated the predissociation lifetime variation of a quasi-bound IBr excited state yielding ground I and Br (only channel open at 569 nm) when the intensity of a IR control laser and the delay time with respect to the pump pulse were modified.[17] García-Vela has shown recently how the lifetime of an excited state can be controlled using a second, time delayed pump pulse, resonant with another excited state close to the first one. The Br_2–Ne complex was proposed for an experimental demonstration.[18]

In this work, strong field control over the CH_3I predissociation dynamics at the origin of the *B*-band has been explored. A moderately intense ($<5 \times 10^{12}$ W cm^{-2}), near-infrared (NIR) control pulse in the picosecond range has been used to alter the potential energy surfaces involved in *B*-band absorption and subsequent dynamics. The effect of the Stark field on the absorption step has been directly measured and calibrated as a function of laser intensity. Together with the Stark effect affecting absorption, the presence of this field induces new processes and it will be shown that both the lifetime of the excited state and the branching ratio of the reaction can be significantly, and, to an extent, controllably altered.

CH_3I predissociation of the bound $^3R_1$ Rydberg state is an interesting topic in itself due to the strikingly diverse lifetimes for the different vibrational levels, and has been the subject of recent work by the authors.[19,20] Due to the discrete spectrum, it is experimentally possible to have state selectivity for the initial state, and the dynamics then proceeds after non-adiabatic crossing at the conical intersection with the valence dissociative $^3A_1$ surface (see Fig. 1), which yields CH_3 with different degrees of vibrational excitation in the C–H stretch (ν_1) and umbrella (ν_2) modes in coincidence with spin–orbit excited I*($^2P_{1/2}$). For the band origin (transition to the $\nu = 0$ level of the Rydberg state), a quantum yield $\Phi^* = 1$ and an excited state level lifetime of 1.5 ps have been found. The present work shows how the absorption spectrum, the product branching ratio and the excited state lifetime can be significantly altered in the presence of a moderately strong NIR control laser pulse.

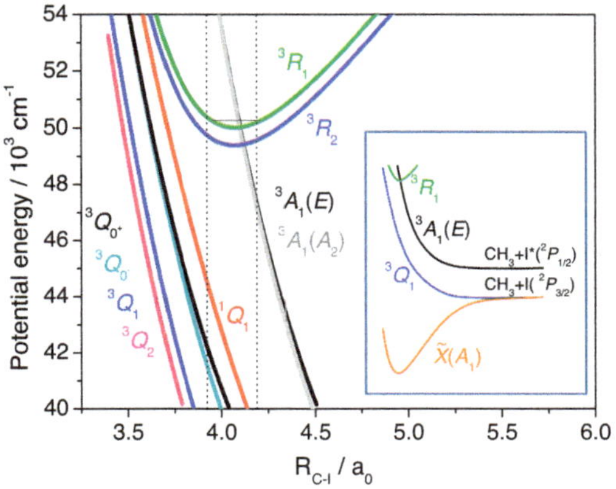

Fig. 1 Potential energy curves of the CH_3I molecule in the C–I distance coordinate (in atomic units, a_0). *A*-band, composed by a set of dissociative states, and *B*-band, formed by the dark $^3R_2$ and bright $^3R_1$ Rydberg states, are shown for short internuclear distances. Dotted lines indicate the Franck–Condon region and the black horizontal line corresponds to the vibrationless level of the $^3R_1$ state. Population in this state advances through the conical intersection near the bottom of the curve with the $^3A_1$ (*E*) and $^3A_1$ (A_2) dissociative curves from the *A*-band. Inset presents surfaces of interest for the present study for longer C–I distances, showing the products correlating with them. Adapted from ref. 28.

2 Experiment

The experimental setup has been described previously[19,20] and only some details relevant for the current work will be given here. Three laser pulses are required for the experiment, all of them generated by modifying the output of an amplified femtosecond laser system, 3.5 mJ pulses of 50 fs duration at 1 kHz repetition rate with a central wavelength of 804 nm.

The first arm (the "pump pulse") is generated by frequency quadrupling part of the fundamental beam in a device consisting of a tripling unit followed by sum-frequency mixing between the third harmonic and the fundamental. This device allows some tunability in a narrow range around 200 nm with a ~0.3 nm bandwidth and a pulse energy of about 0.5 μJ. The pump pulse linear polarization can be rotated from vertical to horizontal using a half wave plate, and an adjustable telescope allows the control of the propagation conditions. This laser pulse is used to pump the CH_3I molecules to an excited state when its central wavelength is tuned to the transition between the vibrationless ground state and the vibrationless excited Rydberg $^3R_1$ bound state.

The second arm (the "probe pulse") comes from an optical parametric amplifier (OPA) pumped by the main fundamental laser output. The 1333.8 nm signal pulse is twice frequency doubled to produce 333.45 nm-centered pulses with a typical pulse energy of around 5 μJ. This pulse is used to probe CH_3 fragments that appear as a result of CH_3I dissociation. The pulse spectral width of 2.5 nm FWHM allows (2 + 1) resonance enhanced multiphoton ionization (REMPI) of the CH_3 product in different vibrational states: $\nu_1 = 0, 1, 2$; $\nu_2 = 0$.[21,22]

The description of the predissociation dynamics of CH_3I in the second absorption *B*-band, using the pump–probe scheme described above has been the subject of recent work by the authors.[19,20]

A third laser pulse (the "control pulse") is added to those described above. The control laser pulses, at the fundamental laser frequency, are generated by temporarily stretching the pulses from a third arm of the main laser output with a pair of diffraction gratings. They are characterized using a chirp-scan type measurement, giving a value of 75000 fs^2 for the linear chirp, which corresponds to 4 ps FWHM pulses. Pulse energies of up to 400 μJ corresponding, after focusing the beam, to an intensity of 8 TW cm^{-2} can be reached, but it is possible to attenuate it in order to use lower intensities. For some measurements, the linear polarization is rotated to vertical with a tunable waveplate set for half wave retardation.

The delay times between the pump and probe pulses and between the pump and control pulses are controlled with two motorized delay stages placed at the pump and control laser arms that allow time steps as low as 1 fs.

All beams are focused with a 25 cm focal length lens onto a pulsed molecular beam inside a differentially pumped vacuum chamber. The molecular beam is generated by expansion of 0 °C CH_3I, seeded in He, at a total pressure of 1.5–2 bars, through a 0.5 mm nozzle diameter, 1 kHz homemade cantilever piezoelectric pulsed valve.[23] The molecular beam goes through a 1 mm diameter skimmer from the expansion chamber to the ionization chamber, where it interacts with the laser pulses. Generated ions are extracted by a set of electrostatic lenses working in velocity mapping configuration,[24] and mass separated in a 50 cm field-free region time-of-flight (TOF) mass spectrometer.

Ion spheres are projected onto a detector consisting of a dual microchannel plate (MCP) in Chevron configuration coupled to a phosphor screen and a Peltier-cooled 12-bit charge-coupled device camera. The front MCP plate is gain gated in order to achieve mass selection. Kinetic energy-radius calibration is done by measuring resonantly ionized CH_3 ($\nu = 0$) fragments produced after photodissociation of CH_3I with a 268 nm pump laser pulse, taking advantage of the well known kinetic energy release (KER) for both $I^*(^2P_{1/2})$ and $I(^2P_{3/2})$ yielding channels.

Ion images are Abel inverted using the polar basis set expansion (pBASEX) method,[25] and a Levenberg–Marquardt squared residuals minimization algorithm[20] is used for non-linear fits to the data.

The instrument temporal response time, considered as the temporal cross-correlation of the pump and probe beams, was measured through multiphoton ionization of Xe, obtaining a value of ≈400 fs.

3 Results

An intense picosecond 804 nm laser pulse has been used to explore the effects of a strong optical field on the photodissociation of CH_3I in the *B*-band. Fig. 2 shows a series of CH_3^+ images obtained for a case where a 5.5 TW cm^{-2}, 4 ps duration 804 nm control pulse is introduced in a typical pump–probe experiment that employs a 201.2 nm pulse to initiate photodissociation (origin of the *B*-band) and a 333.5 nm pulse for (2 + 1) REMPI detection of the $CH_3(v_1 = 0,1,2;v_2 = 0)$ fragments. A long pump–probe delay time (~30 ps) was fixed in all cases, and images were taken for different pump–control (PC) delay times as indicated in the Figure. Zero PC delay time is chosen at an arbitrary position of overlap between the fs pump and ps control pulses.

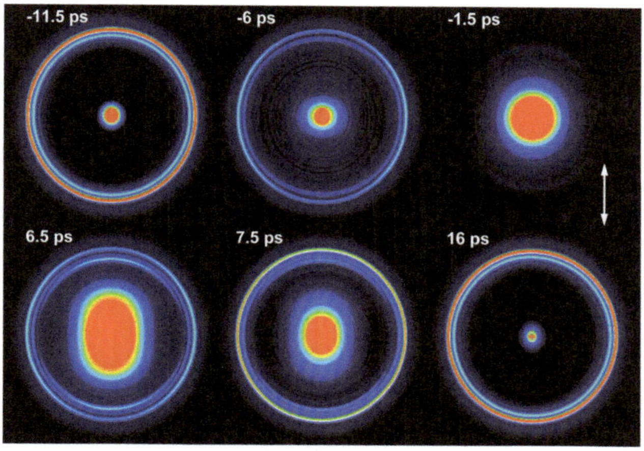

Fig. 2 Set of pBASEX inverted images of CH_3^+ taken for different pump–control delay times at a control laser intensity of 5.5 TW cm^{-2}. A pump–probe delay time of ~30 ps is long enough to detect all the CH_3 fragments generated in the process. The double arrow shows the direction of the polarization of the three laser pulses. See text for more details.

In the first image, acquired at −11.5 ps PC delay time, three almost isotropic rings are present. For this delay time, the control pulse finishes long before the pump pulse arrives, so it has no effect on the photodissociation process. Hence, image interpretation can be made as in the control field free case:[19] predissociation from the $^3R_1$ state occurs with a 1.5 ps lifetime because of a crossing with the dissociative $^3A_1$ surface that leads to fast photofragmentation, yielding CH_3 + I*($^2P_{1/2}$) fragments. The outer ring corresponds to vibrationless CH_3(v = 0) and its radius corresponds to a kinetic energy of 2.51 eV. The inner rings correspond to CH_3(v_1 = 1) and CH_3(v_1 = 2) fragments in correlation with I*($^2P_{1/2}$).[19]

When the control pulse is delayed towards the pump pulse, the signal from the above-mentioned rings starts to decrease, as can be seen for the −6 ps PC delay time, until it vanishes entirely for −1.5 ps. As will be discussed later, this is due to absorption suppression induced by dynamic Stark shift (DSS) of the $^3R_1$ surface. It can also be seen that the signal from the centre of the images increases as the temporal overlap between the control and pump pulses increases. This enhancement of the central part of the image is attributed to dissociative ionization processes that require temporal overlap between the pump and control pulses, and that produce low kinetic energy fragments.

For a 6.5 ps PC delay time, the signal in the rings reappears, although not with full intensity. It is important to note that a new ring arises, that corresponds to CH_3 fragments with 1.95 eV of kinetic energy. This channel shows parallel character, with the intensity concentrated near the poles of the ion sphere. This observation is compatible with a pump–dump channel consisting of 804 nm pulse stimulated emission that carries population from the $^3R_1$ surface to the valence $^3Q_1$ surface after absorption of a pump photon. The molecule then undergoes ultrafast photofragmentation yielding CH_3 + I($^2P_{3/2}$). This dissociation pathway will be explored and discussed in more detail below.

When the PC delay time is further increased, the pump–dump ring decreases in intensity, and the signal in the outer rings regains intensity until, for very long delay times, images look again the same as without the control pulse. In the last situation, taken at 16 ps delay time, the control pulse arrives after predissociation has already finished and thus it has no repercussions on the photodissociation dynamics.

Fig. 3a shows fragment kinetic energy distributions, obtained by angular integration of the pBASEX inverted images, for different PC delay times. The signal from predissociation is suppressed for delay times between about −5 and 9 ps, as can be better seen in Fig. 3b, which shows energy integration of the signal

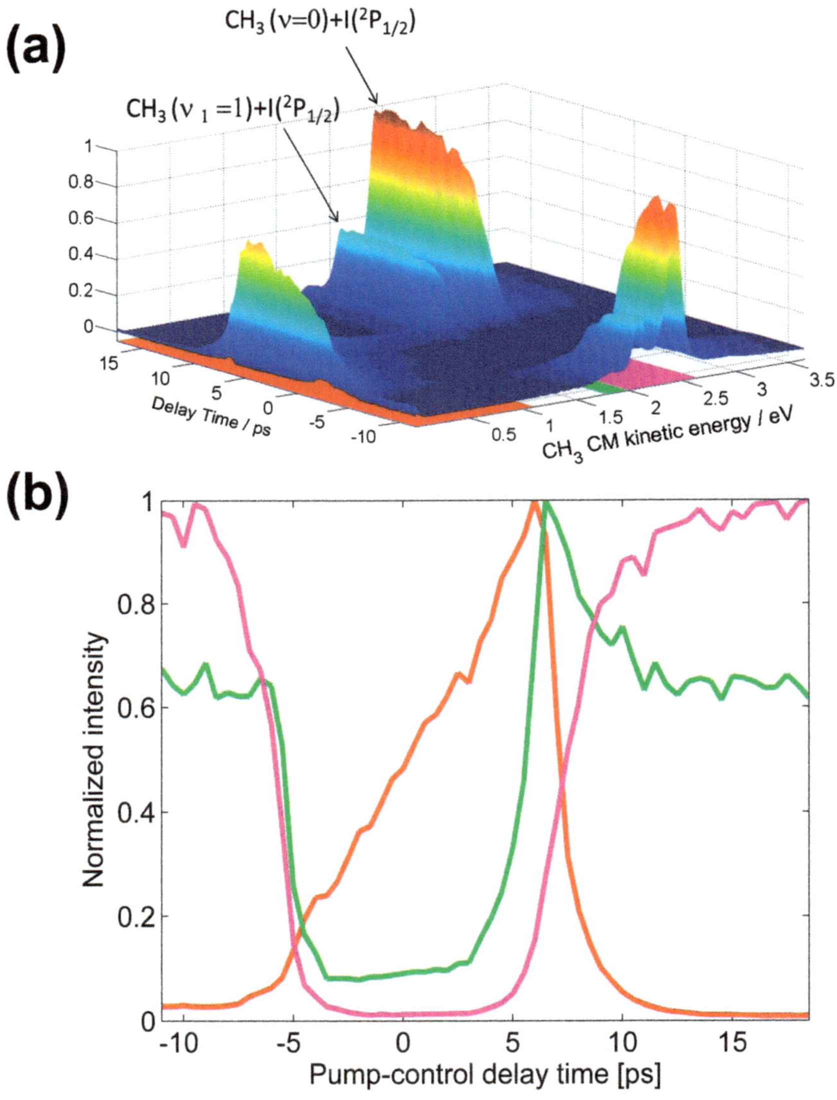

Fig. 3 (a) 3D plot of the CH_3 fragment kinetic energy distributions as a function of the pump (201.2 nm)–control (804 nm) delay time. Control laser intensity is 5.5 TW cm^{-2} and pump–probe delay time is long enough (∼30 ps) to detect all the CH_3 fragments produced in the photodissociation events. Colored areas below the surface represent the regions shown in (b), after integration across the energy coordinate. The three regions correspond to the conical intersection predissociation (magenta), pump–dump (green), and dissociative ionization (red) channels.

corresponding to each of the channels. It can also be seen how the dissociative ionization channel is enhanced for delay times where the pump and control pulses are overlapped and is higher for high delay times, *i.e.*, when the pump pulse arrives at the beginning of the control pulse. Finally, the pump–dump channel is only clear from a delay time of 6 ps and vanishes, disappearing completely at about 8 ps. For delay times shorter than 6 ps, its reduced intensity is the result of the suppressed absorption of the pump pulse. It should be noted that, on Fig. 3b, the signal corresponding to the pump–dump channel is contaminated by the weak $CH_3(v_1 = 2) + I^*(^2P_{1/2})$ predissociation channel due to their similar kinetic energy. This is the reason why the green curve is not zero for delay times outside pump–control overlap.

In principle, cancellation of the *B*-band rings near pump–control overlap could be just a consequence of faster competing processes. It is clear that this is not the case, however, because the rings can be found again if the wavelength of the pump photon is tuned slightly to the blue in the presence of the control field. This is proof of a DSS effect that produces a blue shift in the discrete absorption spectrum of the *B*-band. A calibration of the blue shift has been performed, with the result shown in Fig. 4. For this measurement, the magnitude of the pump photon energy shift required to recover the resonance between the vibrationless ground and $^3R_1$ states (diagnosed by the appearance of the CH_3^+ rings) was measured for different intensities of the control laser. As expected,[26] the relationship between the Stark shift and the intensity of the field is found to be linear. The energy of the $^3R_1$ state shifts by up to 13 meV for the highest intensity of 0.64 TW cm^{-2}, as shown in the Figure. For these measurements, the PC delay time was fixed at −1 ps, that is, the pump pulse was close to the maximum of the control pulse.

As was mentioned above, in the presence of the strong control field, but after retuning the pump wavelength – and always maintaining the probe pulse at a sufficiently long pump–probe delay time – the signal from the rings is recovered, but its intensity is not as high as without the control laser. However, the total ion signal is about the same due to the enhancement of the dissociative ionization channel and the appearance of the pump–dump channel. Fig. 5a shows a raw image taken in these conditions, and Fig. 5c contains its pBASEX inversion. The dump process is assigned to stimulated emission carrying population from the

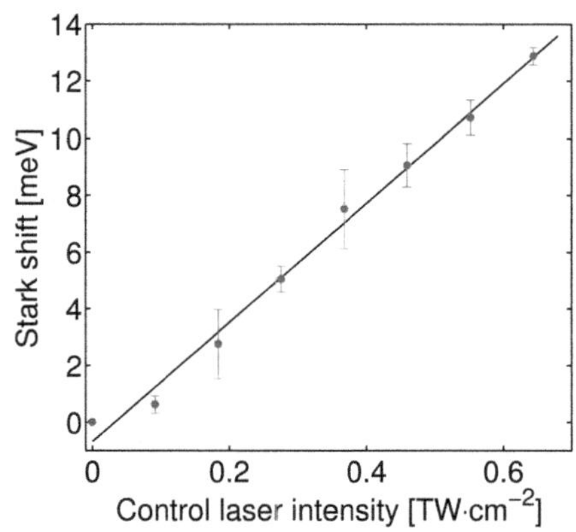

Fig. 4 Measured Stark shift as a function of the intensity of the Stark field and linear regression of the data (solid line). Error bars correspond to the standard deviation.

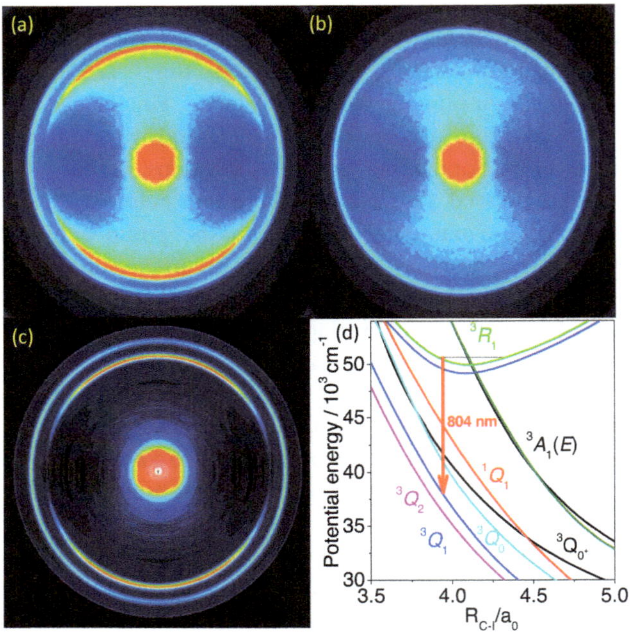

Fig. 5 (a) Raw CH_3^+ image acquired for a long (30 ps) pump–probe delay time and a −1 ps pump–control delay time, with all laser polarizations horizontal and after retuning the pump laser to a wavelength resonant with the transition between the vibrationless ground state and the "dressed" vibrationless $^3R_1$ state. (b) Raw (non Abel-invertible) image taken in the same conditions as (a) but with vertical polarization of the control laser. (c) pBASEX inversion of (a). (d) Detail of the potential energy curves of CH_3I along the C–I coordinate showing the transition assigned to the pump–dump channel.

Rydberg $^3R_1$ state to the dissociative $^3Q_1$ state, as schematically shown in Fig. 5d. This assignment is based on the parallel anisotropy of the pump–dump ring, and will be discussed in Section 4.

Fig. 5b contains a raw image acquired with the linear polarization of the control pulse rotated 90°; *i.e.* perpendicular to the detector plane, and it is not, therefore, Abel invertible. As can be observed, the pump–dump ring is not visible. However, total ion count remains the same, which is an indication that the ions produced through the pump–dump channel are in this case concentrated around the poles of a sphere whose north–south axis is defined by the polarization of the control laser field. This confirms the parallel character of this transition, most probably to the $^3Q_1$ repulsive surface considering the energy of the 804 nm photon and the symmetry of the involved states in the transition (see Fig. 5d).

A delay scan of the control pulse (5.5 TW cm^{-2}) with respect to the blue-shifted (200.5 nm) pump pulse while maintaining an asymptotic probe pulse was carried out. Fig. 6a shows the 2D map constructed from the CH_3^+ kinetic energy distributions obtained from this experiment, after pBASEX inversion and angular integration of the images. In this delay scan, excitation is low for delays outside the pump–control temporal overlap because the pump photon energy is detuned with respect to the transition from the vibrationless ground state to the $\nu = 0$ level of the undressed $^3R_1$ surface. It can be noted, however, that even in these conditions, absorption is not totally suppressed. This is probably related to the asymmetric shape of the absorption line.

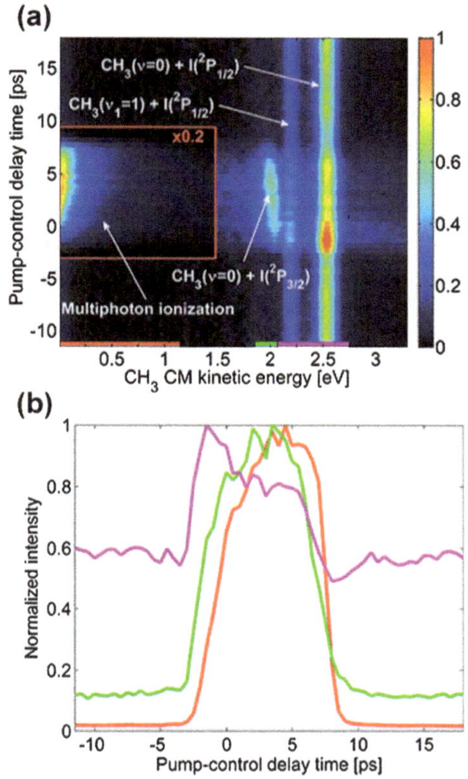

Fig. 6 (a) 2D intensity graph of the CH_3 fragment kinetic energy distribution as a function of the pump (200.5 nm)–control (804 nm) delay time. Note that the pump photon energy has been retuned to be in resonance with the transition between the ground state and the $\nu = 0$ level of the "dressed" $^3R_1$ state. Control laser intensity is 5.5 TW cm^{-2} and a pump–probe delay time of ~30 ps is long enough to detect all the CH_3 fragments. Colored segments on the energy axis represent the energy limits used for integration of the signal, which are represented on the bottom panel. (b) CH_3^+ signal originating from the three channels marked on panel (a), that is, conical intersection predissociation (magenta), pump–dump (green), and dissociative ionization (red) channels.

For delays where pump and control laser pulses are temporally overlapped, the resonance is recovered, since the blue shift is compensated by the shorter pump wavelength. For these delay times, an enhancement of the dissociative ionization channel is observed, due to a two-color processes. Interestingly, a chirped contribution for the pump–dump channel is observed for overlap times. The observed dependence of the CH_3^+ kinetic energy distribution with time can be attributed to the inherent chirp of the control laser pulse. This maps time to energy, showing that for negative delays, when the pump pulse is overlapped with the beginning of the control pulse, kinetic energy of the fragments is higher than for positive delays, when the pump pulse is overlapped with the end of the control pulse. The magnitude of the chirp qualitatively agrees with the measured positive chirp of 75 000 fs^2 for the control pulse, *i.e.*, low frequencies arriving earlier than high frequencies. However, a measurable difference is detected between the chirp of the laser (97 meV ps^{-1}) and the chirped pump–dump signal (76 meV ps^{-1}). It is possible that the spectral probability distribution of the dump process acts as a spectral filter that creates a bias in the observed chirp, giving rise to a reduced chirp.

One of the main aims of this work has been the strong field control of predissociation dynamics (lifetimes) using the *B*-band of CH_3I as a test case. In order to measure directly the effect of the presence of the control field on the lifetime of the excited $^3R_1$ state, a delay scan of the 333.45 nm probe pulse with respect to the pump pulse, in the presence of the control field, was carried out. The method consisted of acquiring a set of images for different pump–probe delay times with a fixed −1 ps pump–control delay time. The reason for using this specific pump–control delay time is to ensure that photodissociation occurs during the flattest part of the control pulse, since its intrinsic lifetime is 1.5 ps. Each image was then pBASEX inverted and kinetic energy distributions were obtained by angular integration. Further integration across the energy range of each of the channels produced the transients that are shown in Fig. 7. Different control laser intensities (0.18, 0.46 and 0.55 TW cm^{-2}) were used to investigate its influence on predissociation dynamics. These relatively low control laser intensities do not significantly shift the transition to the $^3R_1$ state and no pump laser wavelength retuning is required.

Fig. 7 shows transients for the pump–dump and $CH_3(v = 0) + I^*(^2P_{1/2})$ predissociation channels corresponding to experiments without a control pulse and with 0.18, 0.46 and 0.55 TW cm^{-2} control laser intensities. As can be seen, the lifetimes for the pump–dump and $CH_3(\nu = 0) + I^*(^2P_{1/2})$ predissociation channels for a given control intensity are compatible, within error. Furthermore, a reduction of the lifetime with increasing control laser intensity can be observed. The lifetime is reduced from 1.5 ± 0.1 ps for the no control-laser to 1.07 ± 0.07 ps for 0.18 TW cm^{-2}, 0.85 ± 0.06 ps for 0.46 TW cm^{-2} and 0.65 ± 0.04 ps for 0.55 TW cm^{-2}. The temporal behavior does not change when polarization of the control laser is rotated (not shown), with a 0.60 ± 0.04 ps lifetime measured for the

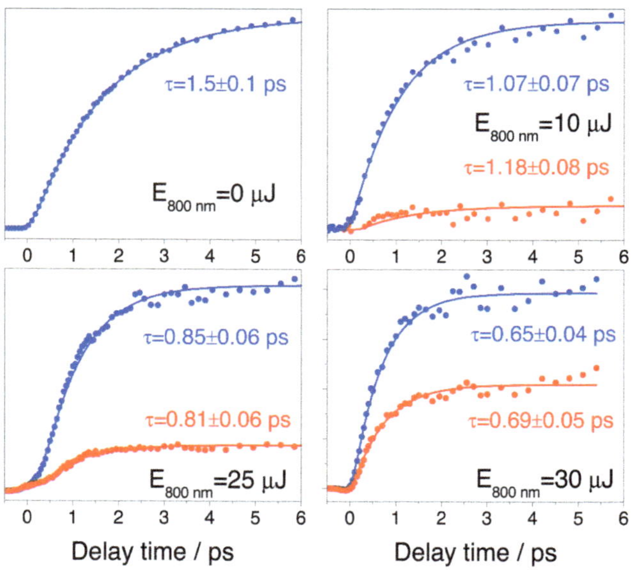

Fig. 7 Experimental transients of the $CH_3(\nu = 0) + I^*(^2P_{1/2})$ (blue) and pump–dump (red) channels as a function of the pump–probe delay time without control laser (a) and with three different intensities of the control laser (b,c,d). The delay time between the pump and control lasers is fixed at −1 ps, so that the pump pulse arrives near the beginning of the control pulse. Fits are also included (solid lines).

0.55 TW cm^{-2} control laser intensity. This fact supports the idea that, upon control laser polarization rotation, the pump–dump channel angular distribution rotates but the channel has similar probability.

It is important to note that, because of the presence of the new pump–dump channel, the $I^*(^2P_{1/2})$ quantum yield ($\Phi^* = [I^*(^2P_{1/2})]/\{[I(^2P_{3/2})] + [I^*(^2P_{1/2})]\}$) is also modified. As has been shown in the literature,[19] for standard, field-free conditions Φ^* has a value of 1. Since the pump–dump channel yields ground-state iodine, $I(^2P_{3/2})$, the global quantum yield is found to decrease as a function of laser intensity. It is not possible to determine meaningful values for the quantum yield from the data reported in the present work because only CH_3 fragments in certain vibrational states ($\nu_1 = 0,1,2$ $\nu_2 = 0$) are measured through resonant probing. As a guide, however, the quantum yield obtained from the results of the measured CH_3 fragments are 0.90, 0.83 and 0.65 for 0.18, 0.46 and 0.55 TW cm^{-2} control laser intensities. This constitutes a very significant control of the quantum yield across this explored intensity range.

4 Discussion

Three main results concerning the strong field laser control on CH_3I photodissociation dynamics upon excitation to the *B*-band will be discussed in this section: the observed Stark shift of the discrete resonances, the modification of the branching ratio of the reaction and the change in the excited state lifetime induced by the presence of a strong NIR laser field.

For relatively high NIR field intensity, it was found that the transition was so significantly shifted to the blue that resonance conditions were lost, even for the broadband fs pump pulses. Notably, it was possible to calibrate the magnitude of the effect as a function of laser intensity by finding the optimum frequency retuning of the pump laser wavelength to the dressed resonance.

As was described in the Results section, the Φ^* quantum yield, which is unity in field-free conditions, has been found to be significantly less than one even when moderate intensities of the control field are employed. This is related to the appearance of the additional so-called pump–dump channel, which yields ground state iodine. The assignment of the inner ring in the CH_3 images to a pump–dump channel is based on energetic considerations as well as on the analysis of the angular distributions of the CH_3 fragments. For this kind of process, if the counterpart fragment has no internal energy, the maximum kinetic energy release (KER) of the CH_3 fragments can be calculated as KER $= (E_{\text{pump}} - D_0 - E_{\text{dump}})\cdot(m_I/m_{CH_3I})$, where D_0 is the dissociation energy of the molecule, E_{pump} and E_{dump} are the energies of the pump and control laser pulses, respectively, and m_I and m_{CH_3I} are masses. The calculated value of 1.97 eV for the 201.2 nm pump and 804 nm probe is consistent with the measured value of 1.95 eV, within the energy resolution of the system, of about 50 meV for this region. Parallel anisotropy of the pump–dump ring shows that the dump process is produced with conservation of the projection of the total angular momentum over the internuclear axis of the pseudo-diatomic CH_3I molecule ($\Delta\Omega = 0$). Taking this into account, as well as the fact that the final state has been found to correlate with $I(^2P_{3/2})$, and considering that only a few states are near resonance from $^3R_1$ with a 800 nm photon, it is deduced that dumping carries population to the $^3Q_1$ surface. Therefore, the new ring that appears when the NIR control laser is applied is assigned to photodissociation through the $^3Q_1$ dissociative surface after absorption of a pump photon and subsequent dump from the $^3R_1$ Rydberg state by the control laser, yielding $CH_3(\nu = 0)$ and ground state $I(^2P_{3/2})$.

To rationalize the results that show a lowered lifetime for the photodissociation under the effect of a strong NIR field, two effects have to be considered. The first one is DSS; as it has been shown in the Results section, the control pulse shifts the transition to the $^3R_1$ state towards higher energies. It is likely that this alters the overlap between the wavefunction in the bound state and the dissociative continuum $^3A_1$ state and consequently the probability of predissociation varies. This would produce a modification of the excited state lifetime.

The second effect that can change the lifetime is the presence of the two new channels opened due to the application of the control laser pulse: the pump–dump channel, which brings population from the $^3R_1$ state to the dissociative $^3Q_1$ state, and the dissociative ionization channel after two-color ionization of the excited CH_3I molecule. In order to determine quantitatively the variation of the lifetime that this second mechanism would produce, a simple kinetic model, neglecting coherence effects, has been applied. In the model, it will be considered that, after absorption of a pump photon, population in the $^3R_1$ state undergoes photodissociation by three different channels. The first one is the well known predissociation, after conical intersection to the $^3A_1$ surface, that has a lifetime of 1.5 ps. The others, pump–dump and dissociative ionization, are opened due to the presence of the control laser. The depopulation of the Rydberg state would proceed as follows:

$$\frac{dP}{dt} = -(k_{CI} + k_{dump} + k_{DI})P$$

where k_{CI}, k_{dump} and k_{DI} are the rate coefficients representing the probability of the conical intersection, pump–dump and dissociative ionization channels, respectively. The excited state lifetime when all channels are open is, therefore, $\tau = 1/k$, with $k = k_{CI} + k_{dump} + k_{DI}$. Since both k_{dump} and k_{DI} are dependent on the control laser intensity, the lifetime can be user-tuned, always to shorter values than the field-free lifetime.

The experiment can be carried out in field-free conditions, for which $k_{CI} = 1/\tau_{CI}$ with $\tau_{CI} = 1.5$ ps, since only the conical intersection dissociation channel is open. Contrarily, k_{dump} and k_{DI} are intensity dependent, and it is not possible to experimentally isolate the corresponding processes. It is possible, however, to estimate them for a given intensity through the analysis of the measured kinetic energy distributions (for asymptotic pump–probe delay times). This allows quantification of the population branching into the three channels, since the corresponding CH_3 fragments appear with different kinetic energies, and the final branching ratio can be extracted through partial integration. By assuming that the rate coefficients (k) of each channel are proportional to the magnitude of the signal, and taking the known $k_{CI} = 1/\tau_{CI}$ as a reference, k_{dump} and k_{DI} can be calculated for a given intensity.

In the absence of other effects, this strategy would allow the prediction of the measured lifetime τ simply by inspection of the asymptotic image. Indeed, a good agreement is found, with 1.07 ± 0.07, 0.85 ± 0.06 and 0.65 ± 0.04 ps experimental lifetimes obtained at 0.18, 0.46 and 0.55 TW cm^{-2} control laser intensities compatible within errors with the predicted 1.1, 0.8 and 0.6 ps lifetimes. This result suggests that the first mechanism initially proposed for the modification of the lifetime through DSS has a negligible effect in this case, and can therefore be neglected. This is not entirely surprising if the shape of the molecular wave function in the vibrationless level of the excited CH_3I molecule is considered. Indeed, the region of the conical intersection is situated near the maximum of the wave function amplitude, and calculations[27] show that a very large Stark shift would be required in order to alter significantly its overlap with the dissociative continuum function in the repulsive state.

The situation would be different for a vibrational level wave function with a steeper slope along the reaction coordinate near the conical intersection. In this case, moderate Stark shift could have a stronger influence on the coupling. A possible candidate could be the $\nu_3 = 1$ level (C–I stretching mode) of the $^3R_1$ state for the CH_3I molecule.

5 Conclusions

In this work, it has been demonstrated how strong NIR laser pulses can be used as a tool to control photodissociation dynamics of CH_3I at the origin of the B absorption band. The magnitude of the dynamic Stark shift (DSS) of the $^3R_1$ Rydberg state produced by a NIR strong laser pulse has been measured for different control laser intensities by quantifying the pump wavelength retuning needed to recover resonance conditions, altered by the control field. The strategy of applying a moderately strong control laser field near the temporal overlap with the pump pulse has been shown to dramatically alter some of the properties of the process. In particular, the branching ratio between the channels yielding I($^2P_{3/2}$) and I*($^2P_{1/2}$) has been quantified for different intensity control laser pulses. It has been found that the production of I($^2P_{3/2}$), negligible in field-free conditions, increases with control laser intensity. This is due to the presence of a pump–dump channel that yields I($^2P_{3/2}$). As a final important result, the lifetime of the excited state is dramatically reduced with moderate control laser intensities. Based on the predictions of a simple kinetic model, we believe that the lifetime change is not mainly due to DSS, but to the simultaneous occurrence of additional channels in the presence of the control field. Strategies for the optimization of the contrast obtained for the DSS-related effects have been suggested.

Acknowledgements

M. E. C. acknowledges financial support from Spanish MECD through a FPU fellowship. V. L. is grateful for financial support from the JAE-DOC program of CSIC. This work has been financed by the Spanish MICINN through Grant No. CTQ2008-02578, the Consolider program "Science and Applications of Ultrafast Ultraintense Lasers", Grant No. CSD2007-00013, and by the European Union ITN network "Ultrafast control of quantum systems by strong laser fields-FAST-QUAST" (Grant No. PITN-GA- 2008-214962). This research has been performed within the Unidad Asociada "Química Física Molecular" between Departamento de Química Física of UCM and CSIC. The facilities provided by the Centro de Láseres Ultrarrápidos (UCM) are gratefully acknowledged.

References

1 C. Brif, R. Chakrabarti and H. Rabitz, *New J. Phys.*, 2010, **12**, 075008.
2 R. S. Judson and H. Rabitz, *Phys. Rev. Lett.*, 1992, **68**, 1500.

3 A. P. Peirce, M. A. Dahleh and H. Rabitz, *Phys. Rev. A: At., Mol., Opt. Phys.*, 1988, **37**, 4950.
4 R. Kosloff, S. A. Rice, P. Gaspard, S. Tersigni and D. J. Tannor, *Chem. Phys.*, 1989, **139**, 201.
5 P. Brumer and M. Shapiro, *Chem. Phys. Lett.*, 1986, **126**, 541.
6 M. Shapiro, J. W. Hepburn and P. Brumer, *Chem. Phys. Lett.*, 1988, **149**, 451.
7 D. J. Tannor and S. A. Rice, *J. Chem. Phys.*, 1985, **83**, 5013.
8 D. J. Tannor, R. Kosloff and S. A. Rice, *J. Chem. Phys.*, 1986, **85**, 5805.
9 U. Gaubatz, P. Rudecki, M. Becker, S. Schiemann, M. Külz and K. Bergmann, *Chem. Phys. Lett.*, 1988, **149**, 463.
10 G. G. Balint-Kurti, S. Zou and A. Brown, *Adv. Chem. Phys.*, 2008, **138**, 43.
11 T. Baumert, M. Grosser, R. Thalweiser and G. Gerber, *Phys. Rev. Lett.*, 1991, **67**, 3753.
12 E. D. Potter, J. L. Herek, S. Pedersen, Q. Liu and A. H. Zewail, *Nature*, 1992, **355**, 66.
13 J. L. Herek, A. Materny and A. H. Zewail, *Chem. Phys. Lett.*, 1994, **228**, 15.
14 G. Zgrablić, A. M. Novello and F. Parmigiani, *J. Am. Chem. Soc.*, 2012, **134**, 955.
15 H. Chen, V. Tagliamonti and G. N. Gibson, *Phys. Rev. Lett.*, 2012, **109**, 193002.
16 B. J. Sussman, D. Townsend, M. Y. Ivanov and A. Stolow, *Science*, 2006, **314**, 278.
17 B. J. Sussman, M. Y. Ivanov and A. Stolow, *Phys. Rev. A: At., Mol., Opt. Phys.*, 2005, **71**, 051401.
18 A. García-Vela, *J. Phys. Chem. Lett.*, 2012, **3**, 1941.
19 G. Gitzinger, M. E. Corrales, V. Loriot, G. A. Amaral, R. de Nalda and L. Bañares, *J. Chem. Phys.*, 2010, **132**, 234313.
20 G. Gitzinger, M. E. Corrales, V. Loriot, R. de Nalda and L. Bañares, *J. Chem. Phys.*, 2012, **136**, 074303.
21 A. T. J. B. Eppink and D. H. Parker, *J. Chem. Phys.*, 1998, **109**, 4758.
22 A. T. J. B. Eppink and D. H. Parker, *J. Chem. Phys.*, 1999, **110**, 832.
23 R. de Nalda, J. Durá, A. García-Vela, J. G. Izquierdo, J. González-Vázquez and L. Bañares, *J. Chem. Phys.*, 2008, **128**, 244309.
24 A. T. J. B. Eppink and D. H. Parker, *Rev. Sci. Instrum.*, 1997, **68**, 3477.
25 G. A. Garcia, L. Nahon and I. Powis, *Rev. Sci. Instrum.*, 2004, **75**, 4989.
26 B. J. Sussman, *Am. J. Phys.*, 2011, **79**, 477.
27 G. Balerdi, M. E. Corrales, G. Gitzinger, J. González-Vázquez, I. R. Solá, V. Loriot, R. de Nalda and L. Bañares, EPJ Web of Conferences, 2013, 41, 02035.
28 A. B. Alekseyev, H.-P. Liebermann and R. J. Buenker, *J. Chem. Phys.*, 2011, **134**, 044303.

Faraday Discussions RSC Publishing

PAPER

Formation of fragment ions (H^+, H_3^+, CH_3^+) from ethane in intense femtosecond laser fields – from understanding to control

Nora Schirmel, Nicola Reusch, Philipp Horsch and Karl-Michael Weitzel*

Received 16th December 2012, Accepted 17th January 2013
DOI: 10.1039/c3fd20152j

The dissociative ionization of ethane in intense femtosecond laser fields has been investigated as a function of the laser pulse shape by systematically varying the quadratic spectral phase, *i.e.* the linear chirp. A very pronounced effect of the sign of the chirp is observed for the parent ion and all fragment ion yields, all ion yields being strongly favored by negative chirp of the laser field. The ratio of the H_3^+ ion yield to H^+ ion yield can also be manipulated by changing the linear chirp, the maximum being observed for a significantly smaller chirp value than that for the individual ion yields. Since the H^+ ions and the H_3^+ ions predominantly originate from the dication of ethane, this indicates control of fragmentation within one charge state of the ethane. Additional experiments performed with d_3-ethane demonstrate that the control is operative prior to the statistical scrambling of hydrogen atoms, further supporting the concept of intra-charge-state control. In the case of formation of CH_3^+ ions two different ensembles occur, one from the monocation, another from the dication. The ratio of these ensembles can again be controlled by means of the linear chirp parameter implying control between the two different charge states (inter-charge-state control).

Introduction

The dissociative ionization of small organic molecules has attracted considerable interest in recent years.[1–4] Depending on the laser pulse parameters, in general, formation of monocations but also multiple ionization as well as excessive fragmentation can be observed. A well investigated molecular system in this respect is the ethane molecule.[4–8] The dissociative ionization of ethane leads *e.g.* to the formation of H^+, H_2^+, $CH_n^+ (n = 0–3)$ and $C_2H_m^+ (m = 0–5)$. Yet, much of the attention dedicated to the dissociative ionization of ethane and, in fact, our own interest are connected to the formation of a rather intriguing fragment ion, the H_3^+. Its equilibrium geometry is that of an equilateral triangle, the binding energy

---

Philipps-Universität Marburg, Fachbereich Chemie, Hans-Meerwein-Str., 35032 Marburg, Germany. E-mail: weitzel@chemie.uni-marburg.de; Fax: +49 (0)64212822360; Tel: (0)64212822527

with respect to $H_2 + H^+$ is in the order of 4.6 eV.[9] The H_3^+ ion is observed in the dissociative ionization of ethane as well as several other organic molecules. There, its formation involves not only the breaking of three C–H-bonds and the formation of one three-center-two-electron bond but also ultrafast hydrogen migration.[6–8,10,11] H_3^+ itself has first been reported by J. J. Thomson in a discharge experiment.[12] Today H_3^+ receives considerable attention in fields connected, *e.g.*, to kinetics[13–15] and interstellar chemistry.[16,17] In fs-laser ionization the first observation of H_3^+ was most likely accomplished by Xia *et al.*[5] In that work signal at mass-to-charge ratio $m/z = 3$ was assigned to C^{4+}, but in the light of today's knowledge it probably was rather H_3^+.

In the following we first review past attempts to understand the molecular dynamics involved in the dissociative ionization of ethane and some related molecules, before turning our attention to the goal of the current work, *i.e.* the attempt to control these dynamics.

The importance of ultrafast hydrogen migration in femtosecond laser ionization processes was emphasized in a systematic study about different hydrocarbons with and without methyl groups[7] as well as in H_3^+ formation experiments looking at methanol and partially deuterated methanol (CD_3OH).[10,18] To the best of our knowledge the first complete reaction path for the formation of H_3^+ in the dissociative fs-laser ionization of ethane was reported by Kraus *et al.* in a combined theoretical and experimental study.[6] In this study, quantum chemical calculations and supporting measurements of kinetic energy releases (KER) allowed to identify a reaction path, which involves a doubly charged $C_2H_6^{2+}$ ion evolving over a transition state (TS) leading to the formation of $H_3^+ + C_2H_3^+$. The most stable structure of $C_2H_6^{2+}$ can be described as a donor–acceptor complex of a planar methane dication (CH_4^{2+}) and a carbene (CH_2) which is consistent with the results of previous studies.[19] The calculated structure of the transition state (TS) consists of a neutral H_2 molecule which is attached to a $C_2H_4^{2+}$ entity, nevertheless this TS does not correlate to the $H_2^+ + C_2H_4^+$ or $H_2 + C_2H_4^{2+}$ channel but exclusively to $H_3^+ + C_2H_3^+$. A similar TS ($C_3H_2^{2+}\cdots H_2$) was proposed theoretically for the dissociation of the allene dication.[20] There, a hydrogen molecule was discussed to roam around the $C_3H_2^{2+}$ dication consequently leading either to the loss of a neutral H_2 or to the formation of H_3^+, the latter, however, *via* another TS.

The pathway indicated above as well as the kinetic energy releases were later confirmed in a photoion–photoion coincidence (PIPICO) experiment detecting coincidences between H_3^+ and $C_2H_3^+$ from ethane (CH_3CH_3).[8] Additional experiments performed for partially deuterated ethane (CD_3CH_3) allowed to analyze the yield ratios of $H_3^+ : H_2D^+ : HD_2^+ : D_3^+$ as $(8.01 \pm 1.20) : (43.0 \pm 5.4) : (43.1 \pm 0.3) : (5.89 \pm 0.10)$, which clearly show significant, almost statistical H-D scrambling through hydrogen migration.[8]

In the current work we present an attempt to go beyond understanding the dissociative ionization of ethane toward control of ion yields in this dissociative ionization. The control of fragmentation ratios in optically driven molecular dynamics has received enormous attention in the last decades and is in fact also the topic of the current Faraday Discussion meeting.[21] While the overall state of this field may be regarded as mature, several specific challenges remain. A common challenge is connected to the question in which charge state of the molecular system the control is operative. The other challenge pertains to the approach, with which the control is achieved. There are examples where control has been achieved by employing genetic algorithms,[22] but also examples where the shape of the laser pulses has been systematically varied (*e.g.*, linear chirp).[23] The question of which of these approaches is to be preferred is almost philosophical in nature. The former in general leads to the "optimum" field for maximizing the fitness of a specific objective (*e.g.*, the ratio of two product yields). The latter in general leads to a local maximum of that objective in parameter space which may be more descriptive and easier to interpret.

In this contribution we describe the control of fragmentation channels employing a systematic variation of the laser pulse shape. Recent work employing the same concept includes systematic variation of the linear chirp[24–27] as well as binary pulse shaping.[28,29] Itakura *et al.*[24] reported on the dissociative ionization of ethanol in chirped laser fields finding that nuclear dynamics can be controlled on the singly charged state. However, positive and negative chirp led to the same fragment ratios, *i.e.* the control originated from laser intensity variation. Marthur *et al.*[25] investigated the dissociative ionization of methane in chirped laser fields reporting fragment ion yields as a function of the linear chirp parameter α. The results indicated a predominantly intensity-driven ion yield, superimposed on which a small sign effect of the chirp parameter was observed. Ultimately, the yields of H^+ and H_2^+ – presumably originating from dissociation of dications – were favored by a positive chirp. Our group has reported the control of the dissociative ionization of DCl by linearly chirped fs-laser pulses, where a positive chirp favored the formation of D^+ while a negative chirp favored the formation of Cl^+.[26] On one hand, it appears important to add further experimental studies hoping to lay a foundation for future theoretical work. On the other hand the observation of linear chirp effects in dissociative ionization has obvious applications in chemical analytics, that is, it is possible to distinguish structural isomers, *e.g.* *o*-xylene and *p*-xylene by ionization employing either linearly chirped[27] or binary shaped[28] fs-laser pulses.

The current work will concentrate on the dissociative ionization of ethane with particular attention on the formation of H^+, H_3^+ and CH_3^+. We will attempt to control total ion yields as well as ratios of ion yields by means of linearly chirping fs-laser fields. From previous work we know that formation of H_3^+ proceeds on the dicationic potential energy surface (PES). The H^+ formation most likely proceeds from a different conformational region of the same dicationic PES. These data will be complemented by a study of CH_3^+ formation, which may either occur on the monocationic or the dicationic PES. A major part of the studies will deal with normal ethane (C_2H_6). Additional experiments performed for d_3-ethane (CD_3CH_3) will shed light on the question whether control is operative prior to or after hydrogen scrambling.

Experimental

The dissociative ionization of ethane has been investigated *via* systematic variation of the linear chirp of fs-laser pulses. These fs-laser pulses were generated in a chirped pulse amplifier (ODIN, Quantronix, repetition rate 1 kHz) seeded by a Ti:Sapphire oscillator (SYNERGY, Femtolasers). The pulses were phase corrected using a spatial light modulator (SLM-S640, Jenoptik) inside a 4f-shaper setup, resulting in nearly Fourier-limited pulses with a typical pulse duration of 45 fs, a central wavelength of 810 nm and a typical pulse energy up to 140 μJ. These pulses were linearly chirped by application of a quadratic spectral phase onto the spatial light modulator. We have also studied quadratically chirped laser pulses by applying a cubic spectral phase to the SLM, however this will not be at the focus of the current work. The corresponding Taylor parameter is $\alpha = 1/2!\cdot\partial^2\varphi/\partial\omega^2$ for linear chirp ($\beta = 1/3!\cdot\partial^3\varphi/\partial\omega^3$ for quadratic chirp), where φ is the spectral phase and ω is the central angular frequency. Note that imprinting a linear chirp onto the laser field is equivalent to stretching the pulse in the time domain according to the relation $\tau^2 = \tau_0^2 + \left(8\ln 2\,\alpha/\tau_0\right)^2$,[30] with τ_0 being the pulse duration of an unchirped laser pulse.

The laser pulses were characterized using a frequency-resolved optical gating technique (GRENOUILLE, Swamp Optics). After pulse-shaping the beam is transmitted through two 1 mm thick optics; one being a beam splitter to couple into the GRENOUILLE and one being a quartz glass window of a vacuum chamber. The shaped and characterized laser pulses were focused *via* a cylindrical mirror ($f = 75$ mm) into the ion source of a home build Wiley McLaren time-of-flight mass spectrometer leading to the ionization of ethane under space focus conditions.[31] Ethane or d^3-ethane was effusively introduced into the reaction chamber at typical pressure values of 3×10^{-5} mbar. Ethane pressure tins with 99.95% purity were purchased from Fluka and used without further purification. Deuterated ethane was purchased from CDN Isotopes (99.3atom% D).

Ion yields of the parent and fragment ions were recorded as a function of the chirp parameter of interest. The experiments were performed for a range of laser pulse energies between 15 μJ and about 130 μJ. The corresponding peak intensities are 6.6×10^{12} W cm^{-2} and 5.7×10^{13} W cm^{-2}, respectively, both numbers referring to the shortest laser pulses employed, *i.e.* about 45 fs. Typically 10 000 mass spectra were averaged for each set of laser pulse parameters.

Results and discussion

In the following we will first present typical time of flight mass spectra (TOF-MS) of normal h_6-ethane as well as d_3-ethane. We will then discuss the ion yields as well as ratios of ion yields as a function of the linear chirp parameter α for h_6-ethane. In a third section we will demonstrate the influence of the laser pulse energy on the ability to control molecular dynamics. Finally, we will present the corresponding data for d_3-ethane.

Time of flight mass spectra

The TOF spectra of normal h_6-ethane and deuterated d_3-ethane have been recorded as a function of the linear chirp of the fs-laser pulses. Typical TOF spectra for both ethane species are presented in Fig. 1. Major signals arise from the C_2H_m (m = 0–6) group and the CH_n^+ (n = 0–3) group for ethane and the corresponding $C_2H_pD_q^+$ (p,q = 0–3; $p + q \leq 6$) and $CH_pD_q^+$ (p,q = 0–3; $p + q \leq 3$) groups for d_3-ethane. Of particular interest in the current study is the formation of the hydrogen ions H^+, H_2^+ and H_3^+ in the case of ethane. The corresponding ion signals appearing at m/z equal to 1, 2, and 3, respectively, are unique under all conditions investigated. The corresponding hydrogen ions in the case of d_3-ethane are H_3^+, H_2D^+, HD_2^+ and D_3^+. In that case ion signals are observed

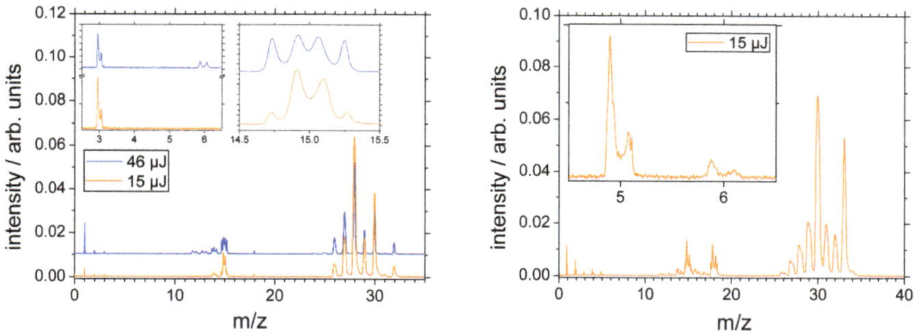

Fig. 1 TOF-MS of $C_2H_6$for 46 μJ (blue) and 15 μJ (orange) at 6.0 × 10^{-6} mbar and 3.0 × 10^{-5} mbar, respectively (left). TOF-MS of $CH_3CD_3$for 15 μJ (right). All mass spectra were recorded for negatively chirped laser pulses.

at $m/z = 1$ (H^+), 2 (H_2^+, D^+), 3 (H_3^+, HD^+), 4 (H_2D^+, D_2^+), 5 (HD_2^+) and 6 (D_3^+), as has been reported in previous studies.[8] Thus, the ion signals at m/z equal to 2, 3 and 4 are not unique, but those at m/z equal to 5 and 6 are, provided that contributions from the doubly charged carbon ion C^{2+} ($m/z = 6$) can be ruled out. As Fig. 1 shows, normal ethane does not show any signal at $m/z = 6$ at laser pulse energies of 15 μJ. Consequently, we assume that d_3-ethane experiments performed at the same laser pulse energy should also be free from contributions of C^{2+}. Note that normal ethane experiments can still be analyzed at considerably higher pulse energies, as long as C^{4+} ($m/z = 3$) can be excluded, which is the case for all pulse energies discussed here. The mass spectra in general contain contaminations from oxygen ($m/z = 32$) and water ($m/z = 18$), however they do not interfere with any of the interpretations.

Control of ion yields and ratios of ion yields from ethane *via* linear chirping

In Fig. 2 (left) the ion yields of the parent ion and the fragment ions CH_3^+, H_3^+, and H^+ from ethane are shown as a function of the pulse duration τ (lower axis) and the corresponding linear chirp parameter α (upper axis) at a pulse energy of

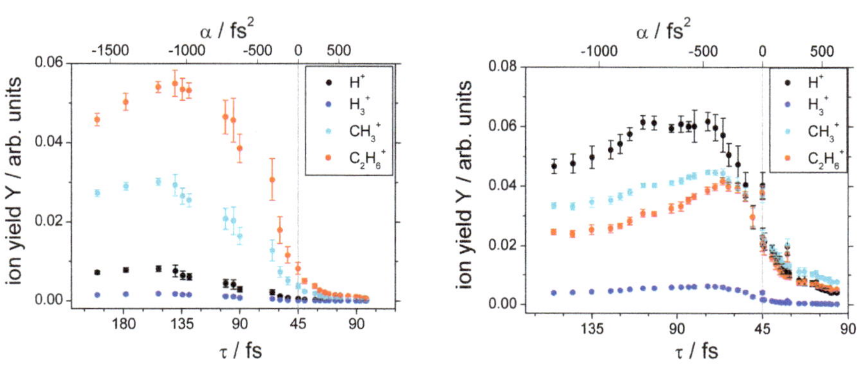

Fig. 2 Ion yields Y of the parent ion ($C_2H_6^+$), as well as the fragment ions CH_3^+, H_3^+ and H^+ from ethane as a function of the pulse length (FWHM, τ) and the linear chirp parameter α for a pulse energy of 15 μJ (left) and 46 μJ (right).

15 μJ. Evidently, all ion yields, including the parent ion yield are rather small for positive chirp, but increase significantly with increasing negative chirp. The ion yields appear to reach a maximum for negative chirp parameters around −1200 fs^2 before decreasing again with decreasing α. Thus, these data demonstrate a highly pronounced asymmetry of the ion yields for different signs of α, *i.e.* the formation of ions is not simply driven by the laser peak intensity. Even the yield of the parent ion is not maximized for the shortest pulses ($\tau = 45$ fs, $\alpha = 0$ fs^2), but rather for pulse durations of approximately 110–153 fs and negative α. This observation holds true for the total ion yield (not shown), which is maximized for ($\tau = 153$ fs, $\alpha = -1186$ fs^2).

In Fig. 2 (right) the corresponding ion yields are shown at a slightly higher laser pulse energy of 46 μJ. Several features deserve to be emphasized. First, the most intense ion (among the four ions shown) is now the H^+ ion. This indicates that the amount of fragmentation increases with increasing laser pulse energy. Second, the overall sign effect of the chirp control persists. However, the chirp parameter for which specific ion yields are maximized is clearly shifted to smaller α values. While at 15 μJ for positive chirp basically all ion signals vanish – making any further analysis numerically difficult – the data at 46 μJ have the advantage that finite signal intensities are also observed for positive chirp values. We will come back to these aspects below.

The fact that the yield of a parent ion is maximized for a pronounced negative chirp, implying a rather long laser pulse, is quite remarkable as the parent ions of methane[25] or ethanol[24] and several aromatic molecules[23,24,27] were shown to be maximized for the shortest laser pulse. To the best of our knowledge, this is the first observation of such a pronounced influence of the sign of the chirp on the ionization efficiency of an organic compound such as ethane. One important difference between the current work and the studies cited above (apart from that for xylene[27]) lies in the laser pulse intensity. While the literature work cited has been performed in the regime of 10^{15} to 10^{16} W cm^{-2}[23–25] our intensity was on the order of 10^{13} W cm^{-2}. As we will show below – in particular for the H_3^+ formation – most of the control effects depending on the sign of α tend to be smeared out at higher laser pulse energies.

The very pronounced increase of the parent ion yield for negatively chirped pulses suggests that we are dealing with a control of ionization efficiency for ethane and d_3-ethane *via* the laser field, more specifically the linear chirp parameter. In order to gain further insight at which point of the dissociative ionization the control is operative we have analyzed various ratios of ion signals, in particular those involving fragment ions. To this end, Fig. 3 addresses the ratio of the parent ion yield to the total ion yield. While the parent ion yield is not maximized for the shortest pulses the ratio of parent to total ion yield shows a maximum for the shortest pulse. Since the ratio decreases with increasing chirp, we conclude that fragmentation is enhanced for longer pulses, which is in agreement with literature reports.[23,24] The data shown in Fig. 3 suggest that there might be an effect of the sign of the chirp on the amount of fragmentation. However, the data for positive chirp have large error bars, since ion signals are vanishing there.

Clearly, both the enhancement of ionization efficiency and the enhancement of fragmentation with increasing negative chirp parameter are operative. The question remains: which part is more important for which molecular process. If

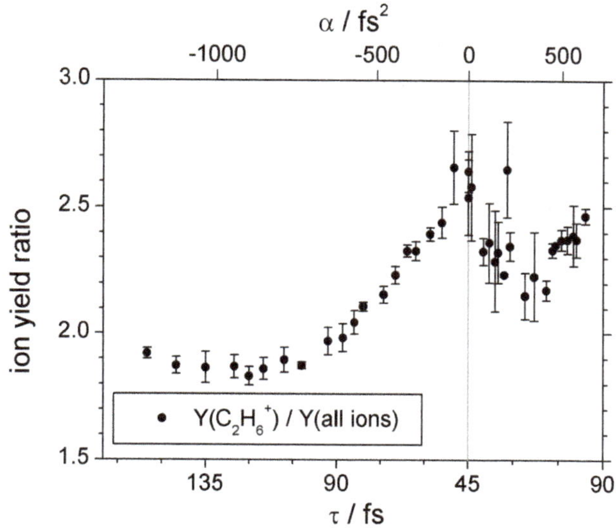

Fig. 3 Ratio of the parent ion yield to the total ion yield of ethane at 46 μJ. All fragments have been taken into account.

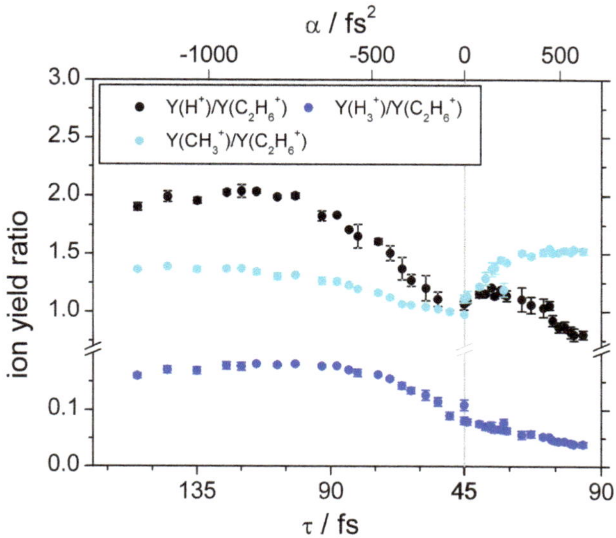

Fig. 4 Several fragment to parent ion yield ratios as a function of pulse duration and corresponding linear chirp parameter. Data taken at 46 μJ.

we assumed the fragmentation to be only dependent on the pulse duration, we would expect no dependency on the sign of the linear chirp parameter for a plot of a specific fragment ion yield normalized to the parent ion yield. As shown in Fig. 4 for different fragment ion to parent ion yield ratios this is not the case. Evidently, all normalized fragment ion yields shown (H^+, H_3^+, CH_3^+) exhibit pronounced sign effects of the linear chirp. The formations of H^+, H_3^+ and CH_3^+ all appear to be favored by negative chirp, the formation of CH_3^+ is also favored by positive chirp, at least at 46 μJ. Since large chirp parameters α correlate with increased laser pulse duration, one might consider the possibility that enhanced ion yields might be influenced by charge resonance enhanced ionization (CREI).[32] On one hand, we do not easily see that CREI would depend on the sign of the chirp parameter. On the other hand, *e.g.* for the formation of H_3^+ on the dicationic PES there is no corresponding accessible region on the monocationic PES from which CREI could effectively start.

Note that there are literature reports of influence of the sign of the linear chirp for the generation of fragments, *e.g.* for NO_2^+ ions ($m/z = 46$) from *p*-nitrotoluene[23] and for the H_2^+ formation from methane,[25] all resulting from highly charged precursors. Further, the fluorescence intensity of N_2^+ ions can be enhanced by applying a negative chirp in the multiphoton ionization of N_2.[33] The selective population of dressed electronic states of an atom can also be controlled by means of linear chirp.[34]

In the following we discuss further characteristics of these ion yields, in particular ratios of fragment ion yields as a function of the linear chirp parameter. First, we show the ratio of ion yields $Y(H_3^+)$ / $Y(H^+)$ as a function of the linear chirp parameter and the corresponding pulse duration in Fig. 5. Evidently, the ratio $Y(H_3^+)$ / $Y(H^+)$ goes through a maximum in the region of negative chirp for both laser pulse energies displayed. The maximum in the yield ratio occurs around $\alpha = -750$ fs^2 for 15 μJ and is shifted toward $\alpha = -300$ fs^2 at 46 μJ. With further increasing negative chirp the ratio $Y(H_3^+)$ / $Y(H^+)$ decreases, *i.e.* the H^+ channel is favored. For positive chirp – as mentioned above – the ion signals are so small at 15 μJ that it is impossible to extract reliable information from the ion yield ratios due to the large error bars. In fact, this is part of the reason why also data for a larger pulse energy are presented. Here, very clearly the error bars are reasonable and we conclude that the ratio of $Y(H_3^+)$ / $Y(H^+)$ decreases in going to positive chirp parameters.

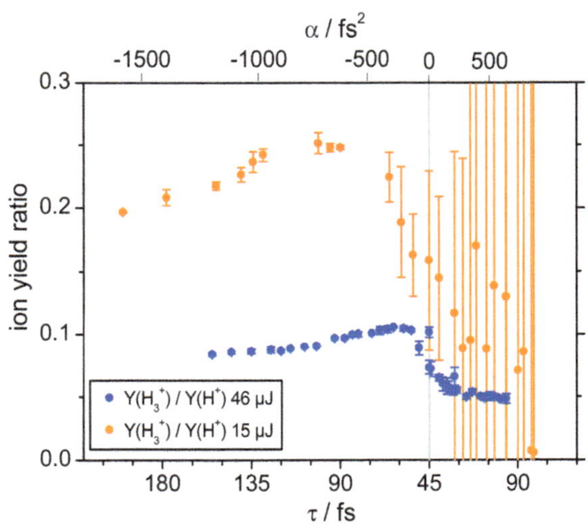

Fig. 5 Ion yield ratio of $Y(H_3^+)$ / $Y(H^+)$ ion as a function of pulse duration and the corresponding linear chirp. Data taken at 15 μJ and 46 μJ.

While the origin of the H_3^+ is clear from previous work – it originates from a decay of the ethane dication[8] – the origin of the H^+ requires additional discussion. Assuming this H^+ might also originate from the dication one would first expect coincidences between H^+ and $C_2H_5^+$. However, Kanya *et al.* observed very few coincidences between H^+ and $C_2H_5^+$.[8] The angular distribution for the H^+ coincidence signal was strongly anisotropic, implying that its precursor has a lifetime shorter than the rotational period of the precursor (about 700 fs). The angular distribution of the H_3^+ coincidences on the other hand was isotropic implying that the lifetime of the precursor was larger than the rotational period. Yet, another puzzling detail was that the yield of H^+ was significantly larger than that of H_3^+. Further in the current work we find that the average kinetic energy of the H^+ is on the order of 6 eV at a pulse energy of 15 μJ, not compatible with the decay of a monocation, but only with Coulomb explosion. A possible solution to this mystery is offered in a paper by Eland, where it was shown that a significant amount of H^+ signal is observed in coincidence with the $C_2H_3^+$ ion![35] Possible mechanisms are a concerted three-body dissociation of $C_2H_6^{2+}$ or a consecutive mechanism in which either the H^+ comes off first with subsequent decay of the $C_2H_5^+$ into $C_2H_3^+ + H_2$, or the neutral H_2 comes off first with the intermediate $C_2H_4^{2+}$ ion subsequently breaking up into $H^+ + C_2H_3^+$. The analysis by Eland suggests that the latter process dominates. We note the average kinetic energy of the H^+ ion is on the order of 6 eV which is compatible with the conclusion above. Kraus *et al.* reported that the Coulomb energy expected from the dissociation of the dication in the eclipsed conformation of neutral ethane is 6.4 eV.[6] In fact, there are different conformations representing local minima on the dicationic PES that do not only differ in nuclear coordinates but also in electron density. Thus, the control of H_3^+ *versus* H^+ yields contains elements of electron density control. The results presented and discussed above provide evidence that we are able to control the fragmentation ratio of two reaction channels taking place on the same dicationic PES.

We now turn to the CH_3^+ ion signal, which exhibits a very peculiar TOF structure (*cf.* Fig. 1). Clearly, there are two different ensembles of CH_3^+ ions formed, a slow one with an average energy of 0.7 eV (inner two peaks) and a fast one with 2.7 eV (outer two peaks). Most likely the former ensemble originates from a decay of the monocation, the latter is due to the Coulomb explosion of the dication leading to $CH_3^+ + CH_3^+$. As shown in the inset of Fig. 1 (left) the relative amount of the slow and the fast CH_3^+ ions changes significantly as the laser pulse energy is increased, the fast component being favored by higher pulse energy. However, the relative amount of slow and fast CH_3^+ ions also depends on the linear chirp parameter. We demonstrate this in Fig. 6 by plotting the ratio of the slow to the fast CH_3^+ ions as a function of the linear chirp parameter α for laser pulse energies of 15 μJ and 46 μJ. At 15 μJ the relative fraction of slow CH_3^+ is significantly higher than at 46 μJ. At both pulse energies this ratio exhibits an S-shaped characteristic. At 15 μJ a maximum occurs around -500 fs^2, a minimum in the region of the transform limited pulse. At 46 μJ a maximum occurs around -500 fs^2, but the minimum is now shifted toward positive α around +250 fs^2. Concentrating on the negative chirp regime, where the signals are strongest, we conclude that increasing negative chirp appears to favor the formation of dications *versus* that of monocations. We interpret this as an example of inter-charge-state control.

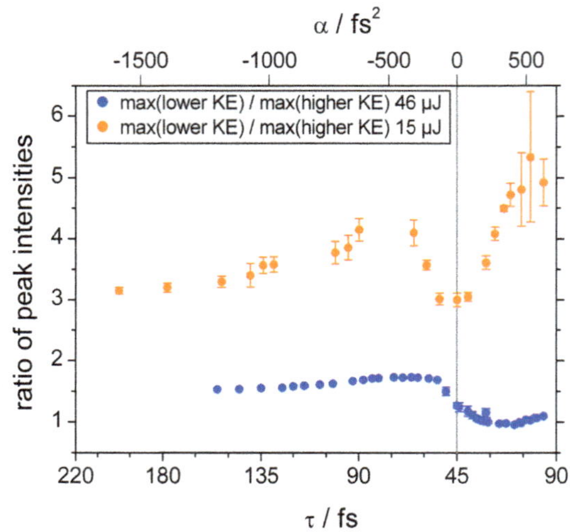

Fig. 6 Ratio of the peak intensities for the slow CH_3^+ divided by that of the fast CH_3^+ as a function of the pulse duration, respectively the chirp parameter.

The role of the laser pulse energy

In the following we discuss the H_3^+ formation from ethane as a function of the laser pulse energy. In Fig. 7 (left) normalized ion yields of H_3^+ are shown as a function of α and τ for different laser pulse energies. With higher laser pulse energies the value of α for the largest H_3^+ yield decreases. Consequently, for the highest pulse energy of 130 μJ we observe the highest ion yields for a vanishing chirp. Still, the ion yields depend on the sign of the chirp, and are higher for negatively chirped pulses as compared to positively chirped pulses.

On the right part of Fig. 7 the α values for which maximized H_3^+ yields are observed are shown as a function of laser pulse energy. This again serves to illustrate that this $\alpha(Y(H_3^+)_{max})$ decreases with increasing pulse energy. Note that

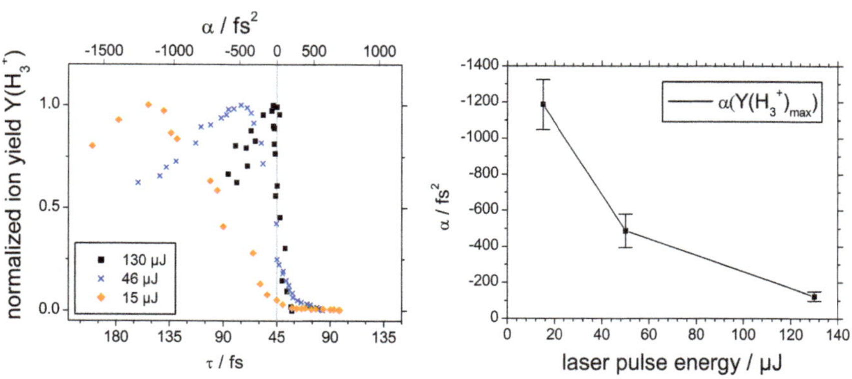

Fig. 7 Normalized ion yields of H_3^+ for different laser pulse energies from 130 to 15 μJ (left); α values, for which a maximum of H_3^+ ion yield is observed, as a function of the laser pulse energy (right). The connecting line serves to guide the eye.

in several chirp dependent experiments the relevant chirp effects observed were more pronounced at lower laser pulse energies.[27,36]

Formation of HD_2^+ and D_3^+ from d_3-ethane

We have demonstrated above that in the case of the CH_3^+, control of population is possible between different charge states (inter-charge-state control). For the ratio of H_3^+ to H^+ we demonstrated the possibility to control ion yield ratios within the dicationic state (intra-charge-state control). At this point the question remains whether the control affects the hydrogen scrambling known to dominate the statistical part of the dissociation dynamics leading to the formation of H_3^+ and $C_2H_3^+$.[8] Kanya *et al.* employed deuterium labeling of the ethane in order to resolve the issue of hydrogen scrambling. d_3-ethane with all three deuterons connected to the same carbon atom (CD_3CH_3) form four different $(H,D)_3^+$ species, *i.e.* H_3^+, H_2D^+, HD_2^+, and D_3^+ with a ratio of $8.01 \pm 1.20 : 43.0 \pm 5.4 : 43.1 \pm 0.3 : 5.89 \pm 0.10$, resembling almost statistical hydrogen scrambling prior to passing the relevant TS for $(H,D)_3^+$ formation. Here, we have also performed systematic chirp studies on d_3-ethane. Fig. 8 illustrates the impact of the linear chirp on the parent ion of d_3-ethane as well as fragment ions at $m/z = 1$ (H^+), 3 (H_3^+, HD^+), 4 (H_2D^+, D_2^+), 5 (HD_2^+), and 6 (D_3^+). The linear chirp dependence is very similar to normal ethane for all ions (Fig. 2). As indicated ion signals at $m/z = 3$ and 4 are not unique because they can originate from a three-atomic and a two-atomic species. However, the ion signals at $m/z = 5$ and 6 are unique. If the chirp dependent control influences the hydrogen scrambling, then the ratio of HD_3^+ yield to D_3^+ yield should exhibit marked variations.

As shown in Fig. 9 the ratio $Y(HD_2^+) / Y(D_3^+)$ is in fact independent of the linear chirp parameter between values of -1600 fs^2 and about -250 fs^2. The value of this ratio is about 6 : 1, very close to the experimental ratio reported by Kanya *et al.*[8] Thus, we conclude that the control between H_3^+ and H^+ discussed above for normal ethane is operative prior to the hydrogen scrambling. In fact, this is not surprising, since we would not expect coherent control for a statistical process.

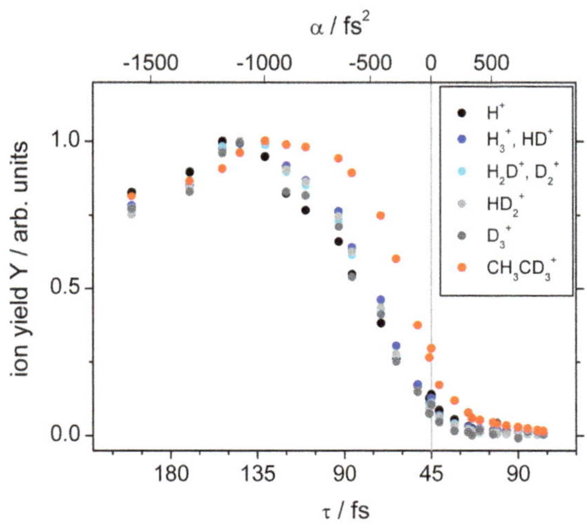

Fig. 8 Normalized ion yields of several ions generated form d_3-ethane at 15 μJ. All ion yields are normalized to their maximum value.

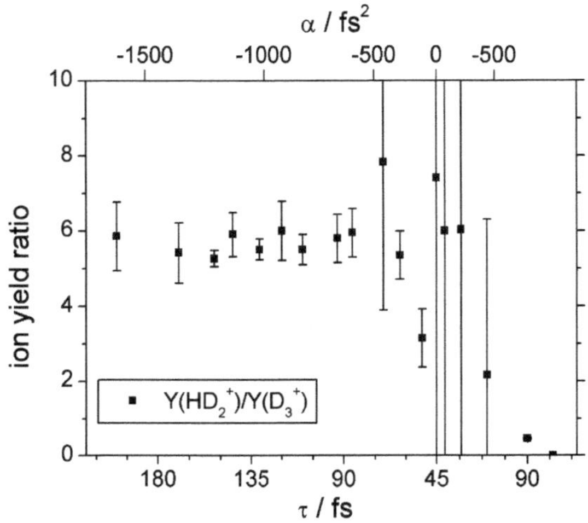

Fig. 9 Dependence of the ion yield ratio $Y(HD_2^+)$ / $Y(D_3^+)$ for d_3-ethane.

Summary and conclusions

We have investigated the dissociative ionization of h_6-ethane and d_3-ethane by means of shaped fs-laser pulses. Particular attention was given to the yields of the monocationic parent ion and the fragment ions H^+, H_3^+ and CH_3^+. Their yields have been recorded as a function of the linear chirp parameter α. A very pronounced sign effect of the chirp parameter is observed for all ion yields, a maximum of the yields being observed for negative chirp around -1000 fs^2. The position of this maximum shifts toward $\alpha = 0$ as the laser intensity is increased. Thus, for larger laser pulse energies the dynamics become more and more intensity driven.

The ratio of H_3^+ ion yields to H^+ ion yields can be controlled by means of the linear chirp parameter. Since there is evidence that both fragmentation channels occur on the dicationic PES this represents a case of control of dissociation dynamics within one charge state. The results shed new light on the complex PES of dicationic ethane with conformations not only differing in the nuclear coordinates but most of all also in the electron density. Ultimately, the ability to control the ratio H_3^+ to H^+ may be connected to the control of electron density.

On the other hand, the TOF spectra for CH_3^+ ions clearly indicate two different ensembles, one originating from the monocationic, the other from the dicationic channel. The ratio of these two channel yields can be manipulated by changing the laser pulse energy. This might be considered a trivial case of control. More interestingly, the ratio of the two CH_3^+ ensembles can also be controlled by the linear chirp parameter. There again a sign effect of α is observed implying that the control is not merely based on laser intensity variation. These results demonstrate that we are able to control the formation of different charge states *via* the linear chirp parameter α.

The current work appears to be one of the rare cases where very pronounced sign effects for parent ion and fragment ion yields are reported in dissociative fs-laser ionization. Lozovoy *et al.* reported on systematic chirp investigations for a number of organic molecules.[23] In most cases no significant sign effects were observed, the exemption being the formation of NO_2^+ from *p*-nitrotoluene, which was also strongly favored by negative chirp.

For the future we hope that the current experimental results stimulate further theoretical work which would have to take into account strong field ionization dynamics as well as dissociation dynamics of a monocation and a dication within the same conceptual approach. This still appears to be a challenge for the time being.

Acknowledgement

We gratefully acknowledge technical assistance from Kerstin Bücker.

References

1 S. A. Buzza, E. M. Snyder, D. A. Card, D. E. Folmer and A. W. Castleman, *J. Chem. Phys.*, 1996, **105**(17), 7425–7431.
2 M. J. Dewitt, D. W. Peters and R. J. Levis, *Chem. Phys.*, 1997, **218**(1–2), 211–223.
3 C. Kosmidis, P. Tzallas, K. W. D. Ledingham, T. McCanny, R. P. Singhal, P. F. Taday and A. J. Langley, *J. Phys. Chem. A*, 1999, **103**(35), 6950–6955.
4 K. Hoshina, Y. Furukawa, T. Okino and K. Yamanouchi, *J. Chem. Phys.*, 2008, **129**(22), 104302-1-6.
5 L. Xia, H. Z. Ren, M. Ri, J. X. Chen, Y. Hong and Q. H. Gong, *Chin. Phys.*, 2004, **13**(9), 1564–1568.
6 P. M. Kraus, M. C. Schwarzer, N. Schirmel, G. Urbasch, G. Frenking and K.-M. Weitzel, *J. Chem. Phys.*, 2011, **134**(11), 114302.
7 K. Hoshina, H. Kawamura, M. Tsuge, M. Tamiya and M. Ishiguro, *J. Chem. Phys.*, 2011, **134**(6), 064324.
8 R. Kanya, T. Kudou, N. Schirmel, S. Miura, K.-M. Weitzel, K. Hoshina and K. Yamanouchi, *J. Chem. Phys.*, 2012, **136**(20), 204309.
9 D. Talbi and R. P. Saxon, *J. Chem. Phys.*, 1988, **89**(4), 2235–2241.
10 T. Okino, Y. Furukawa, P. Liu, T. Ichikawa, R. Itakura, K. Hoshina, K. Yamanouchi and H. Nakano, *J. Phys. B: At., Mol. Opt. Phys.*, 2006, **39**(13), S515–S521.
11 D. Mathur and F. A. Rajgara, *Laser Phys.*, 2009, **19**(8), 1686–1690.
12 J. J. Thomson, *Philos. Mag., Ser. 5*, 1912, **24**(140), 209–254.
13 S. Laube, A. Le Padellec, O. Sidko, C. Rebrion-Rowe, J. B. A. Mitchell and B. R. Rowe, *J. Phys. B: At., Mol. Opt. Phys.*, 1998, **31**(9), 2111–2128.
14 O. Pysanenko, O. Novotny, P. Zakouril, R. Plasil, V. Poterya and J. Glosik, *Czech. J. Phys.*, 2002, **52**, 681–694.
15 J. Tabet, S. Eden, F. Gobet, B. Farizon, M. Farizon, S. Ouaskit, P. Scheier and T. D. Mark, *Int. J. Mass Spectrom.*, 2008, **272**(1), 48–56.
16 T. Oka, *Rev. Mod. Phys.*, 1992, **64**(4), 1141–1149.
17 D. Smith and P. Spanel, *Int. J. Mass Spectrom. Ion Processes*, 1993, **129**, 163–182.
18 T. Okino, Y. Furukawa, P. Liu, T. Ichikawa, R. Itakura, K. Hoshina, K. Yamanouchi and H. Nakano, *Chem. Phys. Lett.*, 2006, **423**(13), 220–224.
19 W. Koch, G. Frenking, J. Gauss and D. Cremer, *J. Am. Chem. Soc.*, 1986, **108**(19), 5808–5817.
20 A. M. Mebel and A. D. Bandrauk, *J. Chem. Phys.*, 2008, **129**(22), 224311.
21 P. W. Brumer and M. Shapiro, *Principles of the Quantum Control of Molecular Processes*, Wiley, Hoboken, 2003.
22 A. Assion, T. Baumert, M. Bergt, T. Brixner, B. Kiefer, V. Seyfried, M. Strehle and G. Gerber, *Science*, 1998, **282**(5390), 919–922.
23 V. V. Lozovoy, X. Zhu, T. C. Gunaratne, D. A. Harris, J. C. Shane and M. Dantus, *J. Phys. Chem. A*, 2008, **112**(17), 3789–3812.
24 R. Itakura, K. Yamanouchi, T. Tanabe, T. Okamoto and F. Kannari, *J. Chem. Phys.*, 2003, **119**(8), 4179–4186.
25 D. Mathur and F. A. Rajgara, *J. Chem. Phys.*, 2004, **120**(12), 5616–5623.
26 H. G. Breunig, A. Lauer and K.-M. Weitzel, *J. Phys. Chem. A*, 2006, **110**(20), 6395–6398.
27 G. Urbasch, H. G. Breunig and K.-M. Weitzel, *ChemPhysChem*, 2007, **8**(15), 2185–2188.
28 J. M. Dela Cruz, V. V. Lozovoy and M. Dantus, *J. Phys. Chem. A*, 2005, **109**(38), 8447–8450.
29 J. M. Dela Cruz, V. V. Lozovoy and M. Dantus, *J. Mass Spectrom.*, 2007, **42**(2), 178–186.

30 P. Balling, D. J. Maas and L. D. Noordam, *Phys. Rev. A: At., Mol., Opt. Phys.*, 1994, **50**(5), 4276–4285.
31 W. C. Wiley and I. H. Mclaren, *Rev. Sci. Instrum.*, 1955, **26**(12), 1150–1157.
32 T. Zuo and A. D. Bandrauk, *Phys. Rev. A: At., Mol., Opt. Phys.*, 1995, **52**(4), R2511–R2514.
33 J. Plenge, A. Wirsing, C. Raschpichler, M. Meyer and E. Rühl, *J. Chem. Phys.*, 2009, **130**(24), 244313.
34 M. Wollenhaupt, A. Prakelt, C. Sarpe-Tudoran, D. Liese and T. Baumert, *Appl. Phys. B: Lasers Opt.*, 2006, **82**(2), 183–188.
35 J. H. D. Eland, *Rapid Commun. Mass Spectrom.*, 1996, **10**(12), 1560–1562.
36 P. Horsch, G. Urbasch, K.-M. Weitzel and D. Kröner, *Phys. Chem. Chem. Phys.*, 2011, **13**(6), 2378–2386.

Faraday Discussions RSC Publishing

PAPER

On the possibility of enhanced multiple ionization near conical intersections

P. H. Bucksbaum*[a] and V. Petrovic*[b]

Received 18th December 2012, Accepted 1st February 2013
DOI: 10.1039/c3fd20154f

Recent measurements of the transient laser-induced fragmentation of photoexcited 1,3-cyclohexadiene undergoing isomerization have seen that some of the fragments have higher velocity at the pump–probe delay corresponding to the time of passage through a conical intersection (CoIn). (http://arxiv.org/abs/1212.3728) Here we explore several possible explanations related to the electronic and ionic structure of molecules in the vicinity of CoIns and suggest some experimental tests that could illuminate the mechanism and ultimately control this behaviour.

1 Introduction

A recent paper has reported a new observation that may yield valuable insights concerning the laser-induced control of unimolecular isomerization or dissociation.[1] The experiments employed ultrafast strong field fragmentation to probe the photoinitiated ring opening of 1,3-cyclohexadiene (CHD) to form 1,3,5-hexatriene (HT).[2] Strong-field photofragmentation, also called "Coulomb explosion", is a well-established method to study transient structural changes,[3] and in particular it has been used extensively in previous experiments that studied this reaction in the gas phase.[4–7]

The new observation augmented more conventional fragment analysis based on the charge-to-mass ratio of the fragments with velocity map imaging (VMI) of the fragment kinetic energies.[8] VMI experimental studies of the Coulomb explosion effect on molecules have been reported previously, and are often employed to study the angular and energy distribution of electrons coming from ultrafast strong field ionization.[9–11] When they have been used for ions, they can measure detailed dissociation kinetic energies of simple molecular systems to view different dissociation channels or capture more exotic phenomena such as coherent wave packet motion.[12–16] When combined with mass filtering, they are capable of investigating complex fragmentation patterns in polyatomic molecules.

[a]*Stanford PULSE Institute and the Departments of Physics, Applied Physics, and Photon Science, Stanford University and SLAC National Accelerator Laboratory, 2575 Sand Hill Road, Menlo Park, CA 94025, USA. E-mail: phbuck@stanford.edu; Tel: +01-926-5337*

[b]*Physics Department, Stanford University, Stanford, CA 94305, USA. E-mail: petrovic@stanford.edu; Tel: +01-725-3306*

In the new experiments discussed here, analysis of the fragment kinetic energies showed a dramatic increase in many fragment velocities. The authors concluded that multiple ionization was responsible for this, and speculated that the timing was consistent with the passage through the conical intersection (CoIn) that is thought to be responsible for the non-radiative relaxation into the HT final state.[17]

These new observations, and similar findings in other molecular systems,[13] motivate the main topic of discussion in this paper: What properties of the transient molecule can give rise to a propensity for multiple ionization? Is this mainly due to the electronic degeneracy that occurs at a CoIn? Perhaps this behaviour is more closely related to the molecular distortion, which may form a particularly favourable geometry for multiple electron ejection. Or, perhaps there are large hidden transient resonances, so that the laser field comes into resonance with one or a series of transitions that can lead to rapid absorption of energy from the laser.

1A Conical intersections

Conical intersections are seams of degeneracy in the Born–Oppenheimer potential energy landscapes of polyatomic molecules. They are natural violations of the no-crossing theorem[18] which are present whenever the number of independent degrees of freedom on the B–O surface exceed the capacity of independent non-adiabatic coupling parameters to lift the degeneracy.[19] In the most common case where the degeneracy occurs between only two B–O energy surfaces, the essential Hamiltonian matrix involves only these two states. Then Hermiticity demands that there is only a single off-diagonal coupling, and therefore the $3N$-6 internal degrees of freedom in an N-atom molecule must remain degenerate over a smaller manifold of dimension $3N$-8.

Excited state populations in molecules may relax to the ground state by traversing these seams of degeracy, without having to exchange energy through radiation or collisions. Since the crossing is not impeded by any gap, the natural time scale for the transition is a small fraction of a vibrational period along the reaction coordinate, which means that the transition must happen in 100 fs, or

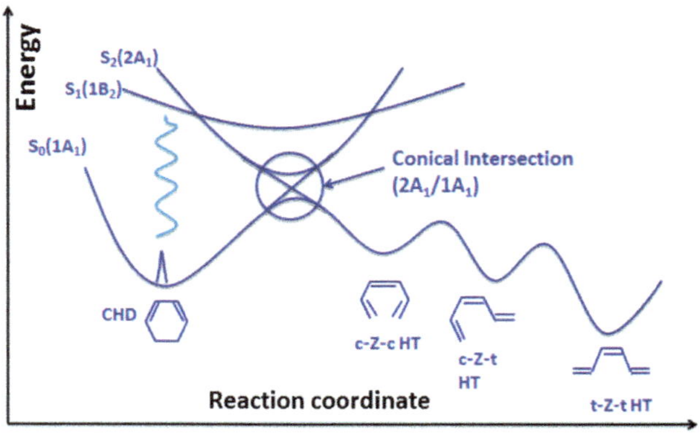

Fig. 1 Simplified reaction path for the photoisomerization of 1,3-cyclohexadiene (CHD) to form hexatriene (HT). Neither energies nor distances are to scale.

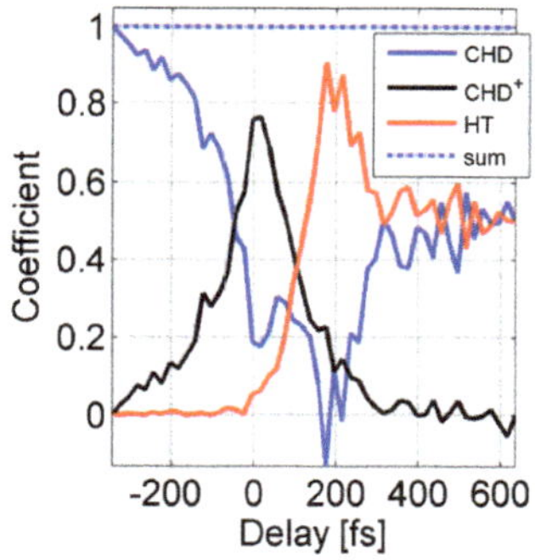

Fig. 2 Relative abundance of the mass fragment TOF spectral signatures associated with CHD and HT following photoexcitation. Reprinted from White, *et al.*,[7] with permission. The mass spectrum at each time delay was decomposed using Principal Component Analysis into a basis of fingerprint spectra. Three were required to account for most of the data. The blue curve corresponds to the parent CHD, which dominates at negative time delays, and still makes up half of the total at long positive delays. The red curve corresponds to HT, which is the other half of the final spectrum at long time delays, but appears to peak at about 200 fs delay. The third is a spectral species that evidently requires the simultaneous presence of the UV pump and the strong infrared probe, because it is only present when they are overlapped. The authors assigned this to the fragmentation of the cation CHD^+, and indeed it is primarily made of this fragment.

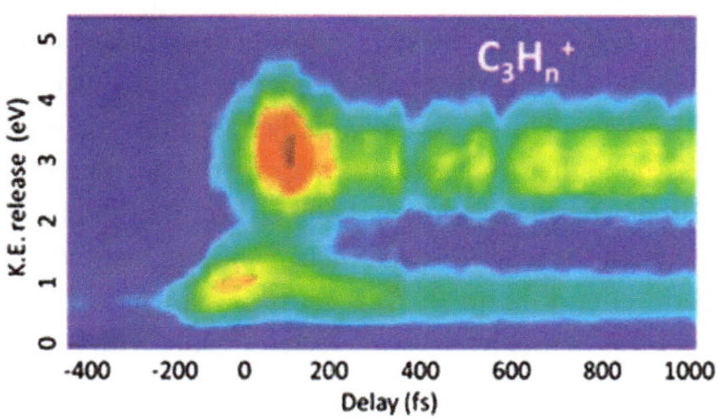

Fig. 3 Detail of the fragment kinetic energies for the C_6H_n ion fragments created by Coulomb explosion of CHD during photo-induced ring opening as described in a recent paper by Petrovic *et al.*[1] The time axis shows the relative delay between the ultraviolet pulse that initiates the ring opening and a strong infrared laser that probes the molecule by Coulomb explosion. A sharp peak in high kinetic energy fragments is seen approximately 150 fs after photoexcitaition. Note that following the peak, this fragment continues to contribute to the signal, suggesting that similar dissociative ionization pathways occur following passage through the vicinity of the CoIn, but have less strength.

faster, for many common cases. It is now established that nonradiative electronic relaxation in polyatomic molecules generally proceeds *via* conical intersections.[20]

2 Summary of the experiments showing enhanced energy fragments

The light-induced ring opening reaction of CHD to form HT can be detected by ion fragment time-of-flight mass spectroscopy. The parent CHD and daughter HT have significantly different fragmentation patterns.[4,7,21] Over a broad range of intensities the fragmentation of CHD by ultrashort 800 nm pulses is dominated by the parent cation, while the daughter displays a prominent mass 1 peak (hydrogen). Other fragments also have characteristic ratios so that a fingerprint spectrum may be developed to differentiate between these species.

2A Enhanced HT spectral signatures

Spectral unmixing and other similar techniques can be employed to isolate the spectral signatures associated with different chemical species. When this analysis was performed, a clear pattern emerged showing that isomerization occurs within approximately 200 fs following the photoexcitation of CHD by 266 nm light pulses.

It is known from this and previous experiments that the isomerization branching percentage for this reaction is approximately 50%, and that only 0.5 ps after photoexcitation this ratio is clearly evident in the TOF data. However, a remarkable increase in the HT component of the TOF spectrum and corresponding decrease in the CHD component occurs in the vicinity of 200 fs. A simple interpretation suggests that the HT channel peaks at 200 fs and then decreases back to its asymptotic ratio of 50%. This marked effect has been noted but not explained.[7]

A significant new piece of evidence has come from the newly reported VMI experiments on the same system, which now show that the excess portion of the fragmentation signature seen at approximately 150 fs delay consists of molecular fragments with higher kinetic energies.[1] This suggests that they come from more highly charged parent molecules. The VMI experiments show that fragment kinetic energies typically occupy two distinct bands of energies. The lower energy fragments and higher energy fragment bands have distinctly different time evolution. In some channels such as H, the higher KE fragments also are the prominent features at later delays, while for other channels, such as $C_4H_n^+$, the high kinetic energy channel persists at later times, but it reaches a distinct maximum around 150 fs.

This high kinetic energy band of fragments is typically separated from the lower kinetic energy fragments by several electron volts, which is consistent with originating in a parent with one higher net charge contributing several eV of additional potential energy. The time delay is characteristic of the passage through the CoIn responsible for the CHD-HT isomerization. These observations led the authors to suggest that the CoIn region has enhanced ionization in the presence of the fragmentation field.

These two observations, enhanced HT signature in White *et al.*,[7] and enhanced kinetic energy in Petrovic *et al.*,[1] seem likely to be related. Although the time delays are different, 200 fs for the HT signature in the White paper, *vs.* 150 fs kinetic energy enhancement in the new Petrovic report, it is not known whether this difference denotes a separate process, or merely a process that has a dependence on parameters such as intensity and time that were different in the two experiments. In the remainder of this paper we will investigate this conjecture.

3 Discussion: strong field methods for enhanced ionization

We wish to investigate the conjecture that multiple ionization is enhanced at or near the vicinity of conical intersections. Although this has not been predicted as an explicit feature of CoIns, nonetheless there are a number of hints and previous physical theories consistent with this idea. Here we list several of the ideas, with the hope that this will guide the Faraday Discussion that accompanies this paper. Each of these will then be described.

1. The *sine qua non* of CoIn physics is the degeneracy of Born Oppenheimer potential energy surfaces. This degeneracy may itself be an important source of enhanced ionization, as we will describe.

2. Enhanced ionization can be induced by strong-field molecular distortion. This has been discussed widely in the context of dissociative ionization, and implies a time delay due to the necessity of molecular motion.

3. Recent strong field control experiments have suggested the importance of a quantum resonance ring or light-induced conical intersection (LICI) that may enhance energy transfer between the field and the system.

4. Finally, recent calculations show that multiple ionizations in strong fields might not be a tunnelling phenomenon at all, but rather are due to the transient localization of charge near an ion in the molecule. This leads to attosecond bursts of multiple ionization.

3A Degeneracy leads to enhanced tunnel ionization

The defining characteristic of a conical intersection is a seam of degeneracy, between two or more potential energy surfaces, and in this section we review calculations and experiments that have suggested that this degeneracy may be responsible for enhanced ionization in molecules. A key feature in enhanced ionization is that the potential energy surfaces that come close to degeneracy involve electronic states of different general shapes. Even in situations where the molecule is so distorted in the process of isomerization or dissociation that no rotational or inversion symmetry remains, the electronic states associated with intersecting surfaces still have orthogonal shapes that are sufficiently different that their change in energy with respect to geometrical coordinates near the conical intersection are quite different, and often even differ in sign.

An instructive simple model of such degeneracy between dissimilar electronic states is the degeneracy between atomic states with the same principal quantum number but different angular momemtum in atomic hydrogen, as shown in Fig. 4. Although this example does not even contain a Born–Oppenheimer potential energy surface, even here there are startling changes in the structure of the electronic wave function induced by a strong field and these properties are intimately tied to the degeneracy of states with different symmetry. In hydrogen, each energy level n has n^2 degenerate orthogonal states, with quantum numbers $(l;m) = (0,\ldots n-1; -l,\ldots,l)$. Note that according to this labelling scheme, there are n-m states with the same quantum number m, but either even or odd parity. When a hydrogen atom is placed in an electric field the levels undergo a linear Stark shift. The states with quantum number m re-order into equally spaced Stark states which are superpositions of the original (n,l,m) eigenfunctions. A new "parabolic" quantum number k labels these states according to their shift. The new states no longer have any definite parity, and are displaced from the nucleus, so that the atom acquires a static dipole moment. The situation is shown in Fig. 4, which plots the probability density of the $(n = 4, k = 3)$ state. These asymmetric charge distributions are known to affect the ionization threshold and the field ionization rates.[22]

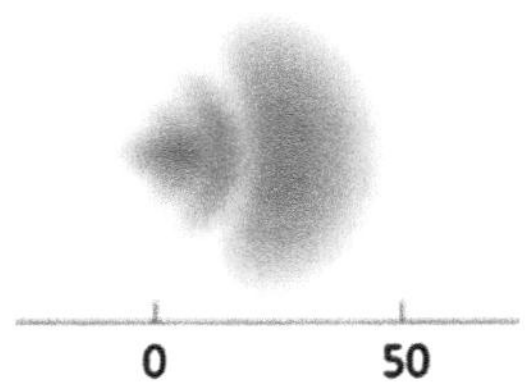

Fig. 4 Stark mixing of degenerate states in hydrogen yield atomic eigenfunctions that are offset from the nucleus. The degree of asymmetry depends on the parabolic quantum number k. Here we show the electron probability density for the most extreme example in the states with $n = 4$, $k = 3$ and $m = 0$.

The physics described in the previous paragraph is quite general. Any degeneracy between states of different symmetry will lead to a strong dipole coupling and linear Stark shift. The situation of strong-field ionization of molecules is much richer, but the presence of a degeneracy is still expected to have an important influence on field ionization if the two states are coupled by the field. The figure of merit here is the relative size of the energy splitting and the field-induced dipole energy $F\langle z\rangle$, where F is the field strength. This has been studied in detail for the simplest molecule H_2^+, where direct TDSE calculations show that the tunnel ionization rate peaks if the protons are displaced enough to bring the two ground states $1s\sigma_g$ and $2p\pi_u$ within this energy separation[23] and experiments appear to confirm this prediction.[24] In polyatomic molecules this phenomenon has been generalized as a resonant tunnel ionization mechanism that proceeds through a light-induced structure formed out of a combination of the molecular potential and the external quasi-static field.[25–27]

3B Enhanced ionization *via* strong field distortion-induced resonances

It has long been known that the threshold for ultrafast laser-induced strong-field ionization is lower in molecules than in atoms when they have the same ionization potential.[28–33] Evidently the propensity towards field ionization depends not only on energy, but also on the shape of the bound state potential. A simple explanation was put forward that when a strong field distorts the molecular potential, then bound electrons that are localized on the side of the molecule that is energetically "uphill" feel both the force of the external field and the force of the internal Coulomb attraction of the downhill atoms.

Model calculations have attempted to give this idea a firmer theoretical basis and predict how the effect could be enhanced.[34] A powerful idea has emerged that there are two main contributors to this phenomenon: localization of charge on the uphill side of the molecule; and a critical bond distance between the charge centers, which is generally longer than the ground state interatomic spacing, but may be achieved during the early stages of dissociation.[31,34–37]

So much experimental and theoretical work was done on this problem twenty years ago that it is fair to say that the principle of enhanced ionization is well established. It is not so clear how this might be contributing to enhanced ionization near CoIns. It is possible that the presence of the CoIn at a geometry where enhanced ionization occurs may simply be a coincidence. But closer examination suggests that the conditions for enhanced ionization and for electronic state degeneracy are quite similar, because both require the removal of a geometrical barrier in the molecule. Tunneling is enhanced because bound states localized on a single uphill charge can easily move downhill once the potential gradient is sufficiently strong to reduce the potential barrier between charges to the bound state energy. A conical intersection is also a point in the geometry of the molecule where an electron can move from one shape to another without a barrier.

Before leaving this subject we should point out that a closely related idea has been proposed by which ionization can be enhanced due to delocalization, particularly in alkanes or other short polymers. In this case the strong field does not induce transient localization, but instead the long potential well has a larger energy change from the uphill to the downhill end, and this enhances the ionization for a delocalized electron.[38]

3C Quantum resonance ring

The concepts expressed in section 3A concern the influence of static fields on field ionization when two states are degenerate. A strong oscillating field may also exert a comparably strong effect on the two potential energy surfaces, not at degeneracy but rather at points of resonance, when the photon energy is equal to the splitting. In strong fields this can also occur at multiphoton resonances. In atoms these "Freeman resonances" are known to dominate the strong field ionization rate for some rare gas systems.[39]

The "dressed state" or "Floquet" picture can be used to analyse how states couple to the continuum, and just as in the case of atoms one finds that the ionization rate is generally enhanced at crossings of the quasi-energies in the dressed state picture.

The connection between laser-induced dressed states and CoIns is more complicated, and has been considered in the context of quantum control of transitions involving the CoIn. In recent work involving the same CHD-HT isomerization considered in this paper, it was shown that a strong dressing laser field can substantially alter the isomerization reaction.[6] The dynamics were investigated by incorporating an *ab initio* spawning method on quasi-energy potential surfaces, as shown in Fig. 5. In this picture, the resonances take the form of crossings, and the first order crossings must occur over a locus of points surrounding the CoIn where the two potential energy surfaces are one photon apart. These points must form a ring in the two-dimensional space in which the

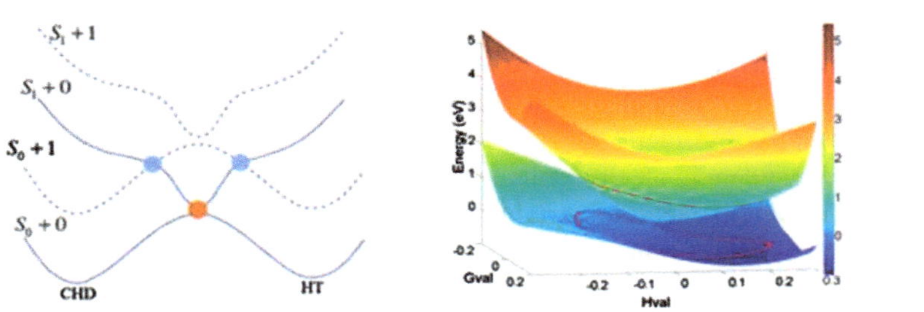

Fig. 5 Left: The dressed state picture of the CoIn region of Fig. 1 is shown schematically. The S0 and S1 states are each repeated at intervals corresponding to the energy of one photon. Thus the S0 state intersects the S1 state on either side of the CoIn point of degeneracy. Right: When viewed in the 2-dimensional subspace where the CoIn appears as a cone, the LICI points appear as a ring of resonance surrounding the CoIn[6] [from Kim *et al.*, with permission].

CoIn appears as a cone; and so this has been called a "quantum resonance ring". This has also been referred to as a "laser-induced conical intersection" or LICI.[40] Ionization activity as the molecule passes through the LICI could be enhanced by the same kinds of mechanisms described in 3A, except the static field is replaced by a resonant field. An important difference is that the form of the enhancement might be expected to depend on the laser wavelength as well as field strength and polarization. In addition, since the resonance ring is not at the same geometry as the CoIn, there could be a slight time difference between passage through the CoIn and passage through the LICI, which could be related to the 50 fs time difference observed in the White and Petrovic reports.[1,7]

3D Non-adiabatic attosecond explanations

Recent full calculations of the strong-field ionization phenomena in very simple systems such as H_2^+ have suggested that the tunnelling paradigm may be incorrect. These results show that ionization is not governed by the tail of the coulomb wave function at the tunnel barrier, but rather caused by attosecond bursts that accompany the transient localization of the electron on the atomic nuclei. This idea is still fairly new, and the connection between attosecond burst ionization and tunnel ionization is not yet fully explored; but we might consider whether there could be any connection to the presence of a CoIn. Just as in the mechanism considered in 3B, attosecond bursts depend on the details of the wave function for the electron, and particularly how it responds to the applied field and the internal motion of the molecule.

In his paper on this subject Becker points out that the bursts of multiple ionization are related to the strong coupling between charge resonant states, which might be the case for the two states in the CoIn since, as we have already pointed out, such states must have shapes that drive their energies to evolve in opposite directions as the geometry of the molecule changes.[41,42] This could lead to the multiple ionization bursts described by Becker.

The calculations involving attosecond dynamics here are quite new. Most previous work in this extremely short time domain have not been explored extensively. There are a few other calculations that make an attempt to explore the emergence of transient localization in strong laser fields on attosecond time scales.[43,44]

4 Possible experimental tests

Future experimental programs could help resolve how each of the mechanisms described above may contribute to the multiple ionization and corresponding high kinetic energy fragments observed in the vicinity of CoIn in the CHD to HT isomerization experiments.

The dependence of the enhanced high momentum fragments on laser parameters could yield important clues about the origin of the effect. For example, if the multiple ionization is dominated by multiphoton resonant pathways, then there could be a distinct non-monotonic wavelength dependence in order to satisfy a resonance requirement $n\hbar\omega = \Delta E$. On the other hand, if the resonant pathway is reached in the strong field limit, then it will tend to scale as the ponderomotive potential $U_p = I\lambda^2$, where I is the laser intensity.

The different time delays observed for the enhancement in the two experiments (150 fs *vs.* 200 fs) could also be investigated further. Varying the field parameters of intensity, wavelength, or polarization while monitoring the delay at which the onset of multiple ionization appears would be informative. Varying wavelength against time delay would test the hypotheses that a Freeman resonance or LICI are involved.

Intensity scans are always very difficult in Coulomb explosion experiments because of the high degree of nonlinearity and problems of saturation that affect secondary dissociation processes differently. Nonetheless, a careful intensity scan might help provide data that would test the distorted-geometry possibility described in section 3B.

The spectral unmixing analyses performed by White, *et al.*,[7] shows the detailed evolution of all of the ion-ion coincidences as a function of time delay. Further examination of this could give additional information about the mechanism by showing, for example, if different states are populated in the dication prior to dissociatoin. In addition, the energy and correlated angular distribution of the multiple electron emission could be informative, although it is technically more difficult.

Finally, there could be valuable information from similar measurements in other systems both with and without CoIns, to test whether the specific properties of the CoIn influence this process. Enhanced ionization has been studied extensively in diatomic molecules, which do not possess CoIns.[30,31,36] Similar experimental tests should be compared from other systems with CoIns.

5 Conclusions

Recent observations of higher levels of multiple ionizations during the transient passage in the vicinity of a conical intersection suggests that there may be some connections between the properties of the CoIn and the multiple ionization phenomenon.[1] We show that several known strong field effects could contribute to this, including stark-induced polarization, transient electronic state resonances, distortion-induced localized charge transfer resonances, and non-adiabatic ionization bursts. We propose that more study of these connections could lead to new insights about the connections between these quite different strong-field phenomena. Ultimately this may lead to new protocols to control molecular isomerization and dissociation through conical intersections.

Acknowledgements

We particularly wish to acknowledge useful discussions with Jaehee Kim, Adi Natan, and Andreas Becker in the preparation of this work for Faraday Discussions. This work was supported by the National Science Foundation.

6 References

1 V. S. Petrovic, S. Schorb, J. Kim, J. White, J. Cryan, M. Glownia, L. Zipp, D. Broege, S. Miyabe, H. Tao, T. Martinez, M. Guehr and P. H. Bucksbaum, http://arxiv.org/abs/1212.3728, 2012.
2 S. Deb and P. Weber, *Annu. Rev. Phys. Chem.*, 2011, **62**, 19–39.
3 C. Ellert, H. Stapelfeldt, E. Constant, H. Sakai, J. Wright, D. Rayner and P. Corkum, *Philos. Trans. R. Soc. London, Ser. A*, 1998, **356**, 329–344.
4 K. Kosma, S. Trushin, W. Fuß and W. Schmid, *Phys. Chem. Chem. Phys.*, 2009, **11**, 172–181.
5 M. Kotur, T. Weinacht, B. J. Pearson and S. Matsika, *J. Chem. Phys.*, 2009, **130**.

6 J. Kim, H. Tao, J. L. White, V. S. Petrovic, T. J. Martinez and P. H. Bucksbaum, *Journal of Physical Chemistry A*, 2011.
7 J. L. White, J. Kim, V. S. Petrović and P. H. Bucksbaum, *J. Chem. Phys.*, 2012, **136**, 054303.
8 A. T. J. B. Eppink and D. H. Parker, *Rev. Sci. Instrum.*, 1997, **68**, 3477–3484.
9 Y. Huismans, A. Rouzée, A. Gijsbertsen, J. H. Jungmann, A. S. Smolkowska, P. S. W. M. Logman, F. Lépine, C. Cauchy, S. Zamith, T. Marchenko, J. M. Bakker, G. Berden, B. Redlich, A. F. G. van der Meer, H. G. Muller, W. Vermin, K. J. Schafer, M. Spanner, M. Y. Ivanov, O. Smirnova, D. Bauer, S. V. Popruzhenko and M. J. J. Vrakking, *Science*, 2011, **331**, 61–64.
10 M. F. Kling, J. Rauschenberger, A. J. Verhoef, E. Hasovic, T. Uphues, D. B. Milosevic, H. G. Muller and M. J. J. Vrakking, *New J. Phys.*, 2008, **10**.
11 S. Unny, Y. Du, L. C. Zhu, K. Truhins, R. J. Gordon, A. Sugita, M. Kawasaki, Y. Matsumi, R. Delmdahl, D. H. Parker and A. Berces, *J. Phys. Chem. A*, 2001, **105**, 2270–2280.
12 F. Aguirre and S. T. Pratt, *J. Chem. Phys.*, 2003, **118**, 1175–1183.
13 W. G. Roeterdink and M. H. M. Janssen, *Phys. Chem. Chem. Phys.*, 2002, **4**, 601–612.
14 F. Rosca-Pruna and M. J. J. Vrakking, *J. Chem. Phys.*, 2002, **116**, 6567–6578.
15 J. Wei, A. Kuczmann, J. Riedel, F. Renth and F. Temps, *Phys. Chem. Chem. Phys.*, 2003, **5**, 315–320.
16 J. Wei, J. Riedel, A. Kuczmann, F. Renth and F. Temps, *Faraday Discuss.*, 2004, **127**, 267–282.
17 M. Garavelli, P. Celani, M. Fato, M. J. Bearpark, B. R. Smith, M. Olivucci and M. A. Robb, *J. Phys. Chem. A*, 1997, **101**, 2023–2032.
18 J. von Neumann and E. P. Wigner, *Z. Physik*, 1929, **30**, 467.
19 E. Teller, *J. Phys. Chem.*, 1937, **41**, 109–116.
20 W. Domcke, D. Yarkony and H. Köppel, *Conical Intersections: Electronic Structure, Dynamics & Spectroscopy*, World Scientific, 2004.
21 S. Trushin, W. Fuß, T. Schikarski, W. Schmid and K. Kompa, *J. Chem. Phys.*, 1997, **106**, 9386.
22 T. F. Gallagher, *Rep. Prog. Phys.*, 1988, **51**, 143.
23 Z. Mulyukov, M. Pont and R. Shakeshaft, *Phys. Rev. A: At., Mol., Opt. Phys.*, 1996, **54**, 4299–4308.
24 G. N. Gibson, M. Li, C. Guo and J. Neira, *Phys. Rev. Lett.*, 1997, **79**, 2022–2025.
25 E. Lötstedt, T. Kato and K. Yamanouchi, *Phys. Rev. A: At., Mol., Opt. Phys.*, 2012, **85**, 041402.
26 E. Lötstedt, T. Kato and K. Yamanouchi, *Phys. Rev. A: At., Mol., Opt. Phys.*, 2012, **86**, 023401.
27 W.-C. Jiang, O. I. Tolstikhin, L.-Y. Peng and Q. Gong, *Phys. Rev. A: At., Mol., Opt. Phys.*, 2012, **85**, 023404.
28 E. Constant, H. Stapelfeldt and P. B. Corkum, *Phys. Rev. Lett.*, 1996, **76**, 4140–4143.
29 M. Schmidt, D. Normand and C. Cornaggia, *Phys. Rev. A: At., Mol., Opt. Phys.*, 1994, **50**, 5037–5045.
30 W. T. Hill, III, J. Zhu, D. L. Hatten, Y. Cui, J. Goldhar and S. Yang, *Phys. Rev. Lett.*, 1992, **69**, 2646–2649.
31 K. Codling, L. J. Frasinski and P. A. Hatherly, *J. Phys. B: At., Mol. Opt. Phys.*, 1989, **22**, L321.
32 L. J. Frasinski, K. Codling, P. Hatherly, J. Barr, I. N. Ross and W. T. Toner, *Phys. Rev. Lett.*, 1987, **58**, 2424–2427.
33 J. Wu, M. Meckel, L. P. H. Schmidt, M. Kunitski, S. Voss, H. Sann, H. Kim, T. Jahnke, A. Czasch and R. Dörner, *Nat. Commun.*, 2012, **3**, 1113.
34 T. Seideman, M. Y. Ivanov and P. B. Corkum, *Phys. Rev. Lett.*, 1995, **75**, 2819–2822.
35 A. N. Markevitch, D. A. Romanov, S. M. Smith and R. J. Levis, *Phys. Rev. Lett.*, 2004, **92**, 063001.
36 T. Zuo and A. D. Bandrauk, *Phys. Rev. A: At., Mol., Opt. Phys.*, 1995, **52**, R2511–R2514.
37 S. Chelkowski and A. D. Bandrauk, *J. Phys. B: At., Mol. Opt. Phys.*, 1995, **28**, L723.
38 M. J. DeWitt and R. J. Levis, *J. Chem. Phys.*, 1998, **108**, 7045–7048.
39 R. R. Freeman and P. H. Bucksbaum, *J. Phys. B: At., Mol. Opt. Phys.*, 1991, **24**, 325–347.
40 R. N. Coffee, L. Fang and G. N. Gibson, *Phys. Rev. A: At., Mol., Opt. Phys.*, 2006, **73**, 043417.
41 N. Takemoto and A. Becker, *Phys. Rev. Lett.*, 2010, **105**, 203004.
42 N. Takemoto and A. Becker, *Phys. Rev. A: At., Mol., Opt. Phys.*, 2011, **84**, 023401.
43 A. Schiffrin, T. Paasch-Colberg, N. Karpowicz, V. Apalkov, D. Gerster, S. Muhlbrandt, M. Korbman, J. Reichert, M. Schultze, S. Holzner, J. V. Barth, R. Kienberger, R. Ernstorfer, V. S. Yakovlev, M. I. Stockman and F. Krausz, *Nature*, 2013, **493**, 70–74.
44 M. Schultze, E. M. Bothschafter, A. Sommer, S. Holzner, W. Schweinberger, M. Fiess, M. Hofstetter, R. Kienberger, V. Apalkov, V. S. Yakovlev, M. I. Stockman and F. Krausz, *Nature*, 2013, **493**, 75–78.

Faraday Discussions RSC Publishing

PAPER

Coherent one-photon phase control in closed and open quantum systems: A general master equation approach

Leonardo A. Pachón,[ab] Li Yu[bc] and Paul Brumer*[b]

Received 6th December 2012, Accepted 17th January 2013
DOI: 10.1039/c3fd20144a

The underlying mechanisms for one photon phase control are revealed through a master equation approach. Specifically, two mechanisms are identified, one operating on the laser time scale and the other on the time scale of the system–bath interaction. The effects of the secular and non-secular Markovian approximations are carefully examined.

1 Introduction

The possibility of manipulating and controlling quantum systems by using quantum features is the goal of many modern control protocols.[1] An unusual type of control, one photon phase control (OPPC), has been the subject of recent attention and discussion (see, *e.g.*, *Faraday Discussions* 153[2]). In this case, control takes place by varying the relative *phase* of components of a weak pulse, while keeping the power spectrum of the light source fixed. A seminal proof[3] showed that one-photon phase control was not possible for isolated molecular systems in which control is over products in the continuum. A subsequent experiment on control of retinal isomerization in bacteriorhodopsin[4] motivated controversy[5–8] and the need for clarification of conditions under which such control was possible. This clarification, provided in ref. 9 showed that control was *possible* for both closed systems and for open quantum systems under well defined conditions. Specifically, we showed the following for control of an observable $\hat{O}$ in a system defined by Hamiltonian $\hat{H}_{\mathrm{M}}$:

(a) If the system is closed (*i.e.*, not coupled to an environment), then one-photon phase control is possible if $[\hat{H}_{\mathrm{M}},\hat{O}] \neq 0$. For example, control over

[a]Instituto de Física, Universidad de Antioquia, AA 1226, Medellín, Colombia. E-mail: leonardo.pachon@fisica.udea.edu.co; Tel: +57 4 219 6436

[b]Chemical Physics Theory Group, Department of Chemistry and Center for Quantum Information and Quantum Control, University of Toronto, Toronto, Canada M5S 3H6. E-mail: pbrumer@chem.utoronto.ca; Fax: +1 416 978 5325; Tel: +1 416 978 7044

[c]Department of Physics, Harvard University, Cambridge, Massachusetts 02138, USA. E-mail: lyu@fas.harvard.edu

isomerization in an isolated molecule is possible since the probability of observing an isomer (*e.g.*, *cis* or *trans*) is an observable that does not commute with $\hat{H}_M$. The situation is similar for control over product formation in radiationless transitions, such as intersystem crossing and internal conversion.[10]

However, if $[\hat{H}_M,\hat{O}] = 0$, then one-photon phase control is not possible in an isolated system.

(b) If the system is coupled to an environment then, as above, control is possible if $[\hat{H}_M,\hat{O}] \neq 0$. However, even in the case where $[\hat{H}_M,\hat{O}] = 0$ control may still be possible, in which case it is *environmentally assisted*.

Ref. 9 provided these general rules, derived in a somewhat different way below, but did not provide any insight into the way in which environmental assistance works, or the conditions under which such control is appreciable. This is the subject of this paper. Specifically, we demonstrate the significant role of non-Markovian dynamics and strong system–bath coupling in enhancing the extent of one-photon phase control. To do so we utilize a master equation based approach to the open system case, which, as seen below, proves enlightening. The net result of this paper is the demonstration of the underlying mechanism responsible for environmentally assisted one-photon phase control.

Note, also, as an aside, that we have also computationally demonstrated[11] one-photon phase control in a model retinal system, making a connection with the original experiments on related systems.

2 One-photon phase control and closed quantum systems

Let us consider a molecular system M described by the Hamiltonian $\hat{H}_M$ with $\hat{H}_M|a\rangle = E_a|a\rangle$. If the system state is described by the density operator $\hat{\rho}_M$, the expectation value of a given observable $\hat{O}$ of system M at time t can be calculated as

$$\langle \hat{O}(t) \rangle = \sum_{a,b} \langle a|\hat{O}|b\rangle \langle b|\hat{\rho}_M|a\rangle. \tag{1}$$

If $[\hat{O},\hat{H}_M] = 0$, we simply get

$$\langle \hat{O}(t) \rangle = \sum_{a} \langle a|\hat{O}|a\rangle \langle a|\hat{\rho}_M|a\rangle. \tag{2}$$

For this case, only diagonal terms contribute to the time evolution of the expectation value of the observable $\hat{O}$.

The essence of phase control is trying to manipulate the expectation of $\hat{O}$ by changing the *phase* of an applied laser pulse. When the laser pulse is taking into account, the Hamiltonian can be re-written as $\hat{H}_0 = \hat{H}_M + \hat{H}^{MR} = \hat{H}_M - \hat{\mathbf{d}}\cdot\mathbf{E}(z,t)$, with $\mathbf{E}(z,t) = \boldsymbol{\varepsilon}\mathscr{E}(z,t)$ where

$$\mathscr{E}(z,t) = \int_{-\infty}^{\infty} d\omega \varepsilon(\omega) \exp\left[-i\omega\left(t - \frac{z}{c}\right)\right], \tag{3}$$

is the amplitude of the field with polarization vector $\boldsymbol{\varepsilon}$, which can be expressed in terms of frequency spectrum $\varepsilon(\omega) = |\varepsilon(\omega)|\exp[i\phi(\omega)]$ and, $\hat{\mathbf{d}}$ is the transition-dipole operator. We are interested in manipulating the time evolution of the observable $\hat{O}$ by means of varying the spectral phase $\phi(\omega)$ only.

In order to clearly understand the role of the environment in OPPC, it is illustrative to analyze in detail the circumstances under which[9] phase control is not possible. The evolution of the density matrix, for unitary time-evolution, is governed by

$$\dot{\rho}_{ab} = -\mathrm{i}\omega_{ab}\rho_{ab} + \frac{\mathrm{i}}{\hbar}\sum_{c}\left[H_{ac}^{\mathrm{MR}}(t)\rho_{cb} - \rho_{ac}H_{cb}^{\mathrm{MR}}(t)\right] \tag{4}$$

where $\rho_{ab} = \langle a|\hat{\rho}_{\mathrm{M}}|b\rangle$, $\hbar\omega_{ab} = E_a - E_b$ and $H_{ab}^{\mathrm{MR}}(t) = \mathrm{d}_{ab}\mathscr{E}(z,t) = \langle a|\varepsilon\cdot\hat{\mathbf{d}}|b\rangle\mathscr{E}(z,t)$. *Due the unitary character of the dynamics*, we can integrate eqn (4) by means of the following ansatz for the wavefunction

$$|\Psi\rangle = \sum_{n} b_n(t)|E_n\rangle\exp\left(-\frac{\mathrm{i}}{\hbar}E_n t\right), \tag{5}$$

with $\hat{\rho}_{\mathrm{M}}(t) = |\Psi(t)\rangle\langle\Psi(t)|$.

Let us consider, *e.g.*, that the molecule is initially $(t = -\infty)$ in the pure state $|E_1\rangle$, so we have $b_1(t = -\infty) = 1$ and $b_k(t = -\infty) = 0$ for $k \neq 1$. After formally integrating eqn (5) for b_m, we get

$$b_m(t) = \int_{-\infty}^{t} \mathrm{d}s b_1(s)\exp(\mathrm{i}\omega_{m,1}s)H_{m1}^{\mathrm{MR}}(s). \tag{6}$$

If the perturbation is weak, for unitary evolution, one can assume that $b_1(t) = 1$ at all times, thus

$$b_m(t) = \frac{\mathrm{i}}{\hbar}\mathrm{d}_{m1}\int_{-\infty}^{\infty}\mathrm{d}\omega\bar{\varepsilon}(\omega)\int_{-\infty}^{t}\mathrm{d}s\,\exp\left[\mathrm{i}(\omega_{m,1} - \omega)s\right], \tag{7}$$

where we have made use of eqn (3) and defined $\bar{\varepsilon}(\omega) = \varepsilon(\omega)\exp(\mathrm{i}\omega z/c)$. Since our interest is in the long time regime, we extend the upper integration limit to infinity and obtain $2\pi\delta(\omega_{m1} - \omega)$ for the integral over s. Finally, we have

$$b_m(\infty) = 2\pi\frac{\mathrm{i}}{\hbar}\mathrm{d}_{m1}|\varepsilon(\omega_{m1})|\exp\left[\mathrm{i}\left(\phi(\omega_{m1}) + \frac{1}{c}\omega_{m1}z\right)\right], \tag{8}$$

which yields

$$\begin{aligned}\rho_{ab}(t) = &\frac{4\pi}{\hbar^2}\mathrm{d}_{a1}\mathrm{d}_{b1}^{*}|\varepsilon(\omega_{a1})||\varepsilon(\omega_{b1})|\exp(-\mathrm{i}\omega_{ab}t)\\ &\times\exp\left\{\mathrm{i}\left[\phi(\omega_{a1}) - \phi(\omega_{b1}) + \frac{z}{c}(\omega_{a1} - \omega_{b1})\right]\right\}.\end{aligned} \tag{9}$$

From here, it is clear that no spectral phase information is present in the diagonal terms, $a = b$, and therefore it is impossible to manipulate the time evolution of the observable $\hat{O}$ in eqn (2).

Based on eqn (9), it is easy to track the vanishing of the spectral phase contribution to the fact that $\hat{\rho}_{\mathrm{M}}$ was constructed as the projector onto the state $|\Psi(t)\rangle$, *i.e.*, to the fact that the evolution is unitary. This can be understood as follows. Under unitary time evolution, the dynamics is generated by the unitary time-evolution operator $\hat{U}(t) = \hat{\mathscr{T}}\exp\left(-\mathrm{i}\int_{t_0}^{t}\mathrm{d}s\hat{H}(s)/\hbar\right)$, by means of the relation $\hat{\rho}_{\mathrm{M}}(t) = \hat{U}(t)\hat{\rho}_{\mathrm{M}}^{(0)}\hat{U}^{\dagger}(t)$. If $\hat{\rho}_{\mathrm{M}}^{(0)}$ represents the density operator of a pure state $|\Psi\rangle$, then $\hat{U}(t)$ evolves $|\Psi\rangle$ (in the direct Hilbert space) whereas $\hat{U}^{\dagger}(t)$ evolves $\langle\Psi|$ (in the dual Hilbert space). Since the generator in the dual space, $\hat{U}^{\dagger}(t)$, is just the time-reversed version of the generator in the direct space, $\hat{U}(t)$, then the evolution in the two spaces interferes in the density operator in such a way that no phase information is encoded in the diagonal elements [see eqn (9)]. This will no longer be true in the case of non-unitary dynamics (*e.g.*, for open quantum systems) because of time-symmetry breaking. This fact can be identified as a first contribution from the environment to OPPC (see below).

In what follows, we explore in detail the possible ways in which the environment can support the generation of OPPC.

3 One-photon phase control and open quantum systems

The study of open-quantum-systems dynamics is a demanding and intricate task due to the various time/energy and coupling strength scales involved. In the case of molecular systems, it is common that one component of the system plays the role of the environment. For example, if one is interested in electronic dynamics, then the vibrational and rotational modes of the same system comprise the environment. In this situation, one expects that the coupling to the environment is strong and the dynamics are non-local in time, *i.e.*, it depends on the previous conformational states of the molecular system (non-Markovian dynamics).[12,13] This non-local character of the dynamics has been proven to be vital in physical systems under the influence of external fields[14,15] Additionally, there is no apparent justification for neglecting the initial correlations between the part we name the system and the part we name the environment. All these features prevent us from applying standard approximation schemes and somehow hide the underlying physical processes.

In order to identify the contribution from the environment to the one-photon phase control, we consider the following situation: a molecular system M is at equilibrium with its environment, a stationary/incoherent situation that is subsequently irradiated by a coherent laser pulse. In the first part of our analysis, section 3.1, we neglect the initial correlation between the system and the environment and discuss how the non-unitary dynamics induced by the environment could assist OPPC. In the second part, section 3.2, we include the initial correlations with the environment and show how they could enhance the role of the environment in assisting OPPC.

3.1 Influence of the non-unitary dynamics

In order to consider the influence of the environment on the system in a consistent way, we start by considering the total dynamics of the two parts by means of the total Hamiltonian

$$\hat{H} = \hat{H}_{\rm M} + \hat{H}^{\rm ME} + \hat{H}_{\rm E} + \hat{H}^{\rm MR}, \tag{10}$$

where $\hat{H}^{\rm ME}$ denotes the interaction between the system and the environment and $\hat{H}_{\rm E}$ characterizes the environment. Let us consider that the interaction between the system and the bath is of the general form

$$\hat{H}^{\rm ME} = \sum_u \hat{K}^u \otimes \hat{\Phi}^u, \tag{11}$$

where $\hat{K}^u$ and $\hat{\Phi}^u$ are operators of the system and the environment, respectively. Note that $\hat{K}^u$ and $\hat{\Phi}^u$ could be non-Hermitian as long as $\hat{H}^{\mathrm{ME}}$ is Hermitian. At this point, we assume that the initial density operator $\hat{\rho}_{\mathrm{ME}}$ of the system + bath factorizes at $t = t_0$ into a product of the system part $\hat{\rho}_{\mathrm{M}}$ and the environment part $\hat{\rho}_{\mathrm{E}}$, that is as $\hat{\rho}_{\mathrm{ME}}(t_0) = \hat{\rho}_{\mathrm{M}}(t_0) \otimes \hat{\rho}_{\mathrm{E}}(t_0)$.

For this case, the non-Markovian evolution of the system density matrix elements can be cast in the form (*cf.* chapter 3 in ref. 16)

$$\dot{\rho}_{ab} = -\mathrm{i}\omega_{ab}\rho_{ab} + \frac{\mathrm{i}}{\hbar}\sum_{c}\left(H^{\mathrm{MR}}_{ac}\rho_{cb} - \rho_{ac}H^{\mathrm{MR}}_{cb}\right) - \sum_{c,d}\int_0^{t-t_0}\mathrm{d}\tau\Big\{\mathsf{M}_{cd,db}(-\tau)\mathrm{e}^{\mathrm{i}\omega_{da}\tau}\rho_{ac}(t-\tau)$$
$$+ \mathsf{M}_{ac,cd}(\tau)\mathrm{e}^{\mathrm{i}\omega_{bc}\tau}\rho_{db}(t-\tau) - \left[\mathsf{M}_{db,ac}(-\tau)\mathrm{e}^{\mathrm{i}\omega_{bc}\tau} + \mathsf{M}_{db,ac}(\tau)\mathrm{e}^{\mathrm{i}\omega_{da}\tau}\right]\rho_{cd}(t-\tau)\Big\}, \tag{12}$$

where $\mathsf{M}_{ab,cd}(t)$ determines the time span for correlations and are called the memory matrix elements (*cf.* chapter 3 in ref. 16). They are defined by

$$\mathsf{M}_{ab,cd}(t) = \sum_{u,v} C_{uv}(t)K^u_{ab}K^v_{cd}, \tag{13}$$

with $K^u_{ab} = \langle a|\hat{K}^u|b\rangle$, and satisfy the relation $\mathsf{M}^*_{ab,cd}(t) = \mathsf{M}_{dc,ba}(-t)$. The reservoir correlation function $C_{uv}(t)$ is given by

$$C_{uv}(t) = \frac{1}{\hbar^2}\langle\hat{\Phi}^u(t)\hat{\Phi}^v(0)\rangle_{\mathrm{R}} - \frac{1}{\hbar^2}\langle\hat{\Phi}^u\rangle_{\mathrm{R}}\langle\hat{\Phi}^v\rangle_{\mathrm{R}}, \tag{14}$$

where $\langle\hat{\Phi}^u(t)\hat{\Phi}^v(0)\rangle_{\mathrm{R}} = \mathrm{tr}[\hat{\rho}_{\mathrm{E,equ}}\hat{\Phi}^u(t)\hat{\Phi}^v(0)]$. The first two terms in the first line of eqn (12) describe the unitary evolution and is completely equivalent to eqn (4). Hence, if no phase control can be achieved from eqn (4), no phase control can be achieved from this part of the evolution in eqn (12). The remaining terms describe the non-unitary non-Markovian dynamics, characterized by the non-local evolution implied by the integration from 0 to $t - t_0$. The most relevant feature of this second part of eqn (12) is the fact that the off-diagonal elements affect the dynamics of the diagonal terms, despite the weak field condition.

Consider then the case where the system is initially in equilibrium with the bath and the excitation follows. For convenience, let us write eqn (12) as $\dot{\hat{\rho}}_{\mathrm{M}} = -\mathrm{i}\hat{\mathscr{L}}_{\mathrm{M}}\hat{\rho}_{\mathrm{M}} - \hat{\mathscr{D}}\hat{\rho}_{\mathrm{M}} - \mathrm{i}\hat{\mathscr{L}}_{\mathrm{laser}}\hat{\rho}_{\mathrm{M}}$, where we have introduced the superoperators $\hat{\mathscr{L}}_{\mathrm{M}}\cdot = [\hat{H}_{\mathrm{M}},\cdot]/\hbar$, $\hat{\mathscr{L}}_{\mathrm{laser}}\cdot = [\hat{H}_{\mathrm{laser}},\cdot]/\hbar$ and the so-called dissipative or relaxation superoperator $\hat{\mathscr{D}}$, which cannot be given simply *via* a Liouville superoperator of a simple commutator. In the absence of the laser field, the solution to eqn (12) can then be written as $\hat{\rho}_{\mathrm{M}}(t) = \hat{\mathscr{T}}\exp\left[-\mathrm{i}\int_{t_0}^{t}\mathrm{d}s\left(\hat{\mathscr{L}}_{\mathrm{M}}(s) - \mathrm{i}\hat{\mathscr{D}}(s)\right)\right]\hat{\rho}_{\mathrm{M}}(t_0)$. Thus, in this representation, our initial state is an eigenstate of the superoperator $\hat{\mathscr{T}}\exp\left[-\mathrm{i}\int_{t_0}^{t}\mathrm{d}s\left(\hat{\mathscr{L}}_{\mathrm{M}}(s) - \mathrm{i}\hat{\mathscr{D}}(s)\right)\right]$, *i.e.*, $\hat{\mathscr{T}}\exp\left[-\mathrm{i}\int_{t_0}^{t}\mathrm{d}s\left(\hat{\mathscr{L}}_{\mathrm{M}}(s) - \mathrm{i}\hat{\mathscr{D}}(s)\right)\right]\hat{\rho}_{\mathrm{M}} = \lambda_{\hat{\rho}_{\mathrm{M}}}\hat{\rho}_{\mathrm{M}}$. For high enough temperature and weak coupling to the bath [*cf.* ref. 17 and references therein], one can show that in the long time regime, and in the absence of the laser, the density operator of the system is well characterized by the canonical distribution $\hat{\rho}_{\mathrm{M,can}}$, *i.e.*,

$$\hat{\rho}_{\mathrm{M,can}} = \lim_{t\to\infty}\hat{\rho}_{\mathrm{M}}(t) = \frac{1}{Z}\mathrm{e}^{-\hat{H}_{\mathrm{M}}\beta} = \frac{1}{Z}\sum_a \mathrm{e}^{-E_a\beta}|a\rangle\langle a|, \tag{15}$$

where $\{|a\rangle\}$ are eigenstates of $\hat{H}_{\rm M}$, $Z = {\rm tr}(e^{-\hat{H}_{\rm M}\beta})$ is the partition function and $\beta = 1/k_{\rm B}T$. Thus, $\hat{\rho}_{\rm M,can}$ is an eigenstate of non-driven dynamics with eigenvalue 1, *i.e.*, $\hat{\mathscr{T}}\exp\left[-{\rm i}\int_{t_0}^{t}{\rm d}s\left(\hat{\mathscr{L}}_{\rm M}(s)-{\rm i}\hat{\mathscr{D}}(s)\right)\right]\hat{\rho}_{\rm M,can}=\hat{\rho}_{\rm M,can}$ or equivalently, $\hat{\rho}_{\rm M,can}$ is a zero mode of the generator $\hat{\mathscr{L}}_{\rm M}(s)-{\rm i}\hat{\mathscr{D}}(s)$. Thus, in this description the laser excitation finds the system in an incoherent superposition of energy eigenstates.

For this initial state, under the weak field assumption, we can solve eqn (12) perturbatively in the laser term,

$$\dot{\hat{\rho}}_{\rm M}^{(0)}(t) = -{\rm i}\left(\hat{\mathscr{L}}_{\rm M}-{\rm i}\hat{\mathscr{D}}\right)\hat{\rho}_{\rm M,can}, \tag{16}$$

$$\dot{\hat{\rho}}_{\rm M}^{(1)}(t) = -{\rm i}\left(\hat{\mathscr{L}}_{\rm M}-{\rm i}\hat{\mathscr{D}}\right)\hat{\rho}_{\rm M}^{(0)}(t)-{\rm i}\hat{\mathscr{L}}_{\rm laser}\hat{\rho}_{\rm M}^{(0)}(t), \tag{17}$$

$$\dot{\hat{\rho}}_{\rm M}^{(2)}(t) = -{\rm i}\left(\hat{\mathscr{L}}_{\rm M}-{\rm i}\hat{\mathscr{D}}\right)\hat{\rho}_{\rm M}^{(1)}(t)-{\rm i}\hat{\mathscr{L}}_{\rm laser}\hat{\rho}_{\rm M}^{(1)}(t). \tag{18}$$

The zeroth-order term $\hat{\rho}_{\rm M}^{(0)}(t)$ describes the dynamics in the absence of the laser and based on the description above, can be trivially obtained as $\hat{\rho}_{\rm M}^{(0)}(t) = \hat{\rho}_{\rm M,can}$. The first-order term $\hat{\rho}_{\rm M}^{(1)}(t)$ accounts for the dynamics induced by the laser pulse on the incoherent superposition of eigenstates $\hat{\rho}_{\rm M,can}$. Since $-{\rm i}(\hat{\mathscr{L}}_{\rm M}-{\rm i}\hat{\mathscr{D}})\hat{\rho}^{(0)}(t) = -{\rm i}(\hat{\mathscr{L}}_{\rm M}-{\rm i}\hat{\mathscr{D}})\hat{\rho}_{\rm M,can}=0$, the system-bath contribution is absent here.

In deriving eqn (9), we had assumed that $\hat{\rho}_{\rm M}^{(0)} = |1\rangle\langle 1|$, whereas in our case we have an incoherent superposition of eigenstates (15). Hence, by virtue of the linearity of the master eqn (12), we can infer directly that

$$\begin{aligned}\rho_{ab}^{(1)}(t) = &\frac{1}{Z}\frac{4\pi}{\hbar^2}\sum_n {\rm e}^{-E_n\beta}{\rm d}_{an}{\rm d}_{bn}^{*}|\varepsilon(\omega_{an})||\varepsilon(\omega_{bn})|\exp(-{\rm i}\omega_{ab}t)\\ &\times\exp\left\{{\rm i}\left[\phi(\omega_{an})-\phi(\omega_{bn})+\frac{z}{c}(\omega_{an}-\omega_{bn})\right]\right\}.\end{aligned} \tag{19}$$

Therefore, it is clear from eqn (19) that no phase information is encoded in the populations, $a = b$, of $\hat{\rho}_{\rm M}^{(1)}(t)$.

The second-order term $\hat{\rho}_{\rm M}^{(2)}(t)$ accounts for the evolution of the state prepared by the laser excitation in the presence of the system-bath dynamics. For the populations of $\hat{\rho}_{\rm M}^{(2)}$ we have

$$\begin{aligned}\dot{\rho}_{aa}^{(2)} = &\frac{\rm i}{\hbar}\sum_c\left(H_{ac}^{\rm MR}\rho_{ca}^{(1)}-\rho_{ac}^{(1)}H_{ca}^{\rm MR}\right)-\sum_{c,d}\int_0^{t-t_0}{\rm d}\tau\Big\{{\sf M}_{cd,da}(-\tau){\rm e}^{{\rm i}\omega_{da}\tau}\rho_{ac}^{(1)}(t-\tau)\\ &+{\sf M}_{ac,cd}(\tau){\rm e}^{{\rm i}\omega_{ac}\tau}\rho_{da}^{(1)}(t-\tau)-\left[{\sf M}_{da,ac}(-\tau){\rm e}^{{\rm i}\omega_{ac}\tau}+{\sf M}_{da,ac}(\tau){\rm e}^{{\rm i}\omega_{da}\tau}\right]\rho_{cd}^{(1)}(t-\tau)\Big\},\end{aligned} \tag{20}$$

Explicitly integrating eqn (20) requires that we specify the system–bath interactions [eqn (11)] and the bath nature. For real environments this integration is not possible and one has to appeal to specific models; in the Appendix we discuss the most relevant features of the commonly used bosonic bath. Despite the sheer complexity of integrating eqn (20), one can still extract some of the relevant physical contributions directly from the master equation (as an example, see ref. 18 for a specific implementation). By inserting eqn (19) into the right-hand side of eqn (20), *it is clear that phase control of the population will result from the system–bath interaction, which couples* $\rho_{ab}^{(1)}$ *to* $\rho_{aa}^{(2)}$. This phase control is of second order in both the radiation field and in the system–bath coupling strength. Note that the radiation field (the first two terms) also couples $\rho_{ab}^{(1)}$ to $\rho_{aa}^{(2)}$. This contribution, however, is third order in the radiation field and thus is negligible in our current considerations of second order effects.

The time dependence of the M matrix elements signifies the non-Markovian character of the dynamics, which can be associated with the environment self-correlations. These self-correlations allows for a back and forth information flow between the system and the bath.[19] If the dynamics is Markovian (see next case below), the information flow is unidirectional: from the system to the bath. This information flow can be understood as an entropy flux and in the case of Markovian dynamics results in an always increasing entropy of the system M. In the presence of external driving fields, non-Markovian terms are essential.[20] In our case, this back and forth flow allows for an *efficient* transfer of phase information *via* the bath. Additionally, the non-Markovian dynamics is characterized by slower correlations decay-rate,[21] allowing thus the presence of phase control for longer times.[18]

3.1.1 Markovian approximation. If the environment quickly "forgets" any internal self-correlations during its interaction with the system, that is, if any dynamically established quantum correlations between parts of the environment are destroyed on a time scale much shorter than the characteristic time scale of the interaction-picture reduced density operator of the system, the Markovian approximation can be invoked. Mathematically, this means that the correlations $C_{uv}(\tau)$, and hence the M matrix elements, die off quickly as $\tau > 0$, and thus the τ integral in eqn (20) picks up contribution only from $\rho_{ab}^{(1)}(t)$, rendering the dynamics time-local. Moreover, it does not make a difference if we extend the upper limit of integration to infinity, in which case the matrix elements $\Gamma_{ab,cd}$, defined below, become constant. As a result, a Markovian master equation is obtained:

$$\dot{\rho}_{aa}^{(2)} = \frac{\mathrm{i}}{\hbar}\sum_{c}\left(H_{ac}^{\mathrm{MR}}\rho_{ca}^{(1)} - \rho_{ac}^{(1)}H_{ca}^{\mathrm{MR}}\right) - \sum_{c,d}\Big\{\Gamma_{ad,dc}(\omega_{cd})\rho_{ac}^{(1)}(t) + \Gamma_{ac,cd}(\omega_{dc})\rho_{da}^{(1)}(t)$$
$$- [\Gamma_{ca,ad}(\omega_{da}) + \Gamma_{da,ac}(\omega_{ca})]\rho_{cd}^{(1)}(t)\Big\}, \tag{21}$$

where

$$\Gamma_{ab,cd}(\omega) = \Re\int_0^{\infty}\mathrm{d}\tau e^{\mathrm{i}\omega\tau}\mathrm{M}_{ab,cd}(\tau) = \Re\sum_{u,v}K_{ab}^{u}K_{cd}^{v}\int_0^{\infty}\mathrm{d}\tau e^{\mathrm{i}\omega\tau}C_{uv}(\tau). \tag{22}$$

This master equation retains the essential feature of the coupling of $\rho_{ab}^{(1)}$ to $\rho_{aa}^{(2)}$ as in eqn (20). *As a result, even given the Markovian approximation, the population will show phase dependence of second order in the radiation field and of second order in the system–bath coupling.* The Markovian approximation, which results in the time locality of the master equation, does not undermine the essential feature of phase dependence in the population. However, Markovian dynamics results, in general, in a stronger correlations decay-rate[21] and therefore, the phase information will rapidly flow back to the bath. *Hence, phase control may not be noticeable when the bath relaxation is very fast.*

3.1.2 Markovian secular approximation. If, in addition, one now applies the secular approximation in eqn (21), we get

$$\dot{\rho}_{aa}^{(2)} = \frac{i}{\hbar}\sum_c \left(H_{ac}^{MR}\rho_{ca}^{(1)} - \rho_{ac}^{(1)}H_{ca}^{MR}\right) - 2\sum_c \left[\Gamma_{ac,ca}(\omega_{ac})\rho_{aa}^{(1)}(t) - \Gamma_{ca,ac}(\omega_{ca})\rho_{cc}^{(1)}(t)\right]. \tag{23}$$

This approximation decouples the evolution of the diagonal elements from that of the off-diagonal terms *via* the bath. Hence no phase control in the population can be observed, due to the artificial truncation of the off-diagonal terms. Thus, the secular approximation is not appropriate for the study of environmentally assisted phase control. *Alternatively, from a physics perspective, if the secular approximation is indeed valid for some system of interest, then our analysis shows that OPPC is not achievable in such a system.*

3.2 The influence of the initial system-bath correlations

The presence of initial correlations between the system and bath, $\hat{\rho}_{ME}(t_0) \neq \hat{\rho}_M(t_0) \otimes \hat{\rho}_E(t_0)$, introduces an important new contribution in the master equation [eqn (12)]. For completeness we write the entire expression,

$$\begin{aligned}\dot{\rho}_{ab} = &-i\omega_{ab}\rho_{ab} + \frac{i}{\hbar}\sum_c \left(H_{ac}^{MR}\rho_{cb} - \rho_{ac}H_{cb}^{MR}\right) - \frac{i}{\hbar}\sum_u \Big\langle a\Big|\mathrm{tr}_E\left(\left[\hat{K}^u\hat{\Phi}^u, \hat{\rho}_{ME}(t_0)\right]\right.\\ &\left.-\left[\hat{K}^u\hat{\Phi}^u, \hat{\rho}_E(t_0)\hat{\rho}_M(t_0)\right]\right)\Big|b\Big\rangle - \sum_{c,d}\int_0^{t-t_0} d\tau\Big\{\mathsf{M}_{cd,db}(-\tau)e^{i\omega_{da}\tau}\rho_{ac}(t-\tau)\\ &+\mathsf{M}_{ac,cd}(-\tau)e^{i\omega_{bc}\tau}\rho_{db}(t-\tau) - \left[\mathsf{M}_{db,ac}(-\tau)e^{i\omega_{bc}\tau} + \mathsf{M}_{db,ac}(-\tau)e^{i\omega_{da}\tau}\right]\rho_{cd}(t-\tau)\Big\},\end{aligned} \tag{24}$$

where $\hat{\rho}_E(t_0) = \mathrm{tr}_S\hat{\rho}_{ME}(t_0)$ and $\hat{\rho}_M(t_0) = \mathrm{tr}_E\hat{\rho}_{ME}(t_0)$. The new term in the dynamics exposes the initial correlations between the system and the environment. Note that this is an extension beyond standard master equations, which do not usually include the state of the environment explicitly.

If $\hat{\rho}_{ME}(t_0)$ factorizes into system and bath terms, then it is clear that the terms in the second line in eqn (24), the ones which account for the initial correlations, vanish. In the opposite case, this would imply that the initial stationary state should be re-derived taking into account this new driving term, *i.e.*, the new stationary state will not be diagonal in the energy eigenbasis. The immediate consequence of this fact is that the initial state, before the coherent excitation takes place, will contain stationary coherences,[15] *i.e.* non-zero time independent off-diagonal elements of the system density matrix in the energy representation, which will allow for phase control.

As in the previous case, we can write eqn (24) as $\dot{\hat{\rho}}_M = -i\hat{\mathscr{L}}_M\hat{\rho}_M - i\hat{\mathscr{L}}_0\hat{\rho}_M - \hat{\mathscr{D}}\hat{\rho}_M - i\hat{\mathscr{L}}_{laser}\hat{\rho}_M$ and assume that the initial state is an eigenstate of the superoperator $\hat{\mathscr{T}}\exp\left[-i\int_{t_0}^t ds\left(\hat{\mathscr{L}}_M(s) + \hat{\mathscr{L}}_0(s) - i\hat{\mathscr{D}}(s)\right)\right]$. Here $\hat{\mathscr{L}}_M$, $\hat{\mathscr{D}}$, $\hat{\mathscr{L}}_{laser}$ are defined as before and $\hat{\mathscr{L}}_0$ denotes the contribution from the initial correlations given in the second line of eqn (24). Due to the initial correlations between the system and the bath, it is clear that the eigenstates of the super operator $\hat{\mathscr{T}}\exp\left[-\mathrm{i}\int_{t_0}^{t}\mathrm{d}s\left(\hat{\mathscr{L}}_{\mathrm{M}}(s)+\hat{\mathscr{L}}_0(s)-\mathrm{i}\hat{\mathscr{D}}(s)\right)\right]$ are no longer diagonal in the eigenbasis of $\hat{H}_{\mathrm{M}}$. For this case,

$$\hat{\rho}_{\mathrm{M,eq}}=\lim_{t\to\infty}\hat{\rho}_{\mathrm{M}}(t)=\frac{1}{Z'}\mathrm{tr}_{\mathrm{E}}\,\mathrm{e}^{-\left(\hat{H}_{\mathrm{M}}+\hat{H}^{\mathrm{ME}}+\hat{H}_{\mathrm{E}}\right)\beta},\tag{25}$$

where $Z'=\mathrm{tr}\left(\mathrm{e}^{-(\hat{H}_{\mathrm{M}}+\hat{H}^{\mathrm{ME}}+\hat{H}_{\mathrm{E}})\beta}\right)$ is the partition function. Thus, in this case the laser excitation finds the system with stationary coherences of energy eigenstates, a concept that we have discussed in ref. 15. *The presence of these off-diagonal elements guarantees that the phases of the laser pulse contribute to the dynamics of the populations.*

Following a perturbative approach similar to that in section 3.1, we can study the dynamics by means of the sequence

$$\dot{\hat{\rho}}_{\mathrm{M}}^{(0)}(t)=-\mathrm{i}\left(\hat{\mathscr{L}}_{\mathrm{M}}+\hat{\mathscr{L}}_0-\mathrm{i}\hat{\mathscr{D}}\right)\hat{\rho}_{\mathrm{M,eq}},\tag{26}$$

$$\dot{\hat{\rho}}_{\mathrm{M}}^{(1)}(t)=-\mathrm{i}\left(\hat{\mathscr{L}}_{\mathrm{M}}+\hat{\mathscr{L}}_0-\mathrm{i}\hat{\mathscr{D}}\right)\hat{\rho}_{\mathrm{M}}^{(0)}(t)-\mathrm{i}\hat{\mathscr{L}}_{\mathrm{laser}}\hat{\rho}_{\mathrm{M}}^{(0)}(t),\tag{27}$$

$$\dot{\hat{\rho}}_{\mathrm{M}}^{(2)}(t)=-\mathrm{i}\left(\hat{\mathscr{L}}_{\mathrm{M}}+\hat{\mathscr{L}}_0-\mathrm{i}\hat{\mathscr{D}}\right)\hat{\rho}_{\mathrm{M}}^{(1)}(t)-\mathrm{i}\hat{\mathscr{L}}_{\mathrm{laser}}\hat{\rho}_{\mathrm{M}}^{(1)}(t).\tag{28}$$

Since the off-diagonal elements in the system energy eigenbasis are present from the beginning, phase control of the population will take place at the time scale of the radiation field. In other words, the phase information can "flow" directly into the population *via* the radiation, without having to wait for the system–bath interaction to first establish phase dependence in the off-diagonal elements.

4 Concluding remarks

Environmentally assisted one-photon phase control arises in open systems when control is achieved over system properties that commute with the system Hamiltonian. In such cases, phase control is not possible unless the environment participates in the dynamics. In this paper we have analyzed the origins of such control and shown that the effects of molecule-bath coupling on phase control are two-fold and take place on two different time scales. First, the thermalization of the coupled system-bath gives rise to non-zero off-diagonal elements in the initial molecular state, which can result in a direct phase control of the population; this control takes place at the timescale of the radiation field. Second, through the molecule-bath coupled dynamics, as modeled by eqn (20) or eqn (21), the phase information contained in the off-diagonals can "flow" through the bath into the diagonal elements, resulting in an indirect phase control; this control takes place on the timescale of the molecule-bath interaction. The different timescales provide useful indicators to distinguish the two kinds of phase control in experiments.

One final note is in order. Environmentally assisted one-photon phase control depends on the system-bath interaction. In the case where the bath is in equilibrium, as could be the case of a solvated molecule, this interaction will cause, long after the pulse is over, the system to relax to a laser-phase-independent equilibrium state as well.[18] As such, and under these circumstances, environmentally assisted one-photon phase control will only survive for a well defined time after the laser excitation is over. By contrast, if the bath comprises part of the same system as, *e.g.*, in macromolecules, then the induced OPPC may survive for longer times assisted by the complex structure of the bath and the strong coupling to it.[11] In general, rates and time scales over which this control persists is dependent upon, and requires numerical studies of, individual systems of interest.

Appendix: Harmonic oscillator bath

Since the explicit calculation of the correlation functions C_{uv} in eqn (14) is not feasible for real baths such as solvents, one has to appeal to specific models of the bath and the system–bath interaction. The most common approximation is the normal mode approximation. In this case one assumes that, *e.g.*, the vibrational modes can be described by small oscillations around the equilibrium point, thus allowing the use of the harmonic approximation. In this case, we replace the interaction term $\hat{H}^{\rm ME}$ in eqn (11) by

$$\hat{H}^{\rm ME} = \hat{K}(s)\hat{\Phi}(\mathbf{Z}) = \hat{K}(s)\sum_{\xi}\hbar\gamma_{\xi}\hat{Z}_{\xi}, \tag{29}$$

where s denotes the system coordinates, γ_ξ denote the system-environment coupling constants and $\hat{\mathbf{Z}} = \{\hat{Z}_\xi\}$ denotes the environment normal mode coordinates. For simplicity, we have assumed that $\hat{H}^{\rm ME}$ contains a single term and have dropped the index u. Note that since $\langle\hat{Z}_\xi\rangle_{\rm R} = 0$, the expression for C_{uv} in eqn (14) reduces to the two-point correlation function of the bath operator. Thus, we have for $C_{uv}(t) = C(t)$

$$C(t) = \int_0^{\infty} d\omega\left[\cos(\omega t)\coth\left(\frac{\hbar\omega}{2k_{\rm B}T}\right) - i\sin(\omega t)\right]\omega^2 J(\omega), \tag{30}$$

where $J(\omega)$ is the spectral density,

$$J(\omega) = \sum_{\xi} g_{\xi}^2\delta(\omega_{\xi} - \omega). \tag{31}$$

Some typical spectral densities are, *cf.* ref. 12, $\omega^2 J(\omega) = \theta(\omega) j_0 \omega^p e^{-\omega/\omega_c}$ or $\omega^2 J(\omega) = \theta(\omega)\dfrac{j_0\omega}{\omega^2 + {\omega_D}^2}$, where ω_c and ω_D both denote a cut off frequency.

Acknowledgements

This work was supported by NSERC and by the US Air Force Office of Scientific Research under contract number FA9550-10-1-0260, by *Comité para el Desarrollo de la Investigación* –CODI– of Universidad de Antioquia, Colombia under contract number E01651 and under the *Estrategia de Sostenibilidad 2013-2014* and by the *Colombian Institute for Science and Technology Development* –COLCIENCIAS– under contract number 111556934912.

References

1 M. Shapiro and P. Brumer, *Quantum Control of Molecular Processes*, Wiley-VCH, Weinheim, 2nd edn, 2012.
2 Coherence and Control in Chemistry, *Disc. Far. Soc.*, 2011, **153**, pp. 1–428.
3 P. Brumer and M. Shapiro, *Chem. Phys.*, 1989, **139**, 221–228.

4 V. Prokhorenko, A. Nagy, S. A. Waschuk, L. S. Brown, R. R. Birge and R. J. D. Miller, *Science*, 2006, **313**, 1257–1261.
5 M. Joffre, *Science*, 2007, **317**, 453.
6 V. Prokhorenko, A. Nagy, S. A. Waschuk, L. S. Brown, R. R. Birge and R. J. D. Miller, *Science*, 2007, **317**, 453.
7 C. Florean, D. Cardoza, J. L. White, J. K. Lanyi, R. J. Sension and P. H. Bucksbaum, *Proc. Natl. Acad. Sci. U. S. A.*, 2009, **106**, 10896.
8 V. I. Prokhorenko, A. Halpin, P. J. M. Johnson, R. Miller and L. S. Brown, *J. Chem. Phys.*, 2011, **134**, 085105.
9 M. Spanner, C. A. Arango and P. Brumer, *J. Chem. Phys.*, 2010, **133**, 151101.
10 T. Grinev, M. Shapiro and P. Brumer, *J. Chem. Phys.*, 2013, **138**, 044306.
11 C. A. Arango and P. Brumer, *J. Chem. Phys.*, 2013, **138**, 071104.
12 L. A. Pachón and P. Brumer, *Phys. Chem. Chem. Phys.*, 2012, **14**, 10094.
13 L. A. Pachón and P. Brumer, *J. Phys. Chem. Lett.*, 2011, **2**, 2728.
14 L. A. Pachón and P. Brumer, *Phys. Rev. E (submitted)*, 2013.
15 L. A. Pachón and P. Brumer, *Phys. Rev. A*, 2013, **87**, 022106.
16 V. May and O. Kühn, *Charge and Energy Transfer Dynamics in Molecular Systems*, Berlin, Wiley, 2001.
17 J. F. Triana, L. A. Pachón and P. Brumer, *Canonical Typicality Deviations at Low Temperature*, 2013, In preparation.
18 L. A. Pachón and P. Brumer, *Environmentally Assisted One-photon Phase Control: The Role of Stationary Coherences*, 2013, In preparation.
19 H.-P. Breuer, E.-M. Laine and J. Piilo, *Phys. Rev. Lett.*, 2009, **103**, 210401.
20 R. Schmidt, A. Negretti, J. Ankerhold, T. Calarco and J. T. Stockburger, *Phys. Rev. Lett.*, 2011, **107**, 130404.
21 A. F. Estrada, D. Zueco and L. A. Pachón, *Out-of-Equilibrium Non-Markovian Quantum Limit*, 2013, In preparation.

PAPER

Photophysics of fulvene under the non-resonant stark effect. Shaping the conical intersection seam

Sergi Ruiz-Barragan and Lluís Blancafort*

Received 20th December 2012, Accepted 17th January 2013
DOI: 10.1039/c3fd20155d

We introduce a mechanistic strategy to control the excited state lifetime of fulvene based on shaping the topography of an extended seam of intersection with the non-resonant dynamic Stark effect. Fulvene has a very short excited state lifetime due to an energetically accessible seam of intersection which lies along the methylene torsion coordinate, and the initial decay occurs at the seam segment around the planar conical intersection structure. We have followed a three-step approach to simulate the control. First, we have calculated the effect of a non-resonant electric field on the potential energy surface at the *ab initio* level, including the field in a self-consistent way. The relative energy of the planar segment of the seam is increased by the non-resonant field. In the second step we simulate the control carrying out MCTDH quantum dynamics propagations under a static non-resonant field to derive the main control mechanisms. At moderately intense fields ($\varepsilon \leq 0.03$ a.u.) the decay is faster as compared to the field free case because the vibrational overlap between the excited and ground state vibrational functions is increased. However, at more intense fields ($\varepsilon = 0.04$ a.u.) the planar conical intersection is energetically inaccessible and the decay occurs at a slower time scale, at the segment of the seam with more twisted geometries. In the third step, the control over the dynamics is exerted with a non-resonant dynamic field. The acceleration of the decay due to the improved vibrational overlap does not occur, but the decay can be made slower with a dynamic field of 0.08 a.u. The results show the viability of our approach to control the photophysics shaping the topology of the conical intersection seam, and they prove that the extended nature of the seam is crucial for simulating and understanding the control.

1 Introduction

The control of chemical processes with light is a long pursued goal of chemists.[1–3] Thanks to the development of laser technology, in recent years it has become a realistic goal, which can be applied at many different levels. Control

Institut de Química Computacional and Departament de Química, Universitat de Girona, Campus de Montilivi, 17071 Girona, Spain

has been achieved, for example, over the translational and rotational degrees of freedom of molecules, orienting them with the help of external electric fields.[4] It has also become possible to control chemical processes with properly shaped resonant pulses that guide the wave packet on the potential energy surface.[5–7] Usually these pulses have complicated shapes or consist of so-called 'trains of pulses' which are developed using optimal and local control theories.[8–11] While these approaches have been successful in a number of cases, one of their drawbacks is that they remain notoriously difficult to interpret.

In recent years, an alternative approach has been developed where the control is achieved by means of a non-resonant field. The field shifts the potential energy surfaces through the Stark effect.[12,13] Intense fields are required to achieve a chemically relevant effect, and in turn this requires the use of dynamic fields. In this case one speaks of the non-resonant dynamic Stark effect (NRDSE). Control by means of the NRDSE is a very appealing approach because the picture of shifting potential energy surfaces is more accessible to interpretation and even prediction than the optimal and local control schemes.

Experimentally, the most successful example of control with the NRDSE is the photodissociation of IBr, where the reaction channel probabilities have been modified by tuning the potential energy barriers with infrared laser pulses.[12] From the point of view of theory, several studies have addressed the control of the photochemistry of this and other diatomic molecules with the NRDSE.[14–17] In most of these cases, nonadiabatic crossings play a key role in the control, and the control of the dynamics can be understood by changes in the position of the crossings and the couplings between the surfaces. More recently, the NRDSE has been used to simulate optical control in polyatomic molecules.[18–20] Here the crossings have the form of conical intersections, which appear as the ideal target to exert control. In fact, during the last decades theoreticians have developed a very detailed picture of conical intersections and their role in photochemistry and photophysics.[21–25] In this context, the challenge is to achieve control by combining this mechanistic knowledge with the ability to tune the potential energy surfaces, and develop what one can call *mechanistic control* schemes.

One important feature of conical intersections that has not been considered in previous NRDSE studies is the fact that they are not isolated points on the potential energy surface, but form part of multidimensional seams of intersection.[26–29] The multidimensional nature is important in photophysics and photochemistry and has to be included in the control simulations of polyatomic molecules. Thus, the multi-dimensionality facilitates the traversal of the intersection and has a strong influence on the excited state lifetimes. For photochemical problems, we have also introduced the notion of the extended seam of intersection, where the seam space is composed of different segments associated to different reaction paths on the surface, leading to different products.[30–34] The traversal of the wave packet through the different seam segments determines the chemical outcome of the excitation. Based on this idea, the possibility of control at the seam has been explored guiding semiclassical dynamics trajectories to the different seam segments by adding momentum in different directions.[31,35–37] Here we propose to follow a similar idea in the context of the NRDSE. Thus, the extended seam should be the ideal playground for the NRDSE, since by shifting the potential energy surface one may favour one or the other associated path. Therefore, our central aim is to use the NRDSE to develop mechanistic control strategies centred on the traversal of the wave packet through the extended seam.

As a first attempt to develop such a strategy, we consider the photophysics of fulvene, a non-fluorescent hydrocarbon.[38–40] We aim to simulate the control of its excited state lifetime by combining resonant and non-resonant laser pulses, as shown in the inset of Fig. 1. The photophysics of fulvene have been studied extensively, and the non-emitting nature is due to an energetically accessible conical intersection seam between the ground and excited state, where the excited state decays after the excitation.[41] The relevant segment of the seam of intersection lies along two nuclear coordinates, one corresponding to a bond-alternation mode, and the other to the torsion of the methylene group (Fig. 1).[34] Every point on the seam is a conical intersection when plotted along the proper coordinates (the branching space vectors, see below). We have recently simulated the excited state dynamics of fulvene with quantum dynamics calculations using the multiconfigurational time-dependent Hartree approach (MCTDH)[42–45] and a four-dimensional model surface that includes the extended seam and its connection with the Franck–Condon structure.[46] In the calculations, the wave packet decays initially along the bond-alternation coordinate and reaches the seam at a segment with a planar geometry, centred around structure **CI$_{plan}$**. This corresponds to path (1) in Fig. 1. Most of the excited state population follows this mechanism and returns to the ground state in less than 10 fs after the excitation pulse. For the fraction of the wave packet that remains in the excited state, the torsional mode is activated after 40–100 fs, depending on the

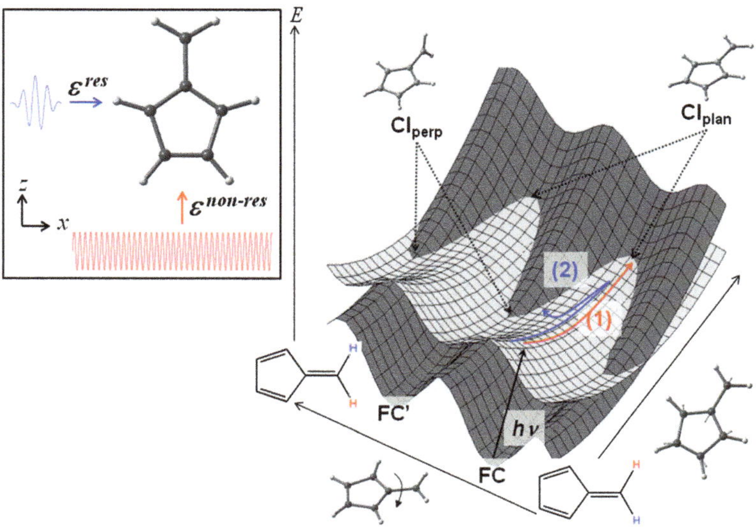

Fig. 1 Two-dimensional potential energy surface for the decay of singlet excited fulvene along the torsional and symmetric bond alternation coordinates, displaying the path for decay at (1) planar and (2) twisted conical intersection geometries. The equivalent ground state minima, related by double bond torsion, are labelled **FC** and **FC′**. Structures **CI$_{plan}$** and **CI$_{perp}$** correspond to the conical intersection stationary points on the seam where the methylene group lies in the plane of the ring and is perpendicular to the plane, respectively. The inset shows the combination of resonant pump pulse and non-resonant control pulse used in the propagations.

intramolecular vibrational energy redistribution (IVR) rate from the initially activated bond stretching mode. As a result, the wave packet samples the other regions of the seam and decays following path (2) in Fig. 1.

In this contribution we have used the NRDSE to control the excited state lifetime by modulating the topography of the seam. Thus, potential energy surface calculations show that the planar segment of the seam associated to the early decay can be made inaccessible at sufficiently intense fields. This suggests that the excited state lifetime can be prolonged under these conditions. To prove this hypothesis, we have added the non-resonant field to our four-dimensional model potential, and we have carried out MCTDH propagations with static and dynamic non-resonant strong fields. With a static field of 0.04 au (2.1×10^{10} V m^{-1} or 1.1×10^{14} W cm^{-2}), the lifetime is extended to about 40 fs, and a similar effect can be obtained with a dynamic field of 0.08 a.u (4.1×10^{10} V m^{-1} or 4.5×10^{14} W cm^{-2}). We also discuss the main practical limitation of the approach for the present fulvene case. Thus, our model only considers two electronic states, but at high field strengths the S_2 excited state interferes with S_0 and S_1. This indicates that more states should be included in the dynamics to make the treatment more accurate.

2 Four-dimensional model for fulvene

To describe the photophysics of fulvene with quantum dynamics it is necessary to construct a model potential of reduced dimension that reproduces the relevant regions of the potential energy surface. These are the low energy regions of the seam and the minimum energy path from the Franck–Condon structure, which is where the excited state dynamics starts, to the seam. Such a model surface, in four dimensions, has been described in detail in ref. 46. Here we summarize its main features.

The model is based on electronic structure calculations at the complete active space self-consistent field (CASSCF) level of theory, and a detailed characterization of the seam of intersection. The four coordinates are shown in Fig. 2. The initial decay

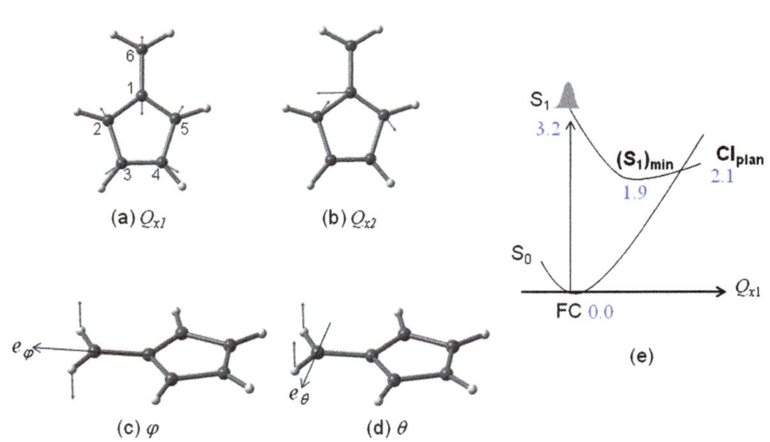

Fig. 2 Coordinates of the four dimensional model: (a) totally symmetric bond stretch, (b) non-totally symmetric bond stretch, (c) methylene torsion, and (d) methylene pyramidalization; (e) energy profile along the Q_{x1} coordinate at planar geometries (model energies in eV). e_φ and e_θ in (c) and (d) are the rotation axes of the rigid CH_2 fragment for φ and θ, respectively (see text).

takes place along the bond alternation coordinate Q_{x1} (Fig. 2a), and the corresponding energy profile is shown in Fig. 2e. The decay along Q_{x1} leads to a minimum on the excited state surface, $(S_1)_{min}$, and further to a conical intersection with the ground state, $\mathbf{CI_{plan}}$. $\mathbf{CI_{plan}}$ belongs to an extended seam with two low energy segments along the CH_2 torsion and pyramidalization. These segments are described in our model by coordinates φ and θ, respectively. The global minimum energy CI is found along the torsion coordinate, at $\varphi = 63°$, approximately. The model also contains the coupling coordinate Q_{x2}. Q_{x2} is a non-totally symmetric bond stretch coordinate approximately equal to the interstate coupling vector **ic** along the seam. This vector is one of the branching space vectors that lift the degeneracy linearly at the intersection, together with the gradient difference **gd** (see eqn (1)):

$$\mathbf{gd} = \nabla_{\mathbf{Q}}(E_2 - E_1) \tag{1a}$$

$$\mathbf{ic} = \langle \Phi_1 | \nabla_{\mathbf{Q}} \hat{H}_{ele} | \Phi_2 \rangle \tag{1b}$$

Φ_i is the adiabatic electronic wave function for state i. The four coordinate model captures the essential details of the early excited state dynamics, since it connects the Franck–Condon structure with the seam and reproduces the main features of the seam. Thus, by including coordinates φ and θ we include the low energy seam segments. Our model also reproduces the branching space along these segments, since the **gd** vector corresponds to combinations of Q_{x1} with φ or θ, while Q_{x2} is a good approximation to the **ic** vector along the whole seam.

Using this model, the dynamics are carried out solving the time-dependent Schrödinger equation (eqn (6)):

$$i\hbar \frac{\partial}{\partial t} \begin{pmatrix} |\Psi_1^d(t)\rangle \\ |\Psi_2^d(t)\rangle \end{pmatrix} = (\mathbf{T} + \mathbf{W} + \mathbf{H}_{int}) \begin{pmatrix} |\Psi_1^d(t)\rangle \\ |\Psi_2^d(t)\rangle \end{pmatrix} \tag{2}$$

The calculations are carried out using the regularized diabatic states approach, where the couplings are expressed through off-diagonal potential terms of the potential matrix **W** and the kinetic energy operator (KEO) $\hat{T}$ is diagonal.[47] Q_{x1} and Q_{x2} are treated as mass-weighted Cartesian displacements, whereas φ and θ are curvilinear coordinates. The elements of the KEO are given by eqn (3).

$$\hat{T} = \sum_{i=1,2} \hat{T}_{Q_{xi}} + \hat{T}_{\varphi} + \hat{T}_{\theta} \tag{3a}$$

$$\hat{T}_{Q_{xi}} = -\frac{1}{2\mu_i} \frac{\partial^2}{\partial Q_{x_i}^2} \quad (i = 1, 2) \tag{3b}$$

$$\hat{T}_{\xi} = -\frac{1}{2I_{\xi}} \frac{\partial^2}{\partial \xi^2} \quad (\xi = \varphi, \theta) \tag{3c}$$

where μ_i is the reduced mass of Q_{xi}, φ and θ correspond to rotations of the rigid CH_2 fragment around the e_{φ} and e_{θ} axes shown in Fig. 2c and 2d, respectively, and I_{ξ} is the moment of inertia corresponding to coordinate ξ. The details of the KEO and the approximations involved are discussed further in ref. 46.

$\mathbf{H}_{int}$ describes the interaction between the wave packet and the external electric field. For the parametrization of the surface, it is convenient to separate $\mathbf{H}_{int}$ in a resonant and a non-resonant term (eqn (4)):

$$\mathbf{H}_{\text{int}} = \mathbf{H}_{\text{int}}^{\text{res}} + \mathbf{H}_{\text{int}}^{\text{non-res}} \tag{4a}$$

$$\mathbf{H}_{\text{int}}^{\text{res}} = \begin{pmatrix} 0 & \mu_{12}\varepsilon^{\text{res}}(t) \\ \mu_{12}\varepsilon^{\text{res}}(t) & 0 \end{pmatrix} \tag{4b}$$

$$\mathbf{H}_{\text{int}}^{\text{non-res}} = \begin{pmatrix} \mu_1(\mathbf{Q})\varepsilon^{\text{non-res}}(t) + \frac{\alpha_1(\mathbf{Q})}{2}\left(\varepsilon^{\text{non-res}}(t)\right)^2 & 0 \\ 0 & \mu_2(\mathbf{Q})\varepsilon^{\text{non-res}}(t) + \frac{\alpha_2(\mathbf{Q})}{2}\left(\varepsilon^{\text{non-res}}(t)\right)^2 \end{pmatrix} \tag{4c}$$

ε^{res} and $\varepsilon^{\text{non-res}}$ are the resonant and non-resonant field components, respectively. In eqn (4b), μ_{12} is the transition dipole moment between states 1 and 2, while μ_i and α_i in eqn (4c) are the dipole moments and polarizabilities of state i, respectively. $\mathbf{H}_{\text{int}}$ has a simple physical interpretation: Thus, the resonant field interaction that pumps the population from the ground to the excited state works through the transition dipole moment between the states, while the non-resonant interaction that shifts the potentials takes place through the dipole moment and the polarizability of the states. In our model we assume that the molecules are aligned in space during the propagation, and in this case the two components of $\mathbf{H}_{\text{int}}$ correspond also to different spatial orientations of the pulse. Thus, in fulvene, μ_{12} is oriented along the x axis (see Fig. 1), whereas the dipole moments are oriented along the z axis. Therefore, in our model the resonant and non-resonant pulses are perpendicular to each other, as shown in Fig. 1. The model only includes the zz components of the polarizability tensor and neglects the Stark shift due to the resonant pulse, since this pulse is only applied during a short pumping period of 5 fs.

3 Parametrization of the field-dependent potential

To carry out the dynamics it is necessary to obtain the parameters of the potential matrix $\mathbf{W}$ and the dipole moments and polarizabilities from the non-resonant field matrix $\mathbf{H}_{\text{int}}^{\text{non-res}}$. The parametrization is carried out against *ab initio* calculations. The dipole moments and polarizabilities in eqn (4c) are coordinate dependent and are not calculated explicitly. Instead, the combined set of parameters of both matrices are obtained from a single fit. To this end, we start by defining a coordinate- and field-dependent potential $\mathbf{V}$ of the form:

$$\mathbf{V}(\mathbf{Q},\varepsilon) = \mathbf{W}(\mathbf{Q}) + \mathbf{H}_{\text{int}}^{\text{non-res}}(\mathbf{Q},\varepsilon) = V^0(\mathbf{Q},\varepsilon)\mathbf{I} + \begin{pmatrix} V_{11}(\mathbf{Q},\varepsilon) & V_{12}(\mathbf{Q}) \\ V_{12}(\mathbf{Q}) & V_{22}(\mathbf{Q},\varepsilon) \end{pmatrix} \tag{5}$$

The terms of $\mathbf{V}$ have a similar form to the field free potential $\mathbf{W}$ from our previous study. They form a Taylor expansion along Q_{x1}, Q_{x2}, $\sin(\varphi)$ and θ, centred at the planar conical intersection $\mathbf{CI_{plan}}$:

$$\begin{aligned} V^0(\mathbf{Q},\varepsilon) &= V^0(\varepsilon) + \left(\lambda_{x1}(\varepsilon) + \lambda_{x_1,\varphi}(\varepsilon)\sin^2(\varphi)\right)Q_{x_1} \\ &+ \frac{1}{2}\left(\omega_{x_1}(\varepsilon) + \omega_{x_1,\varphi}(\varepsilon)\sin^2(\varphi)\right)Q_{x_1}^2 + \frac{1}{2}\omega_\varphi(\varepsilon)\sin^2(\varphi) \\ &+ \frac{1}{2}\varsigma_\varphi\sin^4(\varphi) + \left(\frac{1}{2}\omega_\theta + \frac{1}{2}\omega_{x_1,\theta}Q_{x_1} + \frac{1}{2}\omega_{\varphi,\theta}\sin^2(\varphi)\right)\theta^2 \\ &+ \left(\frac{1}{2}\varsigma_\theta + \frac{1}{2}\varsigma_{x_1,\theta}Q_{x_1} + \frac{1}{2}\varsigma_{\varphi,\theta}\sin^2(\varphi)\right)\theta^4 + \frac{1}{2}\omega_{x_2}Q_{x_2}^2 \end{aligned} \tag{6a}$$

$$V_{11}(\mathbf{Q},\varepsilon) = -\frac{1}{2}\begin{bmatrix} \delta V(\varepsilon) + \left[\delta\kappa_{x1}(\varepsilon) + \delta\kappa_{x_1,\varphi}(\varepsilon)\sin^2(\varphi)\right]Q_{x_1} \\ +\frac{1}{2}\left[\delta\gamma_{x_1}(\varepsilon) + \delta\gamma_{x_1,\varphi}(\varepsilon)\sin^2(\varphi)\right]Q_{x_1}^2 + \frac{1}{2}\delta\gamma_{\varphi}(\varepsilon)\sin^2(\varphi) \\ +\frac{1}{2}\chi_{\varphi}\sin^4(\varphi) + \left(\frac{1}{2}\delta\gamma_{\theta} + \frac{1}{2}\delta\gamma_{x_1,\theta}Q_{x_1} + \frac{1}{2}\delta\gamma_{\varphi,\theta}\sin^2(\varphi)\right)\theta^2 \\ +\left(\frac{1}{2}\chi_{\theta} + \frac{1}{2}\chi_{x_1,\theta}Q_{x_1} + \frac{1}{2}\chi_{\varphi,\theta}\sin^2(\varphi)\right)\theta^4 + \frac{1}{2}\delta\gamma_{x_2}Q_{x_2}^2 \end{bmatrix} \tag{6b}$$

$$V_{22}(\mathbf{Q},\varepsilon) = -V_{11}(\mathbf{Q},\varepsilon) \tag{6c}$$

$$V_{12}(\mathbf{Q}) = \left(\kappa_{\varphi,\theta}^{AB}\sin^2(\varphi) + \kappa_{\phi,\theta 2}^{AB}\sin^4(\varphi)\right)\theta + \kappa_{x_2}^{AB}Q_{x_2} \tag{6d}$$

The field dependence is introduced making some of the parameters in eqn (6a) and (6b) depend on $\varepsilon^{\text{non-res}}$ (the superscript is left out for simplicity). These parameters are second-order expansions of ε, and the parameters of this expansion correspond to the dipole and polarizability terms from eqn (4c). For example, the first term of the right hand side of eqn (6a) is:

$$V^0(\varepsilon) = V_0^0 + \mu^0\varepsilon + \frac{\alpha^0}{2}\varepsilon^2 \tag{7}$$

where V_0^0 is zero because the energy at the centre of the expansion is zero. The same applies for the $\delta V(\varepsilon)$ term of eqn (6b). Similarly, the second term has the form:

$$\lambda_{x1}(\varepsilon) = \lambda_{x1,0} + \mu_{\lambda,x1}\varepsilon + \frac{\alpha_{\lambda,x1}}{2}\varepsilon^2 \tag{8}$$

The remaining terms have a similar form. This model has several simplifications with respect to the way the field is included. First, the field dependent terms along Q_{x2} and θ are neglected for simplicity. In practice, this implies that the non-resonant field produces a constant Stark shift along these coordinates. In addition, the Stark effect on the parameters of the coupling element V_{12} and the transition dipole moment μ_{12} is also neglected. Finally, we only consider the *zz* component of the polarizability and we neglect the Stark shift caused by the resonant pulse along the *x* axes, which would correspond to the *xx* and *xz* polarizability components.

The field-dependent parameters are fitted to *ab initio* calculations in the subspace formed by Q_{x1} and φ. The energy is calculated for several values of ε, where the external field along the *z* axis is included in a self-consistent way. In practice, this means fitting the terms of V to a three-dimensional grid of points along Q_{x1}, φ and ε. The calculations are carried out at the CASSCF/cc-pvdz level of theory, using an active space of six electrons in six orbitals, with the Gaussian program package.[48] This level of theory does not include dynamic electronic correlation. However, the comparison of the CASSCF data with complete active space second order perturbation CASPT2 calculations shows that the effect of dynamic correlation is small and can be neglected for the present purposes.[46]

In the space formed by Q_{x1} and φ, the electronic states S_0 and S_1 have *A* and *B* symmetry. Because of the different symmetry, the *ab initio* potential energy surface is equal to the diabatic potential, and the *ab initio* data can be used directly for the fit. The elements of V^0 are fitted to the average of the S_1 and S_0

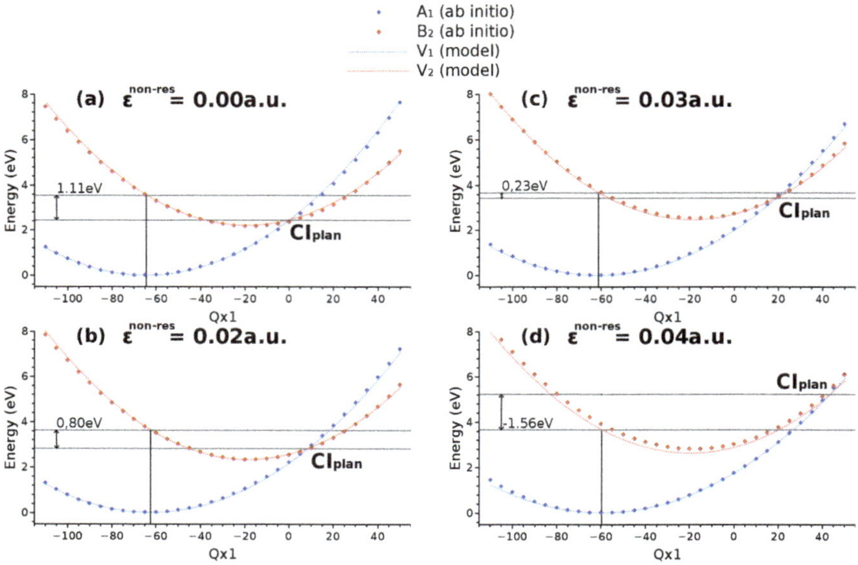

Fig. 3 Energy profile along Q_{x1} at different non-resonant field intensities (cuts for $(Q_{x2},\varphi,\theta) = (0,0,0)$), including the energy difference between the vertical excitation and $\mathbf{CI_{plan}}$. Blue and red diamonds: *ab initio* energies of the A_1 and B_2 states, respectively. Blue and red dashed lines: V_1 and V_2 model energies, respectively. Q_{x1} displacements in mass-weighted atomic units, energies in eV.

ab initio energies, while the elements of V_{11} are fitted to the difference between the energies of the two states. In the *ab initio* calculations, the electric field is included over a range of ± 0.04 a.u. However, at field strengths of -0.02 a.u. and lower, S_2 (of *A* symmetry) interferes with the S_0 and S_1 states at regions with values of φ close to 90°. At these geometries, S_2 has a large dipole moment because of a charge transfer from the methylene group to the ring, and the Stark shift is more pronounced than for the lowest pair of states. The interference of S_2 leads to discontinuities in the potential, and to facilitate the fit the corresponding points are discarded. This is the case for 805 *ab initio* points out of a total of 14190. The set of 34 parameters that depend on Q_{x1}, φ and ε are then fitted to the remaining 13 385 points using the nonlinear least-squares Marquartdt–Levenberg algorithm implemented in Gnuplot.[49] The root mean square of residuals is 0.0024 Hartree for V_0 and 0.0032 for V_{11}. The good agreement between the model and the *ab initio* is shown below (Fig. 3) for energy cuts along Q_{x1} at different field strengths. The remaining 19 parameters of the model are taken over from ref. 46.

4 Stark effect on the potential energy surface

We start our approach to the control with an analysis of the Stark effect on the potential energy surface, focusing on the changes along the decay path. This analysis gives a preliminary idea of the field strengths that will be required to achieve a significant extension of the excited state lifetime in the propagations, and it also shows the physical origin of the shifts that will allow for the control. In Fig. 3 we plot the energy profile along Q_{x1} at different field strengths, 0.00 a.u., -0.02 a.u., -0.03 a.u. and -0.04 a.u. As the field is applied, $\mathbf{CI_{plan}}$ is shifted along

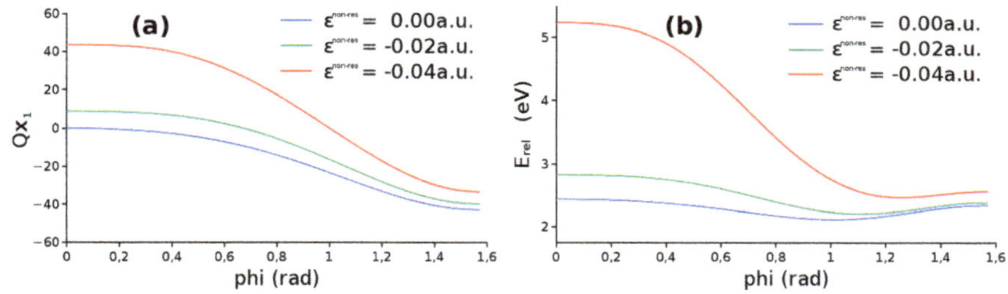

Fig. 4 (a) Locus of the V_2/V_1 seam in the (Q_{x1},φ) plane (projections for $(Q_{x2},\theta) = (0,0)$) and (b) seam energy along φ, at different field strengths. $\mathbf{CI_{plan}}$ corresponds to $\varphi = 0$ rad, and $\mathbf{CI_{perp}}$ to $\varphi = \pi$ rad.

Q_{x1} and its energy relative to the S_1 vertical excitation energy (marked with dotted lines) increases. Thus, $\mathbf{CI_{plan}}$ lies well below the vertical excitation with no field and with a -0.02 a.u. field (1.11 and 0.80 eV, respectively). At -0.03 a.u., $\mathbf{CI_{plan}}$ lies only 0.23 eV lower than the vertical excitation, and at -0.04 a.u. it lies 1.56 eV higher in energy. The effect of the field on the topography of the extended seam is shown in Fig. 4, where we display the projection of the seam in the (Q_{x1},φ) space. Thus, Fig. 4a shows the locus of the seam in this two-dimensional space, and Fig. 4b shows its energy. Both plots show that the field mainly changes the seam at planar structures, shifting the position and energy of $\mathbf{CI_{plan}}$ as discussed for Fig. 3. In contrast to this, the changes in the seam around $\mathbf{CI_{perp}}$ are small.

Overall, our results show that the potential does not respond linearly to the external field. Instead, the main changes are due to the second order terms with respect to the field, *i.e.* the polarizability. This is due to the fact that the two states are only weakly polar. This is important for our purposes because in practice the control has to be exerted with a dynamic, oscillating pulse, and in this case the dominant effect will be due to the polarizability. The energy profiles in Fig. 3 also show that the changes in the energy and position of the CI are due to the relative shifts of the two states, combined with changes in the shape of the surface. These changes are especially important for the ground state, which becomes flatter as the field is increased. This reflects the changes in the dipole moment and the polarizability along Q_{x1} and justifies the inclusion of the coordinate dependence in the dipole and polarizability parameters of our model (eqn (4)). On the basis of these results we can also expect that a field strength of -0.04 a.u. will be required to extend the lifetime of the excited state, since at this intensity the energy of the CI lies notably higher than the vertical excitation.

5 Excited state dynamics under a non-resonant field

The quantum dynamics propagations controlled by the non-resonant field have been carried out with the MCTDH approach. We have run two sets of calculations with static and dynamic non-resonant electric fields. In this section we describe the details of the propagations and the results for the two sets.

5.1 MCTDH propagation details

The MCTDH dynamics were carried with the Heidelberg MCTDH package.[45] For Q_{x1} and Q_{x2} we use a primitive basis of 48 and 25 harmonic oscillator functions, respectively, for φ a fast Fourier transform collocation of 128 functions, and for θ a primitive basis of 25 sine functions. For the single particle basis, we have used a combined mode for Q_{x1} and φ (50 single particle functions (SPF) per diabatic state), as explained in section 4.5 of ref. 42, and 25 and 20 SPF per state for Q_{x2} and θ, respectively. The propagations, which were run over 150 fs, are well converged with respect to the number of SPF, since in all runs the maxima over time of the lowest natural weight for all modes did not exceed 0.001. The propagations were initiated relaxing the wave packet in the electronic and vibrational ground state, and exciting it to the electronic excited state with a single laser pulse of the form:

$$\varepsilon^{\mathrm{res}}(t) = \varepsilon_0^{\mathrm{res}} \cos\left(\omega^{\mathrm{res}} t\right) \sin^2\left(\frac{\pi}{\tau} t\right) \tag{9}$$

In eqn (7), $\varepsilon_0^{\mathrm{res}}$ is the maximum intensity of the resonant pulse, which was set to 0.075 a.u. (3.9×10^{10} V m^{-1} or 3.9×10^{14} W cm^{-2}) for all runs except for the run with a non-resonant dynamic pulse with maximum intensity 0.08 a.u., where $\varepsilon_0^{\mathrm{res}}$ was set to 0.10 a.u. to achieve an initial V_2 population of 0.4. Further, $\hbar\omega^{\mathrm{res}}$ is the excitation energy (3.6 eV), and τ the pulse duration (5 fs). The pulse was set to zero outside the $[0,\tau]$ interval. For the propagations with a dynamic non-resonant field, the field has a simple sinusoidal shape with a carrier frequency $\hbar\omega^{\mathrm{non-res}}$ of 1 eV (eqn (8)), which corresponds to a wave length of 1239 nm:

$$\varepsilon^{\mathrm{non-res}}(t) = -\varepsilon_0^{\mathrm{non-res}} \sin(\omega^{\mathrm{non-res}} t) \tag{10}$$

5.2 Propagations under static non-resonant fields

To understand the effect of the off-resonant pulse on the dynamics, we start with a set of propagations with a constant (static) field of various strengths. In Fig. 5a we show the evolution of the excited state populations (V_2 diabatic state) during the first 50 fs of three propagations with fields of 0.00, −0.02 and −0.03 a.u. In all cases, the resonant pulse takes place during the first 5 fs of the propagation (eqn (7)). During this time, the V_2 population rises from an initial value of 0.0 to a maximum value of approximately 0.4. Due to the very favourable energy profile (Fig. 3), the wave packet decays quickly to the ground state and there is a sharp drop of the population in the first 20 fs. However there are some differences between the propagations with the different fields. In the field free case, the population decreases to approximately 0.15 after a total of 20 fs (blue line). When the field is applied, the initial decay becomes

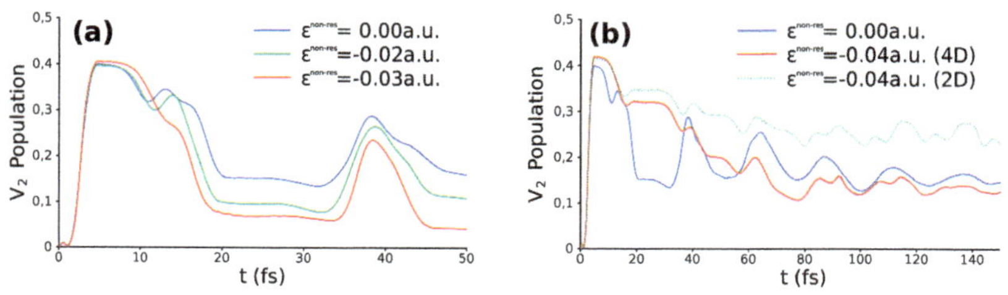

Fig. 5 V_2 population for propagations with static non-resonant fields of different intensities. (a) Non-resonant field intensity 0.00 a.u. (blue profile), −0.02 a.u. (green profile), and −0.03 a.u. (red profile). (b) Non-resonant field intensity 0.00 a.u. (blue profile), −0.04 a.u. (red profile) and −0.04 a.u. on a two-dimensional reduced model (green hashed profile, see text). Note the different time axes for (a) and (b).

more efficient, and the populations after 20 fs are approximately 0.10 and 0.08 with fields of −0.02 and −0.03 a.u., respectively. This result is somewhat counterintuitive, since the field shifts the planar conical intersection up in energy (Fig. 3), in which case a less efficient decay might be expected. However the result can be understood in terms of overlap between vibrational levels. According to the Franck–Condon principle, the V_2 vibrational level that absorbs most of the excited state population during the pump pulse will be that corresponding to the vertical excitation. Then, the $V_2 \rightarrow V_1$ population transfer rate depends in first instance on the overlap between the vibrational function of this level and that of the degenerate V_1 vibrational level. The overlap increases as the vertical excitation and conical intersection energies get closer, and it becomes maximal when both energies are equal, which occurs in our model for a field of approximately −0.032 a.u. This explanation agrees with the result that the most efficient decay is observed at a field of −0.03 a.u.

With a stronger field of −0.04 a.u., the situation is reversed and the decay is slower (Fig. 5b). Thus, the population that remains in V_2 after 20 fs is approximately 0.30, and it takes approximately 45 fs to reach a value of 0.20 (red line of Fig. 5b). This is due to the fact that the planar conical intersection lies substantially higher in energy than the vertical excitation. To illustrate this point, we compare the evolution of the wave packet on V_2 with fields of 0.00 a.u. and −0.04 a.u. for several snapshots (Fig. 6). The wave packet isodensity and the seam are plotted in red and black, respectively. The time and V_2 population of the snapshot are also displayed. Note that the wave packet is split in two halves because φ is a periodic coordinate, and the wave packet is centred initially around $\varphi = 0°$. In the field free case, after 20 fs, the wave packet has traversed the seam at the planar region ($\varphi \approx 0°$), and there is a substantial population decay from a maximum of 0.40 to 0.16. In contrast to this, at a field of −0.04 a.u. the wave packet does not reach the planar seam region in early times. The population decay is slower, and the population after 20 fs is 0.32. In this propagation, the wave packet has to spread along φ to reach the lower energy regions of the seam and decay to V_1 efficiently. This happens after approximately 36 fs, and at 44 fs the V_2 population is reduced to 0.21. Once the wave packet spreads along φ and accesses the seam, the control effect is lost, and the populations in the two propagations are similar. The time that the wave packet requires to spread corresponds to IVR from the initially activated Q_{x1} mode to the torsion coordinate φ, and the present value of approximately 36 fs is in good agreement with our previous study on fulvene excited state dynamics. In that case IVR took place after about 40 fs with an excitation energy of 3.5 eV.

To assess this point further, we have repeated the propagation with constant −0.04 a.u. field strength on a reduced two-dimensional potential surface, where the rotation and pyramidalization coordinates φ and θ are not included. This excludes the possibility of decay along the extended seam, since the CI is reduced to a point. The decay of the V_2 population is slower than in the four dimensional case (green line of Fig. 5b), and the population decrease at around 40 fs due to access to the twisted regions of the seam is not observed. This confirms that the main fraction of the wave packet decays at the twisted segment of the seam in the propagation under the −0.04 a.u. static field. In more general terms, it illustrates the need to include the extended seam to carry out reliable simulations of the decay.

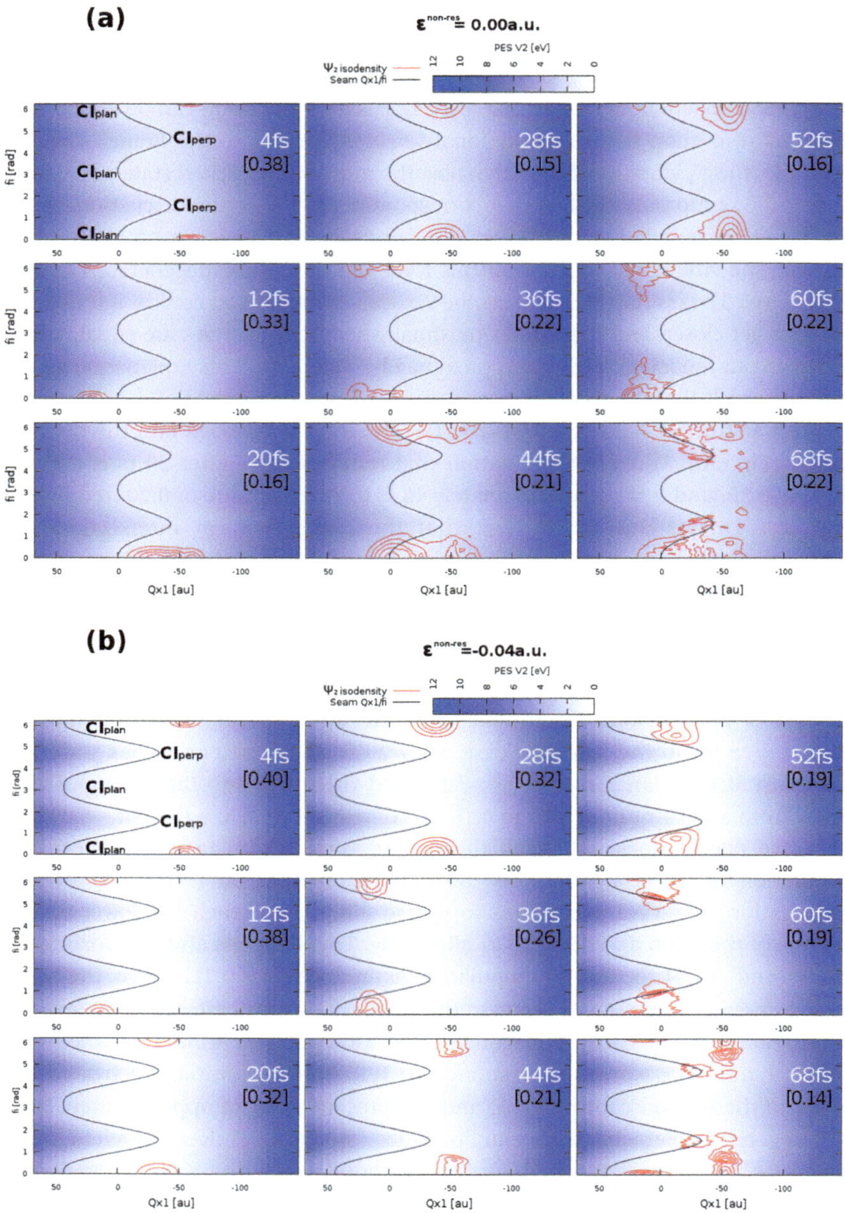

Fig. 6 Snapshots of the wave packet propagations on the V_2 model surface along Q_{x1} and φ (integrated density along Q_{x2} and θ; coordinates in mass-weighted atomic displacements and radians, respectively), displaying the time of the snapshot and the V_2 population. Propagations carried out (a) without non-resonant field and (b) with a static non-resonant field of −0.04 a.u. Energy scale shown in the top of the figure. Wave packet density plotted in red and seam position in black. The wave packet is split in two halves because it is centred initially around $\varphi = 0°$ in the periodic coordinate φ.

5.3 Propagations with dynamic non-resonant field

A realistic simulation requires the use of a dynamic non-resonant field to achieve the necessary field strengths. Thus, we have assessed the viability of our control strategy carrying out propagations with a dynamic field and a carrier wave length

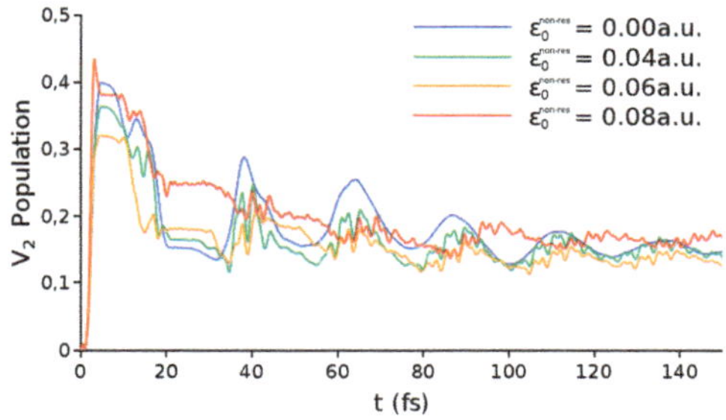

Fig. 7 V_2 population for propagations with dynamic non-resonant fields of different intensities. Maximum field intensity $\varepsilon_0^{\text{non-res}}$: 0.00 a.u. (blue profile), 0.04 a.u. (green profile), 0.06 a.u. (yellow profile), and 0.08 a.u. (red profile).

of 1239 nm, which is close to the value of 1700 nm used in experiments.[12] This corresponds to a non-resonant pulse period of approximately 4 fs. We assume that the non-resonant pulse is switched on before the resonant one and is not modified during the whole propagation. As a rule of thumb, one can estimate that the average effect on the potential of a dynamic field with maximum strength $\varepsilon_0^{\text{non-res}}$ is that of a static field with half the strength, because both have the same intensities integrated over time. Thus, we have carried out propagations using dynamic pulses with $\varepsilon_0^{\text{non-res}}$ equal to 0.04 a.u., 0.06 a.u. and 0.08 a.u. and compared them with the propagations using static fields of −0.02 a.u., −0.03 a.u. and −0.04 a.u. The results are shown in Fig. 7. With dynamic non-resonant fields of 0.04 a.u. and 0.06 a.u. maximum intensity, the V_2 populations have a similar evolution to the propagation without resonant field. Moreover, the acceleration of the decay seen with static fields of −0.02 a.u. and −0.03 a.u. is not seen with dynamic fields. This happens because the improved vibrational overlap that causes the faster decay in the static case is not a linear function of the field, and the average effect of the dynamic field on the rate is not the same as the effect caused by a static field with the same intensity integrated over time. In contrast to this, the delay of the decay observed for a dynamic field of 0.08 a.u. is similar to that observed with a static −0.04 a.u. field, although it is less pronounced. These examples show that the 'rule of thumb' given above to estimate the effect of a dynamic field on the basis of the static field results can indeed provide the correct general picture, since it has allowed us to predict that a significant delay of the decay would be observed with a dynamic field of maximal −0.08 a.u. intensity. However, this approximation also fails to detect subtle effects such as the acceleration due to improved vibrational overlap observed for static fields of −0.02 a.u. and −0.03 a.u.

6 Conclusions

Our work shows that it is possible to simulate the control of the photophysics of an organic molecule by shaping the topography of an extended conical intersection seam with the NRDSE. We have followed a three step approach, where we have first characterized the Stark effect on the topography of the seam. *Ab initio* calculations suggest that a field strength of 0.04 a.u. is required to increase the excited state lifetime significantly. Second, we have simulated the control with static non-resonant pulses and find that the lifetime can be reduced with fields up to 0.03 a.u. because the overlap between the degenerate electronic excited and ground state vibrational functions is increased. In contrast to this, a field of 0.04 a.u. increases the excited state lifetime from less than 10 fs in the field free case to approximately 40 fs. Finally, in the third step we have shown that a similar reduction of the decay rate can be achieved with a dynamic, sinusoidal non-resonant pulse of 0.08 a.u. maximal strength.

In fulvene, the decay along the non-planar segment of the seam is associated with rotation of the methylene group and isomerization of the exocyclic double bond. The present results raise the question if the isomerization process can also be controlled with the NRDSE. Preliminary analyses indicate that the NRDSE on the isomerization is not so substantial, *i.e.* the fraction of fulvene that undergoes double bond isomerization before the decay is not modified much by the non-resonant field. This happens presumably because the changes on the topography of the seam at twisted methylene angles are not so pronounced than at planar geometries. Moreover, it seems that the IVR rate which induces the methylene torsion is not affected by the topological changes induced by the Stark effect. This issue will be considered in more detail in future work, where we will also consider the NRDSE on propagations with different vertical excitation energies, where the IVR rate can be modulated.

Our approach also has some limitations. The first one is the dimensionality of the reduced model surface. Our surface captures the main features of the decay because we are considering a very fast process where the excitation energy is concentrated in only a few modes. In addition, the choice of the coordinates is based on a careful analysis of the extended seam of intersection responsible for the photophysics. However, as discussed in ref. 46, the inclusion of more modes should improve the agreement of the model with the full-dimensional *ab initio* surface. It would also improve the description of the NRDSE. Thus, our model assumes implicitly that the energy shift due to the non-resonant pulse will be constant along all coordinates except Q_{x1} and φ. However, the Stark shift does change significantly along Q_{x1} and φ. This may also occur along other coordinates, which should be taken into account in a more refined model. Another limitation is due to the interference of the second excited state, S_2, at fields of 0.02 intensity and more. At geometries where the methylene group is twisted, this state is more polar than the ground and the excited state, and becomes lower in energy. More states should be included in the simulation to take this effect into account and make the simulations more realistic. In addition, field strengths of up to 0.08 a.u., which corresponds to a power density of 4.5×10^{14} W cm^{-2}, were needed to achieve the effects presented here. In practice, such power densities are likely to cause ionization, which makes the experimental implementation of an NRDSE based control scheme in fulvene challenging. To carry out the control experimentally, it is necessary to find molecules where the ground and excited state dipole moments and polarizabilities differ more substantially than in fulvene.

In more general terms, our work has provided us with a new insight into the relationship between the intersection topography and the dynamics. Thus, by modifying the topography with the external field we have shown that the most efficient decay occurs when the conical intersection is energetically degenerate with the Franck–Condon point. The work is also important because it shows the effect of the non-resonant field on the extended seam. The Stark shift of the potential energy surface is not homogeneous, and as a result the topography of the seam undergoes different changes at different segments. In the present case, the largest effects are seen at planar structures. This makes it possible to control the excited state lifetime during the first 40 fs of the simulation, approximately, but the control is lost at longer times because the seam segment at twisted geometries is still accessible energetically. This demonstrates that the extended nature of the seam has to be taken into account in optical control simulations of polyatomic molecules, not only with the NRDSE but probably also with other strategies.

Acknowledgements

This work has been supported by grants CTQ2008-06696 and CTQ2011-26573 from the Spanish Ministerio de Ciencia e Innovación (MICINN) and Ministerio de Economía y Competitividad (MINECO), respectively, SGR0528 from the Catalan Agència de Gestió d'Ajuts Universitaris i de Recerca (AGAUR), UNGI08-4E-003 from MICINN and the European Fund for Regional Development, and the Xarxa de Referència en Química Teòrica i Computacional de Catalunya from AGAUR. S.R.-B. thanks the MICINN for grant BES-2009-029177.

References

1 C. Brif, R. Chakrabarti and H. Rabitz, *New J. Phys.*, 2010, **12**, 075008.
2 C. Brif, R. Chakrabarti and H. Rabitz, *Adv. Chem. Phys.*, 2012, **148**, 1–76.
3 A. H. Zewail, in *Femtochemistry*, ed. F. C. De Schryver, S. De Feyter and G. Schweitzer, Wiley-VCH, Weinheim, Germany, 2001, pp. 1–86.
4 H. Stapelfeldt and T. Seideman, *Rev. Mod. Phys.*, 2003, **75**, 543–557.
5 T. Brixner and G. Gerber, *ChemPhysChem*, 2003, **4**, 418–438.
6 M. Dantus and V. V. Lozovoy, *Chem. Rev.*, 2004, **104**, 1813–1859.
7 V. V. Lozovoy and M. Dantus, *ChemPhysChem*, 2005, **6**, 1970–2000.
8 G. G. Balint-Kurti, S. Zou and A. Brown, *Adv. Chem. Phys.*, 2008, **138**, 43–94.
9 V. Engel, C. Meier and D. J. Tannor, in *Advances in Chemical Physics, Vol 141*, ed. S. A. Rice, 2009, vol. 141, pp. 29–101.
10 R. J. Levis, G. M. Menkir and H. Rabitz, *Science*, 2001, **292**, 709–713.
11 G. A. Worth and C. S. Sanz, *Phys. Chem. Chem. Phys.*, 2010, **12**, 15570–15579.
12 B. J. Sussman, D. Townsend, M. Y. Ivanov and A. Stolow, *Science*, 2006, **314**, 278–281.
13 D. Townsend, B. J. Sussman and A. Stolow, *J. Phys. Chem. A*, 2011, **115**, 357–373.
14 B. Y. Chang, H. Choi, S. Shin, S. Lee and I. R. Sola, *J. Mod. Opt.*, 2009, **56**, 811–821.
15 H. Choi, W.-J. Son, S. Shin, B. Y. Chang and I. R. Sola, *J. Chem. Phys.*, 2008, 128.
16 P. Marquetand and V. Engel, *Phys. Chem. Chem. Phys.*, 2005, **7**, 469–474.
17 S. Scheit, Y. Arasaki and K. Takatsuka, *J. Phys. Chem. A*, 2012, **116**, 2644–2653.
18 S. Belz, S. Zilberg, M. Berg, T. Grohmann and M. Leibscher, *J. Phys. Chem. A*, 2012, **116**, 11189–11198.
19 D. Kinzel, P. Marquetand and L. Gonzalez, *J. Phys. Chem. A*, 2012, **116**, 2743–2749.
20 G. W. Richings and G. A. Worth, *J. Phys. Chem. A*, 2012, **116**, 11228–11240.
21 F. Bernardi, M. Olivucci and M. A. Robb, *Chem. Soc. Rev.*, 1996, **25**, 321–328.
22 W. Domcke, D. R. Yarkony and H. Köppel, ed., *Conical Intersections: Electronic Structure, Dynamics & Spectroscopy*, World Scientific, Singapore, 2004.
23 W. Domcke, D. R. Yarkony and H. Köppel, ed., *Conical Intersections: Theory, Computation and Experiment*, World Scientific, Singapore, 2011.
24 M. Klessinger and J. Michl, *Excited States and Photochemistry of Organic Molecules*, VCH Publishers, Inc., New York, 1995.
25 D. R. Yarkony, *Rev. Mod. Phys.*, 1996, **68**, 985–1013.
26 G. J. Atchity, S. S. Xantheas and K. Ruedenberg, *J. Chem. Phys.*, 1991, **95**, 1862–1876.
27 A. Migani, M. A. Robb and M. Olivucci, *J. Am. Chem. Soc.*, 2003, **125**, 2804–2808.
28 D. R. Yarkony, *J. Chem. Phys.*, 2005, **123**, 204101.
29 L. Blancafort, B. Lasorne, M. J. Bearpark, G. A. Worth and M. A. Robb, in *The Jahn–Teller Effect: Fundamentals and Implications for Physics and Chemistry*, ed. H. Köppel, D. R. Yarkony and H. Barentzen, Springer, Heidelberg, Berlin, 2009, pp. 169–200.

30 M. Kobylecka, A. Migani, D. Asturiol, J. Rak and L. Blancafort, *J. Phys. Chem. A*, 2009, **113**, 5489–5495.
31 B. Lasorne, M. J. Bearpark, M. A. Robb and G. A. Worth, *J. Phys. Chem. A*, 2008, **112**, 13017–13027.
32 Q. Li, A. Migani and L. Blancafort, *J. Phys. Chem. Lett.*, 2012, **3**, 1056–1061.
33 A. Migani, L. Blancafort, M. A. Robb and A. D. Debellis, *J. Am. Chem. Soc.*, 2008, **130**, 6932–6933.
34 F. Sicilia, M. J. Bearpark, L. Blancafort and M. A. Robb, *Theor. Chem. Acc.*, 2007, **118**, 241–251.
35 C. S. M. Allan, B. Lasorne, G. A. Worth and M. A. Robb, *J. Phys. Chem. A*, 2010, **114**, 8713–8729.
36 D. Asturiol, B. Lasorne, G. A. Worth, M. A. Robb and L. Blancafort, *Phys. Chem. Chem. Phys.*, 2010, **12**, 4949–4958.
37 D. Mendive-Tapia, B. Lasorne, G. A. Worth, M. J. Bearpark and M. A. Robb, *Phys. Chem. Chem. Phys.*, 2010, **12**, 15725–15733.
38 K. R. Asmis, M. Allan, O. Schafer and M. Fulscher, *J. Phys. Chem. A*, 1997, **101**, 2089–2095.
39 P. J. Domaille, J. E. Kent and M. F. Odwyer, *Chem. Phys.*, 1974, **6**, 66–75.
40 J. E. Kent, P. J. Harman and M. F. Odwyer, *J. Phys. Chem.*, 1981, **85**, 2726–2730.
41 M. J. Bearpark, F. Bernardi, M. Olivucci, M. A. Robb and B. R. Smith, *J. Am. Chem. Soc.*, 1996, **118**, 5254–5260.
42 M. H. Beck, A. Jackle, G. A. Worth and H. D. Meyer, *Phys. Rep.*, 2000, **324**, 1–105.
43 H. D. Meyer, U. Manthe and L. S. Cederbaum, *Chem. Phys. Lett.*, 1990, **165**, 73–78.
44 H.-D. Meyer, F. Gatti and G. A. Worth, ed., *Multidimensional Quantum Dynamics: MCTDH Theory and Applications*, WILEY-VCH, Weinheim, Germany, 2009.
45 G. A. Worth, M. H. Beck, A. Jackle and H. D. Meyer, *The MCTDH Package, Version 8.2*(2000), University of Heidelberg, Heidelberg, Germany. H.-D. Meyer, Version 8.3(2002), Version 8.4 (2007). See http://mctdh.uni-hd.de.
46 L. Blancafort, F. Gatti and H. D. Meyer, *J. Chem. Phys.*, 2011, **135**, 134303.
47 H. Köppel, J. Gronki and S. Mahapatra, *J. Chem. Phys.*, 2001, **115**, 2377–2388.
48 M. J. Frisch, G. W. Trucks, H. B. Schlegel, G. E. Scuseria, M. A. Robb, J. R. Cheeseman, G. Scalmani, V. Barone, B. Mennucci, G. A. Petersson, H. Nakatsuji, M. Caricato, X. Li, H. P. Hratchian, A. F. Izmaylov, J. Bloino, G. Zheng, J. L. Sonnenberg, M. Hada, M. Ehara, K. Toyota, R. Fukuda, J. Hasegawa, M. Ishida, T. Nakajima, Y. Honda, O. Kitao, H. Nakai, T. Vreven, J. A. Montgomery Jr., J. E. Peralta, F. Ogliaro, M. J. Bearpark, J. J. Heyd, E. Brothers, K. N. Kudin, V. N. Staroverov, T. Keith, R. Kobayashi, J. Normand, K. Rhagavachari, A. Rendell, J. C. Burant, S. S. Iyengar, J. Tomasi, M. Cossi, N. Rega, J. M. Millam, M. Klene, J. E. Knox, J. B. Cross, V. Bakken, C. Adamo, J. Jaramillo, R. Gomperts, R. E. Stratmann, O. Yazyev, A. J. Austin, R. Cammi, C. Pomelli, J. W. Ochterski, R. L. Martin, K. Morokuma, G. Zakrzewski, G. A. Voth, P. Salvador, J. J. Dannenberg, S. Dapprich, A. D. Daniels, O. Farkas, J. B. Foresman, J. V. Ortiz, J. Cioslowski and D. J. Fox, *Gaussian 09, Revision C.01*; Gaussian, Inc.: Wallingford, CT, 2010.
49 T. Williams, C. Kelley, H.-B. Bröker and E. A. Merritt, *Gnuplot, Version 4.2*, 2007.

DISCUSSIONS

General discussion

DOI: 10.1039/c3fd90020g

Professor Whitaker opened the discussion of the paper by Professor Herek by communicating: I would just like to make a quick observation and then I have a more substantive question. What you call "Passive Control" seems to me to be just a fancy way of saying "Chemistry". I'm a little unclear if your aim to use coherent control to inform chemistry or chemistry to inform control strategies? Perhaps both?

My real question concerns Fig. 8, particularly the bottom right panel. As you write in the paper the shape of the response as a function of the wavelength of the phase-flip is remarkably reminiscent of Meshulach and Silberberg's classic *Phys. Rev. A* paper (your ref. 11) in which they examined a two photon response in atomic Cs vapour. You take this as supporting evidence for a multiphoton control mechanism. But the constructive interference observed by them for a 180 degree phase-flip in the centre of the excitation spectrum is the consequence of non-resonant two-photon (more generally N-photon) excitation, and this seems to contradict your chirp data, which would suggest to me a wavepacket following mechanism *via* a detuned resonance. Can you rationalise these two seemingly opposing views?

Professor Herek communicated in reply: Professor Whitaker's remark is correct; in this case, passive control refers to chemical modifications on the periphery of the molecule that affect the binding interaction and thereby the chemical reactivity. In particular, these modifications may result in substantial changes in the forward and backward electron injection rates thereby providing an effective mechanism of passively tuning the functional efficiency.

In general, passive control refers to all of the modifications that are made to the system or environment (*i.e.* temperature, pH, synthetic design of reactants, *etc.*) at the onset in an attempt to modify the outcome of the reaction, whereas active control implies engaging with the system during the reaction itself so as to influence directly the evolving dynamics. Coherent control is a subset of active control in which quantum coherence effects play a role. These are generated by tuning laser pulse parameters such as the phase, amplitude and polarization of individual frequency components to manipulate the generation and evolution of the excited state wavepacket.

Overall, our aim is to use both active and passive control strategies to gain fundamental insight into excited state dynamics of molecules, and the underlying factors that govern these.

Regarding the "w" shape observed in the π-phase-step scan, we were indeed inspired by the classic work of Meshulach and Silberberg[1] in suggesting that this behavior reflects a two-photon (or potentially N-photon) control mechanism. Likewise, the chirp scan results also imply a two-photon pump-dump control mechanism, analogous to that explored in early experiments by Bardeen and Wilson.[2] The former is generally a non-resonant interaction, while the pump-dump process tracks a wavepacket on an excited state potential energy surface. Note that for both the π-step and chirp scans (see Fig. 8), the control yield was much smaller than was found in the blind optimization. For such a complex system of a large organic molecule in the condensed phase, it is likely that many different control mechanisms are at play. The systematic open-loop scans presented here suggest two of them.

1. D. Meshulach and Y. Silberberg, *Nature*, 1998, **396**, 239.
2. A. H. Buist *et al.*, *Opt. Lett.*, 1999, **24**, 244.

Professor Fielding communicated: Professor Herek has presented a beautiful illustration of both active and passive control. Would she be able to comment on the possibilities for exploring links between the two methods, *i.e.* learning from active control experiments how and what passive control methods might be realised to achieve a particular goal?

Professor Herek communicated in reply: I appreciate this comment and the excellent question, if not proposition, to further explore the powerful combination of active and passive control strategies as a means for obtaining deeper understanding of photophysical processes and chemical reaction mechanisms. Insights gleaned from active control experiments, in particular open-loop studies in which specific pulse shapes (*i.e.* chirps, pulse trains, π-steps) are scanned, can reflect features of the potential energy landscape or identify key vibrational modes. These are precisely the sorts of parameters than can be fine-tuned by passive control strategies involving chemical modification.

Professor Field communicated: Spin-orbit is one of my favorite topics. Owing to its exceptionally atom-localized character, spin-orbit is useful in formulating an atom-in-molecule picture of electronic structure or in devising numerically specific modifications of dynamical processes. For example, by replacing the Zn atom in your ZnPc molecule, you could scale the spin-orbit interaction strength up or down by a known amount (see Table 5.6 on pages 316–317 of "The Spectra and Dynamics of Diatomic Molecules," H. Lefebvre-Brion and R. W. Field, Elsevier, 2004). This would allow you to modify the rate and mechanism of Inter-System Crossing. My question concerns how ISC is exploited in your active control schemes. If the spin-orbit constant is small, then the existence of spin-mixed states will depend on accidental degeneracies between vibronic levels of singlet and triplet states. Most eigenstates will be nearly pure singlet or nearly pure triplet, and population transfer between singlet and triplet will occur only *via* the small number of spin-mixed levels. It seems to me that this is a good situation for cleverly designed control schemes. If the spin-orbit constant is large, then nearly all vibronic states will be of spin-mixed character. In such a case, population will flow between nominal singlet and triplet levels promiscuously. There will be little opportunity for control schemes to modify the rate or efficiency of singlet to triplet population transfer. It would be very interesting if you could compare active control schemes for ZnPc to those for CdPc and CaPc.

Professor Herek communicated in reply: This is an excellent suggestion for a passive control scheme that exploits spin–orbit coupling to modify reaction dynamics, and presents a nice case for combining active and passive control strategies. At this point in our studies, we don't really know how ISC is exploited in the active control scheme, if at all! After all, ISC occurs on a very slow time scale in ZnPc; our feedback window for this path is some 20 ns after the excitation pulse. Comparing the control yield in ZnPc with its Cd- and Ca- siblings, could be very insightful. It would also be interesting to study systems which exhibit ultrafast spin dynamics, on a comparable timescale as the vibrational dynamics.

Professor Buma communicated: In your experiments you observe after optimization a certain ratio for the T/S ratio of ZnPc. Can you say *a priori* anything about what values should theoretically be possible and what the limits would be ? Or, put in a different way, how does one know that the ratio that you have obtained is the best one possible ? Would it be, for example, possible that if you would have started with a completely different set of phase and intensity values that you would have ended ten times as high?

Professor Herek communicated in reply: It's unlikely that the result of a blind optimization experiment is "the best one possible", and even more so that starting with a different set of random phase and intensity values would yield an order of magnitude difference in the control yield. Most likely, the real optimum lies somewhere in between. Not only do learning algorithms need to be sufficiently smart in efficiently sampling the enormous multidimensional parameter space, but limitations of the spatial light modulator preclude many solutions. Furthermore, as discussed before, there may be many possible control mechanisms. Hence any result can be a good result, if we can discern how the corresponding pulse shape acts on the system to induce a change in the reaction outcome.

Mr Arruda communicated: In Fig. 5, you show some X-FROG traces of optimized pulses with strong features that span several hundred femtoseconds, and up to a picosecond in Fig. 5d. How do you determine your time resolution and effective *t*-zero position? Does it even matter to you in the context of these measurements?

Professor Herek communicated in reply: The X-FROG traces are made by cross-correlation of the shaped pump pulse and a well-characterized transform limited unshaped reference pulse containing the same power spectrum and with duration of 25 fs. The reference pulse traverses the same path length as the supercontinuum probe pulse, and is overlapped with the pump in a nonlinear crystal placed at the sample position, hence the effective time zero is sufficiently well defined (especially in the context of the measurements, where the control effect is probed at times of 25 ps and 20 ns after the shaped excitation pulse).

The temporal resolution is determined by the FWHM of the cross-correlation signal at time zero.

Professor Riedle communicated: How far into saturation is your experiment performed? The reason this seems most important to me is the following: if you go close to saturation, any extra pump energy will not increase the excitation level. This is at least true if your pulses are longer than the electronic dephasing time. However, if the excited state vibronic wavepacket moves away from the Franck–Condon point sufficiently to loose resonance with the pump pulse, you basically start your excitation process again from a pure ground state - yet diminished - sample and can propagate further population into the excited electronic state. Could such a scenario explain your measurements? Is the spacing between sub-pulses appropriate and/or the chirp counterproductive for one sign and productive for the other? Such a chirp dependence has been utilized in early experiments of Bardeen and Wilson.

Professor Herek communicated in reply: The experiments are performed with an excitation intensity corresponding to the onset of saturation, where the scenario described by Prof. Riedle is indeed occurring in our sample. The strongest evidence for this lies in the initial 6% jump in the T/S ratio seen in Generation 1 of our blind optimization experiments (Fig. 5). This step-wise increase of the fitness value at the onset of the learning curve occurs concomitant with a 50% loss in actinic energy, and reflects a spreading of the pump intensity to allow for multiple excitation processes. However, beyond this initial (trivial) control mechanism, the fitness continues to climb even though the actinic energy does not vary. This further tweaking of pulse shape implies the involvement of more complicated mechanisms than just signal saturation.

Professor Reid communicated: If it were possible to achieve substantial active control of the photodynamics in zinc phthalocyanine, would it be possible for this to be used in practice in photodynamic therapy?

Professor Herek communicated in reply: I guess this would depend on how substantial the control, and how simple the optimal pulse shape. At the moment, our aim is to extract insight on the mechanisms of photodynamic therapy, to identify the factors that determine efficiency and eventually enhance these by, *e.g.* chemical modification. We're not looking to bring shaped pulses into the clinic. But perhaps that day is not so far off. Mosk and co-workers are developing other shaping strategies–sculpting the wavefront of propagating light–to overcome the problems of scattering in turbid media, that may revolutionize medical optics. Adding spectral shaping will open even more possibilities.

To note, ZnPc is already being used in clinic trials of photodynamic therapy, as it has a high absorption cross-section in the red/near-infrared region where human tissue is most penetrable. The efficiency, however, is limited by an ISC yield of only 60%.

Professor Meier communicated: Do you expect the ISC to be favoured when exciting to higher lying states?

Professor Herek communicated in reply: I expect that by engaging higher excited states, we have access to more pathways, including some that will ultimately lead to ISC. Hence I think that by accessing these states *via* multiphoton excitation, we can enhance the ISC yield.

Professor Riedle communicated: In the experiments reported by Prof. Herek, the aim is to improve the singlet oxygen yield. As desirable as this might be, I'm wondering whether this type of control experiment is not going in a direction that does not make best use of the technical capabilities. One tries to optimize a quantity that is already produced in a fairly large amount by a Fourier-transform-limited pulse. As the shaper diminishes the pulse energy considerably, it would be much easier to increase the total yield of the experiment by just using the unshaped pulse.

A different case would be given if the photochemical reaction under investigation has two more or less equally likely outcomes. If one of them is highly undesirable, coherent control should try to minimize it. In this way a much larger effect could be reached. To find proper examples, one can simply consult the large body of photochemical literature and look at those examples that have a good total quantum yield, but not the selective outcome desired. Experienced photochemists will readily be able to point out such "misbehaving" systems.

Dr Blancafort communicated in reply: Our contribution on the control of fulvene photophysics goes in the direction pointed out by Prof. Riedle. It is a molecule where the decay to the ground state can take place with or without rotation of the double bond. Under normal conditions, the decay without rotation is favored, and our goal is to simulate conditions where we force the rotation of the double bond to be the favored process.

Moreover, what makes fulvene an interesting model system is that the competition between two photochemical processes takes place along a seam of intersection, and we try to use this mechanistic knowledge to derive a control scheme. This seems to be quite a general situation. For example, we find that the competition between the stepwise and concerted paths in the light-induced Wolff rearrangement also takes place along a seam [Q. Li, A. Migani, L. Blancafort, *J. Phys. Chem. Lett.*, 2012, **3**, 1056–1061]. Therefore, understanding the dynamics at the seam is essential to derive viable control schemes for photochemical reactions.

Professor Stolow communicated: Space-time coupling at the focus of an electromagnetic field is an unavoidable consequence of using a spatial light modulator, such as in a typical '4f' arrangement. How does one ensure that such effects do not alter the interpretation of a given control mechanism?

Professor Herek communicated in reply: I appreciate this question, which highlights a serious consideration for this sort of experiment. I think Prof. Stolow answered it best himself in his recent publication on the topic.[1] We recognize that the applied field of the shaped pulses can vary spatially in the focal plane. Even in the absence of pulse shaping, a misaligned NOPA can create significant spatial chirp. As such, different target molecules can experience different excitation fields, thereby smearing the control effect, and the observed control effect can depend on the overlap of the probe laser. For this reason, we focus the probe laser to a much smaller diameter than the pump, and optimal pulse shapes are tested for robustness by repeated scans, sometimes days/weeks after the original optimization was performed. Further, we monitor how the fitness level varies when tweaking the pump–probe overlap, to confirm that no significant spatial variations. The space-time coupling can be experimentally determined from the spatial intensity distribution (*e.g.*, with a CCD camera) and the spectrally resolved intensity distribution (by scanning the entrance slit of a spectrometer across the image of the focal plane).

1. B. J. Sussman, R. Lausten and A. Stolow, *Phys. Rev. A.*, 2008, 77, 043416.

Professor Temps opened the discussion of the paper by Professor Bañares by communicating: Repulsive potential energy surfaces may be hard to investigate by direct measurements starting in the electronic ground state. Could your experiments provide a general methodology to characterize elusive repulsive potential energy surfaces? In particular, you could also change polarizations to observe anisotropies, which would tell you about the symmetries of unknown repulsive states.

Professor Bañares communicated in reply: In principle, we are limited to study one-photon optically allowed transitions as those typical in methyl iodide to the A-band ($^3Q_0$, $^1Q_1$ and $^3Q_0$ repulsive states) which have enough oscillator strength. Of course, we may access other one-photon dark electronic surfaces by, for instance, two-photon absorption if that is optically allowed, and I am thinking in the $3p_z$ or $4p_z$ Rydberg states of the methyl radical. Changing polarization using velocity map imaging would not help because doing so you loose the cylindrical symmetry in the experiment, necessary to perform Abel inversion.

Professor Fielding commented: I am interested to hear about the H-atom transfer Professor Bañares observes and wonder if he would be able to provide some information about the timescale of the process.

Professor Dr Bañares responded: In the experiments we have carried out a study of the excited state hydrogen transfer in pyrrol–ammonia clusters, we have employed pump–probe laser pulses in the nanosecond timescale. We have not obtained information about the timescales of the H-atom transfer process within the clusters.

Professor Neumark communicated: How are you able to calibrate the Stark shift, when the total shift of 13 meV in Fig. 4 is considerably less than your laser linewidth (around 40 meV)?

Professor Bañares communicated in reply: In the intensity range of the control laser where the Stark shift calibration is shown in Figure 4 of the paper, the signal from the rings in the images does not dissapear completely but it reduces with respect to the situation when there is no control laser. Then, the pump laser is slightly retuned to the blue to recover the full signal and this is the way we obtain the Stark shift as a function on the control laser intensity. Of course, at higher intensities of the control laser for which the Stark shift is larger, the signal from the rings totally disappear, as depicted in Fig. 2 of the paper.

Professor Domcke enquired: You observed and discussed the excitation of one quantum of the symmetric stretching mode of the methyl radical. However, the methyl group in methyliodide is pyramidal, while it is planar in the free methyl radical. Why didn't you observe a strong excitation of the umbrella mode of the methyl radical?

Professor Bañares responded: We are selectively detecting the vibrational ground state (ν=0) of the CH_3 radical by tuning our probe laser to 333.5 nm, by 2+1 REMPI on the Q branch of the 0^0_0 transition through the $3p_z$ Rydberg state (resonant by two photons). Interestingly, given that we are using a broadband femtosecond laser pulse as probe, we are capable of ionizing the CH_3 radical produced with one quantum of excitation in the symmetric stretch (ν_1=1), whose Q branch of the 1^1_1 transition is quite close in wavelength and within the bandwidth of the probe laser pulse. Actually, at variance with photodissociation of CH_3I in the A-band, where vibrational activity into the symmetric stretch of CH_3 is small, in the B-band predissociation, the vibrational activity in this mode is important. We can detect CH_3 radicals with one or two quanta of excitation into the umbrella mode (ν_2=1 and ν_2=2) just by tuning our probe laser to the blue. The Q branches of the 2^1_1 and 2^2_2 transitions are at 329.6 nm and 325.7 nm, respectively. We have studied the vibrational activity in both the stretching and umbrella modes when exciting CH_3I to the 0^0_0, 2^1_0 and 3^1_0 vibronic bands of the $^3R_1$ Rydberg state of the B-band in separate publications and deduce vibrational populations.[1,2] There is indeed substantial vibrational activity in the CH_3 umbrella mode. However, the present experiment is focussed in the strong field control of predissociation from the 0^0_0 band of the $^3R_1$ Rydberg state of the B-band yielding CH_3 in ν=0 and ν_1=1. We have performed similar experiments when detecting CH_3 in ν_2=1 and ν_2=2, and qualitatively similar results have been obtained regarding Stark shift, dump and multiphoton dissociative ionization processes.

1. G. Gitzinger, M. E. Corrales, V. Loriot, G. A. Amaral, R. de Nalda, L. Bañares, *J. Chem. Phys.*, 2010, **132**, 234313.
2. G. Gitzinger, M. E. Corrales, V. Loriot, R. de Nalda, L. Bañares, *J. Chem. Phys.*, 2012, **135**, 074303.

Professor Weitzel communicated: The methyl radical is known to have a planar equilibrium geometry. The formation of methyl radicals by photodissociation from methyliodide implies that it should be formed in an excited quantum state of the umbrella mode. The population in the ν=1 state of this umbrella mode can easily be probed by REMPI at about 330 nm (see *e.g.* Luther, Troe and Weitzel, *J. Phys. Chem.*, 1990, **94**, 6316–6320. This wavelength appears to be just barely outside the band width of your current fs-REMPI. Have you made any attempts to observe the population of excited umbrella states by shifting the central wavelength of your laser field?

Professor Bañares communicated in reply: As mentioned in my previous answer, we have centered our probe laser at 333.5 nm to resonantly multiphoton ionize methyl fragments in the ground vibrational state and, given the bandwith of the probe laser pulse, also with one quantum of excitation into the symmetric stretch mode. Of course, we can tune our probe laser down to 329.6 nm to probe methyl fragments with one quantum of excitation in the umbrella mode. In our previous works,[1,2] we were able of measuring vibrational populations of the methyl fragment into the different vibrational modes. Regarding the strong field control experiments, we have performed similar experiments when detecting methyl with one or two quanta of excitation in the umbrella mode, and qualitatively similar results have been obtained.

1. G. Gitzinger, M. E. Corrales, V. Loriot, G. A. Amaral, R. de Nalda, L. Bañares, *J. Chem. Phys.*, 2010, **132**, 234313.
2. G. Gitzinger, M. E. Corrales, V. Loriot, R. de Nalda, L. Bañares, *J. Chem. Phys.*, 2012, **135**, 074303.

Professor Riedle asked: The bond cleavage of a methylhalide transforms the central carbon from a sp^3 hybridization to a sp^2 one. This means that the fragment has to planarize, a motion that takes some time and possibly leads to vibrational ringing. This has recently been analyzed and demonstrated for the related diphenylmethylchloride.[1] In this system a rather universal 300 fs time constant is found at which the fragment transient signal increases largely due to a change in transition strength with the planarization. How can you tell that in your measurements the sub-ps times are not largely influenced by such a nuclear motion? Is the eigenstate picture really justified or shouldn't one rather think in terms of vibronic wavepackets? Wouldn't this behavior also influence the interaction with the control pulse?

1. B. P. Fingerhut, C. F. Sailer, J. Ammer, E. Riedle and R. de Vivie-Riedle, *J. Phys. Chem. A*, 2012, **116**, 11064.

Professor Bañaress replied: It is of course correct that the methyl group undergoes planarization in the course of iodine elimination in the dissociation of methyl iodide. Two situations must be discriminated, though: in the rapid dissociation that follows after the transition to the first absorption band (A-band) of methyl iodide, centered around 260 nm, a vibrational wavepacket is formed, mainly in the umbrella mode associated to this geometrical change. Resonant detection of methyl occurs after approximately 100 fs.[1,2] In the case studied here, however, excitation of the molecule occurs to the second absorption band (B-band) around 200 nm, where bound Rydberg states exist. Discrete lines constitute this section of the absorption spectrum, which are the result of vibronic transitions to specific vibrational levels of the excited electronic state.[3] Although they are lifetime broadened, due to the interaction with dissociative surfaces, these transitions are spectrally separated further than the laser bandwidth. Therefore, a molecular vibronic wavepacket cannot be created upon absorption under these conditions, and an eigenstate picture, although not perfect, is most useful. It has been described in the literature that the lifetime of such excited states is strongly dependent on the vibrational level of the Rydberg state,[3,4] with values ranging from 800 fs to 4 ps (significantly longer than the planarization times found in the paper you cited). Interaction with the dissociative continuum and subsequent dissociation does of course create a vibrational wavepacket, which correlates with an important degree of vibrational excitation in methyl, both in the umbrella and the symmetric stretch modes.[4]

1. R. de Nalda, J.G. Izquierdo, J. Durá, L. Bañares, *J. Chem. Phys.*, 2007, **126**, 021101.
2. R. de Nalda, J. Durá, A. García-Vela, J.G. Izquierdo, J. González-Vázquez, L. Bañares, *J. Chem. Phys.*, 2008, **128**, 244309.
3. A. P. Baronavski, J. C. Owrustky, *J. Chem. Phys.*, 1998, **108**, 3445.
4. G. Gitzinger, M. E. Corrales, V. Loriot, R. de Nalda, L. Bañares, *J. Chem. Phys.*, 2012, **136**, 074303.

Professor Duxbury commented: Recently Thire *et al.*, *PCCP*, 2011, **13**, 18485 measured time resolved predissociation of the vibrationless level of the B state of

CH_3I. In their model one of the routes for predissociation is by spin–orbit interaction with the continuum of the anti-bonding lowest unoccupied molecular orbital localised on the C–I band.

In your study of strong field control of predissociation dynamics of CH3I, what role do you envisage for spin–orbit interaction given that the atomic spin–orbit coupling of Iodine is very large?

Professor Bañares responded: Spin–orbit coupling is crucial for this system due to the heavy iodine atom both in the A and B bands. Without considering the spin–orbit interaction no crossing between the bound Rydberg state and the valence repulsive state would be produced and no predissociation would take place. No special effect on the predissociation dynamics is expected due to the existence of the strong spin–orbit interaction when the control laser field is present for this molecule.

Professor Meier communicated: How does the observed change in the lifetime compare with theory?

Professor Bañares communicated in reply: There is no theory available to reproduce the present experimental results thus far. A theoretical simulation of this experiment would be very demanding. High level multidimensional *ab initio* calculations of electronically excited states would be required before any dynamics is calculated, and this is a difficult system to be studied at such high level, considering the heavy iodine atom and the strong spin–orbit coupling present in the molecule

Professor Ashfold communicated: Corrales *et al.*[1] offer persuasive evidence that the increased rate of decay from the $^3R_1$ Rydberg state of CH_3I observed when the intense 800 nm 'control' pulse overlaps (in time) with their 200 nm pump pulse is due to control-laser-induced depopulation routes (*viz.* a resonant one photon dump transition from the $^3R_1$ state to the dissociative $^3Q_1$ valence state that correlates to the ground state dissociation limit, and multiphoton dissociative ionisation pathways) rather than a consequence of a dynamic Stark shift. The authors suggest that a simple rate equation model can account for the measured >2-fold decrease in excited state lifetime at the highest control laser intensities. To what extent can this same approach account for the deduced variation in the quantum yield of CH_3 + I* products over the same range of control laser intensities?

1. M. E. Corrales, *et al.*, *Faraday Discuss.*, 2013, **163**, DOI:10.1039/C2FD20143G .

Professor Bañares communicated in reply: The quantum yield of CH_3+I* products (ϕ*) is significantly affected by the opening of the one-photon dump transition from the $^3R_1$ state to the dissociative $^3Q_1$ valence state. This dissociative state correlates with CH_3+I products and, thus, the dump process, which is control laser intensity dependent, provides a way to decrease ϕ* signficantly. The answer to your question is yes, the same rate equation model can account for the variation (decrease) of the ϕ* quantum yield as a function of the control laser intensity.

Professor Meier asked: The best control seems to be achieved with a pump/control delay which is not zero. Is this delay experimentally significant? If yes, is there a specific reason for this delay?

Professor Bañares replied: The control laser pulse has 4 picosecond duration whereas the pump laser is about 300 fs duration. This means that there is no a clear way to define time zero between such pulses. In our experiments, the zero delay time between the pump and the control laser pulses has been chosen at an arbitrary position of overlap between the fs pump and ps probe pulses. For the experiments controlling the lifetime, a pump-control delay time of -1 ps (in our arbitrary scale) has been chosen to ensure that predissociation occurs during the flattest part of the control pulse, since the intrinsic predissociation lifetime is 1.5 ps and the control laser pulse is 4 ps in duration.

Professor Buma communicated: In your paper you show that your idea to control the photodissociation of CH_3I at the origin of the B absorption band is compromised by a pump-dump channel that yields $I(^2P_{3/2})$. In this respect control *via* higher-lying excited states such as the C state or even higher lying states might be energetically more favourable. One-photon excitation of these states would not be possible, but previous REMPI-PES studies[1] have shown that they can be excited without any problem using two-, three- or even four-photon excitation.

1. M. R. Dobber, W. J. Buma, C. A. de Lange, *J. Chem. Phys.*, 1993, **99**, 836.

Professor Bañares communicated in reply: You are right that it would be interesting to try this control strategy in the C state. However, to reach the C state you need a 180 nm photon, which is far beyond our experimental capabilities. A two-photon absorption using a 360 nm pump pulse would be worth to try. However, you would really need high intensity and thus, using femtosecond pulses, more than two photons could be absorbed and ionize the molecule, having dissociative ionization. I see this as a difficult experiment since you cannot control easily the number of photons absorbed by the molecule when using femtosecond pulses.

Professor Neumark opened the discussion of the paper by Professor Weitzel by communicating: In describing the isotope effects in your paper, you conclude that "control between H_3^+ and H^+ ... is operative prior to the hydrogen scrambling." Can you explain in more detail what is going on here in terms of isomerization and dissociation barriers?

Professor Weitzel communicated in reply: All the evidence we have suggests that there is a barrier for hydrogen scrambling in the ethane dication lying energetically below the transition state for breaking into H_3^+ and $C_2H_3^+$. This barrier for scrambling connects the isomeric structures $CH_3DCD_2^{++}$, $CH_2D_2CDH^{++}$ and so on.

Professor Field enquired: In your paper you mentioned different effectiveness of upward *vs.* downward linear chirps. Have you attempted downward followed by upward "V" chirps or upward followed by downward "lambda" chirps?

Professor Weitzel responded: This is an interesting question. Chirping femtosecond laser pulses is a systematic way of shaping the pulse form.

In the current work we have considered the role of linear chirp. We have also investigated the role of quadratic chirp (third order dispersion), but basically all ion yields exhibit maxima for quadratic chirp parameters β close to zero (work to be published). Interestingly for certain objectives, *e.g.* maximizing the H_3^+ ion yield, optimizing the laser field by means of a genetic algorithm leads to a pulse shape consistent with the dominance of a linear chirp (work to be published). Another way of systematically shaping the laser field is represented by femtosecond interferometry (*Chem. Phys. Lett.*, 2010, **487**, 209–213 and *J. Chem. Phys.*, 2008, **128**, 121104). The V-chirp (or lambda chirp), suggested by you, is yet another concept for systematically shaping fs-laser pulses. We have not performed experiments employing this concept yet, but will consider it in future work.

Professor Whitaker addressed Professor Field and Professor Weitzel: I would just like to add to Professor Field's question that some time ago we investigated chirp control of the (slightly off resonant) three photon pumping of an ion pair state in molecular iodine using visible light (N. T. Form, B. J. Whitaker and C. Meier, *J. Phys. B.: At. Mol. Phys.*, 2008, **41**, 074011). Similarly to the results just presented by Professor Weitzel we also found that the transform limited pulse was a relatively inefficient means of pumping the target state and that higher efficiencies (by a factor of three or more) could be achieved by applying linear chirp. But we also investigated the effects of higher order chirp and made systematic maps (see the figure below) of the pumping efficiency as a function of group delay and third order dispersion (linear and quadratic chirp). We found that the addition of time ordering (quadratic chirp) to the excitation pulse had a significant extra enhancing effect. This was attributed to creating a double pulse temporal structure (an effect that could also be generated by applying a pi-phase flip in the middle of the excitation spectrum) which we showed was effective through a time delayed resonance in which the first pulse created population on an intermediate state which then had to propagate (as wavepacket) to larger interatomic distances before it could be efficiently transferred to the final state by the second temporal component. The fact that in ethane no effect is observed for third order dispersion may indicate that here the

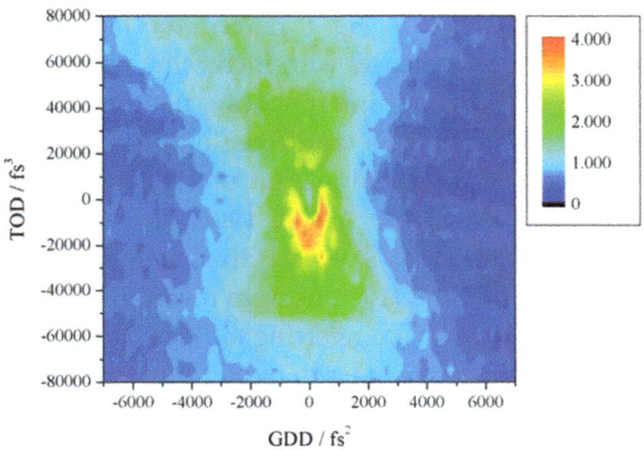

Fig. 1 Pumping efficiency as a function of group delay and third order dispersion.

fragmentation channels are reached by a final photon only connecting two states through a wavepacket following mechanism.

Professor Reid communicated: Do you have any physical insight into why changing the sign of the chirp changes the branching ratio of the photofragments?

Professor Weitzel communicated in reply: Clearly a profound answer would require a thorough theoretical calculation taking into account both ionization dynamics as well as wavepacket dynamics (for the nuclei and for the electrons). We are not in the position to provide this. Thus we can only offer qualitative arguments. In the dissociative ionization of DCl we observed that the formation of Cl^+ from the electronic ground state potential of the ion was favored over the formation of D^+ from the first excited electronic state of the ion for the application of negative linear chirp, but not for positive linear chirp. We interpreted this as an indication that climbing the anharmonic ladder of vibrational states is favored by a negative linear chirp (*J. Chem. Phys. A*, 2006, **110**, 6395–6398). On the other hand theoretical wavepacket work suggested that in addition to vibrational ladder climbing effects also electronic resonances can become important. Here, we recall that the energetic distance between electronic states in general depends on the reaction coordinate (in particular for bound-free transitions) and consequently may change during the chemical process of interest. To our knowledge there is no unique answer as to when the "jump" to another electronic state is favored by negative or positive chirp.

Professor Duxbury communicated: In Figure 8, ion yields from d3 ethane, one curve stands out as having a much larger ion yield over a longer time period than the other 5 examples, it would be interesting to have an explanation for this.

In 2011, in the Faraday Discussion 150, Frontiers of Spectroscopy, Oka gave an interesting introduction on "Spectroscopy and astronomy: H_3 + from the laboratory to the Galactic center. In Figure 11 of this discussion The observed spectra lines observed go well above the barrier to linearity well into the visible region. It would be interesting to reconcile the behaviour of the triatomic ions which your group is studying with the highly excited H_3+ ions produced by a very different series of methods.

Professor Weitzel communicated in reply: The particular ion exhibiting the largest yield for the shortest pulse duration up to about 120 fs pulse duration is the parent ion. The data shown in Fig. 8 reflects the fact that the ratio of all fragment ions to parent ions increases with increasing negative chirp value (see also Fig. 4) This again supports our interpretation that ionization is followed by dissociation. As I discuss in my general discussion of papers 18–20 the photodissociation of H_3^+ leading to the formation of H_2^+ + H has been observed experimentally (Bae and Cosby, *Phys. Rev. A*, 1990, **41**, 1741). The threshold for observing H_2^+ was around 2.5 eV (!). Since the dissociation energy is about 4.9 eV this immediately implied that the H_3^+ ions were rovibrationally hot, presumably above the barrier for linearity you are referring to. This barrier for linearity lies at 9913 cm^{-1} (the same source as the one you quote). Now the interesting question is, whether one will be able to photoinduce the formation of H^+ + H_2 on the electronic ground state potential. The dissociation limit of the latter is 2 eV below that for H_2^+ formation.

Professor Meier asked: The yield of all ions seems to be favoured by a negative chirp. Is this an indication that the chirp effect occurs on the very first ionisation step ?

Professor Weitzel replied: There is most likely more than one chirp effect in this work. Quite obviously there is a chirp effect on the formation of the parent ion, thus in the first ionisation step. But there are also chirp effects in what we call inter-charge-state-control (control between fragments originating from the monocation and from the dication) and intra-charge-state-control (control beween fragments originating from the same charge state).

Professor Whitaker commented: If we think back to Gustav Gerber's early evolutionary control experiments, *e.g.* M. Bergt, T. Brixner, B. Kiefer, M. Strehle and G. Gerber, *J. Phys. Chem. A*, 1999, **103**, 10381, the usual way to provide a feedback signal to the control loop is to monitor two or more ion products, because ions are easy to detect, and to generate a fitness parameter from the ratio of the signals, much as in your experiments. Although clear control was demonstrated, the mechanism in polyatomic molecules was as clear as mud. A particularly striking example was the experiment by Levis and coworkers (R. J. Levis, G. M. Menkir and Herschel Rabitz, *Science*, 2001, **292**, 709) in which they observed the formation of toluene and carbon monoxide (ions) with highly structured electric fields centred around 800 nm applied to acetophenone. If the dynamics occurred on exited state surfaces of the neutral molecule this would be a remarkable result (requiring as it does the concerted breaking of two bonds and the formation of a third). It seems much more likely that the dynamics are in fact occurring on the ionic surfaces, similarly to McLafferty rearrangement reactions. Since at least 4 or 5 near IR photons are required to ionize ethane do your chirp results (which implicitly imply some kind of wavepacket following mechanism between two near resonantly coupled surfaces) not strongly suggest that in your results too it is the ionic PESs that play the crucial role in directing the fragmentation patterns?

Professor Weitzel commented in reply: As we tried to demonstrate by kinetic energy analysis the fragment ions discussed in the current work are mainly formed on the dicationic potential energy surface (PES), some on the monocationic PES, thus implying an "ionization followed by dissociation" mechanism.

Professor Meier communicated: What are the challenges, from a theoretical point of view, to fully simulate the observed chirp and intensity dependent results?

Professor Weitzel communicated in reply: The main theoretical challenge for simulating experiments of this kind (starting from the neutral molecule) from my point of view is that you need to combine state-of-the-art theory for ionization dynamics with state-of-the-art theory for dissociation dynamics. We have done both separately, *e.g.* Korolkov and Weitzel, *Chem. Phys. Lett.*, 2010, **487**, 209–213, for the dissociation dynamics and Znakovskaya *et al.*, *PCCP*, 2011, **13**, 8653–8658, for ionization dynamics, but not combined in one calculation. The combination of the two is currently perhaps only tractable for molecular hydrogen and a few other small molecules. There is an intriguing advantage of looking at the photodissociation dynamics of the ion in the first place. Typically the next higher charge-state is energetically far away. Thus, you do not have to worry about that. The separation of electronic states in the ion (mono-cation) is typically smaller than in the neutral. In the end, this is the reason, why I have suggested the H_3^+ experiment (see the general comment below).

Professor Weitzel communicated: A general comment in the discussion of papers 18–20. The H_3^+ ions, the formation of which we describe in our paper, is not only intriguing because it involves the breaking of three C–H bonds and the formation of three H–H bonds (it's a cyclic species). It is most of all the top molecular system for studying the possibilities and the limits of photochemical control schemes. There are several reasons for this claim:

i) There is no neutral analogue to the H_3^+ ion. Thus, the photodynamics of this system can be investigated free from any contamination of neutrals.

ii) The H_3^+ ion has two dissociation limits, the formation of H_2^+ + H and the formation of H^+ + H_2. These two limits are separated by about 2 eV (the difference in I.E.s of H_2 and H). I would like to suggest attempting to control the yields of these two reaction channels by applying any of the experimental control schemes currently available to H_3^+ ions prepared in an ion trap.

iii) The H_3^+ ion consists of three protons and two electrons. It is the smallest molecular system for which we can envisage the competition between two reaction channels easily distinguishable in any mass spectrometer. It appears conceivable to test the experimental data against the highest levels of theory available to date.

Professor Whitaker communicated in reply: It's known that the vertical excitation energy from the $(1)^1A'$ to $(2)^1A'$ state in H_3^+ is 20 eV or so. The latter state, which correlates with the H_2^+ plus H dissociation channel, is purely dissociative and never crosses the ground state. So I'm not sure that I understand your proposed experiment. How will it be possible to control the branching ratio between the H_2 and H_2^+ channels?

Professor Weitzel commented further: One does not have to rely on the vertical excitation from the rovibrational ground state of H_3^+ (1) $^1A'$ all the way to the (2) $^1A'$ state. The photodissociation of H_3^+ has been described *e.g.* by Bae and Cosby in 1990 (Y. K. Bae and P. C. Cosby, *Physical Review A*, 1990, **41**, 1741) above 2.5 eV starting from hot H_3^+. More recent work is by Petriginani *et al.*, *Journal of Physical Chemistry*, 2010, **114**, 4864. I am convinced that chirped pulse excitation of cold H_3^+ will eventually bring you to the same "region", where one or two additional photons at 800 nm can decide between the H_2^+ and the H^+ channel. I would speculate that down chirp leads may lead to H^+ on the ground state surface, and up chirp leads may lead to H_2^+ by jumping to the upper surface (in the last moment). We believe we have seen something similar in HCl (H. G. Breunig, A. Lauer, and K. M. Weitzel, *Journal of Physical Chemistry A*, 2006, **110**, 6395), however there we started from the neutral, and we were not able to prove that the control is in the ion. That would be the advantage of the H_3^+ experiment.

Professor Whitaker responded: Thank you for the clarification, but given that the energetic gap between the (1) ^{1}A' to (2) ^{1}A' states converges along the dissociation co-ordinate perhaps a down chirped pulse might be more successful in promoting the molecular ion channel. I look forward to someone trying the experiment to see who is right!

Mr Lehmann asked: As a question addressed to Professor Weitzel, we would like to ask the following. Do you think that it would be useful for the interpretation of your chirped pulse shaping experiments on ethane, and similar molecules, at these fluence levels of about 10^{13} Watt/cm^2 to extent the ion-TOF data with more advanced three-dimensional momentum imaging, and in particular to combine this with coincidence electron imaging?[1,2]

In this respect we (C. S. Lehmann, N. B. Ram, I. Powis and M. H. M. Janssen) would like to provide the following comment. Recently, we have introduced a novel technique, multi-photon photoelectron circular dichroism (MP-PECD) to discriminate the enantiomers of chiral molecules with mass-selectivity using photoelectron-photoion coincidence imaging [Fig. 2 and Ref. 3,4]. We employ femtosecond multi-photon ionization with circular polarized light and depending on the wavelength and laser fluence we also observe ion fragmentation.[4] The forward-backward asymmetry in the angular distribution of photoelectrons is measured in coincidence with an ion. The coincidence technique makes it possible to measure and analyse this asymmetry in correlation with the observed mass of the ion. Furthermore, the photoelectron spectra measured provide information about the mechanism that leads to the various fragmentation channels. For the chiral molecule Camphor, the asymmetry observed for excitation near 400 nm is 8%.[3] Because the ionization is due to a three-photon excitation we can expand the angular asymmetry in Legendre polynomials up to order

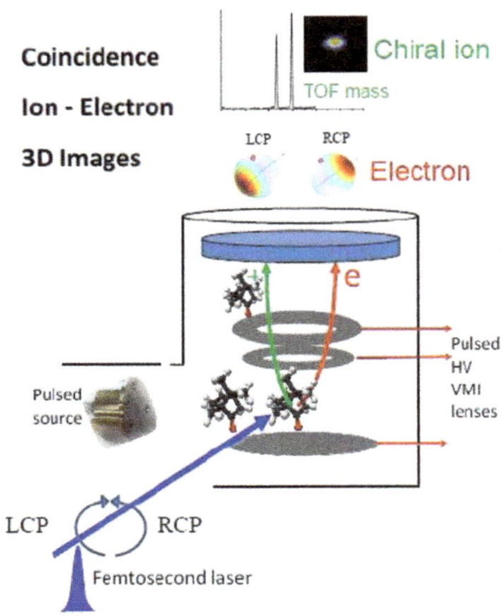

Fig. 2 Schematic illustration of the coincident measurement of the forward-backward asymmetry in the electron distribution from femtosecond multi-photon ionization of a chiral molecule with circular polarized light.

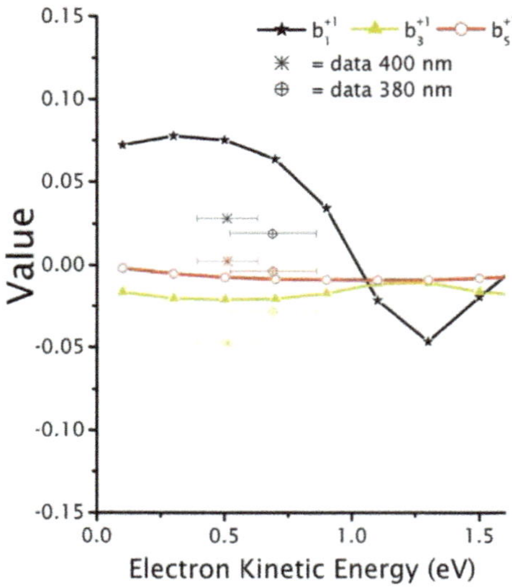

Fig. 3 Comparison of the measured and calculated odd Legendre moments of the electron angular distribution of Camphor.[4]

six. In collaboration with Professor Powis we have attempted to get a better theoretical understanding of the mechanism and the magnitude of the Multi Photon PECD effect.[4] The calculated Legendre coefficients of the photoelectron angular distribution parameters (b_1 – b_6), see Fig. 3, can be compared with the experimental values. It is very encouraging that there is good qualitative agreement regarding the sign of the coefficients and to some extent quantitative agreement. Further theoretical work is in progress, as well as experiments on other molecules.

1. C. S. Lehmann, N. B. Ram, D. Irimia and M. H. M. Janssen, *Faraday Discussions*, 2011, **153**, 173.
2. C. S. Lehmann, N. B. Ram and M. H. M. Janssen, *Rev. Sci. Instrum.*, 2012, **83**, 093103.
3. N. B. Ram, C. S. Lehmann and M. H. M. Janssen, *European Physical Journal, Web of Conferences*, 2013, **41**, 02029.
4. C. S. Lehmann, PhD Thesis, VU University Amsterdam, ISBN 978-90-6464-660-7 (2013).

Professor Weitzel replied: I agree that coincidence studies may add to a better understanding of the ionization /dissociation dynamics. In fact, there was a time when I was concentrating on electron-ion-coincidence experiments, which finally led to my Habilitation (K.-M. Weitzel, "Energetik, Kinetik, und Mechanismus von unimolekularen Elementarreaktionen in Molekuel- und Cluster-Ionen." Shaker-Verlag, 1998, Aachen, ISBN 3-8265-3376-3). More recently we have employed the ion-ion-coincidence imaging technique to improve our understanding of the ethane dication dissociation (Kanya *et al.*, *J. Chem. Phys.*, 2012, **136**, 204309). We certainly would like to continue performing coincidence imaging experiments in the future. However I would like to comment on the relation of imaging experiments *versus* straight time-of-flight measurements: the latter will normally provide a higher kinetic energy resolution. This holds true for the ions as well as for the electrons. It depends on your scientific question whether the imaging or the ToF concept is preferable.

I am pleased to finally see the measurement of a circular dichroism in the photoelectron angular distribution (PECD /PAD) in coincidence with the ion signal. As you know, we have (again) concentrated on the ion in related studies. To the best of my knowledge we were the first to demonstrate the measurement of a circular dichroism in femtosecond laser ionization of a chiral molecule four years ago (Breunig *et al.*, *ChemPhysChem*, 2009, **8**, 1199–1202). While our CD effect originated mainly from the magnetic dipole transition moment, the PECD effect you show is electrically dipole allowed. It would be intriguing if you could perform an experiment on the 3-methyl-cyclopentanone around 310 nm, where we have observed CD values as large as 15% in total ion yields (angle integrated!). Your experiment should then demonstrate whether the contributions from the different transition moments add up. Finally I would like to point out that in my view the prospect for applying femtosecond laser pulses for chirality analysis is in the non-resonant ionization, since there you can detect any molecule. Again we were the first to demonstrate the viablity of this concept (Horsch *et al.*, *Chirality*, 2012, **24**, 684–690). Here, my question would be, will it be possible to extend your fs-PECD measurement to the non-resonant multi-photon-ionization, where we talk about 8 photons or more, or will your PECD effect be killed by the higher order Legendre polynomials?

Mr Lehmann communicated: We (C. S. Lehmann and M. H. M. Janssen) would like to address the following question to Professor Weitzel. In Fig. 2, 3 you show that the influence of the linear chirp is not symmetric, but that only negative chirp enhances the fragmentation. Therefore, it is concluded that it is not a time duration effect that is responsible for the fragmentation dynamics. In our group in Amsterdam we have reported previously pulse shaping studies in combination with electron and ion imaging detection.[1,2] In these experiments we used femtosecond multi-photon excitation at visible wavelengths around 520 nm. However, we observed that the effects of, for instance chirp on fragmentation dynamics in molecules like CF_3I and CH_2BrCl, were very dependent on the central wavelength. In Table 1 we have summarized some more recent data measured with our coincidence set-up at 2 different wavelengths and quite different effects are observed regarding the effect of chirp on the fragmentation ratios. Do you think that your results on ethane are dependent on the exact central wavelength

Table 1 Single color multi-photon ionization of CH_2BrCl. The laser pulses (repetition rate 1 kHz) were centered at 521 and 526 nm, had a bandwidth of 20 nm and a pulse energy of 10 μJ. Experimental data by C. S. Lehmann and M. H. M. Janssen (unpublished results Amsterdam).

	$\phi'' = -200\ fs^2$	$\phi'' = 0\ fs^2$	$\phi'' = +200\ fs^2$	$\phi'' = +400\ fs^2$
521 nm				
Total ions (per sec)	0.81	4.63	3.63	0.49
CH_2Cl^+/CH_2BrCl^+	0.34	0.37	0.51	0.65
CH_2Cl^+/all ions	0.71	0.68	0.62	0.57
CH_2BrCl^+/all ions	0.24	0.25	0.32	0.37
526 nm				
Total ions (per sec)	2.48	4.04	3.94	1.94
CH_2Cl^+/CH_2BrCl^+	0.34	0.33	0.36	0.41
CH_2Cl^+/all ions	0.24	0.23	0.25	0.28
CH_2BrCl^+/all ions	0.70	0.71	0.70	0.68

used, and would it be useful to change the central wavelength to other wavelengths and measure the effect of the linear chirp? Do you expect changes with wavelength?

1. D. Irimia, and M. H. M. Janssen, *J. Chem. Phys.*, 2010, **132**, 234302.
2. C. S. Lehmann, N. B. Ram, D. Irimia and M. H. M. Janssen, *Faraday Discussions*, 2011, **153**, 173

Professor Weitzel communicated in reply: Your CF_3I experiments performed between 520 nm and 550 nm are basically two-photon resonant with the first electronically excited A state, compare *e.g. J. Chem. Soc., Faraday Trans.*, 1997, **93**, 2839–2846. You are probably influenced by several additional resonances at the three photon level and higher. Most likely all your very interesting observations are the result of resonances and moreover interferences between resonances. The situation may be very similar for the CH_2BrCl The situation is fundamentally different for the current ethane study, where the first very weak optical absorption is centered at 8.7 eV (143 nm) and the first strong absorption occurs at about 9.4 eV (132 nm), see *Chemical Physics*, 1993, **173**, 209–239. Our experiment has been performed at 800 nm. Thus it would take 6 photons to reach the first excited electronic state in a resonance picture. Very likely the fs-laser ionization of ethane at 800 nm is non-resonant. Changing the wavelength is not expected to alter our observations.

Professor Burgdoerfer opened the discussion of the paper by Professor Bucksbaum by communicating: You discussed two somewhat complementary models for the enhanced multiple ionization near conical intersections. Would it be possible to distinguish between the two, for example by varying the frequency?

Professor Bucksbaum communicated in reply: Yes, varying the frequency might reveal valuable distinguishing characteristics. We already pointed out that a feature of the conical intersection region is a "resonant ring" surrounding the singularity in the g-h plane. The laser frequency controls the size and shape of this ring. An equally important question in the case of double ionization is whether the cation *also* contains a conical intersection or resonances in the vicinity that could facilitate removal of the second electron.

Dr Kirrander communicated: Your observation of enhanced multiple ionization near conical intersections is intriguing: High harmonic generation (HHG) spectroscopy is being actively explored as a tool for studying molecular dynamics in the vicinity of conical intersections[1]. Crucially, the HHG process relies on *single* electron ionization and recombination. If enhanced *multiple* ionization is a common feature at conical intersections, do you believe it will constitute a problem for this otherwise promising set of techniques?

Also, do you have any plans to examine the reported phenomenon in simpler molecules, where it may be easier to characterize the states and the structure of the molecule at the conical intersection?

1. H. J. Wörner, J. B. Bertrand, B. Fabre, J. Higuet, H. Ruf, A. Dubrouil, S. Patchkovskii, M. Spanner, Y. Mairesse, V. Blanchet, E. Mevel, E. Constant, P. B. Corkum, and D. M. Villeneuve, *Science*, 2011, **334**, 208–212.

Professor Bucksbaum communicated in reply: Dr Kirrander has identified some important questions regarding the implications of our results. If multiple ionization occurs, this may suppress HHG, both through poorer phase matching effects in the resulting plasma and by limiting the neutral recombination probability for the molecule. Future experiments will investigate this possibility. We do have plans to pursue this research with simpler molecules. Even diatomic molecules, which do not have conical intersections, might have enhanced multiple ionizaiton if the enhancement is aided by laser-induced resonant phenomena.

Professor Stolow communicated: Great progress in strong field ionization (SFI) obtained from use of the quasi-static approximation wherein a single driven electron responds adiabatically to the applied field, allowing development of a well known tunnelling picture of SFI. This simple model has worked exceedingly well in atomic physics, giving birth to the entire field of Attosecond Science. However, the interaction of strong laser fields with polyatomic molecules can be a complicated situation in which both the adiabatic[1,2] and single active electron approximations [3,4] can fail dramatically. The latter can be somewhat naively thought of as a consequence of the non-zero low frequency polarizability of the core (*i.e.* dynamic screening). The former, the failure of the adiabatic approximation, is a consequence of inverse electronic level spacings being comparable to time scale of the strong field dipole coupling. These two situations are not wholly independent, as the polarizability (a derived quantity) is a sum over dipole matrix elements divided by electronic energy level spacings. In situations where these approximations fail, it has been shown that multiple ionization and extensive fragmentation of polyatomic molecules can result.[1]

In a dynamic, photoinduced process such as passage through a conical intersection, I would expect that both electronic level spacing and polarizabilities would vary as a function of internal vibrational coordinates. A conical intersection is, in essence, a 'transition state' in an electronically excited state rather that the ground state. In ground state reaction dynamics, I would expect the 'transition state' to be a point of maximum polarizability. Intuitively, this could be understood from the fact that, at the 'transition state', the valence electrons are most delocalized, being by definition somewhere between the old bonds and the news ones. A bit more quantitatively, barriers on ground state potential energy surfaces often arise from very strongly avoided crossings with far away electronically excited states. In this situation, the relevant excited state will have a minimum at geometries corresponding to that of the ground state transition state.

The strong variation of both polarizabilities and electronic level spacing near conical intersections could have important consequences for SFI. Particularly relevant is that the SFI molecular response could change from adiabatic to non-adiabatic upon passage through a conical intersection. The consequences of this could certainly be enhanced ionization and fragmentation.[1] I propose that the variation of the SFI response seen in your results may be due to such effects. If this can be confirmed, then I would say that this method could be used to perhaps locate or determine passage through conical intersections.

1. M. Lezius, V. Blanchet, M. Yu. Ivanov and A. Stolow, Polyatomic molecules in strong laser fields: Nonadiabatic multielectron dynamics, *Journal of Chemical Physics*, 2002, **117**, 1575.
2. M. Lezius, V. Blanchet, D. M. Rayner, D. M. Villeneuve, A. Stolow and M. Yu. Ivanov, Non-adiabatic multi-electron dynamics in strong field molecular ionization, *Physical Review Letters*, 2001, **86**, 51.

3. M. Smits, C. A. de Lange, A. Stolow and D. M. Rayner, Absolute ionization rates of multielectron transition metal atoms in strong infrared laser fields, *Physical Review Letters*, 2004, **93**, 213003.
4. M. Smits, C. A. de Lange, A. Stolow and D. M. Rayner, Dynamic polarization in the strong field ionization of small metal clusters, *Physical Review Letters*, 2004, **93**, 203402

Professor Neumark asked: Can you elaborate on the data in Fig. 2? In particular, how do you use mass spectroscopy to distinguish among the three species responsible for the curves in Fig. 2 given that all three have the same parent mass?

Professor Bucksbaum responded: Here Professor Neumark is referring to the following figure:

This figure, as described in the caption, is from the paper of White *et al.* [J. L. White, J. Kim, V. S. Petrović and P. H. Bucksbaum, *The Journal of Chemical Physics*, 2012, **136**, 054303]. The White paper uses the method of spectral unmixing to discover the linearly independent mass spectral signatures of each of the three channels, corresponding to fragmentation of CHD, CHD+, and HT in the figure.

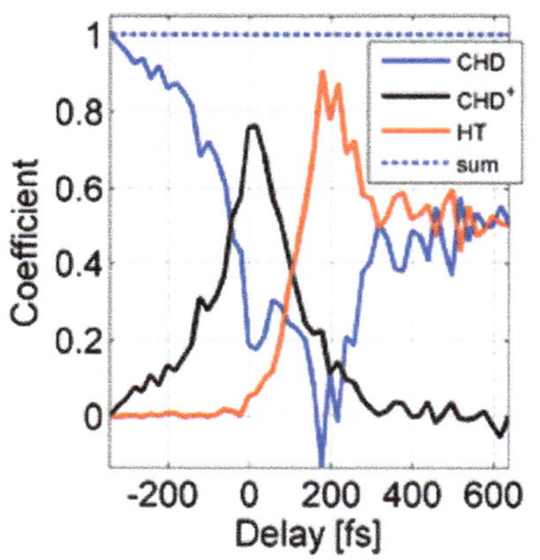

Fig. 4 Relative abundance of the mass fragment TOF spectral signatures associated with CHD and HT following photoexcitation. Reprinted from White *et al.*, with permission.

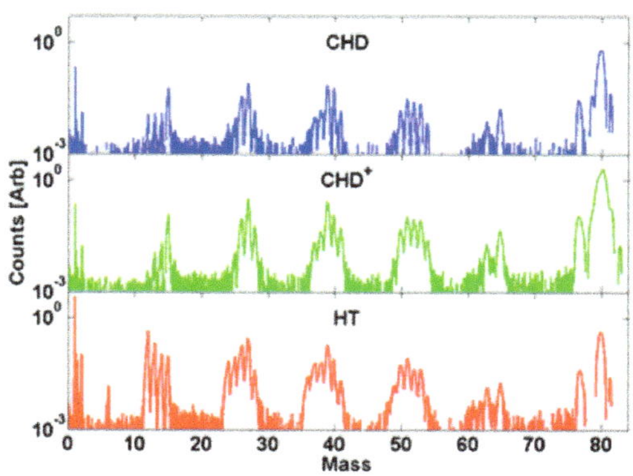

Fig. 5 Spectral amplitudes of the data projected into the three dimensional subspace. Reprinted from White *et al.*, with permission.

Although these species all have the same mass, their fragmentation patterns are quite distinct. Here is another figure from White *et al.* that shows that:

These so-called "endmember" spectra represent the best approximation from the data of the pure spectral signatures of the three species. Then the data are projected into the three dimensional subspace of these three channels, and the spectral amplitudes of these projections are depicted in the figure.

Professor Domcke communicated: There are various ways how quasidegeneracies of potential-energy surfaces can arise in polyatomic molecules. What is unique about conical intersections is the Longuet-Higgins phase, also known as the Berry phase. Does this phase play a role in your double-ionization mechanism *via* conical intersections?

Professor Bucksbaum communicated in reply: The data presented here do not help to answer this question, so I can only speculate about it. Double ionization suggests either ionization energy suppression or intermediate resonances in the cation states of the molecule in this geometry, and of course the Berry phase is also a manifestation of topology. If the geometry-induced CI energy degeneracy between pairs of states with different symmetry persists in both the neutral and the cation, then enhanced ionization through the dipole interaction of these two states might also be present in both. I stress that this is only speculation, but calculations of the location of the CI seams in the cation might help support this line of reasoning.

Dr Mendive-Tapia opened the discussion of the paper by Professor Brumer: Besides the expressions derived for the density matrix, I was wondering if you have searched for further insights by looking at relaxation time-dependent rate constants. As said in the last sentence of the paper, this could be done numerically for individual systems, or perhaps in a complementary way, analytically for a specific form of the Hamiltonian (*i.e.* linear vibronic coupling Hamiltonian).[1–6]

1. M. H. Cho, R. J. Silbey, *J. Chem. Phys.*, 1995, **103**, 595.
2. R. D. Coalson, D. G. Evans, A. Nitzan, *J. Chem. Phys.*, 1994, **101**, 436.
3. A. F. Izmaylov, D. Mendive-Tapia, M. J. Bearpark, M. A. Robb, J. C. Tully, M. J. Frisch, *J. Chem. Phys.* 2011, 135.
4. S. Jang, Y. J. Jung, R. J. Silbey, *J. Chem. Phys.*, 2002, **275**, 319.
5. S. Mukamel, V. Rupasov, *Chem. Phys. Lett.* 1995, **242**, 17.
6. E. Neria, A. Nitzan, *J. Chem. Phys.*, 1993, **99**, 1109.

Professor Brumer communicated in reply: Dr Leonardo Pachon and I have indeed examined this in greater detail in order to gain insight into the origins of the phase control, and its possible magnitude.

Rather than using the approaches that you suggest, we studied a exactly soluble model using an exact influence functional method. Results will be submitted shortly for publication.

Professor Meier enquired: In a publication by M. Joffre (*Science*, 2007, **317**, 453b), very general rules on when a phase control should or should not be present are established. In the examples shown when a one-photon control is observed, which of his general rules are violated in the different cases ?

Professor Brumer communicated in reply: As you know, Miller and co-workers addressed this issue in a reply to the Joffre comment. They suggested that the difficulty with his argument related to the issue of time translation symmetry. Further, do note that a subsequent analysis of the general situation [Spanner, Arango and Brumer, *J. Chem. Phys.*, 2010, **133**, 151101] we showed that phase control is allowed for the measurement of observables that do not commute with the system Hamiltonian. This is indeed the case for the property that Miller *et al.* measured, *i.e.* the control of isomerization. In this case the environment is necessary in order to prevent the system from going periodically between the *cis* and the *trans* configurations. However, it is not necessary for phase control to be permitted. Indeed, we recently demonstrated [Arango and Brumer, *J. Chem. Phys.*, 2013, **139**, 071104] one-photon phase control in a multidimensional model of retinal in rhodopsin, a case related to the original experiment.

Dr Santoro communicated: I really enjoyed reading your elegant paper and I think that the establishment of a formal condition that indicates when a one-photon phase control is possible is a very nice result. You state in the introduction that in a closed system one can achieve a one-photon phase control when the observable one tries to control does not commute with the molecular Hamiltonian. Control of a photoisomerization is possible since the probability to find the system in one of the two isomers does not commute with the Hamiltonian. In other terms, if I understand properly, the control is possible because the vibrational states in the *cis* and *trans* wells of the potential energy surface (PES) are not eigenstates of the molecular Hamiltonian. To fix the ideas, let us say that at time $t = 0$ we know that the system is in the *cis* isomer. This also means, if I am correct, that one can create an excess of population of the *trans* isomer at a given time but, waiting for a sufficiently long period, the system will be back in the *cis* isomer, if it is completely isolated. In my understanding, in all practical cases this does not happen because interaction with the environment introduces dephasing mechanisms, so that the population of the *trans* isomer does not revert spontaneously to zero. Maybe this question is a bit philosophical but: can therefore we say that also in the case of an isolated system one needs to take into account the coupling with an "environment" to explain why the population of the reaction product is permanent? Probably everything rings around the definition of "system" and "environment" but, if the role of the "environment" is played by a bath of secondary modes of the system itself, does this conceptually change the picture? In an open system it is possible to control also an observable that commutes with the system Hamiltonian and, as you clearly state in Ref. 1, this is connected with the fact that the observable does not commute with the Hamiltonian of the "super-system" comprising the system and the environment. These considerations help me to introduce the following very qualitative questions:

(i) can we say that one of the results of your work is that in most of the problems of chemical interest one-photon phase control is possible because the system will always interact with an environment?

(ii) any computational simulation of a control experiment adopts a model of the real system, neglecting therefore some parts that can play the role of an "environment". In this context, does your work in some sense support the idea that a control scheme able to focus, at a given time, a large excess of population in a desired reaction channel, or in a isomer well (even if this is a transient population for the "isolated model system") can provide an effective route for control, if chemical intuition suggests that interaction with the neglected environment will make the "reaction" not reversible?

Clearly your analysis allows getting a much deeper understanding of the control mechanism. For example you show that the environment-assisted control acts on two different timescales, the timescale of the radiation field and the timescale of the system-bath interaction. Is it possible to make some general, qualitative, statement on which of the two timescales is expected to be more "crucial" for the control?

1. M. Spanner, C. A. Arango and P. Brumer, *J. Chem. Phys.*, 2010, **133**, 151101.

Professor Brumer communicated in reply: My co-author, Leonardo Pachon, and I appreciate your questions and answer them below.

Control over isomerization in an isolated molecule is indeed possible since the probability of finding the system in one of the two isomers (e.g., *cis* or *trans*) is an observable that does not commute with the molecular Hamiltonian. However, as pointed out in the question, if one waits long enough, the a flow back of population should be observed, and it is in this particular aspect that the environment plays a major role, i.e. in "stabilizing the product".

In the case where "environment" is comprised of secondary modes of the system itself, we break the time reversal symmetry over an effective time scale which depends upon the dimensionality of the "environment". According to the Poincare recurrence theorem, that time scale grows exponentially with the number of degrees of freedom of the system. In the case discussed in retinal isomerization by Arango and Brumer, *J. Chem. Phys.*, 2013, **138**, 071104, the 24 modes of vibration of the system are able to prevent, on an effective time scale, the back and forth transfer of population between *cis* and *trans*.

Your comment that phase control should then be possible for most of problems of chemical interest since the system will always interact with an environment, is correct. However, our general results show only the existence of that effect. The magnitude of the effect, as we discuss in the end of our paper, will depend on the nature of the system-environment interaction. In a forthcoming paper, we explore in a quantitative way the magnitude of that effect, showing that it depends on the field characteristics, such as strength and frequency, as well as on environment properties such as temperature and memory.

The next question asks whether our work supports the idea that a control scheme able to focus a large excess of population in a desired reaction channel can provide an effective route to control, if chemical intuition suggests that the interaction with the neglected environment will make a process non-reversible.

Our results do suggest this. However, we think that control of such a process could be challenging because it would imply the control over, in principle, several degree of freedoms. However, one could try to "engineer" the environment in order to get a desired result. For example, one could ask whether, e.g., coherently pumping the vibrational modes in order to transfer energy from one electronic freedom to another or to break an electronic bonding is a realistic approach. This could provide an alternative, indirect, means of environmentally assisted coherent control.

Finally, you ask if it is possible to make some general qualitative arguments as to which of the two time scales is expected to be more crucial for the control. Since one is interested in control in the long time regime, one expects that this limit is dictated by the system-environment interaction. Hence, we anticipate that the contribution taking place in the radiation field timescale, mediated by the off-diagonal elements, is less "crucial", in the long time regime, than the contribution coming from the system-environment interaction itself. For a detailed study of a specific model we refer to a forthcoming publication by Pachon and Brumer.

Miss Perveaux opened the discussion of the paper by Dr Blancafort by communicating: When the external field is on, the second excited state comes very close to the states you are studying. You may thus want to include this second excited state in your model.

Do you know if your choice of active coordinates is suited to study this state or should you consider some other coordinates? Can you trust calculations at the CASSCF level of theory for this state, or do you need to include dynamic electron correlation?

Dr Blancafort communicated in reply: In our work we assume that the active coordinates derived for the field free situation are also suitable when the non-resonant field is active, in other words, we assume that the position of the critical points on the surface does not depend on the non-resonant field. One may question this assumption already for the first excited state and the ground state. However, we believe that the changes will be small, and since the present study is more of a proof of principle, we have not taken this into account yet. It is something that we will test in the future. Of course, if one wants to include more states in the dynamics, the number of relevant coordinates will increase because other regions of the surface may become relevant.

Regarding dynamic correlation, we know that CASSCF gives a good description of S_0 and S_1, but we have not checked the importance of dynamic correlation for S_2.

Professor Domcke communicated: My question refers to Fig. 7 in your paper. While the oscillatory structure at short times is strongly affected by the control field, the long-time limit of the population of the V state seems not to be affected. What is the physical reason for this?

Dr Blancafort communicated in reply: This is due to the extended nature of the seam, and it is one of the punchlines of our work. Initially the decay follows the totally symmetric stretch coordinate, and the seam of intersection at planar geometries is energetically not accessible. After approximately 50 fs, intramolecular vibrational energy redistribution activates the torsional mode, and the region of the seam at twisted geometries can be accessed. Clearly, such an effect only appears if the extended nature of the seam is explicitely taken into account in the model.

Dr Santoro communicated: The decay of your system is ultrafast, so there is only a very short time-window to try to control it. You mention in fact that most of the photoexcited population returns to the ground state in less than 10 fs. I imagine therefore that you used an ultrashort resonant pump laser pulse in your simulations, and this seems coherent with what is seen in Fig. 5 of your paper, where the excited state population rises in about 5 fs. Then you use a second dynamic non-resonant "control" field that oscillates with a period of 4 fs, assuming that it is switched on before the resonant one. In such a situation, does the control efficiency depend on the "phase" of the non-resonant field, or in other terms on the delay between the starting of the two fields?

Dr Blancafort communicated in reply: You're right, the time window to control the decay is very short, and therefore we consider it a success to be able to delay it by a few tens of femtoseconds. Regarding your question, it is an interesting point, since the duration of the resonant and non-resonant pulses is similar, as you point out. We will check the influence of the phase and the delay in future work.

Mr Binder asked: Usually, the Multiconfigurational Time-Dependent Hartree (MCTDH) formalism is quite robust, *i.e.* a moderate number of configurations in most cases yields an at least qualitatively correct result. However, in your work you employed an unusually high number of configurations. Was this high number necessary to catch the qualitatively correct dynamics of the system or did you opt for this high number *a priori* to avoid convergence assessment? Furthermore, can you give a rough estimate, when to find most (say = 95%) of the population back in the ground state?

Dr Blancafort replied: We employed a large number of single particle functions in the combined Q_{x1}/φ mode because the surface along the φ coordinate is very flat, and the wave packet spreads along that coordinate. This makes the large basis necessary.

Concerning the time required to transfer most of the population to the ground state, we cannot give an estimate with the present model, because more modes would be required to dissipate the vibrational energy and avoid the recrossings to the excited state.

Professor Whitaker communicated: When one looks at Fig. 3 in your paper, and the population decay traces in Fig. 5, there is a striking similarity with the pictures presented to describe Marcus theory on moving from the inverted to normal regions, and I wonder to what extent a laser dressed state resembles a solvent shift and what further analogies can be drawn.

Dr Blancafort communicated in reply: Yes, there is a clear analogy with the Marcus picture and the passage from the inverted to the normal region. The behaviour is also similar to what Fermi's golden rule would predict for internal conversion, since the fastest decay rate is observed when the overlap between the S_1 vibrational level accessed during the excitation and the ground state degenerate level is optimized.

In a first approximation, the laser dressed state resembles a solvent shift because both 'environments' (if we consider the non-resonant pulse as such an environment) act on the dipole moment of the molecule. However one should note that the laser induced shift depends more on the polarizability than on the dipole moment.

Dr Malhado addressed Professor Whitaker and Dr Blancafort: Since Professor Whitaker raised the question about possible analogies with the role of solvent polarity in this type of non-adiabatic processes, I would like to mention how in the context of a simple model[1–3] it is possible to show the important effect of a polar solvent on the potential energy surfaces, changing the position and topography of the conical intersection seam and inducing different reaction paths and product yields, in a very similar way to the effect of the applied external field shown in Dr Blancafort's article. Because upon electronic excitation the surrounding solvent is out of equilibrium with the solute charge distribution, I would like to highlight the importance of the solvent dynamics in such processes, as it occurs in a similar time scale as these ultra-fast reactions, and the above mentioned calculations indicate that different solvent dynamics *alone* can lead to different product yields. While the active control of the solvent environment does not seem *a priori* the simplest way to determine the outcome of a chemical reaction, in the context of "passive control" discussed at this conference, it has been known by chemists throughout history the importance of choosing the right solvent to optimize a certain product yield.

1. I. Burghardt, L. S. Cederbaum, J. T. Hynes, *Faraday Discuss.*, 2004, **127**, 395–411.
2. J. P. Malhado, J. T. Hynes, *J. Chem. Phys.*, 2012, **137**, 22A543.
3. J. P. Malhado, R. Spezia, J. T. Hynes, *Int. J. Quantum Chem.*, 2013, **113**, 296–305.

Dr Blancafort communicated in reply: Clearly, the position and topography of the seam are very important for the photochemistry, and I agree that the described solvent effects are a nice analogy to the non-resonant electric field effect.

Dr Santoro communicated: You mentioned that the solvent can control a nonadiabatic process altering the position and energy of a conical intersection (similarly to what an external field can do). I would like to add that this is exactly what, in our opinion, happens in nucleobases. Considering the $\pi\pi^*$->$n\pi^*$ internal conversion in uracil derivatives, our results[1] suggest that, moving from acetonitrile to water, the destabilization of the $n\pi^*$ due to the loss of an hydrogen bond causes a shift of the conical intersection with the $\pi\pi^*$ state, making it accessible or not from the Franck Condon region, depending on the substituent in position 5.

1. R. Improta, V. Barone, A. Lami and F. Santoro, *J. Phys. Chem. B*, 2009, **113**, 14491.

Professor Duxbury opened further discussion of the paper by Professor Bucksbaum by communicating: In neutral and ionised open shell triatomic molecules there are many examples of unusual intersections where three states are involved, and where two states are degenerate when the triatomic molecule becomes linear. Three series of molecules which this occurs are the CH_2, SiH_2, GeH_2 family, the NH_2, PH_2, AsH_2 family, and the H_2O^+, H_2S^+ and H_2Se^+ family. The states which become degenerate when linear are a Renner–Teller pair, for example the A1 and B1 States of CH_2, and the other interacting state is triplet CH_2. The triplet state of CH_2 can only interact with the lower singlet state of the pair of singlet states *via* an electronic Coriolis coupling between the Renner–Teller pair. As one proceeds from the first row of the periodic table to the second row the atomic spin–orbit constant of the heavy atom becomes much larger, and predissociation may occur so that breaking off of emission can occur, as in the hydrogen sulphide ion, or rapidly increasing broadening, as in AsH_2.

Although these are not strictly conical intersections, these effects do occur in some of the most commonly occurring small molecules, for example ionised water. They have also been studied by velocity map imaging, such as H_2S^+ by Mike Ashfold and his group. There have also been doubly ionised examples such as those studied by John Eland.

Professor Bucksbaum communicated in reply: Professor Duxbury brings up an interesting point. Excited systems such as the A1 excited state in the water ion undergo rapid bending distortion and become degenerate with the ionic ground state as the molecule becomes linear. This is a degeneracy, but not a conical intersection. Therefore a fragmentation study of this molecule could shed some light on the necessary conditions for enhanced multiple ionization.

Professor Weitzel communicated: You have described an intriguing analysis of enhanced multiple ionization near conical intersections. Can you possibly relate your findings to the observation that multiple ionization in chirped femtosecond laser fields is capable of imaging local electron density onto fragment formation in a way which is very sensitive to structural isomerism (see *e.g.* G. Urbasch, H.G. Breunig, K.-M. Weitzel, *ChemPhysChem*, 2007, **8**, 2185–2188)?

Professor Bucksbaum communicated in reply: Professor Weitzel here refers to the paper by Urbasch *et al.*, which reported that the presence of C^{2+} in the photofragmentation spectrum of xylenes could be a high-contrast indicator of the isomer present. In particular, for the right pulse energies and chirp, para-xylene fragments included distinct amounts of C^{2+}, and *ortho*-xylene did not. This might perhaps be related to the effect reported here, but to study this possible connection, it would be highly desirable to obtain not only the fragment charge to mass ratios, but also their fragmentation momenta. This would make it easier to determine if multiple ionization was occurring in the parent prior to dissociation, as observed in our report, or rather is a "final state effect" where the laser ionizes a fragment after dissociation. In our work the parent became multiply charged, leading to high kinetic energy fragments.

As an aside, White *et al.* [*J. Chem Phys.*, 2012, **136**, 054303] did observe some C^{2+} fragmentation in the product isomer 1,3,5-hexatriene, in the spectral unmixing study reported above (see the reply following the comment by Professor Neumark, earlier in this discussion). We may therefore suppose that it may also be present in the experiment of Petrovic *et al.* However, this fragment channel was not studied by Petrovic

Dr Hockett gave further comment on the paper by Dr Blancafort: It occurred to me that, since the discussion this week has touched a number of times on the closer integration of experiment and theory, I might show some recent work in order to provide a concrete example to frame this discussion, which was so far very general. This work also serves as an illustration of another, related, topic which came up in several discussions: the quantitative analysis of data from complex dynamical systems, and the utility – or otherwise – of time constants drawn from kinetic models as a metric for such systems.

Fig. 6 shows some recent results from time-resolved photoelectron spectroscopy (TRPES) of butadiene. In this work we were using 200nm to pump, and 266nm to probe; the duration of both pulses was on the order of 40 fs. Detection was *via* a magnetic bottle spectrometer. The spectrum shows a clear stripe of photoelectron signal with "chirp" - a delayed onset of signal as a function of energy. If we look at the same data, normalized for each energy slice, it is clear that the peak shapes are approximately Gaussian for all photoelectron energies, with some broadening towards the lower energy part of the spectrum.

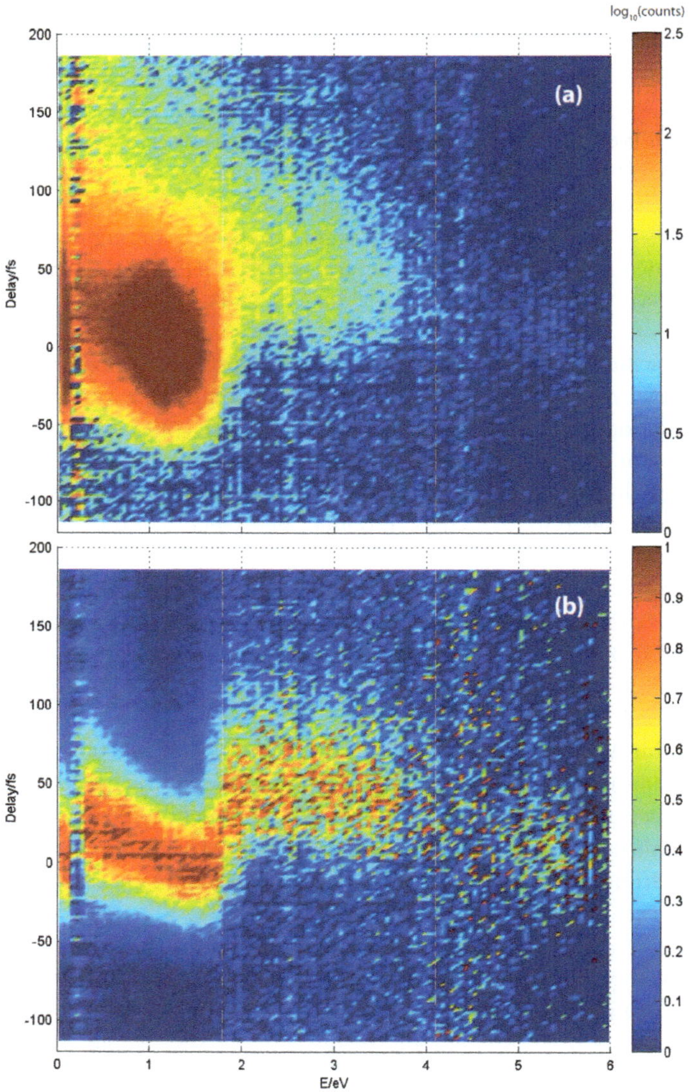

Fig. 6 (a) time-resolved photoelectron spectrum, colour bar shows log(counts); (b) same data as (a), but normalized to peak of signal for each energy slice (0.1 eV slices), colour bar shows normalized counts. The 1-photon cut-off is at ~1.8eV, 2-photon cut-off at ~4.1eV.

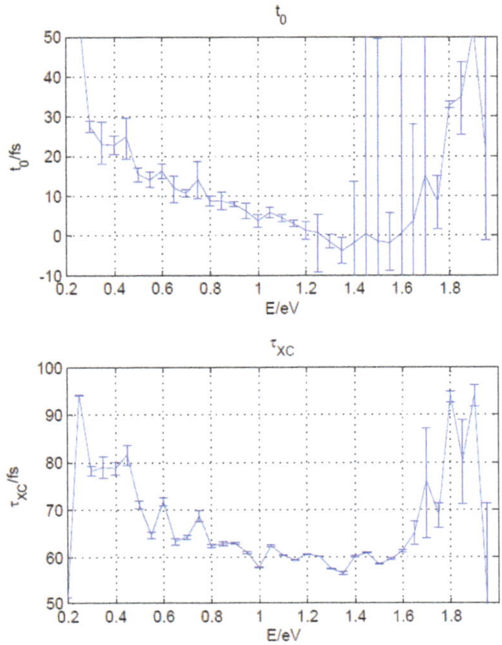

Fig. 7 Parameters for Gaussian temporal fits, performed independently for each energy slice. Fits were weighted to the rising edge of the signal, and by experimental uncertainties (statistical). The Gaussian centre is defined by t_0, and full-width half-max by τ_{XC}. Regions with large error bounds indicate low signal or non-Gaussian line-shapes.

It is clear that "standard" kinetic models, based on two or more analytic temporal functions (usually exponential delays convoluted with a Gaussian cross-correlation feature) with globally fitted time constants and spectral components, aren't really appropriate for this kind of complex behaviour. In practice a reasonable fit might be found with enough temporal functions, but there is no clear physical significance or import to such a fit (which essentially originates from a statistical picture of the dynamics) – beyond the use of such an analysis as an empirical metric. Therefore, an alternative empirical metric might be considered in order to quantify the observed spectrum, and potentially provide

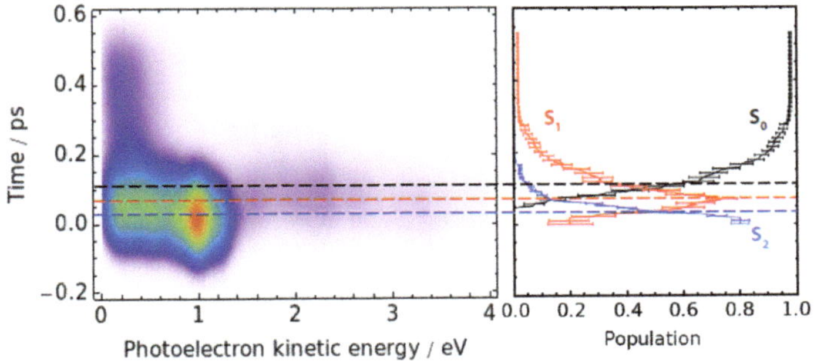

Fig. 8 Calculated time-resolved ionization potential (TRIP) from full-dimensional wavepacket calculations and electronic state populations on S_2 (bright state), S_1 and S_0. Calculations and figure courtesy of M. S. Schuurman (2012, unpublished).

deeper insight into the underlying dynamics in the case where phenomenological inferences may be drawn from such a metric.

Fig. 7 shows the results of Gaussian temporal fits, performed independently for each energy slice. This provides a metric to quantify both the chirp of the spectrum and the broadening observed – hence a way to clock the observed wavepacket dynamics. Here the fitted parameters, as a function of energy, are approximately linear over the observed spectral region, down to around E=0.3 eV where there is significant contribution from background signal. This suggests a

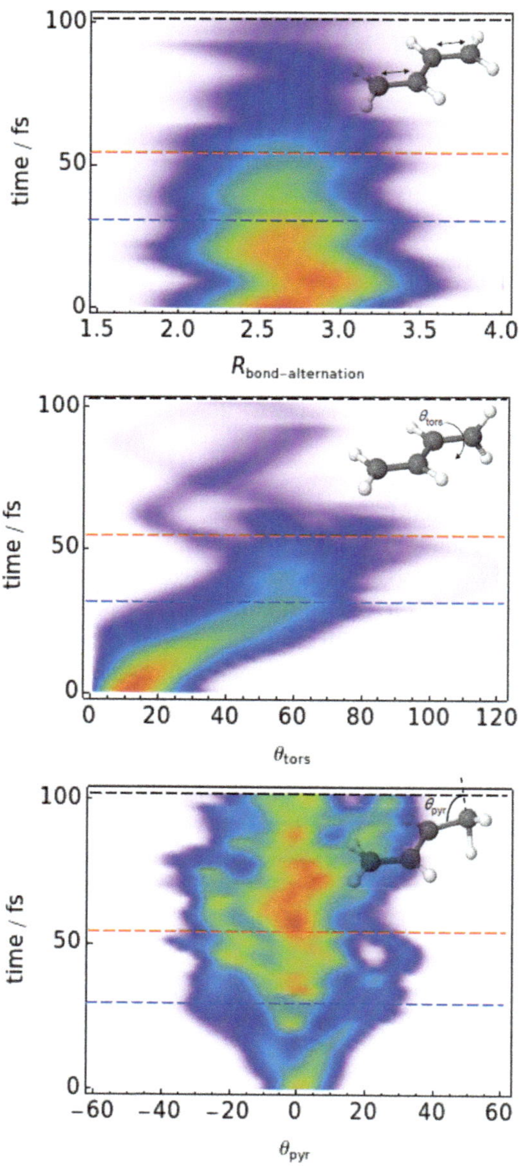

Fig. 9 Wavepacket dynamics projected onto (a) bond-alternation coordinate, (b) terminal methyl torsion, (c) terminal methyl pyramidalization. Calculations and figure courtesy of M. S. Schuurman (2012, unpublished).

phenomenological picture of the initial wavepacket on the excited state moving rapidly (and with little dispersion) along a steep gradient(s) on the potential energy surface, with this motion mapped primarily to the instantaneous ionization potential. It is important to note that this is a very different picture from that of a kinetic model fit, which would provide two (or more) time constants, often interpreted as indicative of wavepacket dephasing or excited state lifetimes. In this case, such time constants would indicate only the decay of the signal from a given region of the spectrum.

Fig. 8 shows some examples of the results of recent *ab initio* multiple spawning (AIMS) calculations, performed by Michael Schuurman at NRC; these are state-of-the-art, full-dimensionality wavepacket calculations. From these results we can compare directly the experimental TRPES and calculated TRPES or, in this case, the calculated TRIP (time-resolved ionization potential). The TRIP represents a reasonable simplification to obviate calculation of the full ionization matrix elements, and may be a good approximation in many cases. In this case there is, visually, good agreement between theory and experiment. Furthermore, by fitting the calculated TRIP to the models discussed above, a more quantitative comparison can be made, revealing good agreement between the extracted metrics from the calculated and experimental data.

Of course, the benefit of theory is that we can now look behind the curtain at the underlying wavepacket dynamics which give rise to the observable. As was already discussed a little this morning, the understanding, mapping and visualization of multi-dimensional spaces is a hard problem, so we show here just a couple of example projections along key vibrational coordinates. In particular the torsional coordinate shows rapid and irreversible motion, while the carbon backbone shows damped oscillatory behaviour. In this case the steep gradient along the torsional coordinate is primarily responsible for the rapid wavepacket motion observed experimentally, and maps essentially to the IP of the system. Other observables may be sensitive to different aspects/dimensions of the wavepacket motion – for example the photoelectron angular distributions show fast oscillations which appear to map some of the carbon backbone dynamics.[1]

In conclusion, this joint experimental and theoretical study indicates both some of the important issues in moving forward with time-resolved measurements of polyatomic systems – in particular the way we might move towards using more detailed metrics for clocking wavepacket motions – and the benefit of close collaboration between experiment and theory.

1. P. Hockett, E. Ripani, A. Rytwinski and A. Stolow, *Journal of Modern Optics*, 2013, DOI:10.1080/09500340.2013.801525.

Faraday Discussions RSCPublishing

PAPER

Photoinitiated quantum molecular dynamics: Concluding Remarks

Michael N. R. Ashfold

Received 14th May 2013, Accepted 14th May 2013
DOI: 10.1039/c3fd00090g

Photochemistry and molecular photophysics have long been very active fields of research. Recent advances in both experimental technology and computational methodology have triggered further spectacular progress in these areas—exploring, understanding and even controlling photochemistry and photophysics at the quantum level, and greatly expanding the range and complexity of photochemical systems amenable to study. The frequency of recent *Faraday Discussions* on related themes, *e.g. Molecular Reaction Dynamics in Gases, Liquids and Interfaces* (FD157, Assisi, 6/2012), *Coherence and Control* (FD153, Leeds, 7/2011), *Frontiers in Spectroscopy* (FD150, Basel) and *Cold and Ultracold Molecules* (FD142, Durham, 4/2009), is testimony to this burgeoning interest.

Faraday Discussions addressing photochemical processes have a long and distinguished tradition. The first, on *Photochemical Reactions in Liquids and Gases*, was held at Jesus College, Oxford, in 1925, and published in the *Transactions of the Faraday Society* in 1926.[1] That Discussion featured two introductory addresses, by Allmand (on Einstein's law of photochemical equivalence) and by Bodenstein (on the mechanisms of photochemical reactions) and a summary presentation by Rideal, which opened with the sentence 'It must be admitted that there is more divergence in opinion than unanimity in views as to the mode of operation of photochemical processes'.

The field had advanced considerably by the time of the 1931 Discussion on *Photochemical Processes*, held in Liverpool.[2] The programme of this Discussion sub-divided into four parts, with sessions on Molecular spectra in relation to photochemical change, Photochemical kinetics in gaseous systems, Photochemical change in liquids and solids and Photosynthesis. Mecke gave the opening lecture in the first session. His paper contains sentences like 'But we can go still further now and make the general statement, that the appearance of a continuous, *i.e.* non-quantised, absorption spectrum of a gas is always a sign of photochemical dissociation taking place in the gas' and 'It still remains to explain why in one spectrum (*e.g.*, the ultraviolet absorption spectrum of oxygen) the convergence limit is directly observed, whilst in another absorption spectrum of the same gas (*i.e.*, red absorption spectrum of O_2) it is missing. The explanation is

School of Chemistry, University of Bristol, Bristol, BS8 1TS, UK

due to Condon-Franck'—whose ideas he then proceeded to describe. Such interpretations still feature prominently in undergraduate textbooks of today.

One other Discussion certainly deserves reference in any retrospective section such as this—the 1962 Discussion at St Andrews University in Dundee on *The Structure of Electronically Excited Species in the Gas Phase*.[3] Herzberg presented the Introductory Lecture, and the list of those contributing papers (*e.g.* Buckingham, Coulson, Dixon, Douglas, Kuppermann, Linnett, Longuet-Higgins, Price, Ramsay, Thrush, van der Waals, Walsh, Watson, *etc*) reads like a who's who of the leading proponents of molecular spectroscopy, structure and bonding at that time. This 1962 Discussion featured the seminal paper by Herzberg and Longuet-Higgins describing the intersection of potential energy surfaces (PESs) in polyatomic molecules[4] that was highlighted in the Introductory Lecture to the present Discussion.[5] As with the 1931 Discussion, concepts aired and honed in the 1962 Discussion now feature in contemporary textbooks. In concluding this retrospective, it is marvellous to note the sense of continuity provided by Dr Mike Hollas—a young contributor at the 1962 Discussion[6] and a participant at the present Discussion.

FD163 has also focussed on core fundamental science, but the potential impact of photochemistry and molecular photophysics now stretch much more widely. There are myriad reasons for developing a better understanding of the coupling and flow of energy between light, electrons and chemical bonds in molecules, not least to enable this community to contribute to the challenges of designing light harvesting systems for clean energy generation, and to diverse fields like photocatalysis, the design of efficient light-driven molecular devices for data storage and processing, and photomedicine. Four over-arching themes were identified when planning the present Discussion: Single molecules: photochemistry and photophysics in isolated molecular systems; Extended systems: photochemistry and photophysics of chromophores in proteins, solution or clusters; Controlling molecular dynamics: controlling photochemistry using sequences of light pulses, shaped light pulses or bond selection prior to photoexcitation; and Applications of molecular dynamics to global challenges: photovoltaic cells, photodynamic therapy, imaging. It is too early to predict the likely lasting legacy of the work presented at this Discussion, and I do not intend to provide a blow-by-blow summary of all the presentations and the accompanying debate. This can be gained by reading the Discussion volume. I am very happy to offer a few personal observations however.

The Introductory Lecturer (Professor Stolow, NRC Ottawa) provided a masterly overview of the current state-of-the-art couched in terms of 'three pillars' of light-matter interaction: structure, function *and dynamics*, viewed from the perspectives of energy/time, phase/coherence and intensity.[5] Stolow posed and offered answers to many of the current 'hot' issues in the field, and set the tone for much of the ensuing discussion. For example, are there compelling reasons why most natural photoprocesses (*e.g.* vision, photostability and/or photoprotection mechanisms in DNA and DNA bases, *etc*) occur on ultrafast timescales? How applicable is the quasi-static model (developed to account for strong field ionisation in atoms) to polyatomic molecules? What are the relevant timescales when thinking about electron dynamics in regions of conical intersection (CI) between PESs? The contributed papers that followed spanned all of the four target themes though, perhaps inevitably, those addressing applications of molecular dynamics to global challenges like light harvesting, phototherapy, *etc.* could only offer a taste of (selected aspects of) the current state-of-the-art in these fields; each could readily form the basis of a *Faraday Discussion* in their own right.

Were there topics that I had expected to see represented more strongly? Yes, I had anticipated more contributions addressing quantum coherence effects,[7] and one or more papers reporting dynamical studies initiated by short pulse XUV radiation. In the event, the late-breaking contribution from Professor Scholes (Toronto) describing 2-D spectroscopy studies of two phycobiliproteins (both of which act as successful light harvesting complexes in nature, but only one of which is found to show measurable coherences between exciton states) served to provoke much interesting discussion of quantum coherence effects on the final morning. I am also aware that the Faraday Standing Committee on Conferences is already addressing my other perceived 'omission': FD171 (Sheffield, July 2014) will be devoted to *Emerging photon technologies for chemical dynamics.*

As at any Discussion, the contributed papers triggered many complementary short presentations and a wealth of specific questions and comments, amongst which I identified several recurring issues. The transient spectra obtained in condensed phase pump–probe experiments (even those involving relatively small molecules) are generally comprised of many broad, overlapped features. Interpreting such data usually involves some form of global analysis, but is made challenging by our imperfect knowledge of the number, nature and time evolution of the constituent basis functions used in the decomposition. Riedle and coworkers[8] showed the clear benefits of recording such broadband spectra over the widest possible range of probe wavelengths (which, ideally, would stretch into the infrared (IR) fundamental region also), but rigorous analysis and interpretation of such transient spectra remains a challenge. Challenges are by no means limited to the experimental domain, however! The photochemistry/physics community is attempting to tackle ever bigger, more complex systems, and the theorists in the audience were repeatedly quizzed regarding the most appropriate form of theory to address a particular problem, and what confidence any non-expert should have in the results of different calculation types. The relentless progress in experiment and theory is enabling study of an ever greater range of systems; indeed, the range of chemistry now accessible to the photophysics community far outstrips that which can realistically be investigated. Hence the suggestion of a consortium approach, wherein a limited number of key systems are identified for detailed investigation using the full armoury of available methods (experimental and theoretical)—a concept pioneered to great effect by Delbrück's promotion of the Phage Group and the exploration of genetics through researches on bacterial viruses.

Two further themes of the Discussion relate closely to recent work in our own group, and I claim summariser's prerogative to conclude this presentation by highlighting some such links. One discussant enquired 'passive control—is that not just chemistry?' to which the answer is surely yes. This Discussion provides several illustrations of just how effective chemistry can be at controlling photophysical outcomes. For example, Rothlisberger and coworkers[9] provide a theoretical rationale for the observation[10] that C–NH_3^+ bond fission following UV excitation of the tryptophan chromophore in a 'dry' Lys^+-Trp pair can be completely suppressed by solvating the amino group with just two water molecules. They also find that this photo-protection mechanism (which can be traced to a relative increase in the energy of the dissociative states formed by $\sigma_{CN}^* \leftarrow \pi$ excitation) is maintained under fully solvated conditions. In a similar vein, Meech and co-workers[11] find that the excited state photophysics of both the neutral and anionic states of the chromophores of green fluorescent protein (GFP) and the kindling fluorescent protein in methanol solution are very similar, yet their photophysical behaviour in the two proteins (*i.e.* in the presence of the respective protein scaffolds) are strikingly different.

These examples highlight the potential sensitivity of excited states, and their photophysics, to intermolecular interactions with the local environment. Photophysics can also be 'tuned' by intramolecular modifications – as illustrated in Professor Stolow's overview of the contrasting near UV photochemistry of acrolein and other α,β-enone isomers[5] and in the paper by Sension and co-workers on the photo-induced ring-opening of various substituted 1,3-cyclohexadienes.[12] The data shown in Fig. 1 provide another example. Excitation to the $S_1(^1\pi\pi^*)$ state of phenol in the gas phase results in O–H bond fission, by tunnelling through the barrier under the CI between the $^1\pi\pi^*$ and the dissociative $^1\pi\sigma^*$ PESs—as discussed by Ramesh and Domcke in this Discussion.[13] 4-substitution affects the relative energies of the two PESs, and thus the magnitude of the barrier under the CI and the tunnelling probability from the $^1\pi\pi^*$ state. Substituting a strong electron withdrawing group like CN completely suppresses this O–H bond fission channel.[14] The choice of substitution site can also be important. O–H bond fission is just one of several possible decay routes from the $^1\pi\pi^*$ state of phenol – others include internal conversion (IC, to the ground (S_0) state) and intersystem crossing. Each of these non-radiative processes must be relatively slow in phenol, given the measured fluorescence lifetime of the $^1\pi\pi^*$ state ($\tau \sim 2.2$ ns).[15] The

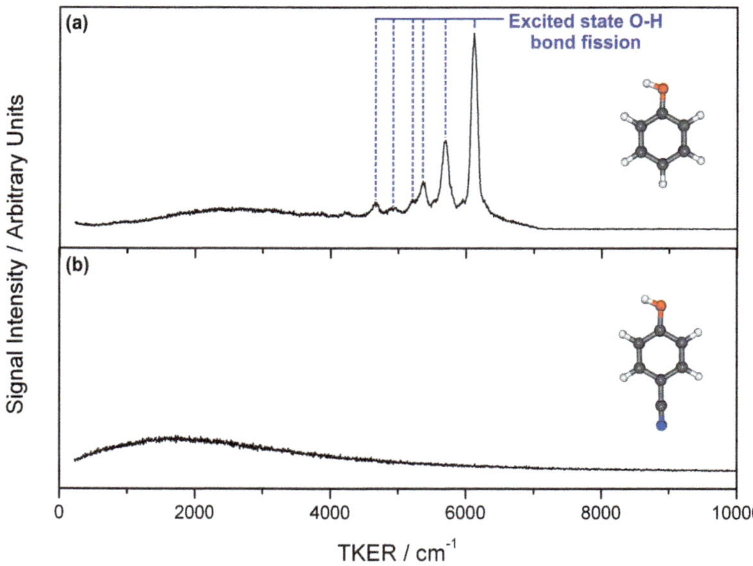

Fig. 1 Total kinetic energy release (TKER) spectra obtained from time-of-flight spectra of H atoms formed in the photolysis of (a) phenol at $\lambda = 275.113$ nm and (b) 4-cyanophenol at $\lambda = 283.998$ nm. The peaks in (a) are attributable to formation of phenoxyl radicals, in selected vibrational levels of the ground electronic state, as a result of O–H bond fission by tunnelling through the barrier under the CI between the $^1\pi\pi^*$ and $^1\pi\sigma^*$ PESs.

corresponding $^1\pi\pi^*$ state of 4-chlorophenol has a similar lifetime ($\tau \sim 1.6$ ns[16]) yet, as Fig. 2 shows, recent pump–probe studies confirm earlier suggestions[17] that the same $^1\pi\pi^*$ state in 2-chlorophenol decays on a sub-ps timescale. This orders of magnitude increase in loss rate may be attributable to a much enhanced IC rate, as a result of a reduction in the energy of the CI between the $^1\pi\pi^*$ state and the lowest $^1\pi\sigma^*$ potential in the C–Cl stretch coordinate, which offers another radiationless pathway to the S_0 state.[18]

Substitution can also affect the branching into different product channels. Consider thiophenol. The PESs for the S_0, $^1\pi\pi^*$ and $^1\pi\sigma^*$ states are qualitatively similar to those for phenol, but the barrier to S–H bond fission by tunnelling under the $^1\pi\pi^*/^1\pi\sigma^*$ CI is much smaller. Thus the H atom products, even those formed when exciting at the $^1\pi\pi^*$–S_0 origin, display anisotropic recoil velocity distributions; *i.e.* the timescale for fragmentation is shorter than the parent rotational period.[19] The $^1\pi\sigma^*$ PES of thiophenol (and phenol) exhibits another CI, with the S_0 PES, at larger S–H(O–H) bond lengths. The energy separation between the ground ($\tilde{X}$) and first excited ($\tilde{A}$) electronic states of the resulting radicals is much smaller in the case of thiophenoxyl, and the radical products from near UV photolysis of thiophenol are formed in both of these states. 4-Substitution can have a dramatic effect on the $\tilde{A}/\tilde{X}$ product branching ratio, which decreases from ~0.8 to ~0.6 as the excitation wavelength is tuned from that of the $^1\pi\pi^*$–S_0 origin to ~250 nm in the case of 4-methylthiophenol, but increases from ~1.3 to >3 (*i.e.* a population inversion throughout) across the same wavelength range when the methyl is replaced by a methoxy substituent. Branching is established in the vicinity of the $^1\pi\sigma^*/S_0$ CI, but is largely controlled by the orientation of the S–H

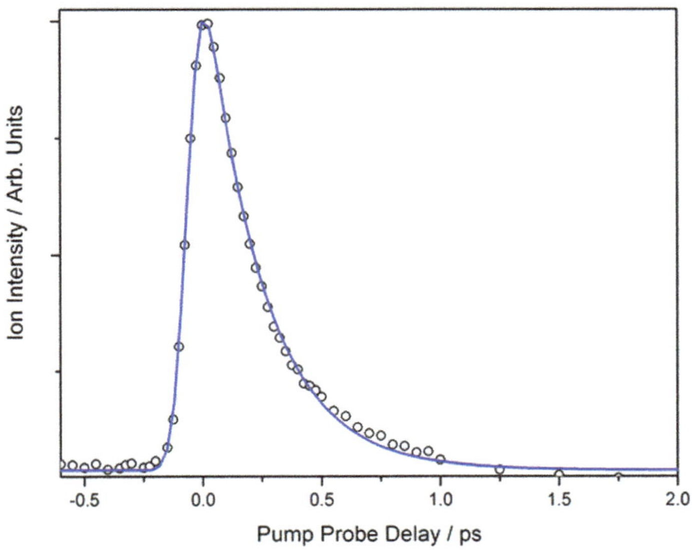

Fig. 2 Time-resolved parent ion signal obtained following excitation of 2-chlorophenol at the origin of the $^1\pi\pi^*$–S_0 transition of the *cis*-isomer and probing (ionization) at 243.1 nm. 'One colour' contributions from the pump and probe lasers alone have been subtracted prior to display. The solid curve (obtained by convoluting a 120 fs (Gaussian) instrument response function and a single exponential decay with time constant 260 fs) is intended simply to illustrate the rapidity of the excited state decay (*cf.* 1.6 ns for the corresponding state of 4-chlorophenol).

bond (relative to the ring plane) as the dissociating molecules approach this CI.[19] Such 'passive' control may well be 'just chemistry' but, as with so much of chemistry, it can be very influential!

We end by revisiting the question 'to what extent can dynamical insights gleaned from detailed studies of a photo-induced process in the gas phase inform our understanding of the equivalent process in solution?' As noted above, Rothlisberger and coworkers[9] find that the excited state photophysics displayed by the Lys^+-Trp pair in the gas phase and in aqueous solution are very different, but the discussion of the paper by Bochenkova and Andersen noted evident similarities in the decay kinetics measured for the anionic GFP chromophore in the gas and solution phases.[20,21] Much of our recent activity has focussed on comparing selected benchmark photodissociation processes and bimolecular reactions in the gas and solution phases. For example, investigations of the near UV photodissociation of phenol and of various thiophenols and thioanisoles, in the gas phase and in weakly interacting solvents like acetonitrile and cyclohexane, show that the gas phase results provide an excellent guide to the early time dynamics prevailing in solution though, clearly, additional processes (*e.g.* vibrational relaxation, geminate recombination, adduct formation, *etc*) specific to the condensed phase reveal themselves on longer timescales.[15,22,23] Similar conclusions have been reached in photo-initiated studies of F and Cl atoms, and of CN radicals, with alkanes. For example, transient broadband IR probe measurements of the HCN products formed in the reaction of CN radicals with cyclohexane in a range of chlorinated organic solvents reveal preferential excitation of one quantum of the C–H stretch and up to two quanta in the bending mode, which is progressively quenched through solvent interaction on the longer timescale.[24] Such state-specific vibrational energy disposal at early times parallels that found in similar CN + alkane reactions in the gas phase,[25] though the extent of vibrational relaxation is less—consistent with partial damping of the developing HCN vibrational motion in the post-transition state region.[26] Many more such comparisons involving larger, more complex systems and a wider range of solvents (particularly water) can be anticipated in the coming months and years, which might well form the basis for a future Faraday Discussion!

I conclude by congratulating, and thanking, the organizers of FD163—Helen Fielding, Steve Leone, Christoph Meier, Andrew Orr-Ewing, Katharine Reid and Graham Worth—for proposing and coordinating such an interesting and thought provoking meeting.

References

1 *Trans. Faraday Soc.*, 1926, **21**, p. 437.
2 *Trans. Faraday Soc.*, 1931, **27**, p. 357.
3 *Discuss. Faraday Soc.*, 1963, **35**, p. 1.
4 G. Herzberg and H. C. Longuet-Higgins, *Discuss. Faraday Soc.*, 1963, **35**, 77.
5 A. Stolow, paper 1 of FD163.
6 J. M. Hollas, *Discuss. Faraday Soc.*, 1963, **35**, 233.
7 See, for example, G. D. Scholes, G. R. Fleming, A. Olaya-Castro and R. von Grondelle, *Nat. Chem.*, 2011, **3**, 763 and references therein.
8 E. Riedle, M. Bradler, M. Wenninger, C. F. Sailer and I. Pugliesi, *Faraday Discuss.*, 2013, DOI: 10.1039/C3FD00010A.
9 M. Guglielmi, M. Doemer, I. Tavernelli and U. Rothlisberger, *Faraday Discuss.*, 2013, DOI: 10.1039/C3FD00037K.
10 S. Mercier, O. V. Boyarkin, A. Kamariotis, M. Guglielmi, I. Tavernelli, M. Cascella, U. Rothlisberger and T. R. Rizzo, *J. Am. Chem. Soc.*, 2006, **128**, 16938.

11 K. Addison, I. A. Heisler, J. Conyard, T. Dixon, P. C. Bulman Page and S. R. Meech, *Faraday Discuss.*, 2013, DOI: 10.1039/C3FD00019B.
12 B. C. Arruda, B. Smith, K. G. Spears and R. J. Sension, *Faraday Discuss.*, 2013, DOI: 10.1039/C3FD20149J.
13 S. G. Ramesh and W. Domcke, *Faraday Discuss.*, 2013, DOI: 10.1039/C3FD00006K.
14 T. N. V. Karsili, A. M. Wenge, S. J. Harris, D. Murdock, J. N. Harvey, R. N. Dixon and M. N. R. Ashfold, *Chem. Sci.*, 2013, **4**, 2434.
15 S. J. Harris, D. Murdock, Y. Zhang, T. A. A. Oliver, M. P. Grubb, A. J. Orr-Ewing, G. M. Greetham, I. P. Grant, M. Towrie, S. E. Bradforth and M. N. R. Ashfold, *Phys. Chem. Chem. Phys.*, 2013, **15**, 6567.
16 M. Böhm, C. Ratzer and M. Schmitt, *J. Mol. Struct.*, 2006, **800**, 55.
17 S. Yamamoto, T. Ebata and M. Ito, *J. Phys. Chem.*, 1989, **93**, 6340.
18 S. J. Harris, T. N. V. Karsili, D. J. Hadden, G. M. Roberts, V. G. Stavros and M. N. R. Ashfold, unpublished results.
19 T. A. A. Oliver, G. A. King, D. P. Tew, R. N. Dixon and M. N. R. Ashfold, *J. Phys. Chem. A*, 2012, **116**, 12444.
20 A. V. Bochenkova and L. H. Andersen, *Faraday Discuss.*, 2013, **163**, DOI: 10.1039/C3FD20150C.
21 C. R. S. Mooney, D. A. Horke, A. S. Chatterley, A. Simperler, H. H. Fielding and J. R. R. Verlet, *Chem. Sci.*, 2013, **4**, 921.
22 T. A. A. Oliver, Y. Zhang, M. N. R. Ashfold and S. E. Bradforth, *Farad. Disc. Chem. Soc.*, 2011, **150**, 439.
23 Y. Zhang, T. A. A. Oliver, M. N. R. Ashfold and S. E. Bradforth, *Farad. Disc. Chem. Soc.*, 2012, **157**, 141.
24 S. J. Greaves, R. A. Rose, T. A. A. Oliver, D. R. Glowacki, M. N. R. Ashfold, J. N. Harvey, I. P. Clark, G. M. Greetham, A. W. Parker, M. Towrie and A. J. Orr-Ewing, *Science*, 2011, **331**, 1423.
25 C. Huang, W. Li, A. D. Estillore and A. G. Suits, *J. Chem. Phys.*, 2008, **129**, 074301 and references therein.
26 D. R. Glowacki, R. A. Rose, S. J. Greaves, A. J. Orr-Ewing and J. N. Harvey, *Nat. Chem.*, 2011, **3**, 850.

Poster titles

Excited potential curves and vertical excitations using state-averaged Monte Carlo configuration interaction, **J. P. Coe and M. J. Paterson**, *Heriot-Watt University, UK*

Controlling the photocycle of a plant photoreceptor using structural analogues of FMN, **A. Wood and G. Richter**, *Cardiff University, UK*

Elucidating the mechanism of a unidirectional molecular motor, **S. Amirjalayer, W. R. Browne, B. L. Feringa, W. J. Buma and S. Woutersen**, *University of Amsterdam, The Netherlands*

Watching the ultrafast photodynamics of azobenzene in the frequency domain, **E. M. M. Tan, S. Amirjalayer, E. Maltseva, S. Smolarek and W. J. Buma**, *University of Amsterdam, The Netherlands*

Laser control in open quantum systems: a Cope rearrangement, **A. Chenel, M. Ndong, G. Dive, C. Meier and M. Desouter-Lecomte**, *Universite Paris-Sud, France*

Electronic spectroscopy of 1,3-cyclohexadiene derivatives, **B. Arruda, B. Smith, K. G. Spears and R. J. Sension**, *University of Michigan, USA*

Simulation of laser-induced quantum dynamics of the electronic and nuclear motion in the ozone molecule on the attosecond time scale, **A. Perveaux, G. J. Halász, B. Lasorne, M. A. Robb, F. Gatti and Á. Vibók**, *Universite Paris-Sud, France*

Full dimensional quantum description (48D) of the ring-opening dynamics of benzopyran through a conical intersection with ML-MCTDH, **A. Perveaux, L. Joubert-Doriol, D. Lauvergnat, H.-D. Meyer, B. Lasorne and F. Gatti**, *Universite Paris-Sud, France*

Correlated electron dynamics in the multi-ionization of atoms and molecules, **H. Price, C. Lazarou, V. Hakobyan, A. Wallis, A. Emmanouilidou**, *University College London, UK*

Investigating the role of intramolecular vibrations in EET using *ab initio* quantum classical dynamics: the BChl dimer, **C. G. Gillis and G. A. Jones**, *University of East Anglia, UK*

Vibronic energy localization in weakly coupled small molecular aggregates, **J. Wehner, A. Schubert, C. Brüning and V. Engel**, *Universität Würzburg, Germany*

A simple model for the relaxation dynamics in perylene-bisimide dimers excited by femtosecond laser pulses, **M. Kess, A. Schubert, V. Settels, B. Engels,**

F. Würthner, S. Lochbrunner, C. Meier and V. Engel, *Universität Würzburg, Germany*

Solving the coupled-channels Schroedinger equation for N_2 photodissociation, **A. N. Heays, B. R. Lewis and S. T. Gibson**, *Leiden University, The Netherlands*

Ultrafast exciton quenching upon geometry deformation in molecular aggregates, **A. Schubert, V. Settels, W.-L. Liu, F. Würthner, C. Meier, R. Fink, S. Schindlbeck, S. Lochbrunner, B. Engels and V. Engel**, *Université Toulouse III, France*

The Jahn–Teller effect in the ultrafast dissociation of binary transition metal carbonyl complexes, **R. G. McKinlay and M. J. Paterson**, *Heriot-Watt University, UK*

Challenges for wavefunction and density functional response methods in benchmarking the electronic excited states of 1st - row transition metal complexes, **N. Almeida, R. G. Mckinlay and M. J. Paterson**, *Heriot-Watt University, UK*

Elucidating the nonadiabatic processes in short model peptides, **M. Mališ, I. Ljubić, N. Došlić, Y. Loquais, E. Gloaguen, H. S Biswal, F. Piuzzi, B. Tardivel, V. Brenner, M. Broquier, C. Jouvet and M. Mons**, *Ruđer Bošković Institute, Croatia*

Probing ultrafast dynamics with time-resolved multi-dimensional coincidence imaging: butadiene, **P. Hockett, E. Ripani, A. Rytwinski and A. Stolow**, *National Research Council of Canada, Canada*

Ab initio ultrafast diffraction from photoexcited atoms and molecules, **A. Kirrander, T. Northey, K. Saita and D. V. Shalashilin**, *University of Edinburgh, UK*

Modelling excited state relaxation channels in macrocyclic chromophores of use in photodynamic therapy, **L. T. Bergendahl and M. J. Paterson**, *Heriot-Watt University, UK*

Non-Markovian excitionic dynamics in presence of strong fields, **A. Debnath and C. Meier**, *Universite Paul Sabatier, France*

Computational studies of excited states and photochemistry of phytochromes, **P. Repiščák and M. J. Paterson**, *Heriot-Watt University, UK*

Theoretical description of inter-atomic Coulombic decay, **P. Kolorenč, N. Sisourat, N. Kryzhevoi and L. S. Cederbaum**, *Charles University in Prague, Czech Republic*

Ultrafast relaxation dynamics of excited argon clusters, **A. Lietard, G. Piani, L. Poisson, M.-A. Gaveau, J.-M. Mestdagh and B. Soep**, *CNRS, IRAMIS, SPAM, France*

Multiconfigurational quantum dynamics of ultrafast energy and charge transfer in organic polymers, **J. Wahl, R. Binder, Sarah Römer and I. Burghardt**, *Goethe University Frankfurt, Germany*

A generalized Hückel type procedure for the diabatization of *ab initio* potential energy surfaces of semiconducting polymers, **R. Binder, S. Römer and I. Burghardt**, *Goethe University Frankfurt, Germany*

Excited-state dynamics and intersystem crossing in rhenium(I)–halide complexes, **J. Eng, C. Gourlaouen, C. Daniel and E. Gindensperger**, *Universite de Strasbourg - CNRS, France*

The pixel imaging mass spectrometry (PImMS) sensor applied to imaging the molecular structure and torsional motion of a chiral molecule, **A. Lauer, C. S. Slater, S. Blake, L. Christensen, J. H. Nielsen, M. Brouard and H. Stapelfeldt**, *University of Oxford, UK*

Excited state dynamics of acetylene: Vibronic coupling 8 state model, **C. Robertson and G. Worth**, *University of Birmingham, UK*

XUV pulses for photoelectron spectroscopy and studies of autoionisation dynamics, **E. Springate, C. M. Cacho, I. C. E. Turcu, R. Chapman, R. S. Minns, S. Weber, M. Siano, F. Frank, J. Marangos, D. Holland, R. Spesyvstev, O. Kirkby, H. Fielding, D. Hadden, G. Roberts, V. Stavros, A. Bainbridge and W. A. Bryan**, *STFC Rutherford Appleton Laboratory, UK*

Two dimensional electronic-vibrational spectroscopy of dye molecules in solution, **T. A. A. Oliver, N. H. C. Lewis and G. R. Fleming**, *University of California, Berkeley, USA*

Vibronic coupling and quenching of excitonic splitting in the *o*-cyanophenol dimer, **S. Kopec, P. Ottiger, S. Leutwyler and H. Köppel**, *University of Heidelberg, Germany*

Coherent state trains for quantum molecular dynamics, **D. V. Makhov and D. V. Shalashilin**, *University of Leeds, UK*

Calculation of the time-resolved photoelectron spectrum of toluene using first-order perturbation theory, **G. W. Richings and G. A. Worth**, *University of Birmingham, UK*

Light harvesting dendrimers containing 1,8-naphthalimide and thieno[3,4-c]-pyrrole-4,6(5H)-dione derivatives: Synthesis and optoelectronic properties, **A. El-Betany, A. Stocken, A. Davies and H. Tai**, *Bangor University, UK*

Towards a three-coordinate multi-surface potential for H atom photodissociation in phenol, **S. G. Ramesh and W. Domcke**, *Indian Institute of Science, India*

The role of $\pi\sigma^*$ states in the ultrafast relaxation of pyrrole: excitation between 249.5 nm and 200 nm, **O. M. Kirkby and H. H. Fielding**, *University College London, UK*

The space-saving bicycle-pedal photoisomerization mechanism in conjugated protonated Schiff bases, **J. P. Malhado, Y. Orozco-Gonzalez, K. Coutinho and S. Canuto**, *Universidade de São Paulo, Brazil*

Ab initio multiconfigurational Ehrenfest simulation of ultrafast photodynamics of pyrrole, **K. Saita, M. G. D. Nix and D. V. Shalashilin**, *University of Leeds, UK*

The influence of electronic coupling strength on transport activated by high-energy quantized molecular vibrations, **E. J. O'Reilly, F. Fassioli and A. Olaya-Castro**, *University College London, UK*

Charge transfer in photosynthetic reaction centres and the role of the coupling to coherent vibrational modes, **F. Fassioli, R. Stones and A. Olaya-Castro**, *University College London, UK*

Transient absorption studies of the photo-Claisen and photo-Fries rearrangement, **S. J. Harris, D. Murdock, M. P. Grubb, G. M. Greetham, I. P. Clark, M. Towrie and M. N. R. Ashfold**, *University of Bristol, UK*

Mass-selective discrimination of chirality by femtosecond coincidence imaging, **C. S. Lehmann, N. B. Ram, M. Rafiee-Fanood, H. Alvaro Galue, I. Powis and M. H. M. Janssen**, *VU University of Amsterdam, The Netherlands*

Ultraviolet photodissociation action spectroscopy of pyridinium derivatives, **C. S. Hansen, B. B. Kirk, S. J. Blanksby and A. J. Trevitt**, *University of Wollongong, Australia*

Photophysics of fulvene under the non-resonant Stark effect. Shaping the conical intersection seam, **S. Ruiz-Barragan and Ll. Blancafort**, *Institut de Química Computacional i Catàlisi and Dpt. De Química, Universitat de Girona, Spain*

The Skinner Prize for the best poster was awarded to Edward J. O'Reilly of University College London, UK, for his poster on the influence of electronic coupling strength on transport activated by high-energy quantized molecular vibrations.

List of participants

Miss Kiri Addison, *University of East Anglia, United Kingdom*
Dr Salih Al-Jabour, *Alquds University, Palestine*
Mr Nuno Almeida, *Heriot Watt University, United Kingdom*
Dr Hector Alvaro Galue, *Vrije Universitiet Amsterdam, The Netherlands*
Dr Saeed Amirjalayer, *Van 't Hoff Institute for Molecular Sciences University of Amsterdam, The Netherlands*
Professor Lars Andersen, *Aarhus University, Denmark*
Ms Anna Andrejeva, *University of Nottingham, United Kingdom*
Mr Brenden Arruda, *The University of Michigan, U.S.A.*
Professor Michael Ashfold, *University of Bristol, United Kingdom*
Professor Dr Luis Bañares, *Universidad Complutense de Madrid, Spain*
Dr Michael Bearpark, *Imperial College London, United Kingdom*
Dr Rudi Berera, *VU University Amsterdam, The Netherlands*
Miss Lisa Therese Bergendahl, *Heriot Watt University, United Kingdom*
Mr Robert Binder, *Goethe University, Germany*
Dr Lluís Blancafort, *Institute of Computational Chemistry, Universitat de Girona, Spain*
Dr Anastasia Bochenkova, *Department of Physics and Astronomy, Aarhus University, Denmark*
Professor Paul Brumer, *University of Toronto, Canada*
Professor Dr Philip Bucksbaum, *Stanford University - PULSE Institute, U.S.A.*
Professor Dr Wybren Jan Buma, *University of Amsterdam, The Netherlands*
Professor Dr Joachim Burgdörfer, *Vienna University of Technology, Austria*
Professor Irene Burghardt, *Göthe University Frankfurt, Germany*
Mr Allan Chappell, *None, United Kingdom*
Mr Adam Chatterley, *Durham and Warwick Universities, United Kingdom*
Miss Aurélie Chenel, *Laboratoire de Chimie Physique d'Orsay and Laboratoire Collisions, Agrégats, Réactivité (LCAR), Toulouse, France*
Dr Jeremy Coe, *Heriot-Watt University, United Kingdom*
Dr Rebeca de Nalda, *CSIC, Spain*
Mr Arunangshu Debnath, *Université Paul Sabatier, France*
Professor Wolfgang Domcke, *Technische Universität München, Germany*
Dr Christof Drechsel-Grau, *Ruhr-Universität Bochum, Germany*
Dr Bo Durbeej, *Linköping University, Sweden*
Emeritus Professor Geoffrey Duxbury, *University of Strathclyde, United Kingdom*
Ms Agapi Emmanouilidou, *University College London, United Kingdom*
Mr Clyde Fare, *Imperial College London, United Kingdom*
Dr Francesca Fassioli, *University College London, United Kingdom*
Professor Robert Field, *Massachusetts Institute of Technology, U.S.A.*
Professor Helen Fielding, *University College London, United Kingdom*
Professor Graham Fleming, *University of California, Berkeley, USA*
Mr Colm Gillis, *UEA, United Kingdom*

Dr Etienne Gindensperger, *Institut de Chimie UMR7177 - Université de Strasbourg - CNRS, France*
Dr Ilya Glebov, *Moscow State University, Russian Federation*
Dr Scott Habershon, *University of Warwick, United Kingdom*
Miss Stephanie Harris, *University of Bristol, United Kingdom*
Dr Alan Heays, *Leiden University / Sterrewacht, The Netherlands*
Professor Dr Jennifer Herek, *University of Twente, The Netherlands*
Professor Jonathan Hirst, *University of Nottingham, United Kingdom*
Dr Paul Hockett, *National Research Council of Canada, Canada*
Dr Michael Hollas, *University of Reading, United Kingdom*
Mr Martin Kess, *Universität Würzburg, Germany*
Mr Hjalte Vinther Kiefer, *Aarhus University, Denmark*
Mr Oliver Kirkby, *University College London, United Kingdom*
Dr Adam Kirrander, *University of Edinburgh, United Kingdom*
Dr Premysl Kolorenc, *Charles University in Prague, Czech Republic*
Ms Sabine Kopec, *University of Heidelberg, Germany*
Dr Alexandra Lauer, *University of Oxford, United Kingdom*
Mr C. Stefan Lehmann, *Vrije Universiteit Amsterdam, The Netherlands*
Professor Dr Kuiying Li, *Yanshan University, China*
Dr Dmitry Makhov, *University of Leeds, United Kingdom*
Dr João Pedro Malhado, *Imperial College London, United Kingdom*
Mr Momir Malis, *Rudjer Boskovic Institute, Croatia*
Dr Russell McKinlay, *Heriot-Watt University, United Kingdom*
Professor Steve Meech, *UEA, United Kingdom*
Professor Dr Christoph Meier, *Université Paul Sabatier, France*
Dr David Mendive-Tapia, *Imperial College London, United Kingdom*
Dr Daniel Murdock, *University of Bristol, United Kingdom*
Professor Daniel Neumark, *University of California, Berkeley, USA*
Mr Edward O'Reilly, *University College London, United Kingdom*
Dr Tom Oliver, *UC Berkeley, United States*
Professor Andrew Orr-Ewing, *University of Bristol, United Kingdom*
Miss Aurélie Perveaux, *Université Montpellier 8, France*
Mr David Picconi, *Technische Universität München, Germany*
Dr Lionel Poisson, *Laboratoire Francis Perrin - CEA-CNRS, France*
Dr Sai Ramesh, *IPC Department, IISc-Bangalore, India*
Professor Katharine Reid, *University of Nottingham, United Kingdom*
Mr Peter Repiscak, *Heriot-Watt University, United Kingdom*
Dr Gareth Richings, *University of Birmingham, United Kingdom*
Professor Dr Eberhard Riedle, *LMU München, Germany*
Dr Gareth Roberts, *University of Warwick, United Kingdom*
Mr Christopher Robertson, *University of Birmingham, United Kingdom*
Professor Ursula Röthlisberger, *Ecole Polytechnique Fédérale de Lausanne, Switzerland*
Mr Sergi Ruiz-Barragan, *Universitat de Girona, Spain*
Dr Kenichiro Saita, *University of Leeds, UK*
Miss Valentina Santolini, *Imperial College of London, United Kingdom*
Dr Fabrizio Santoro, *ICCOM-CNR, Italy*
Professor Gregory Scholes, *University of Toronto, Canada*
Dr Alexander Schubert, *Université Toulouse III - Paul Sabatier, France*
Professor Roseanne Sension, *University of Michigan, United States*
Mr Craig Slater, *University of Oxford, United Kingdom*
Dr Emma Springate, *STFC Central Laser Facility, United Kingdom*
Dr Vasilios Stavros, *University of Warwick, United Kingdom*

Professor Albert Stolow, *National Research Council*, *Canada*
Miss Mayra Stuhldreier, *University of Kiel*, *Germany*
Dr Hongyun Tai, *Bangor University*, *United Kingdom*
Mr Eric Meng Meng Tan, *University of Amsterdam*, *The Netherlands*
Professor Dr Friedrich Temps, *Christian-Albrechts-University Kiel*, *Germany*
Mr Lee Thompson, *Imperial College London*, *United Kingdom*
Dr Adam Trevitt, *University of Wollongong*, *Australia*
Professor Alessandro Troisi, *University of Warwick*, *United Kingdom*
Miss Morgane Vacher, *Imperial College London*, *United Kingdom*
Dr Jan Verlet, *Durham University*, *United Kingdom*
Mr Jan Wahl, *Göthe University*, *Germany*
Dr Derek Wann, *University of Edinburgh*, *United Kingdom*
Mr Johannes Wehner, *Universität Würzburg*, *Germany*
Professor Dr Karl-Michael Weitzel, *Philipps-Universität Marburg*, *Germany*
Mr Chris West, *Durham University*, *United Kingdom*
Professor Benjamin Whitaker, *University of Leeds*, *United Kingdom*
Miss Victoria Wilton, *University of Nottingham*, *United Kingdom*
Mr Andrew Wood, *Cardiff University*, *United Kingdom*
Dr Graham Worth, *University of Birmingham*, *United Kingdom*
Professor Timothy Wright, *University of Nottingham*, *United Kingdom*

Index of contributors*

*The page numbers in **bold** type indicate papers submitted for discussions.

